OXYGEN TRANSPORT
TO TISSUE XIX

ADVANCES IN EXPERIMENTAL MEDICINE AND BIOLOGY

Recent Volumes in this Series

OXYGEN TRANSPORT TO TISSUE XIX

Edited by

David K. Harrison

Ninewells Hospital and Medical School
Dundee, Scotland

and

David T. Delpy

University College London
London, England

SPRINGER SCIENCE+BUSINESS MEDIA, LLC

Library of Congress Cataloging-in-Publication Data

Oxygen transport to tissue XIX / edited by David K. Harrison and David
T. Delpy.
        p.   cm. -- (Advances in experimental medicine and biology :
    428)
    Includes bibliographical references and index.
    ISBN 978-0-306-45711-1   ISBN 978-1-4615-5399-1 (eBook)
    DOI 10.1007/978-1-4615-5399-1
    1. Tissue respiration--Congresses. 2. Oxygen--Physiological
transport--Congresses. 3. Oxygen--Pathophysiology--Congresses.
I. Harrison, D. K. (David Keith), 1951-   . II. Delpy, David T.
III. Series.
QP121.0885 1998
612.2'6--dc21
                                                        97-31219
                                                             CIP

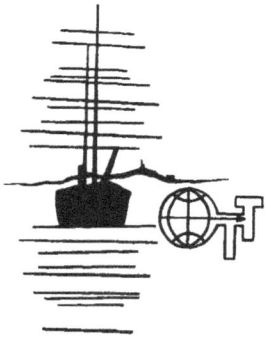

Proceedings of the 24th Annual Meeting of the International Society on Oxygen Transport to Tissue,
held August 19 – 23, 1996, in West Park Centre, Dundee, Scotland

ISBN 978-0-306-45711-1

© 1997 Springer Science+Business Media New York
Originally published by Plenum Press, New York in 1997

http://www.plenum.com

10 9 8 7 6 5 4 3 2 1

INTERNATIONAL SOCIETY ON OXYGEN TRANSPORT TO TISSUE

1996

Officers

President:	David K. Harrison (UK)
Past President:	Edwin M. Nemoto (USA)
President-Elect:	Antal G. Hudetz (USA)
Secretary:	David F. Wilson (USA)
Treasurer:	Peter E. Keipert (USA)
Award Committee:	Duane F. Bruley (USA)

Executive Committee

Sanjay Batra (USA)
Thomas E. Gayeski (USA)
Karlfried Groebe (Germany)
Jens Höper (Germany)
Kyung Kang (USA)

Joseph C. LaManna (USA)
Roland N. Pittman (USA)
Koen van Rossem (Belgium)
Peter W. Vaupel (Germany)

Scientific Programme Committee

Local

Dr. D. K. Harrison
Professor J. J. F. Belch
Dr. R. A. Lerski
Mr. P. T. McCollum
Professor M. J. Rennie

ISOTT

Professor D. T. Delpy (UK)
Dr. A. Eke (Hungary)
Dr. J. Moravec (France)
Dr. E. M. Nemoto (USA)

Local Organising Team

Jennie Bryce
Barbara Clark
David Harrison (President, ISOTT)
Eileen Harrison

Faisel Khan
David Newton
Kelly O'Brien
Zahid Raza

SPONSORS

ISOTT is a scientific society and as such is most grateful for the generosity of the following sponsoring organisations. However, it does not endorse any of their products or services. This applies also to any material which may be distributed in mailings or at its meetings.

Sponsoring Scientific Organisations

- British Microcirculation Society
- Institute of Physics and Engineering in Medicine
- 6th World Congress on Microcirculation

Principal Sponsors

- Alliance Pharmaceuticals
- Baxter Healthcare Corporation (Blood Substitutes Division)
- Dundee City Council
- Dundee Teaching Hospitals
- Scottish Enterprise Tayside

Exhibitors

- Cambridge Medical Books
- Eppendorf-Netheler-Hinz GmbH
- KK Technology
- Moor Instruments Ltd
- Oxford Optronix Ltd
- Perimed UK Ltd
- Radiometer UK Ltd

Sponsors of Symposia

- Hamamatsu UK
- Oxygen Sensors Inc
- Pfizer UK
- Pharmacia and Upjohn
- University of Dundee

PREFACE

In 1996, for its 24th scientific meeting, the International Society on Oxygen Transport to Tissue made its third visit to the United Kingdom. The previous two meetings were held in Cambridge in 1977 and 1986, but this was the first meeting to be held "north of the border" in Scotland. It was attended by some 186 delegates and accompanying persons and there were 128 presentations. The venue was the West Park Centre, the University of Dundee's residential conference centre, and ISOTT was only the second major meeting to be held there using the new Villa accommodation.

Dundee's slogan is "City of Discovery" since it became the permanent home of the Royal Research Ship *Discovery* which was built in the city and was used by Captain Scott on his first expedition to the Antarctic. The ISOTT meeting also fulfilled its promise of being a meeting of discovery with sessions on all aspects of oxygen transport to tissue. The inclusion of a session on oxygen transport in vascular disease reflected the interests of the local participants.

All of the manuscripts were reviewed both for their scientific and editorial acceptability and in some 50% of cases, revisions were requested from the authors. Some manuscripts were ultimately rejected. However, in view of the importance of producing the Proceedings as quickly as possible it is possible that some minor errors may have slipped through, for which the editors apologise.

We wish to acknowledge the enormous hard work that has been put in by Laraine Visser-Isles in copy-editing this volume. Kelly O'Brien and Eileen Harrison have also contributed significantly in the organisation of the manuscripts and the preparation of the Proceedings. The editors, on behalf of ISOTT, also acknowledge the help of all (named and unnamed in this volume) who helped to make the 24th meeting possible.

We congratulate Sergei Vinogradov and Resit Demir respectively for being selected as the Mevin H Knisely and Dietrich W Lübbers Award winners of 1996 based on the scientific excellence of their submitted manuscripts.

We hope that the memories which remain with the many members of the ISOTT "family" who attended the Dundee Conference will be those of the high scientific standard of the presentations, lively debates and a thoroughly enjoyable Scottish social programme.

Finally, I (DKH) wish to dedicate this volume to all of the staff of Dundee Teaching Hospitals and Edinburgh Royal Infirmary NHS Trusts who played a part in my treatment for and recovery from cancer in 1994–95, and to Eileen, Natalie and Francesca who supported me not only throughout that time, but also during the organisation of the meeting.

David K. Harrison
David T. Delpy

CONTENTS

O₂ in Vascular Disease

O$_2$ Transport in Tumours

Hyperoxia and Hypoxia

O_2 Transport in Organs: Brain

O_2 Transport in Organs: Heart and Lung

O_2 Transport in Organs: Skeletal Muscle

Systemic O₂ Transport

Local Regulation of O_2 Supply

Cellular Metabolic State *in Vivo*

Fluorocarbons

Functional Imaging

Perfusion and Diffusion Imaging and Spectroscopy

Methods and Instrumentation

Theoretical Models

LEUKOCYTE ACTIVATION AND TOXIC OXYGEN METABOLITES/FREE RADICALS

Jill J. F. Belch and Paul Hickman

University Department of Medicine
Ninewells Hospital and Medical School
Dundee, Scotland, United Kingdom

1. INTRODUCTION

The transport of oxygen to tissues is crucial for tissue viability. Much research has been directed towards enhancing oxygen transport but there are situations in which oxygen, through the production of toxic oxygen metabolites (TOMs), can be detrimental. This is particularly so in the case of diseases affecting the blood vessels. Whilst the red blood cell is conventionally associated with oxygen carriage, the white blood cell is associated with the generation of TOMs or free radicals (FRs) which are generated as required to mediate the inflammatory and immunological responses. Increasing evidence now shows that another key event may also take place, namely the obstruction of the microcirculation by activated leucocytes and subsequent tissue injury by FRs released from the white blood cell (WBC).[1]

2. SUPPORTING EVIDENCE

There is no doubt that the WBC count is a powerful predictor of vascular events. This is true in terms of myocardial infarction,[2] stroke,[3] and critical limb ischaemia.[4] Initial studies suggested that a white cell count in the high normal range could predict further vascular events after the initial myocardial infarction[5] or stroke. It was thought that this was merely a reflection of tissue injury; damaged tissue producing an inflammatory response and a corresponding increase in white cell count. It has become apparent however that the WBC count in the high normal range is a predictor of first vascular events also. Moreover, a lower WBC count appears to reduce the risk of vascular events. A fall of 10^9 WBC/l was associated with a 14% fall in the risk of developing coronary artery disease.[6] Thus there is good supporting evidence for the role of the WBC in the production of vascular damage.

Oxygen Transport to Tissue XIX, edited by Harrison and Delpy
Plenum Press, New York, 1997

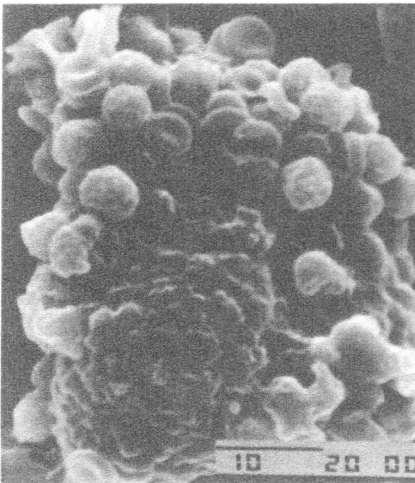

Figure 1. Adherent white cells with red cells and platelet clumps forming vessel occlusive thrombus.

3. VASCULAR DAMAGE – HOW IS IT MEDIATED?

Leucocytes are able to cause vascular damage in a number of different ways through both physical obstruction and the release of noxious chemicals. The WBC is a large cell that is poorly deformable and these hard cells have been found in patients with vascular diseases such as myocardial infarction,[7] acute cerebral infarction,[8] chronic cerebral vascular disease,[9] and peripheral arterial obstructive disease.[10]

Another property of the WBC that affects flow is leukocyte adhesion to the endothelium or to other blood cells where it is known as aggregation[11] (See figure 1). Increased WBC adhesion/aggregation is found in patients with risk factors for vascular disease such as cigarette smoking[12] and hyperlipidaemia. WBC adhesion may have a link with vascular stenoses, particularly those occurring following angioplasty.[13]

Once the WBC is adherent either to other blood cells or to the endothelium, it is able to release its noxious chemicals. There are a number of chemicals released from WBC when it is activated. These include biologically active lipids, neutral hydrolases and toxic oxygen metabolites.

4. OXYGEN DERIVED FREE RADICALS AND REACTIVE OXYGEN SPECIES

Free radicals such as the superoxide anion and hydroxyl radical interact with nucleic acids, other proteins and lipids. The oxidation of membrane proteins adversely affects membrane function as can damage to the cell membrane lipid by peroxidation. Free radicals can directly damage the endothelial cells lining the blood vessels thus increasing the likelihood of thrombosis and subsequent atheroma formation. These prothrombotic effects are enhanced by the ability of the products of free radical oxidation, lipid peroxides, to alter the prostacyclin-thromboxane balance in favour of vasoconstriction and platelet aggregation.[14,15]

Lipid peroxides increase the cholesterol donating ability of low density lipoprotein and inhibit the cholesterol accepting capacity of high density lipoprotein.[16] This promotes cholesterol accumulation in the blood vessels augmenting the development of atherosclerosis. Blood coagulation factors in the plasma are also affected by TOMs. Thrombin generation is increased by lipid peroxides acting on lipoproteins and antithrombin III.[17]

Although the most important source of FRs is likely to be the leukocyte, FRs can also be produced via other mechanisms such as the generation of prostaglandins, catecholamine metabolism and normal cellular respiration. Reperfusion of tissue following obstruction of blood supply to that tissue is another important source of FR generation and occurs following clinical situations such as reperfusion following thrombolysis,[18] and deflation of the balloon during angioplasty.[13]

Many studies have demonstrated increased products of free radical generation in diseases associated with vascular damage. These include heart disease,[19] peripheral arterial obstructive disease,[20] stroke and Raynaud's phenomenon.[21]

As one would expect, the body has a number of defence mechanisms at its disposal for protection against free radical attack. These include the formation of chemicals such as superoxide dismutase and catalase and utilisation of the cell antioxidants such as vitamin C and vitamin E. A circadian rhythm is found in the physiological production and activity of radical scavengers[22] which matches the diurnal variation of the development of thrombotic vascular disease. Patients with both peripheral arterial disease and coronary artery disease have low levels of scavengers such as the superoxide dismutases and plasma thiols.[23] The antioxidant vitamins are also important with low levels of vitamins C and E being found in patients with angina and peripheral arterial disease.[24]

It is interesting to note that subjects with risk factors for the development of arterial disease also have increased free radical products within their blood. Smoking increases lipid peroxide levels[25] as does hyperlipidaemia and diabetes mellitus.[26]

5. THERAPEUTIC CONSIDERATIONS

It is now clear that the WBC can contribute to vascular disease both through rheological and secretory changes in the cell. Thus it might be suggested that therapies which decrease leucocyte activation and free radical production might be beneficial. A large amount of animal work has already taken place in this area with the production of encouraging results.[27,28] Studies are now moving into the human arena with antioxidants being given directly to patients with vascular disease and assessment of the anti WBC, antioxidant properties of conventional and new drugs. Caution however should be exercised to ensure that these therapies that are able to modulate WBC function do not impair behaviour essential to health such as tissue repair, control of infection and the destruction of cancerous cells.

6. CONCLUSION

Blood flow and thus tissue nutrition is affected by both functional and rheological changes within the WBC. Patients with risk factors for vascular disease and those with established vascular disease have evidence of enhanced toxic oxygen metabolite generation and decreased scavenging ability. Thus treatments that inhibit WBC activation in the early stages of ischaemia may have beneficial effects. It should be noted however that normal

WBC response may be essential for the later stages of tissue healing. A number of published studies have shown abnormal WBC function in arterial disease. We are now beginning to see the results of interventional studies, in which drugs which alter WBC behaviour are introduced into patients with vascular disease. This area will continue to be a challenge for the future.

REFERENCES

1. Belch JJF. The relationship between white blood cells and arterial disease. Curr Opinion Lipidol 1994; 5: 440–446.
2. Friedman GD, Klatsky AL, Siegelaub AB. The leukocyte count as a predictor of myocardial infarction. N Eng J Med 1974; 290: 1275–1278.
3. Prentice RL, Szatrowski TP, Kato H, Mason MW. Leukocyte counts and cerebrovascular disease. J Chronic Dis 1982; 35: 703–714.
4. Belch JJF, Frings M, Voleske P, Sohngen W. Neutrophil count and amputation in critical limb ischaemia (Abstract). Crit Limb Ischaemia 1994; 13: 6.
5. Schlant RC, Forman S, Stamler V, Canner PL. The natural history of coronary heart disease: prognostic factors after recovery from myocardial infarction in 2789 men. Circulation 1982; 66: 401–408.
6. Shoenfeld Y, Pinkhas J. Leukopenia and low incidence of myocardial infarction. New Eng J Med 1984; 305: 1606.
7. Nash GB, Christopher B, Morris AJR, Dormandy JA. Changes in the flow properties of white blood cells after acute myocardial infarction. Br Heart J 1989; 62: 329–334.
8. Mercuri M, Ciuffetti G, Robinson M, Toole J. Cell rheology in acute cerebral infarction. Stroke 1989; 20: 959–962.
9. Vermes I, Strik F. Altered leukocyte rheology in patients with chronic cerebrovascular disease. Stroke 1988; 19: 631–633.
10. Ciuffetti G, Mercuri M, Lombardini R, Maragoni G, Santambrogio L, Mannarino E. Leucocyte behaviour in controlled ischaemia of the calves. J Clin Pathol 1989; 42: 1083–1087.
11. Fisher TC, Belch JJF, Barbenel, Fisher AC. Human whole blood granulocyte aggregation in vitro. Clin Sci 1989; 76: 183–187.
12. Bridges AB, Hill A, Belch JJF. Cigarette smoking increases white blood cell aggregation in whole blood. J R Soc Med 1993; 86: 139–140.
13. Lau CS, Scott N, Brown JE, Shaw W, Belch JJF. Increased activity of oxygen free radicals during reperfusion in patients undergoing percutaneous peripheral artery balloon angioplasty. Int Angiol 1991; 10: 244–246.
14. Moncada S, Gryglewski RJ, Bunting S, Vane JR. A lipid peroxide inhibits the enzyme in blood vessel microsomes that generate from prostaglandin enderoperoxides the substance (prostaglandin X) which prevents platelet aggregation. Prostaglandins 1976; 12: 715–737.
15. Tate RM, Morris HG, Schroeder WR, Repine JE. Oxygen metabolites stimulate thromboxane production and vasoconstriction in isolated saline perfused rabbit lung. J Clin Invest 1984; 74: 608–613.
16. Azizova OA, Panasenko OM, Vol'Nova TV, Vladimirov YA. Free radical lipid oxidation affects cholesterol transfer between lipoproteins and erythrocytes. Free Rad Biol 1989; 7: 251–257.
17. Barrowcliffe TW, Gray E, Kerry PJ, Gutteridge JMC. Lipid peroxides, lipoproteins and thrombosis. Life Reproduction 1985; 3: 174–188.
18. Young IS, Purvis JA, Lightbody JH, Adgey AAJ, Trimble ER. Lipid peroxidation and antioxidant status following thrombolytic therapy for acute myocardial infarction. Eur Heart J 1993; 14: 1027–1033.
19. McMurray J, McLay J, Chopra M, Bridges A, Belch JJF. Evidence for enhanced free radical activity in chronic congestive heart failure secondary to coronary artery disease. Am J Cardiol 1990; 65: 1261–1262.
20. Reaven PD, Khouw A, Beltz WF, Parthasarathy S, Witztum JL. Effect of dietary antioxidant combinations in humans. Protection of LDL by Vitamin E but not by β-Carotene. Arterioscler Thromb 1993; 13: 590–600.
21. Lau CS, O'Dowd A, Belch JJF. White blood cell activation in Raynaud's phenomenon of systemic sclerosis and vibration induced white finger syndrome. Ann Rheum Dis 1992; 51: 249–252.
22. Bridges AB, McLaren M, Scott NA, Pringle TH, McNeill GP, Belch JJF. Circadian variation in white blood cell aggregation and free radical indices in men with stable ischaemic heart disease. Eur Heart J 1992; 13: 1632–1636.

23. Bridges AB, McNeill GP, Pringle TH, Niblock A, Belch JJF. Free radical markers in patients with angina pectoris and normal coronary angiograms. Cardiology 1993; 82: 7–11.
24. Riemersma RA, Wood DA, MacIntyre CCA, Elton RA, Gey KF, Oliver MF. Risk of angina pectoris and plasma concentrations of vitamins A, C and E and Carotene. Lancet 1991; 337: 1–5.
25. Bridges AB, Hill A, Belch JJF. Cigarette smoking increases white blood cell aggregation in whole blood. J R Soc Med 1993; 86: 139–140.
26. Jennings PE, Jones AF, Florkowski CM, Lunec J, Barnett AH. Increased diene conjugate in diabetic subjects with microangiopathy. Diabet Med 1987; 4: 452–456.
27. Laurindo FR, de Almeida Pedro M, Barbeiro HV, Pileggi F, Carvalho MHC, Augusto O, Lemas de Luz P. Vascular free radical release. *Ex vivo* and *In vivo* evidence for a flow-dependent endothelial mechanism. Circ Res 1994; 74: 700–709.
28. McCord JM. Oxygen-derived free radicals in post-ischemic tissue injury. N Engl J Med 1985; 312: 159–163.

2

IMPAIRED HEMORHEOLOGY IN THE AGED ASSOCIATED WITH OXIDATIVE STRESS

Joseph M. Rifkind, Ranjeet S. Ajmani, and Jane Heim

National Institutes of Health
National Institute on Aging
Laboratory of Cellular and Molecular Biology
Molecular Dynamics Section
4940 Eastern Avenue, Baltimore, Maryland 21224

1. INTRODUCTION

Oxygen transport to the tissues depends on the ability of blood cells to flow through the circulatory system. The flow of blood in the circulatory network depends upon many factors. The viscosity which measures the resistance to flow is a major factor in determining the hemorheological properties of blood. A number of studies (Torre la de et al.; 1993, Paul et al., 1996; Sharp et al., 1996) indicate that the hemorheological properties, which determine blood flow, are impaired in the aged. Thus, aging has been shown to be associated with an increase in fibrinogen, plasma viscosity, whole blood viscosity and whole blood filterability (Avellone et al., 1993, Lechner et al., 1987).

Fibrinogen, which increases during aging (Hager et al., 1994, Barasch et al., 1995) plays a significant role in producing impaired hemorheology. Fibrinogen increases the plasma viscosity and facilitates aggregation of red blood cells, which influence whole blood viscosity. There is also evidence for increased oxidative stress in the aged (Pierrefiche et al., 1995). Recent studies suggest a relationship between oxidative stress and impaired red cell deformability. It has thus been reported (Snyder et al., 1985) that the addition of hydrogen peroxide to red cells results in cross linking of hemoglobin to spectrin which impairs deformability.

One proposed mechanism predicts an increased oxidative stress under the partially hypoxic conditions expected to exist in the aged, both as the result of impaired hemorheological properties and the vascular problems prevalent in the aged. This mechanism is associated with the higher rates for autoxidation of the partially oxygenated hemoglobin intermediates present under hypoxic conditions (Abugo et al., 1994; Balagopalakrishnan et al.,1996). Associated with the autoxidation is the production of superoxide (Misra and Fridovich, 1972; Rifkind et al., 1989). The generation of oxyradicals in hypoxic erythrocytes has been shown *in vitro* to result in membrane damage as indicated by a higher rate of lysis and the cross linking of membrane proteins similar to that found by the addition of hydrogen peroxide (Rifkind and Abugo , 1994). The associated membrane damage impairs erythrocyte deformability, which will affect the ability of the erythrocytes to pass through narrow pores. These changes in deformability may further exacerbate impaired oxygen transport in the aged.

In this paper we have studied the hemorheological changes taking place during aging and investigated the possibility that some of these hemorheological changes may originate from oxidative stress.

2. MATERIALS AND METHODS

Blood was obtained from participants of the Baltimore Longitudinal Study on Aging. The blood was collected in sterile vacutainers containing EDTA.

2.1. Whole Blood Viscosity

Whole blood viscosity was measured using a Brookfield DV-III cone and plate viscometer. Blood viscosity was measured at shear rates of 450, 225, 90, 45 and 22.5 /sec. Samples were maintained at 37^0C throughout the measurement. Since viscosity is a function of the hematocrit, the results for all samples were corrected to a hematocrit of 45%.

2.2. Red Cell Rigidity

Whole blood was centrifuged at 3000 rpm for 10 minutes and the plasma was removed. These cells were resuspended in PBS and the centrifugation procedure was repeated twice to remove any residual plasma. These plasma free cells were then suspended in Eagle's buffer at a hematocrit of 33%. The viscosity of this suspension was measured at a shear rate of 450 /sec with the results corrected to a hematocrit of 40%. At a high shear rate in the absence of fibrinogen or other plasma factors the viscosity is determined by the red cell rigidity.

2.3. Lipid Peroxidation

A lipid extract of the red blood cells was obtained by the method of Rose and Okleander(1965) using isopropyl alcohol and chloroform. Fluorescent adducts associated with lipid peroxidation (Jain 1989) were measured using an excitation wavelength of 400 nm and an emission wavelength of 460 nm. The measurements were made on a Perkin-Elmer fluorescence spectrophotometer model 650-40.

3. RESULTS AND DISCUSSION

3.1. Blood Viscosity and Shear Rate

Blood is a non-Newtonian fluid with its viscosity dependent on shear rate. At low shear rate red cell aggregation is largely responsible for an increase in viscosity. As the shear rate increases the aggregates tend to be disrupted and the viscosity decreases. The lower viscosity at high shear rates are determined predominantly by the properties of isolated red blood cells.

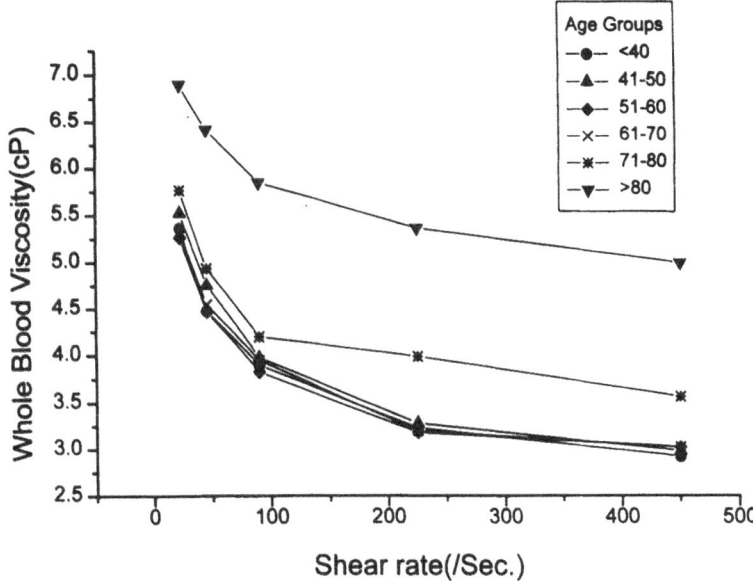

Figure 1. Shear rate vs. blood viscosity for different age groups.

Fibrinogen present in plasma is largely responsible for the aggregation of red blood cells. Even at high shear rate there is a residual contribution of fibrinogen and aggregation to the whole blood viscosity. This factor is eliminated by removing the plasma and replacing it with Eagles buffer containing albumin. The measured viscosity at high shear rate then reflects the red cell rigidity.

3.2. Changes in Blood Viscosity during Aging

Figure 1 shows the shear dependence of whole blood viscosity for six different age groups with the youngest group <40 years of age (33.92±1.46; n = 9) and the oldest group >80 years of age (86.53±1.99; n = 9).

This data shows that there is almost no effect of age below 70 years of age. For the 71–80 year old age group there are marked increases in viscosity at the two highest shear rates (225&450/sec),and for the oldest age group (>80) there is a very marked increase in viscosity at all shear rates.

Previous studies (Avellone et al., 1993; Shirakura et al., 1993) on the change in whole blood viscosity as a function of age have found a trend towards increased viscosity in older subjects which was not statistically significant. By separating the data into different age groups we find that no significant changes are found for the younger groups, but a very significant increase in viscosity is found when comparing the subjects below 70 years of age with those over 70 years of age (p<0.0003).

Figure 2 shows the age dependence for both whole blood viscosity (2a) and red cell rigidity (2b) at a shear rate of 450/sec. The red cell rigidity indicates very small changes in the youngest groups (<60 years of age) and a marked increase in older subjects with a very significant difference between the values obtained above and below 60 years of age (p<0.009). Although the rigidity increases in the sixth decade instead of the seventh decade as found for whole blood viscosity, in both cases a clearly non-linear increase with respect to age is found.

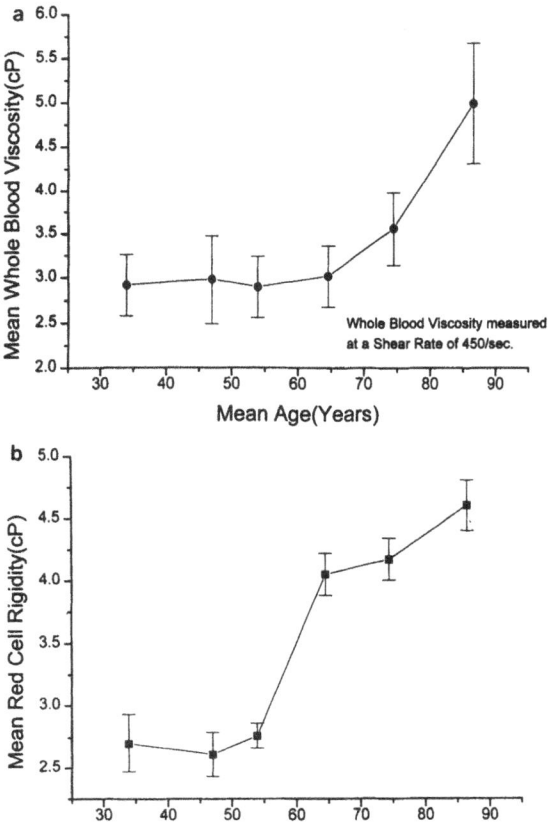

Figure 2. (a) Dependence of whole blood viscosity at a shear rate of 450/sec. on age. (b) Dependence of red cell rigidity on age.

3.3. Relationship between Red Cell Rigidity and Lipid Peroxidation

Figure 3 shows the scatter plot illustrating the dependence of red cell rigidity and lipid peroxidation on age. Individual points are indicated for each of the subjects studied. As indicated in this figure a very significant linear correlation with respect to age is found for both red cell rigidity (R=0.8,p<0.0003) and lipid peroxidation (r=0.477,p<0.0004). As indicated in the analysis for which the subjects were separated into different age groups (Figure 2b), there is a non-linear increase in red cell rigidity as a function of age. In the scatter plot this is indicated by systematic deviations from the linear fit to the data, particularly for young subjects. As indicated in the figure, fitting with a second order polynomial improves the fit.

For the lipid peroxidation deviations from the linear plot are even more pronounced with a number of young subjects falling below the line and a number of old subjects falling above the line. The second order polynomial for lipid peroxidation produces a very marked improvement in the fit with the r value increasing from 0.477 to 0.546.

The correlation of both red cell rigidity and lipid peroxidation with age suggests a possible relationship between these two parameters. The non-linear dependence with the dominant increases occurring for older subjects further supports the hypothesis that the factors responsible for increased lipid peroxidation also influence red cell rigidity. Figure 4a shows a scatter plot relating red cell rigidity to lipid peroxidation. The elimination of the systematic deviations found for lipid peroxidation versus age in Figure 3b and the

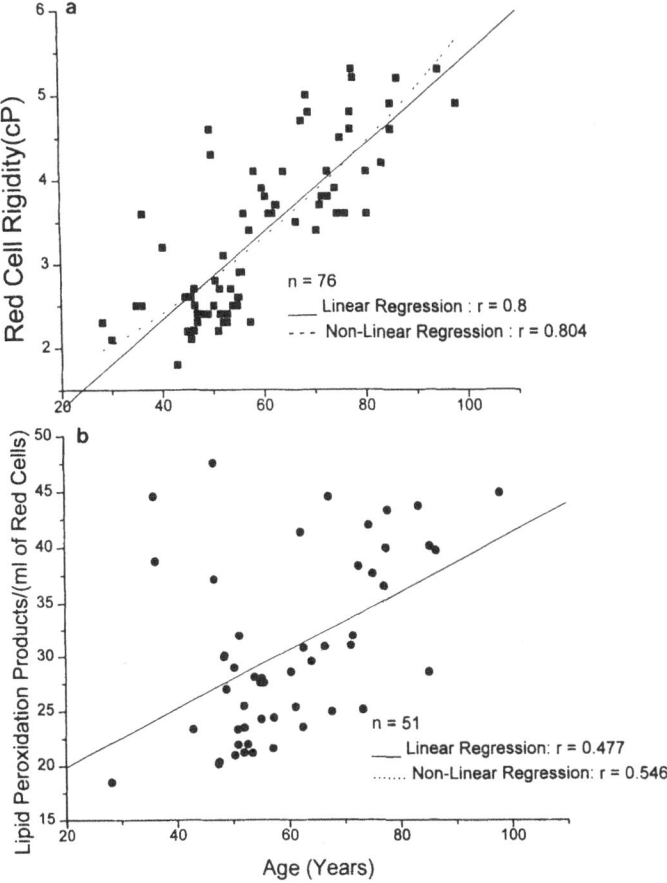

Figure 3. (a)Linear and nonlinear relationship between age and red cell rigidity. (b)Linear and nonlinear relationship between age and red cell lipid peroxidation.

increase in the r value from 0.477 to 0.508 suggests that the non-linear effects in both red cell rigidity and lipid peroxidation have a common origin. Furthermore, the correlation between whole blood viscosity and lipid peroxidation (Figure 4b) implies that the changes in red cell properties resulting from oxidative stress contribute to the overall rheological properties of the blood even with plasma and fibrinogen present.

Lipid peroxidation is a direct measure of oxyradical damage. Therefore, the findings that lipid peroxidation increases with age in a non-linear fashion, and the correlation of this pattern of increase with red cell rigidity and whole blood viscosity, implies that oxyradical damage contributes to impaired hemorheology during aging.

3.4. Oxyradical Damage and Impaired Hemorheology

Lipid peroxidation influences membrane fluidity which is indicative of a more rigid membrane structure. However, oxidative stress which produces lipid peroxidation also produces other changes in membrane properties. It has thus been reported (Snyder et al., 1985) that oxidative stress causes cross linking of hemoglobin to the spectrin cytoskeleton. This protein cross linking has been directly implicated in a loss in red cell deformability and increased membrane rigidity.

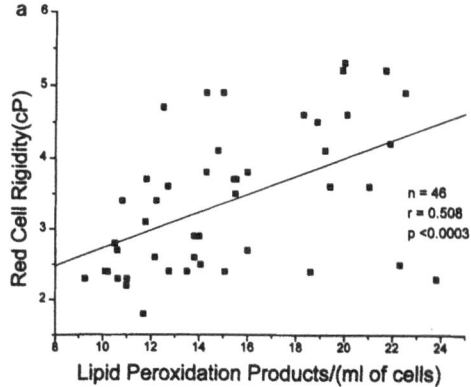

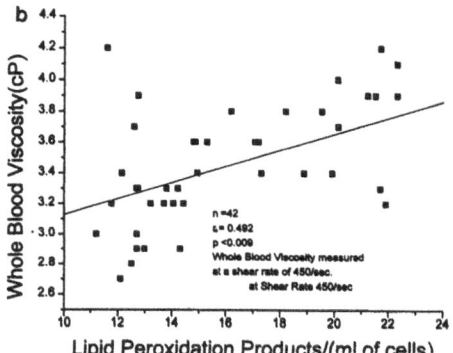

Figure 4. (a) Cross correlation between red cell lipid peroxidation and red cell rigidity. (b) Cross correlation between red cell lipid peroxidation and whole blood viscosity at a shear rate of 450/sec.

3.5. The Origin of Oxyradicals which Damage the Red Blood Cells during Aging

Oxyradicals released into the circulation by neutrophils, endothelial cells, monocytes or erythrocytes can damage the red cell membrane. Less effective antioxidant protective mechanisms can lead to enhanced lipid peroxidation during aging.

A particularly effective means to damage the red cell membrane involves the oxyradicals produced in the red cell during hemoglobin autoxidation (Rifkind et al., 1991). Under hypoxic conditions autoxidation is enhanced because of the particular susceptibility of the partially oxygenated intermediates to autoxidation (Balagopalkrishnan et al., 1996). Hypoxia also enhances the binding of hemoglobin to band 3 of the red cell membrane (Rifkind et al., 1992). This mechanism generates oxyradicals at the cytoplasmic surface of the membrane facilitating membrane damage and bypassing the cytoplasmic antioxidant capabilities.

Aging is associated with impaired blood flow, both because of vascular changes, increased erythrocyte mean cell volume (Araki et al., 1980; Helman et al., 1975) and increased fibrinogen (Bauer et al., 1987). These changes cause hypoxia which is expected to increase red cell oxyradical production. These oxyradicals may be the source for increased lipid peroxidation and possible membrane proteins cross linking, which exacerbate the hemorheological decrement.

REFERENCES

Abugo O., Rifkind J.M. Oxidation of hemoglobin and the enhancement produced by Nitroblue Tetrazolium. Journal of Biochemistry; 269: 24845–24853, 1994.

Araki K, Rifkind J.M. Age dependent changes in osmotic hemolysis of human erythrocytes. Journal of Gerontology. 35; 499–505, 1980.

Avellone G, Garbo D, Panno AV, R Cordova, Alletto G, Raneli G, De Simone R, Strano A, Bompani GD. Haemorheological components in the pre-geriatric and geriatric age range in a randomly selected western Sicily population sample (Casteldaccia Study). Clinical Hemorheology 13: 83–92. 1993.

Balagopalkrishnan C., Manoharan P. T., Abugo O. O., Rifkind J.M. Production of superoxide from hemoglobin-bound oxygen under hypoxic conditions. Biochemistry; 35: 6393–6398, 1996.

Barasch E, Benderly M, Graff E, Behar S, Reicher-Reiss H., Caspi A, Pelled B, Reisin L, Roguin N, Goldbourt U. Plasma fibrinogen levels and their correlate in 6457 coronary heart diseases patients. The Bezafibrate Infarction Prevention (BIP) study. J Clin Epidemiol. 48: 757–765, 1995.

Bauer K.A., Weiss L. M., Sparrow D., Vokonas P.S., Rosenberg R.D. Aging associated changes in indices of thrombin generation and protein C activation in humans. Normative aging study. Journal of Clinical Investigation; 80: 1527–1534, 1987.

Hager K, Felicetti M, Seefrid G, Platt D. Fibrinogen and aging. Aging Clinical Exp Res. 6:133–138, 1994.

Helman N, Lawrence S. The effects of age, sex and smoking on erythrocytes and leukocytes. American Journal of Cardiology; 63: 35–44, 1975.

Jain S. K. Hyperglycemia can cause membranes lipid peroxidation and osmotic fragility in human red blood cells. Journal of Biochemistry; 264:21340–21345, 1989.

Lechner H, Ott E, Schmidt R: Present state of hemorheology. Gerontology, 33:259–264, 1987.

Misra H.P., Fridovich I. The generation of superoxide radical during the autoxidation of hemoglobin. Journal of Biological Chemistry; 247: 6960–6962, 1972.

Paul SD, O'Gara PT, Mahjoub ZA, DiSalvo TG, O'Donnell CJ, Newell JB, Levy GV, Smith AJC, Kondo NI, Cararach M, Ferrer L, Eagle KA. Geriatric patients with acute myocardial infarction: Cardiac risk profiles, presentation, thrombolysis, coronary interventions and prognosis. American Heart Journal. 131: 710–715, 1996.

Pierrefiche G, Laborit H. Oxygen free radicals, melatonin and, aging. Experimental Gerontology; 30: 213–227, 1995.

Rifkind J.M., Zhang L., Heim J.M., Levy A. The role of hemoglobin in generating oxyradicals. Oxygen Radicals in Biology and Medicine, 157–162, 1989.

Rifkind J. M., Zhang L., Levy A., Manoharan P.T. The hypoxic stress on erythrocytes associated with superoxide formation. Free Radical research Communication.12–13:645–652, 1991.

Rifkind J. M., Abugo O., Levy A., Monticone R., Heim J. Formation of free radicals under hypoxia. In : Surviving Hypoxia : Mechanism of control and adaptation (ed: Hochachka PW, Lutz PL, Sick T, Rosenthal M, Thillart G. Van den) CRC Press. 1992:510–521.

Rifkind J.M., Abugo O. Alteration in erythrocyte deformability under hypoxia : implication for impaired oxygen transport. Oxygen Transport to Tissue XVI, 345–351, 1994.

Rose H.G, Okleander M. Improved procedure for the extraction of lipids from human erythrocytes. Journal of Lipid Research; 6: 428–431, 1965.

Sharp DS, Abbott RD, Burchifel CM, Rodriguez BL, Tracy RP, Yano K, Curb DJ. Plasma fibrinogen and coronary heart disease in elderly Japanese-American men. Aretrioscler Thromb Vasc Biol. 16: 262–268, 1996.

Shirakura T, Kubota K, Tamura K. Blood viscosity and cerebral blood flow. Jpn J Geriat. 30: 174–181, 1993.

Snyder L.M., Fortier N.L., Trainor J., Jacobs J., Leb L., Dubin D., Shohet S., Moandas N. Effect of hydrogen peroxide on erythrocyte membrane cross linking. Journal of Clinical Investigation; 76:1971–1977, 1985.

Torre la de J., Mussivand T: Can disturbed brain microcirculation cause Alzheimer's disease?. Neruol Res. 15:146–153, 1993.

MEASUREMENTS OF OXYGEN SATURATION IN NORMAL HUMAN SURAL NERVE BY MICROLIGHTGUIDE SPECTROPHOTOMETRY AND ENDOSCOPY

N. D. Harris,[1] S. Ibrahim,[2] M. Radatz,[3] and J. D. Ward[2]

[1]Department of Medical Physics
Royal Hallamshire Hospital
Sheffield, United Kingdom
[2]Diabetes Research Unit
Royal Hallamshire Hospital
Sheffield, United Kingdom
[3]Department of Neurosurgery
Royal Hallamshire Hospital
Sheffield, United Kingdom

Micro-lightguide spectrophotometry is a new technique which allows continuous determination of the intracapillary haemoglobin oxygen saturation (HbO2%) from the remission spectra of small volumes of tissue. We have used the Erlangen Micro-lightguide Spectrophotometer (EMPHO II) to measure the HbO_2% in human sural nerve. The measurements are made 0.5–1 mm from the surface of the nerve through a small 10mm inscision under local anaesthetic. The lightguide, which consists of a single 250 µm illuminating and six receiving fibres, can be attached to the eyepiece of a small (1.9 mm diameter) endoscope via a beam splitter so the exact site of measurement can be visualised.

One hundred HbO_2% measurements (frames) were taken at each of 3 sites directly with the fibre optic lightguide and then through the endoscope. There was a reduction in the amplitude of the remission spectra through the endoscope and an increase in the range of HbO_2% measurements. Nine healthy control subjects were studied, the mean (SD) HbO_2 was 73.8 (3.8)% with the fibre and 76.7 (5.9)% with the endoscope. Nerve capillary blood flow was assessed using spectrophotometric measurements of nerve fluorescence following IV injection of sodium fluorescein. There was a significant negative correlation between the nerve intravascular oxygen and the fluorescein rise time suggesting that nerve oxygen levels are closely related to the local nerve blood flow. There were no correlations between any of the measurements and patient age, haemoglobin level or tissue temperature. This is the first time that MLS has been used to measure oxygen saturation and blood flow in human sural nerve.

Oxygen Transport to Tissue XIX, edited by Harrison and Delpy
Plenum Press, New York, 1997

1. INTRODUCTION

Micro-lightguide spectrophotometry is a technique pioneered by Kessler [1991] and Frank [1991]. The technique uses a scanning spectrophotometer to analyse the spectra of light reflected back (remission) from the surface of body tissues [Frank 1991]. Light absorption in the tissues is predominantly due to oxygenated and de-oxygenated haemoglobin. Thus the intracapillary haemoglobin oxygen saturation can be calculated by analysing the spectral composition of the remitted light. The instrument covers a range of wavelengths and can therefore be used as a fluorimeter to monitor the concentration of contrast agents as a means of determining local tissue perfusion. The aim of the study was to develop the technique for the measurement of intravscular oxygen saturation and blood flow in human peripheral nerve and to establish a range of normal values in control subjects.

Our group has previously made measurements of sural nerve oxygen and blood flow [Newrick 1986, Tesfaye 1993]. However, these are technically difficult and require exposure of the nerve through a relatively large 3 cm incision. There is therefore a need for more robust minimally invasive procedures which will allow routine clinical measurements of nerve blood flow and oxygenation in patients with neuropathy. In this study measurements of intravascular oxygen and blood flow have been made through a small 1cm incision using micro-lightguide spectrophotometry. The micro-endoscopic technique was previously used by Hoper for measurements in the rabbit eye [Hoper 1994] and has the advantage that the measurement site can be directly visualised.

2. METHODS

2.1. Equipment

The Erlangen micro-lightguide spectrophotometer II (EMPHO II) was used in this study (Bodenseewerk Geratetechnick GmbH, Germany). Light from a 75-W Xenon lamp is carried by a single fibre (diameter 250 μm) to the surface of the tissue. The reflected light is transfered by six fibres surrounding the transmitting fibre to a rotating interference filter disk. The light is monochromatized in a range between 502–628 nm, detected by a photomultiplier and the electrical signal digitised for processing by computer. The shape of the measured haemoglobin spectra is compared to reference haemoglobin spectra and the haemoglobin oxygen saturation calculated by an on-line program as described by Frank [Frank et al. 1991].

A rigid endoscope with an outer diameter of 1.9 mm (Storz, Germany, Type 28301B) was attached via a 80/20 beam splitter (Storz, Type R5199) and f30 mm lens (Type 43830), to a monochrome video camera (Hitachi KP-M1). The light guide was held in the central focal plane of the beam splitter eyepiece by means of a purpose built adaptor so that the EMPHO receives light from the central third of the endoscope field of view. For these endoscopic measurements a separate 250W Halogen light source (Storz Cold Light Fountain 485 BF) was used at the lowest power setting, to illuminate the tissue. At the start of each study a 'white balance' calibration with a surface coated mirror was used to correct for differences in the spectral composition of the halogen light source.

The EMPHO is able to collect approximately 70 spectra/second, although the signal to noise ratio is improved by averaging a number of spectra. Using the light-guide 5 spectra were averaged for each measurement (or frame). However, the combination of the lightguide and the endoscope causes a 4-fold reduction intensity due to light loss through the lenses. To compensate for this, the photomultiplier operating voltage was increased (HV=2) and the number of spectra averaged per frame was increased from 5 to 25 to improve the signal to noise ratio. During the procedure temperature measurements were made by placing a needle thermocouple (C-P Instruments, Herts, UK) into the tissue alongside the nerve.

Nerve fluorescence was measured using 10 spectra averaged at a rate of 4.8 frames/sec with the EMPHO (HV setting 3) and endoscope combination. Fluorescein exitation was illicited using the Storz light source (maximum intensity setting) fitted with a blue barrier filter. Nerve fluorescence was quantified directly from the uncorrected spectra by subtracting the baseline intensity values and integrating between 500 and 560 nm. The change in fluorescence intensity is biphasic with an initial 'fast phase' as fluorescein enters the nerve microvasculature, followed by a second 'slow phase' as the dye diffuses into the endoneurium. We used a manual least squares curve stripping technique [Franklin 1973] to remove the slow phase component and then calculated the time to first apperance of fluorescense (fluorescein apperance time - FAT) and the time between 10 and 90% maximum intensity (fluorecein rise time - FRT) of the first component as shown in figure 2.

2.2. Patient Studies

All invasive procedures were approved by the local Medical Ethics Committee and were performed by an experienced neurosurgeon in the hospital day case surgery unit. Measurements were carried out on 6 male and 3 female subjects, there was one smoker, no patients had a history of respiratory or cardiovascular disease. After 2% lignocaine skin infiltration, a small 10 mm incision was made above the right lateral maleolus and using blunt dissection, the surrounding tissues were parted and the sural nerve exposed. The exact position of the nerve had previously been identified and marked using transcutaneous nerve stimulation with small bipolar electrodes.

One hundred individual $HbO_2\%$ measurements were taken at each of 3 sites directly with the fibre optic light-guide and then through the endoscope as previously described. After each series of measurements the tip of the light-guide and exposed nerve were washed with warm saline and dried. Measurements were taken by placing the fibre / endoscope (held by a micro-manipulator) approximately 0.5 mm away from the surface of the nerve.

Nerve fluorescence was measured over the central portion of the exposed nerve. The tip of the endoscope was placed 1mm from the surface and the patient given 3.5 mg/kg Sodium Fluorecein (Martindale Pharmacetical, UK) as a bolus intravenously into the antecubital vein. Fluorescence within the nerve was measured for 5 minutes, during this time the position of the endoscope over the nerve was monitored via the camera and recorded on Sony U-matic video recorder. Cases in which there were significant changes in the position of the nerve under the endoscope were excluded.

3. RESULTS

The measurement of intravascular oxygen saturation using the light-guide and endoscope are summarised in Table 1 together with the values from the fluorescein blood flow measurements.

Nerve intravascular oxygen saturation levels are relatively high, typically between 70 and 80%. There is relatively small variation in the values during each measurement pe-

Table 1. Sural nerve intravascular haemoglobin saturation (HbO$_2$) and blood flow

Subject	Age (years)	Temp. (°C)	HbO$_2$(%) ±SD bare fibre	HbO$_2$(%) ±SD endoscope	FAT (seconds)	RT (seconds)
AT (F)	45	32.1	70.9 ± 14.7	79.5 ± 5.7	38	10.9
TP (F)	52	30.1	76.9 ± 4.7	71.8 ± 4.2	28	21.4
KP (S)	51	31.3	74.1 ± 2.8	68.7 ± 6.7	49	25.5
JV	67	29.3	68.4 ± 1.4	76.8 ± 4.1	—	—
MB(F)	50	31.4	76.7 ± 3.8	87.7 ± 6.6	—	—
DC	42	30.1	79.7 ± 7.8	76.4 ± 5.8	34	8.6
JD	56	33.3	70.1 ± 3.6	79.9 ± 4.5	22	6.9
JS	44	—	71.5 ± 2.7	72.4 ± 3.6	—	—
LN	50	30.2	75.8 ± 2.3	—	35	10.9
Mean ± SD	50.7 ± 7.5	30.9 ± 1.3	73.8 ± 3.76	76.7 ± 5.9	36.1 ± 8.1	14.03 ± 7.66

Key: F - Female S - Smoker

riod. Figure 1 shows a histogram depicting the range of values measured with the fibre and endoscope over two regions of the nerve in the same subject. Using the combination of the light-guide and endoscope there is an increase in the variability of the SaO2 measurements resulting in approximately a twofold increase in the standard. In the group as a whole there was no significant difference in the oxygen levels measured using the light-guide and the endoscope 73.8 (3.76)% and 76.7 (5.92)%, mean (SD), respectively.

Due to technical difficulties, fluorescein blood flow measurements could only be made in 6 subjects. This was primarily due to large fluctuations in the fluorescense signal caused by patient movements. An example of the change in intensity with time is shown in figure 2. There was a significant negative correlation between the nerve intravascular oxygen measured by the endoscope and the FRT (2-tailed Spearman r=-.9 p<.04 n=5). These data suggest that nerve oxygen levels are related to the local nerve blood flow. There were no correlations between any of the measurements and patient age, haemoglobin level or tissue temperature.

4. DISCUSSION

In common with other neural tissues, peripheral nerve has a high metabolic rate and well developed vascular supply which feeds into the nerve at regular intervals. These ves-

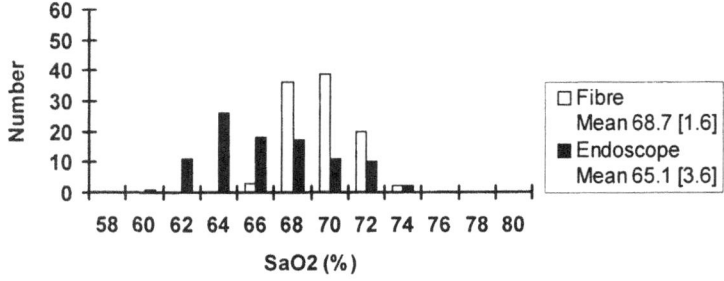

Figure 1. Histograms showing the oxygen saturation values measured in proximal regions of the sural nerve with the fibre optic light-guide and the light-guide plus endoscope.

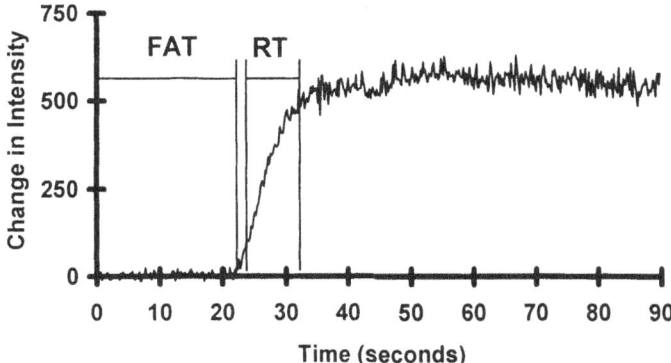

Figure 2. The change in fluorescence intensisty (500–560nm) in the sural nerve following injection of IV Sodium fluorescein showing the measurement of the fluorescein appearance time (FAT) and 10 to 90% rise time (RT).

sels run along the outer surface (perineurium) with numerous small branches into the body of the nerve (endoneurium) and go on into the individual fascicles. This high degree of vascularisation is reflected in the high intravascular oxygen saturation levels. Similar spectrophotometric measurements on the surface of the retina in the rabbit gave values between 57 and 92% [Hoper 1994], measurements on the outer surface of the brain were lower with values between 39 and 56% [Hoper and Gaab 1994]. Physiologically the nerve may be closer to the retina than the brain since nerve vascular supply has limited autoregulatory capacity [Low 1984].

The technique assumes that in the microcirculation there is an equilibrium between mixed arteriolar and venous intravascular oxygen and nerve metabolic activity. Therefore intravascular oxygen saturation reflects the level of nerve blood flow and the rate of oxygen extraction by tissues. The only other comparable human nerve oxygen measurements were carried out on non-neuropathic diabetic subjects using polargraphic microelectrodes [Newrick 1986]. Nerve oxygen levels were between 6.1 and 7.1 kPa which would be equivalent to a saturation of around 80%. The correspondence between nerve microvascular oxygen and nerve PO_2 is good despite differences in the methodology and possible measurement depth. The microelectrode measurements were made by stripping away the epineurium and inserting the needle into the perineurium of a superficial fascicle. The spectrophotometer measures back scattered light from vessels within the first 500 μm, this will include the perineurium and fascicles around the edge of the nerve.

Overall there was no significant difference in the oxygen levels measured using the light-guide fibre and the endoscope. The principal advantages of the endoscope are the smaller measurement area and the ability to visualise directly the area of measurement. The fibre collects light from the central third of the endoscope field of view, corresponding to an area of around 0.3 mm diameter, compared with an area of 2 mm diameter. The difference between the two measurements is generally around 5%, although it can be as high as 10%. However, this is similar to the variability seen along the length of the exposed nerve and may be attributed to differences in the exact placement of fibre and endoscope over each regions.

The fluorescence measurements provide a means of assessing nerve blood flow. We have previously carried out similar measurements using video microscopy with digital image analysis of intensity [Tesfaye 1993]. Due to the low light levels involved it would be difficult to carry out this type of analysis through the endoscope. The spectrophotometer provides a more sensitive method and gives a more precise measurement. The appearance time is a measure of the transit time of the fluorescein in the large vessels and may be influenced by factors such as cardiac output or peripheral vascular resistance. The rise time has previously been used to derive an absolute measure of blood flow in intestinal capillaries [Purbeck 1985]. However, this requires precise knowledge of dose and detector sensitivity and so in this study we have used the rise time as a relative measure. In spite of the small number of subjects there was a significant correlation between the rise time and nerve intravascular oxygen level indicating that under these resting conditions nerve oxygen levels are closely related to the local nerve blood flow. Clearly further studies are required to validate the nature of this relationship.

In summary, the combination of microlight-guide spectrophotometry and micro-endoscopy provides a valuable technique for clinical investigation of microvascular oxygen and blood flow. Although a skin incision is required to gain access, this is a minor surgical procedure which is readily carried out under local anaesthetic. The measurement does not require insertion of needles into the tissue of interest and is therefore minimally invasive, offering the possibility of repeated measurements. We are currently using the technique to continue our investigations into pathogenisis and treatment of diabetic peripheral neuropathy.

ACKNOWLEDGMENTS

This work is supported by a British Diabetic Association Project Grant. The support of Scotia Pharmaceuticals Ltd for provision of the EMPHO spectrophotometer.

We are very grateful to all the volunteers who took part in the study and to staff of the Day Care Unit at the Royal Hallamshire Hospital.

REFERENCES

Frank KH et al (1991) The Erlangen micro-lightguide spectrophotometer 1 Physics in Med. and Biol. 34: 1883

Franklin DA and Newman GB (1973) Computational methods and curve fitting: A guide to medical mathematics Blackwell Scientific Publications Chapter 11, 323–342

Hoper J and Gaab MR (1994) Effect of arterial PCO_2 on local HbO_2 and relative Hb concentration in the human brain. Physiol Meas; 15: 107–113

Hoper J and Funk R (1994) A combination of microendoscopy and spectrophotometry allowing real-time analysis of microcirculatory parameters during in vivo observations. Physiol. Meas.; 15:333–337.

Kessler M, Frank K, Hoper J, Tauschek D and Zundorf J (1991) Reflection Spectrophotometry In 'Oxygen Transport in Tissue XIV' Ed. Erdmann W.

Low PA and Tuck RR (1984) Effect of changes in blood pressure, respiratory acidosis and hypoxia on blood flow in the sciatic nerve of the rat. J. Physiol. 347, 513–524

Newrick PG, Wilson AJ, Jakubowski J, Boulton AJM, Ward JD (1986) Sural nerve oxygen tension in diabetes B.M.J. 293: 1053–54

Purbeck L, Lund F, Svensson L, Thulin L (1985) Fluorescein flowmetry: a method for measuring relative capillary blood flow in the intestine. Clin. Physiol. 5: 281–292

Tesfaye S, Harris ND, Jakobowiki S, Mody C, Ward JD (1993) Impaired blood flow and arterio-venous shunting in human diabetic neuropathy. Diabetologia 36,1266–1274.

4

MICROVASCULAR BLOOD FLOW AND OXYGEN SUPPLY IN ULCERATED SKIN OF THE LOWER LIMB

D. J. Newton,[1] D. K. Harrison,[1] G. B. Hanna,[2] C. J. A. Thompson,[3] J. J. F. Belch,[3] and P. T. McCollum[2]

Vascular Laboratory
[1]Department of Medical Physics
[2]Department of Surgery
[3]Department of Medicine
Ninewells Hospital, Dundee, Scotland

1. INTRODUCTION

Successful cutaneous wound healing requires an inflammatory response, the formation of new epithelial tissue and the remodelling of that tissue (Clark, 1993). After haemostasis has been achieved, inflammation is the necessary first step, in which local microvascular blood flow increases in order to aid the recruitment of leukocytes to the area. A sufficient rate of oxygen delivery is essential to supply these infiltrating cells and the newly forming granulation tissue, and hence ensure the healing process.

Ulceration develops when that process is impaired and tissue begins to break down, and the presence of peripheral vascular disease is a predisposing factor for these problems in the lower limb. The features and pathogenesis of leg ulcers are characterised by the nature of the disease and the involvement of macro- or microvascular abnormalities. Numerous hypotheses have been presented in explanation of their development, particularly in patients with venous insufficiency (Loosemoore & Dormandy, 1993) and diabetes mellitus (Flynn & Tooke, 1992), but the existence of hypoxia at the surface of the skin is a common finding. This suggests that oxygen delivery may in some way be impaired, either through insufficient perfusion or diffusion.

Microvascular abnormalities in chronic inflammation of the skin have been studied with a variety of non-invasive techniques, using the tuberculin reaction as a model (Harrison et al, 1992). A consistent finding has been lower measurements of transcutaneous oxygen tension at the surface of the inflamed skin, compared with normal skin, but in the presence of a high microvascular flow of blood which is well oxygenated. The main conclusion that has been drawn from this work is that the rate of oxygen extraction from the erythrocytes is reduced, which may be due to some impairment in its diffusion to the sur-

Oxygen Transport to Tissue XIX, edited by Harrison and Delpy
Plenum Press, New York, 1997

face of the skin, exacerbated by a reduced blood cell transit time which may limit the amount of oxygen off-loaded.

The aim of this pilot study was to investigate whether similar findings may be made in the inflamed skin around leg ulcers, which may illuminate the cause of hypoxia in this situation. The same measurement techniques of laser Doppler flowmetry, lightguide spectrophotometry and transcutaneous oximetry were used to assess microvascular oxygen transport in the skin. Ulcers of a variety of origins were studied, with the aim of concentrating on specific types in later work, subject to the results obtained here.

2. MATERIALS AND METHODS

2.1. Patient Groups

Twelve patients with leg ulcers were recruited to the study, which had been approved by the Tayside Committee on Medical Ethics. The ulcers were of varying aetiology and their positions on the lower limb were largely characteristic of the underlying disease. Three were secondary to peripheral arterial occlusive disease indicated by reduced peripheral pressures measured with Doppler ultrasound, and were found around the shin region of the leg. These patients all ultimately progressed to amputation.

Another three patients had undergone recent, successful bypass procedures but with no apparent, immediate effect on healing, possibly because of the development of microvascular disease, due to vasculitis in two cases and diabetes mellitus in the other. These ulcers were on the shin and the side of the foot.

Two patients with diabetes mellitus and associated peripheral neuropathy had ulcers on the soles of their feet, at the base of the metatarsal heads. The remaining four ulcers were on the gaiter region of the legs of patients with deep venous insufficiency, with decreased venous refilling times, as assessed with photoplethysmography. The patient group overall had a mean age of 61 years.

Twenty control subjects with no evidence of vascular disease were recruited from patients undergoing day-case surgery at the hospital and from members of staff. The mean age was 41 years.

2.2. Protocol

Each subject was rested supine for 15 minutes in a temperature-controlled laboratory at $25 \pm 1°C$. Microvascular blood flow and oxygenation were assessed non-invasively using the techniques of laser Doppler flowmetry, lightguide spectrophotometry and transcutaneous oximetry. In the patient group, steady state measurements of each parameter were made in the skin bordering the ulcer and on the dorsum of the same foot, in unaffected skin. In the control group, measurements were made on the medial aspect of the gaiter region of the leg, a comparable site with most of the ulcers studied here.

2.3. Laser Doppler Flowmetry

This is a well established technique for assessing microvascular blood flow in the skin, and other organs. Light from a 2 mW helium-neon laser is directed to the skin through optical fibres, and that which is backscattered from the tissue is picked up by similar fibres and passed to a photodetector. The resulting signal is processed to obtain a value of laser Doppler flux (LDF), which is a function of the Doppler shifts in frequency of the backscattered light, related to the velocities of the red blood cells in the catchment volume of the instrument (estimated to be of the order of 1 mm^3). LDF is the product of the average velocity and total concentration of moving blood cells. It is a relative measure and generally expressed in arbitrary units, in this case volts. The instrument used for this work was the PF2b flowmeter (Perimed, Stockholm, Sweden) with the following settings: gain = 10, time constant = 0.2 s and bandwidth = 12 kHz. A standard probe, with 2.4 mm diameter fibre, was attached to the measurement site with an adhesive probe holder. The signal was monitored until it had stabilised, and then recorded for two minutes, from which the average flux was calculated.

2.4. Lightguide Spectrophotometry

This method makes use of the characteristic light absorption properties of haemoglobin in the visible wavelength range. The absorption spectrum of light remitted from a tissue sample varies with haemoglobin concentration and saturation (SO$_2$); other chromophores have less effect. A typical spectrophotometer for this application consists of a light source with optical fibres for light transmission, some method of spectral filtering and photodetection to pick up the remitted light, and computer software to analyse recorded spectra and calculate values of SO$_2$.

Two systems were used in this work, each with different catchment volumes. The Photal MCPD-1000 (Otsuka Electronics, Osaka, Japan) uses 18 quartz fibres of internal diameter 200μm to transmit light to the skin and 12 similar fibres to pick up the remitted light. Analysis was performed using an algorithm developed in this centre (Harrison et al, 1992) which uses the deoxygenated peak and isosbestic points of each spectrum to calculate SO$_2$. The size and spacing of the optical fibres mean that the catchment volume of the instrument in this configuration is estimated to be similar to that for laser Doppler flowmetry (that is, 1 mm^3). The average of three measurements at each site was taken.

The EMPHO II (BGT, Nuremburg, Germany) was the other spectrophotometer used here (Frank et al, 1989), and has optical fibres of 250μm diameter; one transmitting and six receiving. The penetration depth is therefore much less than the MCPD and most of its signal is estimated to be from the capillary layer of vasculature in the skin. Measurements are made by scanning the lightguide over the area of interest, so building up a picture of the distribution of SO$_2$ of which, in this case, the mean was taken. For this study, 50 measurements were made at each site; about 1 per second. Analysis is performed in real time by fitting standard spectra to the measured spectra. The two methods of analysis employed by these systems have been found to be comparable in physiological ranges (Newton et al, 1994).

2.5. Transcutaneous Oximetry

The TCM3 oximeter (Radiometer, Copenhagen, Denmark) was used to obtain measurements of transcutaneous oxygen tension (tcpO$_2$) in the skin. A heater set at 44°C dilates the skin microcirculation and aids oxygen diffusion to the head of the electrode on the surface of the skin, where it is reduced at the cathode to produce a current proportional to the partial pressure of oxygen at the sensor. In normal human skin this value approximates arterial pO$_2$ in the skin. A measurement was read from the digital display of the instrument after the value had stabilised; about 20 minutes.

2.6. Statistical Analysis

Paired-sample t-tests were used to analyse differences between measured parameters in the skin bordering the ulcers and on the dorsum of the foot of the same subject. Independent-sample t-tests were employed to look at the differences between the patient and control groups. All parameters were normally distributed. Practical limitations meant that measurements of $tcpO_2$ were only made in five of the patients under study. Statistical analysis of these results was therefore not considered appropriate.

3. RESULTS

A summary of the results is presented in Table 1. Measurements of LDF were found to be significantly higher in skin bordering ulcers than in unaffected skin on the dorsum of the foot ($p < 0.05$) and in normal limbs ($p < 0.01$). Similarly SO_2, measured using the MCPD system tended to be higher around ulcers than in dorsal skin (although not statistically significant), and was significantly higher than in the control group ($p < 0.05$). SO_2, measured with the EMPHO instrument was also higher around the ulcers than in dorsal ($p < 0.05$) or control skin ($p < 0.01$).

By contrast, measurements of $tcpO_2$ tended to be lower in skin bordering ulceration than in dorsal or control skin.

4. CONCLUSIONS

Transcutaneous oximetry has proved to be a valuable technique in the assessment of various types of vascular disease. Low values of $tcpO_2$ are generally measured in regions of slow-healing ulceration (Pecoraro et al, 1991; Mani et al, 1986), and reflect the poor nutritional state of the tissue in these areas. This parameter, however, is influenced by a number of other factors, apart from blood oxygenation, in adult skin, including vascular responsiveness (Cheatle et al, 1991; Hanna et al, 1995) and, as suggested by our own work in inflammation (Harrison et al, 1992), oxygen diffusion. It seems, therefore, that the value of this technique lies partly in its sensitivity to abnormalities in any of these elements, each of which may affect tissue health.

Previous work, looking at microvascular oxygen supply in chronic inflammation, has employed lightguide spectrophotometry to measure values of SO_2 in the skin (Harrison et al, 1992; Newton et al, 1994). This technique has the advantage that it can be used to directly assess levels of oxygen in the blood, and the results of this work indicate high microvascular SO_2 in areas of high blood flow in inflammation. This is in contrast to low levels of $tcpO_2$ at the surface of the skin. This suggests that oxygen is not being extracted

Table 1. Summary of results (mean ± standard deviation)

	LDF (V)	SO_2 MCPD (%)	SO_2 EMPHO (%)	$tcpO_2$ (mmHg)
Ulcer	4.7 ± 2.7	58 ± 25	47 ± 7	22 ± 26
Dorsum	2.3 ± 1.5	41 ± 20	35 ± 17	47 ± 33
Control	1.2 ± 0.4	33 ± 17	31 ± 13	70 ± 12

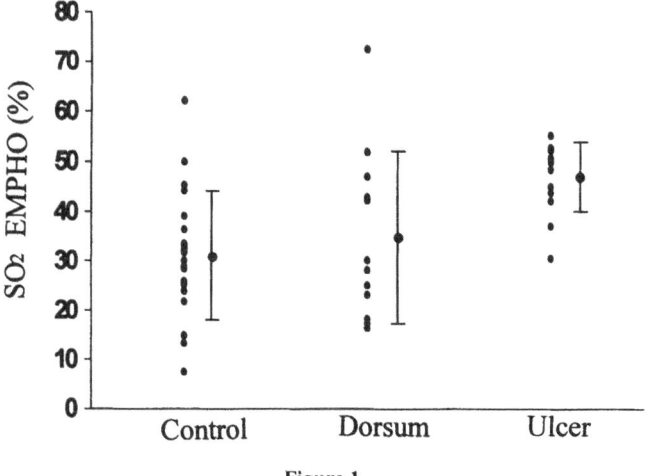

Figure 1.

at an adequate rate, and this may be due to some impairment in diffusion to the skin's surface. A critical transit time for offloading of oxygen has also been postulated.

The aim of this study was to investigate whether a similar situation might exist in the inflamed site of leg ulcers, and the preliminary results presented here confirm this. Microvascular blood flow is increased by a factor of four from normal skin, and is double that in unaffected skin of the same limb. This hyperaemia is consistent with the inflammatory response to injury. Figure 1 shows that this blood is adequately oxygenated, and in fact SO_2 is raised around the ulcer site. The EMPHO system interrogates mainly the capillary level of vasculature, and this parameter seems to be somewhat more sensitive than that measured with the MCPD instrument, which is looking at a larger volume of tissue (Newton et al, 1994). Measurements of $tcpO_2$ on the whole tended to be lower around the ulcers, and in some cases were very low indeed, as shown in Figure 2. Although only a few values were possible, the results here are generally quite clear, and this finding has been verified in other work (Pecoraro et al, 1991; Mani et al, 1986).

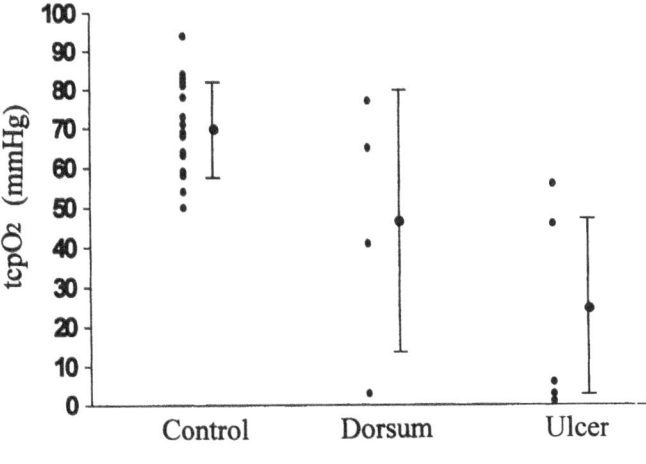

Figure 2.

These results, therefore, suggest that capillary oxygen supply to the skin does appear to be sufficient, but it is not being extracted from the haemoglobin because of poor diffusion conditions to the surface of the skin. This results in low tcpO$_2$ values and poor tissue oxygenation which impairs the healing process. The existence of oedema alone does not seem to affect oxygen transport (Michel, 1990), and so some additional factor may have some influence. The infiltrating cells may present some barrier, the diffusion distances will be longer, and the effect of fibrin cuffs around the capillaries has been suggested in some cases (Loosemoore & Dormandy, 1993). This may have some significance when considering therapy, especially hyperbaric oxygen treatment (Harrison et al, 1994). There is also evidence that reduced red cell transit times may also have some effect on the poor extraction of oxygen (Newton et al, 1996).

Lightguide spectrophotometry has great value in the investigation of tissue oxygen supply because of its direct nature and ease of use. Of particular interest may be the distribution of SO$_2$ over a site. More detailed investigations, differentiating between various types of ulcers and also looking at the underlying disease, are intended to follow up this initial work.

ACKNOWLEDGMENTS

This work was supported by a grant awarded by the Medical Research Council, short project no. G9327691.

REFERENCES

Cheatle TR, Stibe ECL, Shami SK, Coleridge Smith PD. Vasodilatory capacity of the skin in venous disease and its relationship to transcutaneous oxygen tension. *Br J Surg* **78**: 607–10, 1991.

Clark RAF. Mechanisms of cutaneous wound repair. In: Westerhof W, ed. *Leg ulcers - diagnosis and treatment* pp 29–50. Elsevier, Amsterdam, 1993

Flynn MD, Tooke JE. Aetiology of diabetic foot ulceration: a rôle for the microcirculation? *Diabetic Med* **8**: 320–9, 1992.

Frank KH, Kessler M, Appelbaum K, Dümmler W. The Erlangen micro-lightguide spectrophotometer (EMPHO I). *Phys Med Biol* **34**: 1883–1900, 1989.

Hanna GB, Newton DJ, Harrison DK, Belch JJF, McCollum PT. Use of lightguide spectrophotometry to quantify skin oxygenation in a variable model of venous hypertension. *Br J Surg* **82**: 1352–6., 1995.

Harrison DK, Evans SD, Abbott NC, Beck JS, McCollum PT. Spectrophotometric measurements of haemoglobin saturation and concentration in skin during the tuberculin reaction in normal human subjects. *Clin Phys Physiol Meas* **13**: 349–63, 1992.

Harrison DK, Abbott NC, Carnochan FMT, Beck JS, James PB, McCollum PT. Protective regulation of oxygen uptake as a result of reduced oxygen extraction during chronic inflammation. *Adv Exp Med Biol* **245**: 789–96, 1994.

Loosemoore TM, Dormandy JA. Pathophysiology of venous ulceration. *Vasc Med Rev* **4**: 49–57, 1993.

Mani R, Gorman FW, White JE. Transcutaneous measurements of oxygen tension at edges of leg ulcers: preliminary communication. *J R Soc Med* **79**: 650–4, 1986.

Michel CG. Diffusion in oedematous tissue and through pericapillary cuffs. *Phlebology* **5**: 223–30, 1990.

Newton DJ, Harrison DK, Delaney CJ, Beck JS, McCollum PT. Comparison of macro- and micro-lightguide spectrophotometric measurements of microvascular haemoglobin saturation in the tuberculin reaction in normal human skin. *Physiol Meas* **15**: 115–28, 1994

Newton DJ, Harrison DK, McCollum PT. Oxygen extraction rates in inflamed human skin using the tuberculin reaction as a model. *Int J Microcirc* **16**: 118–23, 1996.

Pecoraro RE, Ahroni JH, Boyko EJ, Stensel VL. Chronology and determinants of tissue repair in diabetic lower-extremity ulcers. *Diabetes* **40**: 1305–13, 1991.

5

SKIN OXYGENATION FOLLOWING GROIN INCISION FOR FEMORO-POPLITEAL BYPASS SURGERY

Z. Raza,[1] D. J. Newton,[2] D. K. Harrison,[2] P. T. McCollum,[1] and P. A. Stonebridge

Vascular Laboratory
[1]Department of Surgery
[2]Department of Medical Physics
Ninewells Hospital and Medical School
Dundee, Scotland

1. INTRODUCTION

In all forms of infra-inguinal surgery, the femoral artery is commonly exposed with a longitudinal incision made directly over the vessel (Figure 1). A prosthetic or vein graft acts as the conduit for revascularisation of the affected leg in a bypass procedure. However, wound infection in this area is a fairly common finding in the post-operative period. Up to 15% of all groin wounds become infected, ranging from a straight forward superficial cellulitis to a suppurative infection with wound dehiscence, often necessitating removal of the graft (Newington et al, 1991). In contrast, more distal wounds at the popliteal or tibial vessels have a much lower incidence of infection.

The skin of the groin is likely to be supplied mainly from the branches of the femoral artery. The superficial circumflex artery has probably the majority of blood supply to the groin as this is the pedicle upon which a free groin flap is raised during plastic surgical procedures (McGregor and Jackson, 1972). The groin area upon which a flap is raised has its pedicle situated at its lateral side. This means that a large proportion of the blood flow to the medial aspect of the flap is supplied from this lateral pedicle.

Following a longitudinal incision for femoral artery exposure, dermal blood flow may be disrupted to the extent that a gradient develops between the medial and lateral sides of the wound. Should this be the case, the medial and lateral aspects of the wound will show notable differences in blood flow post-operatively.

A study was performed to establish if there was a difference in skin oxygenation between the medial and lateral sides of the wound and, if so, in which direction this exists. Finally we were also interested to see whether a groin wound that went on to become infected had any special distinguishing features in terms of its skin oxygenation, compared to those wounds that healed.

Oxygen Transport to Tissue XIX, edited by Harrison and Delpy
Plenum Press, New York, 1997

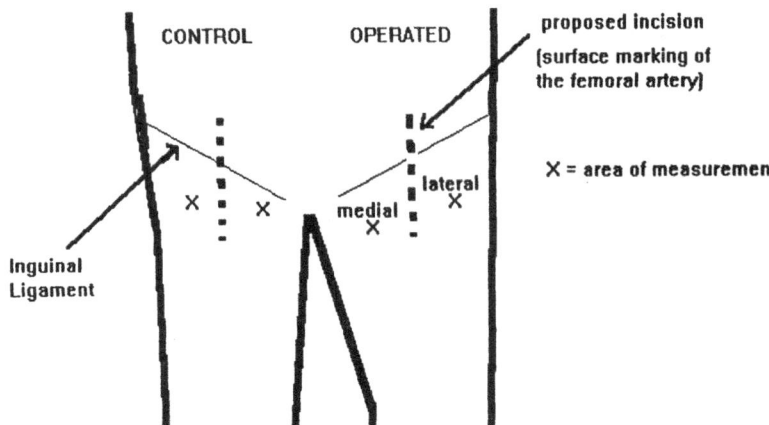

Figure 1. A diagrammatic representation of the surface marking of the femoral artery where measurements are taken either side of the proposed incision.

2. MATERIALS AND METHODS

2.1. Patient Groups

Following approval by the Tayside Ethical Committee, 15 consecutive patients were invited to take part in the study. All patients were elective admissions, scheduled to undergo a femoro-popliteal or femoro-distal bypass procedure. There were 9 males and 6 females. All patients had clinically proven peripheral vascular disease with Doppler or toe pressures which confirmed this. Thirteen patients were operated on because of disabling claudication (Fontaine IIb) and the other two underwent surgery for rest pain with critical limb ischaemia. There were 4 patients who were known diabetics. The mean age was 69.2 ± 10 years and all patients had a history of tobacco intake with 5 patients who were active smokers.

2.2. Protocol

Each patient was allowed to rest supine for 10 minutes in a thermostatically controlled room (21 ± 2 °C). Measurements of haemoglobin saturation were made using the Erlangen Micro-lightguide Spectrophotometer (EMPHO II, BGT GmbH, Germany). Baseline measurements were made pre-operatively. Fifty spectra were recorded from the skin on each side of the proposed groin incision, moving the lightguide from the mid-in-

Table 1. SO_2 (mean ± standard deviation) in the control and operated groins pre-op and at 2 and 7 days post-op

	Operated groin		Control groin	
	Medial	Lateral	Medial	Lateral
Pre-op	24.1 ± 10.6	22.2 ± 11.9	26.3 ± 11.9	23.8 ± 12.1
2 days	33.4 ± 11.7	47.1 ± 12.0	17.5 ± 8.3	15.0 ± 7.8
7 days	25.2 ± 14.2	32.8 ± 17.6	14.7 ± 10.3	11.3 ± 9.8

guinal point (surface marking of the common femoral artery) to a point 10 centimetres lateral, and similarly on the medial side. The mean SO_2 was calculated from each set of recordings. This was repeated on the contra-lateral side for a control.

These measurements were repeated 2 and 7 days post-operation, on either side of the subsequent wound and in the control groin.

3. RESULTS

Table 1 shows the mean SO_2 recorded on the medial and lateral aspects of the groin on the control and operated lower limbs, pre-operatively and at 2 and 7 days post-operation. All the statistics used were analysis of variance for repeated measures.

Post-operatively, groin SO_2 changed from baseline values ($p < 0.01$), but this change was different in the control and operated limbs ($p < 0.001$). In the non-operated (control) groin, SO_2 decreased at 2 days post-op and had not returned to its normal value 1 week following surgery. By contrast SO_2 in the operated groin increased at day 2 and was returning back to normal at day 7.

The increase in SO_2 in the operated groin was significantly greater on the lateral aspect of the wound compared with the medial side (Figure 2) and this was highly significant ($p < 0.001$) and persisted even at 7 days. There was no such difference between the medial and lateral aspects of the control groin.

4. DISCUSSION

There has been very little work to establish possible contributing factors relating to groin wound infection. Oxygen supply to the tissue of a wound is by far one of the most important determinants of successful wound healing (Forrester 1988). Obviously previous surgery, contamination of the operative site, handling of the tissue and the nutritional state of the patient are also important aspects to determine healing of wounds in general.

Following a longitudinal incision to expose the common femoral artery, a blood flow gradient develops across the wound. This gradient shows that there is a greater oxygenation on the lateral side of the wound compared with the medial side. This difference persists at 7 days, although to a lesser extent. All of our groin wounds healed and there-

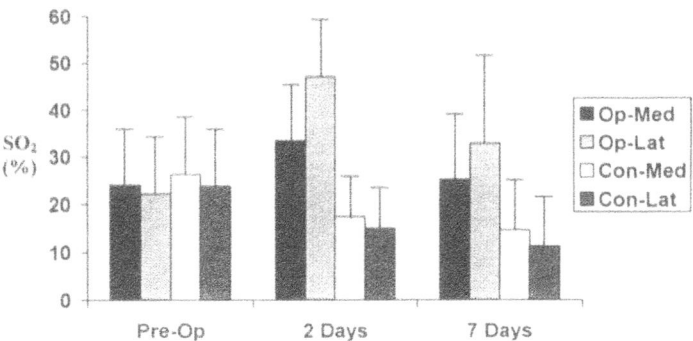

Figure 2. Groin SO_2 prior to and following femoro-popliteal bypass.

fore it is not possible to assess differences between patients who go on to heal and the ones who have wound problems. The fact that a gradient is developing following surgery may be a contributing factor to the increased incidence of groin wound infections compared with other sites.

It is interesting to note that the blood supply appears to flow from the lateral aspect of the groin in a medial direction. The higher oxygenation of the lateral side of the wound compared to the medial side in our study, following a groin incision would also support this.

The important message of note from our study is that a significant skin oxygenation gradient has developed across the line of incision, which can only be explained by a disruption of the blood supply. The relative oxygenation values are high because of the surrounding erythema from the acute inflammatory reaction following a surgical incision. Wound infection may occur when the gradient between either side of the wound is much greater than we have recorded. Clearly only a controlled prospective study will be able to confirm this hypothesis. A further study is warranted to establish what pattern of oxygenation occurs in groins of patients whose wounds become infected.

REFERENCES

Forrester JC. Wounds and their management. In: Cuschieri A, Giles GR and Moussa AR. (Editors). *Essential Surgical Practice*. 2nd Edition pp. 3–15. *Butterworth - Heinemann Ltd, UK. 1988.*

McGregor IA and Jackson IT. The groin flap. *Br J Plas Surg.* 1972; **25**: 3–16.

Newington DP, Houghton PWJ, Baird RN and Horrocks M. Groin wound infection after arterial surgery. *Br J Surg.* 1991; **78**: 617–619.

6

NEAR INFRARED TIME RESOLVED SPECTROSCOPY FOR THE DETECTION OF DEEP VEIN THROMBOSIS WITHIN THE HUMAN LEG

Jamie T. Barnett, Mark W. Hemelt, Duane F. Bruley, and Kyung A. Kang

Department of Chemical and Biochemical Engineering
College of Engineering
University of Maryland Baltimore County (UMBC)
5401 Wilkens Ave., ECS 101, Baltimore, Maryland 21228

1. INTRODUCTION

It is intended to locate deep vein thrombosis (DVT) inside the human leg utilizing near infrared (NIR) time resolved spectroscopy (TRS).

Deep vein thrombosis is one of the initial symptoms of a thrombo embolic disorder caused by the deficiency of anti-coagulants such as protein C and protein S, or by the activated protein C resistant symptom. During DVT, anoxia and hypoxia occur due to the blockage in the vein. It is dangerous and can lead to death if not detected and treated immediately. It is estimated that over 5 % of the total population may have this condition.

Currently, DVT in the leg is detected by the injection of radiolabeled fluid into the heel of the patient. Then, the radiolabeled fluid is traced to the clot using X-rays. This method is invasive, costly, and painful to the patient. Ultrasound is also used for DVT diagnosis, however, it has a limitation in penetration depth.

Our method of detection involves the use of near infrared light. Near infrared light includes wavelengths starting from the end of the visible light to the beginning of the infrared. At these wavelengths, the light does not effect the biological system as do X-rays. Once the method for detecting DVT by NIR-TRS is realized in real time, the patient will have a non-invasive, low cost, painless, portable alternative.

2. THEORY

Frequency response analysis theory has been implemented in identifying chemical systems for several decades.[1,3,4,10,11] Recently, this analysis has been used for NIR-TRS spectra to identify and characterize biological systems.[6,7,8] Time resolved spectroscopy (TRS) takes a series of single photons of near infrared (NIR) light and converts them to a pulse by way of a time to amplification converter (TAC), which emulates pulse testing.[9] From a single pulse, magnitude ratio (MR), phase shift (ϕ), steady state gain (K), time constant (τ), and system order (n) can be obtained by frequency response analysis.[6,7,8] The theory of Filon's quadrature is used by way of a data reduction program[2,5] to obtain the five engineering parameters.[2,5]

A system has a unique transfer function which identifies and characterizes it from other systems. The transfer function is the ratio of the Laplace transforms of the output, [Y(s)], and input, [X(s)].

$$G(s) = \frac{Y(s)}{X(s)} \tag{1}$$

where the 's' represents Laplace variables which are complex numbers. By substituting 'jω' for 's' the transfer function changes to an equation in the frequency domain, the Fourier Transform G(ω):

$$G(\omega) = \frac{Y(\omega)}{X(\omega)} = \frac{\int_0^{T_y} y(t)\, e^{-j\omega t}\, dt}{\int_0^{T_x} x(t)\, e^{-j\omega t}\, dt} \tag{2}$$

where j indicates an imaginary number and ω represents frequency. The variables T_x and T_y are the times when the input and output pulses become zero, respectively.

The ratio of the output magnitude to the input in the frequency domain is the magnitude ratio (MR). The output wave typically lags behind the input wave by a phase shift (ϕ), which is also a function of frequency ω. Using the real, Re(ω), and imaginary, Im(ω), parts of the transfer function, magnitude ratio (MR) and phase shift (ϕ) values can be calculated:

$$MR(\omega) = |G(\omega)| = \sqrt{\mathrm{Re}^2(\omega) + \mathrm{Im}^2(\omega)} \tag{3}$$

$$\phi(\omega) = \phi\big|_{G(\omega)} = \tan^{-1}\left[\frac{\mathrm{Im}(\omega)}{\mathrm{Re}(\omega)}\right] = \phi\big|_{Y(\omega)} - \phi\big|_{X(\omega)} \tag{4}$$

The MR, in decibels, and ϕ, in radians, are calculated over a range of frequencies (Eq. 3 and Eq. 4, respectively). Semilog Bode plots are constructed for MR and ϕ versus modulation frequency [Fig. 1].

In addition to MR and ϕ at various modulation frequencies, three additional parameters can be obtained from Bode plots. The steady state gain (K) is the MR value when the frequency (ω) is 0 and can be used to determine the system linearity. The time constant (τ) is the system response time and is the inverse of the break frequency (ω_b), which is the frequency at which the steady state gain and decay slope intersect. The slope of the magnitude ratio decay plot at higher frequencies is often referred to as the system order (n).

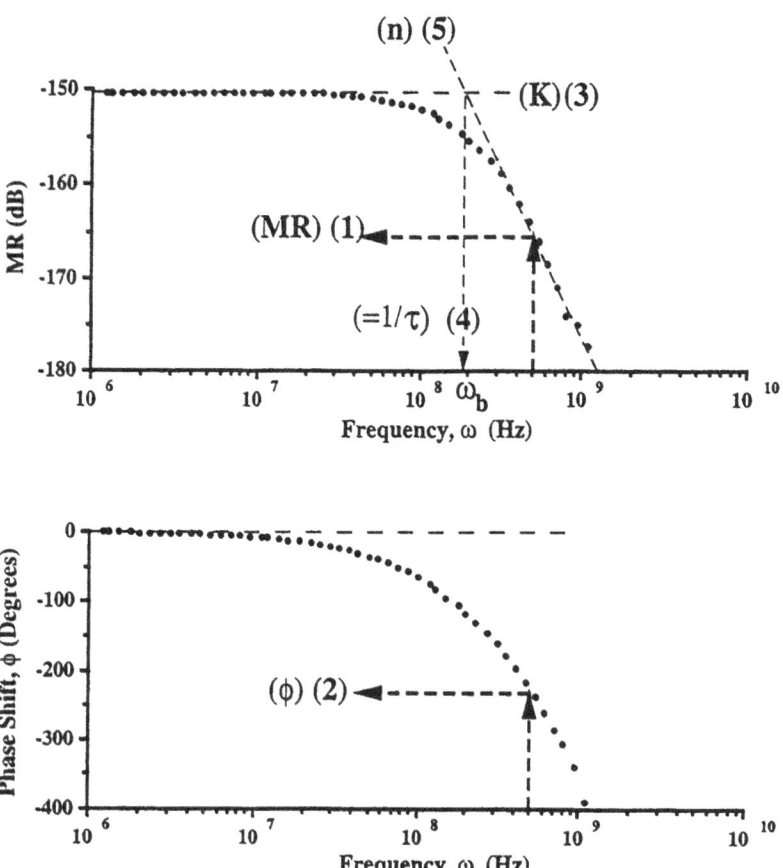

Figure 1. Frequency response analysis of a TRS spectrum. (1) Magnitude ratio (MR), (2) Phase shift (ϕ), (3) Steady state gain (K), (4) Time constant (τ), and (5) System order (n).

The normalized frequency content (NFC) is a useful indicator of how meaningful each data point is for a specific frequency. The values of the NFC for normal pulse functions range from 0.0 to 1.0. Frequencies where the NFC values are below 0.3 are considered to be inaccurate.

3. INSTRUMENTS AND MATERIALS

The following instruments and materials were used to perform the experiments:

1. Time resolved spectroscope (TRS): The equipment was manufactured by Hamamatsu Photonics, KK. (Hamamatsu, Japan). The wavelength used for experimentation was 780 nm. The TRS equipment was a single photon counting system which has a pulse repetition rate of 5 MHz. The detector system was a multi-microchannel plate photomultiplier tube (MCP-PMT).

2. Two optical fibers: For this experiment, two optical fibers with diameter 250 μm and 2.5 mm were used to deliver the source light and to collect the out signal, respectively.

3. Agar (Sigma Chemical, St. Louis, MO): Agar was used for the experimental model preparation.
4. Non fat dried milk powder (Giant Co., Landover, MD): The milk was used as the scatterer once it has been mixed with the agar.
5. Black absorber: Black rubber was cut to obtain a spherical absorber size of 1.0 cm in diameter.
6. India Ink (Eberhard Faber, Lewisburg, TN): The ink was used as an absorber for homogeneous and heterogeneous models.
7. Neutral density filter (Kodak Co., New York, NY): To attenuate the input light intensity, neutral density gelatin filters were used for the instrument function measurement, x(t). The filters were inserted between the source and the detector so that the input pulse to the system could be measured.
8. Plastic container (13 x 17 cm): The agar and milk mixture were poured into the container to produce the leg phantoms.
9. Bovine hemoglobin (Sigma Chemical, St. Louis, MO): The hemoglobin was used to simulate a thrombus.

4. METHODS

4.1. Instrument Linearity Test

To test the instrument linearity, multiple TRS spectra are taken at different input intensities, in the range of use for the experiments. The system is linear if the steady state gain is constant over the range of intensities. A TRS spectrum was taken with the source and detector positioned on a homogeneous scattering model. Then, the source and detector were removed from the surface of the model and an instrument function was taken as described in section 4.5. Then, the source and detector were placed at the same position on the surface of the gel, the intensity was increased and, again, the spectrum was recorded and the instrument function was taken at the increased intensity. This process was repeated for 5 different input intensities in the photon count experiments range of our experiments. The steady state gain (K) is obtained and plotted with respect to the intensity.

4.2. Homogeneous Model Study

Homogeneous experiments were completed using a 0.5 L (13x11x4.5 cm) rectangular container. These tests were designed so that trends of five parameters could be found for changes in the source-detector distance. TRS spectra for a source and detector separation of 3 cm, 4 cm, 5 cm, 6 cm, and 7 cm was taken in the middle of the phantom to avoid boundary effects. Phantoms were prepared with 80 g of milk powder and 200 μL of 10 % ink solution. These amounts were used so that the absorbtion (μ_a) and scattering (μ_s) coefficients of the model were in the range of human tissue. The phantoms were scanned to find trends in 4 of the 5 engineering parameters (MR, ϕ, τ, n) for different source-detector (S-D) distances.

4.3. Heterogeneous Model with a Spherical Absorber

To investigate the sensitivity of four parameters with a change in S-D separation in detecting an absorber, the area of interest (i.e. around the absorber) was scanned at S-D separa-

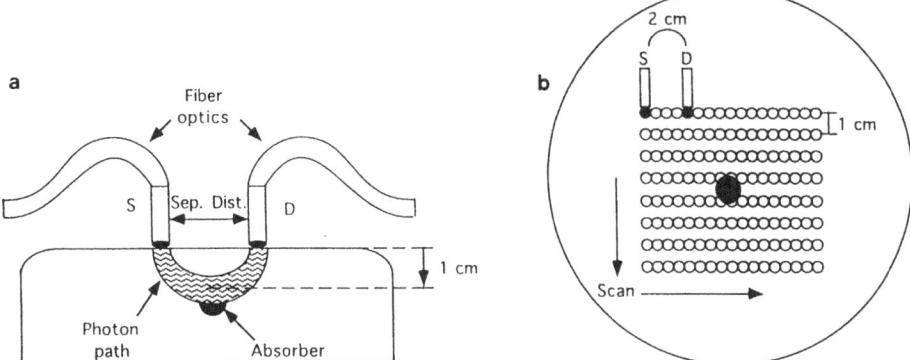

Figure 2. A heterogeneous model used to study the correlation of the S-D distance and the absorber location. (a) Side view of the phantom with a 1 cm black rubber absorber, 1 cm from the surface. (b) The top view of the phantom. The phantom was scanned in increments of 1 cm in x and 0.5 cm in y at a constant S-D separation.

tions of 2, 3, and 4 cm. Heterogeneous phantoms were made in a 3 L (height = 13 cm, diameter = 17 cm) cylindrical plastic container [Fig. 2]. The mixture of 18.6 g of agarose, 20 g of milk powder, and 1000 mL of water was heated then poured into the plastic container. A black sphere with a diameter of 1 cm was suspended 1 cm from the bottom of the container (from the surface of the sphere to the surface of the model). The rubber was attached to a thin metal wire so that the gel could be solidified around it. After a few hours, the solution was solidified and the wire was removed from the rubber. Then, the bottom of the model was scanned at the increment of 1 cm in the x-axis and 0.5 cm in the y-axis.

4.4. Leg Model with a DV Thrombus

A phantom was developed with a hemoglobin absorber to simulate a leg with a thrombus. To simulate a thrombus, 4.2 g of bovine hemoglobin was dissolved in 10 mL of water. The normal hemoglobin concentration in humans is 0.14 g of hemoglobin per 1 mL blood and the hemoglobin concentration used was 3 times the normal concentration. A transparent plastic straw (0.6 cm in diameter and 4 cm in length) was used to hold the hemoglobin solution. Using a hot glue gun, the straw was sealed at both ends with 1 mL of the hemoglobin solution in the center [Fig. 3 (a)]. Absorption coefficient (μ_a) and effective scattering (μ_s') values for each of the homogeneous models were compared with actual values obtained from human legs. After obtaining proper absorption and scattering coefficients of legs, 300 g of milk powder, 1200 μL of ink solution (10 % ink), 58.8 g of agar and 3 L of water were heated and mixed in a 3 L container. The simulated DVT clot was, then, suspended 1 cm from the outer perimeter of the container and the solution was refrigerated. Once solidified, the round side of the phantom was scanned with a source detector separation of 3 cm.

4.5. Instrument Function

The instrument function or input, x(t), was measured by taking a TRS spectrum when the source and detector fibers were facing each other. Since the TRS spectrum intensity was too high, neutral density gelatin filters (with a known N.D. number) were placed between the source and detector to reduce the photon intensity to the MCP-PMT.

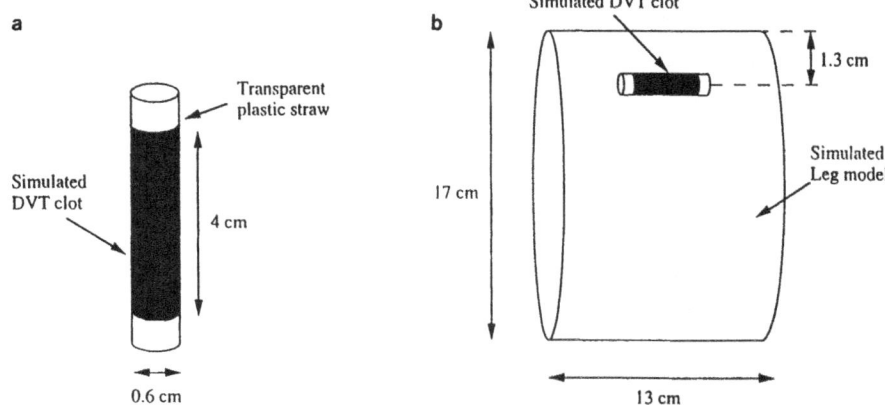

Figure 3. Experimental leg model with a thrombus. (a) Simulated DV thrombus using bovine hemoglobin. (b) Leg phantom with simulated thrombus 1.3 cm from surface (center of absorber to surface of phantom).

5. RESULTS AND DISCUSSION

Note: The term "phase shift", in engineering, always has a negative value. In this paper, the term "phase" represents the absolute value of the phase shift.

5.1. System Linearity Test

Time resolved spectroscopy (TRS) applies the single photon counting system where the time for each individual photon to reach the detector is measured. Then, a pulse is generated by a time to amplification converter (TAC). Theoretically, the system should be linear for the single photon counting system. However, to test the linearity of the entire system as a whole unit, including the photon multiplier tube (PMT), time to amplification converter (TAC), constant flux discriminator (CFD), etc., the steady state gain (K) was

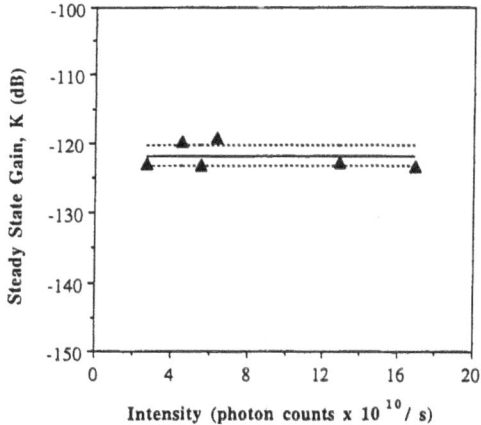

Figure 4. Linearity test for 5 different intensities of NIR-TRS system. The values of K are constant over the photon count range.

obtained for five different input intensities (photon counts/s). The results [Fig. 4] show that the TRS instrument used in these experiments behaves linearly in the input intensity range used for our experiments.

5.2. Homogeneous Model Study

The magnitude ratio (MR) at various frequencies was plotted for source and detector separations of 3, 4, 5, 6, 7 cm in [Fig. 5(a)]. The figures show that an increase in the S-D separation yields linearly decreasing MR values, in dB, for all frequencies. Theoretically, the light intensity (MR) decreases exponentially with the increase in S-D distance. The natural log of MR is taken to obtain values in decibels (dB). After this function is performed, the resulting values yield a linear plot versus increasing S-D separation.

The phase shift, ϕ, at various frequencies was also plotted at various S-D separations in [Fig. 5 (b)]. From the figure, it is shown that the phase increases (phase shift (ϕ) decreases) as the S-D separation increases. The increase in the phase is due to the photons traveling a longer path when the S-D separation is increased.

The time constant (τ) and order of the system (n) were plotted for changes in S-D separation in [Fig. 5 (c), 5 (d)]. When the S-D separation increases, the time constant (τ) increases because the photons are traveling a longer distance and the system responds slowly. The order tends to increase linearly as S-D separation increases.

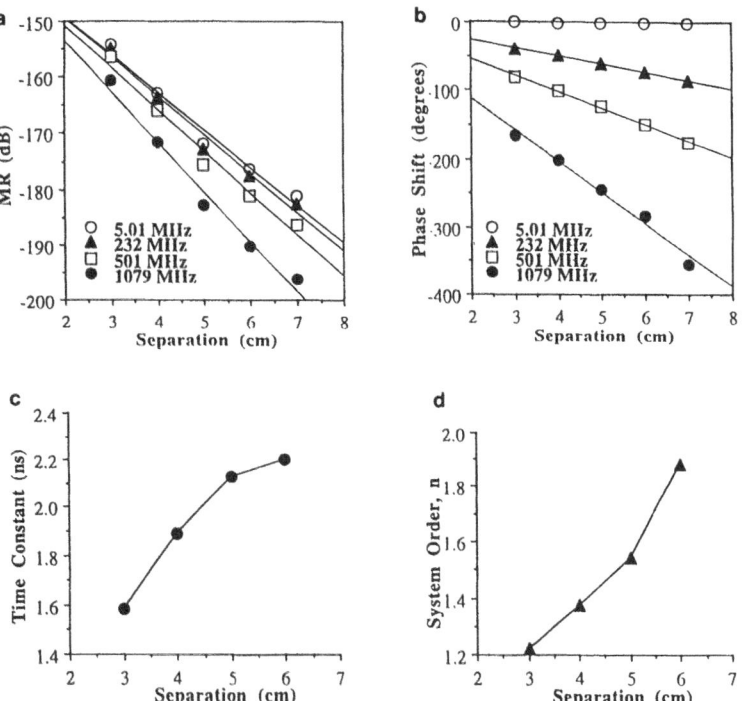

Figure 5. Four engineering parameters vs. S-D separation. (a) Magnitude ratio (MR) vs. S-D separation at 4 different modulation frequencies. The MR decreases linearly with S-D separation. (b) Phase shift (ϕ) vs. S-D separa-

5.3. Heterogeneous Model with a Spherical Absorber

These initial heterogeneous experiments determine whether an absorber can be located using the 4 parameters (MR, ϕ, τ, n). A black sphere of 1 cm diameter at a depth of 1 cm (surface to surface) was used as an absorber in the initial heterogeneous models [Fig. 2 (a)]. The area of interest was then scanned at separation distances of 2 cm, 3 cm, and 4 cm to study the relation between the absorber location and the S-D separation.

A three-dimensional plot of the magnitude ratio (MR) was constructed for 2, 3, 4 cm S-D separations at an arbitrary chosen frequency (681 MHz) [Fig. 6 (a), 7 (a), 8 (a)]. The z axis shows the difference in MR from a reference point, ΔMR ($MR_{measured}$ - $MR_{reference}$). The reference point was chosen to be the average of the MR values for the points on the outside of the scan. This was done because this would show no effect on the plot itself, however, make it easy to see the value change. As the S-D separation increases the sensitivity of the magnitude ratio increases. The ΔMR peak height is the greatest as the S-D separation is the greatest. For increasing S-D separation, the penetration depth of the photons is increased. The photon density in the path varies inversely with the length. For example, the photon density in the shorter length path is higher than that of the longer path. As the lower density path hits the absorber its effect will not be as great on the MR as the higher density path. Since the penetration of the photons is smaller with a shorter S-D separation, the ΔMR values will be less because only the low density path is being absorbed [Fig. 6

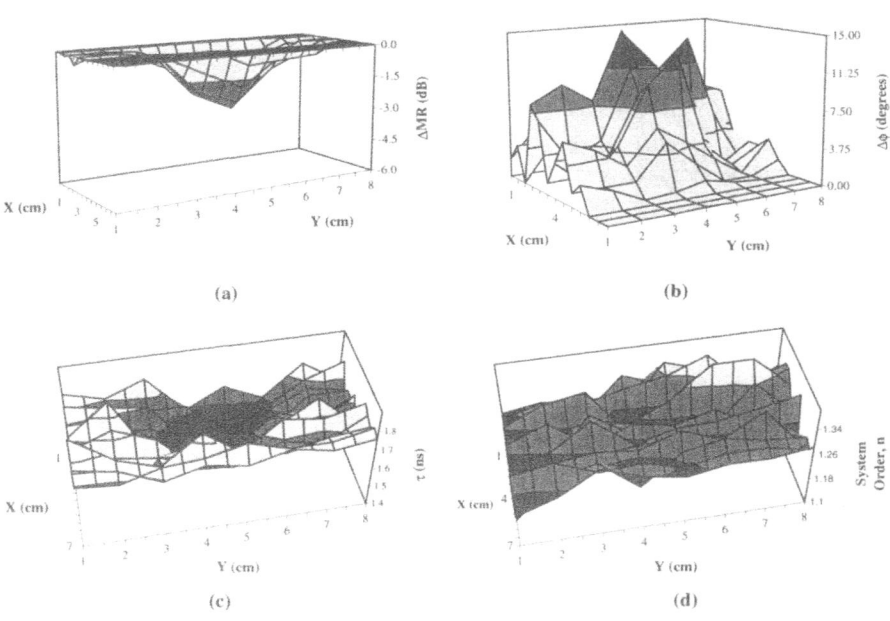

(a)   (b)

(c)   (d)

Figure 6. 3-D plots of four engineering parameters at 2 cm S-D separation for an absorber localization. (a) ΔMR vs. S-D position, where the reference MR value was subtracted from all values. (b) Δ Phase shift (ϕ) vs. S-D position, where the reference phase shift value was subtracted from all values. (c) Time constant (τ) vs. S-D position. (d) System order (n) vs. S-D position.

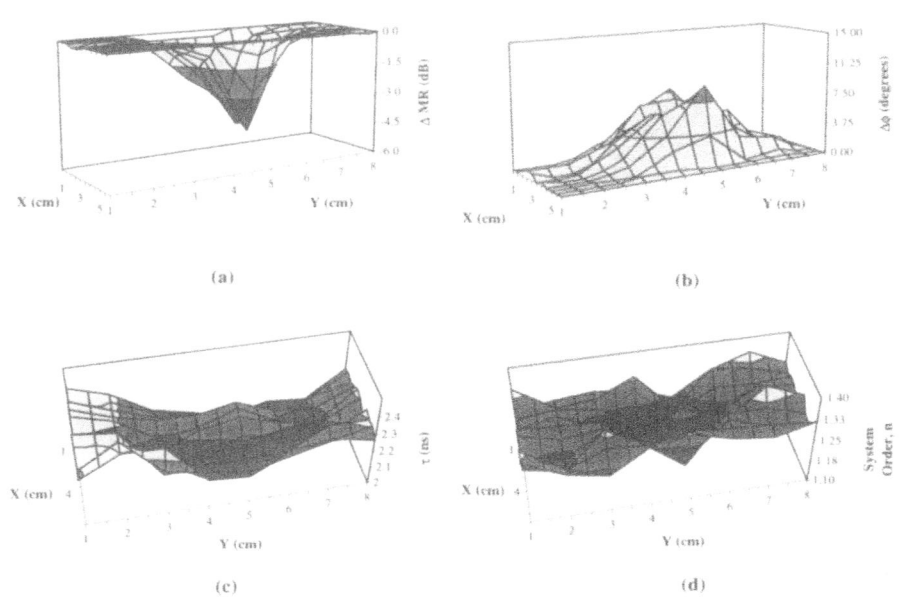

Figure 7. 3-D plots of four engineering parameters for 3 cm S-D separation for an absorber localization. (a) ΔMR vs. S-D position. (b) Δ Phase shift (ϕ) vs. S-D position. (c) Time constant (τ) vs. S-D position. (d) System order (n) vs. S-D position.

(a)]. In the larger S-D separation, the higher density path will be absorbed which explains the higher drop in photon count and thus the ΔMR values [Fig. 7 (a), 8 (a)]. The location of the absorber was determined and compared with the actual absorber location. The two dimensional location of the absorber can be determined using MR.

The $\Delta\phi$ ($\phi_{measured}$ - $\phi_{reference}$) values for 2, 3, 4 cm S-D separations are plotted to determine how the $\Delta\phi$ changes with the increase in separation [Fig. 6 (b), 7 (b), 8 (b)]. As the S-D position approaches the absorber, $\Delta\phi$ increases. This can be explained by the fact that the photons that travel the longer paths will most likely be absorbed, or leave the phantom, while the photons that travel the shorter paths will reach the detector [Fig. 2 (a)].

The $\Delta\phi$ is less sensitive when the S-D separation increases. As previously stated, the increase in S-D separation will allow the higher density path of photons (shorter pathlength) to be absorbed. This results in a greater $\Delta\phi$ value when the S-D separation is shorter because only the shorter path photons will reach the detector. Further experiments with ϕ could give information in absorber depth.

A plot of time constant (τ) at each position was constructed for the three S-D separations [Fig. 6 (c), 7 (c), 8 (c)]. The mean time constant increased for increasing S-D separations as in the homogeneous experiments and for the same reasons. The τ decreased as the S-D position approached the absorber for 2 cm separation [Fig. 6 (c)], however, it increased for 3, 4 cm S-D separations [Fig. 7 (c), 8 (c)].

System Order (n) was plotted at each separation distance [Fig. 6 (d), 7 (d), 8 (d)]. The trend in the system order (n) plots is a general increase where the absorber is located.

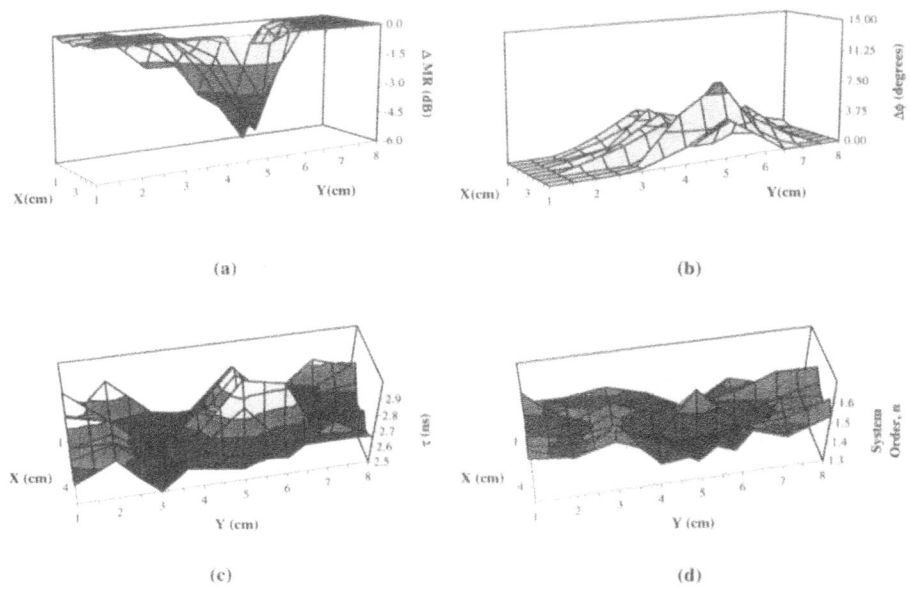

Figure 8. 3-D plots of four engineering parameters for 4 cm S-D separation for an absorber localization. (a) ΔMR vs. S-D position. As the S-D separation increased to 3, 4 cm [Fig. 7 (a), 8 (a)], the MR was more sensitive. (b) Δ Phase shift (φ) vs. S-D position. As the S-D separation increased to 3, 4 cm [Fig. 7 (b), 8 (b)], the phase shift was less sensitive. (c) Time constant (τ) vs. S-D position. (d) System order (n) vs. S-D position.

This trend is magnified with an increase in S-D separation as can be seen from a comparison of the 2, 3, and 4 cm plots.[Fig. 6 (d), 7 (d), 8 (d)].

Correlations between absorber location and sensitivity will be studied further.

5.4. Heterogeneous Leg Model

Our overall objective was to use the 4 engineering parameters (MR, φ, τ, n) to locate a thrombus in the human leg. A leg phantom was constructed to simulate a human leg with a thrombus 1 cm from the surface [Fig. 3 (b)]. The distance from the surface of the phantom to the center of the thrombus was 1.3 cm. To accurately simulate the leg, absorption (μ_a) and scattering (μ_s') coefficients were measured from human legs and correlated with values obtained from homogeneous experiments. Then, a phantom and a hemoglobin absorber, simulating a DVT clot was prepared [Fig. 3]. The (μ_a) and (μ_s') values of the simulated leg model were 0.12 cm^{-1} and 3.3 cm^{-1}, respectively. The area of interest (around the thrombus) on the phantom was scanned at a constant S-D separation of 3 cm.

A 3-dimensional plot of the MR, at each X and Y position at a specific frequency (681 MHz) is shown in [Fig. 10 (a)]. The peak where the MR is the lowest shows the location of the hemoglobin absorber. [Fig. 9 (b)] is a contour plot of MR where the dashed white line represents the actual position of the thrombus in the phantom.

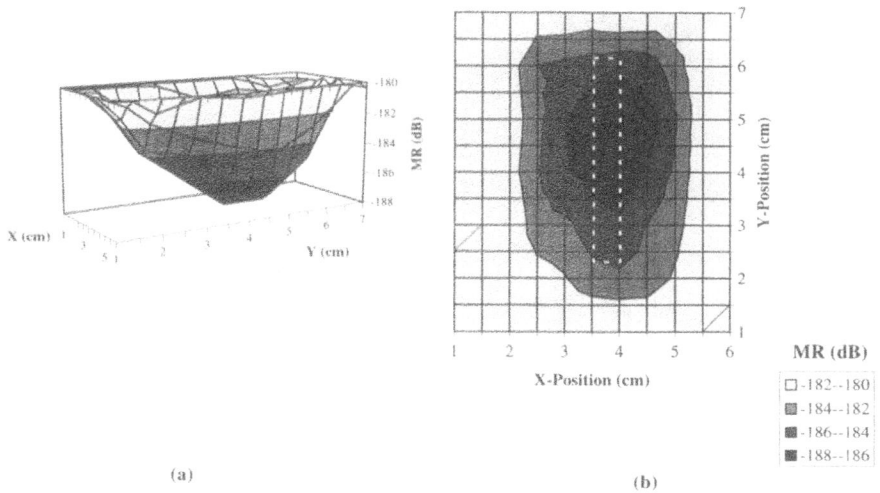

(a)   (b)

Figure 9. (a) A 3-D plot of MR vs. S-D position for 3 cm S-D separation scan of leg model with simulated DV thrombus (b) A contour plot of MR vs. S-D position. The dashed line represents the actual location of the hemoglobin absorber.

Phase Shift (ϕ) vs. S-D position at a specific frequency (681 MHz) is plotted in [Fig. 10 (b)]. As the source and detector approach the absorber, the phase decreases because the photons taking the longer path are absorbed by the stimulated DVT clot.

Time constant (τ) and system order (n) are shown in Figures 10 (c) and 10 (d). The results from the simulated thrombus [Fig. 10] and the experiments using the black rubber absorber [Fig. 6, 7, 8] show similar trends in time constant and system order. The τ increases as the S-D position approaches the absorber. The n decreases as the S-D position approaches the absorber.

The trends for order and time constant will be studied further. In future experiments, we plan to vary the size, depth, and absorbance of the DVT clot and analyze the four parameters from the TRS spectra, to locate a DVT clot in three dimensions.

6. CONCLUSION

It appears that TRS can be an effective, non-invasive, low cost way of detecting DVT in the human leg. These initial experiments are intended to show 3-dimensional localization of a thrombus so that correlations between S-D separation, depth, and thrombus size may be developed. The magnitude ratio and phase shift data were successful in locating the position of the DVT clot. The time constant (τ) and order show trends that will be analyzed in future experiments to locate a DVT clot in three dimensions. Also, the engi-

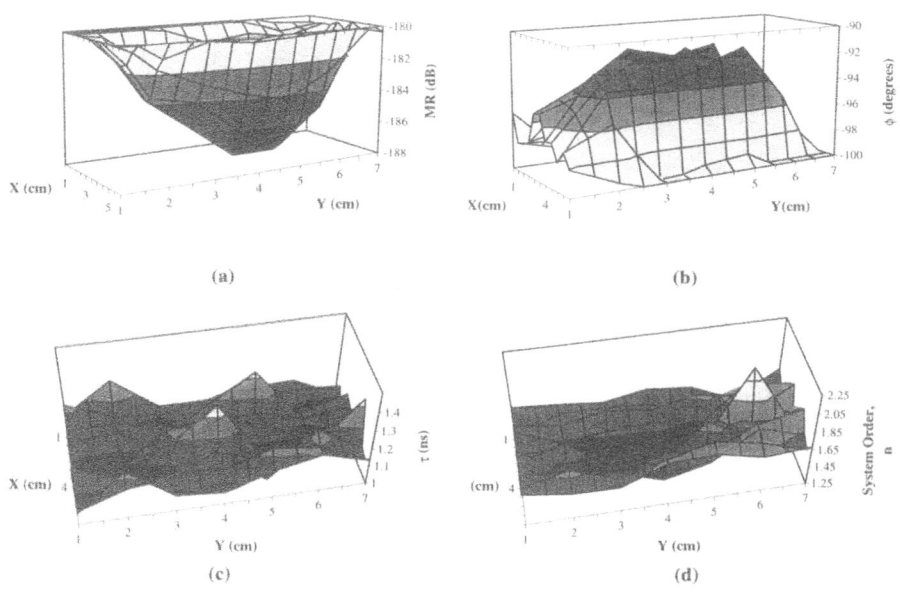

(a)   (b)

(c)   (d)

Figure 10. 3-D plots of four engineering parameters for 3 cm S-D separation scan of simulated leg phantom (a) MR vs. S-D position. (b) Phase shift (ϕ) vs. S-D position. (c) Time constant (τ) vs. S-D position. (d) System order (n) vs. S-D position.

neering parameters may be used to characterize the DVT clot. With the combination of these variables, the 3-D localization could be greatly enhanced. This technique might be useful in locating heterogeneities in many other physiological systems, such as tumor localization.

ACKNOWLEDGMENT

This research is supported by The Whitaker Foundation, Biomedical Engineering Research Grant. The authors would like to thank Dr. Britton Chance and other researchers in his lab at the University of Pennsylvania for their technical support and the TRS instrument.

REFERENCES

1. Bruley, D.F. and Prados, J.W., "The frequency response analysis of a wetted wall adiabatic humidifier", *AIChE Journal*, 10(5):612 (1964).
2. Bruley. D.F., Pulse reduction code written for process identification (1974).
3. Clements, W.C., Jr. and Schelle, K.B., "Pulse testing for dynamic analysis", *I & EC Process Design and Development*. 2(2);94 (1963).
4. Coughanowr, D.R., in: *Process systems analysis and Control*, McGraw-Hill, New York (1991).

5. Filon, L.N.G., "On a quadrature Formula for Trigonometric Integrals", *Proc. of the Royal Soc. Edinburgh*, 49:38 (1928).

6. Kang, K.A., Bruley, D.F., Londono, J.M., and Chance, B., "Frequency Response by Pulse Reduction for the Analysis of TRS Spectra", ISOTT XVI, *Advances in Experimental Medicine and Biology* Plenum, New York (1993).

7. Kang, K.A., Bruley, D.F., Londono, J.M., and Chance, B., "Highly Scattering Optical System Identification via Frequency Response Analysis of Optical NIR-TRS Spectra", *Annals of Biomedical Engineering*, 22, 241–253, (1994).

8. Kang, K.A., Bruley, D.F., Kitai, M., and Chance, B., "System Parameter Analysis of NIR-TRS Spectra from Homogeneous Media with and without an Absorbing Boundary and Heterogeneous Media with a Single Absorber", ISOTT XVII, *Advances in Experimental Medicine and Biology*, Plenum, New York, (1996).

9. Koyama, K. and Fatlowitz, D., "Application of MCP-PMTs to time correlated single photon counting and related procedures", Hamamatsu technical information, No. ET-03/OCT (1987).

10. Lewis, C.I., Jr., Bruley, D.F., and Hunt, D.H., "Evaluation of temperature pulse characteristics and pulse testing for thermal dynamic analysis ", *I&EC Process Design and Development*, 6(3):281 (1967).

11. Luyben W.L., *Process Modeling Simulation and Control for Chemical Engineers*, 2nd edition, McGraw-Hill , New York (1990).

7

INCIDENCE OF ASYMPTOMATIC PERIPHERAL ARTERIAL OCCLUSIVE DISEASE IN DIABETIC PATIENTS ATTENDING A HOSPITAL CLINIC

T. A. Elhadd,[1] R. T. Jung,[1] R. W. Newton,[1] P. A. Stonebridge,[2] and J. J. F. Belch[1]

[1]University Department of Medicine
[2]University Department of Surgery
Ninewells Hospital and Medical School
Dundee, Scotland

1. INTRODUCTION

Vascular events are the major cause of morbidity and mortality in patients with diabetes mellitus, with an overall 50% mortality after 40 years duration of diabetes (Deckert 1978). Peripheral arterial occlusive disease (PAOD) develops prematurely, more frequently, rapidly and extensively than in non-diabetic subjects. Peripheral atherosclerosis in patients with diabetes has a particular predilection for the peroneal and tibial arteries between the knees and ankles. Aorto-iliac disease occurs less commonly, whereas disease in the femoro-popliteal area has an incidence similar to that of non-diabetic subjects (Banga 1994). The incidence of clinically symptomatic PAOD in patients with diabetes is estimated to be two to five fold compared with the general population. Ten percent of patients have evidence of PAOD at the time of diagnosis (UKPDS 1991) and the cumulative incidence rises to 45% after 20 years duration (Janka et al 1980). The incidence of asymptomatic PAOD in diabetes has not yet been studied.

Symptomatic and asymptomatic PAOD are strong predictors of high mortality from cardiovascular causes within five years (Hiatt and Sussman 1994). In one study, 67% of patients dying from cardiovascular events had PAOD compared to 15% of survivors (Janka et al 1980). Morbidity is very high in such a group, with a 20 fold risk of ischaemic gangrene and a four fold risk of amputation. One in every two amputees is diabetic, and diabetic foot ulceration is a leading cause for long term hospitalisation. The incidence of asymptomatic PAOD in diabetes is not known; therefore this pilot study was conducted to establish the actual incidence in the diabetic population attending our diabetic clinic.

Table 1. Demographic features of the study groups

	Peripheral pulse absent	Peripheral pulse present
Number of patients	20	10
Sex (male / female)	7/13	5/5
Age (years)	66 (36-80)	61 (40-70)
Duration of diabetes(years).	10 (1-24)	9 (1-22)
Body mass index (m2/kg)	30 (18-40)	29 (23-40)
Smoking (Y/N/Ex)	3/9/8	4/3/3
Hypertension	12	4
Dyslipidaemia	4	0
HbA1c *(mean±SD)	7.45±1.07	6.89±1.13
IDDM	3	1
NIDDM	17	9

Key: IDDM: insulin dependent diabetes mellitus. NIDDM: non-insulin dependent diabetes mellitus.
* normal <5.8%.

2. PATIENTS AND METHODS

Patients with both type 1 and type 2 diabetes mellitus were identified from the computerised register of Ninewells Hospital at Dundee satisfying the following criteria: 1. No history of cardiovascular disease, in particular no angina or intermittent claudication; 2. Absent peripheral arterial pulse (both dorsalis pedis and posterior tibial arteries); and 3. Not taking aspirin. Twenty patients agreed to participate in the study. Ten additional patients with the same inclusion criteria but with a palpable pulse were included to serve as controls. Ethical committee approval was granted for the study. All patients gave their written consent to participate in the study. All patients wereseen over a period of 4 months.

Ankle brachial pressure index (ABPI) was estimated by Doppler ultrasound (Model 811-B, Parks Medical Elect. Inc., Aloha, Oregon, USA). An ABPI value of ≤ 0.9 was taken as indicative of occlusive disease, and a value of >0.9 was taken as normal. The various clinical characteristics of the study groups is shown in Table 1. All variables were expressed as median, range unless stated otherwise.

Of the control group, four were smokers, three were non-smokers and three were ex-smokers, four had hypertension, and none had any lipid abnormality. In the other group, three were smokers, nine were non-smokers and eight were ex-smokers, twelve patients were known to be hypertensive and only four were known to have lipid abnormality.

3. RESULTS

Of all study patients, 23% had evidence of asymptomatic PAOD. All members of the control group had a ABPI > 1.0 . In the other group, 10 patients had an adequate peripheral arterial circulation with a mean ABPI of 1.05 ± 0.09 (mean $\pm$ SD), and the remaining 10 patients had evidence of severe peripheral arterial occlusive disease with a mean ABPI of 0.69 ± 0.13. The results from the study groups are shown in Figure 1. The toe pressure measurements did not differ from the ABPI results in the entire study group. In the group with low ABPI, 70% of the patients were hypertensive, were mostly ex-smokers, and predominantly female. They were also significantly older when compared to the group with normal ABPI, p <0.05 Student

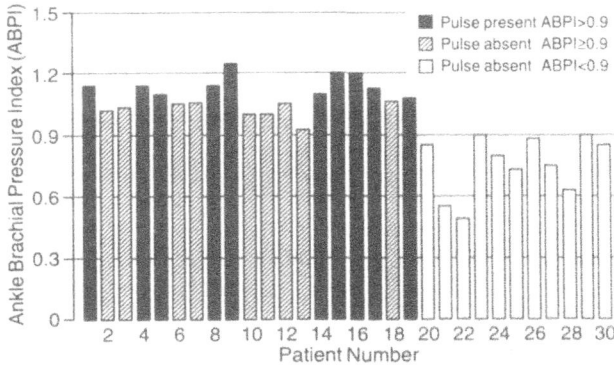

Figure 1. ABPI in the study groups.

t-test. Otherwise, the two groups did not differ in their smoking habit, BMI, lipid profile, glycaemic control, or duration of diabetes.

4. DISCUSSION

Despite the relatively small number of patients studied, these preliminary results show a 23% incidence of PAOD among asymptomatic diabetic patients, with no history of cardiovascular disease.Walters et al (1992) found a prevalence of 23.5% among NIDDM and 8.7% among IDDM, versus 16.4% reported in the general population (Fowkes et al 1991). This also indicates that clinical examination alone is not a sensitive method of detecting PAOD, as it had only 50% accuracy for the actual presence of peripheral arterial insufficiency in asymptomatic diabetic patients. Because we measured APBI at rest, a few patients with PAOD may have remained undetected. It might be expected that toe pressure measurement will enhance the pick up rate of patients with low APBI; however, we did not find this. Due to the small cohort, it might prove more useful in a larger study, particularly in patients with medial wall calcification denoted by an artificially high ABPI (Vincent et al 1983).

There was a high prevalence of hypertension in patients with low ABPI. As hypertension has been shown by others (Palumbo et al 1991) to be the most important factor in both development and progression of PAOD in diabetic patients, we recommend that this should be treated early and aggressively in those at risk.We also recommend that in the presence of other risk factors for foot ulceration, such as foot deformity or neuropathy, such patients should have the peripheral circulation assessed by the more accurate measurement of ABPI using Doppler ultrasound, rather than palpation of pulses. ABPI measurement is an inexpensive, precise, and reproducible technique.

We conclude that the natural history of asymptomatic peripheral arterial disease in diabetic patients needs to be evaluated in a longitudinal study, and the value of a simple prophylactic drug, such as aspirin (Yudkin,1995), be evaluated as a protection against future cardiovascular events in an at risk cohort.

ACKNOWLEDGMENT

T. A. El-hadd is supported by a grant from the Medical Research Council, U K.

REFERENCES

Banga JD. Lower extremity arterial disease in diabetes mellitus. Diab Rev Int 1994; 3 (4) : 7–11.

Deckert T. Prognosis of diabetics with diabetes onset before age 31. Diabetologia 1978; 14: 463–477.

Fowkes FGR, Housley E, Cawood EHH et al . Edinburgh Artery Study : Prevalence of asymptomatic and sympto-matic peripheral arterial disease in the general population. Intern J Epid 1991; 2: 384- 391.

Hiatt WR, Sussman KE. Peripheral arterial disease in patients with diabetes mellitus. Can we make the difference? Diabet Res Clin Pract 1994; 24: 65–67.

Janka HU, Standle E, Mehnert H. Peripheral vascular disease in diabetes and its relation to cardiovascular risk fac-tors: screening with Doppler ultrasonic technique. Diabetes Care 1980; 3: 207–213.

Palumbo PJ, Michael O'Fallon M, Osmondson PJ et al. Progression of peripheral occlusive arterial disease in dia-betes mellitus. Arch Intern Med 1991; 151: 717–721.

United Kingdom Prospective Diabetes Study Group.UK prospective diabetes study (UKPDS) VII: Study design, progress and performance. Diabetologia 1991; 34:877–90.

Vincent DG, Salles-Cunha SX, Bernhard VM, Towne JB. Non-invasive assessment of toe systolic pressure with special reference to diabetes mellitus. J Cardiovas Surg 1983; 24(1):22–28.

Walters DP, Gatling W, Mullee MA, Hill RD. The prevalence, detection and epidemiological correlates of periph-eral vascular disease: A comparison of diabetic and non-diabetic subjects in an English community. Diab Med 1992 9: 710–715.

Yudkin JS. Which diabetic patients should take aspirin? Br Med J 1995; 311: 641–642.

8

CUTANEOUS MICROVASCULAR RESPONSES ARE IMPROVED AFTER CHOLESTEROL-LOWERING IN PATIENTS WITH PERIPHERAL VASCULAR DISEASE AND HYPERCHOLESTEROLAEMIA

Faisel Khan, Stuart J. Litchfield, and Jill J. F. Belch

University Department of Medicine
Ninewells Hospital and Medical School
Dundee, DD1 9SY, Scotland

1. INTRODUCTION

High blood cholesterol concentrations and the risk of coronary heart disease are related. It is now known that lowering cholesterol reduces the incidence of ischaemic cardiac events[1]. The reasons for this improvement in cardiac function, however, are not clear. Angiographic assessment of coronary blood vessels reveals only slight regression of large vessel atherosclerosis with lipid-lowering therapy.[2] Studies in vivo have shown that the endothelium is abnormal in the early preclinical stages of atherosclerosis, even before anatomical evidence of plaque formation is present[3]. Damage to the endothelium predisposes to thrombosis, leucocyte adhesion, and proliferation of smooth muscle cells in the arterial wall. The endothelium play an important role is the controlling of vascular tone, and it has been suggested that the beneficial effects of cholesterol-lowering may be related to an improvement in endothelial cell function. Indeed, studies in human coronary blood vessels show lowering cholesterol improves endothelial cell function.[4]

Most studies investigating endothelial cell function in hypercholesterolaemia use intra-coronary or intra-brachial artery infusions of substances, namely acetylcholine, that stimulate production of endothelium-derived relaxing factor (EDRF). If, however, assessment of endothelial cell function is to be used routinely to provide direction for therapeutic intervention or outcome, then non-invasive evaluation is preferred. Endothelial dysfunction has been reported in the forearm muscle circulation of patients with hypercholesterolaemia,[5] but not in the skin circulation. Assessment of the blood vessels in the skin of the forearm may, therefore, provide an accessible site for such investigations. In the present study, we have used the non-invasive techniques of iontophoresis and laser Doppler flowmetry to examine the effects of lowering of cholesterol on endothelial cell

Oxygen Transport to Tissue XIX, edited by Harrison and Delpy
Plenum Press, New York, 1997

function in forearm skin of patients with hypercholesterolaemia and peripheral arterial occlusive disease (PAOD).

2. METHODS

2.1. Patient Selection

Eight patients (5 male, 3 female) with an average age (±SEM) of 64.3±2.7 years were selected from their documented history of lower limb PAOD. Those patients with a raised total serum cholesterol (≥ 6.5 mmol/L) received dietary counselling and returned in 7 weeks for a fasting blood sample. Following dietary intervention, those patients who still had an elevated LDL cholesterol (≥ 4.1 mmol/L) that justified the introduction of a lipid-lowering agent were recruited into the study. Lipid-lowering therapy was with fluvastatin (40 mg once daily) (Sandoz Pharmaceuticals) for 24 weeks. No patient had a history of previous angioplasty or prior therapy with cholesterol-lowering agents. Patients were receiving no medication and non-steroidal anti-inflammatory drugs, including aspirin, were avoided for at least two weeks before any vascular assessment. The study was approved by the Hospital Ethics Committee. Written informed consent was obtained from all patients.

2.2. Lipid Measurements

Blood samples were obtained after a 12 hour fast. Plasma total cholesterol was measured on a Hitachi 704/717 analyser using the cholesterol oxidase/PAP method (Boehringer Mannheim Diagnostics). Plasma high density lipoprotein (HDL) cholesterol was isolated using heparin manganese and the cholesterol content measured as for total cholesterol. Plasma triglycerides were measured using enzymatic hydrolysis of triglycerides with subsequent enzymatic determination of the liberated glycerol.

2.3. Experimental Protocol

Measurements of skin blood flow were performed in a temperature-controlled laboratory (23±1°C). Patients wore light clothing and their arms were supported at heart level. Forearm skin blood flow was measured using a laser Doppler flowmeter. Assessments of endothelium-dependent and -independent vascular responses were made during iontophoresis of acetylcholine chloride and sodium nitroprusside, respectively. The Moor Instrument combined iontophoresis (MIC1) and laser Doppler probe (MBF3/D) system was used.

Iontophoresis provides a convenient method for producing a quantifiable stimulus to the microvasculature and can be used effectively to transport drugs in solution across intact skin.[6] The technology is based on the principle that an electrical potential difference will cause ions in solution to migrate according to their electrical charge. To transfer a drug into the skin, the polarity of the active electrode has to have the same charge as the active ions of the drug. For a given drug, the quantity delivered is directly proportional to the total charge that migrates through the skin surface. The total charge can be increased by either increasing current or time.

The forearm skin was cleaned with alcohol and a probe holder was attached approximately midway on the volar aspect of the forearm using double sided adhesive tape. This holder kept the laser probe in position and also contained a well around the probe into

Table 1. Lipid measurements in 8 patients before and after
cholesterol-lowering

	Before fluvastatin	After fluvastatin
Total cholesterol (mmol/l)	6.96 ± 0.28	5.57 ± 0.34[†]
LDL cholesterol (mmol/l)	4.97 ± 0.33	3.75 ± 0.31[*]
HDL cholesterol (mmol/l)	1.13 ± 0.08	1.33 ± 0.15[*]
Triglyceride (mmol/l)	1.54 ± 0.15	1.21 ± 0.23

[*]p<0.02, [†]p<0.002.

which solutions (0.5 ml) were applied for iontophoresing at the test site. An additional probe holder was attached 20 mm distal to the test site and served as the control.

Acetylcholine was made up to a 1% solution in de-ionized water and iontophoresed for 10 seconds at a current strength of 0.2 mA (giving a charge of 2 milli Coulombs (mC)). To increase the dose, acetylcholine was iontophoresed for 20 and 40 seconds. Vascular responses to each dose were measured for 4 minutes. At a different site sodium nitroprusside was iontophoresed using similar doses to those for acetylcholine (i.e 2, 4 and 8 mC). Blood flow responses to acetylcholine and sodium nitroprusside were repeated after 24 weeks of lipid-lowering therapy.

2.4. Statistics

Blood flow is expressed in arbitrary units (AU). The area under the curve (AUC) was calculated for a 4 minute basal period and for each 4 minute period following iontophoresis of drug. The percentage change in AUC from baseline was calculated for the test site and expressed as a ratio of the corresponding change at the control site. This ensures that only the direct effects of local drug administration are taken into account. Differences in these ratios and in plasma cholesterol, before and after lipid-lowering, were compared using Student's t-test.

3. RESULTS

Lipid-lowering therapy for 24 weeks produced significant reductions in total and LDL cholesterol of 20% (p<0.002) and 25% (p<0.02), respectively (Table 1). HDL cholesterol increased by 25% (p<0.02) resulting in a change in the ratio of HDL/total cholesterol from 16% to 24%. Triglycerides did not change significantly over the 24 week period (Table 1).

Cutaneous vascular response to acetylcholine were improved significantly after cholesterol-lowering (Fig 1a). This augmentation in endothelium-dependent vasodilatation was significant at the two higher doses of acetylcholine. The blood flow ratio changed from 1.99 to 6.46, p<0.05 and from 3.32 to 8.29, p<0.01, respectively. In contrast, Fig 1b shows that there was no significant improvement in the cutaneous vascular response to sodium nitroprusside after cholesterol-lowering, although a trend towards an increase was noted at the highest dose (p=0.06).

4. DISCUSSION

The results of the present study show that lowering cholesterol for 24 weeks improves cutaneous vascular responses to acetylcholine, an endothelium-dependent vasodila-

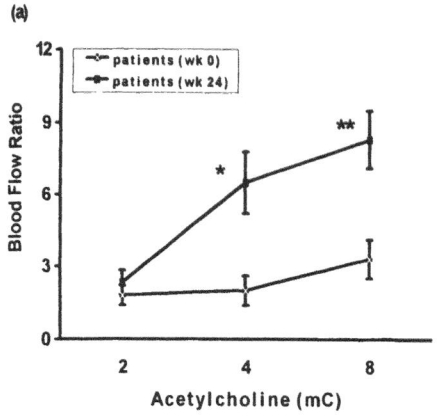
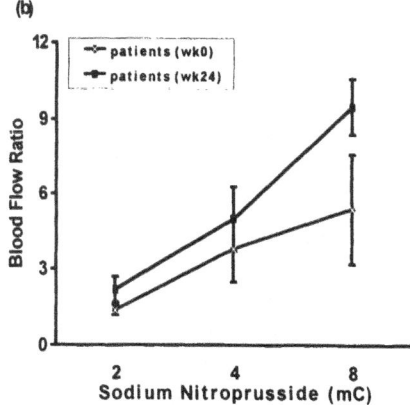

Figure 1.

tor. In contrast, lowering of cholesterol had no significant effect on the vascular response to sodium nitroprusside, an endothelium-independent vasodilator that mimics the action of endothelium-derived NO. The beneficial effects of cholesterol-lowering, therefore, may be related to an improvement in endothelial cell function.[4]

Most studies investigating the integrity of endothelial cell function use the invasive procedure of intra-arterial infusions of substances that produce vascular relaxation via release of EDRF. We have used the non-invasive techniques of iontophoresis and laser Doppler flowmetry to measure blood flow responses in the skin of the forearm, a site that is readily accessible. Additionally, while venous occlusion plethysmography has been used in other studies to measure the total blood flow in a limb, which includes flow to muscle and skin, we have assessed endothelial cell function in the resistance vessels of the skin only. It is important to distinguish between these two vascular beds because of the different mechanisms controlling blood flow through them. Indeed, we have shown that intra-arterial infusions of a muscarinic agonist produce a comparable vasodilator response in the forearm of patients with Raynaud's phenomenon and control subjects when using venous occlusion plethysmography to measure total forearm blood flow.[7] Subsequent measurements of skin blood flow, however, using laser Doppler flowmetry, show a significant difference between the two groups, localizing a defect in the cutaneous vasculature which is masked by a normal muscle blood flow response.

Endothelial cell dysfunction in the coronary and forearm circulations have been shown to be correlated closely.[8] If, therefore, alterations in forearm skin blood flow, associated with lipid-lowering therapy, are representative of a more generalized improvement in endothelial cell function, then the application of the techniques of iontophoresis and laser Doppler flowmetry to the forearm skin could be used to provide important markers for the integrity of endothelial cell function and its modulation by therapy.

Endothelial cells release a potent vasodilator, EDRF, in response to muscarinic stimulation. Bossaller et al[9] suggested that hypercholesterolaemia might lead to a selective impairment in endothelial muscarinic receptor function, although more recent work by Casino et al.[10] shows that impaired endothelium-dependent vasodilatation extends beyond the muscarinic receptor, perhaps related to a defect in the pathway that regulates the endothelial modulation of vascular tone. Several ideas have been proposed for explaining the impairment of endothelial cell function by hypercholesterolaemia, including decreased availability of L-arginine,[11] the precursor of endothelium-derived NO, decreased bioactivity of NO,[12] or increased inactivation of NO by production of superoxide anions from the endothelium.[13] We cannot determine the precise location of this defect from the current investigation. One possibility for the improvement in endothelium-dependent vasodilatation associated with lowering of lipids is a reduction in the production of oxygen-derived free radicals. Ohara et al[13] demonstrated an increased production of superoxide anion from the endothelium of hypercholesterolaemic blood vessels, which may subsequently be responsible for an increased degradation of NO. In fact, it has been proposed that hypercholesterolaemia produces an increase in NO synthase activity but its action is reduced as a result of oxidative degradation.[14] A reduction in serum cholesterol produces normalization of oxygen-derived free radical production[13] and a combination of cholesterol-lowering with antioxidant therapy has been shown to have a beneficial effect on coronary vasomotion.[15]

Cholesterol-lowering produced a trend towards an augmented vasodilator response to sodium nitroprusside, especially at the highest concentration (Fig 1b), although this increase was not significantly different from baseline values. Since sodium nitroprusside causes vasodilataion through the production of NO, and a lowering of cholesterol reduces the production of oxygen-derived free radicals, the tendency towards an improved response to this agent may reflect the decreased degradation of NO. A possible reason for the lack of significance was the relatively large variability in the vascular response to sodium nitroprusside and a small sample size (n=8).

A limitation of our study is that it was not placebo-controlled. Thus, we do not know whether the improvement in endothelium-dependent vasodilatation is due to cholesterol-lowering or to drug effect. Treasure et al,[4] however, found in their study that placebo treatment only produced a 1% change in the coronary artery response to the peak dose of acetylcholine in comparison to a 16% improvement in the group receiving lovastatin.

In conclusion, we have shown an improvement in endothelium-dependent vasodilatation after cholesterol-lowering therapy in the forearm cutaneous circulation of patients with hypercholesterolaemia and PAOD. The application of the non-invasive techniques of iontophoresis and laser Doppler flowmetry may provide useful tools for early detection of endothelial dysfunction, thereby enabling steps to be taken towards preventing the possible later, harmful ischaemic events. The utility of these techniques may also be valuable for assessing the efficacy of drug therapy targeted at improving endothelial cell function.

ACKNOWLEDGMENTS

This study was supported by funds from Sandoz Pharmaceuticals and the Sir Jules Thorn Charitable Trust. We thank Sisters Jean Bancroft and Rosalind Robb for technical assistance.

REFERENCES

1. Levine GN, Keaney JF, Vita JA. Cholesterol reduction in cardiovascular disease. *N Eng J Med.* 1995; 332: 512–521.
2. Brown G, Albers JJ, Fisher LD, Schaeffer SM, Lin JT, Kaplan C, Zhao XQ, Bisson BD, Fitzpatrick VF, Dodge HT. Regression of coronary artery disease as a result of intensive lipid-lowering therapy in men with high levels of apolipoprotein B. *N Eng J Med.* 1990; 323: 1289–1298.

3. Celemajer DS, Sorensen KE, Gooch VM, Spiegelhalter DJ, Miller OI, Sullivan ID, Lloyd JK, Deanfield JE. Non-invasive detection of endothelial dysfunction in children and adults at risk of atherosclerosis. *Lancet.* 1992; 340: 111–1115.

4. Treasure CB, Klein JL, Weintraub WS, Talley JD, Stillabower ME, Kosinski AS, Zhang J, Bocuzzi SJ, Cedarholm JC, Alexander RW. Beneficial effects of cholesterol-lowering therapy on the coronary endothelium in patients with coronary artery disease. *N Eng J Med.* 1995; 332:481–487.

5. Chowienczyk PJ, Watts GF, Cockcroft JR, Ritter JM. Impaired endothelium-dependent vasodilation of forearm resistance vessels in hypercholesterolaemia. *Lancet.* 1992; 340: 1430–1432.

6. Sloan JB, Soltani K. Iontophoresis in dermatology. *J Am Acad Derm.* 1986; 15(4): 671–684.

7. Khan F, Coffman JD. Enhanced cholinergic cutaneous vasodilation in Raynauds phenomenon. *Circulation.* 1994; 89: 1183–1188.

8. Uehata A, Gerhard MD, Meredith IT, Lieberman EL, Selwyn AP, Creager M, Polak J, Ganz P, Yeung AL, Anderson TJ. Close relationship of endothelial dysfunction in coronary and brachial artery. *Circulation.* 1993; No 4, part 2: I-618.

9. Bossaller C, Habib GB, Yamamoto H, Williams C, Wells S, Henry PD. Impaired muscarine endothelium dependent relaxation and cyclic guanosine 5'- monophosphate formation in artherosclerotic human coronary artery and rabbit aorta. *J Clin Invest.* 1987; 79: 170–174.

10. Casino PR, Kilcoyne CM, Cannon III RO, Quyyumi AA, Panza JA. Impaired endothelium-dependent vascular relaxation in patients with hypercholesterolemia extends beyond the muscarinic receptor. *Am J Cardiol.* 1995; 75: 40–44.

11. Creager MA, Gallagher SJ, Girerd XJ, Coleman SM, Dzau VJ, Cooke JP. L-Arginine improves endothelium-dependent vasodilation in hypercholesterolemic humans. *J Clin Invest.* 1992; 90: 1248–1253.

12. Casino PR, Kilcoyne CM, Quyyumi AA, Hoeg JM, Panza JA. The role of nitic oxide in endothelium-dependent vasodilation of hypercholesterolemic patients. *Circulation.*1993; 88: 2541–2547.

13. Ohara Y, Peterson TE, Harrison DG. Hypercholesterolemia increases endothelial superoxide anion production. *J Clin Invest.* 1993;91: 2546–2551.

14. Minor RLJ, Myers PR Guerra R, Bates JN, Harrison DG. Diet-induced atherosclerosis increases the release of nitrogen oxides from rabbit aorta. *J Clin Invest.* 1990; 86: 2109–2116.

15. Anderson TJ, Meredith IT, Yeung AC, Frei B, Selwyn AP, Ganz P. The effect of cholesterol-lowering and antioxidant therapy on endothelium-dependent coronary vasomotion. *N Eng J Med.* 1995; 332: 488–493.

9

OXYGEN TENSION IN PRIMARY TUMOURS OF THE UTERINE CERVIX AND LYMPH NODE METASTASES OF THE HEAD AND NECK

Heidi Lyng,[1] Kolbein Sundför,[2] Gunnar Tanum,[3] and Einar K. Rofstad[1]

[1]Department of Biophysics
Institute for Cancer Research
[2]Department of Gynecology
[3]Department of Oncology
The Norwegian Radium Hospital
Oslo, Norway

1. INTRODUCTION

Tumours often show low oxygen concentrations due to abnormal delivery and metabolism of oxygen (Vaupel *et al.*, 1989; Gulledge and Dewhirst, 1996). The vascular network is highly irregular with large intercapillary distances leading to impaired blood supply. The capacity of the blood to carry oxygen is reduced because of low extracellular pH. Moreover, the oxygen demand is elevated due to the high proliferation activity of tumour cells. These characteristics lead to the development of hypoxic and necrotic areas within tumours. However, the oxygen concentration can differ considerably among different tumours, even among tumours of the same histological type.

Knowledge about the oxygen concentration in human tumours is useful because oxygen is important for the growth and development of cancer and for the response to cancer treatment. The growth of solid tumours beyond a size of about 1–2 mm necessitates invasion of blood vessels for the delivery of oxygen and nutrients (Folkman, 1990). Moreover, low oxygen concentration may induce gene amplification and enhanced metastatic potential (Hill, 1990). The radiotherapeutic response is often impaired in poorly oxygenated tumours (Durand, 1991; Stone *et al.*, 1993) and the toxicity of several chemotherapeutic agents depends on the oxygen concentration (Durand, 1991).

Tumour oxygen tension (pO_2) can be measured by use of polarographic needle electrodes (Kallinowski *et al.*, 1990). The method is used to study pO_2 distributions in patients with primary tumours of the uterine cervix and lymph node metastases of the head and neck at The Norwegian Radium Hospital. One aim of the project is to investigate possible differences between the pO_2 distributions of the two tumour types. The present preliminary communication reports measurements performed in 41 patients before the start of treatment.

Oxygen Transport to Tissue XIX, edited by Harrison and Delpy
Plenum Press, New York, 1997

2. MATERIALS AND METHODS

2.1. Patients

Thirty-six patients with squamous cell carcinoma of the uterine cervix (stage Ib, IIb or IIIb, according to the FIGO) and 5 patients with lymph node metastases from squamous cell carcinoma of the head and neck were included in the study. The largest tumour diameter was 2 cm or more. Informed consent was obtained from all patients.

2.2. Measurement of pO_2

Oxygen tension was measured in tumour tissue and in subcutaneous tissue in the pubic area before the start of treatment. General anaesthesia (Propofol, *i.v.*) was used in 9 patients with cervix carcinoma, otherwise no anaesthetic was used. Studies performed in our institution have shown that Propofol has no significant influence on the pO_2 distribution in cervix tumours.

Measurements were performed by use of sterile polarographic needle electrodes with a shaft diameter of 300 μm (Eppendorf pO_2 Histograph 6650) (Lyng *et al.*, 1996). A sterile venflon (20 G) was used to guide the oxygen electrode into the tissue. The electrode was moved automatically through the tissue in preset steps of either 1 mm or 0.7 mm. Each forward step was followed by a backward step of 0.3 mm, leading to a distance of either 0.7 mm or 0.4 mm between each pO_2 reading. In each tumour, a total of 57 to 252 readings were performed in 2 to 6 different tracks. The length of each track was determined from the size of the tumour. In normal tissue, a total of 30 to 64 readings were performed in a single track.

2.3. Data Analysis

A pO_2 histogram was generated for each tumour and normal tissue. Six pO_2 parameters; *i.e.*, the median pO_2, the 10th and 90th percentiles and the fraction of pO_2 readings below 2.5 mmHg, 5 mmHg and 10 mmHg, were calculated from each histogram. An analysis of variance on ranks and a Dunn's test were applied to search for differences between three pO_2 histograms. A Mann-Whitney rank sum test was used to search for differences between two pO_2 histograms and to investigate whether the pO_2 parameters differed between the cervix tumours and the head and neck tumours. A significance level of $P = 0.05$ was used throughout.

3. RESULTS

Histograms of pO_2 values measured in normal subcutaneous pubic tissues, cervix tumours and head and neck tumours are shown in Fig. 1. The three histograms were significantly different ($P < 0.05$). The highest pO_2 values were measured in the normal tissues, whereas the lowest values were measured in the cervix tumours. Most tumours showed large heterogeneity in pO_2 compared to subcutaneous pubic tissue.

Differences in pO_2 were also found among tumours. Typical pO_2 histograms for two different cervix tumours, one with low and the other with high pO_2, are shown in Figs. 2 a and b ($P < 0.05$), whereas histograms for two head and neck tumours with different pO_2

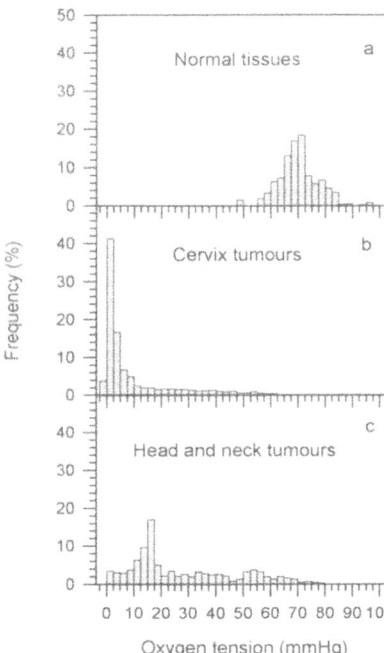

Figure 1. Frequency distributions of pO_2 measured in subcutaneous pubic tissue (a), cervix carcinoma (b) and lymph node metastases of head and neck carcinoma (c) in humans. The histograms are based on 929 measurements in 21 patients (a), 5434 measurements in 36 cervix tumours (b) and 717 measurements in 5 head and neck tumours (c).

are shown in Figs. 2 c and d ($P < 0.05$). Although most cervix tumours had low pO_2, a few tumours with relatively high pO_2 were also seen.

Cumulative frequency diagrams of median pO_2 and fraction of pO_2 readings below 5 mmHg are shown in Fig. 3 for cervix tumours and head and neck tumours. Median pO_2 ranged from 1 mmHg to 25 mmHg for the cervix tumours and from 13 mmHg to 40 mmHg

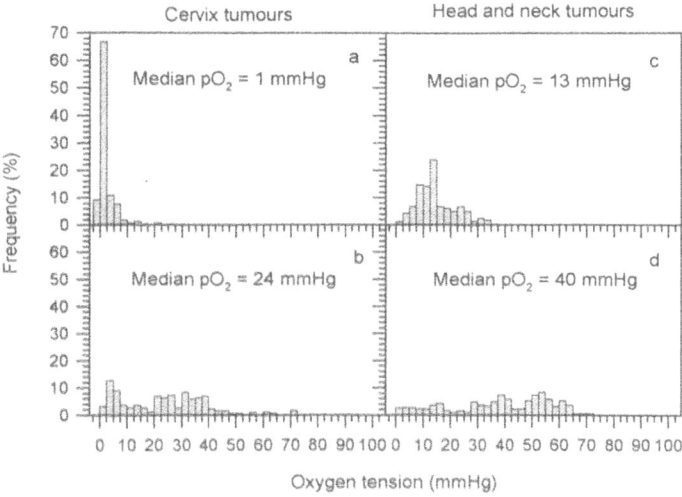

Figure 2. Frequency distributions of pO_2 for a human cervix carcinoma with low pO_2 (a), a human cervix carcinoma with high pO_2 (b), a lymph node metastasis of human head and neck carcinoma with low pO_2 (c) and a lymph node metastasis of human head and neck carcinoma with high pO_2 (d). The histograms are based on 238 measurements (a), 189 measurements (b), 163 measurements (c) and 189 measurements (d).

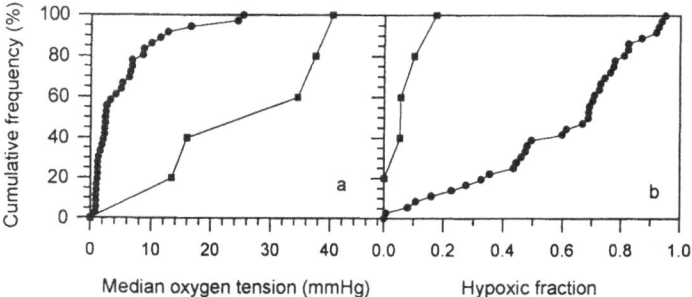

Figure 3. Cumulative frequency distributions of median pO_2 (a) and hypoxic fraction; *i.e.*, fraction of pO_2 readings below 5 mmHg (b) for human cervix carcinoma (circles) and lymph node metastases of human head and neck carcinoma (squares).

for the head and neck tumours. Moreover, fraction of pO_2 readings below 5 mmHg ranged from 0.01 to 0.95 for the cervix tumours and from 0.00 to 0.17 for the head and neck tumours. The median pO_2 and the 10th and 90th percentiles of the pO_2 histograms were significantly lower for the cervix tumours than for the head and neck tumours ($P < 0.05$), whereas the fractions of pO_2 readings below 2.5 mmHg, 5 mmHg and 10 mmHg were significantly higher for the cervix tumours than for the head and neck tumours ($P < 0.05$).

4. DISCUSSION

Frequency distributions of pO_2 values measured in cervix carcinoma and lymph node metastases of head and neck carcinoma were compared in the present work. Both tumour types showed highly heterogeneous pO_2 distributions, consistent with previously published data on pO_2 in human tumours (Vaupel *et al.*, 1989). The pO_2 values were within the ranges of those reported earlier for cervix tumours (Höckel *et al.*, 1993; Brizel *et al.*, 1995) and head and neck tumours (Gatenby *et al.*, 1988; Martin *et al.*, 1993; Nordsmark *et al.*, 1994; Brizel *et al.*, 1995). Although only a limited number of patients with head and neck tumours were included in the present study, it was found that these tumours had significantly higher pO_2 than the cervix tumours. Low pO_2 can occur if the measurements are performed in necrotic tumour tissue. However, necrosis was scarcely found in biopsies taken from the measurement tracks in the cervix tumours studied here (Lyng *et al.*, 1995). Presence of necrosis was therefore not the reason why pO_2 was measured to be lower in cervix tumours than in head and neck tumours. A difference in pO_2 between tumours can be caused by a difference in oxygen supply and/or oxygen consumption.

Tumour cells are supplied with oxygen through the ingrowth of vessels from the surrounding normal tissues (Folkman, 1985). Experimental studies have shown that the vascularization of the surrounding normal tissues has significant influence on the oxygen supply of tumours; *e.g.*, tumours with intramuscular growth have a higher blood flow than tumours of the same type growing intradermally or intra-abdominally (Hirst *et al.*, 1991). The normal tissues surrounding cervix tumours are poorly vascularized, especially in parous women (Höckel *et al*, 1991). Head and neck tumours, on the other hand, are surrounded by well vascularized normal tissues. This difference in normal tissue vascularization suggests that the oxygen supply may differ between cervix tumours and head and neck tumours. However, the oxygen consumption may also differ between the two tumour types. Potential doubling times reported for cervix tumours are generally shorter than those reported for head and neck tumours (Wilson, 1991), indicating higher proliferation activity and hence higher oxygen consumption in the former than in the latter tumour type. The difference in pO_2 between cervix tumours and head and neck tumours can therefore probably be caused by a difference in the oxygen supply as well as in the oxygen consumption.

Necrosis was hardly seen in the cervix tumours although the frequency of low pO_2 readings was high. There are several possible explanations for this observation. First, cervix tumours may have a high ability to use anaerobic glycolysis for energy production. Second, the oxygen consumption rate may be reduced significantly at low oxygen tensions in cervix tumours, leading to long oxygen diffusion distances. Large hypoxic areas may therefore develop before oxygen is totally depleted. Third, cervix tumours may have a high disposition for apoptotic response to oxygen and nutrient depletion rather than developing necrosis. Apoptosis can be induced under hypoxic conditions and is an important mechanism for cell death in many tumours (Majno and Joris, 1995; Graeber et al., 1996). Thus, apoptosis has been identified in cervix tumours and has been shown to be correlated with radiation response (Levine et al., 1995; Wheeler et al., 1995).

The present results may have some implications for the treatment of patients presenting with cervix carcinoma or lymph node metastases of head and neck carcinoma. Hypoxia-induced treatment resistance is a major problem in the radiotherapy of both tumour types (Overgaard et al., 1986; Höckel et al., 1993). The low pO_2 of cervix tumours compared to head and neck tumours indicates that the former tumour type contains more radioresistant hypoxic cells than the latter type. Several methods have been used to eliminate the hypoxic cells in these tumour types; e.g., breathing of hyperbaric oxygen, blood transfusion and infusion of nitroimidazole compounds (Glassburn et al., 1977; Bush, 1986; Révész and Balmukhanov, 1987). Our results indicate that improvement of radiation response by eliminating the hypoxic cells will be a more difficult task in cervix carcinoma than in lymph node metastases of head and neck carcinoma.

ACKNOWLEDGMENTS

Financial support from The Norwegian Cancer Society is gratefully acknowledged.

REFERENCES

Brizel, D. M., Rosner, G. L., Prosnitz, L. R., and Dewhirst, M. W., Patterns and variability of tumor oxygenation in human soft tissue sarcomas, cervical carcinomas, and lymph node metastases, Int. J. Radiat. Oncol. Biol. Phys. 32: 1121–1125 (1995).

Bush, R. S., The significance of anemia in clinical radiation therapy, Int. J. Radiat. Oncol. Biol. Phys. 12: 2047–2050 (1986).

Durand, R. E., The influence of microenvironmental factors on the activity of radiation and drugs, Int. J. Radiat. Oncol. Biol. Phys., 20: 253–258 (1991).

Folkman, J., Tumor angiogenesis, Adv. Cancer Res. 43: 175–203 (1985).

Folkman, J., What is the evidence that tumors are angiogenesis dependent?, J. Natl. Cancer Inst. 82: 4–6 (1990).

Gatenby, R. A., Kessler, H. B., Rosenblum, J. S., Coia, L. R., Moldofsky, P. J., Hartz, W. H., and Broder, G. J., Oxygen distribution in squamous cell carcinoma metastases and its relationship to outcome of radiation therapy, Int. J. Radiat. Oncol. Biol. Phys. 14: 831–838 (1988).

Glassburn, J. R., Brady, L. W., and Plenk, H. P., Hyperbaric oxygen in radiation therapy, Cancer 39: 751–765 (1977).

Graeber, T. G., Osmanian, C., Jacks, T., Housman, D. E., Koch, C. J., Lowe, S. W., and Giaccia, A. J., Hypoxia-mediated selection of cells with diminished apoptotic potential in solid tumours, *Nature* 379: 88–91 (1996).

Gulledge, C. J., and M. W. Dewhirst, M. W., Tumor oxygenation: A matter of supply and demand, *Anticancer Res.* 16: 741–750 (1996).

Hill, R. P., Tumor progression: Potential role of unstable genomic changes, *Cancer Metast. Rev.* 9: 137–147 (1990).

Hirst, D. G., Hirst, V. K., Shaffi, K. M., Prise, V. E., and Joiner, B., The influence of vasoactive agents on the perfusion of tumours growing in three sites in the mouse. *Int. J. Radiat. Biol.* 60: 211–218 (1991).

Höckel, M., Knoop, C., Schlenger K., Vorndran, B., Baußmann, E., Mitze, M., Knapstein, P. G., and Vaupel, P., Intratumoral pO_2 predicts survival in advanced cancer of the uterine cervix, *Radiother. Oncol.* 26: 45–50 (1993).

Höckel, M., Schlenger, K., Knoop, C., and Vaupel, P., Oxygenation of carcinomas of the uterine cervix: Evaluation by computerized O_2 tension measurements, *Cancer Res.* 51: 6098–6102 (1991).

Kallinowski, F., Zander, R., Hoeckel, M., and Vaupel, P., Tumor tissue oxygenation as evaluated by computerized-pO_2-histography, *Int. J. Radiat. Oncol. Biol. Phys.* 19: 953–961 (1990).

Levine, E. L., Renehan, A., Gossiel, R., Davidson, S. E., Roberts, S. A., Chadwick, C., Wilks, D. P., Potten, C. S., Hendry, J. H., Hunter, R. D., and West, C. M. L., Apoptosis, intrinsic radiosensitivity and prediction of radiotherapy reponse in cervical carcinoma, *Radiother. Oncol.* 37: 1–9 (1995).

Lyng, H., Sundfør, K., Tropé, C., and Rofstad, E. K., Heterogeneity in pO_2 and histological appearance in human cervix carcinoma, In: Tumor Oxygenation, P. W. Vaupel, D. K. Kelleher, and M. Günderoth (eds), New York: Gustav Fischer Verlag, pp. 249–258 (1995).

Lyng, H., Sundfør, K., Tropé, C., and Rofstad, E. K., Oxygen tension and vascular density in human cervix carcinoma, *Br. J. Cancer* in press (1996).

Majno, G., and Joris, I., Apoptosis, oncosis, and necrosis: An overview of cell death, *Am. J. Pathol.* 146: 3–15 (1995).

Martin, L., Lartigau, E., Weeger, P., Lambin, P., Le Ridant, A. M., Lusinchi, A., Wibault, P., Eschwege, F., Luboinski, B., and Guichard, M., Changes in the oxygenation of head and neck tumors during carbogen breathing, *Radiother. Oncol.* 27: 123–130 (1993).

Nordsmark, M., Bentzen, S. M., and Overgaard, J., Measurement of human tumour oxygenation status by a polarographic needle electrode, *Acta Oncol.* 33: 383–389 (1994).

Overgaard, J., Hansen, H. S., Jørgensen, K., and Hansen, M. H., Primary radiotherapy of larynx and pharynx carcinoma – an analysis of some factors influencing local control and survival, *Int. J. Radiat. Oncol. Biol. Phys.* 12: 515–521 (1986).

Révész, L., and Balmukhanov, S. B., Anaemia as a prognostic factor for the therapeutic effect of radiosensitizers, *Int. J. Radiat. Biol.* 51: 591–595 (1987).

Stone, H. B., Brown, J. M., Phillips, T. L., and Sutherland, R. M., Oxygen in human tumors: Correlations between methods of measurement and response to therapy, *Radiat. Res.* 136: 422–434 (1993).

Sutherland, R. M., Rasey, J. S., and Hill, R. P., Tumor biology, *Am. J. Clin. Oncol.* 11: 253–274 (1988).

Vaupel, P., Kallinowski, F., and Okunieff, P., Blood flow, oxygen and nutrient supply, and metabolic microenvironment of human tumors: A review, *Cancer Res.* 49: 6449–6465 (1989).

Wheeler, J. A., Stephens, L. C., Tornos, C., Eifel, P. J., Ang, K. K., Milas, L., Allen, P. K., and Meyn, R. E., Apoptosis as a predictor of tumor response to radiation in stage 1B cervical carcinoma, *Int. J. Radiat. Oncol. Biol. Phys.* 32: 1487–1493 (1995).

Wilson, G. D., Assessment of human tumour proliferation using bromodeoxyuridine – current status, *Acta Oncol.* 30: 903–910 (1991).

10

MEASUREMENT OF HYPOXIA *IN VIVO* USING A 2-NITROMIDAZOLE (NITP)

R. J. Hodgkiss,[1*] L. Webster,[2] and G. D. Wilson[1]

[1]Gray Laboratory Cancer Research Trust
Mount Vernon Hospital
Northwood, Middlesex HA6 2JR, England
[2]Department of Immunology
The Rayne Institute
St. Thomas' Hospital
Lambeth Palace Road, London SE1 7EH, England

1. INTRODUCTION

The vascularisation of tumours is relatively poor and disordered compared with that of normal tissues, leading to inefficient delivery of oxygen and other nutrients to many tumour cells. As oxygen diffuses from blood capillaries through the mass of surrounding cells it is depleted by normal cell metabolism. In the pioneering work of Thomlinson and Gray an effective diffusion distance of about 150 μm was calculated (Thomlinson and Gray 1955) which closely matched the radii of viable tumour cords. Beyond this distance lie radioresistant hypoxic tumour cells at the boundary between viable and necrotic cells. Transient changes in the perfusion of individual blood vessels may further reduce oxygen transport and change the oxygenation status of some cells near to blood vessels over relatively short time scales (e.g. Chaplin *et al.* 1986). Moderate levels of radiobiological hypoxia are also known to exist in some normal tissues e.g. murine skin (Stewart *et al.* 1982) and liver (Arteel *et al.* 1995).

The bioreductive metabolism of 2-nitroimidazoles proceeds in a series of one-electron reductions. The first one-electron reduction product is very reactive towards oxygen and is oxidised to the parent nitroimidazole so efficiently that there is effectively no substrate for the second bioreductive step, when 2-nitroimidazoles are metabolised in the presence of oxygen (Mason and Holtzman 1975, Wardman and Clarke 1976). This futile metabolism generates toxic oxidising species such as superoxide and this is the basis for the anti-parasitic activity of some 2-nitroimidazoles. Under hypoxic conditions, bioreduc-

---

*Correspondence to: R.J. Hodgkiss, Gray Laboratory Cancer Research Trust, Mount Vernon Hospital, Northwood. Middlesex. HA6 2JR. England. Telephone: 01923-828611; Fax: 01923-835210

tive metabolism leads to binding of portions of the molecule, including side-chains (Raleigh *et al.* 1985), to macromolecular components of cells in tissues and tumours. The oxygen-dependence of 2-nitroimidazoles binding in cells is similar to the dependence of radiosensitivity (Franko *et al.* 1987, Hodgkiss *et al.* 1991).

The hypoxia-specific metabolism and binding of 2-nitroimidazoles has been used to identify hypoxic cells in tumours and some normal tissues, initially using radiolabelled misonidazole (e.g. Chapman *et al.* 1981, Franko and Chapman 1982, Raleigh *et al.* 1985). This approach has been used in the clinic (Urtasun *et al.* 1986a, 1986b), but the relatively large dose of a long-lived radiolabel needed to obtain adequate labelling has limited its application although zones of intense labelling were observed in melanomas and a small cell lung cancer. Some 2-nitroimidazoles labelled with short-lived isotopes may be more suitable for general clinical use as non-invasive hypoxia assays (Rasey *et al.* 1985, 1989, Biskupiak *et al.* 1991, Mannan *et al.* 1992a, 1992b, Linder *et al.* 1994, Rumsey *et al.* 1995) and compounds with fluorinated side-chains can be detected non-invasively by nuclear magnetic resonance (Raleigh *et al.* 1986, Maxwell *et al.* 1987, Li *et al.* 1991, Aboagye *et al.* 1995). Several immunologically detectable 2-nitroimidazoles have been described as hypoxia markers (Raleigh et al 1987, Hodgkiss *et al.* 1991, Cline *et al.* 1994, Hodgkiss *et al.* 1994, Arteel *et al.* 1995, Evans *et al.* 1996) and allow histological or flow cytometric assessment of tumour hypoxia.

Hypoxia and cell proliferation are two microenvironmental factors thought to contribute to the clinical radioresistance of tumours. Although proliferation has been reported to be confined to well-oxygenated regions of tumours around blood vessels (Hirst and Denekamp 1979), simultaneous staining of cells disaggregated from tumours for bound metabolites of an immunologically-identifiable hypoxia marker (NITP) and a proliferation marker (BrdUrd) has shown there to be some S-phase DNA synthesis and cell-cycle movement in the hypoxic sub-compartment of tumours (Webster *et al.* 1994, 1995). However, hypoxia is not evenly distributed through the cell cycle in tumours and a greater proportion of G2 phase cells are hypoxic compared with the other cell cycle phases. This paper extends the analysis of proliferation and hypoxia by examining the dependence of BrdUrd labelling within tumours on the degree of bioreductive labelling with NITP, and therefore on the level of hypoxia at which this labelling occurred.

2. MATERIALS AND METHODS

The murine tumours used (CaNT poorly differentiated mammary adenocarcinoma, Rhodesia (Rh) moderately well-differentiated adenocarcinoma, SaF anaplastic sarcoma) were of spontaneous origin, maintained by serial transplantation and for this work were grown in the dorsal subcutaneous site in syngeneic mice. Tumours were selected for use at mean geometric diameters of 8–10 mm.

A novel 2-nitroimidazole, 7-(4′-(2-nitroimidazol-l-yl)-butyl)-theophylline (NITP), with an immunologically-identifiable side-chain (theophylline) was custom synthesised by Lancaster Synthesis using similar methods to those previously described (Long *et al.* 1991). Use and detection of the NITP bioreductive hypoxia marker and its simultaneous use with the BrdUrd proliferation marker has been described (Hodgkiss *et al.* 1991, Webster *et al.* 1995). Briefly, NITP (0.45 μmol g^{-1}) was administered i.p. dissolved in peanut oil + 10% DMSO. Animals were sacrificed two hours later, tumours excised, weighed and disaggregated into a single cell suspension which was fixed in 70% ethanol. Bound adducts of NITP were identified in cells and normal tissues using a polyclonal antiserum raised against theophylline and BrdUrd-labelled DNA was identified with a monoclonal antibody. In order for the monoclonal antibody to recognise BrdUrd it is necessary to partly denature the DNA and this was accomplished by treating the fixed cells with 0.2 mg/ml pepsin in 2 M HCl for 20 min at room temperature, which also converts the cells to cell nuclei. For flow cytometry indirect immunofluorescent staining with FITC and phyco-erythrin-labelled second antibodies was used to detect bound NITP and BrdUrd respectively and quantitative DNA staining with 7-aminoactinomycin D was used for single-cell discrimination and cell-cycle analysis. Flow cytometry analysis of stained cells was carried out on a Becton-Dickinson FacScan with laser excitation at 488 nm.

3. RESULTS

NITP was well tolerated by experimental animals and had little effect on physiological parameters such as breathing rate, core temperature or relative blood flow in tumours, gastrocnemius muscle, gut, kidney and skin (Hodgkiss *et al.* 1991, Hodgkiss *et al.* 1995). Bound metabolites of the NITP hypoxia marker could be reliably detected in hypoxic cells in tumours using standard immunofluorescent staining of single cells disaggregated from tumours for flow cytometry and the level of binding was greatly reduced by strategies designed to improve the oxygenation of tumours such as allowing the animals to breathe oxygen or carbogen (95% oxygen, 5% carbon dioxide), or treating them with the blood-flow modifier nicotinamide (Hodgkiss and Wardman 1992, Rojas *et al.* 1992, Figure 1).

Converting cells to cell nuclei had little effect on the intensity of fluorescent staining for bound NITP metabolites, despite loss of the cytoplasm. Simultaneous triple-staining for bioreductively bound NITP, incorporated BrdUrd and for DNA content allowed the interdependence of hypoxia, proliferation and cell-cycle stage to be investigated (Figure 2). The aneuploid SaF tumour contained a considerable number of diploid stromal normal cells. A region (R1) was set in Figure 2a to include only 1% of the non-specific background fluorescent staining of aneuploid cells from a tumour not exposed to NITP, and thus cells from tumours exposed to NITP *in vivo*, exhibiting staining within the region R1 must have contained metabolically bound NITP and were therefore hypoxic. Figure 2b shows the pattern of BrdUrd labelling in region R1, with overlap between the hypoxic and proliferating compartments of the tumour (Webster *et al.* 1994), although with a lower labelling index than in the well oxygenated compartment (Figure 2c).

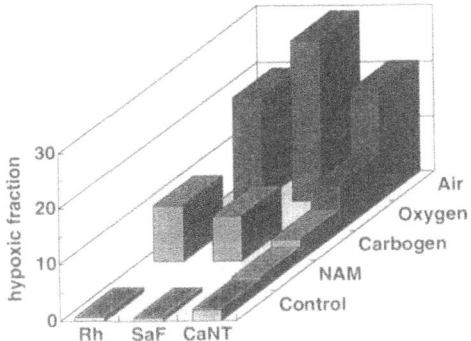

Figure 1. Hypoxic fractions of CaNT, SaF and Rh tumours, measured using NITP, in animals breathing air, oxygen or carbogen or treated with 500 mg/kg nicotinamide in air.

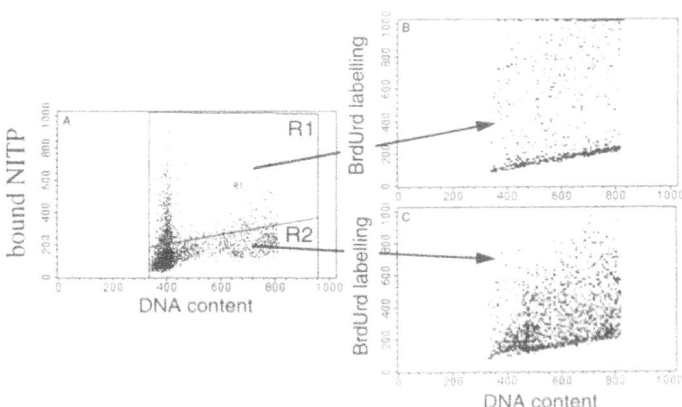

Figure 2. Flow cytometry analysis of single cells disaggregated from an SaF tumour simultaneously stained for bound NITP, BrdUrd labelling and DNA. (a) bound NITP (FITC) *vs.* DNA content (propidium iodide): (b) BrdUrd labelling (R-phycoerythrin) *vs.* DNA content in aneuploid hypoxic tumour cells from region R1: (c) BrdUrd labelling *vs.* DNA content in aneuploid oxic tumour cells from region R2.

To investigate this further, cells from an SaF tumour were simultaneously stained for bound NITP, BrdUrd and DNA and a series of horizontal regions made at different levels of NITP binding (Figure 3a) within which, the amount of cell proliferation (Figure 3b) was determined (Figure 3c). In Figure 3c cells in regions R1 and R2 (well oxygenated cells) have similar background staining intensities to cells from control tumours not exposed to NITP, while the cells in regions R3–R6 have higher staining intensities indicating bound NITP and are therefore classed as hypoxic.

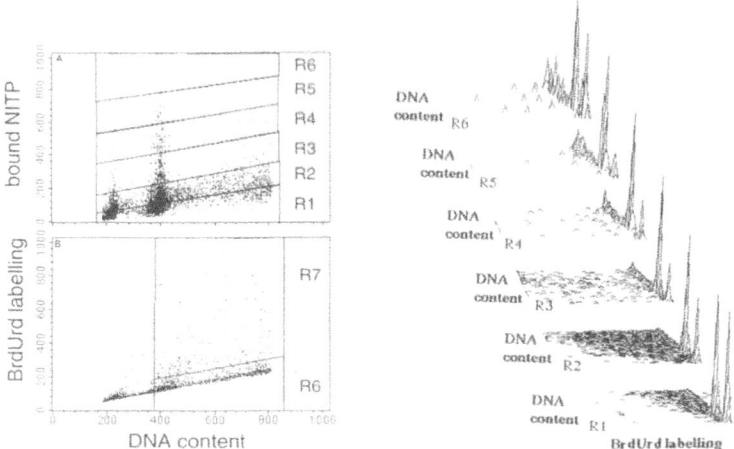

Figure 3. Flow cytometry analysis of single cells disaggregated from an SaF tumour simultaneously stained for bound NITP, BrdUrd labelling and DNA. (a) bound NITP (FITC) *vs.* DNA content (propidium iodide) divided into regions R1 - R6 with increasing NITP binding: (b) BrdUrd labelling (R-phycoerythrin) *vs.* DNA content with regions indicated for proliferating (R7) and non-proliferating cells (R8): (c) BrdUrd labelling of aneuploid cells at different levels of NITP binding in regions R1 - R6 from Figure 3a.

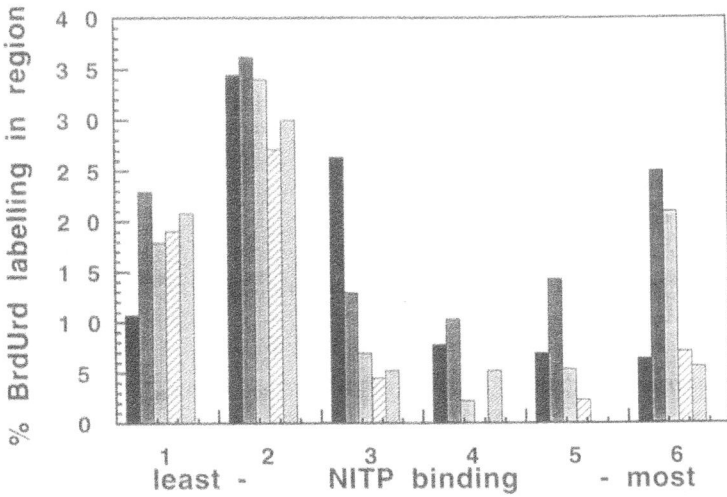

Figure 4. %BrdUrd labelling within each level of NITP binding (Figure 3a, R1 - R6) in cells from 5 separate SaF tumours.

Surprisingly, the population of well oxygenated cells is not homogenous, with different proportions of cells labelling with BrdUrd in regions R1 and R2. Most of the hypoxic cells labelling with BrdUrd are found in the first region above the normoxic cells (R3) with rapidly-decreasing numbers of cells incorporating BrdUrd as the level of NITP binding increases. The BrdUrd labelling of cells from each level of NITP staining is summarised in Figure 4 and shows that a significant proportion of the cells in each region are undergoing S-phase DNA synthesis. However the high proportion of labelled cells observed in some tumours in region 6 represents a tiny fraction of the total cells in the tumour.

4. DISCUSSION

Staining cells simultaneously for multiple parameters provides considerably more information than could be obtained from analysing each parameter individually, and in this case the interplay between hypoxia, proliferation and cell cycle position can be studied. An intercellular range of bound NITP metabolites is observed and is likely to reflect the distribution of oxygen tensions within tumours. Hypoxia occurs in the normal diploid component and throughout the cell cycle of the aneuploid component of these tumours. Proliferating cells are also found in both hypoxic and well oxygenated compartments of the tumour, but most of the hypoxic proliferating cells are found to have bound modest amounts of the hypoxia marker and are therefore probably relatively poorly oxygenated but not completely anoxic. The well oxygenated cells in region R2 have a higher labelling index than the hypoxic population. However, the proportion of hypoxic cells labelling with BrdUrd is not strongly dependent on the extent of their labelling with the hypoxia marker (Figure 4) and therefore the degree of hypoxia experienced.

Acute (perfusion-limited) hypoxia occurs in tumours with changes in oxygen tensions over relatively short time scales, demonstrated by administration of two fluorescent markers separated by a time interval (Chaplin *et al.* 1986). It is therefore possible that cells labelling with both hypoxia and proliferation markers may have experienced fluctuations in oxygen tensions allowing incorporation of the two labels in adjacent time slots. In this case, the moderate amount of bound NITP observed in the majority of hypoxic cells incorporating BrdUrd could reflect a short hypoxic interval rather than an intermediate oxygen tension. Inspection of histological sections of the CaNT corded tumour shows that the majority of cords contain relatively well-oxygenated cells adjacent to blood vessels with bound adducts of NITP occurring 6–8 cell-diameters from the central blood vessel as a result of diffusion-limited hypoxia. However, a minority of tumour cords exhibit NITP binding right up to the central blood vessel, which we interpret as examples of perfusion-limited acute hypoxia. The SaF tumour is not corded and it is harder to assess the relationship between vasculature and hypoxia in this tumour. However, it seems likely that acute hypoxia will contribute to the overlap between hypoxia and proliferation seen in the SaF tumour and experiments are in progress to resolve this issue.

The hypoxia marker NITP is a rapid and convenient method of assessing the occurrence of hypoxia in tumours and normal tissues, and is compatible with simultaneous assessment of proliferation using BrdUrd. Simultaneous assessment of hypoxia, proliferation and cell-cycle phase has already demonstrated that the relationships between these parameters are more complex than expected. NITP is of low toxicity and is currently undergoing formulation studies for trial as a clinical marker of tumour hypoxia.

ACKNOWLEDGMENTS

This work was supported by the Cancer Research Campaign.

REFERENCES

Aboagye, E.O., Lewis, A.D., Johnson, A., Workman, P., Tracy, M. and Huxham, I.M., 1995. The novel fluorinated 2-nitroimidazole hypoxia probe SR-4554: reductive metabolism and semi-quantitative localisation in human ovarian cancer multicellular spheroids as measured by electron energy loss spectroscopic analysis. *Br. J. Cancer,* **72,** 312–318.

Arteel, G.E., Thurman, R.G., Yates, J.M., Raleigh, J.A., 1995. Evidence that hypoxia markers detect oxygen gradients in liver: pimonidazole and retrograde perfusion of rat liver. *Br. J. Cancer,* **72,** 889–895.

Biskupiak, J.E., Grierson, J.R., Rasey, J.E., Martin, G.V. and Krohn, K.A., 1991. Synthesis of an (iodovinyl)misonidazole derivative for hypoxia imaging. *J. Med. Chem.,* **34,** 2165–2168.

Chaplin, D.J., Durand, R.E. and Olive, P.L., 1986. Acute hypoxia in tumours: implications for modifiers of radiation effects. *Int. J. Radiat. Oncol. Biol. Phys.,* **12,** 1279–1282.

Chapman, J.D., Franko, A.J. and Sharplin, J., 1981. A marker for hypoxic cells in tumours with potential clinical applicability. *Br. J. Cancer,* **43,** 546–550.

Cline, J.M., Thrall, D.E., Rosner, G.L. and Raleigh, J.A., 1994. Distribution of the hypoxia marker CCI-103F in canine tumours. *Int. J. Radiat. Oncol. Biol. Phys.,* **28,** 921–933.

Evans, S.M., Jenkins, W.T., Joiner, B., Lord, E.M. and Koch, C.J., 1996. 2-Nitroimidazole (EF5) binding predicts radiation resistance in individual 9L s.c. tumors. *Cancer Res.,* **56,** 405–411.

Franko, A.J. and Chapman, J.D., 1982. Binding of 14C-misonidazole to hypoxic cells in V79 spheroids. *Br. J. Cancer,* **45,** 694–699.

Franko, A.J., Koch, C.J., Garrecht, B.M., Sharplin, J. and Hughes, D., 1987. Oxygen dependence of binding of misonidazole to rodent and human tumors in vitro. *Cancer Res.,* **47,** 5367–5376.

Hirst, D.G. and Denekamp, J., 1979. Tumour cell proliferation in relation to the vasculature. *Cell Tissue Kinet.,* **12,** 31–42.

Hodgkiss, R.J., Jones, G., Long, A., Parrick, J., Smith, K.A., Stratford, M.R.L., and Wilson, G.D., 1991. Flow cytometric evaluation of hypoxic cells in solid experimental tumors using fluorescence immunodetection. *Br. J. Cancer,* **63,** 119–125.

Hodgkiss, R.J. and Wardman, P., 1992. The measurement of hypoxia in tumours. *Br. J. Radiol., Suppl.* **24**, 105–110.

Hodgkiss, R.J., Parrick, J., Porssa, M. and Stratford, M.R.L., 1994. Bioreductive markers for hypoxic cells: 2-nitroimidazoles with biotinylated 1-substituents. *J. Med. Chem.*, **37**, 4352–4356.

Hodgkiss, R.J., Stratford, M.R.L., Dennis, M.F. and Hill, S.A., 1995. Pharmacokinetics and binding of the bioreductive probe for hypoxia, NITP: effect of route of administration. *Br. J. Cancer*, **72**, 1462–1468.

Li, S.J., Jin, G.Y. and Moulder, J.E., 1991. Prediction of tumour radiosensitivity by hexafluoromisonidazole retention monitored by [^{1}H]/[^{19}F] magnetic resonance spectroscopy. *Cancer Comms.*, **3**, 133–139.

Linder, K.E., Chan, Y.W., Cyr, J.E., Malley, M.F., Nowotnik, D.P. and Nunn, A.D., 1994. TcO(PnA.O-1-(2-nitroimidazole)) [BMS-181321], a new technetium-containing nitroimidazole complex for imaging hypoxia: synthesis, characterisation, and xanthine oxidase-catalysed reduction. *J. Med. Chem.*, **37**, 9–17.

A. Long, J. Parrick and R.J. Hodgkiss, 1991. An efficient procedure for the 1-alkylation of 2-nitroimidazoles and the synthesis of a probe for hypoxia in solid tumours. *Synthesis*, **9**, 709–713.

Mannan, R.H., Mercer, J.R., Wiebe, L.I., Kumar, P., Somayaji, V.V. and Chapman, J.D., 1992a. Radioiodinated azomycin pyranoside (IAZP): a novel non-invasive marker for the assessment of tumour hypoxia. *J. Nucl. Biol. Med.*, **36**, 60–67.

Mannan, R.H., Mercer, J.R., Wiebe, L.I., Somayaji, V.V. and Chapman, J.D., 1992b. Radioiodinated 1-(2-fluoro-4-iodo-2,4-dideoxy-beta-L-xylopyranosyl)-2-nitroimidazole: a novel probe for the non-invasive assessment of tumor hypoxia. *Radiat. Res.*, **132**, 368–374.

Mason, R.P. and Holtzman J.L., 1975. The role of catalytic superoxide formation in the O$_2$ inhibition of nitroreductase. *Biochem. Biophys. Comms.* **67**, 1267–1274.

Maxwell, R.J., Workman, P. and Griffiths, J.R., 1987. Demonstration of tumour-selective retention of fluorinated nitroimidazole probes by ^{19}F magnetic resonance spectroscopy. *Int. J. Radiat. Oncol. Biol. Phys.*, **16**, 925–929.

Raleigh, J.A., Franko, A.J., Koch, C.J. and Born, J.L., 1985. Binding of misonidazole to hypoxic cells in monolayer and spheroid culture: evidence that a side-chain label is bound as efficiently as a ring label. *Br. J. Cancer*, **51**, 229–235.

Raleigh, J.A., Franko, A.J., Treiber, E.O., Lunt, J.A. and Allen, P.S., 1986. Covalent binding of a fluorinated 2-nitroimidazole to EMT-6 tumours in BALB/C mice: detection by F-19 nuclear magnetic resonance at 2.35 T. *Int. J. Radiat. Oncol. Biol. Phys.*, **12**, 1243–1245.

Raleigh, J.A., Miller, G.G., Franko, A.J., Koch, C.J., Fuciarelli, A.F. and Kelley, D.A., 1987. Fluorescence immunohistochemical detection of hypoxic cells in spheroids and tumours. *Br. J. Cancer*, **56**, 395–400.

Rasey, J.S., Koh, W.J., Grierson, J.R., Grunbaum, Z. and Krohn, K.A., 1989. Radiolabelled fluoromisonidazole as an imaging agent for tumour hypoxia. *Int. J. Radiat. Oncol. Biol. Phys.*, **17**, 985–991.

Rasey, J.S., Krohn, K.A., Grunbaum, Z., Conroy, P.J., Bauer, K. and Sutherland, R.M., 1985. Further characterisation of 4-bromomisonidazole as a potential detector of hypoxic cells. *Radiat. Res.*, **102**, 76–85.

Rojas, A., Joiner, M.C., Hodgkiss, R.J. Carl, U., Kjellen E. and Wilson, G.D., 1992. Enhancement of tumor radiosensitivity and reduced hypoxia-dependent binding of a 2-nitroimidazole with normobaric oxygen and carbogen: a therapeutic comparison with skin and kidneys. *Int. J. Radiat. Oncol. Biol. Phys.*, **23**, 361–366

Rumsey, W.L., Patel, B. and Linder, K.E., 1995. Effect of graded hypoxia on retention of technetium-99m-nitroheterocycle in perfused rat heart. *J. Nucl. Med.*, **36**, 632–636.

Stewart, F.A., Denekamp, J. and Randhawa, V.S., 1982. Skin sensitisation by misonidazole: a demonstration of uniform mild hypoxia. *Br. J. Cancer*, **45**, 869–877.

Thomlinson and Gray, 1955. The histological structure of some human lung cancers and the possible implications for radiotherapy. *Br. J. Cancer* , **9**, 539–549.

Urtasun, R.C., Chapman, J.D., Raleigh, J.A., Franko, A.J. and Koch, C.J., 1986a. Binding of ^{3}H-misonidazole to solid human tumours as a measure of tumour hypoxia. *Int. J. Radiat. Oncol. Biol. Phys.*, **12**, 1263–1267.

Urtasun, R.C., Koch, C.J., Franko, A.J., Raleigh, J.A. and Chapman, J.D., 1986b. A novel technique for measuring human tissue pO$_2$ at the cellular level. *Br. J .Cancer*, **54**, 453–457.

Wardman, P. and Clarke, E.D., 1976. Oxygen inhibition of nitroreductase electron transfer from nitro radical-anions to oxygen. *Biochem.. Biophys. Res. Com.* **69**, 942–949.

Webster, L., 1994. Hypoxia and proliferation in murine tumour models. PhD Thesis University of London Faculty of Science.

Webster, L., Hodgkiss, R.J. and Wilson, G.D., 1995. Simultaneous triple staining for hypoxia, proliferation and DNA content in solid murine tumors. *Cytometry*, **21**, 344–351.

11

NIR REFLECTION MEASUREMENTS OF HEMOGLOBIN AND CYTOCHROME aa$_3$ IN HEALTHY TISSUE AND TUMORS

Correlations to Oxygen Consumption: Preclinical and Clinical Data

F. Steinberg,[1] H. J. Röhrborn,[1] T. Otto,[2] K. M. Scheufler,[3] and C. Streffer[1]

[1]Institute of Medical Radiation Biology Essen
[2]Urology Essen
[3]Neurosurgery Bonn
45122 Essen, Hufelandstr. 55, Germany

1. SUMMARY

Objectives

A new near infrared reflectance spectroscopy based technology (MULTISCAN OS 10/30) for non-invasive measurements of tissue oxygenation allows detection of absolute tissue hemoglobin concentration and saturation values in real time.

Methods

MULTISCAN OS 10/30 scans a tissue sector of defined geometry at 400 to 1200 nm wavelength with 0.3 nm intervals at a scan rate of up to 400 Hz in reflection mode. The newly developed algorithms are based on the entropy pattern of photons and allow the detection of absolute values of total hemoglobin (tHb), deoxy- (Hb) and oxy-hemoglobin (HbO$_2$) in mg/ml tissue as well as tissue oxygen (saturation = tiSO$_2$) in the range of 0 to 100 % in real time. Cytochrome aa$_3$ (Cyt) can be monitored simultaneously.

Results

Physiological Stimulation. Clamping experiments at the finger of volunteers, in the intestine of a pig, in the kidney of mice and rats showed a very fast, sensitive and tissue specific reaction. Changes in breathing and/or anesthesia conditions were immediately followed by corresponding changes of tissue oxygenation (brain, finger, skin). Simultaneous measurements of oxygen (HbO$_2$), and cytochrome showed a strong correlation between both parameters. For validation purposes parallel polarographic measurements (pO$_2$ measurements) were performed in mice. Furthermore angiographic data of patients as well as NMR/I data were compared with the NIR spectroscopy findings.

Tumor Measurements. Patients with tumors of different origins showed significantly different oxygen values. The lowest level was found in glioblastoma (< 10 % sat, < 1.0 mg/ml tHb), whereas renal carcinoma showed tremendously increased values (> 90 % sat, > 5.0 mg/ml tHb). In contrast the surrounding healthy kidney tissue was significantly *less* oxygenated and perfused compared to the tumors.

Oxygen Consumption Measurements. Oxygen consumption was measured in 16 patients after ligation of the blood supply (A. and V. renalis) to the kidney affected by tumor. All tumors showed a significantly lower consumption rate compared to the healthy tissue. These findings were controlled by animal experiments of human renal carcinomas on nude mice. The same results were obtained under these experimental conditions.

Conclusion

MULTISCAN OS 10/30 is a new and useful tool for in vivo characterization of oxygen and cytochrome in healthy and tumor tissues. In many clinically relevant situations oxygen measurements can be helpful and support the clinical routine diagnostics.

2. MATERIAL AND METHODS

2.1. Patients

Data from 45 patients with solid tumors (glioblastoma, meningeoma) and arteriovenous malformations (AVM) from the departments of neurosurgery and urology (University Essen) and 10 healthy volunteers aged 35 to 59 years (male and female) were included in this study.

2.2. Methods

HbO$_2$, Hb, tHb (in mg/ml tissue) and saturations values (tiSO$_2$ in %) were measured simultaneously in real time, displayed on-line and stored on hard-disc. These parameters defined the individual oxygen status from at least 15 consecutive measurements (1/s) per location. Additionally, blood gas analysis was performed every 30 min. during operation. Measurement protocols were performed within less than 2 min. The measured tissue volume is defined by a cone (diameter of the sensor fiber = 1 mm and depth of 4 mm). The sensor was sterilized with GIGASEPT®. The surface of the tissue must be kept blood free during intraoperative measurements.

MULTISCAN OS 10/30 scans a tissue sector of defined geometry at 400 to 1200 nm wavelength with 0.3 nm intervals at a scan rate of up to 400 Hz in reflection mode (see also Fig 1a). The newly developed algorithms (FOG Theory) are based on the entropy pattern of photons (spectroscopic finger prints) and allow the detection of absolute tissue hemoglobin concentrations and saturation. Lambert Beer's Law is only valid for light in

Table 1. Summary of the differences in components and parameters of the MULTISCAN systems OS 10, OS 20, and OS 30

	OS 10	OS 20	OS 30
Components			
Wavelength	400–1000 nm	500–1000 nm	500–1000 nm
Spectral resolution	> 0.3 nm	> 0.1 nm	> 0.1 nm
Detector	Monochromator	CCD array	CCD array
Fiber	2000 fibers with Bernoulli distribution	Pencil and Pin sensor, Quarz fibers, sterilisable, movable 1 + 6 fibers (125 μm)	Pencil and Pin sensor, Quarz fibers, sterilisable, movable 1 + 6 fibers (125 μm)
Light source	Halogen white light electrical input: 60W(300–1200 nm)	Halogen white light electrical input: 8 W (300–1200 nm)	Halogen white light electrical input: 8 W (300–1200 nm)
Data output (screen)	≈ 1Hz	≈ 1Hz	≈ 1Hz
Data aquisition	1–400 Hz	1–1000 Hz	1–1000 Hz
Parameter			
Saturation	Calculated from HbO$_2$ and Hb	Measured independently from HbO$_2$ and Hb	Measured independently from HbO$_2$ and Hb
Hemoglobin	HbO$_2$, Hb	HbO$_2$, Hb	HbO$_2$, Hb
Cytochrome	—	—	Cytochrome aa$_3$(oxidised or reduced)

homogeneous media and does not account for any kind of scattering effects. Due to the extremely heterogeneous distribution of hemoglobin/cytochrome in tissue and the non-linear relation between absorption and reflection Lambert Beer's Law can not be used in tissues. Based on Lübbers' Axiom (personal communication Lübbers, 1994) it is possible to distinguish between spectral properties, object properties and pathway properties. Thus, the entropy pattern can be measured. The Hopf Theorem (mathematician 1824–1882) allows calculation of interferences of the entropy pattern of back-scattering light. The basis of these spectroscopical finger prints are extremely precise calibration spectra (Siggard-Anderson et al. 1972, Weissbluth, 1974, Lübbers et al. 1965) of hemoglobin and cytochrome in a wide range of 400 up to 1000 nm which were described in 1875 in the thesis

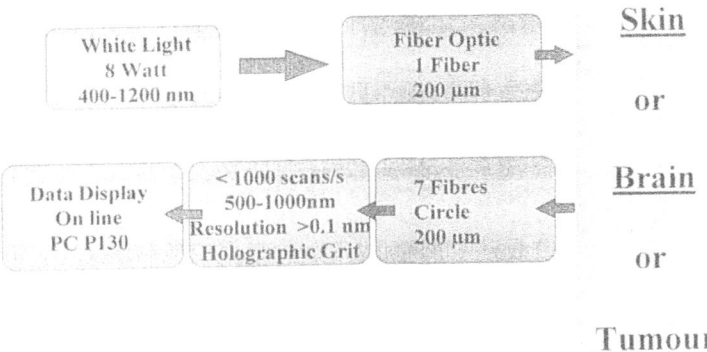

Figure 1. Shows a schematic drawing of the hardware used in MULTISCAN OS 20/30. See Table 1 for a summary of the differences in components and parameters of the MULTISCAN systems OS 10, OS 20 and OS 30.

of Wiskemann for the first time. The hardware components of the systems are shown in a schematic drawing (Fig 1b). The different systems used in the study are described in Fig. 1a.

3. RESULTS

Fig 2 shows measurements of tHb, HbO_2 and $tiSO_2$ taken from the finger tip of a 58-year-old male healthy volunteer. Both A. digitales were clamped (Fig 2a). Zero levels of HbO_2 were reached within 15 to 17 s. tHb decreased down to 0.4 mg/ml. This experiment

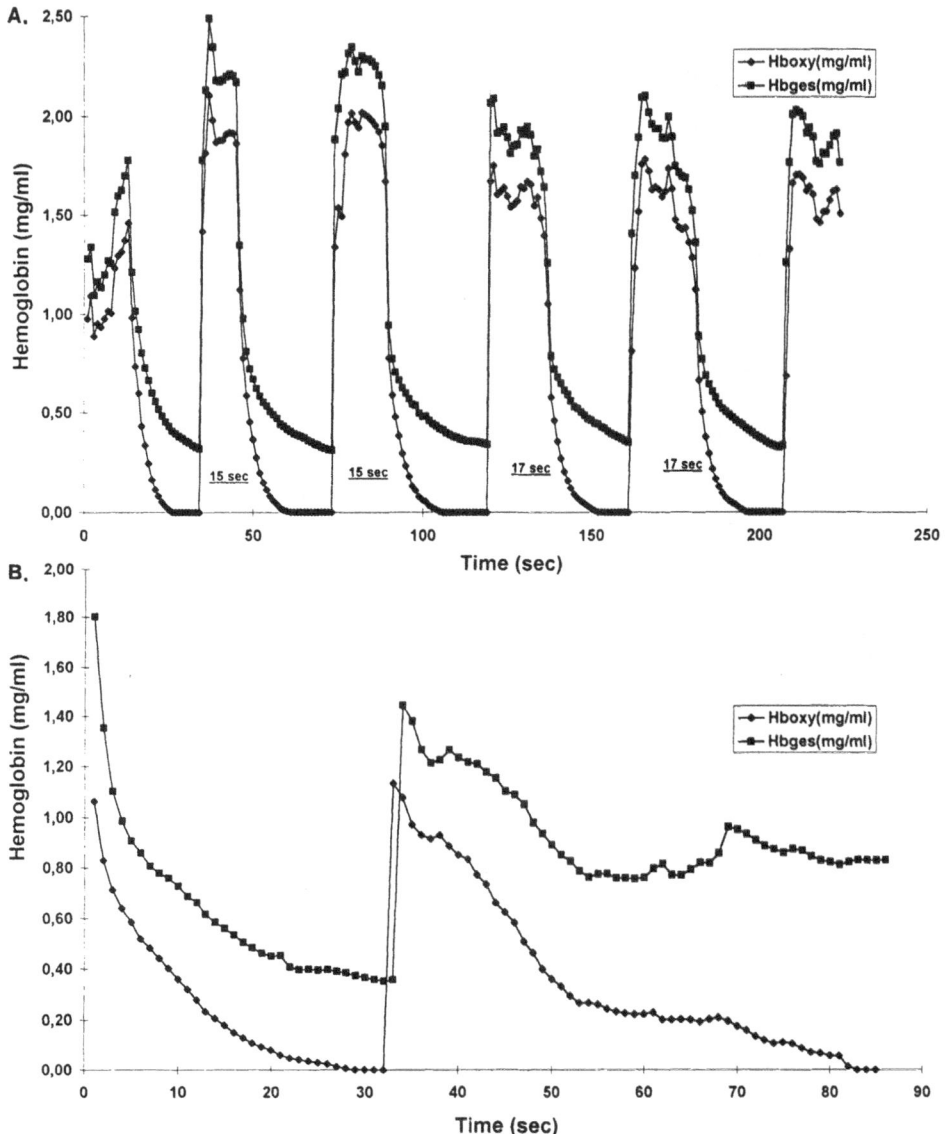

Figure 2. Changing of HbO_2 and tHb after clamping the Arteria digitalis (a) and the Arteria brachialis (b), measured in the fingertip. The zero level (HbO_2) is reached within 15 or 30 sec, respectively.

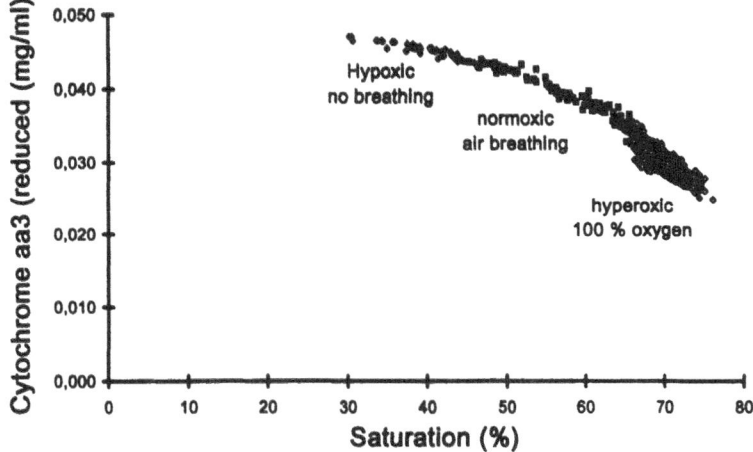

Figure 3. Correlation of cytochrome aa$_3$ (reduced) and saturation in three different breathing situations: air breathing, no breathing (for 90 sec), 100 % oxygen breathing, measured in the finger tip of a volunteer ($r^2 = 0.98$).

was repeated more than 6 times, yielding the same results. A. and V. brachialis were clamped (cuff pressures exceeding 280 mmHg, Fig 2b). The same effect could be observed: within 30 to 40 s HbO$_2$ concentrations decreased to zero (tHb reached the same level as seen at Fig 2a).

Fig 3 shows correlations of tiSO$_2$ and cytochrome aa$_3$ (reduced), measured at the finger tip of a volunteer in three different breathing situations: air breathing (= 21 % O$_2$), 100 % oxygen breathing and breath-holding (approx. 90 s). Graphic display of data as a function of both parameters shows a strong interdependency ($r^2 = 98$).

Table 2 shows data from 13 patients with untreated solid tumors of different origin and metastases. This table demonstrates the high interindividual variability of all three parameters. Similarly, it is important to note that the distribution of individual parameters also showed a wide range. Low saturation values can be seen in combination with both high and low tHb values.

Table 2. 13 patients with solid tumors and metastasis. See the significant differences in oxygenation from one entity to the other and the ranking based on tiSO$_2$ values

	Saturation (%)	HbO$_2$ (mg/ml)	Total Hb (mg/ml)
Astrocytoma	19	0,22	1,10
Astrocytoma	55	1,38	2,52
Meningioma	11	2,54	22,18
Meningioma	15	0,77	4,58
Meningioma	39	3,02	7,65
Brain Metastasis (MammaCa)	19	0,04	0,22
Brain Metastasis (Rectum Ca)	27	0,75	2,79
Kidney Tumour	64	> 4,92	> 7,68
Kidney Tumour	100	3,65	3,65
Kidney Tumour	94	> 4,92	>5,23
Kidney Tumour	98	> 4,92	> 5,02
Testis Tumour	95	4,52	4,59
Liposarcoma	49	> 4,92	> 10,04

Table 3. Simultaneous measurement of the oxygen situation in a
46-year-old male patient with renal cell carcinoma. Tumor was
measured at 3 different locations with significantly higher
oxygenation within the tumor compared to
the healthy tissue of the kidney

	Saturation (%)	HbO$_2$ (mg/ml)	Total Hemoglobin (mg/ml)
Tumor			
1	100	3,65	3,65
2	100	2,59	2,59
3	90	3,54	3,81
Kidney	61	1,22	2,01

Simultaneous measurements were performed in healthy kidneys and renal cell carcinomas intraoperatively in 16 patients. The untreated tumor of a 46-year-old male patient shown in Table 3 was measured in three different positions. tHb, HbO$_2$ and tiSO$_2$ values measured in healthy kidney tissue were significantly lower compared to those measured in the tumor, which was located within the same kidney. The same results were seen in the other patients, too. To prevent vascular metastasis, A. and V. renalis were ligated. Simultaneously, the oxygen consumption was measured in tumor (Fig 4a) as well as in healthy kidney tissue (Fig 4b). All tumors showed a significantly lower consumption rate compared to the healthy tissue as well as a higher initial tiSO$_2$. These findings were controlled by animal studies of human renal cell carcinomas on nude mice. The same results were obtained.

4. DISCUSSION

The clamping experiments showed two unexpected results. 1. The reproducible reduction of tHb: This effect is explained by the measuring technique of our system. The finger tip was measured in reflection mode, i.e., only oxygenation of the surface (maximal 4 mm in depth = capillary region) of the skin was detected. During the clamping period, the blood was ditched into the venous part of the vascular system, which comprises a vascular capacitance system that will pool a considerable volume of blood. 2. The temporal differences in reaching the zero-saturation level: This effect can be explained by the greater arterial blood volume between the pressure cuff and the measured tissue volume at the finger tip. Chance and Nioka (personal communication, 1996) have detected zero-saturation levels after approximately 10 min. in clamping experiments at the forearm. In the muscle, a small oxygen buffer exists (myoglobin). No oxygen pool exists in the finger tip (Grote, 1993).

Cytochrome measurements are of great interest (Chance et al. 1966). Because of technical limitations, spectroscopic systems could not distinguish between the hemoglobin signals and the signals of reduced and oxidized cytochrome. Therefore, most of the data published up to now were performed in blood free perfused animals (Lübbers et al. 1965) or with a cytochrome signal (CtO2 - Ct (oxidized minus reduced cytochrome aa$_3$) which is difficult to interpret (Wollert 1996). This formula cannot be understood in the biochemical meaning of redox state (Chance et al. 1966).

Considerable heterogeneity could be demonstrated in tumors (Fig 4: glioblastoma, astrocytoma, meningioma, renal cell carcinoma, metastasis). This phenomenon has already been described by morphological methods (e.g. corrosion casts; Steinberg et al. 1990), quantification of tumor vessels (Steinberg et al. 1991), blood flow measurements

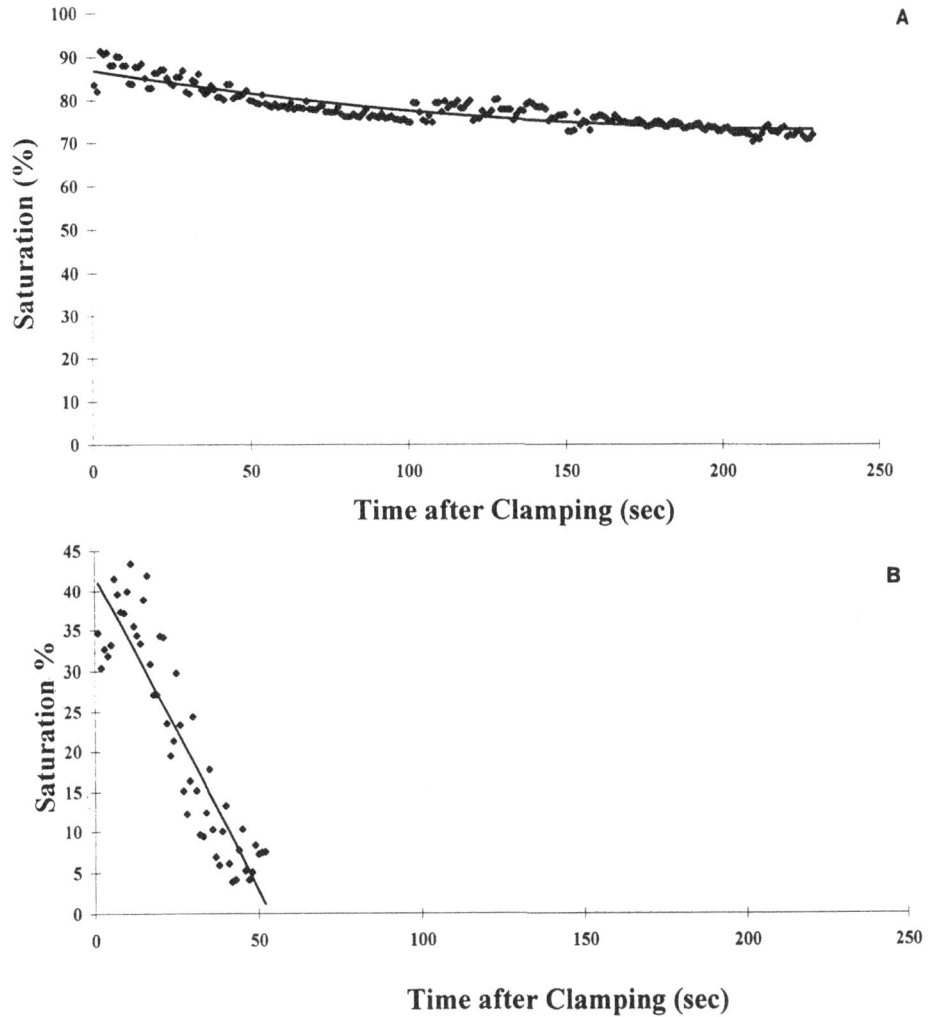

Figure 4. Oxygen consumption measurements in a 46 year old male patient with renal cell carcinoma. The change of saturation was monitored during clamping the Arteria renalis. The neoplastic tissue shows a significantly slower desaturation rate compared to the healthy kidney. This phenomenon can be understood as a reduced oxygen consumption of the tumor.

(Kallinowski et al. 1989), and in correlation to other biological tumor parameters (Steinberg et al. 1994). Polarographic oxygen measurements (pO_2) have shown the same tendency in different tumor entities (Füller et al. 1994, Feldmann et al. 1995, Streffer et al. 1995).

The high tHb values found in renal cell carcinoma were expected because of the angiographic appearance of this tumor entity with a high degree of vascularisation found within the tumor. This can be used for diagnostic purposes. These measurements comprise the first series of extremely high intratumoral oxygenation reported in the literature. The significant difference in oxygen consumption rate is reflected by the ideas of Warburg (1926). He hypothesized, that some tumors are reacting to the needs of their oxidative metabolic needs by increasing glycolysis.

5. CONCLUSION

MULTISCAN OS 10/30 can be used for physiological testing and intraoperative measurements. The routine assessment of oxygen status in tumor patients is possible. It can be used for diagnostic purposes, therapy monitoring and functional testing. The feasibility of measuring cytochrome parameter provides additional information about the intracellular oxygen status in various experimental and clinical settings. Experimental and clinical studies can disclose elementary correlations of physiological parameters, such as the effect of various ventilatory parameters (see Scheufler et al. 1996b) or intracranial hypertension (see Scheufler et al. 1996a) on cerebral oxygenation. Monitoring of intracellular oxygen can also contribute to qualitative and quantitative assessment of diagnostic and therapeutic modalities dealing with (Steinberg et al. 1996) cerebral oxygen parameters. Further studies are required to prove reliability of the system under different pathologic conditions in patients.

ACKNOWLEDGMENT

The authors would like to express their thanks for the technical assistance of Mrs. Hildenhagen and for technical and financial support from NIOS Medizintechnik GmbH/Germany and Krebshilfe Herdecke.

REFERENCES

Chance B, Schoener B, Schidler F. The intracellular oxidation-reduction state. In: Oxygen in the animal organism. Dickens F, Neil E (eds.) Pergamon, p 367, 1966.

Feldmann HJ, M Molls, T Auberger, G Stüben. Oxygenation and perfusion status of recurrent human tumors. In: Funktionsanalyse biologischer Systeme, 24.pp. 319–326. Editors: P Vaupel, DK Kelleher and M Günderoth. Akademie der Wissenschaften und der Literatur, Mainz, 1995.

Füller J, Feldmann HJ, Molls M, Sack H, Untersuchungen zum Sauerstoffpartialdruck im Tumorgewebe unter Radio- und Thermoradiotherapie. Strahlenther Onkol 170, 453–460, 1994.

Grote, J.:Gewebsatmung. In: R.F. Schmidt und G. Thews (Hrsg.) Physiologie des Menschen, 558–572, Springer Verlag Berlin (1993).

Kallinowski F, Schlenger KH, Runkel S, Kloes M, Stohrer M, Okunieff P, Vaupel P, Blood flow, metabolism, cellular microenvironment, and growth rate of human tumor xenografts. Cancer Res 49, 3759–3764, 1989.

Lübbers, D.W., Kessler, M., Scholz, R., Bücher, Th., Cytochrome reflexion spectra and fluorescence of the isolated, perfused, hemoglobin-free rat liver during a cycle of anoxia. Biochem.Zschr, 341, 346–350, 1965

Scheufler K.M, Thees CH, Steinberg F, Zentner J, NIR reflexion spectroscopy based oxygen measurements during intracranial hypertension in rabbits: an experimental study. Adv.Exp..Med.Biol., this issue, 1996a.

Scheufler KM, Thees CH, Steinberg F, Zentner J, Influence of various ventilation parameters on NIR reflection spectroscopy based cerebral oxygen measurements: an experimental animal study. Adv.Exp..Med.Biol., this issue, 1996b.

Siggaard-Andersen O., Norgaard-Pedersen, B., Rem, J., Hemoglobin pigments. Spectrophotometric determination of oxy-, carboxy-, met-, and sulfhemoglobin in capillary blood. Clin. Chim. Acta, 42, 85–100, 1972.

Steinberg F, Konerding MA, Sander A and Streffer C, Vascularisation, proliferation and necrosis in untreated human primary tumors and untreated human xenografts. Int J Rad Biol, 60, 161–168, 1991.

Steinberg F, Konerding MA, Streffer C. The vascular architecture of tumors: Histological, morphometrical and ultrastructural studies. J Canc Res Clin Oncol, 116: 517–524, 1990.

Steinberg F, M Leßmann, C Streffer; Non-invasive infrared-spectroscopic measurements of intravasal oxygen content in xenotransplanted tumors on nude mice - correlation to growth rate. Oxy.Transp.Tissue, 345, 549–555, 1994.

Steinberg F, Röhrborn HJ, Scheufler KM, Asgari S, Trost HA, Seifert V, Stolke D, Streffer C, NIR reflection spectroscopy based oxygen measurements and therapy monitoring in brain tissue and intracranial neoplasms - correlation to MRI and angiography. Adv. Exp.Med.Biol., this issue, 1996.

Streffer C, Steinberg F, Tamulevicius P, Oxygen content and energy metabolism in tumors - are they correlated? . In: Funktionsanalyse biologischer Systeme, 24. pp. 195–204. Editors: P Vaupel, DK Kelleher and M Günderoth. Akademie der Wissenschaften und der Literatur, Mainz, 1995.

Warburg O, Über den Stoffwechsel der Tumoren. Berlin, Springer, 1926.

Weissbluth, M., Hemoglobin. Springer, Berlin - Heidelberg - New York, 1974.

Wiskemann, M., Spectralanalytische Bestimmungen des Haemoglobingehaltes des menschlichen Blutes. Dissertation Universität Freiburg, 1875.

Wollert HG, Eckel L. Letter to the editor: Tissue and cellular oxygenation state measurement during cardiac surgery. Thorac. cardiovasc. 44, 158–160, 1996.

12

TUMOR OXYGENATION UNDER NORMOBARIC AND HYPERBARIC HYPEROXIA

Impact of Various Inspiratory CO_2 Concentrations[*]

O. Thews, D. K. Kelleher, and P. Vaupel

Institute of Physiology and Pathophysiology
University of Mainz
D-55099 Mainz, Germany

1. INTRODUCTION

Tumor hypoxia is an important factor limiting the efficiency of sparsely ionizing radiation and O_2-dependent chemotherapy. Since the tumor pO_2 is the result of a dynamic steady state between oxygen supply and O_2 consumption of the tumor tissue, hypoxia could be reduced either by increasing the O_2-supply or by reducing the O_2 demand of the tumor cells. The O_2 supply can be improved for instance by (i) increasing the arterial oxygen partial pressure, (ii) improving (and homogenizing) the tumor perfusion, or (iii) enhancing the O_2 release from blood into the tissue by right-shifting the HbO_2 dissociation curve. Theoretically, it should also be possible to improve tumor oxygenation by a relatively small decrease in O_2 consumption rate of the tumor cells. However, at present increasing the arterial pO_2 by breathing hyperoxic gas mixtures seems to be the most effective method to improve tumor oxygenation and, thus, to enhance the efficiency of standard radio- and chemotherapy in experimental malignancies [1,7,10,11,15,24] as well as in human tumors [3,9,12,14,22,29,30]. However, since in some tumor entities oxygenation is inadequate and anisotropic [27], normobaric hyperoxia is often not sufficient to completely eradicate tumor hypoxia [6,16,18,26]. In these cases only breathing of hyperoxic gases under hyperbaric conditions may be sufficient to lead to therapeutic results. On the other hand, studies on experimental tumors in animals as well as clinical trials in patients showed non-uniform results concerning the therapeutic benefit of inspiratory hyperoxia ranging from clear improvement of radiosensitivity [3,4,20,30] to no effect on therapeutic outcome [3,4]. Finally, enhancement of tumor growth by hyperbaric oxygena-

* Supported by the Deutsche Krebshilfe (Grant M 40/91 Va 1) and the Dr.med.h.c. Erwin Braun Foundation, Basel (Grant 5.5).

tion has been discussed [13,23]. Several factors, such as tumor entity, site of growth, or tumor vascularisation, seem to influence the therapeutic benefit of hyperbaric hyperoxia [15,17,30]. But also systemic parameters of the tumor-bearing host (e.g., constriction of tumor-feeding vessels, anemia, anesthesia) could also modulate the efficacy of these treatments [5,21]. In particular, the role of varying inspiratory pCO_2 on tumor oxygenation has been discussed controversially. Although carbogen (95% O_2 + 5% CO_2) is used in the clinical setting, it remains unclear whether or not the beneficial therapeutic effects are more pronounced than with pure oxygen. On the other hand, CO_2 has been blamed for inducing a significant constriction in the tumor-feeding vessels reducing tumor perfusion and thus limiting the potential effect of an increased arterial pO_2 resulting in a worsening of tumor pO_2 [2].

In this study, the effect of inspiratory hyperoxia on oxygenation of an experimental sarcoma of the rat has been analyzed under normobaric and hyperbaric conditions with different inspiratory CO_2 concentrations. In order to investigate the mechanism by which the inspiratory CO_2 can influence tumor oxygenation, microregional tumor perfusion was also monitored during breathing of elevated inspiratory CO_2 concentrations.

2. MATERIALS AND METHODS

2.1. Animals and Tumors

Male Sprague-Dawley rats (body weight 239 ± 28 g) were used with DS-sarcomas growing s.c. on the hind foot dorsum after injection of DS-ascites cells (0.4 ml, approx. 10^4 cells/µl). Tumors were used in the experiments when they had reached a volume of between 0.5 and 2.5 ml (approx. 6–10 days after tumor implantation).

2.2. Surgical Procedure

Animals were anesthetized with pentobarbital (40 mg/kg i.p.) and laid supine on a heated operating pad. Mean arterial blood pressure (MABP) was continuously monitored through a catheter placed in the carotid artery and connected to a Statham pressure transducer (type P23 ID, Gould Oxnard, CA, USA).

2.3. Inspiratory Hyperoxia

Animals were allowed to spontaneously breathe one of three inspiratory gases (room air; 100% oxygen; carbogen: 95% oxygen + 5% CO_2) flushed around the tracheotomy tube at a flow rate of 2 l/min. Gassing started 20 to 30 min before pO_2 measurements. Experiments were performed either under normobaric (1 atm) or hyperbaric (2 atm) conditions. Under hyperbaric conditions the CO_2 fraction of the carbogen was reduced to 2.5% resulting in an arterial pCO_2 comparable to the normobaric control.

2.4. Measurement of the Spatial pO_2 Distribution

Tumor oxygen tensions were determined using polarographic needle electrodes (shaft diameter 300 µm) and pO_2-histography (KIMOC-6650, Eppendorf, Hamburg, Germany; for more details see [28]). A small midline incision was made in the skin covering the lower abdomen and the Ag/AgCl reference electrode was placed between the skin and the underlying musculature. For probe insertion, a small incision was made into the skin covering the tumor and the electrode was placed at a depth of approximately 1 mm. The electrode was then automatically advanced through the tissue in pre-set steps of 1 mm immediately followed by a retraction of the electrode by 0.3 mm resulting in an effective step-length of 0.7 mm. Each electrode step was followed by a pO_2 measurement. In this way, 6 to 18 parallel electrode tracks (track-to-track distance of 1.0 to 3.1 mm) were analyzed with the polarographic needle electrode resulting in 78–322 pO_2 measurements per tumor. In most of the tumors the electrode tracks were placed in two horizontal layers though, due to the ellipsoid shape of the tumor, the upper layer was smaller than the lower one. The measured pO_2 distribution within the tumor was described by the mean and median pO_2 as well as by the fraction of pO_2 measurements ≤ 2.5 mmHg and ≤ 5 mmHg.

In order to assess transient changes in tumor pO_2 during changes of the inspiratory pO_2, a catheter pO_2 electrode (LICOX pO_2, GMS, Kiel-Mielkendorf, Germany) was placed in the center of the tumor and pO_2 was continuously monitored. In this experimental series, animals started breathing room air at 1 atm for 30 min after which they received either carbogen or oxygen as inspiratory gas. After 10 min of inspiratory hyperoxia, the environmental pressure was elevated linearly over 10 min up to 2 atm which was retained for 30 min. Subsequently, the pressure was lowered to normobaric conditions within 10 min. After another 10 min of breathing carbogen or 100% oxygen the gassing tube was removed and the animal breathed room air at normobaric conditions again.

2.5. Red Blood Cell (RBC) Flux Measurements

Changes of microregional perfusion during carbogen breathing were assessed using a multi-channel laser Doppler flowmeter (Oxford Array, Oxford Optronics, Oxford, UK) assessing the relative red blood cell flux (RBC flux). Up to 6 light guide needle probes were placed approx. 3–8 mm in depth in peripheral and central locations of the tumor for measurement of the RBC flux in a hemisphere of approximately 1 mm^3 around the probe tip. After a stabilizing period, MABP and RBC flux were continuously recorded for 10 min before and 30 min during carbogen breathing at normobaric conditions.

2.6. Statistical Analysis

All results are presented as mean values $\pm$ SEM. Comparisons between the groups were performed using the two-tailed Wilcoxon rank-sum test for unpaired samples.

3. RESULTS

Since tumor volume is an important factor influencing the oxygenation of the DS-sarcoma, tumors were divided into three groups of different volume ranges (<1.0 ml, 1.0-<1.5 ml, and ≥ 1.5 ml).

3.1. Normobaric Conditions

Under normobaric conditions, either O_2 or carbogen breathing, resulted in a significant improvement of tumor oxygenation compared to air-breathing, although hypoxia could not be eradicated completely with these inspiratory gases (Fig. 1). With all inspiratory gas mixtures, a size dependency of tumor oxygenation was found. In larger tumors

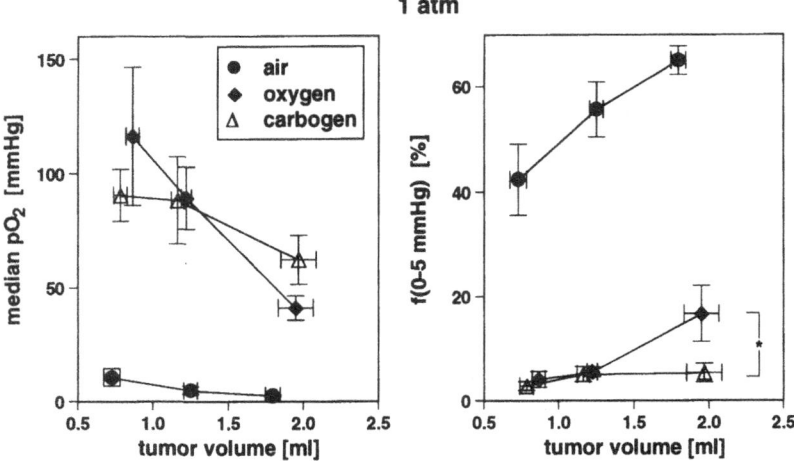

Figure 1. Size dependency of tumor oxygenation (expressed as the median pO_2 and the fraction of pO_2 values ≤ 5 mmHg) while breathing different gas mixtures under normobaric conditions (values are means $\pm$ SEM, * $p<0.05$). Each data point represents a minimum of 8 tumors.

(volumes ≥ 1.5 ml) the fraction of hypoxic pO_2 values was significantly lower with carbogen breathing than with pure O_2 although with carbogen 5% of the pO_2 values were still ≤ 5 mmHg and 3% were ≤ 2.5 mmHg (with pure oxygen: 16% and 12%, respectively). These improvements with carbogen may be the result of a slightly better tumor perfusion as indicated by higher RBC fluxes measured in comparable regions of the tumor tissue parallel to an increase in the mean arterial blood pressure during carbogen breathing (data not shown). Calculating the relative resistance to flow in the tumors by dividing the MABP by the RBC flux shows that the resistance to flow slightly decreases by about 5–10% possibly as a result of a slight dilatation of vessels feeding the tumor.

3.2. Hyperbaric Conditions

During 2 atm pressure the median tumor pO_2 increased to more than three times the baseline value at 1 atm, with tumor hypoxia being almost completely eradicated during O_2 or carbogen breathing (Fig. 2). Under these hyperbaric conditions, a volume dependency of tumor oxygenation could still be found. In large tumors carbogen breathing (as compared to 100% O_2) resulted in a slightly (but not significantly) better oxygenation.

3.3. Transient Changes in Tumor Oxygenation

Under steady state conditions, the effects of pure O_2 and carbogen breathing under elevated pressure on tumor oxygenation were comparable. In contrast, measurements of the transient changes in tumor pO_2 showed differences in the dynamics of tumor oxygenation between the two inspiratory gases. With oxygen the tumor pO_2 increased or decreased parallel to the environmental pressure, reaching a steady state within 3 to 5 min at the final pressure of 2 atm (Fig. 3). With carbogen the maximum tumor pO_2 was reached only after 15 min at elevated pressure (Fig. 4). After reducing the environmental pressure to normal, with carbogen the oxygenation remained at a higher level for more than 30 min whereas with 100% oxygen the tumor pO_2 decreased to baseline values immediately.

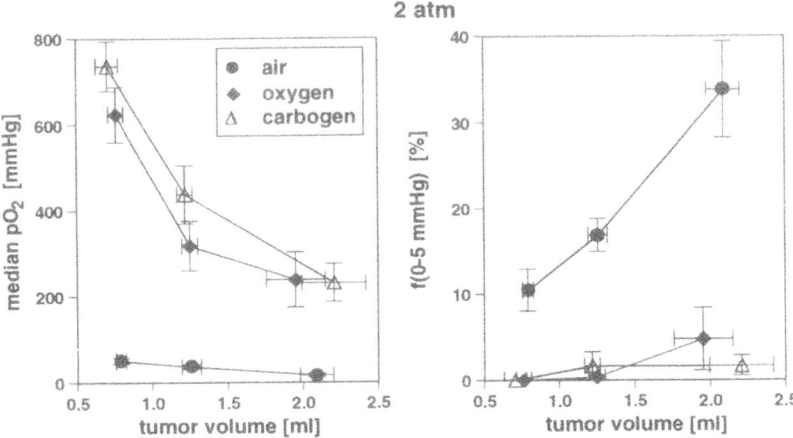

Figure 2. Size dependency of tumor oxygenation (expressed as the median pO₂ and the fraction of pO₂ values ≤5 mmHg) while breathing different gas mixtures under hyperbaric (2 atm) conditions which were maintained for at least 20 min before pO₂ measurement (values are means ± SEM). Each data point represents a minimum of 8 tumors.

4. DISCUSSION

Tumor oxygenation is mainly influenced by the arterial O_2 supply to the tissue, the latter being dependent on the arterial O_2 content (predominantly a function of arterial pO_2 and hemoglobin concentration) and the tissue perfusion. Increasing the arterial pO_2 by inspiratory hyperoxia therefore leads to an improvement of tumor oxygenation (Fig. 1) which is in accordance with previous studies [6,14,16,26]. In the present study, no principal difference was found between the effect of carbogen and pure oxygen on the oxygenation of solid tumors. Only in large tumors carbogen shows a slight but relevant (for the

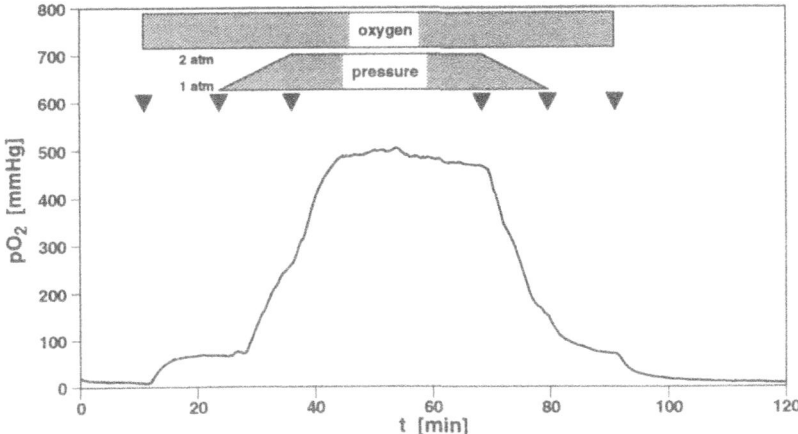

Figure 3. Transient changes in tumor pO₂ during oxygen breathing at normobaric and hyperbaric conditions. The gray bars indicate the period of oxygen breathing and the changes in environmental pressure (n=3).

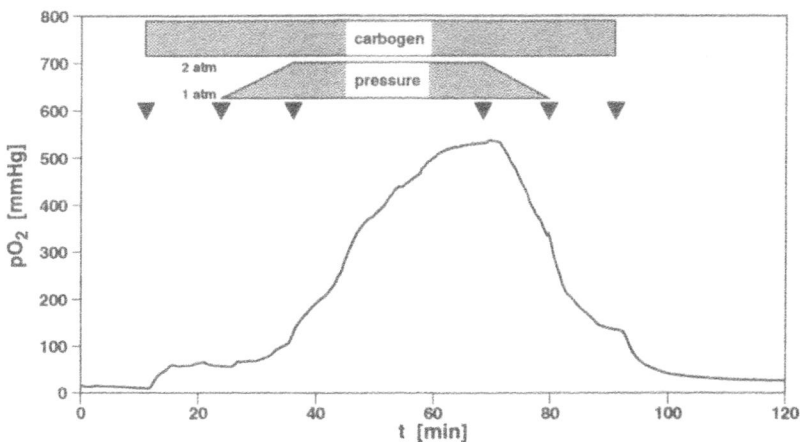

Figure 4. Transient changes in tumor pO_2 during carbogen breathing at normobaric and hyperbaric conditions. The gray bars indicate the period of carbogen breathing and the changes in environmental pressure (n=3).

therapeutic efficiency of sparsely ionizing radiation) improvement in tumor oxygenation as compared to oxygen. The rationale for using carbogen with 5% CO_2 added to oxygen is that increasing the arterial pCO_2 leads to (1) a significant stimulation of ventilation resulting in an improved pulmonary O_2 uptake and (2) a better O_2 delivery from the blood to the tumor cells by a right-shift of the HbO_2-dissociation curve. However, CO_2 has been blamed for inducing a sympathetically induced vasoconstriction reducing tumor perfusion and thus limiting the potential beneficial effect of an increased arterial pO_2. Brizel et al. [2], using a tumor model which showed a similar oxygenation status (median pO_2: 2–8 mmHg; fraction of pO_2 values < 5 mmHg: 49–82%), found that with hyperbaric (3 atm) carbogen breathing the oxygenation is only marginally improved. However, when animals were simultaneously treated with bretylium a significant improvement of tumor oxygenation was found although pO_2 values measured during pure oxygen breathing were not attained. The authors concluded that with carbogen breathing CO_2 induces a severe vasoconstriction which can be counteracted by a "blockade of the sympathicus". However, in the present study this CO_2-induced vasoconstriction was not seen, a finding which is in accordance with previous studies [8]. In contrast, a slight increase in tumor perfusion indicated by an increase in the RBC flux of about 20% was found. Subsequently, pure O_2 as well as carbogen improved tumor oxygenation to approximately the same extent (Figs. 1 & 2). Remarkably, the increase in the median tumor pO_2 during hyperbaric hyperoxia found in the study of Brizel et al. [2] was much smaller (from 8 to 55 mmHg) than in the present study (from 6 to 394 mmHg) although both tumor models initially showed a comparable oxygenation status. Since Brizel et al. measured control values as well as values under hyperbaric hyperoxia in the same tumors with a relatively long time interval between these measurements, the trauma to the tumor tissue resulting in an edema or hemorrhage could lead to an underestimation of the "true" beneficial effect of 100% oxygen or carbogen breathing.

Although the improvement in tumor pO_2 during carbogen and O_2 breathing was found to be comparable in the present study, the dynamic changes in tumor oxygenation during alterations in the environmental pressure were different (Figs. 3 & 4). When breathing room air during changes of the environmental pressure (data not shown) tumor oxygenation increases parallel to pressure similar to the results obtained during O_2 breathing. At present, the reason for these differences between O_2 and carbogen remains unclear. With O_2 breathing the tumor oxygenation seems to be influenced only by the arterial pO_2 whereas with carbogen a delayed increase and a more sustained improvement was found. The dynamic response of the cerebral respiratory control center to the increased arterial pCO_2 possibly influences the pulmonary O_2 uptake and subsequently influences the arterial pO_2 during changes in the environmental pressure. The increased arterial pCO_2 may have a longer lasting vasodilatatory effect on the tumor feeding vessels resulting in an improved tumor perfusion. However, the analysis of changes in tumor perfusion during hyperbaric carbogen breathing will be addressed in further studies.

As shown in Figs. 3 & 4, the maximum tumor pO_2 is reached by 15 minutes under hyperbaric conditions and remained approximately constant during the whole measuring period of 30 min. If the hyperbaric hyperoxic conditions were maintained for longer than 30 min (up to 70 min; data not shown) the tumor oxygenation status did not change significantly either with O_2 or with carbogen breathing. From these results an additional beneficial effect of longer lasting inspiratory hyperbaric hyperoxia on tumor oxygenation is not evident.

The pre-treatment oxygenation status of the DS-sarcoma is quite poor (median pO_2: 6 mmHg; fraction of pO_2 values ≤ 5 mmHg: 53%) resulting in a fraction of cells with less than half-maximal radiosensitivity of about 40 to 50%. Under hyperbaric hyperoxia all of these cells will become maximally radiosensitive and therefore a beneficial effect of the inspiratory hyperoxia on radiotherapy might be expected. However, it should be stated that inspiratory hyperoxia under normobaric as well as under hyperbaric conditions probably only has a positive effect if a tumor is initially hypoxic. Since the radiosensitivity of tissues is half-maximal at a pO_2 of 3–4 mmHg, the effect of sparsely ionizing radiation cannot be improved when no hypoxia is present, as is the case in several human tumor entities. This might explain the limited improvement of therapeutic efficiency with normo- or hyperbaric hyperoxia in earlier clinical trials [3,4,17,30], especially since tumor oxygenation can vary over a wide range depending on the tumor entity, stage, or site of growth. However, in tumors showing severe hypoxia (e.g., in bulky tumors) or in patients with tumor-associated anemia, inspiratory hyperoxia has been found to be very effective in improving the radiotherapeutic outcome [5,30]. In these tumors it seems to be necessary to use hyperbaric conditions to get a maximal therapeutic effect especially when the inspiratory hyperoxia is combined with other radiosensitizers [11,12,22] or oxygen-carrying solutions [19,24,25]. In these cases hyperbaric treatment can eradicate tumor hypoxia completely. Thus a characterization of the oxygenation status of a tumor prior to radiotherapy seems to be necessary for an individualization of treatment modalities and an assessment of the potential of normobaric and hyperbaric oxygen as a treatment modulator.

5. CONCLUSIONS

Oxygen as well as carbogen breathing at 1 atm results in an improvement in tumor oxygenation to approximately the same extent. Hyperbaric breathing (2 atm) increases the tumor pO_2 more than twice the baseline value at 1 atm. The slightly better effect of carbogen on tumor oxygenation could be attributed to an increase in the mean arterial blood pressure resulting in an improved tumor perfusion. However, hyperbaric conditions should be maintained for at least 20 min if maximal effects on tumor oxygenation are to be obtained.

REFERENCES

1. Alagoz T, Buller RE, Anderson B, Terrell KL, Squatrito RC, Niemann TH, Tatman DJ, Jebson P. Evaluation of hyperbaric oxygen as a chemosensitizer in the treatment of epithelial ovarian cancer in xenografts in mice. *Cancer 75*: 2313–2322 (1995).

2. Brizel DM, Lin S, Johnson JL, Brooks J, Dewhirst MW, Piantadosi CA. The mechanisms by which hyperbaric oxygen and carbogen improve tumour oxygenation. *Brit. J. Cancer 72*: 1120–1124 (1995).

3. Denham JW, Yeoh EK, Wittwer G, Ward GG, Ahmad AS, Harvey ND. Radiation therapy in hyperbaric oxygen for head and neck cancer at Royal Adelaide Hospital—1964 to 1980. *Int. J. Radiat. Oncol. Biol. Phys. 13*: 201–208 (1987).

4. Dishe S. What have we learnt from hyperbaric oxygen? *Radiother. Oncol. 20* (Suppl.): 71–74 (1991).

5. Dische S, Anderson PJ, Sealy R, Watson ER. Carcinoma of the cervix—anaemia, radiotherapy and hyperbaric oxygen. *Br. J. Radiol. 56*: 251–255 (1983).

6. Falk SJ, Ward R, Bleehen NM. The influence of carbogen breathing on tumour tissue oxygenation in man evaluated by computerised pO_2 histography. *Br. J. Cancer 66*: 919–924 (1992).

7. Gerweck LE, Hetzel FW. pO_2 in irradiated versus nonirradiated tumors of mice breathing oxygen at normal and elevated pressure. *Int. J. Radiat. Oncol. Biol. Phys. 32*: 695–701 (1995).

8. Grau C, Horsman MR, Overgaard J. Improving the radiation response in a C3H mouse mammary carcinoma by normobaric oxygen or carbogen breathing. *Int. J. Radiat. Oncol. Biol. Phys. 22*: 415–419 (1992).

9. Henk JM. Late results of a trial of hyperbaric oxygen and radiotherapy in head and neck cancer: a rationale for hypoxic cell sensitizers? *Int. J. Radiat. Oncol. Biol. Phys. 12*: 1339–1341 (1986).

10. Horsman MR, Nordsmark M, Khalil AA, Hill SA, Chaplin DJ, Siemann DW, Overgaard J. Reducing acute and chronic hypoxia in tumors by combining nicotinamide with carbogen breathing. *Acta Oncol. 33*: 371–376 (1994).

11. Kjellen E, Joiner MC, Collier JM, Johns H, Rojas A. A therapeutic benefit from combining normobaric carbogen or oxygen with nicotinamide in fractionated X-ray treatments. *Radiother. Oncol. 22*: 81–91 (1991).

12. Laurence VM, Ward R, Dennis IF, Bleehen NM. Carbogen breathing with nicotinamide improves the oxygen status of tumours in patients. *Br. J. Cancer 72*: 198–205 (1995).

13. Margaretten NC, Witschi HP. Effects of hyperoxia on growth characteristics of metastatic murine tumors in the lung. *Cancer Res. 48*: 2779–2783 (1988).

14. Martin L, Lartigau E, Weeger P, Lambin P, Le Ridnat AM, Lusinchi A, Wibault P, Eschwege F, Luboinski B, Guichard M. Changes in the oxygenation of head and neck tumors during carbogen breathing. *Radiother. Oncol. 27*: 123–130 (1993).

15. Martin LM, Thomas CD, Guichard M. Nicotinamide and carbogen: relationship between pO_2 and radiosensitivity in three tumour lines. *Int. J. Radiat. Biol. 65*: 379–386 (1994).

16. Mueller-Klieser W, Vaupel P. Tumor oxygenation under normobaric and hyperbaric conditions. *Br. J. Radiol. 56*: 559–564 (1983).

17. Overgaard J. Sensitization of hypoxic tumour cells—clinical experience. *Int. J. Radiat. Biol. 56*: 801–811 (1989).

18. Perry PM, Nias AH. The optimum pressure of oxygen for radiotherapy of a mouse tumour. *Br. J. Radiol. 65*: 784–786 (1992).

19. Rockwell S, Irvin CG, Kelley M, Hughes CS, Yabuki H, Porter E, Fischer JJ. Effects of hyperbaric oxygen and a perfluorooctylbromide emulsion on the radiation responses of tumors and normal tissues in rodents. *Int. J. Radiat. Oncol. Biol. Phys. 22*: 87–93 (1992).

20. Rojas A, Joiner MC, Hodgkiss RJ, Carl U, Kjellen E, Wilson GD. Enhancement of tumor radiosensitivity and reduced hypoxia-dependent binding of a 2-nitroimidazole with normobaric oxygen and carbogen: a therapeutic comparison with skin and kidneys. *Int. J. Radiat. Oncol. Biol. Phys. 23*: 361–366 (1992).

21. Rojas A, Stewart FA, Smith KA, Soranson JA, Randhawa VS, Stratford MR, Denekamp J. Effect of anemia on tumor radiosensitivity under normo- and hyperbaric conditions. *Int. J. Radiat. Oncol. Biol. Phys. 13*: 1681–1689 (1987).

22. Sealy R, Cridland S, Barry L, Norris R. Irradiation with misonidazole and hyperbaric oxygen: final report on a randomized trial in advanced head and neck cancer. *Int. J. Radiat. Oncol. Biol. Phys. 12*: 1343–1346 (1986).

23. Sklizovic D, Sanger JR, Kindwall EP, Fink JG, Grunert BK, Campbell BH. Hyperbaric oxygen therapy and squamous cell carcinoma cell line growth. *Head Neck 15*: 236–240 (1993).

24. Teicher BA, Sotomayor EA, Robinson MF, Dupuis NP, Schwartz GN, Frei III E. Tumor oxygenation and radiosensitization by pentoxifylline and perflubron emulsion/carbogen breathing. *Int. J. Oncol. 2*: 13–21 (1993).

25. Teicher BA, Holden SA, Cathcart KN, Herman TS. Effect of various oxygenation conditions and fluosol-DA on cytotoxicity and antitumor activity of bleomycin in mice. *J. Natl. Cancer Inst. 80*: 599–603 (1988).

26. Thews O, Kelleher DK, Vaupel PW. Modulation of spatial O_2 tension distribution in experimental tumors by increasing arterial O_2 supply. *Acta Oncol. 34*: 291–295 (1995).

27. Thews O, Kelleher DK, Vaupel PW. pO_2-mapping of experimental rat tumors: visualization and statistical analysis. In: PW Vaupel, DK Kelleher, M Günderoth (eds.), Tumor oxygenation. Gustav Fischer, Stuttgart, 27–38 (1995).

28. Vaupel P, Schlenger K, Knoop C, Höckel M. Oxygenation of human tumors: Evaluation of tissue oxygen distribution in breast cancers by computerized O_2 tension measurements. *Cancer Res. 51*: 3116–3122 (1991).

29. Voute PA, van der Kleij AJ, De Kraker J, Hoefnagel CA, Tiel van Buul MM, Van Gennip H. Clinical experience with radiation enhancement by hyperbaric oxygen in children with recurrent neuroblastoma stage IV. *Eur. J. Cancer 31A*: 596–600 (1995).

30. Whittle RJ, Fuller AP, Foley RR. Glottic cancer: results of treatment with radiotherapy in air and hyperbaric oxygen. *Clin. Oncol. Roy. Coll. Radiol. 2*: 214–219 (1990).

13

BLOOD FLOW AND OXYGENATION STATUS OF HEAD AND NECK CARCINOMAS[*]

Peter Vaupel

Institute of Physiology and Pathophysiology
University of Mainz
D-55099 Mainz, Germany

1. INTRODUCTION

Solid tumors contain a significant fraction of microregions which are chronically or transiently hypoxic. Experimental evidence is growing, showing that hypoxia may have a profound impact on malignant progression and on responsiveness to therapy [1–4]. The clinical relevance of tumor oxygenation in human solid malignancies is under investigation (for a recent review see [5]). In this presentation, relevant clinical findings available to date on blood and oxygen supply of human head and neck carcinomas will be reviewed and emphasis will be given to the relevance of these factors in clinical oncology.

2. BLOOD FLOW OF HEAD AND NECK TUMORS

Blood flow and tumor oxygenation can influence not only the biological behaviour of solid tumors (e.g., invasion and metastasizing potential, growth rate) but also the response to non-surgical therapeutic measures (e.g., pharmacokinetics and -dynamics of anticancer agents, radiotherapy) and the outcome of diagnostic procedures. Knowledge of these relevant physiological parameters before commencement of therapy may allow an individualized tumor therapy which is adapted to suit the biology of a specific malignancy.

Despite this importance only few studies on blood flow through primary and metastatic head and neck carcinomas have been reported so far [6–8]. Considering the presently available data the following (preliminary) conclusions can be drawn when flow data derived from the reports are pooled (Fig. 1).

---

* Supported by the Deutsche Krebshilfe (Grant M 40/91 Va 1).

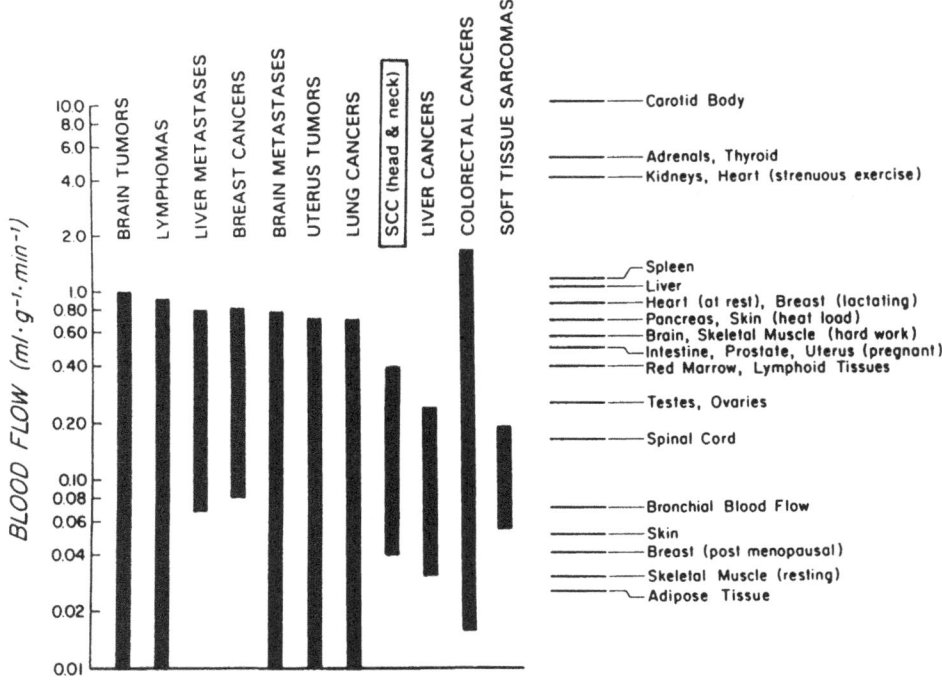

Figure 1. Variability of blood flow in human malignancies (black bars) and mean flow values of normal tissues and organs (pooled data, updated from Ref. 5). SCC = squamous cell carcinomas.

a) Mean (± SE) blood flow through primary head and neck tumors is 0.14 ± 0.07 $\text{ml} \cdot \text{g}^{-1} \cdot \text{min}^{-1}$ [6], with an arteriovenous shunt perfusion of $23 \pm 13\%$ (range 8–43%).

b) Mean blood flow through metastatic lymph node lesions in the head and neck region is 0.15 [7] and 0.24 ± 0.19 $\text{ml} \cdot \text{g}^{-1} \cdot \text{min}^{-1}$ [8], respectively.

c) Blood flow can vary considerably despite similar histological classification and primary site.

d) Tumors can have flow rates similar to those measured in organs with a high metabolic rate, such as heart or brain.

e) Some tumors exhibit flow rates that are as low as those found in tissues with a low metabolic rate, such as skin and resting skeletal muscle.

f) In some tumor entities, blood flow in the periphery is distinctly higher than in the center, whereas in others blood flow is significantly higher at the tumor center compared with the edge of the tumor.

g) There is no association between tumor size and blood flow in most cancers.

3. FLUID PRESSURE AND CONVECTIVE CURRENTS IN THE INTERSTITIAL SPACE

Like all malignant tumors, growing head and neck cancers produce new, often abnormally leaky microvessels but are unable to form their own functioning lymphatic system. As a result of this and due to a large hydraulic conductivity, in solid tumors there is a significant bulk flow of free fluid in the interstitial space. Whereas in the normal tissue convective currents in the interstitial compartment are estimated to be 0.5 - 1% of plasma flow, in xenografted humen cancers interstitial water flux can reach 15% of the respective plasma flow [9].

After seeping copiously out of the highly permeable tumor microvessels, fluid accumulates in the tumor matrix and a high interstitial pressure builds up in solid tumors [10]. Whereas in normal tissue the interstitial fluid pressure is slightly subatmospheric or just above atmospheric values, an interstitial hypertension with mean values of 19 mmHg and higher develops in squamous cell carcinomas of the head and neck region [10]. These increased interstitial fluid pressures, which drop precipitously in the outermost rim of the tumor, can wash out therapeutic agents, especially larger molecules, from the tumor into the surrounding tissue, i.e., fluid is squeezed out from the high- to the low-pressure regions at the tumor/normal tissue interface.

4. OXYGENATION STATUS OF HEAD AND NECK TUMORS

Tumor oxygenation is dependent on the cellular O_2 consumption rate and the O_2 supply to the respiring cells. The latter is determined mainly by the convective transport via the blood and by the diffusional flux from the microvessels to the O_2-consuming sites. Peculiarities of tumor tissue oxygenation can therefore be attributed primarily to characteristic structural and functional abnormalities of the tumor microcirculation (perfusion-limited O_2 delivery) and to a deterioration of the diffusion geometry (diffusion-limited O_2 delivery). As a result of a compromised and anisotropic microcirculation, the O_2 availability to the cancer cells shows a great variability, and many human malignancies reveal hypoxic tissue areas that are heterogeneously distributed within the tumor mass and can be located next to a well-perfused tumor area (intratumor heterogeneity). As a rule, in most solid malignancies the tissue O_2 status is poorer than in normal tissue at the site of tumor growth. This has been shown for a series of solid tumors [5, 9].

The oxygenation status of head and neck tumors has been evaluated using direct or indirect assays. Earlier studies used a cryospectrophotometric ex vivo microtechnique that allows for the measurement of the HbO_2 saturation of individual red blood cells in tumor microvessels [11]. As a rule, the mean HbO_2 values observed in cancers of the oral cavity were distinctly lower than those found in the normal mucosa at the site of tumor growth. The medians of the HbO_2 frequency distributions of the normal oral mucosa and of the tumors decreased from 80 to 49 sat.% and correlated well with changes in vascular density.

In various malignant tumors, considerable inter- and intraindividual differences were observed, even when tumors of the same clinical stage and grade were investigated. Although these investigations have provided a detailed insight into the oxygenation status of human malignancies, the experimental data provided refer only to the O_2 saturation status of red blood cells in microvessels. The O_2 profiles within the cellular and stromal compartment thus must be extrapolated (calculated) from the O_2 loading of the "biological indicator" hemoglobin.

Nuclear medicine assays for the detection of tumor hypoxia have recently been developed and applied to human tumors [12]. With this latter technique, tumor oxygenation has been "indirectly" estimated using ^{123}I-iodoazomycin arabinoside (IAZA) [13] or ^{18}F-fluoromisonidazole (F-MISO) [14] as probes. Correlations between various methods of assessment of the oxygenation status of solid tumors in humans and their response to

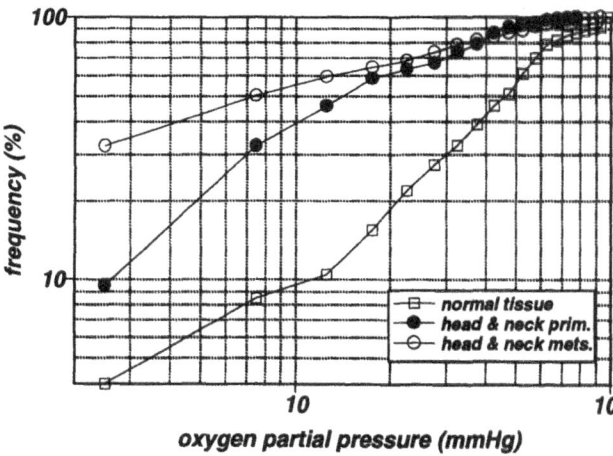

Figure 2. Cumulative frequency distributions of pO$_2$ values in primary head and neck tumors compared to metastatic lesions of this tumor entity together with values obtained in the normal tissue adjacent to the lesions (adapted from Ref. 5).

therapy have recently been evaluated [15]. As a "gold-standard" technology to assess the oxygenation of human malignancies, the polarographic O$_2$ electrode system has been suggested [15].

Oxygen partial pressure distributions in lymph node metastases of head and neck tumors have been described over the last decade. Despite these earlier data, the clinical importance of tumor hypoxia remained uncertain because of the low number of patients involved and limitations of the pO$_2$ determinations [16, 17]. Since about 1990, a clinically applicable standardized procedure has been established that enables the determination of intratumoral O$_2$ tensions in primary tumors and metastatic lesions of patients by use of a computerized polarographic needle electrode system (pO$_2$ histography, Eppendorf, Hamburg, Germany). The following sections summarize currently available pO$_2$ data for head and neck tumors obtained with this reliable system, which is well tolerated by patients.

4.1. Oxygenation Status of Primary Head and Neck Tumors

Oxygenation data derived from primary head and neck tumors are sparse and thus preliminary [11, 18, 19]. In Fig. 2, pooled oxygenation data obtained from primary head and neck tumors using the pO$_2$ histography system are plotted as a cumulative pO$_2$ distribution curve. Mean and median O$_2$ tensions obtained from different clinical stages and histological grades are, on average, distinctly lower than in the normal tissue. Oxygen tensions measured in the subcutis of the head and neck region revealed a median pO$_2$ of approx. 50 mmHg, whereas in primary tumors the median pO$_2$ was 16 mmHg. Thus far, about 10% of the head and neck cancers investigated exhibited pO$_2$ values $\leq$ 5 mmHg.

In Fig. 3, oxygenation data from primary head and neck tumors are compared with relevant pO$_2$ values measured in other tumor entities. From the compilation of pO$_2$ data there is evidence that—according to their O$_2$-status—head and neck tumors can be grouped between breast cancers and glioblastomas.

The vascularity of primary head and neck tumors seems to have a significant impact on the oxygenation status of these malignancies. This statement is supported by the data presented in Fig. 4: In squamous cell carcinomas of the oral cavity, oxyhemoglobin saturation (HbO$_2$) of individual red blood cells in tumor microvessels has been measured in cryobiopsies under ex vivo conditions [11]. From these intravascular HbO$_2$ values, tissue oxygen tensions have been calculated and plotted together with the respective data of the

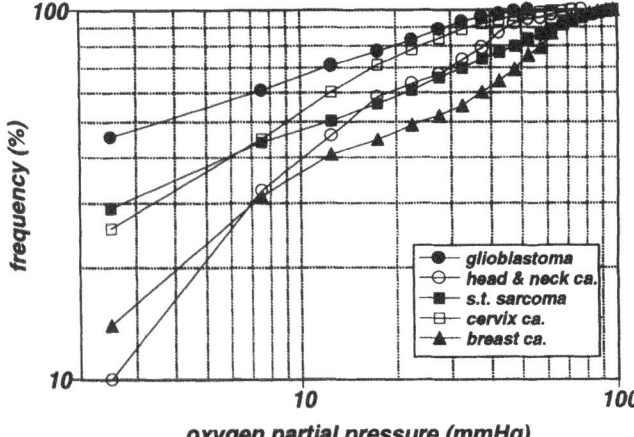

Figure 3. Compilation of cumulative frequency distributions of measured pO$_2$ values in different primary tumors. (Data adapted from Ref. 5).

normal oral mucosa. There is a clear correlation between the oxygenation status and vascular density. Well-vascularized tumors showed a better oxygenation status and less hypoxia than poorly vascularized lesions. Tumors with medium-quality vascularization exhibited an oxygenation status between these two extreme patterns.

4.2. Oxygenation Status of Metastatic Lesions

In contrast to the situation with primary tumors there is a substantial body of information on the pO$_2$ distribution in metastatic lesions of head and neck tumors [20–29]. When the oxygenation status of primary head and neck carcinomas are compared with metastatic lesions of this entity, the fraction of pO$_2$ readings ≤ 5 mmHg is distinctly higher in the latter group (Fig. 2), i.e., metastatic lesions exhibit a poorer oxygenation status than the primaries. Whether this pattern is a characteristic of this entity or a general biological phenomenon has to be elucidated in ongoing studies. In contrast to primary tumors investigated so far, the oxygenation status of metastatic lesions of the head and neck appears to be linked to tumor size, with the lower median pO$_2$ values preferentially occurring in larger nodal sizes [21, 25].

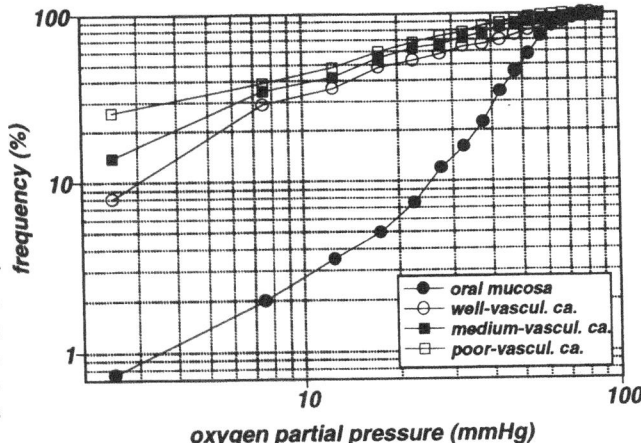

Figure 4. Cumulative frequency distributions of pO$_2$ values in the normal oral mucosa and in squamous cell carcinomas of the oral cavity as a function of vascular density. Data have been calculated from oxyhemoglobin saturation values of individual red blood cells in tumor microvessels measured in cryobiopsies under ex vivo conditions (adapted from Ref. 5).

Recent studies confirmed earlier observations by Gatenby et al. [17] suggesting that the presence of hypoxia corresponded with poor radiation response of metastatic head and neck tumors [28]. Furthermore, information is accumulating, indicating that pretreatment oxygenation in head and neck tumors can predict radiation response in advanced squamous cell carcinomas of the head and neck [28]. Similar observations have been made earlier by Höckel et al. [1] for advanced stage cancers of the uterine cervix.

Summarizing currently available data on pretreatment oxygenation of head and neck tumors, there is evidence that:

a) Tumor oxygenation, like tumor microcirculation, is anisotropic and compromised as compared to normal tissues.

b) Tumor oxygenation is not regulated according to the metabolic demand as is the case in normal tissues.

c) On average, the median pO_2 values in primary and metastatic tumors are lower than in the normal tissue at the site of growth.

d) Many solid tumors contain O_2-depleted areas (about 10% of primary tumors, approx. 30% of metastatic lesions).

e) Tumor-to-tumor variability in oxygenation is significantly greater than intratumor variability.

f) Tumor oxygenation is unpredictable considering clinical stage and grade.

g) Tumor oxygenation is independent of other known oncological parameters.

h) Pretreatment tumor oxygenation can predict the response of advanced squamous cell carcinomas of the head and neck to sparsely ionizing radiation.

REFERENCES

1. M. Höckel, K. Schlenger, M. Mitze, U. Schäffer & P. Vaupel. Hypoxia and radiation response in human tumors. Sem. Radiat. Oncol. *6*, 3–9 (1996)
2. M. Höckel, K. Schlenger, B. Aral, M. Mitze, U. Schäffer & P. Vaupel. Association between tumor hypoxia and malignant progression in advanced cancer of the uterine cervix. Cancer Res., in press (1996)
3. T.G. Graeber, C. Osmanian, T. Jacks, D.E. Housman, C.K. Koch, S.W. Lowe & A.J. Giaccia. Hypoxia mediated selection of cells with diminished apoptotic potential in solid tumors. Nature *379*, 88–91 (1996)
4. D.M. Brizel, S.P. Scully, J.M. Harrelson, L.J. Layfield, J.M. Bean, L.R. Prosnitz & M.W. Dewhirst. Tumor oxygenation predicts the likelihood of distant metastases in human soft tissue sarcoma. Cancer Res. *56*, 441–443 (1996)
5. P.W. Vaupel, O. Thews & M. Höckel. Tumor oxygenation: characterization and clinical implications. In: rhErythropoietin in Cancer Supportive Treatment, J.F. Smyth, M.A. Boogaerts & B.R.-M. Ehmer (eds.). New York, Basel, Hong Kong: Marcel Dekker, 53–85 (1996)
6. R.H. Wheeler, H.A. Ziessman, B.R. Medvec, J.E. Juni, J.H. Thrall, J.W. Keyes, S.R. Pitt & S.R. Baker. Tumor blood flow and systemic shunting in patients receiving intraarterial chemotherapy for head and neck cancer. Cancer Res. *46*, 4200–4204 (1996)
7. F.M. Waterman, D.L. Tupchong, R.E. Nerlinger & J. Matthews. Blood flow in human tumors during local hyperthermia. Int. J. Radiat. Oncol. Biol. Phys. *20*, 1255–1262 (1991)
8. P.H. Wust, H. Strahl, K. Diedemann, S. Scheller, J. Löffel, H. Riess, J. Bier, V. Jahncke & R. Felix. Local hyperthermia of N2/N3 cervical lymph node metastases: correlation of technical/thermal parameters and response. Int. J. Radiol. Oncol. Biol. Phys., in press (1996)
9. P.W. Vaupel. Blood flow, oxygenation, tissue pH distribution, and bioenergetic status of tumors. Lecture *23*. Berlin: Ernst Schering Research Foundation (1994)
10. R. Gutmann, M. Leunig, J. Feyh, A.E. Goetz, K. Messmer, E. Kastenbauer & R.K. Jain. Interstitial hypertension in head and neck tumors in patients: correlation with tumor size. Cancer Res. *52*, 1993–1995 (1992)
11. W. Mueller-Klieser, P. Vaupel, R. Manz & R. Schmidseder. Intracapillary oxyhemoglobin saturation of malignant tumors in humans. Int. J. Radiat. Oncol. Biol. Phys. *7*, 1397–1404 (1981)

12. J.D. Chapman. Measurement of tumor hypoxia by invasive and non-invasive procedures: a review of recent clinical studies. Radiother. Oncol. *20*, 13–19 (1991)

13. D. Groshar, A.J.B. McEwan, M.B. Parliament, R.C. Urtasun, L.E. Goldberg, M. Hoskinson, J.R. Mercer, R.H. Mannan, L.I. Wiebe & J.D. Chapman. Imaging tumor hypoxia and tumor perfusion. J. Nucl. Med. *34*, 885–888 (1993)

14. W.J. Koh, J.S. Rasey, M.L. Evans, J.R. Grierson, T.K. Lewellen, M.M. Graham, K.A. krohn & T.W. Griffin. Imaging of hypoxia in human tumors with [F-18]Fluoromisonidazole. Int. J. Radiat. Oncol. Biol. Phys. *22*, 199–212 (1991)

15. H.B. Stone, J.M. Brown, T.L. Philipps & R.M. Sutherland. Oxygen in human tumors: correlations between methods of measurement and response to therapy. Radiat. Res. *136*, 422–434 (1993)

16. R.A. Gatenby, L.R. Coia, M.P. Richter, H. Katz, P.J. Moldofsky, P. Engstrom, D.Q. Brown, R. Brookland & G.J. Broder. Oxygen tension in human tumours: in vivo mapping using CT-guided probes. Radiology *156*, 211–214 (1985)

17. R.A. Gatenby, H.B. Kessler, J.S. Rosenblum, L.R. Coia, P.J. Moldofsky, W.H. Hartz & G.J. Broder. Oxygen distribution in squamous cell carcinoma metastases and its relationship to outcome of radiation therapy. Int. J. Radiat. Oncol. Biol. Phys. *14*, 831–838 (1988)

18. W. Fleckenstein, J.R. Jungblut, M. Suckfüll, W. Hoppe & C. Weiss. Sauerstoffdruckverteilungen in Zentrum und Peripherie maligner Kopf-Hals-Tumoren. Dtsch. Z. Mund Kiefer Gesichts Chir. *12*, 205–211 (1993)

19. D.M. Saumweber, R.J. Kau & W. Arnold. Tumor tissue oxygenation in primary squamous cell carcinomas of the head and neck - preliminary results. In: Tumor oxygenation, P.W. Vaupel, D.K. Kelleher & M. Günderoth (eds). Stuttgart, Jena, New York: Gustav Fischer Verlag (1995)

20. L. Martin, E. Lartigau, P. Weeger, P. Lambin, A.M. Le Ridant, A. Lusinchi, P. Wibault, F. Eschwege, B. Luboinski & M. Guichard. Changes in oxygenation of head and neck tumours during carbogen breathing. Radiother. Oncol. *27*, 123–130 (1993)

21. E. Lartigau, A.M. Le Ridant, P. Lambin, P. Weeger, L. Martin, R. Sigal, A. Lusinchi, B. Luboinski, F. Eschwege & M. Guichard. Oxygenation of head and neck tumors. Cancer *71*, 2319–2325 (1993)

22. E. Lartigau, H. Randrianarivelo, L. Martin, S. Stern, C.D. Thomas, M. Guichard, P. Weeger, A.M. Le Ridant, B. Luboinski, T. Nguyen, J.-C. Ortoli, F. Grange, M.-F. Avril, A. Lusinchi, P. Wibault, C. Haie-Meder, A. Gerbaulet & F. Eschwege. Oxygen tension measurements in human tumours: The Institut Gustave-Roussy experience. Radiat. Oncol. Invest. *1*, 285–291 (1994)

23. M. Nordsmark, S.M. Bentzen & J. Overgaard. Measurement of human tumour oxygenation status by a polarographic needle electrode. Acta Oncol. *33*, 383–389 (1994)

24. J. Füller, H.J. Feldmann, M. Molls & H. Sack. Untersuchungen zum Sauerstoffpartialdruck im Tumorgewebe unter Radio- und Thermoradiotherapie. Strahlenther. Onkol. *170*, 453–460 (1994)

25. D.J. Terris & E.P. Dunphy. Oxygen tension measurements of head and neck cancers. Arch. Otolaryngol. Head Neck Surg. *120*, 283–287 (1994)

26. D.M. Brizel, G.L. Rosner, L.R. Prosnitz & M.W. Dewhirst. Patterns and variability of tumor oxygenation in human soft tissue sarcomas, cervical carcinomas, and lymph node metastases. Int. J. Radiat. Oncol. Biol. Phys. *32*, 1121–1125 (1995)

27. M.J. Eble, F. Lohr & M. Wannenmacher. Oxygen tension distribution in head and neck carcinomas after peroral oxygen therapy. Onkologie *18*, 136–140 (1995)

28. M. Nordsmark, M. Overgaard & J. Overgaard. Pretreatment oxygenation predicts radiation response in advanced squamous cell carcinoma of the head and neck. Radiother. Oncol., in press (1996)

29. V. Strnad, L. Keilholz, M. Kirschner, M. Meyer & R. Sauer. Sauerstoffdruckverteilung in Lymphknotenmetastasen und die Veränderungen während akuter respiratorischer Hypoxie. Strahlenther. Onkol., in press (1996)

14

INTRATUMORAL pO$_2$ MEASURED USING A NEW OXYGEN SENSITIVE PARAMAGNETIC MATERIAL, GLOXY

Philip E. James, Julia A. O'Hara, Oleg Y. Grinberg, Tomasz Panz, and Harold M. Swartz

Department of Radiology
Dartmouth Medical School
Hanover, New Hampshire 03755

1. INTRODUCTION

There is increasing evidence that tumor oxygenation is clinically important in predicting tumor response to radiation, chemotherapy and/or overall prognosis (Vaupel, et al.,1989; Hall, 1988). More recently, tumor hypoxia has been implicated in promoting survival of tumor cell phenotypes that are more resistant and have lost their apoptotic (self-killing) ability (Giaccia, 1996; Shrieve and Begg, 1985). A valid method for obtaining measurements of intratumor oxygen tensions repeatedly and non-invasively is therefore very desirable. Electron Paramagnetic Resonance (EPR) oximetry is a recently developed technique which has the potential to provide such data. Its full use will be facilitated by the development of oxygen-sensitive materials which can be used under the various circumstances in which repeated measurements of pO$_2$ are desired.

We recently reported on a paramagnetic derivative of coal (termed "gloxy"), whose characteristics suggest that it could be especially useful for making pO$_2$ measurements in tumors where the pO$_2$ is relatively low (James, et al., 1996). The present paper outlines two methods by which we have used gloxy in MTGB and RIF-1 murine tumors to measure tumor pO$_2$ directly, non-invasively and repeatedly by in vivo EPR. A slurry of small particles of gloxy (in saline) was inserted into one or two specific locations by a one-time injection into already established tumors, thereby allowing continuous measurement of pO$_2$ at these locations during growth of the tumors. In other animals, gloxy particles were co-injected with tumor cells at the time of inoculation so that the gloxy was distributed over a broader area spanning (several mm). In these cases, the tumor cells grew in between and surrounding the gloxy particles, allowing tumor pO$_2$ to be monitored throughout the lifetime of these tumors.

The advantages of using each of these techniques to obtain specific information on intratumoral pO$_2$ are discussed, and the potential of using gloxy in combination with L-band EPR for making tumor pO$_2$ measurements assessed.

Oxygen Transport to Tissue XIX, edited by Harrison and Delpy
Plenum Press, New York, 1997

2. METHODS

2.1. Materials

The original coal blend was obtained from Filter Anthracite Ltd (Ammanford, South Wales, UK). This oxygen-sensitive coal was given the name "gloxy" (derived from the Welsh word for coal "glo"). Its EPR signal was found to have some very useful features which give it some advantages over previously used chars or coals. The advantages include a narrow linewidth in nitrogen, a sensitive linewidth dependence with pO_2 (especially from 0 to 40 mm Hg), and high signal intensity. Unlike the oxygen-sensitive chars, gloxy required no preparation prior to use other than grinding to the desired particle size.

The chemical and elemental analysis of gloxy has been carried out (James, et al., 1996).

2.2. EPR Measurements

EPR measurements of pO_2 in mice in vivo were performed using a specifically constructed low-frequency EPR spectrometer (1.2 GHz - L-band) with an external loop for surface detection (Nilges, et al., 1989). A scan range of 10 Gauss and incident microwave power of 12 mW was used in most of these experiments.

2.3. Linewidth Calibration with pO_2

Calibration experiments were performed using a slurry of microscopic particles of gloxy (10 mg ground) suspended in 200 μl of saline. This was filtered through a mesh (50 um), and then 50 μl was drawn into gas-permeable teflon tubing. This was placed inside a 3 mm (internal diameter) quartz tube and placed within the detecting loop. This quartz tube was open at both ends so as to allow gas (of the appropriate mixture) to flow through and equilibrate with the sample. The amount of oxygen in the inflowing gas was varied using a Riteflow flowmeter (Manostat, New York) and the pO_2 was measured by incorporating a Clark-type electrode into the tubing, proximate to the sample.

For each sample, five (30 sec) spectra were recorded at each pO_2. The linewidth was measured as the difference in magnetic field between the maximum and minimum of the first derivative of the EPR signal. This was calculated by computer simulation of each EPR spectrum using an IBM compatible software package (EW Voight, by Scientific Software Services, Urbana, IL.).

2.4. Tumor pO_2

The methodology was essentially the same as that used for incorporating India ink into tumors (Goda, et al., 1995). Briefly, 50 mg of gloxy was ground manually using a 2.5 ml pestle and mortar. This was dissolved in 1 ml sterile saline and filtered though a 20 μm mesh (Small Parts Co.). Two types of tumors were used; a mouse mammary adenocarcinoma, MTG-B (being relatively inhomogeneous) and a radiation-induced fibrosarcoma RIF-1 (being relatively more homogenous). Inoculation of tumors was achieved by injecting 5×10^6 cells sub cutaneously on the rear flank of female C3H HEJ mice.

For EPR measurements of pO_2, un anesthetized animals were placed in a plastic holder with a hole through which the tumor could protrude. The holder was placed on a mechanical micro manipulator device which was secured in the gap of the magnet. This allowed fine adjustment of the animal (tumor) position with respect to the magnetic field gradient, and ensured that the gloxy was located in the center of the field for all measurements. The surface detector was placed over the tumor and 5 spectra (30 sec each) were recorded for each tumor so as to calculate the mean pO$_2$ and allow analysis of variance for each mouse.

We used two variations on the basic methodology:

(a) 50 μl of a filtered slurry of gloxy, was injected directly into 10 day old (250 mm^3) established tumors in female C3H mice. In some cases, two sites were injected with a gloxy slurry within the same tumor and a magnetic field gradient was used to separate the EPR signal arising from each of these sites, allowing simultaneous measurement of the pO$_2$ at each location being probed by the gloxy (see Figure 1).

One of the limitations of in vivo EPR spectroscopy is the limit in depth of detection - in general, the greater the distance between the detecting loop and the paramagnetic species, the smaller the EPR signal detected. (The amplitude of the resultant spectrum also depends on the material in which the measurements are to be made, and the linewidth at that pO$_2$, since, for the same number of spins a narrow linewidth will result in a higher signal amplitude). In order to measure the change in signal amplitude with growth of the tumor (or increased distance between the detecting coil and the gloxy within the tumor), at 24 hour intervals we recorded 5 EPR spectra at each of three positions of the detecting loop above the tumor; firstly, in a position where the entire surface of the tumor was

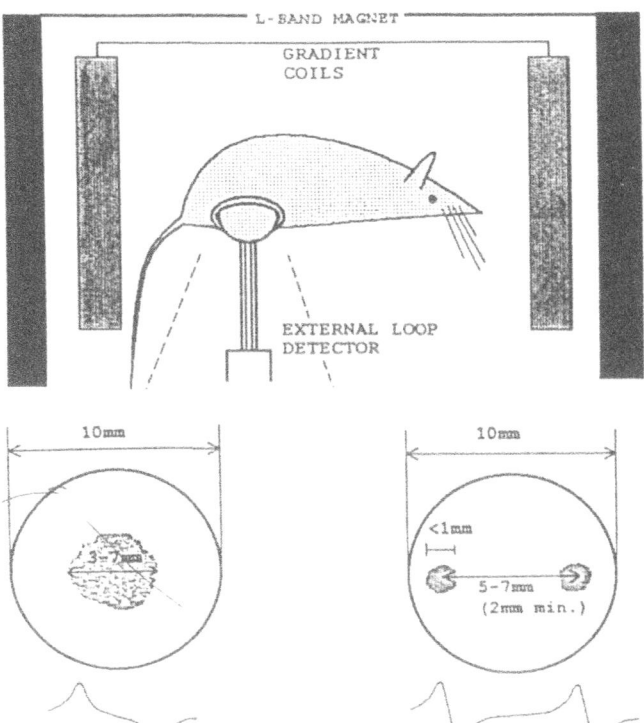

Figure 1. (a) Diagram showing the positioning of the surface detector and animal within the L-band EPR magnet. (b) Where gloxy was co-injected along with tumor cells at the time of tumor inoculation, the gloxy slurry was spread over a large volume, a magnetic field gradient was used to allow measurement of the pO$_2$ at these edges. (c) In cases where a slurry was injected into two sites, the signal arising from each location was separated by using a magnetic field gradient.

within the detecting loop (where the most sensitive region of the loop was toward the rear of the tumor with respect to the animal), secondly, by adjusting the position so that the most sensitive region of the detecting loop was at the apex or most protruding part of the tumor, and then where the loop was toward the head-end of the tumor. Recording conditions (spectrometer settings) were kept as close as possible to those on day 13. In order to compare spectra from day to day, and so that changes in EPR signal intensity could be attributed to depth of detection and not changes in tissue pO_2, we used an approximation to calculate relative signal intensity (RSI) detected:

$$RSI = \text{Signal amplitude x Linewidth}^2$$

(b) In another group of animals, a slurry of gloxy was co-injected along with the tumor cells sub-cutaneously at the time of tumor inoculation. In these mice, tumor pO_2 was monitored after 24 hours and every day throughout the lifetime of the tumors. A magnetic field gradient was also used in these cases to separate the EPR signals arising from gloxy particles in a single plane through the tumor. Using computer simulations of the resulting EPR spectra, we were able to determine the pO_2 at these two edges of the slurry.

3. RESULTS

3.1. Properties of Gloxy

The number of spins/g (i.e. EPR active centers) present in gloxy was found to be 4.6×10^{19}/g (compared to fusinite = 1×10^{19} spin/g, and India ink = 2.5×10^{19} spin/g (James, et al., 1996)). The extent of linewidth change with pO_2 for a slurry of small particles of gloxy can be seen from the data presented in Figure 2 (a), and is compared with samples of fusinite and India ink (at the same concentration and similar instrumental conditions; the modulation amplitude was optimized in each case and was kept to one third of the measured linewidth). The single line of gloxy had high signal amplitude (Figure 2(b)), being ten times greater than fusinite and twenty times that of India ink (recorded in nitrogen), and the spectral linewidth was very sensitive to pO_2 over the physiologically important range (0–40 mm Hg).

3.2. Stability of Gloxy

The number of spins/g and linewidth sensitivity to pO_2 of a slurry of gloxy (stored in saline and incubated at 37^0C and 95% O_2 + 5% CO_2) were monitored over a period of 90 days and we observed no significant change. The spectral shape and intensity also were independent of changes in pH over the range tested (2–11) and observation temperature ($10–45^0$C). The time for equilibration of spectral shape of gloxy (contained within a gas permeable tube) and the perfused gas flowing through the cavity, was relatively fast when the gas was changed from nitrogen to air (>1 min), and slower when the change was from air to nitrogen (10 min).

3.3. Measurement of Tumor pO_2 in Vivo

3.3.1. Gloxy Injected as a Slurry into Established Tumors. Measurements of pO_2 obtained from MTGB tumors into which a single slurry of gloxy was injected (day 10–12),

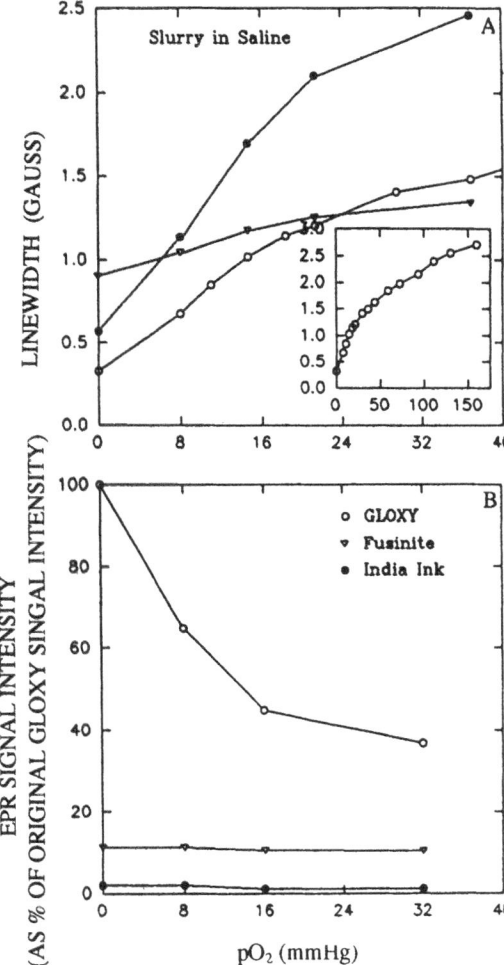

Figure 2. Spectral characteristics of gloxy and other oxygen sensitive paramagnetic materials: a) Linewidth changes with pO₂ for a slurry of particles in saline (prepared by grinding 10 mg each of gloxy or fusinite which was filtered through a 20 um mesh and made up in 200 μl of saline). India ink was used as a 50% solution of the purchased ink (Higgins, Lewisburg, TN). Inset: Changes in linewidth over a broader range of pO₂ (nitrogen to air (160 mm Hg)). (b) Relative Signal Intensity (RSI) over the same pO₂ range. The parameters of linewidth and signal amplitude are normalized for receiver gain of the spectrometer. Modulation amplitude was optimized for each spectrum (and each sample) and is shown as measured.

showed slight variation between individual animals (mean pO₂ was 6.0+/-0.6 mm Hg for a group of 12 animals on day 13 post tumor inoculation), and in all animals there was a decrease in pO₂ with tumor growth (Figure 3).

The measured pO₂ in MTG-B tumors were lower than those of RIF-1 tumors (mean pO₂±s.d: MTG-B = 3.0±0.1 mm Hg; RIF-1 = 5.4±0.4 mm Hg, at day 15). At lower pO₂ values the standard deviation in measurements with gloxy was 4% from the mean pO₂ (compared to >8% in experiments using India ink - data not shown, but also in agreement with Goda, et al., 1995). Subsequent measurement in these same MTG-B tumors at 20 days showed a lower pO₂ value, presumably because the tumors had increased in size (1.3±0.0 mm Hg (error=2.9% of measured pO₂)).

We observed an overall decrease in RSI detectable over the period of investigation, the number on day 24 being 25% of that on day 13 and the signal-to-noise ratio was 11:1 and 5:1, respectively (Figure 4). Following EPR measurement on day 24, this tumor was excised from the mouse and preserved in formalin. Histological sectioning of this tumor showed the upper surface of the gloxy slurry to be 5.4 mm from the tumor capsule (with an additional 0.8–1 mm layer of skin) and was approximately 0.8 mm in diameter. These results were in close agreement with Magnetic Resonance imaging (MRI) of the gloxy

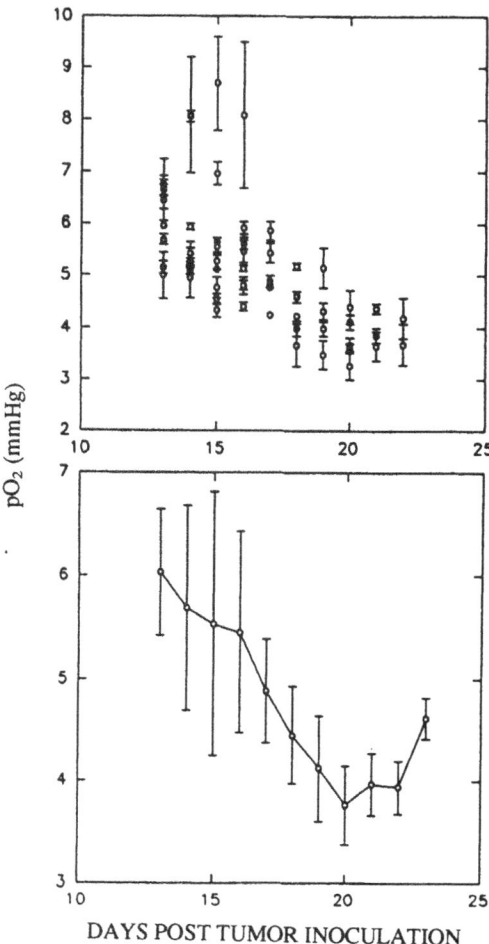

Figure 3. Changes in pO_2 for MTGB tumors during growth (day 13 to 24)(n=12). Gloxy was injected into established tumors at day 10–12. Upper: Error bars represent the standard deviation from the mean for 5 spectra recorded at each time point per mouse (each symbol respresents data from a particular mouse). Below: Average pO_2 values for the group of twelve animals. Error bars represent the standard deviation from the mean of the group on a given day.

within the tumor measured in vivo (refer to Liu, et al., 1995 - we can estimate the size of the slurry in vivo by measuring the susceptibility effect caused by the paramagnetic slurry of gloxy particles in sequential 0.8mm slices through the tumors). In these cases we found the average volume of the slurry to be 2.8+/-0.8% of the total tumor volume.

3.3.2. Gloxy Implanted Along with Tumor Cells at the Time of Inoculation. In these mice, we were able to monitor pO_2 throughout the lifetime of the tumors (as gloxy was co-injected with the tumor cells on the day of tumor inoculation (day zero)). We obtained a profile of pO_2 changes that was characteristic to each tumor (Figure 5) but had features in common; these included an initial increase in tumor pO_2, followed by a plateau which co-incided with the appearance of a palpable tumor, then pO_2 decreased with further increase in tumor volume.

In these same tumors, we also recorded the first derivative of the absorption spectrum in the presence of a magnetic field gradient (1.2 Gauss/cm). As the diameter of the slurry was quite large (histological analysis showed the slurries to be several mm and MRI measurements showed an average slurry volume of 20 +/- 3.7% of the total tumor volume), we detected the first derivative signal arising from each edge of the slurry. This was due to the fact that the absorption spectrum arising from gloxy at each edge gave rise to

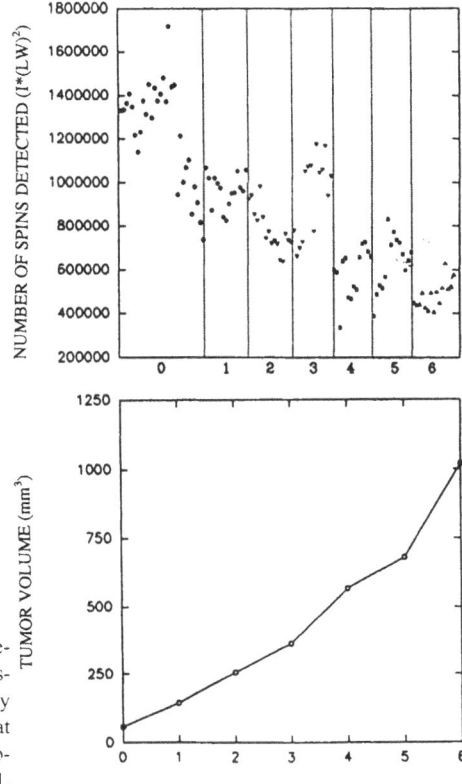

Figure 4. Apparent change in the number of spins detected (relative spin intensity; RSI) due to increased distance between the surface detecting loop and the gloxy within the tumor. Spectra at day 19 (first measurement at day 13 post inoculation, gloxy was injected into established tumors on day 12) had a signal-to-noise ratio of 5:1. Below: Tumor volume over the same investigation period.

DAYS POST GLOXY (13+N FOLLOWING INOCULATION)

one half of the first derivative EPR spectrum (the signal arising from the absorption spectrum of gloxy between these two edges was constant and the resulting first derivative was therefore zero). This is illustrated in Figure 1. The spectra were simulated and fitted assuming 2 centers with variable linewidth and position. As a result, we could measure pO$_2$ at the two edges of the slurry in the plane of the magnetic field gradient (see Figure 6).

Tumors implanted with gloxy (or injected with saline as control) were fixed in formalin prior to embedding in paraffin and 5 μm slices were taken for staining. The injection track into the tissue was visible in some cases, but we saw no evidence of tissue damage or cell death, no edema, and no infiltration of inflammatory cells over the investigation period (refer to O'Hara and James, in this book).

4. DISCUSSION

The choice of oxygen-sensitive materials for measurement of intratumor pO$_2$ by EPR oximetry depends on a number of factors; 1) the intrinsic line width and number of oxygen-sensitive, paramagnetic centers, 2) the sensitivity of the linewidth to pO$_2$ over the range of interest, 3) stability in tissues, 4) occurrence of toxicity, and 5) the conditions under which the experiment will be made (for example, depth of detection, nature of tissue, and motion expected) (Glockner, et al., 1992; Liu, et al., 1993).

The most significant feature of this particular coal derivative is the high EPR signal amplitude. It had a high number of spins/g (compared to previously used probes) and the li-

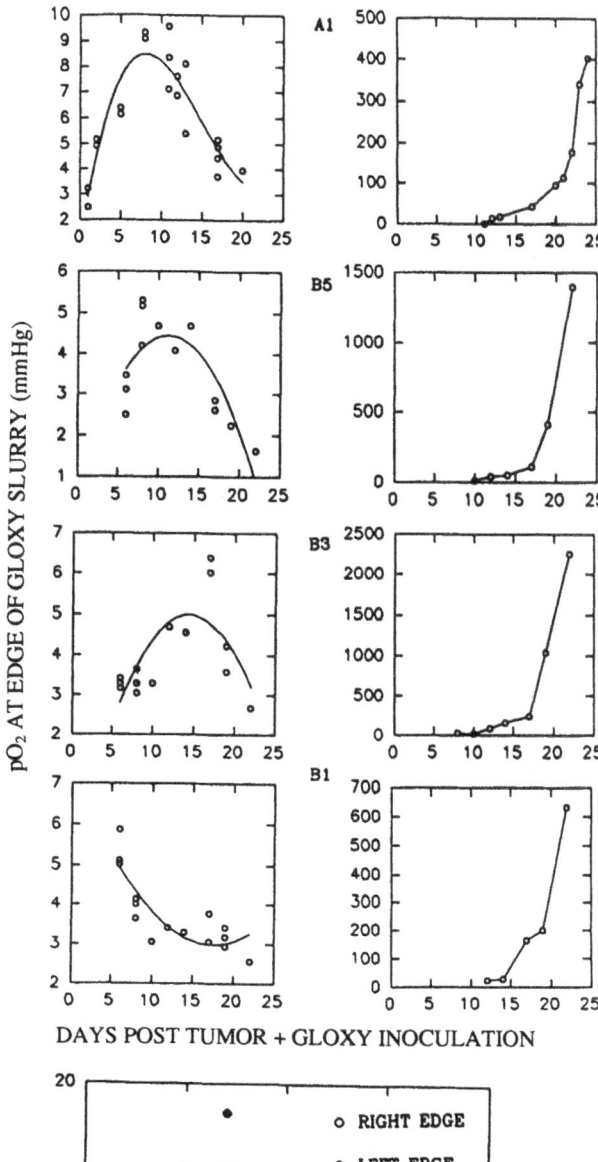

Figure 5. Changes in tumor pO_2 and tumor volume for four individual tumors. In these cases, the gloxy was co-injected along with the tumor cells at the time of tumor inoculation, allowing measurement of tumor pO_2 throughout the lifetime of the tumor.

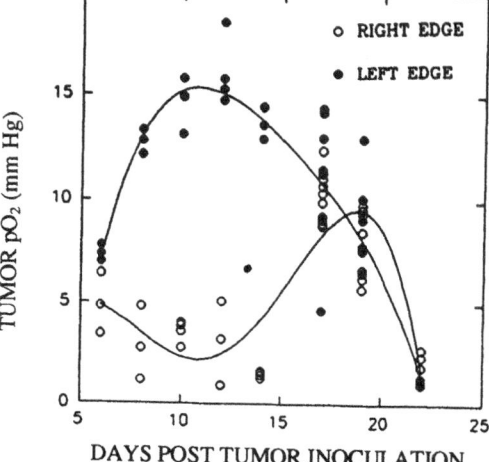

Figure 6. Simultaneous measurement of the pO_2 at two edges of the slurry in a tumor in a mouse. A magnetic field gradient was used during recording of the spectra, and each EPR spectrum was simulated assuming two spin centers with variable linewidth (or pO_2) and intensity.

Table 1. Summary of some strengths and weaknesses of the two methods using
gloxy to measure tumor pO$_2$

TABLE 1	One or two slurries injected into an established tumor	Gloxy seeded with tumor cells at the time of inoculation
POSITIVE ASPECTS OF METHOD	- Multiple site possible - Relatively good control over positioning of slurry - As with "seed" method, the gloxy slurry may be visualized by *in vivo* Magnetic Resonance Imaging (MRI). - Volume of sampling and location of slurry remains the same throughout tumor lifetime - BEST for tissue where there is little inhomogeneity (e.g. RIF-1 tumors)	- One time injection at inoculation - Large volume of sampling (seeded slurry was spread over an area approx. 5-6 mm in diameter in a tumor which was 14 mm, compared to approximate tumor cord at 200 um) - Cells grow mixed with gloxy particles - Determination of pO$_2$ at each edge - Potential future application for determining pO$_2$ composition variation within the volume of tumor containing the slurry - BEST for tissue with relatively large inhomogeneity of histology and pO$_2$ (e.g. MTGB tumors)
NEGATIVE ASPECTS OF METHOD	- Insertion of gloxy requires additional injection - Relatively small volume of measurement - Two center simulation difficult (higher gradient required)	- Little control over positioning of gloxy slurry - Latent period or delay in tumor growth (however, rate of growth of tumor is the same following delay)

neshape at pO$_2$ <40 mm Hg was such that signal amplitude was up to 10 times that of previously used material (for example, fusinite coal (Vahidi, et al., 1994) or India ink (Goda, et al., 1995). This factor and the resulting excellent signal-to-noise were the main reasons for the precision of these tumor pO$_2$ measurements. The linewidth of the EPR spectrum of gloxy was reversibly broadened by oxygen and the response to pO$_2$ was independent of pH, temperature, oxidants, reductants, and the presence of other paramagnetic materials. No additional invasive procedures were required following initial implantation of the gloxy particles through a small bore needle; this feature of repetitive, non-invasive measurements is particularly useful for repeated measurements over a period of time (such as monitoring pO$_2$ in tumors during their development and therapy). Table 1 summarizes some of the strengths and weaknesses of the two methods using gloxy to measure tumor pO$_2$.

Gloxy can be used to measure pO$_2$ at multiple sites simultaneously or particles can be injected along with tumor cells at the time of inoculation in model tumors. We speculate that by co-injecting particles with tumor cells in model tumors, we are able to monitor increasing tumor pO$_2$ with angiogenesis at early stages of tumor growth, and a decrease in tumor pO$_2$ where tumor volume and oxygen demand exceeds supply. This technique also allows measurement of the average pO$_2$ across a larger volume of the tumor, and therefore incorporates the variation in pO$_2$ values across many tumor cords.

REFERENCES

1. P. Vaupel, F. Kallinowski, P. Okunieff. Blood flow, oxygen and nutrient supply, and metabolic microenvironment of human tumors; a review. Can. Res. 49, 6449–6465 (1985).

2. E.J.Hall. Radiobiology for the Radiologist. (J.B.Lippincott ed., Philadelphia), 3rd edition, p 143, (1988).

3. A.J.Giaccia. Hypoxic stress proteins; survival of the fittest. In: J.A.Raleigh (ed) Hypoxia and its clinical significance 6, 46–58 (1996).

4. D.C.Shrieve and A.C.Begg. Cell cycle kinetics of aerated, hypoxic and re-aerated cells in vitro using flow cytometry determination of cellular DNA and incorporated bromodeoxyuridine. Cell Tissue Kinetics 18, 641–651 (1985).

5. P.E.James, O.Y.Grinberg, F.Goda, T.Panz, J.A.O'Hara, H.M.Swartz. Gloxy: An oxygen-sensitive coal for accurate measurement of low oxygen tensions in biological systems. Mag. Reson. Med. - In press (1996).

6. M.J.Nilges, T.Walczak, H.M.Swartz. 1GHz in vivo EPR spectrometer operating with a surface probe. Phys. Med. 2, 195–201 (1989).

7. F.Goda, J.A.O'Hara, E.S.Rhodes, K.J.Liu, J.F.Dunn, G.Bacic, H.M.Swartz. Changes of oxygen tension in experimental tumors after a single dose of x-ray irradiation. Cancer Res. 55, 2249–2252 (1995)

8. K.J.Liu, G.Bacic, P.J.Hoopes, J.Jiang, H.Du, L.C.Ou, J.F.Dunn, H.M.Swartz. Assessment of cerebral pO_2 by EPR oximetry in rodents: effects of anesthesia, ischemia, and breathing gas. Brain Research 685, 91–98 (1995)

9. J.F.Glockner, H.M.Swartz. In vivo EPR oximetry using two novel probes: Fusinite and lithium phthalocyanine, in "Oxygen Transport to Tissue XIV" (W.Erdmann, D.F.Bruley, Eds.), pp 229–245, Plenum Publishing, New York, 1992

10. K.J.Liu, P.Gast, M.Moussavi, S.W.Norby, M.Wu, H.M.Swartz. Lithium phthalocyanine: A new probe for EPR oximetry in viable biological systems. Proc. Natl. Acad. Sci. (USA) 90, 5438–5442 (1993).

11. N.Vahidi, R.B.Clarkson, K.J.Liu, S.W.Norby, M.Wu, H.M.Swartz. In vivo and in vitro EPR oximetry with fusenite: A new coal-based, solid state EPR probe. Magn. Reson. Med. 31, 139–146 (1994).

12. F.Goda, K.J.Liu, T.Walczak, J.A.O'Hara, J.Jiang, H.M.Swartz. In vivo oximetry using EPR and india ink. Magn. Reson. Med. 33, 237–245 (1995)

15

DETERMINING THE ANATOMIC POSITION AND HISTOLOGICAL EFFECTS IN MURINE TUMORS OF GLOXY, AN OXYGEN-SENSITIVE PARAMAGNETIC MATERIAL

Julia A. O'Hara, Philip E. James, Youssef Zaim Wadghiri, Tomasz Panz, Oleg Y. Grinberg, Neelu Jain, Jeffrey F. Dunn, and Harold M. Swartz

Dartmouth Medical School
Department of Diagnostic Radiology
Norris Cotton Cancer Center
Hanover, New Hampshire 03755

1. INTRODUCTION

Oxygen is an important component in tumors that influences the response to radiation therapy and certain forms of chemotherapy. Extensive work has demonstrated that the pO_2 in human tumors cannot be predicted based on size, histological grade or type but must be measured (1). Recent studies have confirmed that tumor pO_2 correlates with the outcome of radiation therapy (2–4). One obvious treatment where individualization of treatment could be of value is during courses of radiation therapy. It has been postulated that fractionated radiation is successful because the tissue reoxygenates after initial doses of radiation and initially hypoxic regions of tumors become more oxygenated and therefore respond better to subsequent doses. Different tumor types and different individual tumors reoxygenate at different times and to different extents. Accurate assessments of tumor pO_2 that reflect the physiological status of the whole tumor could result in better treatments for solid tumors.

Electron paramagnetic resonance (EPR) oximetry using oxygen-sensitive paramagnetic materials implanted directly into a tumor has the potential for providing accurate, easily repeatable measurements of the partial pressure of oxygen (pO_2) (5, 6). Previously we utilized fusinite or India ink as the oxygen sensitive material. In our studies of this technique a new paramagnetic material, gloxy, has been developed that has the potential to provide sensitive, accurate pO_2 measurements in tumors and other tissues (7). Gloxy is used in a similar manner to other carbon-based materials being injected as a slurry of particles in saline directly into an established tumor. In this study we also studied the feasibility and results of co-injecting tumor cells and gloxy slurry so that the pO_2 of the tumor could be monitored throughout the course of its growth.

Oxygen Transport to Tissue XIX, edited by Harrison and Delpy
Plenum Press, New York, 1997

Since EPR oximetry uses a relatively small amount of paramagnetic material into a tumor (compared to the whole tumor volume) it would be of value to establish where the gloxy is located relative to tumor margins and regions of different physiological and structural status in the tumors. The material remains in the tumor over the course of tumor growth and response to treatment, so it is of critical importance to examine whether exposure to gloxy has any effect on the tumor's growth or histological appearance. In this work we have used tumor volume measurements, histology and MRI techniques to evaluate tumor-tissue interactions with implanted gloxy.

2. METHODS

All animal procedures were approved by the Institutional Animal Care and Use Committee of Dartmouth Medical School.

2.1. Animal Tumor Models

These studies were carried out using the radiation induced fibrosarcoma (RIF-1) and mouse mammary adenocarcinoma (MTG-B). All tumors were grown on 6-week old female C3H/HeJ mice (Jackson Laboratories, Bar Harbor, ME) and mice were randomly assigned to treatment protocols at the time of implantation. RIF-1 was grown according to the protocols of Twentyman *et al.*, (8) as previously described (5). The cells were supplied to us by Dr. Janna Wehrle, Johns Hopkins University. An intradermal injection (50 µl) of the tumor cell suspension (2.0×10^5 cells) in the right flank, resulted in tumors of approximately 7 to 8 mm in diameter, 12 days post transplantation.

The MTG-B tumor is a murine adenocarcinoma maintained by serial passage in five to seven week-old female C3H/HeJ mice (THE JACKSON LAB, Bar Harbor, ME). This tumor was initially obtained from a spontaneous C3H mouse mammary tumor originating in the glandular epithelium, and first characterized by Clifton *et al.* (9). It has been used extensively in our laboratory for the study of the effects of radiation and chemotherapy (10, 11). For transplantation, tumors were removed aseptically from mice and minced in minimum essential media (MEM) without serum (GIBCO, Grand Island NY). The minced tumor particles were pressed through a nylon 70 µM cytosieve [FALCON] and centrifuged at 60 x g for 5 minutes. The pellet of tumor cells was resuspended in fresh MEM. A subcutaneous injection of 0.05 ml (10^6 cells) tumor cell suspension in the right flank typically grows to a flattened sphere of approximately 7 to 8 mm within 10 to 12 days.

2.2. Gloxy: Material and Method of Administration

Gloxy, a paramagnetic component of Welsh coal, has EPR features which are especially useful for accurate measurement of low oxygen tensions *in vivo*. Gloxy has a high spin concentration; sensitive linewidth dependence on pO_2, especially from 0 to 40 torr; and a relatively narrow EPR line that is fairly homogeneous in shape. It can be used as single particles of desired size or as a powder or slurry prepared by grinding to the desired particle size. Gloxy is stable in physiological solutions and in tissues and nontoxic to cells in culture. The characteristics of Gloxy have been described in detail (7). Gloxy was inserted into tumors in one of two ways. (1) Twenty-five µl of a suspension of < 50µm particles was sterilized and then injected directly into the tumor using a 250 µl Hamilton syringe and automatic injection device or (2) 50 µl of 50 mg/ml of gloxy was co-injected s.c. into the mouse flank with 50 µl of cell suspension.

2.3. Tumor Growth Analysis

Tumor diameters in three orthogonal directions (d_1, d_2, d_3) were measured daily using calipers. Tumor volumes (V_t) were calculated using the formula: $V_t = \frac{1}{6}(d_1 \ast d_2 \ast d_3)$.

2.4. In Vivo EPR Oximetry

EPR spectra of India ink in tumors were obtained using a 1.2 GHz ("L-band") EPR spectrometer constructed in the EPR Center at Dartmouth. Typical settings for the spectrometer were: magnetic field, 425 Gauss; modulation frequency, 27 kHz; modulation amplitude, 0.2 Gauss; incident microwave power, 15 mW. The mice were restrained in a plastic holder with a hole through which the tumors were exposed. The detecting EPR coil was positioned over the tumor and five spectra were recorded per animal (30 seconds per spectrum, total measurement time was less than 10 min., including positioning of the animal and spectrometer adjustment). Spectral line widths of gloxy were measured and the pO_2 values calculated from a calibration curve of pO_2 vs. line width in the presence of tumor cells using a Bruker, ER 220D-SRC EPR spectrometer (9.6 GHz, X-band) and a commercial oxygen analyzer (Sensor Medics Co., Model OM-11, Anaheim CA). The calibration procedure was the same as described previously, including verification of its validity for 1.2 GHz studies (12, 13).

2.5. Magnetic Resonance Imaging

MRI localization of the EPR oximetry particles and slurries was done in a 7T horizontal bore magnet with a SMIS console. The coils used were a ^{1}H tuned 6 cm birdcage coil or a 2cm surface coil. Mice were anesthetized with ketamine/xylazine (90:9 mg/kg), positioned in the magnet lying on the left side with the tumor uppermost and the tumor was imaged with a gradient echo sequence. Mouse body temperature was monitored with a rectal probe and maintained with a water-circulating heating pad. Techniques related to the imaging of EPR oximetry paramagnetic materials in biological tissues have been previously described (14, 15).

2.6. Histology

Experimental tissues for microscopic examination were fixed in phosphate-buffered formalin, trimmed, and processed for staining/labeling. Sections were routinely stained with hematoxylin and eosin (H&E). The slides were analyzed for determination of the anatomic position of the EPR paramagnetic material and the pathologic tissue reaction to it. Analysis of histology sections was used to verify the placement of the paramagnetic material determination that had been made with MRI.

3. RESULTS AND DISCUSSION

Figure 1 shows tumor growth curves for tumor without gloxy compared to those with gloxy. No difference from controls in the time to appearance of the tumors (latency)

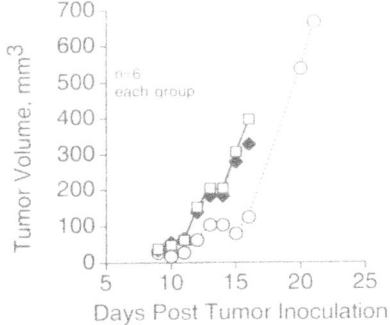

Figure 1. Growth of MTG-B tumors in the absence or presence of Gloxy. Open Squares: Control, 25 µl of saline only; Closed Diamonds: Gloxy injected as a slurry into established tumors on Day 10 after implantation; Open Circles: Gloxy slurry was administered with the tumor cells on Day 0.

or in growth rate was observed when gloxy was injected into an established tumor. When gloxy was co-administered with the tumor cells at the time of tumor implantation there was a delay to the appearance of palpable tumors. Once the tumors reached a detectable size, the rate of growth was not significantly different from tumors with no gloxy or tumors injected with gloxy after the tumor reached 6 mm diameter.

Histological examination of MTG-B tumors resulted in the conclusion that there were no consistent differences between tumors with or without gloxy (Figure 2). Areas of necrosis and viable cells are visible in both sections with no obvious difference in tissue composition and no evidence of increased fibrillar material or excessive macrophage infiltration. In this case gloxy had been present for 8 days after injection into an established tumor. Tumor volume at the time of sacrifice was 500 mm^3. These observations were confirmed in two other tumors, the RIF-1 and C6 rat glioma grown on the flank of athymic nude mice (data not shown). Figure 3 shows a low-magnification H&E section of an MTG-B tumor that had gloxy injected with the tumor cells.

Our observations with Gloxy confirm our previous observations with other paramagnetic materials that there is no undue inflammatory response and no evidence of increased fibrin deposition in either RIF-1 or MTG-B tumors in contact with India ink or fusinite (5, 6, 15). Areas of blood pooling and focal necrosis are seen in all tumors whether or not gloxy was present.

Figure 3 shows a deposition of gloxy in such a seeded tumor. The deposit is in viable tissue, adjacent to a necrotic area. Using this method of gloxy implantation we were able to monitor the pO$_2$ during the full course of tumor growth until sacrifice (about 22 days in untreated tumors). In some cases the gloxy was more diffuse than in other tumors, allowing for a higher percentage of the tumor to be monitored (up to 20% of the tumor).

Magnetic resonance imaging of living animals was used to monitor the tumor volume, the position of the gloxy in tumors and the percentage of the tumor monitored by the gloxy.

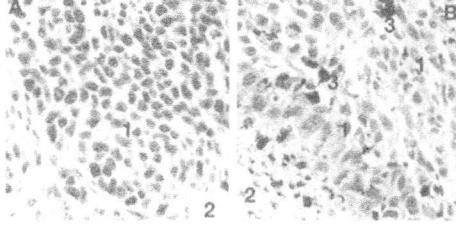

Figure 2. MTG-B tumor without (A) and with (B) gloxy present in the tissue. These are H&E stains of different tumors. Gloxy had been present for 7 days in the tumor. Note that gloxy is in both viable (1) and necrotic (2) regions of the tumor and that there is no obvious tissue reaction to its presence. A necrotic area is also visible in the tumor without gloxy. (1) Viable tumor, (2) Necrosis, (3) Gloxy. The black areas in B are deposits of Gloxy.

Figure 3. Co-Injection of Gloxy with Tumor Cells Resulted in a Focal Deposition of Gloxy. MTG-B tumor cells (approximately 10^6) mixed with particles of gloxy (1 mg) in incomplete medium were injected subcutaneously into the flank of a C3H mouse. Five days after implantation, before tumors were palpable, pO_2 was monitored by EPR. MRI images of slices through the tumor showed that the gloxy particles were sampling a volume which was 3–5% of the tumor. (See Figure 5). On the 22nd day after tumor inoculation, the mouse was sacrificed and the tumor sectioned. Gloxy appears in a viable region adjacent to tumor necrosis. (1) Viable Tumor (2) Necrotic Tumor (3) Gloxy.

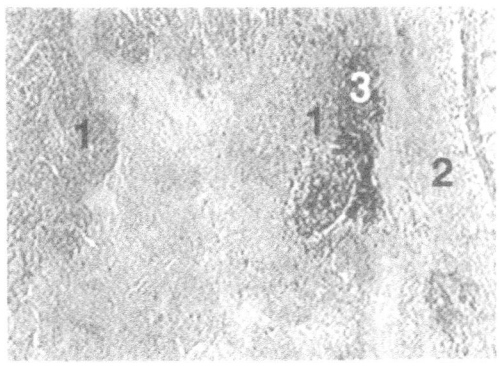

Figure 4 shows an MRI of a 1 cm MTG-B tumor with gloxy implanted as discrete particles or as a slurry when the tumor was about 6mm in diameter. In both cases, the susceptibility effect of the gloxy allows for observation of the position of the material. Using gradient echo images it is possible to discern areas of necrosis and viable tissue, relative to the gloxy. Figure 5 shows a tumor seeded with gloxy and tumor cells at the same time. From images like these the volume of the tumor sampled by the paramagnetic material can be estimated by measuring the total tumor volume (adding the volume of each slice composed of tumor) and the volume of the image represented by the paramagnetic material in each slice of the tumor. Our preliminary results with histology have determined that the hypointense areas are caused by gloxy. We are systematically studying how closely the volume determinations made by MRI agree with those made by histological assessment since the size of the dark areas in MRI slices is influenced by the sequence parameters for image acquisition. Once the relationship between "MRI volume" and "histological volume" is established, the volume sampled by the paramagnetic material can be estimated, by repeating MRI during the course of the tumor growth.

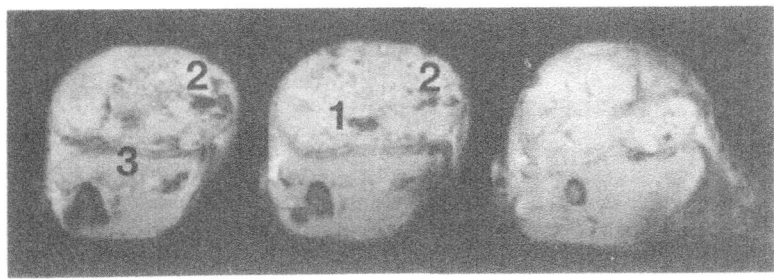

Figure 4. Magnetic Resonance Images of MTG-B Tumors containing Gloxy after injection of gloxy as a slurry and as a large particle. The tumor outlined by its capsule is overlying thigh muscle and the leg bone can be seen as hypointense areas in each of the images. The images were acquired with a 6 cm birdcage coil, FOV= 30mm, Matrix = 256x256, Slice Thickness= 0.7mm, TR/TE/å=800ms/9ms/60°. Sequential slices demonstrate the capability of determining the position of the paramagnetic materials within the tumor. Dark areas near numbers represent (1) Gloxy Particle, (2) Gloxy Slurry (3) Tumor Capsule. Diffuse hypointense areas representing focal necrosis can also be seen in the tumor.

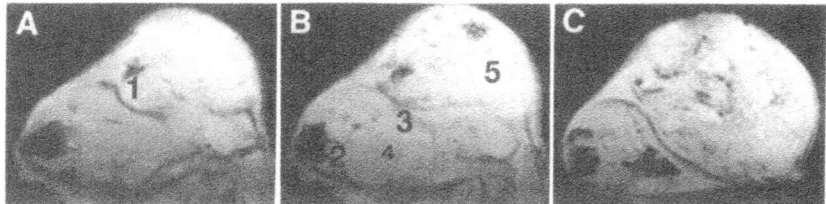

Figure 5. Magnetic Resonance Images of an MTG-B tumor containing gloxy after co-injection of gloxy and tumor cells as in Figure 3. The images were acquired with a surface coil, FOV= 24mm, matrix=128x128, slice thickness= 1.0mm, TR/TE/α=1000ms/9ms/60°.(1) femur, (2) tumor capsule (3) thigh muscle (4) Gloxy slurry. Hypointense areas distributed throughout the tumor represent focal necrosis, not deposits of gloxy. The tumor was 1 cm in diameter.

4. CONCLUSIONS

We have established methodology (MRI and histology) to analyze the site and volume sampled by gloxy during the growth of tumors. Gloxy does not cause either a persistent inflammatory response in the tumor and can be used in conjunction with EPR oximetry to measure the mean pO_2 from a delineated volume of a tumor throughout its growth period, without affecting the rate of growth of the tumor. As with all the materials, a one-time injection is required to implant the particles, but no further invasive procedure is required to monitor the pO_2 during growth of the tumor. We also tested Gloxy by simultaneously injecting the slurry with tumor cells at the time of inoculation (the latter may increase the latent period for development of palpable tumors). The utility of this technique is still being investigated. EPR oximetry has the unique capability of monitoring pO_2 before the tumor becomes palpable. The increased latency and reduced tumor take is a concern and the implications and consistency of these observations are under investigation. Gloxy is a paramagnetic materials with useful properties for EPR oximetry. Ongoing testing of paramagnetic materials will determine the optimal material for determination of pO_2 in tumors in vivo.

ACKNOWLEDGMENTS

This work was supported by ACS Grant BE-186, the Core Grant of the Norris Cotton Cancer Center CA23108, NIH Grant GM51630 and used the facilities of the IERC at Dartmouth supported by NIH grant RR-0811.

REFERENCES

1. P. W. Vaupel, Oxygenation of solid tumors. In *Drug Resistance in Oncology* (B. A. Teicher, Ed.), pp. 53–86. Marcel Dekker, New York, 1993.
2. M. Hockel, C. Knoop, K. Schlenger, B. Vorndran, E. Baussmann, M. Mitze, P. G. Knapstein and P. Vaupel, Intratumoral pO_2 predicts survival in advanced cancer of the uterine cervix. *Radiother Oncol.* **26**, (1), 45–50 (1993).
3. M. Hockel, K. Schlenger, M. Mitze, U. Schaffer and P. Vaupel, Hypoxia and radiation response in human tumors. *Sem in Radiat Oncol.* **6**, 3–9 (1996).
4. P. Okunieff, M. Hoeckel, E. P. Dunphy, K. Schlenger, C. Knoop and P. Vaupel, Oxygen tension distributions are sufficient to explain the local response of human breast tumors treated with radiation alone. *Int J Radiat Oncol Biol Phys.* **26**, (4), 631–6 (1993).

5. F. Goda, J. A. O'Hara, E. S. Rhodes, K. J. Liu, J. F. Dunn, G. Bacic and H. M. Swartz, Changes of oxygen tension in experimental tumors after a single dose of X-ray irradiation. *Cancer Res.* **55**, (11), 2249–52 (1995).

6. J. A. O'Hara, F. Goda, K. J. Liu, G. A. Bacic, P. J. Hoopes and H. M. Swartz, The pO$_2$ in a murine tumor after radiation: an in vivo electron paramagnetic resonance oximetry study. *Radiat Res.* **144**, (2), 222–229 (1995).

7. P. E. James, O. Y. Grinberg, F. Goda, J. A. O'Hara and H. M. Swartz, Gloxy: An oxygen-sensitive coal for accurate measurement of low oxygen tensions in biological systems. *Magn Reson Med.* **38**, 48–58 (1997).

8. P. R. Twentyman, J. M. Brown, J. W. Gray, A. J. Franko, M. A. Scoles and R. F. Kallman, A new mouse tumor model system (RIF-1) for comparison of endpoint studies. *J Natl Canc Inst.* **64**, 595–604 (1980).

9. K. H. Clifton, R. C. Briggs and H. B. Stone, Quantitative radiosensitivity studies of solid carcinomas *in vivo*: methodology and effect of anoxia. *J Natl Canc Inst.* **36**, 965–974 (1966).

10. E. Jones, B. Lyons, E. Douple, A. Filimonov and B. Dain, Response of a brachytherapy model using ^{125}I in a murine tumor system. *Radiat Res.* **118**, 112–130 (1989).

11. E. B. Douple, J. A. O'Hara and E. L. Jones, Paraplatin enhancement of radiation therapy in a murine tumor (MTG-B). In *Anticancer Drug Research* (K. Lapis S. Eckhardt, Ed.), pp. 71–80. Akademiai Kiado, Budapest, 1987.

12. F. Goda, K. J. Liu, T. Walczak, J. A. O'Hara and H. M. Swartz, In vivo oximetry using EPR and India ink. *Magn Reson Med.* **33**, (2), 237–245 (1995).

13. H. M. Swartz, K. J. Liu, F. Goda and T. Walczak, India ink: A potential clinically applicable EPR oximetry probe. *Magn Reson Med.* **31**, (2), 229–232 (1994).

14. J. F. Dunn, S. Ding, J. A. O'Hara, K. J. Liu, E. Rhodes, J. B. Weaver and H. M. Swartz, The apparent diffusion constant measured by MRI correlates with pO2 in a RIF-1 tumor. *Magn Reson Med.* **34**, (4), 515–9 (1995).

15. G. Bacic, K. J. Liu, J. A. O'Hara, R. D. Harris, K. Szybinski, F. Goda and H. M. Swartz, Oxygen tension in a murine tumor: a combined EPR and MRI study. *Magn Reson Med.* **30**, (5), 568–72 (1993).

16

IN VIVO INVESTIGATIONS OF VASCULAR DENSITY AND LOCAL MITOCHONDRIAL METABOLISM AFTER IRRADIATION

Ludwig Plasswilm[1] and Jens Höper[2]

[1]Klinik und Poliklinik für Strahlentherapie
[2]Institut für Physiologie und Kardiologie
Universität Erlangen-Nürnberg, Germany

INTRODUCTION

Angiogenesis, the biological event in which preexisting vessels give rise to new ones, is one of the essential phenomenons under several physiological and pathological conditions. A number of stimulating angiogenic peptides have been identifed, such as fibroblast growth factor, epidermal growth factor and platelet-derived growth factor (Schott and Morrow 1993). Progress in knowledge of the basic mechanisms of primary and secondary angiogenesis appears to be of particular importance in cancer research, since the tumor vascular system plays a decisive role in the genesis and growth as well as in metastasis formation of tumors (Folkman 1990, Weidner et al. 1992, Bikfalvi 1995, Leon et al. 1996). Chick embryos, that have been used in medical research for more than a hundred years, have also been used to study vasculogenesis and angiogenesis as well as the effect of angiogenic factors. Most of these studies focussed on the vascular system of the chorioallantoic membrane (CAM) during the second and third week of incubation of the eggs. Recently, experiments were conducted to clarify the suitability of the vascular system of the yolk sac membrane, which is the earliest extraembryonic blood vessel system appearing already at day 2 of incubation. The yolk sac is an extraembryonic membrane, which surrounds the yolk completely. Its inner layer is formed by the follicle epithelium and its outer layer is a precipitation of mucin from the infundibulum of the oviduct. Huettner (1949) and Rosenbruch (1990) described the structure of the yolk sac membrane centripetally as serosa/ectoderm, somatic mesoderm, visceral mesoderm, and endoderm. The development of blood cells and vessels starts extraembryonically in the visceral mesoderm of the yolk sac membrane (Romanoff 1960, Starck 1975).

With respect to the significance of angiogenesis, the aim of this study was to investigate whether the yolk sac blood vessel system of chick embryos represents a useful model for *in vivo* measurements of angiogenesis in radiobiology.

Oxygen Transport to Tissue XIX, edited by Harrison and Delpy
Plenum Press, New York, 1997

MATERIALS AND METHODS

Eggs

For this study fertilized crossbred 'White-Plymouth-Rocks x Sussex' eggs (Lehr-und Versuchsanstalt für Kleintierzucht, Kitzingen, Germany) were incubated in an upright position in a commercial incubator at $36.8 \pm 0.1°C$ and 60–65 % relative humidity. The daily weight loss of the eggs was 0.3 ± 0.05 g per day which was considered to be within the normal range.

Irradiation

Irradiation was performed on day two of incubation. For this purpose the egg was carefully opened. At this time the sinus terminalis was visible through the inner shell membrane. The position of the area to be radiated was marked on the egg shell.

Radiation was performed with a linear accelerator (Siemens, 2.2 Gy/min) with 100 cm focus-surface distance and a water equivalent bolus of 10 mm. The irradiated part of the *area vasculosa* had a diameter of 10 mm. The eggs received single fraction irradiation with doses from 2 to 10 Gy. After irradiation, the hole was closed again and the eggs were incubated for another 48 hrs.

Vascular Density

On day 4 of incubation, 48 hrs after irradiation, the eggs were opened and the inner shell membrane was removed. Now the vascular system of the *Area vasculosa* was visible. The almost circular, vascularized part of the yolk sac membrane (*Area vasculosa*) was photographed in vivo. Prints of known enlargement were evaluated for density of the blood vessels.

The vascular density was determined according to Höper and Jahn (1995). From all eggs at least two photographs were taken; one from the irradiated and one from a control area. The parameter points of intersection/mm^2 (VIS/mm^2) was compared by Student's t-test. Values with $P < 0.05$ were considered to differ statistically significant from the control.

Mitochondrial Metabolism

For investigations of the local mitochondrial metabolism of the *Area vasculosa* after radiation, a micro-light guide spectrophotometer (EMPHO II, BGT, Überlingen, Germany) with modified light guides was used (Höper, in press) to measure the cleavage of a tetrazolium salt (WST-1, Boehringer-Mannheim, Germany).

Erlanger Micro-Lightguide Spectrophotometer (EMPHO)

The EMPHO (BGT,Überlingen, FRG) consists of four modules: the light source, the micro-lightguide, the detection device and the computing system. Details have been published elsewhere (Frank et al. 1989). Light from a xenon arc lamp (Osram) is parallellised and focused onto the entrance plane of the illuminating micro-lightguide fibre. The light is transmitted to the tissue surface by a central micro-lightguide fibre closely surrounded by a hexagon of six detecting fibres. Each of the fibres has a diameter of 250 µm. Both the illuminating fibre and the detecting micro-lightguides are encased in a flexible rubber tube. At the measuring end they are inserted into a stainless steel cannula (outer diameter 1.6 mm). At the other end they are attached to the optical instrument by means of a special plug. The light transmitted by the detection micro-lightguide passes through a rotating interference bandpass filter disk (502–628 nm, Anders, Nabburg, FRG) and is measured by a photomultiplier. The output is connected to a computer used for signal storage and processing. Before each measurement a response spectrum for the apparatus is recorded. This is used for the evaluation of the tissue spectra. For this purpose the lightguide probe is held above a surface-coated mirror (Spindler & Hoyer, Göttingen, Germany) at a set distance.

Micro-Lightguide

For the in vivo investigation of the cleavage of the tetrazolium salt 4-[3-(4-Iodophenyl)-2-(4-nitrophenyl) -2H-5-tetrazolio]-1,3-benzene disulfonate (WST-1) a special light guide was developed. At the tip of the light guide a microcuvette was fixed. This cuvette was filled with 50 µl of the reagent. At the bottom the cuvette has an opening, which is placed at the surface of the *Area vasculosa*. The WST-1 diffuses into the tissue and is cleaved to formazan. This is accompanied by a change in the absorption, which is measured.

Colorimetric Assay

In recent years different tetrazolium salts have been described which can be used for the in-vitro measurement of cell proliferation and viability (Scudiero et al. 1988, Vistica et al. 1981).

For our investigations we used the WST-1 reagent (Boehringer Mannheim, Germany). WST-1 is the tetrazolium salt (4-[3-(4-Iodophenyl)-2-(4-nitrophenyl) -2H-5-tetrazolio]-1,3-benzene disulfonate) which is cleaved to formazan by the "succinate-tetrazolium reductase" system (EC 1.3.99.1). This system belongs to the respiratory chain of mitochondria and is active only in viable cells (Slater et al. 1963). While WST-1 is slightly red, formazan is dark red. The maximal absorbance for the reaction product formazan is found between 420 and 480 nm. At 502 nm approx. 70% of the maximal absorption is found. Only the signal at 502 nm was evaluated from the spectra recorded. A detailed description of the method will be published (Höper, in press). The measurements were performed 1, 24 and 48 hrs after irradiation. In figure 1 a schematic drawing showing the experimental setup is depicted.

RESULTS

Vascular Density

In fig. 2 and table 1 vascular densities in control eggs (0 Gy) and after irradiation are shown; 48 hours after irradiation with 2 to 8 Gy there is a slight decrease in vascular density. After a single dose of 10 Gy a distinct, statistical significant increase in vascular density was found compared to the value of 4, 6 and 8 Gy ($p < 0.05$). Compared to 0 Gy (control) changes are also obvious but not statistically significant ($p = 0.07$).

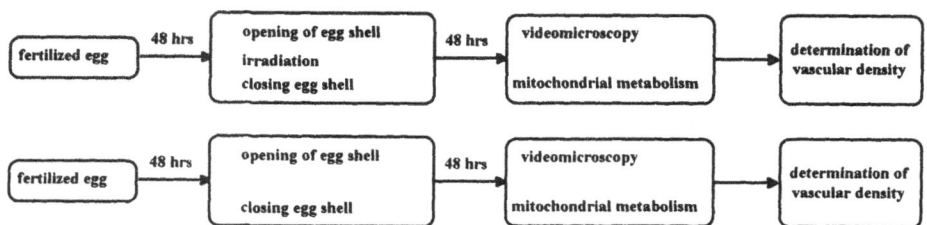

Figure 1. Schematic depiction of the experimental setup. The experimental eggs (upper panel) were irradiated after 48 hrs of incubation. The control eggs (lower panel) were only opened after 48 hrs.

Mitochondrial Metabolism

Figures 3 and 4 show the absorption changes due to the formation of formazan measured in control and irradiated eggs. It is obvious that in the controls there was an increase in absorption from the 4th to the 6th day of incubation.

Compared with this control group there was a significant (p < 0.01) increase in absorption at 1 hour after radiation with 10 Gy (Fig. 3). 24 hours after irradiation no significant difference was seen, whereas at 48 hours a significant decrease (p < 0.005) was observed (Fig. 3). After radiation with 6 Gy there was a distinct, significant increase (p < 0.001) in absorption 1 h after irradiation. Compared to the age matched control eggs no difference was observed 24 and 48 hrs after radiation (Fig. 4).

DISCUSSION

After irradiation an increasing vascular density in the *Area vasculosa* of chick embryos and a change in the cleavage of the tetrazolium salt to formazan occurs. A single radiation dose of 10 Gy leads to a significant increase in vascular density. Despite development dependent changes in the WST-test (Höper, in press) significant alterations have been observed after irradiation. Comparision of these two parameters may lead to a better comprehension of angiogenesis.

Angiogenesis is under the control of different growth factors. Stewart et al. (1989) showed that epidermal growth factor (EGF) promotes extraembryonic angiogenesis in the *Area vasculosa* of 3 day old chicks. Prenatal growth and development of the early embryo is under control of basic fibroblastic growth factor (bFGF, Hill and Han 1993) and insulin-like growth factor (IGF, Milner 1993). bFGF elicits fibrocyte proliferation and minor en-

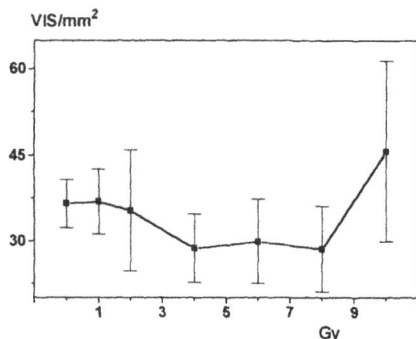

Figure 2. Relation between radiation dose and vascular density (VIS/mm^2). Number of experiments see table 1, mean values ± SD.

Table 1. Vascular density (VIS/mm^2) in control eggs
and after radiation (n = number of experiments)

Gy	VIS/mm^2	SD	n	p
0	36.6	4.1	6	<0.08 vs. 10 Gy
1	36.9	5.7	2	
2	35.3	5.3	4	
4	28.7	6.0	7	
6	29.9	7.4	8	
8	28.6	7.6	7	
10	45.7	15.7	10	<0.05 vs. 4, 6 and 8 Gy

dothelial cell proliferation (Wilting et al. 1993). To our knowledge it has not been shown that these factors stimulate growth and vascularisation of the yolk sac membrane. In the *Area vasculosa* vascular endothelial growth factor (VEGF) is an important mitogen for vasculogenesis as well as for angiogenesis (Flamme et al. 1995).

Vasoproliferative effects of hypoxia have already been described for the chorioallantoic membrane (CAM) of chick embryos (Strick et al. 1991). Furthermore, Hudlicka et al. (1989) investigated angioge-nesis in the CAM after administration of angiogenic factors. The induced new blood vessels showed a 'wheel spokes' pattern on the spot, where the angiogenic stimulus had been administered. Later, Wilting et al. (1993) were able to show that VEGF induces angiogenesis in the regions of precapillary vessels, capillaries and venules, respectively. It has been shown that during hypoxia an up-regulation of the VEGF gene occurs (Wilting et al. 1993, Shweiki et al. 1992, Goldberg and Schneider 1994, Giaccia et al. 1995, Hlatky et al. 1995).

As shown by Höper and Jahn (1995), hypoxia not only induces an increase in vascular density but also induces an enlargement of the *Area vasculosa*. This indicates that hypoxia not only induces proliferation of endothelial cells but also of non-vascular tissue. In the meantime it has been shown that also radiation can induce the expression of bFGF in normal rat astrocytes (Toflion et al. 1995) as well as in breast carcinoma cells (Lee et al. 1995). The proliferative effect of EGF shown by Stewart et al. (1989) may not be confined to the vascular endothelium, but seems to stimulate the development of ecto-, meso- and endoderm of the *Area vasculosa*.

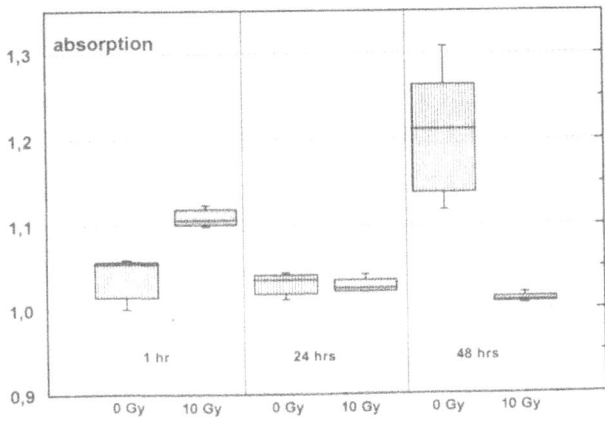

Figure 3. Absorption changes due to formation of formazan. The data are shown as Box-and-Whisker-plots. The line in the box represents the median, in the box are all values between the 25th and 50th percentile, whereas the whiskers correspond to the 5th and 95th percentile (0 Gy n = 7, 10 Gy n = 6). The difference between control (0 Gy) and radiated (10 Gy) eggs is significant after 1 h (p<0.01) and after 48 hrs (p <0.005).

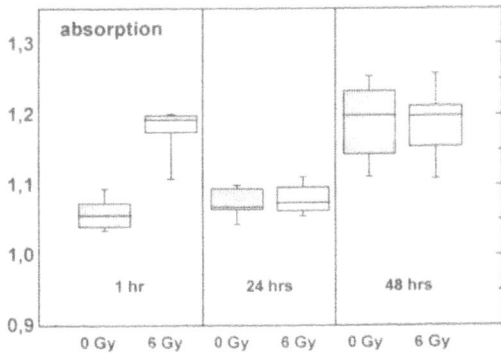

Figure 4. Absorption changes due to formation of formazan. Significant differences (p<0.001) between control (0 Gy) and radiated (6 Gy) eggs were found only 1 h after irradiation.

As shown in figure 2 irradiation with 1–4 Gy was without any effect on vascular density. In contrast, single fraction radiation with 6–8 Gy led to a slight decrease in vascular density. This unexpected result indicates that there is a discrepancy between the enlargement of the area vasulosa and the formation of blood vessels. Macroscopically no unusual enlargement of the *Area vasculosa* was observed, leading to the assumption that the formation of blood vessels is depressed. This could be due to a radiation effect. As indicated by the result of the WST-test there is increase in cellular metabolism. As shown by Gudz et al. (1994) an increase in oxygen uptake rate of thymocytes occurs 1 h after irradiation with single doses of 6 - 10 Gy. The authors showed that the increase in cellular respiration is due to an increased ATP consumption. Krumme et al. (1975) showed already in 1975 that after irradiation a dose-dependent increase in sodium and potassium permeability occurs, indicating disturbances in cellular ionic homeostasis. This in turn will activate the sodium-potassium-pump, thus increasing the ATP consumption. From measurements in local oxygen partial pressure it was concluded that also the local oxygen uptake increases. Longer lasting changes in oxygen uptake rate may induce synthesis and release of angiogenic factors (Brown et al. 1995).

In case of successful repair, oxygen uptake returned to normal values. Therefore 24 and 48 hours after irradiation with 6 Gy the WST-test showed control values.

After radiation with 10 Gy the mitochondrial metabolism increased too, but the reaction was less pronounced compared to 6 Gy (Fig. 4). The reason for this may be acute dose-dependent cellular death. These cells will not participate in oxygen uptake. Radiation with a single dose of 10 Gy causes an acute decrease in the number of cultured fibroblasts by almost 40%. If this also occurs in the *Area vasculosa*, only the remaining cells will participate in the reduction of WST. Therefore, the maximal increase in formazan and thus the absorption change must be lower. Taking this into account, then in the remaining cells an increased WST reduction must take place. These cells will therefore have an increased oxygen need. The same will be true after 24 hrs. This indicates an increase in local oxygen uptake over at least a period of 24 hrs. A further indication for this assumption is given by the result obtained after 48 hrs. The remaining, intact cells return to a normal state, thus the oxygen uptake returns to normal values. Because the number of intact cells is lower than before irradiation, a decreased reduction of WST/tissue volume must occur.

SUMMARY

It is obvious that irradiation increases local oxygen uptake. Because the oxygen supply is not altered, a decrease in local oxygen partial pressure may develop. This state may be called relative hypoxia. If the metabolism is altered for a period of at least 24 hrs an increase in angiogenic mitogens, inducing angiogenesis will occur. Presumably this is the mechanism underlying the therapeutic effect of radiotherapy in benign disease.

REFERENCES

Bikfalvi, A. Significance of Angiogenesis in Tumor Progression and Metastasis. Eur. J. Cancer **31**, 1101–1104, (1995).

Brown, MD, Hudlicka, O, Makki, RF, Weiss, JB. Low-molecular-mass endothelial cell-stimulating angiogenic factor in relation to capillary growth induced in rat skeletal muscle by low-frequency electrical stimulation. Int. J. Microcirc. **15**, 111–116, (1995).

Flamme, I, Breier, G, Risau, W. Vascular endothelial growth factor (VEGF) and VEGF receptor (flk-1) are expressed during vasculogenesis and vascular differentiation in the quail embryo. Dev. Biol. **169**, 699–712, (1995).

Folkman J. What is the evidence that tumors are angiogenesis dependent? J. Natl. Cancer Inst. **82**, 4–6, (1990).

Frank, KH, Kessler, M, Appelbaum, K, Dümmler, W. The Erlangen micro-lightguide spectrophotometer EMPHO I. Phys. Med. Biol. **34**, 1883–1900, (1989).

Giaccia, AJ, Chen, EY, Yeh, P, Waleh, N, Laderoute, KR., Mazure, NM. Regulation of vascular endothelial growth factor by hypoxia. Proc. 10th Int. Congress of Radiation Research (abstract P11–11) p 218, Wuerzburg, Germany, (1995).

Goldberg, MA, Schneider, TJ. Similarities between the oxygen sensing mechanisms regulating the expression of vascular endothelial growth factor and erythropoietin. J. Biol. Chem. **269**, 4355–4359, (1994).

Gudz, TI, Pandelova, IG, Novgorodov, SA. Stimulation of respiration in rat thymocytes induced by ionizing radiation. Radiat. Res. **138**, 114–120, (1994).

Hill, DJ, Han, VK. Cell and Tissue growth. In: Perinatal and Pediatric Pathophysiology (P.D. Gluckmann, M.A. Heymann eds.) pp 155–161, Edward Arnold, (1993).

Hlatky, L, Tsionou, C, Hahnfeldt, P, Coleman, N. VEGF expression in human mammary tumor cells: Upregulation with hypoxia. Proc. 10th Int. Congress of Radiation Research (abstract P11–12) p 218, Wuerzburg, Germany, (1995).

Höper, J, Jahn, H. Influence of environmental oxygen concentration on growth and vascular density of the area vasculosa in chick embryos. Int. J. Microcirc. **15**, 186–192, (1995).

Höper, J. In vivo determination of local mitochondrial metabolism by use of a tetrazolium salt. Physiol. Measure. (Submitted).

Hudlicka, O, West, D, Kumar, S, Khelly, FE, Wright, AJ. Can growth of capillaries in the heart and skeletal muscle be explained by the presence of an angiogenic factor. Br. J. Exp. Path. **70**, 237–246, (1989).

Huettner, AF. Fundamentals of comparative embryology of the vertebrates. Macmillan, New York, (1949).

Krumme, BA, Höper, J, Schönleben, K, Fischedick, O, Kessler, M. Tissue oxygen supply of skeletal muscle and liver after cobalt radiation. Pflügers Arch. **359**, R 40, (1975).

Lee, Y, Berns, C, Galoforo, S, Erdos, G, Corry, P. Ionizing radiation induced AP-1 transcription factor activity and bFGF gene expression. Proc. 10th Int. Congress of Radiation Research abstract P11–20) p 221, Wuerzburg, Germany, (1995).

Leon, SP, Folkerth, RD, Black, PM Microvessel density is a prognostic indicator for patients with astroglial brain tumors. Cancer **77**, 362–372, (1996).

Milner, RDG. Prenatal growth control. In: Perinatal and Pediatric Pathophysiology (P.D. Gluckmann, M.A. Heymann eds.) pp 162–169, Edward Arnold, (1993).

Romanoff, AL. The avian embryo. Structural and functional development. Macmillan, New York 571–600, (1960).

Rosenbruch, M. Toxizitätsuntersuchungen am bebrüteten Hühnerei. Unter besonderer Berücksichtigung der extraembryonalen Gefäßsysteme. Dermatosen **38**, 5–11, (1990).

Schott R J, L A. Morrow. Growth factors and angiogenesis. Cardiovasc. Res. **27**, 1155–1161, (1993).

Scudiero, DA, Shoemaker, RH, Paull, KD, Monks, A, Tierney, S, Nofziger, TH, Currens, MJ, Seniff, D, Boyd, MR. Evaluation of a soluble tetrazolium/formazan assay for cell growth and drug sensitivity in culture using human and other tumor cell lines.Cancer Res. **48**, 4827–4833, (1988).

Shweiki, D, Itin, A, Soffer, D, Keshet, E. Vascular endothelial growth factor induced by hypoxia may mediate hypoxia-initiated angiogenesis. Nature **39**, 843–845, (1992).

Slater, TF, Sawyer, B, Sträuli, U. Studies on succinate-tetrazolium reductase systems. III. Points of coupling of four different tetrazolium salts.Biochem.Biophys. Acta **77**, 383–393, (1963).

Starck, D. Embryologie. 3. Ed. Thieme, Stuttgart, (1975).

Stewart, R, Nelson, J, Wison, DJ. Epidermal growth factor promotes chick embryonic angiogenesis. Cell Biol. Int. Rep. **13**, 957–965, (1989).

Strick, D.M, Waycaster, RL, Montani, JP, Gay, WJ, Adair, TH. Morphometric measurements of chorioallantoic membrane vascularity: effects of hypoxia and hyperoxia. Am. J. Physiol. **260**, H1385-H1389, (1991).

Toflion, PJ, Ijichi, A, Sakuma, S. Radiation induced expression of bFGF in normal rat astrocytes. Proc. 10th Int. Congress of Radiation Research (abstract P11–15) p 219, Wuerzburg, Germany, (1995).

Vistica, DT, Skehan, P, Scudiero, D, Monks, A, Pittman, A, Boyd MR, Tetrazolium-based assays for cellular viability: A critical examination of selected parameters affecting formazan production. Cancer Res. **51**, 2515–2520, (1981).

Weidner, N, Folkman, J, Pozza F, Bevilacqua P, Allred, EN, Moore, DH, Meli, S, Gasparini, G. Tumor angiogenesis: a new significant and independent prognostic indicator in early-stage breast carcinoma. J. Natl. Cancer Inst. **84**, 1875–1887, (1992).

Wilting, J, Christ, B, Bokeloh, M, Weich, HA. In vivo effects of vascular endothelial growth factor on the chicken chorioallantoic membrane. Cell Tissue Res. **274**, 163–172, (1993).

17

EFFECT OF CELL LINE AND DIFFERENTIATION ON THE OXYGENATION STATUS OF EXPERIMENTAL SARCOMAS[*]

O. Thews, D. K. Kelleher, B. Lecher, and P. Vaupel

Institute of Physiology and Pathophysiology
University of Mainz
D-55099 Mainz, Germany

1. INTRODUCTION

Since the measurement of tumor oxygenation status and the fraction of hypoxia within tumor tissue by computerized pO_2-histography [10] has become increasingly important in the clinical setting, many studies on tumor oxygenation in different experimental malignancies as well as in various human tumors have been carried out. Because pO_2-histography is an invasive technique, this method has been used in humans mostly in superficial primary tumors and lymph node metastases [1,3,10] (for a review see [9]). Experimental studies in animals have been carried out on a wide range of isotransplanted rodent sarcomas or carcinomas [2,4,6–8,11] and of xenotransplanted human tumors [6]. The results obtained have however been quite variable, often producing conflicting data which are difficult to interpret, presumably due to the use of a variety of tumor entities, sites of tumor growth, and differentially differentiated tumor cells. However, if therapeutic modulation of the tumor oxygenation is the aim of an experimental study, the pre-treatment oxygenation status is an important factor influencing the experimental results. If, for instance, a study is carried out on the modulatory effect of respiratory hyperoxia on the radiosensitivity of a tumor, the beneficial effect of O_2 breathing depends on the initial tumor pO_2. Since radiosensitivity of tissues is half-maximal at a pO_2 of 3–4 mmHg, the effect of sparsely ionizing radiation cannot be improved when hypoxia is not present in the tumor. This fact might explain the varying results on the therapeutic effect of oxygenation modulators on radiosensitivity of solid tumors.

Since studies on tumor oxygenation are carried out in various experimental malignancies, a characterization of these tumor models concerning their pO_2 distribution is desirable if these studies are to be comparable. Additionally, experimental studies on tumor

---

[*] Supported by the Deutsche Krebshilfe (Grant M 40/91 Va 1).

Oxygen Transport to Tissue XIX, edited by Harrison and Delpy
Plenum Press, New York, 1997

oxygenation have been performed in many different animal models using different anesthesia protocols and experimental designs (e.g., spontaneously breathing vs. artificially ventilated animals) [2,4,6–8,11]. Thus, the aim of the present study was to analyze the impact of different cell lines and differentiation on tumor oxygenation under comparable experimental conditions.

2. MATERIALS AND METHODS

2.1. Animals

Male and female Sprague-Dawley rats, male and female Lewis Han rats, and male WAG/Rij rats with a body weight of 200–300 g were used in this study and were allowed to access to food and acidified water *ad libitum* prior to the experiments. All experiments had been previously approved by the competent animal ethics committee.

2.2. Tumors

Tumors were implanted subcutaneously on the hind foot dorsum. Five experimental sarcoma cell lines of the rat were used:

1. DS-sarcoma,
2. Yoshida-sarcoma,
3. R1H rhabdomyosarcoma,

and two differently differentiated lines of the BA-HAN-1 rhabdomyosarcoma, namely

4. BA-HAN-1C / F1, and
5. BA-HAN-1B / G8.

Cell clone F1 shows morphological features of a highly differentiated rhabdomyosarcoma with myotube-like giant cells and myofilaments whereas in the G8 cell line these signs of differentiation cannot be found.

The DS- and the Yoshida-sarcoma were implanted by subcutaneous injection of ascites cells of the respective cell line (0.4 ml, approx. 10^4 cells/μl). The BA-HAN-1 sarcoma cells were cultured in vitro and injected as a suspension (0.5 ml, approx. $3 \cdot 10^3$ cells/μl). The R1H sarcomas were implanted as solid tumor tissue fragments into the subcutis. Oxygenation was measured when the tumors had reached a volume of between 0.5 and 2.5 ml.

2.3. Surgical Procedure

For experimentation, animals were anesthetized with sodium pentobarbital (40 mg/kg i.p.) and placed supine on a heated operating pad. Pentobarbital was used to avoid the known influence of several inhalative anesthetics on polarographic pO_2 measurement. Throughout all experiments, animals breathed room air spontaneously. Mean arterial blood pressure (MABP) was continually monitored through the connection of a catheter placed in the carotid artery to a Statham pressure transducer (type P23 ID, Gould, Oxnard, CA, USA). Before and after each experiment the arterial blood gas status (pO_2, pCO_2, and pH) was determined from arterial blood samples using a pH/blood gas analyzer (type 178, Ciba Corning, Fernwald, Germany). A second catheter was placed into the external jugular vein for subsequent anesthetic administration.

2.4. pO₂ Measurements

Tumor oxygen tensions were determined using O_2 sensitive, polarographic needle electrodes (shaft diameter 300 μm) and pO_2-histography (KIMOC-6650, Eppendorf, Hamburg, Germany; for more details see [10]). A small midline incision was made in the skin covering the lower abdomen and the Ag/AgCl reference electrode was placed between the skin and the underlying musculature. Calibration was performed in saline solutions equilibrated with either air or pure N_2 immediately before and after tumor pO_2 measurements. For probe insertion, a small incision was made into the skin covering the tumor and the electrode placed at a depth of approximately 1 mm. The electrode was then automatically advanced through the tissue in pre-set steps with an effective step-length of 0.7 mm. These steps consisted of a rapid forward movement of 1.0 mm immediately followed by a retraction of the electrode by 0.3 mm in order to minimize tissue compression artifacts. Each electrode step was followed by a pO_2 measurement. At the end of each track through the tumor, the O_2 probe was automatically removed from the tissue. In this way, at least 5 parallel electrode tracks were analyzed resulting in approximately 100 pO_2 measurements per tumor. pO_2 studies of individual tumors were generally carried out in less than 20 min.

2.5. Statistical Analysis

The oxygenation status of individual tumors was expressed by the mean and median pO_2 values as well as by the fraction of pO_2 readings ≤2.5 mmHg and ≤5 mmHg. For each tumor line the results are presented as means ± SEM (number of tumors indicated in brackets). Comparisons between the lines were performed using the two-tailed Wilcoxon test for unpaired samples and the Kruskal-Wallis test (significance level α=5%).

3. RESULTS

The pO_2 distributions showed pronounced differences in the oxygenation status between the cell lines. When expressed by the mean or median tumor pO_2, Yoshida-sarcomas (median pO_2: 14 mmHg) were best oxygenated, followed by the undifferentiated G8 line of the BA-HAN-1 (7 mmHg), DS-sarcoma (6 mmHg), the highly differentiated F1 line of the BA-HAN-1 (4 mmHg) and the R1H sarcoma (<1 mmHg) (Fig. 1). These differences were highly significant (p<0.0001, Kruskal-Wallis test for overall differences between all five cell lines). Inversely to the mean pO_2, the fraction of low pO_2 values (≤2.5 mmHg and ≤5 mmHg) increased from the Yoshida to the R1H sarcoma which is almost completely hypoxic with more than 90% of the measured values below 2.5 mmHg (Fig. 2).

In most of the sarcoma lines analyzed in this study, tumor oxygenation showed a more or less pronounced dependency on tumor volume. With increasing tumor size the

Figure 1. Mean (left) and median (right) pO_2 values in different experimental sarcomas with approximately the same tumor volume of between 0.5 and 3.0 ml (mean tumor volume 1.25 ml). Data are presented as means ± SEM (number of tumors investigated in brackets).

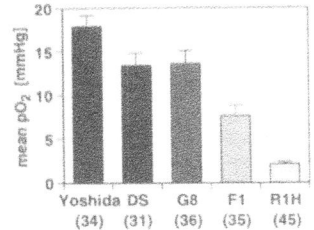
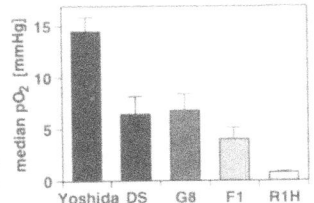

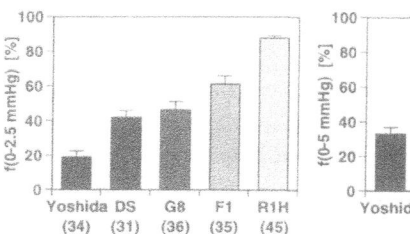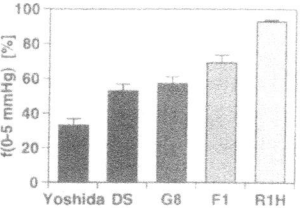

Figure 2. Fraction of hypoxic pO_2 values (≤ 2.5 mmHg and ≤ 5 mmHg) in different experimental sarcomas of approximately the same tumor volume (mean ± SEM, number of tumors in brackets).

oxygenation becomes worse: the median pO_2 decreases (Fig. 3), whereas the fraction of pO_2 measurements ≤ 2.5 mmHg increases with enlarging tumor mass (Fig. 4). Only in the R1H tumors, which is the cell line with the poorest oxygenation in this study, the oxygenation does not change with tumor volume.

The comparison between the two cell lines of the BA-HAN-1 rhabdomyosarcoma (F1 and G8), which both originally stem from the same parental tumor and which differ only in the grade of cell differentiation, show that the undifferentiated G8 line had a slightly better (p<0.03) oxygenation than the more differentiated F1 line (Figs. 1–4).

4. DISCUSSION

The results of this study show, that different cell lines have a strong impact on the oxygenation of sarcomas. The oxygenation status ranges from the well-oxygenated Yoshida-sarcomas, with relatively high median pO_2 (14 mmHg) and a low fraction of hypoxic pO_2 measurements, to the almost completely hypoxic R1H sarcoma with a median pO_2 <1 mmHg. Since tumor growth rate in these tumors is quite different, a poorer oxygenation in faster growing tumors in which oxygen supply to the tissue is limited due to insufficient vascularization might have been expected. However, the results obtained in the present study did not support this hypothesis. The oxygenation status did not correlate with the volume doubling time (VDT) of the tumor entity where the Yoshida- and DS-sarcomas are the fastest growing tumors (VDT: 2.3 d), followed by the F1 line of the BA-HAN-1 (2.5 d), the G8 line (3.0 d) and the slowly growing R1H rhabdomyosarcoma (6.1 d). The fastest growing tumors were those with the better oxygenation and vice versa.

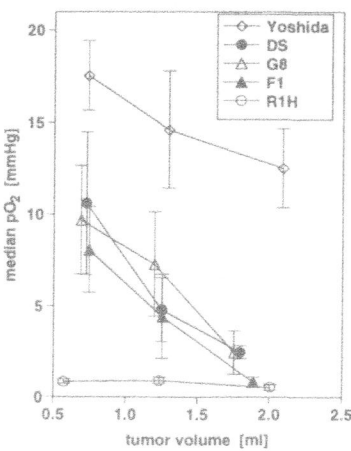

Figure 3. Volume dependency of median pO_2 values in different experimental sarcomas (mean ± SEM). Each data point represents a minimum of 8 tumors.

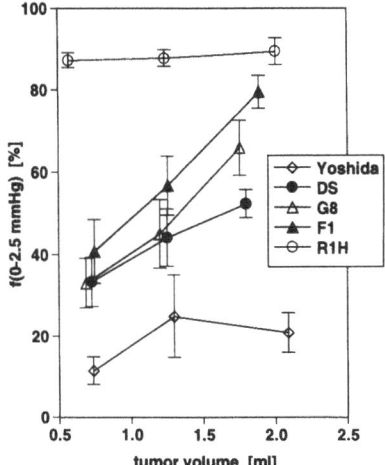

Figure 4. Volume dependency of the fraction of pO$_2$ values ≤2.5 mmHg) in different experimental sarcomas (mean ± SEM). Each data point represents a minimum of 8 tumors.

Besides the tumor entity, the site of tumor growth seems to play an important role for the oxygenation status. Reeker et al. [4] implanted the R1H-sarcoma into the flank of WAG/Rij rats and measured the tumor oxygenation using pO$_2$-histography. They found a median pO$_2$ in these tumors (in rats ventilated with 21% O$_2$) of 10 mmHg which is much higher than that found in the present study (< 1 mmHg) where tumors were growing on the hind foot dorsum.

The grade of tumor cell differentiation also influences the oxygenation status. In the BA-HAN-1 rhabdomyosarcoma the oxygenation is significantly better (p<0.03) in the less differentiated G8 cell line (median pO$_2$: 7 mmHg; fraction of pO$_2$ values ≤5 mmHg: 57%) than in the well-differentiated F1 line (median 4 mmHg; hypoxic fraction 69%). Stier et al. [5] found similar results for the oxygenation of multi-cellular spheroids of these two cell lines in vitro. They described, that low pO$_2$ values in spheroids are associated with a high proportion of highly differentiated myotube-like giant cells in the F1 spheroids. Eble et al. [2] have compared the oxygenation status of two differentially differentiated lines of Dunning prostate tumor R3327 by pO$_2$-histography. In contrast to our data they found that the moderately differentiated HI cell line was much better oxygenated (median pO$_2$ approx. 44 mmHg) than the anaplastic AT1 tumor (median pO$_2$ approx. 7 mmHg). Yeh et al. [11] also found higher pO$_2$ values in the well-differentiated cell line of the R3327 prostate carcinoma (median pO$_2$ 6 mmHg) than in the anaplastic line (median pO$_2$ 2.2 mmHg). It is difficult to compare the results of these authors with the present study since Eble et al. as well as Yeh et al. used bulky tumors (volumes between 1 and 12 ml). Thus, at present it remains unclear, whether cell differentiation in tumors up to a volume of 1% of the body weight (which can be found in tumor patients) can strongly influence the oxygenation status. In the BA-HAN-1 sarcoma, cell differentiation apparently plays only a minor role for tumor oxygenation.

5. CONCLUSIONS

Tumor oxygenation in experimental sarcomas mainly depends on the cell line used in the respective study. The pO$_2$ distributions showed highly significant differences between the tumor lines, although relevant factors influencing oxygenation (e.g., site of tumor growth, tumor volume, anesthesia, mean arterial blood pressure, arterial blood gas status) were identical in all experiments. Thus, it is difficult to compare results between experimental studies using different tumor models and experimental conditions. The site of tumor growth and the cell differentiation also influence tumor oxygenation, although to a minor extent compared to the differences seen between the cell lines.

REFERENCES

1. Brizel DM, Rosner GL, Prosnitz LR, Dewhirst MW. Patterns and variability of tumor oxygenation in human soft tissue sarcomas, cervical carcinomas, and lymph node metastases. *Int. J. Radiat. Oncol. Biol. Phys. 32*:1121–1125 (1995).

2. Eble MJ, Lohr F, Wenz F, Krems B, Bachert P, Peschke P. Tissue oxygen tension distribution in two sublines of the Dunning prostate tumor R3327. In: PW Vaupel, DK Kelleher, M Günderoth (eds.). Tumor oxygenation. Gustav Fischer, Stuttgart, 95–105 (1995).

3. Höckel M, Schlenger K, Knoop C, Vaupel P. Oxygenation of carcinomas of the uterine cervix: Evaluation by computerized O_2 tension measurements. *Cancer Res. 51*: 6098–6102 (1991).

4. Reeker W, Zywietz F, Kocks E. Determination of oxygen partial pressure (pO_2) in a rat rhabdomyosarcoma: a methodical study. In: PW Vaupel, DK Kelleher, M Günderoth (eds.). Tumor oxygenation. Gustav Fischer, Stuttgart, 59–72 (1995).

5. Stier H, Karbach U, Gerharz CD, Gabbert H, Mueller-Klieser W. Oxygenation status of rhabdomyosarcoma spheroids with different stages of differentiation. *Adv. Exp. Med. Biol. 345*: 351–357 (1994).

6. Streffer C, Steinberg F, Tamulevicius P. Oxygen and energy metabolism in tumors: Are they correlated? In: PW Vaupel, DK Kelleher, M Günderoth (eds.). Tumor oxygenation. Gustav Fischer, Stuttgart, 195–204 (1995).

7. Thews O, Kelleher DK, Vaupel PW. Modulation of spatial O_2 tension distribution in experimental tumors by increasing arterial O_2 supply. *Acta Oncol. 34*: 291–295 (1995).

8. Thews O, Kelleher DK, Vaupel PW. pO_2-mapping of experimental rat tumors: visualization and statistical analysis. In: PW Vaupel, DK Kelleher, M Günderoth (eds.). Tumor oxygenation. Gustav Fischer, Stuttgart, 27–38 (1995).

9. Vaupel P, Thews O, Höckel M. Tumor oxygenation: Characterization and clinical implications. In: JF Smyth, MA Boogaerts, BRM Ehmer (eds.). rhErythropoietin in cancer supportive treatment. Marcel Dekker, New York, 205–239 (1996).

10. Vaupel P, Schlenger K, Knoop C, Höckel M. Oxygenation of human tumors: Evaluation of tissue oxygen distribution in breast cancers by computerized O_2 tension measurements. *Cancer Res. 51*: 3116–3122 (1991).

11. Yeh KA, Biade S, Lanciano RM, Brown DQ, Fenning MC, Babb JS, Hanks GE, Chapman JD. Polarographic needle electrode measurements of oxygen in rat prostate carcinomas: Accuracy and reproducibility. *Int. J. Radiat. Oncol. Biol. Phys. 33*: 111–118 (1995).

18

NEW HORIZONS IN HYPERBARIC OXYGENATION

Philip B. James

Wolfson Hyperbaric Medicine Unit
The University of Dundee
Ninewells Medical School
Dundee DD1 95Y, United Kingdom

The use of oxygen at increased atmospheric pressure in the post war years followed clinical research in the 1930's which established the role of hypoxia in a variety of conditions. An increase in the effectiveness of radiation therapy on cultures of cancer cells found using hyperbaric oxygenation was not reflected in clinical experience. The prolongation of the duration of cardiac arrest without brain damage using increased oxygenation became academic with the development of cardiopulmonary bypass procedures. After a period of inactivity there is increasing recognition of the need for hyperbaric oxygenation for conditions not amenable to pharmacological intervention.

BAROMETRIC PRESSURE AND OXYGEN TENSION

The sole determinant of the rate of oxygen transport into tissue is the tension of the gas in plasma. Assuming normal pulmonary function, breathing air at sea level when barometric pressure is a standard atmosphere of 1013 hPa oxygen exerts a pressure of 212 hPa. This gives a plasma tension of about 133 hPa in healthy individuals. Atmospheric pressure varies with the weather and therefore the plasma tension also changes. The lowest pressure recorded at sea level in the UK has been 905 hPa, corresponding to an oxygen partial pressure of 189 hPa. On a high pressure day, where the atmospheric pressure can exceed 1050 hPa more oxygen enters the body and is one factor responsible for the sense of well-being in a period of good weather.

Barometric pressure is not considered in clinical practice, but is of critical importance in aerospace and diving activity. In space exploration the suits used are pressurised to about one third of normal atmospheric pressure and the astronauts breathe pure oxygen. Commercial aircraft are hyperbaric chambers, although at altitude the cabin pressure is subatmospheric causing hypoxia in some passengers.[1] Concorde which flies at altitudes over 60,000 feet has a test pressure in excess of an atmosphere. In diving the World record stands at 523 metres and a breathing mixture consisting of helium, hydrogen and about 1% oxygen was used, graphically illustrating that it is the partial pressure of oxygen that is important not the percentage. Thousands of scientists and engineers have been involved in this technology, but only a small number of physicians. They understand the effects of pressure, the dangers of oxygen deficiency and the use of oxygen at dosages far higher than those normally possible in hospital. The concept of dosage is critical to the acceptance of oxygenation under hyperbaric conditions.

HYPOXIA AND MIXED GAS DIVING

The emergence of the offshore oil and gas industry and the rapid expansion of commercial saturation diving has involved a group of physicians who now have experience of hyperbaric technology and its associated problems. It has brought a new awareness of hyperbaric oxygenation. Whenever artificial breathing mixtures are made there is the possibility that sufficient oxygen may not be included. Over 16 deaths have occurred in commercial diving over the last twenty years[2] because of this problem which leads to anoxic asphyxia. Nothing illustrates the dependency of the brain on the presence of oxygen in the respired gas more than the effect of breathing a pure gas such as helium, but the effect is seen with any gas other than oxygen. Helium is used routinely in commercial diving beyond 50 metres of sea water and is usually supplied premixed with oxygen to offshore diving operations. However gas blenders are also used which allow oxygen to be mixed with helium during diving. If the oxygen supply runs out then pure helium is supplied to a diver and loss of consciousness occurs within a few breaths. This is because the oxygen in the blood of the central venous return supplied to the lungs diffuses out into the respired gas because of the concentration gradient. The lungs can therefore transport oxygen in either direction with equal facility and do not use active transport as originally suggested by JS Haldane.

Acute hypoxaemia is associated with rapid loss of consciousness and if sustained leads to vasodilatation and failure of the energy-dependent endothelial mechanisms responsible for the blood brain barrier, increasing vascular permeability.[3] Cerebral oedema creates further hypoxia and a further increase in permeability and free radicals are released which initiate lipid peroxidation. These effects are initially reversible by oxygen under hyperbaric conditions. In diving using a bell and transfer under pressure to chambers on deck it is possible to give a large dosage of oxygen quickly which may be life saving, but, at the moment few hospitals are equipped with hyperbaric chambers. There is clearly no substitute for oxygen and Taylor[4] has drawn attention to the economics of hypoxia. "Clearly the cost of a few extra days in an ICU to prevent a hypoxic episode would be a fraction of the amount needed for the long-term care of someone brain damaged as a result of an inadequate blood (oxygen) supply."

DECOMPRESSION SICKNESS

Decompression sickness is due to the release of inert gas following a reduction of pressure. It can occur as the result of an ascent to altitude or on decompression from a hyperbaric exposure. The uptake and release of inert gas is not symmetrical and cannot be calculated. Even following conservation decompression procedures gas forms in tissue and some may enter the circulation as bubbles. The conventional view that bubbles cause simple vascular occlusion is no longer tenable. Microbubbles have been found to cause an immediate increase in vascular permeability including the blood-brain barrier.[5] This has been investigated using 15 micron microbubbles injected directly into the carotid artery of the anaesthetised guinea pig. The barrier remains open for between 2 - 3 hours after transit of the bubble as assessed by trypan blue which binds to plasma proteins. The release of protein into tissue initiates an inflammatory response with activation of the complement cascade.

REPERFUSION INJURY

The role of pressure and oxygen in the treatment of decompression sickness has evolved from recompression therapy using compressed air, which was based on the compression of bubbles to re-establish perfusion. The role of oxygen at increased dosage in reperfusion injury has been determined in a rat model using an isolated gracilis muscle preparation.[6] After 4 hours of complete circulatory arrest, the release of the circulation using normally oxygenated blood leads to white cells adhering to the venular endothelium which eventually arrests flow and there is an associated arteriolar constriction. There is evidence that the venular damage is mediated by oxygen free radicals released by leucocytes on reperfusion. The use of hyperbaric oxygenation either in the period prior to, or after, release of the circulation abolishes these effects. The investigators had expected worsening of reperfusion injury using more oxygen, but because of their positive results, they are now using hyperbaric oxygenation routinely and report minimal tissue necrosis and oedema in limbs reimplanted after a delay of 12 hours. These findings are relevant to reperfusion in other organs, especially the heart. Breathing oxygen at 2 atm abs reduces cardiac output by 20%, yet increases tissue oxygen availability including the myocardium. It is difficult to understand how such a major effect, so simply achieved is not used. An excised canine heart perfused with deoxygenated saline maintains contractions for only 10–15 minutes.[7] However, perfused with oxygenated saline a heart will continue to beat for between 2.5 and 11.5 hours, demonstrating the reserves of energy present and the critical need for oxygen.

OXYGEN AND ISCHAEMIC INFARCTION

The first controlled trial of hyperbaric oxygen therapy in myocardial infarction, has shown that the mortality in cardiogenic shock can be halved.[8] Electrocardiograms of the patients breathing oxygen in the chamber show beat by beat improvement, indicating the central role of hypoxia in arrythymias. This is being confirmed in a controlled trial of hyperbaric oxygen therapy combined with thrombolysis.[9] The time taken for the ECG to normalise and the chest pain to be relieved has been halved in the group given hyperbaric oxygenation. A reduction of the release of cytosolic enzymes and an improvement in ejection fraction has also indicated better myocardial function from the additional oxygen. Reperfusion with additional oxygen holds great promise in the management of coronary and cerebral thromboembolism.. In short duration ischaemia 100% oxygen used immediately rather than the 35% normally used could make a substantial improvement to outcome. Infarcts in both the heart and brain are surrounded by a zone where defects in flow, known as the ischaemic penumbra, persist impairing function.[10,11] For highly differentiated cells the level of oxygenation at which function ceases is much greater than the level at which catastrophic membrane failure occurs. The viability of cerebral tissue, which is "not dead but sleeping," has been shown by imaging before and after a single session of hyperbaric oxygenation.[12] Serial sessions benefit by allowing capillaries to grow into damaged, but still viable tissue, to restore function. These techniques would be of value in the assessment and therapy of vegetative coma patients and children with cerebral palsy. A critical issue is how long can oedema persist in a given tissue. There are no a priori reasons to doubt that it can be from birth to grave.

OXYGEN AND HYPERAEMIC HYPOXIA

Ischaemia also occurs in inflammation, when blood flow increases and vessels dilate increasing permeability.[13] Compression of the microcirculation reduces blood flow and oxygen transport. This may be fatal in cerebral oedema, for example, following head injury when "'luxury perfusion" and oedema from damage to the blood-brain barrier elevate intracranial pressure.[14] In tissues with low metabolic activity, such as skin, the temporary reduction of oxygen delivery is not critical, but in tissues such as the brain or heart, some necrosis is inevitable. Inflammation is, of course, usually associated with infection which often complicates wounds. Tissue oxygen measurements have established the level of oxygen necessary for healing to be over 25 mm Hg.[15] Elevation of tissue oxygen tension using serial sessions of hyperbaric oxygenation allows lymphatic drainage of oedema and capillary neogenesis. The new vascular bed improves oxygen delivery restoring macrophage mobility and normal microbial killing from the generation of oxygen free radicals. Fibroblasts also begin to lay down new collagen. The benefit of hyperbaric oxygenation in wound healing, including a controlled trial of venous ulceration,[16] is supported by an extensive literature.[17]

A WAY FORWARD

There is a responsibility for scientists involved in studies of oxygen transport to educate clinicians in the principles involved. High-dosage oxygenation is inexpensive and, although chambers require an investment, their cost can be amortised over many years. Single person chambers, which generally use a pure oxygen atmosphere, can be installed in the space of a bed and should be available in every hospital. Multiplace chambers use masks or hoods and some designs in Japan, Russia and China can accommodate over 30 patients. Implementation of the technology would dramatically improve the quality of outcome in many conditions and reduce the spiralling costs of health care delivery.

REFERENCES

1. James PB. Jet "leg" pulmonary embolism and hypoxia. Lancet 1996;**389**: 1086.
2. James PB, Calder IM. Anoxic asphyxia - a cause of industrial fatalities. JRSM 1991;**84**:493–5.
3. Olesen SP. Rapid increase in blood-brain barrier permeability during severe hypoxia and metabolic inhibition. Brain Res 1986;**368**:24–9.
4. Hills BA, James PB. Microbubble damage to the blood-brain barrier: relevance to decompression sickness. Undersea Biomed Res 1991; **18**:111 - 116.
5. Taylor B. quoted by Bryan J in: Healthcare management 1993;1:30–31.

6. Zamboni WA, Roth AC, Russell RC, et al. Morphologic analysis of the microcirculation during reperfusion of ischaemic skeletal muscle and the effect of hyperbaric oxygen. Plast Recon Surg 1993;**91**:1110–23.

7. Sabiston DC, Talbert JL, Riley LH, Blalock A. Maintenance of the heart beat by perfusion of the coronary circulation with gaseous oxygen. Ann Surg 1959; **150**:361–70.

8. Thurston JGB, Greenwood TW, Bending MR, et al. A controlled investigation into the effects of hyperbaric oxygen on mortality following acute myocardial infarction. Q J Med 1973;**42**:752–770.

9. Ellestad MH et al. Press release. American Heart Association. Oxygen therapy enhances effectiveness of clot-dissolving drugs. 16 November 1992.

10. Braunwald E, Kloner RA. The stunned myocardium: prolonged postischemic ventricular dysfunction. Circulation 1982;**66**:1146–9.

11. Astrup J, Siesjo BK, Simon L. Thresholds in cerebral ischemia - the ischemic penumbra. Stroke 1981; **12**:723–5.

12. Neubauer RA, Gottlieb SF, Kagan RL. Enhancing idling neurones. Lancet 1990;**336**:542.

13. Abbot N, Beck JS, Carnochan FMT, James PB, et al. Effect of hyperoxia at 1 and 2 ata on hypoxia and hypercapnia in human skin during experimental inflammation. J Appl Physiol 1994;**77**:767–73.

14. Obrist WD, Langfitt TW, Jaggi JL, Cruz J, Gennarelli TA. Cerebral blood flow and metabolism in comatose patients with acute head injury. J Neurosurg 1984;**61**:241–53.

15. Hammarlund C, Sundberg T. Hyperbaric oxygen reduced size of leg ulcers: a randomised double-blind study. Plast Recon Surg 1994;**93**:829–33.

16. Uhl E, Sirsjo A, Haapaniemi T, Nilsson G, Nylander G. Hyperbaric oxygen improves wound healing in normal and ischemic skin tissue. Plast Recon Surg 1994;**93**:835–41.

17. Davis JC, Hunt TK. Eds: Problem wounds: the role of oxygen. Elsevier, New York 1988.

19

CLINICAL HYPERBARIC MEDICINE AND THE WWW QUESTION

A. J. van der Kleij

Department of Surgery
Academic Medical Center
University of Amsterdam
Amsterdam, The Netherlands

There is a revival of interest in the literature for clinical hyperbaric medicine[1,2,3,4] and even new horizons are recognized.[5,6] This phenomenon may be a new development, or could it be a matter of history repeating itself? In 1885, E.T. Williams from the Brompton Hospital at London wrote in the *British Medical Journal*: "The use of atmospheric air under different degrees of atmospheric pressure, in the treatment of disease is one of the most important advances in modern medicine, and when we consider the simplicity of the agent, the exact methods by which it may be applied, and the precision with which it can be regulated to the requirements of each individual, we are astonished that in England this method of treatment has been so little used". In that period doctors practised modern medicine and nowadays doctors do claim to do the same. However, there is a difference. Today we know that oxygen is not a simple agent. Furthermore, expected linearity effects of increased ambient pressure and/or FiO_2 are attenuated by regulatory networks within the vascular bed and heterogeneity of perfusion. In fact to unravel the complicated pathway of oxygen from the ambient air to the mitochondrium is one of the fundamental drives for the ISOTT.

Apparently, interest in hyperbaric oxygen therapy has waxed and waned throughout history. In 1993 Kindwall[8] wrote in the same journal an article entitled: "The lack of knowledge of indications for hyperbaric oxygen therapy" reminding the indications and the significance of hyperbaric oxygen therapy for clinical medicine.

Everybody is familiar with the meaning of the acronym WWW (World Wide Web). However, within the context to clinical hyperbaric medicine I would like to propose that the acronym WWW is depicted as: *W*hy is hyperbaric oxygen not in *W*orldwide use to enhance *W*ound healing in problem wounds? We can only speculate to answer this question.

It has been demonstrated that fibroblasts can not synthesize collagen when oxygen availability is deprived.[9] Furthermore, a positive effect of HBO therapy on capillary neoangiogenesis has been observed in chronic hypoxic tissue.[10] Based on these findings HBO is advocated. Maybe it is overemphasised in clinical hyperbaric medicine that the most important factor necessary for a normal wound healing process is the availability of molecular oxygen adjacent in the wound area. Clinical evidence of the positive effect of hyperbaric oxygen therapy on wound healing and/or to prevent major amputation is based on uncontrolled studies and convincing evidence is still lacking.[11,12,13] It is also suggested that hyperbaric oxygen therapy can be used as a possible treatment to prevent major amputations in a subgroup of patients with otherwise untreatable critical leg ischaemia.[14,15] In the last 15 years there are numerous publications of uncontrolled studies using hyperbaric oxygen in the treatment of soft tissue radionecrosis[16,17,18,19,20] and osteoradionecrosis.[21,22,23,24] It has been shown in a prospective randomised clinical trial that hyperbaric oxygen therapy plays an essential role in the size reduction of chronic leg ulcers.[25] However, the clinical endpoint, a healed wound, is not mentioned in this study.

We must admit the existence of a major problem in hyperbaric medicine namely the design of a double-blind controlled clinical trial. Various sham methods have been developed which may vary from 5 to 60 minutes exposure to 0.1 - 2.0 absolute atmospheres and administering normoxic or hypoxic gas mixtures.[26,27] Moreover, clinical trials about wound-healing are difficult to perform because of the lack of objective evaluation methods.[28]

A literature search (Medline) revealed that during the last three decennia the number of articles concerning hyperbaric oxygenation has decreased (Figure 1). Apparently more effort is put into amassing knowledge about wound healing. These numbers could also suggest that the two research/clinician groups are insufficiently informed about each other's data.

Can all patients with problem wounds benefit from the experience with hyperbaric oxygen, in other words is the number of available hyperbaric chambers a limiting factor? Since the early beginnings the number of hyperbaric chambers is increased worldwide. The overall number of the registered multi-place chambers is 601 and of monoplace chambers 478. For the USA these numbers are 266 and 134, for England 8 and 61 and for Italy 7 and 49, respectively. Obviously, these figures do not cover a world-wide need.

What can we learn from the past and our teachers? Jacobson[29] concluded from his lesson of history of hyberbaric therapy that: "If this form of therapy is to achieve a world-

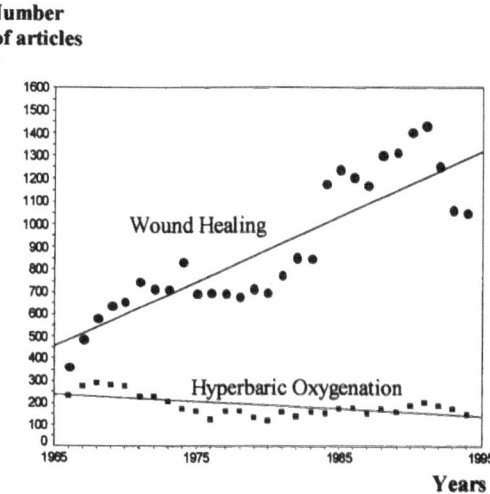

Figure 1. Search in Medline using the keywords hyperbaric oxygenation and wound healing for the period 1965–1995.

wide and lasting place in the medical armentarium, it can only do so on a firm basis of accurate physiological data on the effects of both pressure and oxygen obtained in experiments, as well controlled as clinical medicine permit".

In general, it may be concluded that clinicians who seek for the solution for the WWW question can nowadays use the facilities of the World Wide Web to initiate multicentric controlled clinical trials bearing in mind a strict rationale indication to be present to treat patients with hyperbaric oxygen therapy.[30]

REFERENCES

1. Ashamalla H., Thom SR, Goldwein JW. Hyperbaric oxygen therapy for the treatment of radiation-induced sequelae in children. Cancer 1996; 77:2407–2412.
2. Bevers RFM, Bakker DJ, Kurth KH. Hyperbaric oxygen treatment of 40 patients with hemorrhagic radiation cystitis. Lancet 1995; 346:803–805.
3. Feldmeier JJ, Jelen I, Davolt DA, Valente PT, Meltz ML, Alecu R. Hyperbaric oxygen as aprophylaxis for radiation-induced delayed enteropathy. Radiotherapy and Oncology 1995; 35:138–144.
4. Uhl E, Sirsjö A, Haapaniemi T, Nilson G, Nylander G. Hyperbaric oxygen improves woundhealing in normal and ischemic skin tissue. Plas. Reconstr. Surg. 1994; 93:835–841.
5. Mehdaoui H, Elisabeth L. New Frontiers: Sickle-Cell Anemia. In: Handbook on hyperbaric oxygen therapy. (eds. G. Oriani, a. Marroni, F. Wattel). Springer-verlag. Italia, Milano. 1996; 830–833. (ISBN 3–540–75016–9).
6. Kleij van der AJ, Bakker DJ, Voûte PA. New Frontiers: Tumour oxygenation & radiotherapy. In: Handbook on hyperbaric oxygen Therapy. (eds.. G. Oriani, a. Marroni, F. Wattel). Springer-verlag. Italia, Milano. 1996; 817–829 (ISBN 3–540–75016–9).
7. Williams ET. The compressed air bath and its uses in the treatment of disease. Br. Med J. 1885.
8. Kindwall EP. Hyperbaric oxygen. More indications than doctors realise. Br. Med J. 1993; 307:515 - 516.
9. Hunt TK, Dai MP. The effect of varying ambient oxygen tension on wound metabolism and collagen synthesis. Surg Gynaecol Obstet. 1972; 135:561–567.
10. Knighton DR, Silver IA, Hunt TK. Regulation of wound healing angiogenesis: Effect of oxygen and inspired oxygen concentrations. Surg 1981; 90:262–270.
11. Hauser CJ. Tissue salvage by mapping of skin surface transcutaneous oxygen tension index. Arch Surg. 1987; 122:1128–1130.
12. Mathieu D, Wattel F, Bouachour G, Billard V, Defoin J.F. Post-traumatic limb ischemia: Prediction of final outcome by transcutaneous oxygen measurement in hyperbaric oxygen. J. Trauma. 1990; 30:307–314.
13. Sheffield PJ. Tissue oxygen measurements. In: "Problem Wounds: The Role of Oxygen. JC Davis and T.K. Hunt. (Eds) New York: Elsevier, 1988;17–51.
14. Baroni P, Porro T, Faglia E, Pizzi G, Mastropasqua A, Oriani G, Pedesini G, Favales F. Hyperbaric oxygen in diabetic gangrene treatment. Diab Care 1987: 10:81–86.
15. Cianci P, Hunt TH. In: The Diabetic Foot. 5th ed. ST Louis, Baltimore. Levin ME, O'Neal LW, Bowker JH (Eds). Boston, Chicago, London, Philadelphia, SIdney, Toronto, Mosby Year Book, Adjunctive hyperbaric oxygen therapy in treatment of diabetic wounds. 1993; Ch. 14. 305–319.
16. Hart GB, Mainous EG. The treatment of radiation necrosis with hyperbaric oxygen (OHP). Cancer 1976; 37:2580–2585.
17. Kindwall EP. Hyperbaric oxygen's effect on radiation necrosis. Clin. in Pl. Surg. 1983; 20:473–483.
18. Williams JA, Clarke D, Dennis WA, Dennis EJ, Smith ST. The treatment of pelvic soft tissue radiation necrosis with hyperbaric oxygen. Am. J. Obstet. Gynecol.1992; 167:412–416.
19. Rijkmans BG, Bakker DJ, Dabhoiwala NF, Kurth KH. Successful treatment of radiation cystitis with hyperbaric oxygen.Eur. Urol. 1989; 16:354–356.
20. Weis JP, Mattei DM, Neville EC, Hanno PM. Primary treatment of radiation-induced hemorrhagic cystitis with hyperbaric oxygen: 10-year experience. J. Urol. 1994; 151:1514–1517.
21. Hohn DC, Mackay RD, Halliday BJ. The effect of O_2 tension on microbicidal function of leucocytes in wounds and in vitro. Surg. Forum. 1976; 27:18–20.
22. Mansfield MJ, Sanders DW, Hemback RD. Hyperbaric oxygen as an adjunct in the treatment of osteoradionecrosis of the mandible. J. Oral. Surg. 1981; 39:585–589.
23. Marx RE, Johnson RP, Studies in the radiobiology of osteoradionecrosis and the clinical significance. Oral. Surg. Oral.Med. Oral. Pathol. 1987; 64:379–390.

24. Merkesteyn van JPR, Balm AJ, Bakker DJ, Borgmeyer-Hoelen AM. Hyperbaric oxygen treatment of osteoradionecrosis of the mandible with repeated -pathologic fracture. Report of a case. Oral. Surg. Oral. Med. Oral. Pathol. 1994; 77:461–464.

25. Hammerlund C, Sundberg T. Hyperbaric oxygen reduced size of chronic leg ulcers: A randomized double-blind study. Plas.Recon. Surg. 1994; 93:829–834.

26. Harpur GD, Suke R, Bass BH, Bass MJ, Bull SB, Reese L, Noseworthy JH, Rice GP, Ebers GC. Hyperbaric oxygen therapy in chronic stable mutiple sclerosis: double-blind study. Neurology: 1986; 36:988–991.

27. Nighoghossian N, Trouillas P, Adeleine P, Salord F. Hyperbaric oxygen in the treatment of acute ischaemic stroke: A double blind pilot study. Stroke. 1995; 26: 1369–1372.

28. Mekkes JR, Westerhof W. Image Processing in the study of wound healing. Clinics in Dermatology 1995; 13:401–407.

29. Jacobson JH, Morsch JHC, Rendell-Baker L. Clinical experience and implications of hyperbaric oxygenation. Annals of the New York Academy of Sciences 1965; 117:651–670.

30. Boerema I. The large chamber. Annals of the New York Academy of Sciences 1965; 117:883–887.

20

GAS EXCHANGE AND GAS TRANSPORT IN A WATER BREATHER OF MILLIMETRE SIZE

Ralph Pirow, Frank Wollinger, and Rüdiger J. Paul

Institut für Zoophysiologie
Westfälische Wilhelms-Universität Münster
Hindenburgplatz 55, 48143 Münster, Germany

1. INTRODUCTION

Body size is highly correlated with the structure-function relationship of an animal. Whereas oxygen uptake and distribution occur by simple diffusion in micro-organisms, the development of internal and/or external structures for convective gas transport (circulatory and ventilatory systems) becomes necessary with increasing body size. This constraint is of special importance for the embryonic and larval development of both vertebrates and invertebrates. Water-breathers with a body size of up to 1 mm belong to a group in which oxygen transport via diffusion is sufficient, at least at normoxia, and convection is not yet required. This limitation in body size is derived from the maximum penetration depth of diffusively transported oxygen, which can be calculated by a simplified model (Piiper and Koepchen, 1975). We have focused our interest on a fresh-water invertebrate, the crustacean *Daphnia magna* (Figure 1), which has great ecological importance and is also used as a bioindicating organism for water quality control. This species has a body length of 1–4 mm and possesses an open circulatory system with freely dissolved haemoglobin (Hb). The body appendages (thoracic limbs) are used as a pump to move water through, and retain food by, the filter apparatus.

The importance of these structures for gas exchange, however, is still a matter for discussion. According to Peters (1987) the circulatory system is thought to be ineffective and internal gas transport should occur only by facilitated diffusion using Hb as oxygen carrier. On the other hand *Daphnia* shows - under hypoxic conditions - a compensatory tachycardia (Fig. 2B, Paul et al., 1996), which can be considered as a regulatory response of a convective transport system. This phenomenon is in accordance with the oxyregulating behaviour of the animal (Figure 2A; Kobayashi and Hoshi, 1984; Paul et al., 1996). A gill function attributed to the thoracic limbs is doubtful, because the epithelium of the so-called branchial sacs has a thickness of about 15 to 20 μm, and represents a considerable diffusion barrier (see Peters 1987). No regulatory responses in the frequency of thoracic limb movements were found under hypoxia (Figure 2B; Paul et al., 1996). It was therefore

Oxygen Transport to Tissue XIX, edited by Harrison and Delpy
Plenum Press, New York, 1997

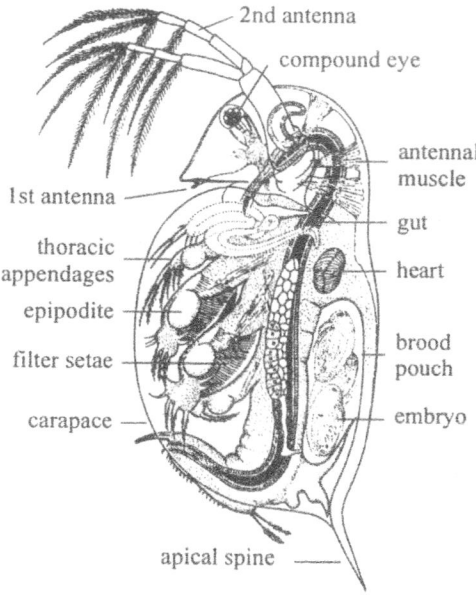

Figure 1. *Daphnia pulex*, lateral view (Modified after Kaestner, 1967).

argued that gas exchange with the environment occurs via diffusion over the entire body surface. The aim of the present study is to determine the nature of internal gas transport as well as the importance of Hb. We also wish to clarify, whether *Daphnia* takes up oxygen from the environment across the entire body surface or whether gas exchange is limited to specialised parts of the body.

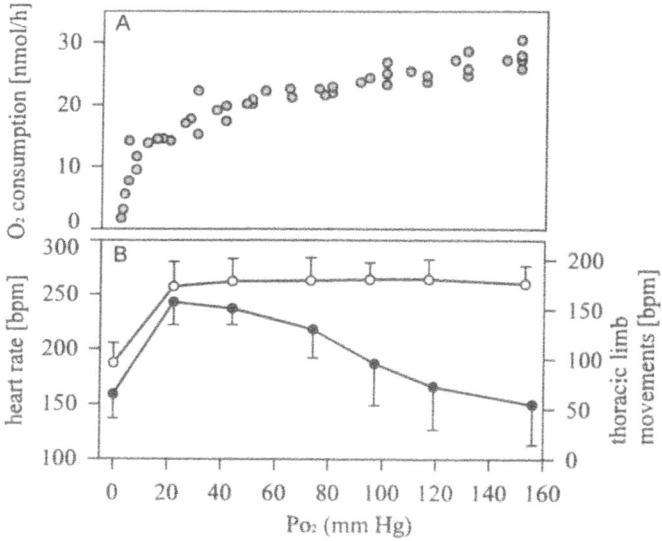

Figure 2. Physiological response of *Daphnia magna* to decreasing ambient oxygen partial pressures at 15°C (Paul et al., 1996). (A) Oxygen consumption, (B) Heart rate (black circles) and thoracic limb movements (white cir-

We used optical techniques - e.g. video spectroscopy, phosphorescence imaging and digital image processing - to be able to measure physiological parameters (Hb, Po_2, ventilatory flow) in small animals, to image physiological functions and to have a better understanding of systemic interactions.

2. MATERIALS AND METHODS

2.1. Animals

Daphnia magna Straus was obtained from the Staatliches Umweltamt Münster/Nordrhein-Westfalen, Germany. The keeping conditions were described elsewhere (Paul et al., 1996).

2.2. Experimental Conditions

All experiments were carried out at 20°C. We used a special flow-through chamber (Paul et al., 1996), which was perfused with medium. The medium—diluted sea water with a conductivity of 3300 µS cm^{-1} (25°C)—was sucked up from a vessel and equilibrated with different O_2/N_2-mixtures (at normocapnic conditions) using gas-mixing pumps (Wösthoff, Bochum, Germany). The animal was fixed by sticking its apical spine with wax to a glass capillary, which was attached to a small plastic block. The block was placed in the flow-through chamber. By turning the block it was possible to observe the animal from dorsal, ventral and lateral.

2.3. Optical pO_2 Measurements

Oxygen profiles outside the animals were measured by an optical method which was based on the determination of the Po_2 dependent phosphorescence quenching (Vanderkooi and Wilson, 1986) of special probes (albumin bound Pd-meso-tetra (4-carboxyphenyl) porphine, Medical System Corp., Greenvale, NY, USA). We built up a phosphorescence microscope (Figure 3) by modifying an inverted microscope (Zeiss Axiovert 100, Carl Zeiss, Oberkochen, Germany). A µs Xenon flash lamp (uF900, Spectrolas GmbH, München, Germany) mounted on the epifluorescence lamp adapter of the microscope was used as excitation source. The reflector slider contained an excitation filter BP 500–580, a dichroic mirror FT 580 and an emission filter LP 590. A photomultiplier (PMT, H5783–01, Hamamatsu Photonics K.K., Japan) picked up the phosphorescence light. A chopper (Mod. 230, HMS Elektronik, Leverkusen, Germany) was placed in front of the PMT to guarantee that only phosphorescence light but no excitation light reached the detector. We reduced the PMT's entrance aperture to 2 mm to analyse only a small region of the microscopic phosphorescence image. The signal of the PMT and the reference signal from the chopper LED was picked up by a DSP lock-in amplifier (SR830, Stanford Research Systems, Sunnyvale, CA, USA), which calculated the phase difference between the reference signal and PMT signal. A computer with a timer circuit card (DAS1602, Keithley Metrabyte, Taunton, MA, USA) checked the chopper status by analysing the reference signal and triggered the flash lamp so that the frequency of the flash lamp was phase-locked to the chopper frequency and the flash only occurred while the emission light path was closed.

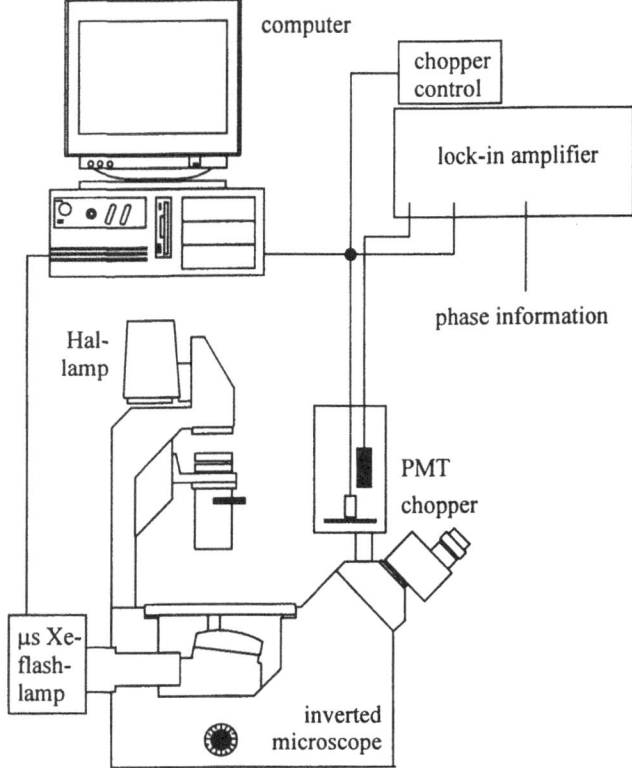

Figure 3. Set-up for pO$_2$ measurement.

The theoretical relationship between phase information (φ) and oxygen partial pressure (Po$_2$) could be derived from the Stern-Vollmer equation and from the Fourier integral of a function consisting of periodic exponential decays:

$$\varphi = \arctan\left(\frac{1 + k_q \tau_0 P_{O_2}}{2\pi f \tau_0}\right)$$

where, for 20°C and pH 7.2, the decay time in absence of quencher τ_0 has a value of 600 μs and the quenching constant k_q is 215 mm Hg^{-1} s^{-1}. We used a chopper frequency f of 90 Hz. This theoretical relationship was experimentally proved by equilibrating a small drop of dye solution (1 mg per ml) with water vapour saturated gas mixtures of different oxygen content. As shown in Figure 4 the experimental relationship differs from the theoretical relationship at Po$_2$-values higher than 20 mm Hg, probably due to the limited maximum sampling rate (102 kHz) of the lock-in amplifier.

2.4. In-Vivo Hb Imaging

The set-up (Figure 5) was built of an inverted microscope (Zeiss Axiovert 100, Germany) combined with a computer-driven scanning-grating monochromator (T.I.L.L. Photonics GmbH, Planegg, Germany) used as illuminating system. This light source was coupled to the transmission lamp adapter of the microscope by a quartz fibre optic light

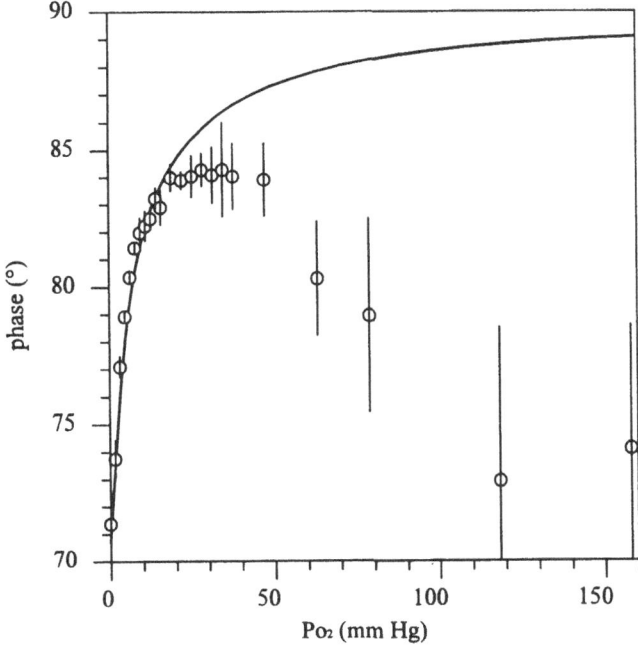

Figure 4. Theoretical and experimentally measured relationship between phase and Po_2.

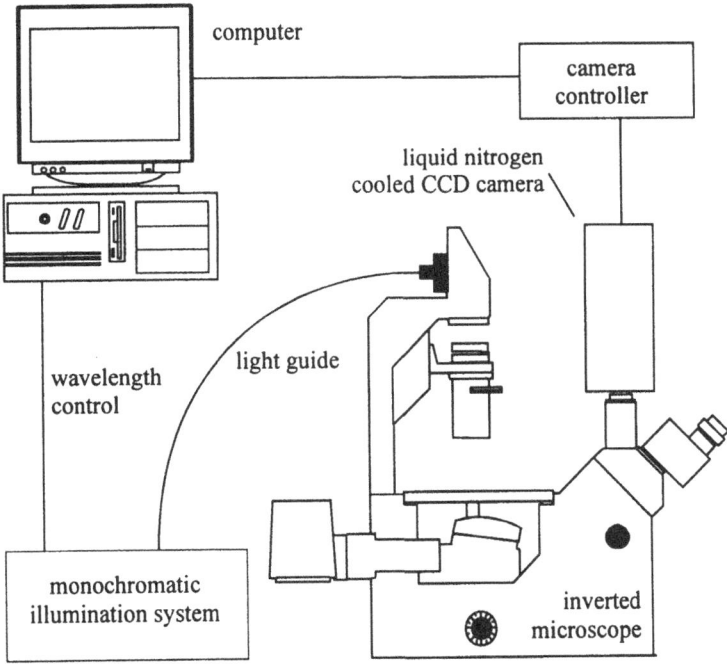

Figure 5. Set-up for Hb measurement.

guide (1.5 mm in diameter). A 16-bit liquid nitrogen cooled CCD detector (LN/CCD-576E, Princeton Instruments, Trenton, NJ, USA) mounted on the camera adapter of the microscope was used to pick up the microscopic images. For the determination of Hb saturation we utilised the Soret-peak that shifts from 415 nm (oxy-Hb) to 430 nm (desoxy-Hb). Using WinView software (Princeton Instruments, Trenton, NJ, USA) a sequence of images was acquired with the CCD camera while varying the illumination wavelength from 380 nm to 448 nm in steps of 2 nm. As a result of this procedure we recorded a spectrum at each pixel of the microscopic image. A small region of the microscopic image outside the specimen was used to acquire a mean reference spectrum, necessary for the calculation of absorption spectra. Before determining Hb saturation, the spectra at each pixel were checked to see whether they were similar to a Hb-typical Soret bond. For all pixels with valid spectra, the peak wavelength was determined.

3. RESULTS

3.1. External Po$_2$ Profiles

We measured the oxygen distribution outside the animal in order to find the site of oxygen uptake. No decrease in Po$_2$ could be found in the dorsal region and on both sides of the carapace (Fig. 6). The measured Po$_2$ values corresponded to pre-set hypoxic values of the medium. Oxygen depletion occurred in the ventral body region. The medium had a lower Po$_2$ after passing the ventral body region.

3.2. Hb Saturation

We have shown that oxygen uptake in *Daphnia* is mainly restricted to the ventral body region. Consequently the oxygenation of the haemolymph should take place at the same site. To demonstrate this we measured Hb saturation in-vivo. Hb was highly oxygenated in the ventral and posterior body region (Figure 7). A lower Hb saturation was found in central regions of the body.

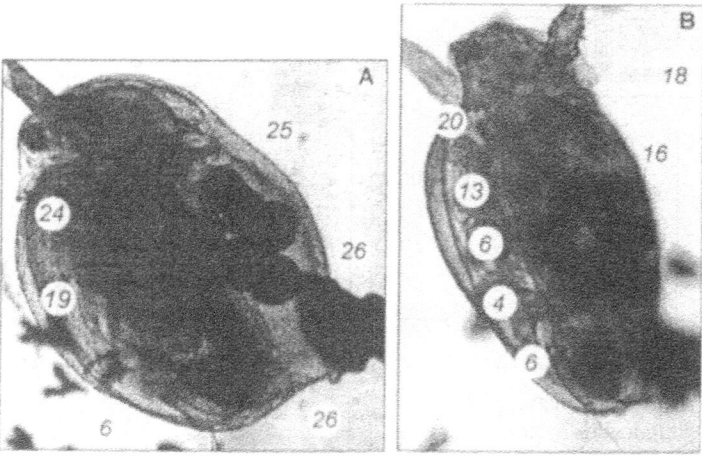

Figure 6. Po$_2$ values outside the animal. The Po2 of the perfusion medium was pre-set to 20 mm Hg.

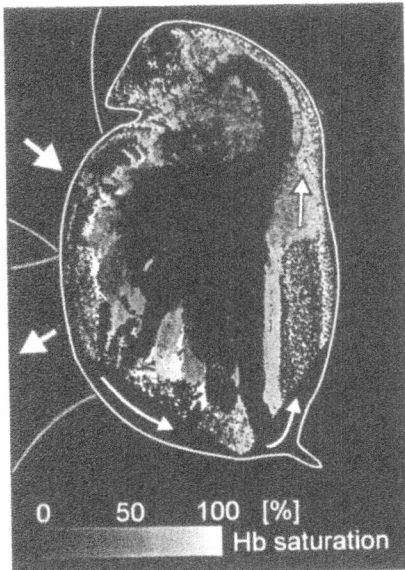

Figure 7. Hb saturation in *Daphnia magna* under normoxic conditions. The arrows mark the direction of ventilatory and circulatory flow.

4. DISCUSSION

Both external convection and internal convection are highly involved in gas transport in *Daphnia*. The thoracic limb movements generate a ventilatory flow and the oxygen uptake is locally restricted to the ventral body region. In-vivo Hb measurements showed, that Hb is also saturated in the ventral body region. As *Daphnia* showed a circulatory but not a ventilatory response to hypoxia (Paul et al., 1996), we conclude that internal convection is more important than external convection in small-sized animals.

ACKNOWLEDGMENTS

This study was supported by the Deutsche Forschungsgemeinschaft (Pa 308/4-1 and Pa 308/4-2).

REFERENCES

Kaestner, A. (1967). Lehrbuch der speziellen Zoologie, Band 1, 2. Teil. Gustav Fischer Verlag, Stuttgart.

Kobayashi, M., Hoshi, T. (1984). Analysis of respiratory role of haemoglobin in *Daphnia magna*. Zool. Sci. (Tokyo) 1: 523–532.

Paul, R.J., Colmorgen, M., Hüller, S., Tyroller, F., Zinkler, D. (1996). Circulation and respiratory control in millimetre-sized animals (*Daphnia magna, Folsomia candida, Tubifex tubifex*) studied by optophysiology. J. Comp. Physiol., in press.

Peters, R.H. (1987). Metabolism in *Daphnia*. In: Peters RH, De Benardi R (eds), Memorie Dell'Istituto Di Idrobiologia Dott. Marco De Marchi. Vol. 45. Verbania Pallanza, 193–243.

Piiper, J., Koepchen, H.P. (1975). Atmung. Urban & Schwarzenberg München.

Vanderkooi, J.M., Wilson, D.F. (1986). A new method for measuring oxygen in biological systems. Adv. Exp. Med. Biol. 200: 189–193.

21

RESPONSE OF PURINE METABOLISM AND CORTICAL OXYGEN PRESSURE TO HYPOXIA AND REOXYGENATION IN NEWBORN PIGLETS

Peter Pastuszko, Peter Marro, Maria Delivoria-Papadopoulos, and
David F. Wilson

Departments of Physiology
Biochemistry and Biophysics and of Pediatrics
Medical School, University of Pennsylvania
Philadelphia, Pennsylvania 19104

INTRODUCTION

It has been proposed that the level of purine metabolites (particularly hypoxanthine) reflects the intracellular energy metabolism and can be used as sensitive, specific markers of response of tissue to hypoxic/ischemic conditions.[26,27] Several studies have shown that hypoxic conditions cause breakdown of cellular nucleotides resulting in accumulation of hypoxanthine and xanthine[4,7,8,28] and these metabolites can serve as oxidizable purine substrates for xanthine dehydrogenase and oxidase. One of the possible mechanisms of tissue damage during reoxygenation following hypoxic/ischemic conditions is through generation of free radicals by the hypoxanthine-xanthine oxidase reactions.[1,5,6,15,16,24,29] Under hypoxic/ischemic conditions the elevated cytosolic calcium concentration can activate a protease, calpain, which converts xanthine dehydrogenase to xanthine oxidase.[32] During posthypoxic reoxygenation xanthine oxidase catalyzes oxidation of xanthine to uric acid with formation of superoxide radical O_2^-.[4,15,28] The superoxide radical can further react with hydrogen peroxide produced in the same reaction and form hydroxyl radicals. Contribution of toxic oxygen metabolites from xanthine oxidase to tissue injury during reoxygenation has been based particularly on studies showing that the inhibitors of xanthine oxidase, such as allopurinol, decrease the postischemic injury of different tissues.[9,19,20,21,24]

It has been also shown that hypoxanthine can affect neuronal metabolism by binding to benzodiazepine receptors[31] inhibiting phosphodiesterase in brain[14] or interacting with neurotransmission.[6,18]

The purpose of this study was to determine the response of purine metabolism in the newborn piglet to hypoxia and reoxygenation. The hypoxia model chosen was stepwise decrease in FiO_2, a model which has been extensively characterized with respect to the relationship of oxygen pressure to neurotransmitter metabolism in the brain.[10,22,23] The degree of brain hypoxia was quantitated by measuring the oxygen pressure in the cortex. Significant increase in plasma level of hypoxanthine and xanthine was not observed until the last step of the hypoxia protocol, which caused cortical oxygen to decrease to about 6 Torr.

MATERIALS AND METHODS

Animal Preparation

Anesthesia was induced with halothane (Halocarbon Laboratories, Augusta, SC; 4% mixed with 96% oxygen) and 1.5% lidocaine-HCl (Abbott Laboratories, North Chicago, IL) was used as a local anesthetic. The halothane was reduced to 0.6–0.8% after tracheotomy and Tubocurine-HCl used for respiratory paralysis (Apothecon, Bristol-Meyers Squibb, Princton, NJ; 1.5 mg/kg), and the femoral artery and femoral vein were cannulated. Once the vessels were catheterized, halothane was withdrawn entirely and Fentanyl-citrate (Elkins-Linn, Inc., Cherry Hill, NJ; 30 µg/kg) was injected intravenously at approximately one hour intervals throughout the experiments. The animals were paralyzed with tubocurarine and mechanically ventilated with a mixture of oxygen and nitrous oxide (in control conditions with 21–22% oxygen and 78–79% N_2O). The head was placed in a Kopf stereotaxic holder and an incision was made along the midline of the scalp. The scalp was removed to expose the skull and a hole approximately 12 mm in diameter was made in the skull over one parietal hemisphere. After the dura was removed, a cranial window for measuring the oxygen pressure in the cortex was placed in the hole and fixed with dental cement. The surface of the brain under the window was flushed with artificial CSF throughout the study.

In all experiments the blood pressure, body temperature, heart rate and end-tidal CO_2 were continuously monitored. Arterial blood samples were taken every 15 min and blood glucose and lactate were measured in a Model 2300 STAT Glucose and L-lactate analyzer (YSI 2300 dual analyzer). In addition, arterial blood pressure and blood pH, $PaCO_2$ and PaO_2 were measured throughout the duration of the experiment using a Model 178 pH/Blood Gas Analyzer (Corning). After the studies were completed, the anesthetized animals were euthanized by intravenous administration of saturated KCl solution.

Experimental Protocol

Hypoxic-hypoxia was obtained by reducing the fraction of inspired oxygen (FiO_2) from normal (21%) to 14%, 11% and 9%, holding at each step for 20 min. At the end of the hypoxic period the FiO_2 was returned to 21% and maintained at that level throughout a 90 minutes period of reoxygenation. This model has been used to determine the relationship between oxygen pressure in the brain and neurotransmitter metabolism.[10,22,23]

Cortical Oxygen Pressure

The cortical oxygen pressure was measured by the oxygen dependent quenching of phosphorescence as previously described by Wilson et al.[35,36]

Purine Metabolite Measurements

The purine metabolites, including inosine, hypoxanthine, xanthine and uric acid were measured in arterial serum samples taken at the end of each decrease in FiO_2 and at

Table 1. Physiological parameters of newborn piglets during graded hypoxia and reoxygenation

Conditions (FiO$_2$)	MAP (Torr)	Blood pH	PaCO$_2$ (Torr)	PaO$_2$ (Torr)	Cortical PO$_2$	Blood Glucose (mg/dL)	Blood Lactate (mmol/L)
Control (21%)	93±12	7.36 ± 0.03	41 ± 3	107 ± 25	33 ± 2	87 ± 21	2.0 ± 0.8
Hypoxia (14%)	97±19	7.35 ± 0.06	38 ± 2	47 ± 9**	24 ± 3**	82 ± 32	2.8 ± 0.8
Hypoxia (11%)	94±23	7.34 ± 0.04	35 ± 4	35 ± 5**	18 ± 4**	78 ± 23	4.14 ± 1.6*
Hypoxia (9%)	82±28	7.23 ± 0.06**	33 ± 3	29 ± 7**	6 ± 3**	92 ± 12	7.2 ± 2.7**
Reoxygenation							
60 minutes	94±20	7.03 ± 0.1**	35 ± 6	102 ± 12	32 ± 12	91 ± 21	10.9 ± 3.2**
90 minutes	90±21	7.09 ± 0.06**	32 ± 3	105 ± 20	32 ± 8	86 ± 18	13.3 ± 2.9**

Data presented are the mean ±SEM for 6 experiments* p < 0.05 **p < 0.01 vs control.

60 and 90 minutes of reoxygenation. The serum samples were immediately frozen and then analyzed for the purine content using a High Pressure Liquid Chromatography System with spectrophotometric detector. Identification and quantitation of the metabolites were conducted by comparison with chromatograms of standard solutions.

Statistical Evaluation

All values are expressed as a means ± SD. The statistical significance of observed changes in the measured parameters was determined using one way analysis of variance with repeated measures by Wilcoxon signed-rank test. p < 0.05 was considered statistically significant.

RESULTS

Physiological Parameters of the Piglets during the Experimental Protocol

The effect of graded hypoxia obtained by reducing the fraction of inspired oxygen on the physiological parameters of newborn piglets are shown in Table 1. As can be seen, hypoxia caused a substantial increase in blood lactate. Stepwise decrease in FiO$_2$ from 21% to 14%, 11% and 9% resulted in an increase in blood lactate from 2 ± 0.8 mmol/l to 2.8 ± 0.8, 4.1 ± 1.6 (p < 0.05) and 7.2 ± 2.7 mmol/l (p <0.01), respectively. During the period of reoxygenation (FiO$_2$ of 21%), the level of lactate steadly increased up to 13.3 ± 3.2 mmol/l at 90 min (p < 0.01). The increase in the level of lactate during the hypoxic period was parallel to the decrease in pH. The pH decreased from a control value of 7.36 ± 0.03 to 7.35 ± 0.05, 7.34 ± 0.04 and 7.23 ± 0.06 (p < 0.01), respectively, as the FiO$_2$ was decreased to 14%, 11%, and 9% . During the 60 min of reoxygenation it further decreased to 7.03 ± 0.1 (p < 0.01). Measurements of the blood glucose levels showed no significant alterations in these parameters during the experiments. Arterial blood pressure was also unaffected.

Effect of Decreased FiO$_2$ on Cortical Oxygen Pressure and the Plasma Levels of Purine Metabolites

Stepwise decreases in the FiO$_2$ from 21% to 14%, 11% and 9% caused progressive decrease in cortical oxygen pressure from 33 ± 2 Torr to 24 ± 3 Torr (p <0.01), 18 ± 4 Torr (p

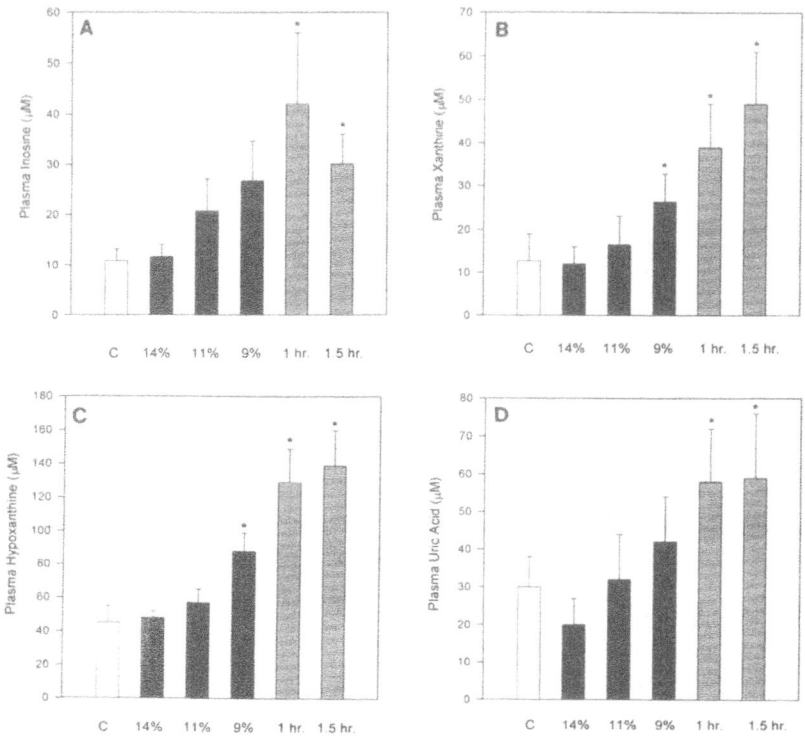

Figure 1. Effect of hypoxia and reoxygenation on plasma levels of purine metabolites. The levels of inosine (A), xanthine (B), hypoxanthine (C) and uric acid (D) were measured in arterial serum by HPLC as described in Materials and Methods. The results are expressed as the means ± SD for 6 experiments. * p < 0.05 for significant difference from baseline values as determined by one way analysis of variance, followed by the Wilcoxon signed-rank test.

<0.01) and 6 ± 5 Torr (p<0.01) (Table 1). Return of FiO$_2$ to control values after these hypoxic conditions resulted in the return of cortical oxygen pressures to control values.

The purine metabolites were measured in arterial serum samples taken at the end of each decrease in FiO$_2$ and at 60 and 90 minutes of reoxygenation. The results are shown in Figure 1 (A-D). Inosine levels increased from 10.8 ± 2.4 µM to 11.6 ± 2.5 µM, 20.7 ± 6.4 µM, 26.7 ± 8 µM, 42 ±14 µM (p < 0.05) and 30 ±6 µM, respectively (Figure 1A). Xanthine levels increased from 12.7 ± 6.2 µM to 12 ± 4 µM, 16.5 ± 6.5 µM, 26.3 ± 6.5 µM (p < 0.05), 39 ± 10 µM (p < 0.05) and 49 ± 12 µM (p < 0.05), respectively (Figure 1B). Hypoxanthine levels increased from 45 ± 7 µM to 48 ± 4 µM, 57 ± 8 µM, 88 ± 11 µM (p < 0.05), 129 ± 20 µM (p < 0.05) and 139 ± 21 µM (p < 0.05), respectively (Figure 1C). Uric acid levels changed from 30 ± 12 µM to 20 ± 7 µM, 32 ± 12 µM, 42 ± 12 µM, 58 ± 19 µM (p < 0.05) and 59 ± 20 µM (p < 0.05), respectively.

4. DISCUSSION

The purpose of this work was to determine the relationship between oxygen delivery to the brain and plasma level of purine metabolites in global hypoxia model in newborn piglets. The plasma levels of the purine metabolites have been reported to increase in hypoxia and these are readily measured (see introduction). The extent to which they might provide a measure of brain hypoxia is of considerable importance.

Global hypoxia produced by decrease of FiO_2 from 21% (control) to 14% and 11%, resulted in the cortical oxygen pressure decreasing significantly from 33 Torr to 24 Torr and 18 Torr as the FiO_2 decreased, respectively. During these hypoxic conditions, no statistically significant changes in plasma purine metabolites were observed. Only further decrease in FiO_2 to 9%, a value which resulted in decrease in cortical oxygen to 6 Torr, resulted in statistically significant increase in plasma levels of xanthine and hypoxanthine. These results suggested that a mild or moderate hypoxia in newborn piglets can not be detected by measuring the level of plasma purine. In contrast, hypoxic conditions resulting in similar values for the cortical oxygen pressure resulted in statistically significant increases in the extracellular levels of dopamine and decreases in DOPAC in the striatum of newborn piglets.[10] Thus, the dopaminergic system in the striatum of newborn is extremely sensitive to decrease in cortical oxygen pressure and changes in extracellular level of dopamine can be used as an indicator of even very mild hypoxia.[22] In contrast, changes in level of plasma purines report only severe hypoxic conditions, when brain oxygen decreases to very low values.

The observed increase in plasma levels of purines metabolites during severe hypoxia, with further increase in posthypoxic period, is generally in agreement with data from other laboratories. In pigs exposed to 6–7% hypoxia, a linear increase of plasma hypoxanthine was reported to occur during hypoxia[29,30] and there was no difference between the levels in arterial and venous plasma. Thiringer et al.[34] also showed that in the exteriorized hypoxic fetal lamb the plasma hypoxanthine level rapidly increased. In the brain of fetal sheep, the same authors[33] reported there was no change in level of hypoxanthine in mild hypoxia but that hypoxanthine did increase in severe hypoxia. This indicated that fetal brain required severely hypoxic conditions before degrading its energy rich intracellular purines. In a model of a hemorrhagic shock, increases in the blood level of hypoxanthine and uric acid have also been reported.[11]

The mechanism of increase of purine metabolites acid may reflect the conversion of xanthine dehydrogenase to xanthine oxidase in ischemic/hypoxic tissue.[32] Conversion of xanthine dehydrogenase to xanthine oxidase has been shown to occur during ischemia in rat liver, kidney, heart and lung[3,37] as well as in ischemic rat brain in a 4-vessel occlusion model.[12]

The question arises, then, as to whether increase of plasma level of purine metabolites during hypoxia and reoxygenation can be used as the index of changes in oxygen pressure in the brain. Our results show that cortical oxygen pressure is much more sensitive to decrease in FiO_2 than the levels of plasma purines metabolites. Equally importantly, the extracellular levels of dopamine in the striatum, an indicator of the metabolically important decrease in oxygen pressure, increase long before there is significant increase of purine metabolites in the plasma. Increased levels in hypoxanthine and xanthine were observed only after the brain was already quite severely hypoxic by the other criteria.

It is possible that the level of purine metabolites in the cerebral spinal fluid more closely represent the extent of brain hypoxia. Meberg and Saugstad[17] found high hypoxanthine concentrations in cerebrospinal fluid of hypoxic newborn babies and in children after convulsions. The blood -brain barrier is considered to be more permeable in newborns than in adults and is reported to be functional only after a few months after birth.[2] The brain of the newborn piglets is histologically comparable to that of the human newborn of 36–38 weeks of gestation.[13] Rootwelt et al.[25] showed that in newborn piglets hypoxanthine can pass from plasma to CSF, such that increased level of hypoxanthine in CSF is not proof of cerebral hypoxia. It is unlikely, that in global ischemia/hypoxia models the amount of increase the purine metabolites generated in the brain will significantly change the total level of purine metabolites in plasma. On the other hand, the possibility that the plasma levels of nucleotide catabolites would provide a useful correlation with the degree of brain hypoxia does not appear justified.

ACKNOWLEDGMENT

This work was supported by NIH grant, NS-31465.

REFERENCES

1. Braunwald E. and Lonar R.A. (1985) Myocardial reperfusion. A double edged sword? *J Clin Invest* **16**, 1713–1720.
2. Cornford E.M. and Oldendorf W.H. (1975) Independent blood brain barrier transport systems for nucleic acid precursors. *Biochem Biophys Acta* **394**, 211–216.
3. Engerson T.D., McKelvey G., Rhyne D.B., Boggio E.B., Synder S.J. and Jones H.P. (1987) Conversion of xanthine dehydrogenase to oxidase in ischemic rat tissues. *J Clin Invest* **79**, 1564–1570.
4. Floyd R.A. (1990) Role of oxygen free radicals in carcinogenesis and brain ischemia. *FASEB J* **4**, 2587–2597.
5. Fridovich I. (1970) Quantitative aspects of the production of superoxide anion radical milk xanthine oxidase. *J Biol Chem* **245**, 4053–4057.
6. Fredholm B.B. and Hedquist P. (1980) Modulation of neurotransmission by purine nucleotides and nucleosides. *Biochem Pharmacol* **29**, 1635–1643.
7. Harkness R.A. (1988) Hypoxanthine, xanthine and uridine in body fluids, indicators of ATP depletion. *J Chromatogr* **429**, 255–278.
8. Harkness R.A. and Lund R.J. (1983) Cerebrospinal fluid concentrations of hypoxanthine, xanthine, uridine and inosine: high concentrations of the ATP metabolite hypoxanthine after hypoxia. *J Clin Pathol* **36**, 1–8.
9. Hearse D.J., Manning A.S., Downey J.M. and Yellon D.M. (1986) Xanthine oxidase: critical mediator of myocardial injury during ischemia and reperfusion. *Acta Physiol Scand* **548**, 65–78.
10. Huang, Ch-Ch., Lajevadi, N.S., Tammela, O., Pastuszko, A., Delivoria-Papadopoulos, M., and Wilson, D.F. (1994) Relationship of extracellular dopamine in striatum of newborn piglets to cortical oxygen pressure. *Neurochemical Research*, **19**, 640–655.
11. Jones C.E., Crowell J.W. and Smith E.E. (1968) Significance of increased blood uric acid following extensive hemorrhage. *Am J Physiol* **124**, 1374–1377.
12. Kinuta Y., Kimura M., Itokawa Y., Ishikawa M. and Kikuchi. H. (1989) Changes in xanthine oxidase in ischemic rat brain. *J Neurosurg* **71**, 417–420.
13. Laptook A. and Stonestreet B.S. (1982) The effects of different rates of plamanate infusion upon brain blood flow after asphyxia and hypotension in newborn piglets. *J Pediatr* **100**, 791–796.
14. Liang C.M., Liu Y.P.and Chabner B.A. (1980) Modes of action of hypoxanthine, inosine and inosine 5-monophosphate from bovine brain. *Biochem Pharmacol* **29**, 277–282.
15. McCord J.M. (1985) Oxygen-derived free radicals in postischemic tissue injury. *N Engl J Med* **312**, 159–163.
16. McCord J.M. and Fridovich I. (1968) The reduction of cytochrome C by milk xanthine oxidase. *J Biol Chem* **243**, 553–560.
17. Meberg A. and Saugstad O.D. (1978) Hypoxanthine in cerebrospinal fluid in children. *Scand J Clin Lab Invest* **38**, 437–440.
18. Niklasson F., Agren H. and Hallgren R. (1983) Purine and monoamine metabolites in cerebrospinal fluid: parallel purinergic and monoaminergic activation in depressive illness? *J Neurol Neurosurg Psychiatry* **46**, 255–260.
19. Parks D.A., Bulkley G.B. and Granger D.N. (1983) Role of oxygen derived free radicals in digestive tract diseases. *Surgery* **94**, 415–421.
20. Parks D.A., Bulkley G.B., Granger D.N., Hamilton S.R. and McCord J.M. (1982) Ischemic injury in the cat small intestine: role of superoxide radicals. *Gastroenterology* **2**, 9–15.

21. Parks D.A. and Granger D.N. (1986) Xanthine oxidase: biochemistry, distribution and physiology. *Acta Physiol Scand* **548**, 87–99.

22. Pastuszko A. (1994) Metabolic response of dopaminergic system during hypoxia-ischemia and reoxygenation in the immature brain. *Biochem. Med. and Metabolic Biology.* **51**, 1–15.

23. Pastuszko, A., Lajevardi, N.S., Chen, J., Tammela, O., Wilson, D.F., and Delivoria-Papadopoulos, M. (1993) The effects of graded levels of tissue oxygen pressure on dopamine metabolism in the striatum of newborn piglets. *J. Neurochem.* **60**, 161–166.

24. Patt A., Harken A.H., Burton L.K., Rodell T.C., Piermattei D., Schorr W.J., Parker N.B., Berger E.M., Horesh L.S., Terada L.S., Linas,S.L., Cheronis J.C. and Repine J.E. (1988) Xanthine oxidase-derived hydrogen peroxide contributes to ischemia reperfusion-induced edema in gerbil brains. *J Clinic Invest* **81**, 1556–1562.

25. Rootwelt T., Oyasaeter S. and Saugstad O.D. (1993) Transport of hypoxanthine from plasma to cerebrospinal fluid and vitreous humor in newborn piglets. *J Perinat Med* **21**, 211–217.

26. Saugstad O.D. (1975) Hypoxanthine as a measurement of hypoxia. *Pediatric Res* **9**, 158–160.

27. Saugstad O.D. (1983) Hypoxanthine and the diagnosis of hypoxia. *Ups J Med Sci* **38**, 29–33.

28. Saugstad O.D. (1988) Hypoxanthine as a indicator of hypoxia: Its role in health and disease through free radical production. *Pediatric Res* **23**, 143–150.

29. Saugstad O.D. and Aasen A.O. (1980) Plasma hypoxanthine levels as a prognostic aid of tissue hypoxia. *Eur Surg Res* **12**, 123–129.

30. Saugstad O.D., Aasen A.O. and Hetland O. (1978) Plasma hypoxanthine concentrations as an indicator of tissue hypoxia in pigs. *Eur Surg Res* **10**, 314–322.

31. Skolnick P., Marangos P.J., Goodwin F.K., Edwards M. and Paul S. (1978) Identification of inosine and hypoxanthine as endogenous inhibitors of (3H) diazepam binding in the central nervous system. *Life Sci* **23**, 1473–1480.

32. Stark K., Seubert P., Lynch G. and Baudry M. (1989) Proteolytic conversion of xanthine dehydrogenase to xanthine oxidase: Evidence against a role for calcium activated protease (calpalin). *Biochem Biophys Res Comm* **165**, 858–864.

33. Thiringer K., Blomstrand S., Hrbek A., Karlsson K. and Kjellmer I. (1982) Cerebral arterio-venous difference for hypoxanthine and lactate during graded asphyxia in the fetal lamb. *Brain Res* **239**, 107–117.

34. Thiringer K., Saugstad O.D. and Kjellmer I. (1980) Hypoxanthine as a measure of tissue hypoxia in acutely exteriorized fetal lamb. *Pediatric Res* . **14**, 905–909.

35. Wilson D.F., Rumsey W.L., Green T.J. and Vanderkooi J.M. (1988) The oxygen dependence of mitochondrial oxidative phosphorylation measured by a new optical method for measuring oxygen concentration. *J Biol Chem* **263**, 2712–2718.

36. Wilson D.F., Pastuszko A., DiGiacomo J.E., Pawlowski M., Schneiderman R. and Delivoria-Papadopoulos M. (1992) Effect of hyperventilation on oxygenation of the brain cortex of newborn piglets. *J App Physiol* **70**, 2691–2696.

37. Zhong Z., Lemasters J.J. and Thurman R.G. (1989) Role of purines and xanthine oxidase in reperfusion injury in perfused rat liver. *J Pharmacol Exp Ther* **250**, 470–475.

22

RAPID ISCHEMIC PRECONDITIONING PROTECTS RATS FROM CEREBRAL ANOXIA/ISCHEMIA

Miguel A. Pérez-Pinzón,* Guang-Ping Xu, Patricia L. Mumford, W. Dalton Dietrich, Myron Rosenthal, and Thomas J. Sick

Department of Neurology
University of Miami School of Medicine
Miami, Florida 33101

1. INTRODUCTION

Mild ischemic insults may limit damage from subsequent ischemic insults in heart and brain (Murry et al., 1986; Kato et al., 1992; Li et al., 1992; Liu et al., 1992; Walker et al., 1993). This phenomenon was defined as ischemic preconditioning (IPC). In brain, for example, 3 min of sublethal ischemia followed by 3 days of reperfusion protected against hippocampal CA1 neuronal damage after 8 min of ischemia (Liu et al., 1992). Preconditioning also improved evoked potential recovery in hippocampal slices (Schurr et al., 1986; Schurr and Rigor, 1987). Studies reported here seek to further characterize IPC in intact brain and to derive insights into its mechanism through studies in hippocampal slices. The rationale for these studies is derived from the belief that understanding the mechanisms of IPC could offer unique insights into basic mechanisms of ischemic injury and into potential therapies to ameliorate the consequences of anoxia or ischemia.

Two factors with important roles in IPC appear to be the intensity of ischemia and the latency between insults. In myocytes, for example, IPC was effective if the latency between insults was 1 hr. In brain IPC was protective against global cerebral ischemia but only if many hours of reperfusion separated the 'conditioning' and 'test' insults (Kato et al., 1992). In intact rat brain, we sought whether IPC could be induced within 30 min as in heart. In parallel studies, we sought mechanisms of IPC by determining: 1) if anoxic preconditioning could be induced in a manner similar to that in intact brain; and 2) whether adenosine plays a role in preconditioning. Emphasis was placed upon adenosine since en-

*Correspondence to: Miguel A. Pérez-Pinzón, Ph.D., Department of Neurology, University of Miami, South Campus, Bldg. B., 12500 SW 152 St., Miami, Fl. 33177. Telephone: (305)254-7178; Fax: (305)254-0661; e-mail: mperez@neuron.med.miami.edu.

dogenous adenosine may mediate IPC in heart (Liu et al., 1991) and in brain (Fitzgibbons et al., 1994; Gidday et al., 1995; Heurteaux et al., 1995).

2. MATERIALS AND METHODS

Methods and protocols for these studies were described previously (Pérez-Pinzón et al., 1996a,b). In brief, studies in intact brain were conducted in fasted Wistar rats anesthetized with 0.5% halothane and 70% nitrous oxide (balance oxygen). Ligatures of polyethylene (PE-10) tubing, contained within a double-lumen silastic tubing, were passed around each carotid artery. Brain temperature was maintained at 36.5 -37 °C throughout the experiment. An extracellular potassium electrode was placed in a burr hole opened left to the midline. Before each ischemic insult, blood was gradually withdrawn from the femoral artery into a heparinized syringe to reduce systemic blood pressure to 40–50 mm Hg. Cerebral ischemia was then produced by tightening the carotid ligatures bilaterally. Ischemia was confirmed by the abrupt increase in extracellular potassium ion activity to 50–70 mM, that is characterized as 'anoxic depolarization' (AD) (Hansen, 1985).

Three days later, the rats were re-anesthetized, perfused for 1 min with physiologic saline, and then perfused with FAM (a mixture of 40 % formaldehyde, glacial acetic acid, and methanol, 1:1:8 by volume) for 19 min. Coronal sections, of 10 μm were stained with hematoxylin and eosin. Ischemic damage by neurons showing moderate to severe shrinkage, eosinophilic cytoplasm, increase nuclear basophilia, and a pyknotic nucleus (Dietrich et al., 1993).

Hippocampal slices were prepared as in prior studies (Pérez-Pinzón et al., 1996a). For measurements of extracellular evoked potentials, slices were placed in an interface recording chamber. Evoked potentials were measured from the stratum pyramidale of the hippocampal field CA1 with standard glass micropipettes filled with 150 mM sodium chloride. Orthodromic field potentials were elicited by stimulating the Schaffer collaterals with bipolar, tungsten electrodes insulated with teflon except at the tip. Slices were made anoxic by switching the gas mixture in the atmosphere above the slice from 95% O_2–5% CO_2 to 95% N_2–5% CO_2.

Adenosine was purchased from Sigma Chemical Company. 2-chloroadenosine (2-CADO) and 8-cyclopentyl-1,3-dipropylxanthine (DPCPX) were purchased from Research Biochemical International.

Statistical significance among groups was established with an ANOVA test. Post-hoc comparisons within groups were carried out with a Scheffe's test. Significance was accepted with $p < 0.05$. All data were expressed as mean ± SEM.

3. RESULTS

In control rats, the 'test' ischemic insult (10 min) produced many ischemic neurons within the CA1 hippocampus. Ischemic neurons were also commonly observed in layers III, IV and V of the cerebral cortex immediately dorsal to the entorhinal fissure. In contrast, rats that underwent the 'preconditioning' insult 30 min prior to the 'test' insult, had fewer ischemic cells within the hippocampus and cerebral cortex (Figure 1).

The ischemic cell counts for rats grouped as controls (n = 11) and IPC-treated (n = 11) averaged from left to right hemispheres are shown in Figure 1. IPC significantly protected by 78, 66 and 54% the lateral, middle and medial subsectors of the hippocampal

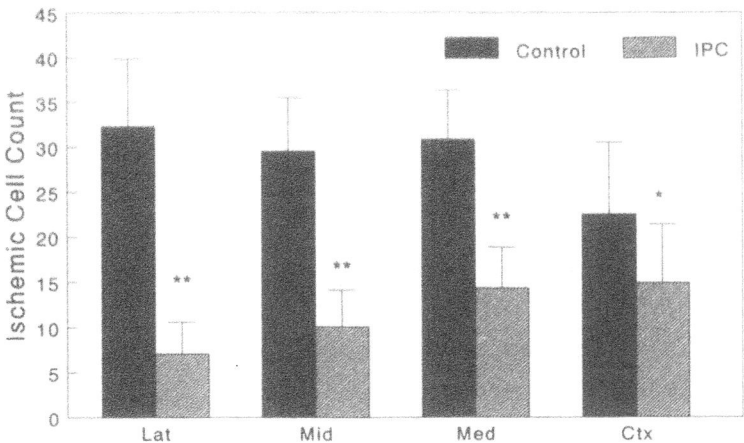

Figure 1. Ischemic preconditioning protects 30 min prior to a 10 min 'test' ischemic insult. Ischemic cell count was significantly reduced in all regions studied in the IPC-treated rats (n = 11) as compared with controls (n = 11). Lat = Lateral subsection of the CA1 subfield of the hippocampus; Mid = Middle subsection of the CA1 subfield of the hippocampus; Med = Medial subsection of the CA1 subfield of the hippocampus; Ctx = Lateral cortex. *p < 0.01; **p < 0.05.

CA1 when compared with control, respectively (Figure 1). In the lateral hippocampus, control ischemic cell count was 32.3 ± 7.5 vs IPC which was 7.1 ± 3.5 (p < 0.01). In the middle hippocampus control ischemic cell count was 29.5 ± 6.0 and IPC was 10.1 ± 4.0 (p < 0.01). In the medial subsector of the hippocampus control ischemic cell count was 30.8 ± 5.5 and IPC was 14.3 ± 4.5 (p < 0.01). IPC protection occurred in the lateral cortex.

No significant differences were found in the levels of $[K^+]_o$ prior to, during and after 'test' ischemia between controls and the IPC group (Figure 2).

Figures 3 and 4 show percent recovery from pre-anoxic evoked potential amplitudes that were recorded for 30 min following anoxia. Anoxia was prolonged to allow for 1

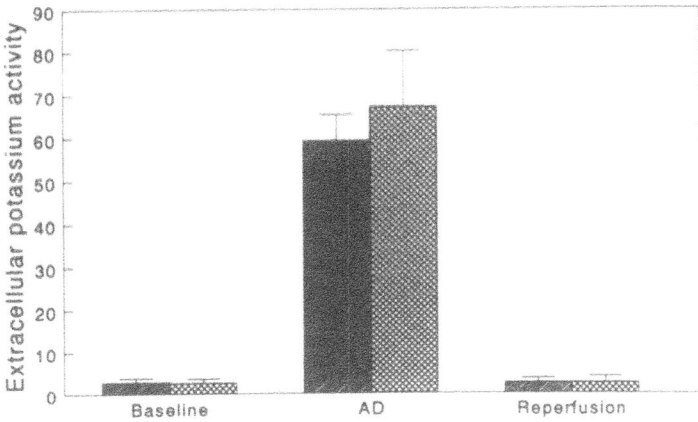

Figure 2. Extracellular potassium activity in controls and preconditioned slices during pre-anoxia, anoxic depolarization (AD) and 30 min of reoxygenation (reperfusion). No significant differences were found between the two groups at any experimental time.

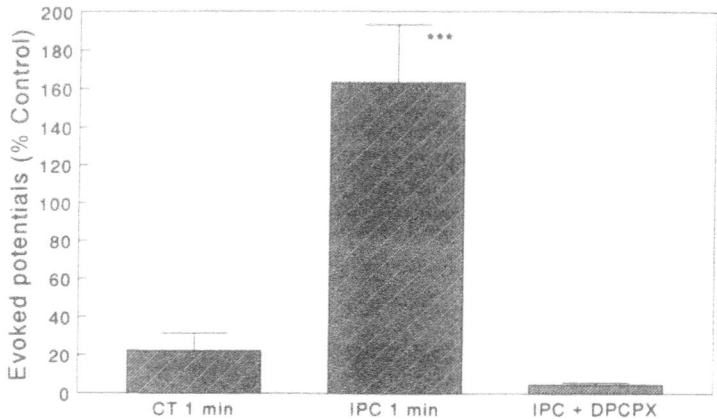

Figure 3. Comparison of average values of evoked potential amplitudes (% from control) 30 min after final re-oxygenation in groups 1, 2 and 4. CT 1 min is control with 1 min AD; IPC 1 min is the preconditioned slice with 1 min AD; and IPC + DPCPX represents group 4, where DPCPX is superfused onto slices during the preconditioning stage. A significant better recovery of evoked potentials was observed in the preconditioned group (IPC 1 min), p < 0.001.

(Figure 3) or 2 (Figure 4) minutes of anoxic depolarization (AD). These EPs were measured in unconditioned slices (Ct 1 min and Ct, Figures 3 and 4, respectively) and in slices that were preconditioned with a period of anoxia that was kept until the onset of AD (IPC, Figures 3 and 4).

In unconditioned slices, EP recovery was poor after anoxic insults which provoked AD for 1 min (22 ± 9% of control, n = 9) (Figure 3). Recovery of evoked potentials was even less (only 10 ± 5% of control; n = 17) when the anoxic insult was prolonged to a duration which produced AD for 2 min (Figure 4).

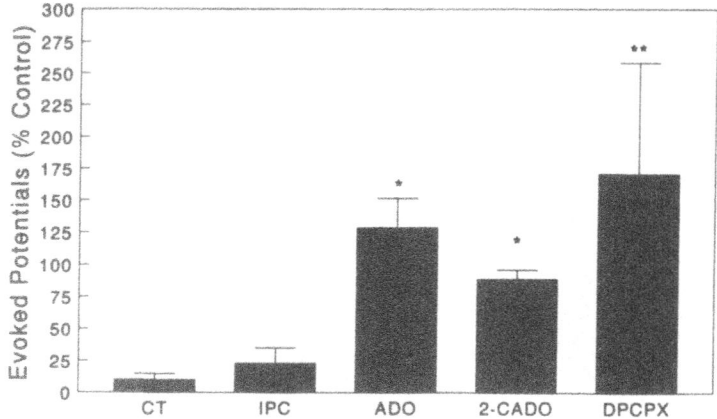

Figure 4. Comparison of average values of evoked potential amplitudes (% from control) 30 min after final re-oxygenation in groups 1, 2 and 3. CT is controls with 2 min AD; IPC is the preconditioned slice with 2 min AD; ADO represents 100 μM adenosine; 2-CADO represents 10 μM 2-chloroadenosine; and DPCPX represents 5 μM DPCPX. Significant improvement in the recovery of evoked potentials was observed in slices superfused with adenosine (p < 0.05), 2-CADO (p < 0.05) and DPCPX (p < 0.01).

Preconditioning significantly improved EP recovery following a 'test' anoxic insult which produced AD for only 1 minute (p< 0.001). In these slices, EP recovery was 163 ± 30% (n = 6) of control (Figure 3). However, preconditioning effects were not significant when the test insult was prolonged to produce AD of 2 min duration (Figure 4). In these slices, EP amplitudes recovered to only 23 ± 12% of the control value (n = 14).

Synaptic activity was protected by 100 μM adenosine or 10 μM 2-CADO when superfused transiently for 10 min onto slices 30 min prior to anoxia. Recovery of evoked potentials was 128 ± 23% (n = 6, p < 0.05) and 88 ± 7% (n = 14, p < 0.05) for adenosine and 2-CADO, respectively.

Further support for the role of adenosine A1 receptors during preconditioning is defined in Figure 3. When 5 μM DPCPX was superfused transiently onto slices during 'conditioning' episodes, the protection afforded by preconditioning was abolished. The recovery of evoked potentials in this group was 4.6 ± 1% (n = 8).

4. DISCUSSION

In brain, IPC provided significant protection but IPC protection was evident only when the latency between 'preconditioning' and 'test' insults was prolonged (Kitagawa et al., 1990; Kato et al., 1991; Liu et al., 1992). This is in contrast to the short latency required for IPC in heart (Murry et al., 1986). Studies in the hippocampal slices suggest that this short latency for IPC may be common to tissues *in vitro* (Schurr et al., 1986; Schurr and Rigor, 1987; Pérez-Pinzón et al., 1996a). To our knowledge, this is the first demonstration that only a short (30 min) reperfusion period between 'conditioning' and 'test' ischemic insults can provide significant protection against acute ischemic neuronal damage in the CA1 hippocampal subfield and neocortex (Pérez-Pinzón et al., 1996b).

One difference in present studies is that the duration of the 'conditioning' and 'test' anoxic/ischemic insults was based on the duration of AD rather than on the latency to reperfusion (Schurr et al., 1986; Schurr and Rigor, 1987; Lin et al., 1992; Lin et al., 1993). Previous studies showed that the AD duration better predicts EP recovery than is the duration of anoxia itself (Pérez-Pinzón et al., 1996a,b).

Since other studies in brain have shown that IPC disappears after 1 h (Kato et al., 1991), we suggest that there are two phases to IPC protection: 1) the one dominated by enhanced glycolysis, suppression of electrical activity, loss of ion homeostasis and depressed ATP consumption. This phase of neuroprotection may disappear rapidly, as in heart (Murry et al., 1986). Another phase of preconditioning may occur later after the preconditioning insult, which may involve expression of specific genes which limit delayed neuronal damage (Liu et al., 1993; Liu et al., 1993; Kato et al., 1994; Kato et al., 1995).

It is likely that IPC involves similar mechanisms in heart and brain (Liu et al., 1991; Alkhulaifi et al., 1993) and that adenosine is involved (Yao and Gross, 1994; Tsuchida et al., 1993; Gruver et al., 1994; Tsuchida et al., 1994). We conclude here that adenosine receptors may mediate APC in hippocampal slices. The adenosine A1 receptor antagonist DPCPX blocked APC while adenosine, or the A1 receptor agonist 2-CADO, induced protection similar to that provided by APC.

A role for adenosine in brain IPC was also supported by findings that the nearly complete protection produced by IPC in the CA1 region of rat hippocampus 3 days after 6 min of global cerebral ischemia (Heurteaux et al., 1995) was abolished by DPCPX. In contrast, infusion of cyclopentyl adenosine (CPA) (1 mg/kg) (an adenosine A1 receptor agonist) 15 min prior to ischemia produced 70% protection of CA1 cells after 3 days of reperfusion. In this latter study, CPA was infused only 15 min prior to the 'test' ischemic insult. This paradigm resembled that employed in hippocampal slices and the *in vivo* ischemia paradigms used in the present study, and supports the hypothesis that activation of the adenosine A1 receptor prior to ischemia can simulate IPC.

Adenosine was also implicated in protection of brain using a model of anoxic preconditioning in neonates. In this study, the adenosine receptor blocker caffeine attenuated protection of neonatal brain afforded by hypoxic preconditioning (Fitzgibbons et al., 1994). In contrast, infusion of the adenosine A1 receptor agonist R-(phenylisopropyl)-adenosine (R-PIA) emulated hypoxic preconditioning protection (Fitzgibbons et al., 1994). Moreover, hypoxic preconditioning was also potentiated by blocking adenosine deaminase (which initiates adenosine catabolism) (Gidday et al., 1995).

Defining the mechanism by which adenosine might mediate IPC is difficult because of the many effects of this neuromodulator but possibilities include: a) increased glycogenolysis (Magistretti et al., 1986) and increased glycolytic rate (Pasteur effect) (Wyatt et al., 1989; Gutierrez-Juarez et al., 1992); and b) increases in the adenylate pool. This latter mechanism is supported by reports that ATP levels increased 85% when 8 mM adenosine was superfused onto slices (Whittingham et al., 1989). Although the adenosine concentration used here was several fold lower than that used by Whittingham et al. (1989), it is possible that adenosine increased the energy charge of hippocampal neurons and thereby prevented the rapid transition to energy failure.

In conclusion, the finding that IPC can be produced within 30 min is important because it suggests that strategies may be available for protection of brain against acute ischemia. The therapeutic window for rescuing neurons may be extended with better understanding of the neuroprotective mechanisms involved in this rapid IPC model. These data also support the hypothesis that adenosine, probably by its activation of A1 receptors, is involved in the neuroprotection afforded by preconditioning.

ACKNOWLEDGMENTS

These studies were supported by PHS grant NS 14325, NS 32167, NS 05820 and Grant in Aid from AHA. We thank Dr. Baowin Lin, Isabel Garcia, and Susan Kraydiseh for technical support.

REFERENCES

Alkhulaifi A. M., Pugsley W. B. and Yellon D. M. (1993) The influence of the time period between preconditioning ischemia and prolonged ischemia on myocardial protection. *Cardioscience* 4(3): 163–169.

Dietrich W. D., Busto R., Alonso O., Globus M. Y. and Ginsberg M. D. (1993) Intraischemic but not postischemic brain hypothermia protects chronically following global forebrain ischemia in rats. *J Cereb Blood Flow Metab* 13(4): 541–549.

Fitzgibbons J. C., Shah A. R., Park T. S. and Gidday J. M. (1994) Caffeine blocks the cerebroprotective actions of preconditioning in the hypoxic-ischemic neonate. *Soc. Neurosci. Abs* 20(1): 618.

Gidday J. M., Fitzgibbons J. C., Maceren R. G., Shah A. R., Shah N. R. and Park T. S. (1995) Inhibition of adenosine deaminase with deoxycoformycin (DCF) potentiates preconditioning cerebroprotection in a perinatal rat model of ischemic tolerance. *Soc. Neurosci. Abs.* 21(1): 513.

Gruver E. J., Toupin D., Smith T. W. and Marsh J. D. (1994) Acadesine improves tolerance to ischemic injury in rat cardiac myocytes. *J Mol Cell Cardiol* 26(9): 1187–1195.

Gutierrez-Juarez R., Castrejon-Sosa M., Martinez-Valdez H., Blancas-Torres P. G., Pina E. and Madrid-Marina V. (1992) Activating effect of adenosine on rat erythrocyte glycolysis. *Int J Biochem* 24(3): 433–436.

Hansen A. J. (1985) Effect of anoxia on ion distribution in the brain. *Physiol Rev* 65(1): 101–148.

Heurteaux C., Lauritzen I., Widmann C. and Lazdunski M. (1995) Essential role of adenosine, adenosine A1 receptors, and ATP-sensitive K+ channels in cerebral ischemic preconditioning. *Proc Natl Acad Sci U S A* 92(10): 4666–4670.

Kato H., Araki T., Murase K. and Kogure K. (1992) Induction of tolerance to ischemia: alterations in second-messenger systems in the gerbil hippocampus. *Brain Res Bull* 29(5): 559–565.

Kato H., Kogure K., Araki T. and Itoyama Y. (1995) Induction of Jun-like immunoreactivity in astrocytes in gerbil hippocampus with ischemic tolerance. *Neurosci Lett* 189(1): 13–16.

Kato H., Liu Y., Araki T. and Kogure K. (1991) Temporal profile of the effects of pretreatment with brief cerebral ischemia on the neuronal damage following secondary ischemic insult in the gerbil: cumulative damage and protective effects. *Brain Res* 553(2): 238–242.

Kato H., Liu Y., Kogure K. and Kato K. (1994) Induction of 27-kDa heat shock protein following cerebral ischemia in a rat model of ischemic tolerance. *Brain Res* 634(2): 235–244.

Kitagawa K., Matsumoto M., Tagaya M., Hata R., Ueda H., Niinobe M., Handa N., Fukunaga R., Kimura K., Mikoshiba K. and et a. (1990) 'Ischemic tolerance' phenomenon found in the brain. *Brain Res* 528(1): 21–24.

Li Y. W., Whittaker P. and Kloner R. A. (1992) The transient nature of the effect of ischemic preconditioning on myocardial infarct size and ventricular arrhythmia. *Am Heart J* 123(2): 346–353.

Lin B., Dietrich W. D., Ginsberg M. D., Globus M. Y. and Busto R. (1993) MK-801 (dizocilpine) protects the brain from repeated normothermic global ischemic insults in the rat. *J Cereb Blood Flow Metab* 13(6): 925–932.

Lin B., Globus M. Y., Dietrich W. D., Busto R., Martinez E. and Ginsberg M. D. (1992) Differing neurochemical and morphological sequelae of global ischemia: comparison of single- and multiple-insult paradigms. *J Neurochem* 59(6): 2213–2223.

Liu G. S., Thornton J., Van Winkle D. M., Stanley A. W., Olsson R. A. and Downey J. M. (1991) Protection against infarction afforded by preconditioning is mediated by A1 adenosine receptors in rabbit heart. *Circulation* 84(1): 350–356.

Liu Y., Kato H., Nakata N. and Kogure K. (1993) Temporal profile of heat shock protein 70 synthesis in ischemic tolerance induced by preconditioning ischemia in rat hippocampus. *Neuroscience* 56(4): 921–927.

Liu Y., Kato H., Nakata N. and Kogure K. (1993) Correlation between induction of ischemic tolerance and expression of heat shock protein-70 in the rat hippocampus. *No To Shinkei* 45(2): 157–162.

Liu Y., Kato H., Nakata N. and Kogure K. (1992) Protection of rat hippocampus against ischemic neuronal damage by pretreatment with sublethal ischemia. *Brain Res* 586(1): 121–124.

Magistretti P. J., Hof P. R. and Martin J. L. (1986) Adenosine stimulates glycogenolysis in mouse cerebral cortex: a possible coupling mechanism between neuronal activity and energy metabolism. *J Neurosci* 6(9): 2558–2562.

Murry C. E., Jennings R. B. and Reimer K. A. (1986) Preconditioning with ischemia: a delay of lethal cell injury in ischemic myocardium. *Circulation* 74(5): 1124–1136.

Pérez-Pinzón M. A., Mumford P. L., Rosenthal M. and Sick T. J. (1996) Anoxic preconditioning in hippocampal slices: role of adenosine. *Neuroscience* : 75(3):687–694.

Pérez-Pinzón M. A., Xu G. P., Dietrich W. D., Rosenthal M. and Sick T. J. (1996) Rapid preconditioning protects rats against ischemic neuronal damage after 3 but not 7 days of reperfusion following global cerebral ischemia. *J Cereb Blood Flow Metab* : In press.

Schurr A., Reid K. H., Tseng M. T., West C. and Rigor B. M. (1986) Adaptation of adult brain tissue to anoxia and hypoxia in vitro. *Brain Res* 374(2): 244–248.

Schurr A. and Rigor B. M. (1987) The mechanism of neuronal resistance and adaptation to hypoxia. *Febs Lett* 224(1): 4–8.

Tsuchida A., Liu G. S., Mullane K. and Downey J. M. (1993) Acadesine lowers temporal threshold for the myocardial infarct size limiting effect of preconditioning. *Cardiovasc Res* 27(1): 116–120.

Tsuchida A., Yang X. M., Burckhartt B., Mullane K. M., Cohen M. V. and Downey J. M. (1994) Acadesine extends the window of protection afforded by ischaemic preconditioning. *Cardiovasc Res* 28(3): 379–383.

Walker D. M., Walker J. M. and Yellon D. M. (1993) Global myocardial ischemia protects the myocardium from subsequent regional ischemia. *Cardioscience* 4(4): 263–266.

Whittingham T. S., Warman E., Assaf H., Sick T. J. and LaManna J. C. (1989) Manipulating the intracellular environment of hippocampal slices: pH and high-energy phosphates. *J Neurosci Methods* 28(1–2): 83–91.

Wyatt D. A., Edmunds M. C., Rubio R., Berne R. M., Lasley R. D. and Mentzer R., Jr. (1989) Adenosine stimulates glycolytic flux in isolated perfused rat hearts by A1-adenosine receptors. *Am J Physiol* 257(6 Pt 2): H1952–1957.

Yao Z. and Gross G. J. (1994) A comparison of adenosine-induced cardioprotection and ischemic preconditioning in dogs. Efficacy, time course, and role of KATP channels. *Circulation* 89(3): 1229–1236.

23

BRAIN METABOLIC AND VASCULAR ADAPTATIONS TO HYPOXIA IN THE RAT

Review and Update

Joseph C. LaManna[1] and Sami I. Harik[2]

[1]Department of Neurology
 Case Western Reserve University School of Medicine
 Cleveland, Ohio
[2]Department of Neurology
 University of Arkansas for Medical Sciences
 Little Rock, Arkansas

1. INTRODUCTION

Prolonged exposure to low oxygen environments provokes systemic adaptations that result in maintained oxygen delivery to the brain, and central cerebrovascular and metabolic adaptations that preserve tissue oxygen and energy supply to support neuronal function. Recently, we have examined a number of aspects of the central nervous system adaptation to mild hypoxia, and though these studies provide a far from complete understanding of the phenomenon, some general conclusions can at least be drawn.

2. MATERIALS AND METHODS

All of the reported studies were carried out using a hypobaric model of mild hypoxia in rats (LaManna et al., 1992). Rats were kept, 3 to cage in chambers constructed to hold 2 to 4 cages, and that could be maintained at 0.5 ATM (380 mmHg) pressure. Rats from the same litter were always used for experimental control purposes. These control, normobaric - normoxic rats were housed in the same room, just outside the hypobaric chambers and were, thus, subject to identical ambient conditions. Rats were exposed to hypoxia for periods up to 3 weeks. Hypoxic exposure was continuous except for up to 1 hour per day to allow cage cleaning, feeding and watering.

Oxygen Transport to Tissue XIX, edited by Harrison and Delpy
Plenum Press, New York, 1997

3. RESULTS

3.1. Systemic Adaptations

At 0.5 ATM, arterial oxygen partial pressure initially falls to about 40 - 45 mmHg, and this hypoxemia must be considered the primary variable responsible for initiating the short and long term physiological adaptations. Mature rats exposed to this level of hypoxia are capable of successful adaptation and are able to survive for periods of at least a few months. The most obvious signs of successful adaptive responses that were observed were hyperventilation, an increase in red blood cell volume fraction, and a decrease in weight gain.

3.1.1. Ventilation. Freely breathing rats exposed to the equivalent of 10% hypoxia hyperventilated, with $PaCO_2$ falling below 20 mmHg (LaManna et al., 1992). The acute respiratory alkalosis was compensated, after several hours, by bicarbonate excretion so that the hypoxic adapted rats maintained arterial blood pH in the normal range despite hypocarbia.

3.1.2. Hematocrit. Hematocrit was elevated to 59 ± 3.5(sd) compared to litermate controls at 52 ± 1 after 4 days. It was significantly higher by 1 week (62 ± 1.5), and was maintained higher than littermate controls at 2 (65 ± 3) and 3 (69 ± 1.5) weeks of hypoxic exposure, recovering to near control (55 ± 1) after three weeks of return to normoxia (Harik et al., 1996). In the rat it was not unusual for hematocrits to reach 60 - 70 or more after 3 weeks of hypoxia (Harik et al., 1996; LaManna et al., 1996; Harik et al., 1995; Harik et al., 1995; Mironov et al., 1994; Harik et al., 1994; LaManna et al., 1992).

3.1.3. Body Weight. Normobaric-normoxic littermate rats beginning at about 300g body weight, gained weight at about 7 - 8 grams/day (2% of body weight increment per day) when allowed food and water ad libitum. In contrast, rats exposed daily to 23 hours of mild hypoxia gained no weight during the 3 weeks of exposure (LaManna et al., 1992). When rats adapted to 3 weeks of hypoxia (344 ± 24 g hypoxic vs 420 ± 13 g controls) were allowed to return to normobaric-normoxic conditions, they demonstrated weight gain similar to littermate controls, recovering to within 95% (446 ± 32 g) of the control (468 ± 32 g) body weight (Harik et al., 1996).

3.2. Vascular Adaptations

3.2.1. Structure.

3.2.1.1. Microstructure. There was a 33% increase in cerebral capillary density beginning at about 1 week of hypoxia (Harik et al., 1996) when bFGF was also elevated (LaManna et al., 1994a), increasing to about 70 % by 3 weeks (LaManna et al., 1992; Harik et al., 1996). Although brain weight was slightly less in the hypoxic exposed rats (Harik et al., 1995a; Harik et al., 1994a), the increased density was apparently due to angiogenesis. There were regional increases (30 - 46 % in the parietal cortex) in the number of capillary segments and in mean segment length (LaManna et al., 1994b); and evidence of capillary sprouting after 1 week exposure (Mironov et al., 1994). It appeared that capillary growth was by early hyperplastic and later hypertrophic mechanisms (Harik et al., 1995a). This was shown by 50% increased microvessel protein by 1 week of hypoxia, with increased microvessel DNA only at 2 and 3 weeks (Harik et al., 1995a). There was an increase by about 30% in GLUT-1 transporters per unit capillary as shown by cytochalasin B binding (Harik et al., 1994), and by the more specific Western blot using mono- and polyclonal GLUT-1 antibodies (Harik et al., 1996).

3.1.1.2. Ultrastructure. There was an 18% increase in capillary diameter (5.52 ± 0.11 µm vs. 4.66 ± 0.34 µm), but no change in wall thickness (0.30 ± 0.01 µm vs. 0.31 ± 0.06 µm), basement membrane thickness (34 ± 1 nm vs. 38 ± 4 nm), or pericyte coverage (23 ± 3 % vs 27 ± 8 %) at 3 weeks of hypoxia (Stewart et al., in press). The number of mito-chondria per endothelial cell profile was unchanged (5 ± 0.2 vs 5 ± 1.7), but the density (% area) of mitochondria in the neuropil was decreased (3.9 ± 0.2 % vs 5.4 ± 0.7 %) by 30% (Stewart et al., in press).

3.2.2. Function. There was an initial 20 -40% increase in regional cerebral cortical blood flow which persisted for at least 3 hours, but returned to baseline before 3 weeks (LaManna et al., 1992). In correspondence with the increased capillary density and GLUT-1 transporter content, there was a more than doubling of the glucose single pass extraction fraction, from about 0.25 to as much as 0.72 in the cerebral cortex and higher in the hippo-campus, cerebellum and striatum (Harik et al., 1994). This resulted in a near doubling of the blood to brain glucose influx rate (Harik et al., 1994).

3.3. Metabolic Adaptations

Hypoxia increased the regional metabolic rate for glucose by 10 - 40 % (Harik et al., 1995b). There were increased concentrations of brain glucose (150% of control) and lac-tate (250% of control), and decreased glycogen (75% of control) content in hypoxic rats, but ATP and phosphocreatine were not different from normoxic controls, and intracellular pH was also unchanged, as would be expected for a successful adaptation (Harik et al., 1995b). Overall, cytochrome oxidase activity, determined by quantitative histochemistry, was decreased by at least 16 - 27% (LaManna et al., 1996).

4. DISCUSSION

4.1. Systemic Adaptations

The systemic adaptation mechanisms elicited by hypoxic exposure in this rat model, especially those controlling ventilation, resemble those of the human adapting to high alti-tude (Lenfant and Sullivan, 1971; Dempsey and Forster, 1982; Monge and León-Velarde, 1991), although the rat hematocrit is much more elevated than the human. The lower $PaCO_2$ results in slightly (~10 - 15 mmHg) higher PaO_2 pressures at the same atmospheric level, and the increased hematocrit allows the normalization of the oxygen-carrying ca-pacity of the hypoxic blood. The effect of hypoxia on body weight is also similar to that observed in human high altitude adaptation (Boyer and Blume, 1984) and is almost cer-tainly due to decreased water (Fregly and Waters, 1966).

4.2. Capillary Density

Increased brain capillary densities in response to longer term hypoxic exposure have been previously reported (Cervós-Navarro et al., 1991; Miller, Jr. and Hale, 1970; Opitz, 1951; Diemer and Henn, 1965). The increased capillary density results in diminished intercapillary distances leading to a real improvement in oxygen diffusing capacity (Hunziker et al., 1974), overcoming the decreased arterial oxygen driving pressure.

4.3. Glucose Metabolism

The only moderately increased glucose metabolism is unlikely to provide a significant source of ATP. But the increased glycolytic acid production could function to balance the decreased tissue CO_2 that would result from hyperventilation, thus maintaining cellular acid-base balance (Lauro and LaManna, in press; Musch et al., 1983). Decreased cytochrome oxidase activity has been reported (Chávez et al., 1995; Dagani et al., 1984). This is consistent with the proposal that brain hypometabolism is a defense mechanism in altitude adaptation (Hochachka et al., 1994), and may occur through hypoxia-induced hypothermia (Wood, 1991) which is protective in acute hypoxia (Carlsson et al., 1976).

5. CONCLUSIONS

The rat brain is apparently capable of structural and functional plasticity to balance energy supply and demand. The primary brain adaptations to hypoxia appear to be accomplished through increased capillary density, and an increased glycolytic, but decreased oxidative metabolism. The cellular sensor and effector mechanisms that are responsible for this plasticity remain unknown.

REFERENCES

Boyer, S.J. and Blume, F.D. 1984, Weight loss and changes in body composition at high altitude, *J. Appl. Physiol.* 57: 1580–1585.

Carlsson, C., Hägerdal, M., and Siesjö, B.K. 1976, Protective effect of hypothermia in cerebral oxygen deficiency caused by arterial hypoxia, *Anesthesiol.* 44: 27–35.

Cervós-Navarro, J., Sampaolo, S., and Hamdorf, G. 1991, Brain changes in experimental chronic hypoxia, *Exp. Pathol.* 42: 205–212.

Chávez, J.C., Pichiule, P., Boero, J., and Arregui, A. 1995, Reduced mitochondrial respiration in mouse cerebral cortex during chronic hypoxia, *Neurosci. Lett.* 193: 169–172.

Dagani, F., Marzatico, F., Curti, D., Zanada, F., and Benzi, G. 1984, Effect of prolonged and

intermittent hypoxia on some cerebral enzymatic activities related to energy transduction, *J. Cereb. Blood Flow Metab.* 4: 615–624.

Dempsey, J.A. and Forster, H.V. 1982, Mediation of ventilatory adaptations, *Physiol. Rev.* 62: 262–346.

Diemer, K. and Henn, R. 1965, Kapillarvermehrung in der hirnrinde der ratte unter chronischem sauerstoffmangel, *Die Natur.* 52: 135–136.

Fregly, M.J. and Waters, I.W. 1966, Posthypoxic drinking response of rats, *Fed. Proc.* 25: 1220–1226.

Harik, N., Harik, S.I., Kuo, N.-T., Sakai, K., Przybylski, R.J., and LaManna, J.C. 1996, Time course and reversibility of the hypoxia-induced alterations in cerebral vascularity and cerebral capillary glucose transporter density, *Brain Res.* 737: 335–338.

Harik, S.I., Behmand, R.A., and LaManna, J.C. 1994, Hypoxia increases glucose transport at blood-brain barrier in rats, *J. Appl. Physiol.* 77: 896–901.

Harik, S.I., Hritz, M.A., and LaManna, J.C. 1995a, Hypoxia-induced brain angiogenesis, *J. Physiol. (Lond.)*, 485.2: 525–530.

Harik, S.I., Lust, W.D., Jones, S.C., Lauro, K.L., Pundik, S., and LaManna, J.C. 1995b, Brain glucose metabolism in hypobaric hypoxia, *J. Appl. Physiol.* 79: 136–140.

Hochachka, P.W., Clark, C.M., Brown, W.D., Stanley, C., Stone, C.K., Nickles, R.J., Zhu, G.G., Allen, P.S.. and Holden, J.E. 1994, The brain at high altitude: hypometabolism as a defense against chronic hypoxia? *J. Cereb. Blood Flow Metab.* 14: 671–679.

Hunziker, O., Frey, H., and Schulz, U. 1974, Morphometric investigations of capillaries in the brain cortex of cats, *Brain Res.* 65: 1–11.

LaManna, J.C., Vendel, L.M., and Farrell, R.M. 1992, Brain adaptation to chronic hypobaric hypoxia in rats, *J. Appl. Physiol.* 72: 2238–2243.

LaManna, J.C., Boehm, K.D., Mironov, V., Hudetz, A.G., Hritz, M.A., Yun, J.K., and Harik, S.I. 1994a, Increased fibroblastic growth factor mRNA in the brains of rats exposed to hypobaric hypoxia, in: "Oxygen Transport to Tissue XVI," M.C. Hogan, O. Mathieu-Costello, D.C. Poole, and P.D. Wagner, eds., pp. 497–502, Plenum Publishing Corp. New York.

LaManna, J.C., Cordisco, B.R., Kneuse, D.E., and Hudetz, A.G. 1994b, Increased capillary segment length in cerebral cortical microvessels of rats exposed to 3 weeks of hypobaric hypoxia, in: "Oxygen Transport To Tissue XV (Advances in Experimental Medicine and Biology, v. 345)," P. Vaupel, R. Zander, and D.F. Bruley, eds., pp. 627–632, Plenum Publishing Corp. New York.

LaManna, J.C., Kutina-Nelson, K.L., Hritz, M.A., Huang, Z., and Wong-Riley, M.T.T. 1996, Decreased rat brain cytochrome oxidase activity after prolonged hypoxia, *Brain Res.* 720: 1–6.

Lauro, K.L. and LaManna, J.C. (In press), Adequacy of cerebral vascular remodeling following three weeks of hypobaric hypoxia examined by an integrated composite analytical model, in: "Oxygen Transport to Tissue XVIII (Advances in Experimental Medicine and Biology)," E.M. Nemoto, ed., Plenum Publishing Corp. New York.

Lenfant, C. and Sullivan, K. 1971, Adaptation to high altitude, *New Engl. J. Med.* 284: 1298–1309.

Miller, A.T., Jr. and Hale, D.M. 1970, Increased vascularity of brain, heart, and skeletal muscle of polycythemic rats, *Am. J. Physiol.* 219: 702–704.

Mironov, V., Hritz, M.A., LaManna, J.C., Hudetz, A.G., and Harik, S.I. 1994, Architectural alterations in rat cerebral microvessels after hypobaric hypoxia, *Brain Res.* 660: 73–80.

Monge, C. and León-Velarde, F. 1991, Physiological adaptation to high altitude: Oxygen transport in mammals and birds, *Physiol. Rev.* 71: 1135–1172.

Musch, T.I., Dempsey, J.A., Smith, C.A., Mitchell, G.S., and Bateman, N.T. 1983, Metabolic acids and [H$^+$] regulation in brain tissue during acclimatization to chronic hypoxia, *J. Appl. Physiol.* 55: 1486–1495.

Opitz, E. 1951, Increased vascularization of the tissue due to acclimatization to high altitude and its significance for the oxygen transport, *Exp. Med. Surg.* 9: 389–403.

Stewart, P.A., Isaacs, H., LaManna, J.C., and Harik, S.I. (In press), Ultrastructural concomitants of hypoxia-induced angiogenesis, *Acta Neuropathol.*:

Wood, S.C. 1991, Interactions between hypoxia and hypothermia, *Ann. Rev. Physiol.* 53: 71–85.

24

INFLUENCE OF ENVIRONMENTAL OXYGEN CONCENTRATION ON ENLARGEMENT AND VASCULAR DENSITY OF THE *AREA VASCULOSA* IN CHICK EMBRYOS

Jens Höper,[1] Herbert Jahn,[1] Resit Demir,[1] and Kirsten Höper[2]

[1]Institut für Physiologie und Kardiologie
[2]Klinik mit Poliklinik für Kinder und Jugendliche
Friedrich-Alexander-Universität
Erlangen-Nürnberg, Germany

INTRODUCTION

Hypoxia is known to induce an upregulation of different genes (Shweiki et al. 1992, Goldberg and Schneider 1994). Among these genes are different ones encoding for growth factors. Thus hypoxia can lead to a consecutive proliferation of e.g. endothelial cells. As shown by Höper and Jahn (1995) in the *Area vasculosa* of the early chick embryo hypoxia not only induces an increase in vascular density but also an enlargement of the vascularized area. Hyperoxia, on the other hand, is followed by a smaller vascularized area but also an increased vascular density. This indicates that vasculogenesis, angiogenesis and growth of the vascularized area are influenced by oxygen in different ways.

In order to find out whether or not oxygen becomes a limiting factor for one of these processes we studied the effect of 100, 21, 10 and 5% oxygen on enlargement of the vascularized area and on vascular density.

METHODS AND MATERIALS

For the investigations fertilized eggs of the crossbreeding White-Plymouth-Rocks X Sussex (Lehr- und Versuchsanstalt für Kleintierzucht, Kitzingen, FRG) were available. The eggs were delivered on the day of collection and after transport they rested in an unheated room at a temperature of 16–18 °C for approx. 24 h. Only eggs with a weight of 52–61 g were used for the experiments. The eggs were put into the incubator in an upright position.

Incubation was performed at a temperature of 36.8 ± 0.1 °C and at a relative humidity of 60–65%. These values were continuously measured by a Technotherm 5400 monitor. Under these conditions the weight loss per day of the single egg was 0.3 ± 0.05 g.

From the first day the eggs were incubated either under conditions of atmospheric oxygen (20.9%), in a mixture of 90% nitrogen and 10% oxygen (mild hypoxia), in a mixture of 95% nitrogen and 5% oxygen (severe hypoxia), or under hyperoxic conditions (100% O_2).

All examinations described in the following were performed on the fourth day of incubation. After opening the egg photographic pictures were taken for later determining the vascular density as well as the diameter of the *Area vasculosa* (Fotomakroskop M 400, Wild, Heerbrugg). The chosen enlargements of the pictures of the yolk sac tissue were evaluated for *area vasculosa* diameter and vascular density as described by Höper and Jahn (1995).

Evaluation of the Data

The mean value of the *Area vasculosa* diameter was taken as 100%, all other values were calculated as percentage of this mean value. The same calculations were made for vascular density.

Total number of blood vessels was calculated from vascularized area and vascular density (VIS/ mm^2).

The measured parameters were grouped for normoxia, mild hypoxia, severe hypoxia and hyperoxia and, as for the diameter of the *Area vasculosa*, were examined as to significant differences by means of Student's t-test.

RESULTS

Figure 1 shows the relative diameter of the *Area vasculosa* after incubation of the eggs in different environmental oxygen concentrations. It is obvious that hyperoxic incubation results in a smaller diameter and thus total vascularized area.

Mild hypoxia (10% oxygen), on the other hand results in an increase in diameter and thus an enlargement of the *Area vasculosa* (Figure 2). Severe hypoxia (5% oxygen) does not cause a further enlargement but is followed by a decrease.

Vascular density has the lowest value at 20% and shows a pronounced increase at 100% oxygen. Mild hypoxia also induces an increase in vascular density but the amount is smaller than in hyperoxia. With severe hypoxia a decrease in vascular density is observed.

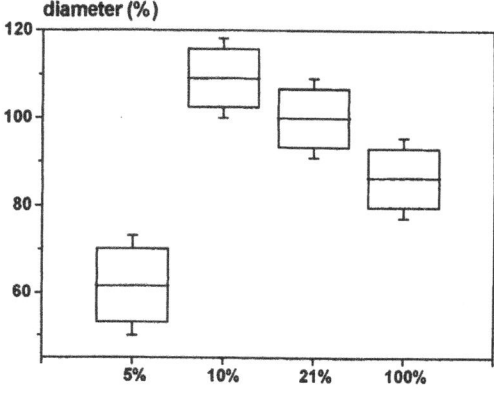

Figure 1. Relative diameter (mean value at 21% oxygen = 100%) after 96 hrs incubation in different oxygen concentrations.

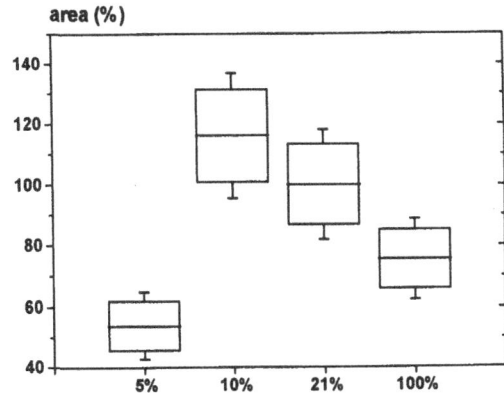

Figure 2. Relative enlargement of the *Area vasculosa* calculated from the diameters shown in figure 1 (mean value at 21% oxygen = 100%) after 96 hrs incubation in different oxygen concentrations.

DISCUSSION

The effect of environmental oxygen on vascularisation and growth of the *Area vasculosa* was studied in early chick embryos.

During mild hypoxia (10% oxygen) an increase in vascular density was observed. Exposure to hyperoxia also caused an increase in vascular density. Severe hypoxia (5% oxygen) leads to a decrease in vascular density. The diameter as well as the enlargement of the *Area vasculosa* also show an oxygen dependency. Compared to normoxia a smaller diameter and thus enlargement was found. Mild hypoxia resulted in an increase in diameter and vascularized area, whereas severe hypoxia resulted in smaller diameter and area. Due to the simultaneous changes in diameter and vascular density compared to normoxia a decrease in the total number of blood vessels was observed in hyperoxia, whereas mild hypoxia caused an increase in the total number of blood vessels (Höper and Jahn 1995). Due to the fact that in severe hypoxia both vascularized area and vascular density were decreased the total number of blood vessels is smaller than in mild hypoxia.

Vasoproliferative effects of hypoxia have already been described for the chorioallantoic membrane (CAM) of chick embryos (Strick et al. 1991). Later, Wilting et al. (1993) were able to show that vascular endothelial growth factor (VEGF) induces angiogenesis in the regions of precapillary vessels, capillaries and venules.

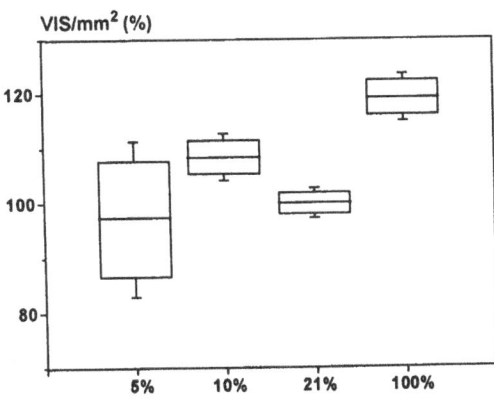

Figure 3. Relative vascular density (VIS/mm^2 = vascular intersections/mm^2, mean value at 21% oxygen = 100%) after 96 hrs incubation in different oxygen concentrations.

It has been shown that during hypoxia an up-regulation of the VEGF gene occurs (Goldberg and Schneider 1994, Shweiki et al. 1992, Ladoux and Frelin 1993). VEGF is not the only factor known to stimulate vasculogenesis and angiogenesis. Stewart et al. (1989) showed that epidermal growth factor (EGF) promotes extraembryonic angiogenesis in the *Area vasculosa* of 3 day old chicks.

bFGF elicits fibrocyte proliferation and minor endothelial cell proliferation (Wilting et al. 1993). The proliferative effect of EGF shown by Stewart et al. (1989) may not be confined to the vascular endothelium, but seems to stimulate the development of ecto-, meso- and endoderm of the *Area vasculosa*.

An increase in basic fibroblastic growth factor (bFGF) and in epidermal growth factor (EGF) primarily will lead to a proliferation of endo-, meso-, and ectoderm and may thus lead to an enlargement of the *Area vasculosa*. This will only occur if a simultaneous proliferation and thus growth of the terminal sinus is induced. Because both bFGF and EGF seem to stimulate endothelial cell proliferation an increase in these mitogens may lead to *Area vasculosa* growth.

Vascular endothelial growth factor (VEGF) only stimulates endothelial proliferation, thus a selective increase in VEGF will lead to an increase in vascular density.

The increase in both, enlargement of the vascularized area and total number of blood vessels observed under conditions of mild hypoxia indicates that under these conditions different growth factors may be up-regulated. During hyperoxia there seems to be a suppression of growth factors, leading to a smaller enlargement of the vascularized area and less blood vessels. The results observed under severe hypoxia indicate that the growth of the vascularized area is more inhibited than vasculogenesis and angiogenesis.

From the data presented two conclusions may be drawn:

1. Proliferation of endothelial- and non-endothelial cells have a different oxygen dependency.
2. Below a critical oxygen supply, both angiogenensis and proliferation of the ecto-, meso- and endoderm become inhibited. Whether or not this is due to less synthesis of the appropriate mitogens or a depressed efficacy cannot yet be answered and will be subject of further experiments.

REFERENCES

Goldberg, MA, Schneider, TJ Similarities between the oxygen sensing mechanisms regulating the expression of vascular endothelial growth factor and erythropoetin. J Biol Chem 269, 4355 - 4359, 1994.

Höper, J, Jahn, H Influence of environmental oxygen concentration on growth and vascular density of the area vasculosa in chick embryos. Int J Microcirc 15, 186 - 192, 1995.

Ladoux, A, Frelin, C Hypoxia is a strong inducer of vascular endothelial growth factor mRNA expression in the heart. Biochem Biophys Res Commun 195, 1005–1010, 1993.

Shweiki, D, Itin, A, Soffer, D, Keshet, E Vascular endothelial growth factor induced by hypoxia may mediate hypoxia-initiated angiogenesis. Nature 39, 843–845, 1992.

Stewart, R, Nelson, J, Wison, DJ Epidermal growth factor promotes chick embryonic angiogenesis. Cell Biol Int Rep 13, 957–965, 1989.

Strick, DM, Waycaster, RL, Montani, J-P, Gay, WJ, Adair, TH Morphometric measurements of chorioallantoic membrane vascularity: effects of hypoxia and hyperoxia. Am. J. Physiol. 260, H1385 - H1389, (Heart Circ. Physiol. 29), 1991.

Wilting, J, Christ, B, Bokeloh, M, Weich HA In vivo effects of vascular endothelial growth factor on the chicken chorioallantoic membrane. Cell Tissue Res 274, 163–172, 1993.

25

NITRIC OXIDE AND ENDOTHELIN IN ACUTE HYPOXIC PULMONARY VASOCONSTRICTION AND ADAPTATION TO CHRONIC HYPOXIA

W. L. Rumsey, B. M. Abbott, D. L. Bertelsen, and M. A. DeSiato

Zeneca Pharmaceuticals
Wilmington, Delaware 19850-5437

1. INTRODUCTION

Hypoxia induced pulmonary vasoconstriction is an important protective mechanism designed to redirect pulmonary blood flow from poorly to well ventilated alveoli in order to maintain adequate ventilation-perfusion throughout the lung and thus preserve oxygenation of arterial blood. The sensing of local oxygen pressure within the lung as well as other organs, and the mechanism(s) by which an alteration in vascular tone is brought about is a complex process that has yet to be fully elucidated. It has become evident, however, that the endothelium lining the vascular wall is a rich source of multiple vascular mediators. These mediators may participate in the oxygen sensing process thereby affecting the balance of vascular tone.

Endothelins have been shown to be some of the most potent mediators of vasomotor tone, in particular, serving to increase resistance to blood flow (Yanagisawa et al., 1988; for review, refer to Masaki, 1993 or Masaki et al., 1992). In the pulmonary circulation, endothelin isopeptides can result in either vasoconstriction or vasodilation (for review refer to Ferro and Webb, 1996). The latter is thought to occur via the endothelin B receptor located on endothelium with subsequent release of prostacyclin and/or endothelium derived relaxant factor (Masaki et al., 1992). Release of nitric oxide, an endothelium derived relaxant factor, potentially can oppose the vasoconstrictive actions of endothelins either under normoxic or hypoxic conditions. Consistent with this notion, pulmonary vasoconstriction stimulated by hypoxia was augmented in the presence of antagonists of endothelium derived relaxant factors (Brashers et al., 1988, Archer et al., 1989).

Chronic exposure to hypoxia converts a "normal" protective process into a pathophysiological state manifested by grossly elevated pulmonary arterial pressure, remodeling of the pulmonary vasculature and right ventricular hypertrophy. Endothelins have been implicated in the pathogenesis of pulmonary hypertension (for review, refer to Barnes, 1994) whereas endothelium dependent relaxation was found to be depressed in rings prepared from pulmonary arteries of patients with chronic obstructive lung disease, a condition that results often in pulmonary hypertension (Dinh-Xuan et al., 1993). Either the increased or decreased synthesis/release of endothelin or nitric oxide may lead, at least in part, to the pathophysiology of pulmonary hypertension. Recently, we (Rumsey et al., 1995b) and others (DiCarlo et al., 1995; Eddahibi et al., 1995) have shown that endothelin antagonists can attenuate pulmonary hypertension in rats induced by hypoxic adaptation. Using animals subjected to acute hypoxia and those adapted to chronic hypoxia, the present report provides preliminary data from studies aimed toward understanding further the role of endothelin and nitric oxide in acute hypoxic pulmonary vasoconstriction and pulmonary hypertension.

2. METHODS

2.1. Animals

Male, pathogen-free, Sprague-Dawley rats were housed either in standard cages in room air (Normoxic) or in Plexiglass boxes containing an atmosphere of 10% oxygen (Hypoxic). In the latter case, no effort was made to control the level of CO_2 (range 0.8–1.8%). The bedding was changed daily (inspired gases were re-equilibrated within 1 hr). Animals were permitted free access to water and standard rat chow.

2.2. Physiological Measurements

After animals were anesthetized (Urethane: 1.5 g /kg IM, followed 45 min later with 0.75 g/kg SC; or Inactin: Na thiobarbital 125 mg/kg IP), catheters were surgically introduced into the left carotid artery for measurement of systemic pressures and for sampling of arterial blood gases. Animals were maintained at 37° C using a thermostatted water blanket. A pressure transducer (2-French, Millar, Inc., Houston, TX) was advanced to the right ventricle via the jugular vein for measurement of intraventricular pressure. The latter served as an alternative measure of pulmonary arterial pressure. The left jugular vein was cannulated for administration of compounds. The animals were tracheotomized but respired freely using room air without mechanical aid. Arterial blood gases were monitored (Radiometer ABL 500, Copenhagen, Denmark) from a small aliquot (0.05 ml) of blood. Data were captured and displayed on a monitor during the experiment using a commercially available data acquisition system (Modular Instruments, Malvern, PA). Upon completion of the experiment, the heart was surgically removed, rinsed in saline, and the ventricles separated. The tissues were weighed and dried overnight at 90° C for determination of dry wt. Wet (or dry) weight of the right ventricle divided by that of the left ventricle plus septum was used as an indicator of hypertrophy.

2.3. General Experimental Protocols

For all experiments, animals were permitted to stabilize following surgical preparation for a period of 15–30 min while respiring on room air. After this period, compounds were administered (refer to figures for details) and physiological parameters were monitored for a period of not less than 60 min.

In some cases, normal rats were subjected to five intermittent bouts of hypoxia (10% O_2: 2–2.5% CO_2 for 3 min). Each hypoxic episode was separated by a 10 min interval of room air. Hypoxic gases were delivered by encapsulating the rat's muzzle within the open barrel portion of a syringe (60 ml) that was connected on its opposite end to a gas intake line (Tygon tubing). Gases from pressurized cylinders containing either pure O_2, CO_2, or N_2 were mixed and adjusted via a flow meter and monitored (Datex Normocap 200, Helsinki, Finland) through a sideport of the syringe used for inhalation. For some experiments, the nitric oxide synthase inhibitor, Nω-nitro-L-arginine methyl ester (L-NAME), was infused constantly beginning 5 min prior to the last bout of hypoxia. The nonselective endothelin (A+B) receptor antagonist, Bosentan, was injected (IV bolus) 3 min prior to administration of L-NAME.

3. RESULTS AND DISCUSSION

The hypoxic environment (10% O_2) used in this study has been shown to increase right intraventricular systolic pressure (RVSP) by about twofold, from 29 ± 2 (age matched normoxic rats; n = 36) to 64 ± 3 (n = 27) mm Hg (Rumsey et al., 1995a). A marked rise of mean arterial pressure was also found, from 79 ± 2 (n = 37) to 99 ± 4 (n = 27) mm Hg. The level of change for either parameter was dependent upon the duration of the period of hypoxic exposure. These measurements were an average of those obtained from animals exposed to hypoxia for 7 to 36 days. Even though a marked rise in RVSP (57 ± 4 mm Hg; n = 8) is evident by 7 days of hypoxia with a concomitant increase in hematocrit (about 29%), these changes were not associated with right ventricular hypertrophy. On the other hand, 14 days of exposure resulted in an elevation, about 69%, of right ventricular mass and a further increase of hematocrit. In normoxic animals, hematocrit levels were typically about 46 ± 0.7% (n = 19) whereas in hypoxic adapted animals these values were about 70 ± 0.8% (n = 29). Consequently, the duration of hypoxic adaptation to produce pulmonary hypertension for the experiments described in the present study was 14 days.

Acute administration of the nonselective endothelin receptor antagonist, Bosentan (10 mg/kg IV bolus), did not alter RVSP in either normal rats or in those adapted to hypoxia for 14 days. Figure 1 shows that RVSP was 32 ± 5 (n = 3) mm Hg and 74 ± 2 (n = 4) mm Hg, respectively, immediately prior to compound delivery and, for the most part, it was maintained at these levels following compound administration for both groups throughout the observation period (150 min). Mean arterial pressure also remained unaltered in normal animals supplied with Bosentan. In hypoxic animals receiving antagonist, mean arterial pressure slowly decreased from 100 ± 6 to 73 ± 10 mm Hg but not in those given its vehicle (117 ± 5 vs 101 ± 9 mm Hg, n = 2). Similar findings have been obtained using ET$_A$ selective antagonists (Rumsey et al., 1995b).

To determine the effect of blockade of nitric oxide synthase on pulmonary pressures, L-NAME was infused (30 mg/kg/min) into the jugular vein of both normal and hypoxic adapted rats anesthetized with Inactin. In normal animals, L-NAME infusion was associated with a rise in RVSP, reaching a maximum level after 10 min of continuous administration, i.e., from 33.7 ± 1.8 to 48.3 ± 3.0 mm Hg. After this peak was reached, pressure subsided. At 30 min of infusion, RVSP was 42 ± 3.8 (n= 6) mm Hg and the infusion was discontinued. Gradually, RVSP returned toward pre-infusion levels (39.8 ± 2.7 mm Hg at 30 min post-infusion). Mean arterial pressure also rose, by about 29%, during L-NAME infusion, although the peak response did not occur for 30 min after infusion had been initiated. The vehicle, saline, infused during the same time frame did not affect either parameter. When Bosentan (10 mg/kg IV) was injected prior to administration of L-NAME, RVSP did not respond to the substrate analog whereas mean arterial pressure increased,

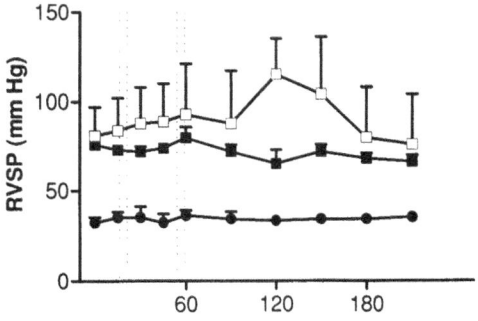

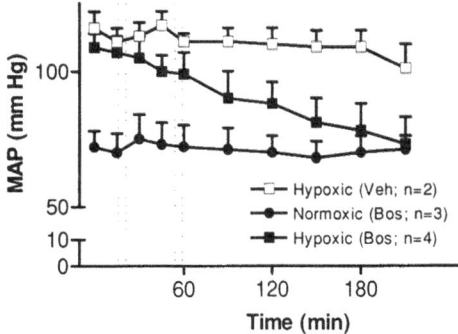

Figure 1. Effect of acute administration of Bosentan on hypoxia induced pulmonary and systemic hypertension. Animals were anesthetized with urethane and prepared as described in METHODS. Bosentan (10 mg/kg) was administered slowly via injection into the jugular vein during the period noted by the second set of dashed lines. The first set of dashed lines denotes injection of its vehicle, 10% Dimethyl Sulfoxide. RVSP = right intra-ventricular systolic pressure and MAP = mean arterial pressure. Values represent means ± SEM.

about 29%, in a manner similar to that obtained in the absence of the endothelin antago-nist. By contrast, in animals adapted to hypoxia for 14 days, administration of L-NAME did not alter RVSP even though mean arterial pressure rose by about 13%, from 117 ± 4 to 132 ± 7 mm Hg.

The data described above prompted a series of experiments designed to explore the potential role of either endothelin or nitric oxide in acute hypoxic pulmonary vasocon-striction. A consecutive series of five intermittent bouts of hypoxia resulted in a progres-sive but moderate rise of RVSP in normal rats anesthetized with urethane. The fourth insult afforded a small elevation of RVSP, about 2 ± 1 (n = 6) mm Hg, above that value obtained immediately prior to this episode of hypoxia (top panel of Figure 2). The fifth bout, however, led to a more robust change, i.e., an increase of 6 ± 2 mm Hg. The fifth hy-poxic episode decreased paO_2 from 96.5 ± 1.5 to 55.3 ± 2.6 (n = 3) mm Hg and $paCO_2$ from 45.5 ± 2.7 to 37.4 ± 1.0 mmHg. Arterial pH rose from 7.403 ± 0.025 to 7.458 ± 0.017. The fall in arterial pCO_2 and the rise in pH suggested a strong ventilatory effort, most likely due to the combined hypoxia and high levels of inhaled CO_2.

When L-NAME was infused before the fifth hypoxic event, RVSP initially did not re-spond. The pre-infusion level of RVSP was 31 ± 2 mm Hg and during infusion it was 32 ± 2 (n=7) mm Hg. Mean arterial pressure, however, rose by 50 ± 5 mm Hg in response to the ana-log. Application of the hypoxic stimulus, however, increased RVSP by 24 ± 3 (n = 7) mm Hg. This change during hypoxia was about fourfold greater than that obtained in the absence of L-NAME. Interestingly, when the fifth hypoxic event was discontinued but infusion of L-NAME was maintained, RVSP returned to its original value, i.e., 30 ± 2 mm Hg.

Administration of Bosentan (10 mg/kg, n = 3) prior to the fifth bout of hypoxia alone (bottom panel of Figure 2) did not alter the response of RVSP to acute oxygen dep-rivation. Hypoxia increased RVSP by about 19 ± 5% above the pre-hypoxic level and, in

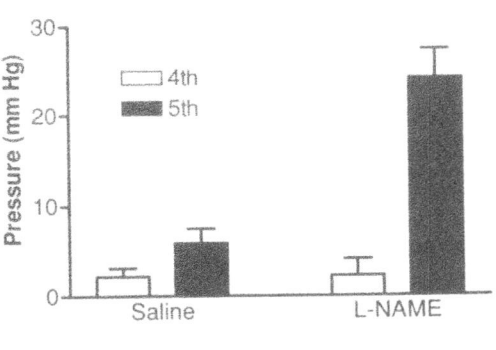

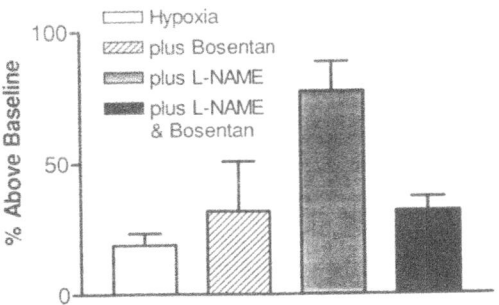

Figure 2. Effect of nitric oxide synthase inhibition and endothelin antagonism on acute hypoxic pulmonary vasoconstriction. Animals were exposed briefly to five intermittent bouts of hypoxia (10% O_2: 2–2.5% CO_2) as described in METHODS. The ordinate of the top panel equals the absolute rise in right intraventricular pressure (RVSP) whereas that in the bottom panel represents the percent change above the value of RVSP obtained immediately prior to the 5th bout of hypoxia. The legend in the top panel refers to 4th and 5th bouts of hypoxia in the same animals receiving either the nitric oxide synthase inhibitor, L-NAME, or its vehicle, saline. The legend in the bottom panel describes changes associated only with the 5th bout of hypoxia in the absence or presence of the nonselective endothelin antagonist, Bosentan, or L-NAME. Values represent means ± SEM (n = 3–7).

the presence of endothelin antagonist, it stimulated a rise of RVSP of 32 ± 19% (n = 3). It can also be seen in Figure 2 that this value was essentially the same as that obtained with hypoxia plus Bosentan during L-NAME infusion, 32 ± 5% (n = 5). This result, however, was markedly less than that produced by hypoxia and L-NAME which provided a 77 ± 11% rise in RVSP.

The findings of the present report, although preliminary, provide some insights about the contribution of endothelins and nitric oxide to pulmonary vascular tone in response to acute or long term exposure to hypoxia. Firstly, using the nonselective endothelin antagonist, Bosentan, acute relief was not obtained from the elevated pulmonary vascular pressures induced either by short or long exposure periods to oxygen deprivation. Secondly, in animals respiring room air, the inhibition of nitric oxide synthesis by L-NAME increased RVSP in normal rats, but not in those adapted to chronic hypoxia. Lastly, administration of L-NAME augments in an additive manner the response of RVSP to acute hypoxia and this augmentation is prevented in the presence of Bosentan.

Results from a number of investigations indicate that secretion of endothelin from tissue increases in response to hypoxia. For example, using primary cultures of human umbilical vein endothelial cells, Kourembanas and colleagues (1991) showed that low oxygen pressure, i.e., 30 Torr, resulted in an increase of endothelin-1 secretion by four to eightfold as compared to cells maintained at ambient oxygen pressures. Secretion was also accompanied by a rise in the transcriptional rate of the preproendothelin gene and in steady state levels of its mRNA. Acute hypoxia has also been reported to stimulate the release of endothelin-1 from resistance vessels perfused in vitro (Rakugi et al., 1990). This enhanced release of peptide (71% in 30 min) was associated with an increase in perfusion pressure. Finally, plasma levels of endothelin-1 were found to be progressively elevated in rats exposed to 24 and 48 hrs of normobaric hypoxia (10% O_2) and endothelin-1 mRNA was increased by twofold in the lungs of these animals as compared to those maintained in room air (Elton et al., 1992).

Given the findings of previous investigators, it was unexpected that Bosentan administered acutely was without effect on RVSP in rats either adapted to hypoxia for 14 days or to those exposed to acute hypoxia. In contrast to our findings, Chen and coworkers (1995) have shown that, in conscious unrestrained rats, Bosentan (100 mg/kg/day PO given once per day for two consecutive days) blocked completely the rise of mean pulmonary artery pressure stimulated by short term hypoxia. It should be pointed out that the animals were challenged on a day following compound administration. These authors also reported that this dose of Bosentan when given two days prior to and concurrent with hypoxic adaptation prevented the development of pulmonary hypertension brought about by two weeks of exposure to normobaric hypoxic environment. Endothelins have a slow rate of dissociation of the receptor-ligand complex and, in rat tissues, binding is essentially irreversible (for review, refer to Sokolovsky, 1995). The half-life of an exogenous bolus of endothelin is about 7 min but the pressor effect is much longer lived (Masaki et al., 1992). Consequently, it may have been difficult to obtain a reduction of ventricular pressure by an endothelin antagonist within the time frame of the present experiments. The latter explanation does not appear to be entirely valid since both Bosentan and BMS182874 (Rumsey et al., 1995b), which is selective for the endothelin A receptor, decreased the enhanced mean arterial pressure of hypoxic adapted rats. It might also be argued that the dose of Bosentan (10 mg/kg IV) used in the present study may not have been adequate to effectively diminish or block the enhanced right ventricular pressures obtained in these animals. On the other hand, Clozel and coworkers (1994) have reported recently that, in pithed rats, this dose of Bosentan significantly reduced, by about 40%, the rise of mean arterial blood pressure stimulated by exogenous administration of endothelin-1 (1 nmol/kg). Given the avid binding and long lived functional response reported for endothelins it seems unlikely that endothelins were primarily responsible for the rise of RVSP during acute hypoxia. Consistent with this notion, the discontinuation of acute hypoxia returned RVSP to levels obtained prior to the hypoxic challenge. Although it cannot be ruled out that endothelins did not contribute to either acute hypoxic pulmonary vasoconstriction or to the maintenance of elevated pulmonary pressures of the hypoxic adapted animals, our data would suggest that factors other than endothelins may also be involved in these processes.

It has been suggested that relaxation due to endothelium derived factors is impaired in rats exposed to chronic hypoxia (Adnot et al., 1991). In the latter study, it was shown using the perfused rat lung that one week of hypoxic exposure reduced responses to acetylcholine or A23187 and by three weeks of hypoxic adaptation these responses were completely abolished. Hypoxia has been shown to inhibit the uptake of L-arginine in pulmonary artery endothelial cells (Block et al., 1995) although L-arginine deficiency is not thought to serve as the mechanism for loss of nitric oxide activity (Dinh-Xuan et al., 1993). Our results obtained in vivo and using L-NAME in hypoxic adapted rats extend these earlier observations and suggest that the nitric oxide system is in some manner dysfunctional in hypoxia induced pulmonary hypertension. The nature of this dysfunction remains to be determined.

Inhibition of nitric oxide synthesis by L-NAME markedly increased the level of RVSP in normal rats and during acute hypoxia suggesting that nitric oxide release to the vasculature may be serving to regulate vascular tone under basal conditions and it may counteract the hypoxic vasoconstrictive stimulus. In rat lung perfused with cell-free media, nitric oxide exhalation was reported to be suppressed during acute hypoxia although its release to the vascular lumen was unchanged (Grimminger et al,. 1995). Brashers and coworkers (1988) using a similar experimental preparation reported that hypoxic pulmonary vasoconstriction is augmented in the presence of antagonists of endothelium derived relaxant factors such as eicosatetraynoic acid or norhydroguaiaretic acid. Moreover, N^G-monomethyl-L-arginine enhanced the hypoxic pressor response by about twofold (Archer et al., 1989). It is interesting that chronic inhibition of nitric oxide synthesis does not result in pulmonary hypertension although systemic hypertension is brought about by this treatment (Hampl et al., 1993). These findings suggest that impairment of nitric oxide dependent relaxation is not necessarily a causative factor in pulmonary hypertension rather it is an undesirable outcome.

The present study also showed that the L-NAME enhancement of RVSP in normal rats and during acute hypoxia could be markedly ameliorated by pretreatment with Bosentan. In the latter case, only that portion of the rise in RVSP attributable to L-NAME was abolished by the endothelin antagonist. These data might suggest that nitric oxide is continuously available and that it represses the vasoconstrictive actions of other mediators. One of these mediators may be endothelin. Clearly, further work is warranted to more effectively define the precise function and activity of these substances.

Finally, the regulation of pulmonary vascular tone during normal conditions and pathophysiological ones is supported by a number of potential mediators such as endothelins and nitric oxide. The availability of new antagonists of these mediators may allow further delineation of this complex process. In particular, the use of inhibitors such as L-NAME afford an opportunity to evaluate this process in a model of greater dynamic range and sensitivity to action by other compounds.

REFERENCES

Adnot, S., B. Raffestin, S. Eddahibi, P. Braquet, and P.-E. Chabrier. Loss of endothelium-derived relaxant activity in the pulmonary circulation of rats exposed to chronic hypoxia. *J. Clin. Invest.* 87: 155–162, 1991.

Archer, S.L., J.P. Tolins, L. Raij, and E.K. Weir. Hypoxic pulmonary vasoconstriction is enhanced by inhibition of the synthesis of an endothelium derived relaxing factor. *Biochem. Biophys. Res. Comm.* 164 (3): 1198–1205, 1989.

Barnes, P.J. Endothelins and pulmonary diseases. *J. Appl. Physiol.* 77(3): 1051–1059, 1994.

Block, E.R., H. Herrera, and M. Couch. Hypoxia inhibits L-arginine uptake by pulmonary artery endothelial cells. *Am. J. Physiol.* 269) *Lung Cell. Mol. Physiol.* 13): L574-L580, 1995.

Brashers, V.L., M.J. Peach, and C.E. Rose, Jr. Augmentation of hypoxic pulmonary vasoconstriction in the isolated perfused rat lung by in vitro antagonists of endothelium-dependent relaxation. *J. Clin. Invest.* 82: 1495–1502, 1988.

Chen, S.-J., Y.-F. Chen, Q.C. Meng, J. Durand, V.S. DiCarlo and S. Oparil. Endothelin-receptor antagonist bosentan prevents and reverses hypoxic pulmonary hypertension in rats. *J. Appl. Physiol.* 79(6): 2122–2131, 1995.

Clozel, M., V. Breu, G.A. Gray, B. Kalina, B.-M. Loffler, K. Burri, J.-M. Cassal, G. Hirth, M. Muller, W. Neidhart and H. Ramuz. Pharmacological characterization of Bosentan, a new potent orally active nonpeptide endothelin receptor antagonist. *J. Pharmaol. Exp. Ther.* 270 (1): 228–235, 1994.

Dinh-Xuan, A.T., J. Pepke-Zaba, A.Y. Butt, G. Cremona and T.W. Higenbottam. Impairment of pulmonary-artery endothelium-dependent relaxation in chronic obstructive lung disease is not due to dysfunction of endothelial cell membrane receptors nor to L-arginine deficiency. *Br. J. Pharmacol.* 109: 587–591, 1993.

DiCarlo, V.S., S.-J. Chen, Q.C. Meng, J. Durand, M. Yano, Y.-F. Chen, and S. Oparil. ETA-receptor antagonist prevents and reverses chronic hypoxia-induced pulmonary hypertension in rat. *Am. J. Physiol.* 269 (*Lung Cell. Mol. Physiol.* 13): L690-L697, 1995.

Eddahibi, S., B. Raffestin, M. Clozel, M. Levame, and S. Adnot. Protection from pulmonary hypertension with an orally active endothelin receptor antagonist in hypoxic rats. *Am. J. Physiol.* 268 (*Heart Circ. Physiol.* 37): H828-H835, 1995.

Elton, T.S., S. Oparil, G.R. Taylor, P.H. Hicks, R.-H. yang, H. Jin, and Y.F. Chen. Normobaric hypoxia stimulates endothelin-1 gene expression in the rat. *Am. J. Physiol.* 263 (*Regulatory Integrative Comp. Physiol.* 32): R1260-R1264, 1992.

Ferro, C.J. and D.J. Webb. The clinical potential of endothelin antagonists in cardiovascular medicine. *Drugs* 51(1): 12–27, 1996.

Grimminger, F., R. Spriestersbach, N. Weissman, D. Walmrath, and W. Seeger. Nitric oxide generation and hypoxic vasoconstriction in buffer perfused rabbit lungs. *J. Appl. Physiol.* 78(4): 1509–1515, 1995.

Hampl, V., S.L. Archer, D.P. Nelson, and E.K. Weir. Chronic EDRF inhibition and hypoxia: effects on pulmonary circulation and systemic pressure. *J. Appl. Physiol.* 75 (4): 1748–1757, 1993.

Kourembanas, S., P.A. Marsden, L.P. McQuillan, and D.V. Faller. Hypoxia induces endothelial gene expression and secretion in cultured human endothelium. *J.Clin. Invest.* 88: 1054–1057, 1991.

Masaki, T. Endothelins: Homeostatic and compensatory actions in the circulatory and endocrine systems. *Endocrine Rev.* 14 (3): 256–268, 1993.

Masaki, T., M. Yanagisawa, and K. Goto. Physiology and pharmacology of endothelins. *Med. Res. Rev.* 12(4): 391–421, 1992.

Mazmanian, G., B. Baudet, C. Brink, J. Cerrina, S. Kirkiacharian, and M. Weiss. *J. Appl. Physiol.* 66 (3): 1040–1045, 1989.

Rakugi, H., Y. Tabuchi, M. Nakamura, M. Nagano, K. Higashimori, H. Mikami, T. Ogihara, and N. Suzuki. Evidence for endothelin-1 release from resistance vessels of rats in response to hypoxia. *Biochem. Biophys. Res. Comm.* 169 (3): 973–977, 1990.

Rumsey, W., B. Abbott, D. Bertelsen, D. Nelson, and M. Erecinska. Hypoxia induced pulmonary hypertension alters myocardial energy metabolism. *J. Mol. Cell. Cardiol.* 27 (5): A33, 1995a.

Rumsey, W.L., D.L. Bertelsen, B. M. Abbott, M. A. Mallamaci, R. A. Bialecki. Pulmonary Hypertension is Decreased by an Orally Active Endothelin-A Antagonist. *Circulation* 92 (8): Suppl I-430, 1995b.

Sokolovsky, M. Endothelin receptor subtypes and their role in transmembrane signaling mechanisms. *Pharmac. Ther.* 68 (3): 435–471, 1995.

Yanagisawa, M., H. Kurihara, S. Kimura, Y. Tomobe, M. Kobayashi, Y. Mitsui, Y. Yazaki, K. Goto, and T. Masaki. A novel potent vasoconstrictor peptide produced by vascular endothelial cells. *Nature* 332: 411–415, 1988.

26

HAEMOGLOBIN OXYGENATION CHANGES DURING VISUAL STIMULATION IN THE OCCIPITAL CORTEX

J. Ruben,[1] R. Wenzel,[1] H. Obrig,[1] K. Villringer,[2] J. Bernarding,[3] C. Hirth,[1] H. Heekeren,[1] U. Dirnagl[1] and A. Villringer[1]

[1]Department of Neurology
 Charité, Humboldt-University, Berlin, Germany
[2]Department of Radiology
[3]Department of Medical Computer Science
 Free University, Berlin, Germany

1.INTRODUCTION

It has been shown that near-infrared spectroscopy (NIRS) permits the assessment of functional brain activation.[8,9,14,21] Four studies were dedicated to the visual system which partly yielded conflicting results.[10,11,13,21] Though all studies showed an increase in [oxy-Hb] during visual stimulation using different stimulation paradigms, the changes observed in [deoxy-Hb] were not uniform. Meek et al.[13] reported on a decrease in [deoxy-Hb] in 4 of 10 subjects but an increase in the other 6 subjects, while Kato et al.[11] and Hoshi et al.[10] using photic stimulation of 8 Hz and 10 Hz found an increase in [deoxy-Hb] in their studies. On the other hand, Villringer et al.[21] observed a decrease in [deoxy-Hb] during photic stimulation and picture observation, but this included only 3 subjects lacking therefore statistical evaluation. All NIRS-investigators[10,11,13,21] localised the optical probes only with respect to external bony landmarks without knowledge of the exact spatial relation of these landmarks to the anatomy of the brain. This could have led to inaccurate positions of the optical probes over the visual cortex as there is a great variability of the calcarine fissure concerning these external landmarks.[18] Localising the optical probes individually according to previously acquired 3D MRI, the aim of the current study was to investigate whether there is a consistent, statistically evident pattern of changes in [oxy-Hb] and [deoxy-Hb] during visual stimulation. Furthermore, we investigated if NIRS is able to detect oxygenation changes due to activation of a secondary or prestriate visual area known from studies both in monkeys and humans as area V5 or MT to be sensitive to the submodality "visual motion".[12,19,22–24]

2. METHODS

We used a NIRO-500 (Hamamatsu Photonics, Japan) to continuously measure changes in [oxy-Hb] and [deoxy-Hb]. The underlying theory of NIRS has been explained elsewhere in detail.[2-4] Seventeen (7 male, 10 female; mean age 26.9 ± 6.2 years) healthy volunteers underwent a visual stimulation consisting of a coloured, randomly moving dodecahedron (speed 6 cm per s, 5 cm in diameter) which was displayed on a computer screen at a distance of 2 m for 30 s alternating with a period of rest (blank dark screen) lasting 30 s as well. In each subject, 10 to 12 cycles were carried out. The optical probes were horizontally placed over the right occipital cortex 1 cm next to the midline on a level with the calcarine fissure according to the individual MRI of the subject **(Position 1)**. The interoptode spacing inter-individually varied from 3.5 to 4 cm. Sampling time was one second.

As a control condition, thirteen (4 male, 9 female; mean age 27.3 ± 7.3 years) of the seventeen subjects also performed a sequential finger opposition task with their dominant right hands. The protocol and the positions of the optical probes were the same as mentioned above.

As changes in the concentrations of the chromophores to functional activation are known to occur with a latency of a few seconds, we determined the state of activation as the mean of a 30 second period beginning 5 s after onset of the stimulus and lasting 5 s beyond the offset of the stimulus. The state of rest was determined as the mean of the remaining 30 s of the 60 s cycle of rest and activation. The changes in [oxy-Hb] and [deoxy-Hb] were determined as the difference between the means of both 30-s-periods characterising the state of stimulation and the state of rest in the average across all subjects and in the single subjects during the visual stimulation. T-statistics for paired samples were used to reveal significant differences between the states of stimulation and rest for each stimulus and between both stimuli. All t-tests were two-tailed except for the one-tailed tests for single subject analysis during visual stimulation at position 1.

In order to include motion-sensitive visual area V5, in a second experiment (n=5; four female, one male; mean age 25.8 ± 3.4 years) the optical probes were placed in a different position **(Position 2)**. This more lateral position was chosen to overlie the "ascending limb of the inferior temporal sulcus" which according to Watson et al.[22] is close to V5. In addition to the moving, coloured dodecahedron a second stationary dodecahedron differing from the first one just in the submodality object motion was displayed. The stimulation paradigms were presented 11 times to each subject in an interleaved fashion for 30 s each alternating with periods of rest of 30 s.

3. RESULTS

3.1. Changes in [Oxy-Hb] and [Deoxy-Hb] at Position 1 during Visual and Motor Stimulation

Figure 1 shows the changes in [oxy-Hb] and [deoxy-Hb] in the average across all subjects for visual stimulation (n=17) and motor control condition (n=13).

During visual stimulation, an increase in [oxy-Hb] can be seen starting with a latency of a few seconds after onset of the visual stimulus. The maximum of the increase in [oxy-Hb] was reached about 14 s after onset of the stimulus. After reaching this maximum, a slight decrease in [oxy-Hb] occurred. Baseline levels were attained about 11 s after the offset of the stimulus. The increase in [oxy-Hb] was accompanied by a decrease in [deoxy-Hb]. As for the

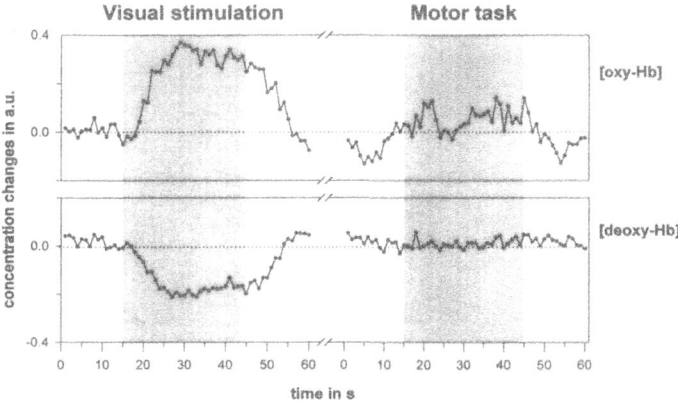

Figure 1. Changes in [oxy-Hb] and [deoxy-Hb] over position 1 in the average across all subjects during visual stimulation (n=17) and during the control condition consisting of the performance of a sequential finger opposition task (n=13), stimulation period grey shaded. The typical response pattern during visual stimulation consisted of an increase in [oxy-Hb] and a decrease in [deoxy-Hb]. T-statistics revealed statistically significant changes between the states of activation and the states of rest for changes in [oxy-Hb] and [deoxy-Hb] to visual stimulation and for the change in [oxy-Hb] during the motor task. The differences between these two stimulation conditions proved to be statistically significant with p=0.019 for [oxy-Hb] and with p<0.001 for [deoxy-Hb].

time course of the changes in [deoxy-Hb], it strongly resembled the one seen for [oxy-Hb]. Comparing the means of the time frames characterising the states of activation and rest, the changes (mean ± standard deviation) were 0.27 ± 0.22 arbitrary units (a.u.) in [oxy-Hb] and -0.18 ± 0.11 a.u. in [deoxy-Hb]. Arbitrary units correspond to μmol/l when assuming a differential pathlength factor (DPF) of 6.26[4]. These changes were statistically significant with p<0.001 each for the rise in [oxy-Hb] and the decrease in [deoxy-Hb].

During the motor control condition, a small increase in [oxy-Hb] of 0.09 ± 0.12 a.u. occurred while [deoxy-Hb] remained constant with 0.00 ± 0.04 a.u.. The rise in [oxy-Hb] nevertheless is not as continuous as during visual stimulation. Though small in comparison to the changes in [oxy-Hb] during visual stimulation, the rise in the [oxy-Hb] during the motor task proved to be statistically significant with p<0.05.

Nevertheless, the comparison between the changes occurring during visual stimulation and motor control task in the thirteen subjects undergoing both stimulation paradigms revealed a statistically significant (p=0.019) stronger increase in [oxy-Hb] of 0.20 ± 0.26 a.u. and a statistically significant (p<0.001) stronger decrease in [deoxy-Hb] of -0.18 ± 0.09 a.u. in favour of the visual stimulation.

As single-subject analysis revealed, this typical response pattern of an increase in [oxy-Hb] and a decrease in [deoxy-Hb] during visual stimulation was seen in 15, regarding the increase in [oxy-Hb], and in all the 17 subjects examined regarding the decrease in [deoxy-Hb]. These changes were statistically significant (p<0.05) in 11 and 14 cases, respectively, according to t-statistics.

3.2. Changes in [Oxy-Hb] and [Deoxy-Hb] at Position 2

The measurements over the lateral occipital cortex revealed a statistically significant difference between the visual stimulation with the moving-coloured and the stationary-

Table 1. Changes in [oxy-Hb] and [deoxy-Hb] in 17 single subjects (10 to 12 cycles of 30 s of stimulation and 30 s of rest) during visual stimulation. Significance levels (one-tailed t-tests) in brackets; * = statistically significant results. As the haemodynamic response is known to occur with a latency of a few seconds, we determined these concentration changes as difference between the mean of a 30 second period beginning 5 s after onset of the stimulus and lasting 5 s beyond the offset of the stimulus and the mean of the remaining 60 s

Subject	Δ[oxy-Hb] ±SD *(Sig.)*	Δ[deoxy-Hb] ±SD *(Sig.)*
1	0.14 ±0.38 *(p=0.082)*	-0.08 ±0.08 *(p=0.001)* *
2	0.38 ±0.10 *(p=0.000)* *	-0.16 ±0.04 *(p=0.000)* *
3	0.35 ±0.57 *(p=0.055)*	-0.08 ±0.18 *(p=0.088)*
4	0.79 ±0.09 *(p=0.000)* *	-0.30 ±0.05 *(p=0.000)* *
5	0.22 ±0.18 *(p=0.000)* *	-0.24 ±0.11 *(p=0.000)* *
6	0.33 ±0.12 *(p=0.000)* *	-0.20 ±0.07 *(p=0.000)* *
7	0.14 ±0.20 *(p=0.020)* *	-0.08 ±0.05 *(p=0.000)* *
8	0.42 ±0.12 *(p=0.000)* *	-0.24 ±0.03 *(p=0.000)* *
9	0.11 ±0.32 *(p=0.147)*	-0.17 ±0.08 *(p=0.000)* *
10	0.35 ±0.28 *(p=0.000)* *	-0.18 ±0.20 *(p=0.005)* *
11	-0.16 ±0.33 *(p=0.083)*	-0.01 ±0.05 *(p=0.364)*
12	-0.02 ±0.26 *(p=0.412)*	-0.05 ±0.12 *(p=0.108)*
13	0.13 ±0.18 *(p=0.025)* *	-0.09 ±0.07 *(p=0.001)* *
14	0.44 ±0.13 *(p=0.000)* *	-0.23 ±0.07 *(p=0.000)* *
15	0.08 ±0.28 *(p=0.173)*	-0.20 ±0.09 *(p=0.000)* *
16	0.32 ±0.12 *(p=0.000)* *	-0.26 ±0.05 *(p=0.000)* *
17	0.54 ±0.30 *(p=0.000)* *	-0.46 ±0.08 *(p=0.000)* *

coloured dodecahedron in the average across all subjects as concerns the changes in [deoxy-Hb]. The stimulation with the moving-coloured stimulus led to a statistically significant (p=0.006) decrease in [deoxy-Hb] of -0.20 ± 0.08 a.u., whereas the stimulation with the stationary stimulus led to a smaller, not statistically significant (p=0.056) decrease in [deoxy-Hb] of -0.09 ± 0.08 a.u. The resulting difference between both paradigms of -0.10 ± 0.03 a.u. in favour of the moving stimulus proved to be statistically significant (p=0.001).

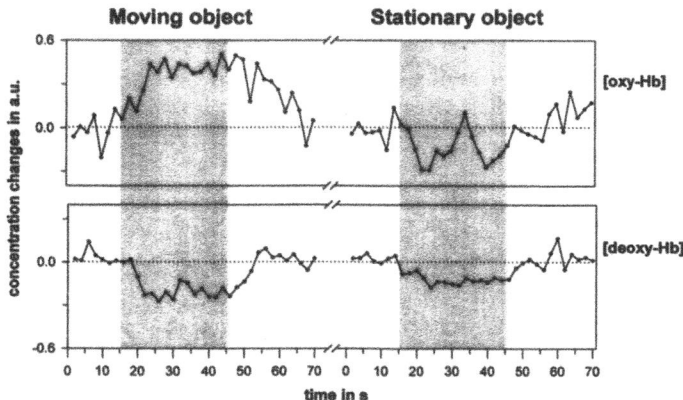

Figure 2. Changes in [oxy-Hb] and [deoxy-Hb] over position 2 in the average (n=11 for each stimulus) across a single subject during observation of the moving and the stationary stimulus are shown. In this subject a statistically significant difference was seen not only for the more pronounced drop in [deoxy-Hb] but also for the changes in [oxy-Hb] during observation of the moving stimulus as compared to the observation of the stationary stimulus (p<0.001 each).

As for [oxy-Hb], in the average across all subjects only small, not statistically significant increases of 0.09 ±0.17 a.u. in this parameter developed during stimulation with the moving dodecahedron and of 0.05 ± 0.20 a.u. during stimulation with the stationary stimulus. The difference between both paradigms of 0.04 ± 0.23 a.u. was not statistically significant.

Figure 2 shows the concentration changes of [oxy-Hb] and [deoxy-Hb] over time in a single subject (female, 24 years). In this subject, an increase in [oxy-Hb] of 0.29 ± 0.14 a.u. and a decrease in [deoxy-Hb] of -0.23 ± 0.04 a.u. developed during the observation of the moving stimulus (p<0.001 each) whereas during observation of the stationary stimulus statistically significant decreases in [oxy-Hb] of -0.13 ± 0.17 a.u. (p=0.032) and [deoxy-Hb] of -0.13 ± 0.04 a.u. (p<0.001) occurred. The differences between the moving and the stationary stimulus were therefore 0.42 ± 0.23 a.u. for [oxy-Hb] and -0.10 ± 0.06 a.u. for [deoxy-Hb]. These differences proved to be statistically significant with p<0.001 each.

DISCUSSION

In this study we have shown a consistent, statistically evident response pattern of NIRS-parameters to visual stimulation when measuring with a probe position close to the calcarine fissure according to previously acquired MRI. This typical response pattern consisted of an increase in [oxy-Hb] and a decrease in [deoxy-Hb]. This response pattern is likely to reflect oxygenation changes to activation of parts of visual areas V1, V2 and V3 and is in agreement with PET-findings indicating a mismatch between regional cerebral blood flow (rCBF) and oxygen consumption[5,6] and fMRI-studies indicating a localised drop in [deoxy-Hb].[1,7,15,16,19,20] Differences with previous NIRS-studies regarding the changes in [deoxy-Hb] may be attributed to differences in the positioning of the optical probes with respect to the visual cortex. Nevertheless, there was also a small increase in [oxy-Hb] during the motor task as control condition. This probably reflects a non-specific global increase in cerebral blood flow. Sitzer et al.[17] reported during a finger opposition task on an increase in the mean cerebral blood flow velocity (CBFV) not only in the middle cerebral artery but also an initial increase in the CBFV in the posterior cerebral artery.

We furthermore demonstrated the possibility of detecting motion-dependent oxygenation changes with NIRS in prestriate, motion sensitive area V5 near the occipito-temporal-parietal junction. Since we found a clearly more pronounced, significant decrease in [deoxy-Hb] during observation of the moving stimulus as compared to the stationary object we conclude that this difference reflects the activation of motion-sensitive visual area V5. These results are in good agreement with anatomical findings using fMRI[12,19] and PET.[22,24]

REFERENCES

1. J.W. Belliveau, K.K. Kwong, D.N. Kennedy, J.R. Baker, C.E. Stern, R. Benson, D.A. Chesler, R.M. Weisskoff, M.S. Cohen, R.B. Tootell, et al, Magnetic resonance imaging mapping of brain function. Human visual cortex, *Invest. Radiol.* 27 Suppl 2:S59–65 (1992).
2. B. Chance, Optical method, *Annu. Rev. Biophys. Biophys. Chem.* 20:1–28 (1991).
3. M. Cope and D.T. Delpy, System for long-term measurement of cerebral blood and tissue oxygenation on newborn infants by near infra-red transillumination, *Med. Biol. Eng. Comput.* 26:289–294 (1988).
4. A. Duncan, J.H. Meek, M. Clemence, C.E. Elwell, L. Tyszczuk, M. Cope, and D.T. Delpy, Optical pathlength measurements on adult head, calf and forearm and the head of the newborn infant using phase resolved optical spectroscopy, *Phys. Med. Biol.* 40:295–304 (1995).
5. P.T. Fox and M.E. Raichle, Focal physiological uncoupling of cerebral blood flow and oxidative metabolism during somatosensory stimulation in human subjects, *Proc. Natl. Acad. Sci. U. S. A.* 83:1140–1144 (1986).
6. P.T. Fox, M.E. Raichle, M.A. Mintun, and C. Dence, Nonoxidative glucose consumption during focal physiologic neural activity, *Science* 241:462–464 (1988).
7. J. Frahm, K.D. Merboldt, and W. Hanicke, Functional MRI of human brain activation at high spatial resolution, *Magn. Reson. Med.* 29:139–144 (1993).
8. C. Hirth, H. Obrig, K. Villringer, A. Thiel, J. Bernarding, W. Mühlnickel, H. Flor, U. Dirnagl, and A. Villringer, Non-invasive functional mapping of the human motor cortex using near-infrared spectroscopy, *NeuroReport* (1996), in press.
9. Y. Hoshi and M. Tamura, Detection of dynamic changes in cerebral oxygenation coupled to neuronal function during mental work in man, *Neurosci. Lett.* 150:5–8 (1993).
10. Y. Hoshi and M. Tamura, Dynamic multichannel near-infrared optical imaging of human brain activity, *J. Appl. Physiol.* 75:1842–1846 (1993).
11. T. Kato, A. Kamei, S. Takashima, and T. Ozaki, Human visual cortical function during photic stimulation monitoring by means of near-infrared spectroscopy, *J. Cereb. Blood Flow Metab.* 13:516–520 (1993).
12. G. McCarthy, M. Spicer, A. Adrignolo, M. Luby, J. Gore, and T. Allison, Brain activation associated with visual motion studied by functional magnetic resonance imaging in humans, *Human Brain Mapping* 2:234–243 (1995).
13. J.H. Meek, C.E. Elwell, M.J. Khan, J. Romaya, J.S. Wyatt, D.T. Delpy, and S. Zeki, Regional changes in cerebral haemodynamics as a result of a visual stimulus measured by near infrared spectroscopy, *Proc. R. Soc. Lond. B* 261:351–356 (1995).
14. H. Obrig, C. Hirth, J. Junge-Hülsing, C. Döge, T. Wolf, U. Dirnagl, and A. Villringer, Cerebral oxygenation changes in response to motor stimulation, *J Appl Physiol* (1996), in press.
15. S. Ogawa, D.W. Tank, R. Menon, J.M. Ellermann, S.G. Kim, H. Merkle, and K. Ugurbil, Intrinsic signal changes accompanying sensory stimulation: functional brain mapping with magnetic resonance imaging, *Proc. Natl. Acad. Sci. U. S. A.* 89:5951–5955 (1992).
16. M.I. Sereno, A.M. Dale, J.B. Reppas, K.K. Kwong, J.W. Belliveau, T.J. Brady, B.R. Rosen, and R.B. Tootell, Borders of multiple visual areas in humans revealed by functional magnetic resonance imaging [see comments], *Science* 268:889–893 (1995).
17. M. Sitzer, U. Knorr, and R.J. Seitz, Cerebral hemodynamics during sensorimotor activation in humans, *J. Appl. Physiol.* 77:2804–2811 (1994).
18. H. Steinmetz, G. Furst, and B.U. Meyer, Craniocerebral topography within the international 10–20 system, *Electroencephalogr. Clin. Neurophysiol.* 72:499–506 (1989).

19. R.B. Tootell, J.B. Reppas, K.K. Kwong, R. Malach, R.T. Born, T.J. Brady, B.R. Rosen, and J.W. Belliveau, Functional analysis of human MT and related visual cortical areas using magnetic resonance imaging, *J. Neurosci.* 15:3215–3230 (1995).

20. R. Turner, P. Jezzard, H. Wen, K.K. Kwong, D. Le Bihan, T. Zeffiro, and R.S. Balaban, Functional mapping of the human visual cortex at 4 and 1.5 tesla using deoxygenation contrast EPI, *Magn. Reson. Med.* 29:277–279 (1993).

21. A. Villringer, J. Planck, C. Hock, L. Schleinkofer, and U. Dirnagl, Near infrared spectroscopy (NIRS): a new tool to study hemodynamic changes during activation of brain function in human adults, *Neurosci. Lett.* 154:101–104 (1993).

22. J.D. Watson, R. Myers, R.S. Frackowiak, J.V. Hajnal, R.P. Woods, J.C. Mazziotta, S. Shipp, and S. Zeki, Area V5 of the human brain: evidence from a combined study using positron emission tomography and magnetic resonance imaging, *Cereb. Cortex.* 3:79–94 (1993).

23. S. Zeki, Functional specialisation in the visual cortex of the rhesus monkey, *Nature* 274:423–428 (1978).

24. S. Zeki, J.D. Watson, C.J. Lueck, K.J. Friston, C. Kennard, and R.S. Frackowiak, A direct demonstration of functional specialization in human visual cortex, *J. Neurosci.* 11:641–649 (1991).

27

MITOCHONDRIAL HYPEROXIDATION AFTER CEREBRAL ANOXIA/ISCHEMIA

Epiphenomenon or Precursor to Residual Damage?

Myron Rosenthal, Patricia L. Mumford, Thomas J. Sick, and
Miguel A. Pérez-Pinzón

Department of Neurology (D4-5)
University of Miami School of Medicine
Miami, Florida 33101

1. INTRODUCTION

Ischemia and reperfusion afterward provoke a characteristic series of events which is common among tissues. Many of the changes produced by ischemia are easily measurable with non-invasive optical or electrode probes in intact or isolated tissues. Among these changes are shifts toward reduction of the electron carriers of the mitochodrial respiratory chain, depletion of tissue oxygen, suppression of the tissue's functional activities and loss of ion homeostasis. Figure 1 is a composite drawing which illustrates some of these typical events in the cerebral cortex of an anesthetized rat.

An advantage of 'on-line' measurements during ischemia is that the intensity of the insult can be defined. For example, we considered ischemia to be 'complete' in rat brain only when: a) declines in tissue oxygen tension, and increases in reduction of NAD or cytochrome a,a_3 were maximal and sustained; b) electrical activity was suppressed; and c) there was loss of ion homeostasis (i.e. anoxic depolarization).

Although much is yet to be learned about events during cerebral ischemia, the events that occur during reperfusion afterward are even less well understood. This latter information is essential since reperfusion and the resultant reoxygenation are prerequisite for recovery of the mitochondrial reduction/oxidation (redox) state and of oxidative energy production. In turn, the latter is essential for recovery of ion transport and electrical activities which provide the first opportunities to reduce or prevent permanent damage and the resulting neurological impairment.

Evidence is increasing, however, that cerebral injury may occur not only during brain anoxia/ischemia but also during (and likely as a consequence of) reoxygenation/reperfusion afterward. Hypotheses have centered upon reperfusion-induced changes which may promote ischemic-induced injury by enhancing lipid peroxidation and oxygen free radical formation.

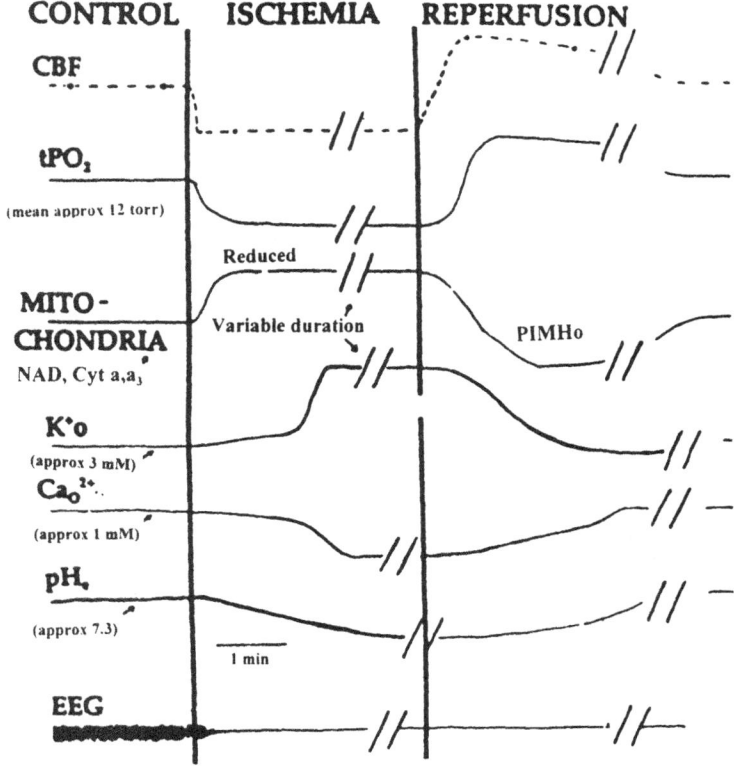

Figure 1. Events during ischemia and reperfusion derived from experiments in rat cerebral cortex which were carried out over many years. Top trace: cerebral blood flow monitored at intervals by microelectrode recording of hydrogen clearance. Second trace: an illustration of typical changes in tissue oxygen tension monitored by micropolarography. Third trace: pictoral representation of the usual changes in the optical signals which indicate redox shifts of NAD (by fluorescence) or cytochrome a,a₃ (by reflection spectrophotometry). Shifts toward reduction are depicted as an upward deflection in this trace. Lower traces: typical changes in extracellular ion activities and EEG during and after cerebral ischemia.

A key to studies reported here is that reperfusion after global ischemia produced characteristic changes which are also measurable with non-invasive probes. Among these changes are shifts in the optical characteristics of rat cerebral cortex (c.f. Figure 1) which were interpreted as indicating hyperoxidation of the mitochondrial respiratory chain components.[3,4,7,8,13,15,16,17] In addition, reperfusion after ischemia usually produced hyperemia, hyperoxygenation, some restoration of the baseline level of potassium ion activity and some recovery of electrical activity. This sequence of changes due to reperfusion occurred following ischemia (4-vessel occlusion model) in most rats tested in our laboratory.[11] In fact, post-ischemic mitochondrial hyperoxidation (PIMHo) was absent only when ischemia was less than 10 min duration, or when reoxidation was incomplete to baseline or absent (occasionally with ischemia of 30–60 min duration; always with 120 min).

Reviewed here are studies aimed toward testing four related hypotheses: 1) that reoxygenation/reperfusion of post-anoxic/ischemic cerebral tissues promotes mitochondrial hyperoxidation; 2) that post-anoxic/ischemic mitochondrial hyperoxidation signals residual intracellular derangements which likely involve limitations to the supply of reducing equivalents to the respiratory chain; 3) that mitochondrial hyperoxidation is promoted by an oxyradical mechanism; and 4) that antioxidants can suppress mitochondrial hyperoxidation and improve electrical recovery after anoxia.

2. RESULTS

2.1. Do the Optical Shifts Seen after Reperfusion Signal Intracellular Events?

Studies were necessary to confirm that the optical changes seen after ischemia signal intracellular events interpretable as shifts in mitochondrial reduction/oxidation (redox) ratios. Such caution was appropriate since spectrophotometric monitoring in intact tissues is influenced also by large changes in hemoglobin volume and saturation which may occur in the optical fields. Although blood flow and tissue oxygenation were usually transiently increased during post-ischemic reperfusion, it was concluded that PIMHo cannot solely be explained by these changes.[11] For example, the amplitude of post-ischemic mitochondrial hyperoxidation (PIMHo) did not increase when cerebral blood flow (CBF) increased above an apparent threshold during reperfusion. Related findings were that tissue hyperoxygenation was not required for PIMHo to occur or to continue. These results suggest that PIMHo is not a function of hyperemia or hyperoxygenation during reperfusion after cerebral ischemia. Rather, PIMHo must be modulated, at least in part, by residual intracellular derangements which limit mitochondrial electron transport

To further test the hypothesis that PIMHo is modulated by intracellular events, optical shifts were recorded in hippocampal slices subjected to anoxia followed by reoxygenation.[11] A key to these studies is that slices must function independent of the cerebrovasculature which eliminated effects due to hemoglobin. Also, optical shifts could not be attributed to tissue hyperoxygenation since post-anoxic oxygen delivery to the tissue could not exceed pre-anoxic levels.

Figure 2 demonstrates that post-anoxic hyperoxidation of certain mitochondrial respiratory chain components (especially hyperoxidation of NAD) occurred in hippocampal slices. This conclusion supports the hypothesis that the optical shifts which we conclude are indicating mitochondrial hyperoxidation occur as a result of intracellular events.

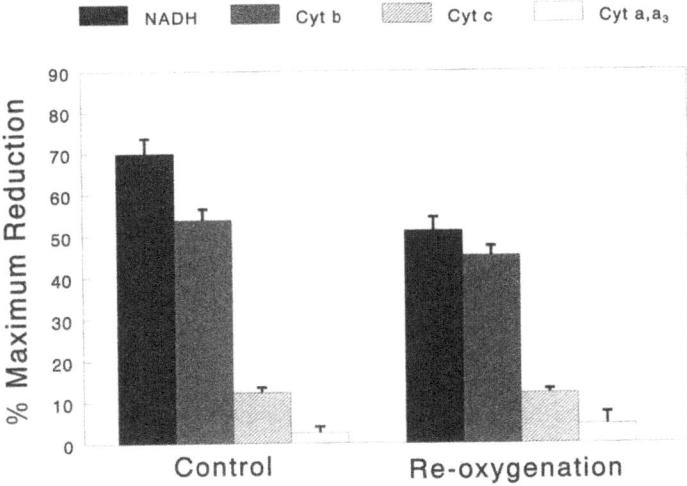

Figure 2. Effects of reoxygenation following anoxia in hippocampal slices. The gradient of reduction from NAD to the cytochromes seen under control conditions is similar to that recorded under 'resting' conditions in isolated mitochondria.[2]

2.2. What Is the Mechanism of PIMHo?

Principles of mitochondrial redox activity were derived from studies of isolated mitochondria[2] where 5 mitochondrial states were described. These mitochondrial states varied with the availability of oxygen, substrate and ADP. With minor exceptions, these principles of electron transport and redox activity that were defined in isolated mitochondrial appear to apply also to intact tissues. In brain, for example, mitochondria became more reduced as the oxygen supply became limited (State 5), while oxidative shifts were recorded in response to increased brain activity (state 3) or decreased substrate availability.[12]

During reperfusion after ischemia, the large oxidative shifts of PIMHo should be similarly defined in terms of the factors which provoked oxidation of mitochondria *in vitro*. Such oxidation could be provoked by: a) blocking electron flow with respiratory inhibitors; b) uncoupling phosphorylation from electron flow; c) decreasing the total electron carrier pool which, in turn, would decrease the capacity of the carriers to absorb light and thereby create the appearance of an oxidative shift; d) increasing the availability of ADP such as by activity-induced increases in ATP use (State 3); or e) decreasing the availability of substrate (State 2).

An early conclusion was that post-ischemic/anoxic mitochondrial hyperoxidation occurs throughout the respiratory chain and does not result from blockade of electron transport within the respiratory chain itself. This is because NAD, at the substrate end of the electron transport chain, and cytochrome a,a_3, at the oxygen end of the chain, both exhibited hyperoxidation after global ischemia. Uncoupling of oxidative phosphorylation is a possible mechanism of PIMHo because uncoupling provoked mitochondrial oxidation *in vitro*. Uncoupling could occur after ischemia due to loss of Ca^{2+} homeostasis or production of free oxygen radicals. However, uncoupling also increased substrate and O_2 consumption by isolated mitochondria, in contrast to our preliminary findings in rat brain which were that glucose was high but oxygen consumption was low during PIMHo. It is possible, however, that uncoupling may underlie secondary deterioration.

PIMHo could be explained by decreased mitochondrial pool size which would be consistent with NAD fluorescence topography studies showing lesser concentrations of reduced NAD in regions of post-ischemic energy failure.[17] However, preliminary studies indicate that transient cerebral anoxia (produced by inspiration of N_2 in anesthetized rat) provoked shifts toward reduction of NAD or cytochrome a,a_3 that reached the same maximal level prior to ischemia or during PIMHo suggesting that pool size is not decreased.

It is also unlikely that increased brain activity, and thereby increased ADP availability, provokes mitochondrial oxidation during the PIMHo period in a manner analogous to the 'state 3' of isolated mitochondria. This is because electrical activity was depressed or suppressed at the time of maximum PIMHo amplitude in post-ischemic rat brain.[11] Also, NAD hyperoxidation was associated with lesser recovery of EEG following middle cerebral artery occlusion in cats;[15] and oxygen consumption was depressed after global ischemia during times corresponding to our observations of PIMHo.[14] This depression in oxygen consumption occurred despite hyperemia.

Research was therefore concentrated on the hypothesis that PIMHo results from limitations in the supply of reducing equivalents to the respiratory chain. This was tested by comparing the effects produced by an inhibitor of glycolysis with those produced by ischemia. Glycolysis was inhibited by superfusion of iodoacetate (IAA) onto rat brain which caused oxidation of cytochrome a,a_3 and increased tPO_2 in a manner analogous to PIMHo[10]. An additional characteristic of the post-ischemic brain was apparent when responses to electrocortical stimulation were compared to those recorded during IAA superfusion. Prior to cerebral ischemia or to inhibition of glycolysis, stimulation provoked increases in extracellular potassium ion activity (K^+o) which recovered to baseline within a few seconds after the cessation of stimulation. Following inhibition of glycolysis, the clearance of the increased K^+o was significantly slower. Slowed K^+o clearance also occurred after ischemia but it was not seen during or after hypoxia. Stimulation prior to ischemia also produced transient oxidative shifts of cytochrome a,a_3. After ischemia, stimulation also provoked oxidative shifts but rates of re-reduction of this cytochrome were slowed. IAA provoked a similar slowing of stimulus-provoked oxidative responses.

These studies support a conclusion that the residual dysfunctions following brain ischemia, manifested as PIMHo and ion transport derangement, result from dysfunction of the glycolytic rather than the oxidative metabolic system.

2.3. Are There Links between Mitochondrial Hyperoxidation, Electrical Recovery, and Oxyradical Mechanisms of Cell Injury?

Links between hyperoxidation and cell injury were suggested by findings that NAD hyperoxidation after focal ischemia was associated with greater histopathology[15] and with incomplete recovery of ATP and elevated lactate.[16] Therefore, this question has been prioritized in studies using intact rat brain and hippocampal slices. In the former model, for example, findings were that respiring animals with a slightly 'hypoxic' gas mixture during reperfusion ameliorated PIMHo. This procedure also promoted recovery of evoked potential activity.[5] When rats were respired with 'hyperoxic' gas mixtures during reperfusion after global ischemia, PIMHo was increased and electrical recovery was lessened.

In hippocampal slices, this question was approached by determining whether antioxidants could suppress mitochondrial hyperoxidation after anoxia and whether this putative suppression of mitochondrial hyperoxidation was linked to improved recovery of evoked potentials.[9] To answer these questions, the slices were made anoxic for a time sufficient to produce 'anoxic depolarization' for 2 minutes. After this time, the slices were re-oxygenated and evoked potential amplitudes compared with control.

Both 500 μM ascorbate and 50 μM glutathione significantly decreased post-anoxic hyperoxidation of NAD (Figure 3). These drugs also improved electrical recovery in hip-

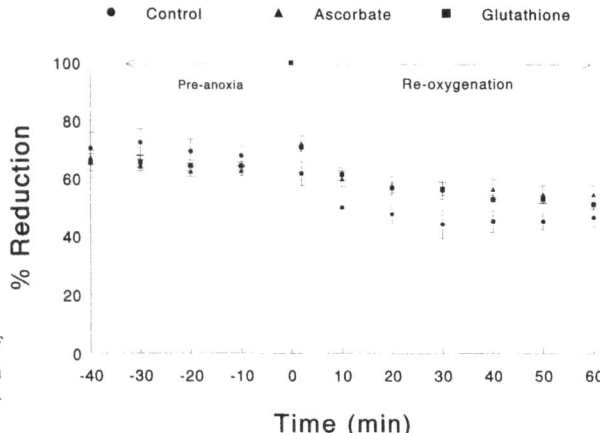

Figure 3. Average percent reduction of NAD prior to and following reoxygenation of hippocampal slices without drugs (control), or with ascorbate or glutathione.

pocampal slices. Such data are compatible with reports that antioxidants protect against electrophysiological damage from brain anoxia or ischemia. What may be more important, however, is that these data support a link between hyperoxidation, oxygen free radicals and post-anoxic (and likely post-ischemic) electrical failure.

3. DISCUSSION /CONCLUSIONS

Evidence presented here clearly demonstrates that reoxygenation following cerebral ischemia or anoxia produced mitochondrial hyperoxidation.This hyperoxidation is not dependent upon systemic influences or hyperemia or tissue hyperoxygenation. Rather, data support that mitochondrial hyperoxidation is due to intracellular events residual to the metabolic insult.

Links between mitochondrial hyperoxidation and cell injury were supported by two findings: 1) manipulation of PIMHo in intact brain, by varying the post-ischemic FiO_2, influenced recovery of evoked potentials. This suggests either a direct link between PIMHo and dysfunction of synaptic transmission or that each is produced by a common mechanism of derangement; 2) Antioxidants (ascorbate or glutathione) suppressed NADH hyperoxidation and improved electrical recovery in hippocampal slices following anoxia.

We propose that mitochondrial hyperoxidation and tissue hyperoxygenation after cerebral ischemia result from imposition of a 'state-2' like limitation in electron availability to the respiratory chain. This may underlie the reported depression in oxygen consumption seen after reperfusion[14] and is consistent with the hypothesis that there may be inhibition of pyruvate dehydrogenase.[1,6] Whether PIMHo and putatively related changes represent residual intracellular metabolic derangements due to ischemia itself or to reperfusion remains to be determined. The antioxidant studies suggest that reactive oxygen species (ROS) may play a role promoting both electrical failure and mitochondrial hyperoxidation, possibly through promoting the putative derangement described above, after cerebral metabolic insults. These data emphasize that defining the mechanism of PIMHo will be important for understanding the physiology and pathophysiology of post-ischemic brain.

REFERENCES

1. Cardell M, Koide T, Wieloch T (1989) Pyruvate dehydrogenase activity in the rat cerebral cortex following cerebral ischemia. J Cerebral Blood Flow Metab 9:350–357
2. Chance B, Williams GR (1956) The respiratory chain and oxidative phosphorylation. Adv Enzymol 17:65–134
3. Dora E, Tanaka K, Greenberg JH, Gonatas NH, Reivich M (1986) Kinetics of microcirculatory, NAD/NADH, and electrocorticographic changes in cat brain cortex during ischemia and recirculation. Ann Neurol 19:536–544
4. Duckrow RB, LaManna JC, Rosenthal M (1981) Disparate recovery of resting and stimulated oxidative metabolism following transient ischemia. Stroke 12:677–686
5. Feng, Z-C, TJ Sick, M Rosenthal, Tissue oxygenation and evoked potential recovery after global cerebral ischemia, Neuroscience Abstracts, 16, 1990
6. Katayama Y, Welsh FA (1989) Effect of dichloroacetate on regional energy metabolites and pyruvate dehydrogenase activity during ischemia and reperfusion in gerbil brain. J Neurochem 52:1817–1822
7. Mayevsky A, Friedli CM, Reivich M (1985) Metabolic, ionic and electrical responses of the gerbil brain to ischemia. Am J Physiol 248:R99-R107
8. Paschen W, Sato M, Pawlik G, Umbach C, Heiss W-D (1985) Neurologic deficit, blood flow and biochemical sequelae of reversible focal cerebral ischemia in cats. J Neurol Sci 68:119–134

9. Perez-Pinzon MA, Mumford PL, Rosenthal M, Sick TJ (submitted) Antioxidants limit mitochondrial hyperoxidation and enhance electrical recovery following anoxia in hippocampal slices

10. Raffin CN, Rosenthal M, Busto R, Sick TJ (1992) Glycolysis, oxidative metabolism and brain potassium ion transport. J Cerebral Blood Flow Metab 12:34–42

11. Rosenthal M, Feng Z-C, Raffin CN, Harrison M, Sick TJ (1995) Mitochondrial hyperoxidation signals residual intracellular dysfunction after global ischemia in rat neocortex. J Cerebral Blood Flow Metab 15:655–665

12. Rosenthal M, Sick TJ (1988) Measurement of metabolic activity associated with ion shifts. In: Neuromethods, vol x: Brain Electrolytes and Water Spaces (eds Boulton A, Baker GB, Walz W eds), Clifton NJ, Humana Press, pp 187–245

13. Rosenthal M, Martel DL, LaManna JC, Jöbsis FF (1976) Oxidative energy metabolism in situ during and following short periods of transient cortical ischemia in cats. Exptl Neurology 50:477–494

14. Singh NC, Kochanek PM, Schiding JK, Melick JA, Nemoto (1992) Uncoupled cerebral blood flow and metabolism after severe global ischemia in rats. J Cerebral Blood Flow Metab 12:802–808

15. Tanaka K, Dora E, Greenberg JH, Reivich M (1986) Cerebral glucose metabolism during the recovery period after ischemia — its relationship to NADH-fluorescence, blood flow, ECG and histology. Stroke 17:994–1004

16. Welsh FA, Marcy VR, Sims R (1991) NADH fluorescence and regional energy metabolites during focal ischemia and reperfusion of rat brain. J Cerebral Blood Flow Metab 11:459–465

17. Welsh FA, O'Connor MJ, Marcy VR, Spatacco AJ, Johns RL (1982) Factors limiting regeneration of ATP following temporary ischemia in cat brain. Stroke 13:234–242

28

RESPONSE OF CORTICAL OXYGEN AND STRIATAL DOPAMINE IN NEWBORN PIGLETS TO ALCOHOL INFUSION UNDER NORMOXIC AND HYPOXIC CONDITIONS

Dekun Song, Stephanie J. Murphy, David F. Wilson, and Anna Pastuszko

Departments of Biochemistry and Biophysics and of Pediatrics
School of Medicine
University of Pennsylvania
Philadelphia, Pennsylvania 19104

INTRODUCTION

Alcohol has a wide range of biochemical and physiological effects on the central nervous system. It has been reported to cause disturbances in cellular metabolism, induce cellular stress as indicated by increased expression of heat shock protein mRNA, inhibit the activities of several enzymes, increase production of free radicals, induce apoptosis, alter the function of a variety of neurotransmitter systems, etc. Several studies have also provided evidence that dopaminergic neurotransmission is affected by alcohol. The effect is dependent on the concentration of alcohol given and the experimental model used. Alcohol, in *in vitro* studies, has been shown to significantly inhibit dopamine release from striatal slices[4,15,17,25] and striatal synaptosomes.[15] Acute doses of alcohol (1–3 g/kg), however, have been reported to increase dopamine release and/or turnover as evaluated by increased level of dihydroxyphenylacetic acid (DOPAC) and homovanillic acid (HVA).[3,7,12,16,20] Imperato and Di Chiara,[11] using *in vivo* microdialysis, demonstrated that alcohol, at 0.5–2.5 g/kg, elevated dopamine release in the nucleus accumbens and to a lesser extent in the caudate. In contrast, in studies on anesthetized rats, addition of alcohol to the dialysate caused a dose dependent increase in dopamine release in both the nucleus accumbens and the caudate but with a greater effect in nucleus accumbens.[26] Several studies also have reported that low doses of alcohol increased functional activity of the dopaminergic system.[22] It has been suggested that this behavioral stimulation is mediated by activation of dopaminergic transmission.[1,5,6,10] The mechanisms by which alcohol alters dopaminergic turnover have not been fully elucidated.

The present study was undertaken to investigate the effect of alcohol on cortical oxygenation and dopamine metabolism in brain of newborn piglets. It is shown that infusion of a low dose of alcohol causes a statistically significant decrease in cortical oxygen pressure and this is additive to mild arterial hypoxia. This alcohol induced decrease in brain oxygenation may be an important mechanism by which alcohol exacerbates injury to the brain in response to hypoxia/ischemic and/or trauma.

MATERIALS AND METHODS

Animal Model

Newborn piglets, age 3–5 days were used for all studies. Anesthesia was induced with halothane (Halocarbon Laboratories, Augusta, SC; 4% mixed with 96% oxygen), and 1.5% lidocaine-HCl (Abbott Laboratories, North Chicago, IL) was used as a local anesthetic. The halothane was reduced to 0.6–0.8% after tracheotomy and Tubocurine-HCl was used to induce respiratory paralysis (1.5 mg/kg), and the femoral artery and femoral vein were cannulated. Once the vessels were catheterized, halothane was withdrawn entirely and Fentanyl-citrate (30 µg/kg) was injected intravenously at approximately one hour intervals throughout the experiments. The animals were paralyzed with tubocurarine and mechanically ventilated with a mixture of oxygen and nitrous oxide (in control conditions with 21–22% oxygen and 78–79% N_2O). The head was placed in a Kopf stereotaxic holder and an incision made along the midline of the scalp. The scalp was removed to expose the skull and a hole approximately 5 mm in diameter made over one parietal hemisphere for measuring the oxygen pressure. Another hole about 4 mm in diameter was drilled contralaterally to the first and the microdialysis probe was implanted into the striatum (coordinates A- 5 mm, L-8 mm and V- 15 mm from brema). In all experiments, the blood pressure, body temperature, and heart rate were continuously monitored. Arterial blood samples were routinely taken and the blood pH, $PaCO_2$ and PaO_2 measured on a Beckman blood gas machine.

All animal use procedures were in strict accordance with the NIH Guide for the Care and Use of Laboratory Animals and were approved by the local Animal Care Committee.

Experimental Protocol

Alcohol Infusion. Following a 2 hr control period, a continuous infusion rate of 225 µl/kg/min, *i.v.* of 50% ethanol (loading dose) was initiated and maintained for 30 min as described by Kim et al.[13] At the end of the 30 min. period, the infusion rate of 50% ethanol was decreased to 22.5 µl/kg/min, *i.v.* (maintenance dose) and continued for the rest of experiment. Controls were sham-operated animals infused with saline at the same rates and times as the ethanol infusions.

Hypoxic Model. Hypoxia was produced by reducing the fraction of inspired oxygen (FiO_2) from normal (22%) to 14% for 30 min. At the end of the hypoxic period, the FiO_2 was returned to 22% and maintained at that level throughout a period of reoxygenation.

Measurement of blood alcohol. Heparinized arterial blood samples (0.5 ml) were taken at 15 min intervals during the first hour and at 30 min intervals during the remainder of the study. The samples were tightly stopped and kept on ice until blood alcohol was determined using a commercial diagnostic kit (Sigma Diagnostics alcohol procedure).

Cortical oxygen pressure. The cortical oxygen pressure was measured by oxygen dependent quenching of phosphorescence.[23,24]

Microdialysis and biochemical measurements. Ringer solution (140 mM NaCl, 2.5 mM KCl, 1.3 mM $CaCl_2$ and 0.9 mM $MgCl_2$) was pumped through the microdialysis probe at 1 µl/min. The microdialysis probes used during present study had a diameter of 0.4 mm and dialysis length at the tip of 3 mm (CMA 12 probes, Bioanalytical System, Inc.). Dialysis samples were collected at 10 min intervals following a 2 hr stabilization period and the samples immediately analyzed for dopamine and its metabolites using a BAS Liquid Chromatography System with electrochemical detection. The column was equilibrated with a mobile phase of 0.1 M monochloroacetic acid (final pH 2.8), containing 0.5 mM EDTA, 0.15 g/l sodium octyl sulfate and 2% acetonitrile. Ten µl of dialysate was directly injected into the microbore column, giving a detection limit of 5–10 fmole/sample. Identification and quantitation of dopamine and its metabolites was by comparison with chromatograms of standard solutions. Retention times for DOPAC, dopamine and HVA were 4.0, 8.3, and 11.2 minutes, respectively.

The efficiency of the microdialysis probe was determined *in vitro* at 37°C for each of the compounds. The relative recoveries were: 16 ± 2% for DA, 19 ± 2% for DOPAC and 20 ± 3% for HVA (means ± SEM for 7 experiments). The values for the levels of different compounds in the dialysate are presented after correction for relative recovery by the microdialysis probe.

Statistical evaluation. The results are expressed as a means ± SEM. Statistical significance of changes in the measured parameters was determined using one way analysis of variance (ANOVA) with post-hoc Dunnett's test. $p < 0.05$ is considered statistically significant.

RESULTS

Blood Alcohol Levels in Newborn Piglets

The effects of continuous intravenous infusion of either saline or ethanol on the blood alcohol levels of newborn piglets are shown in Figure 1. The loading dose infusion

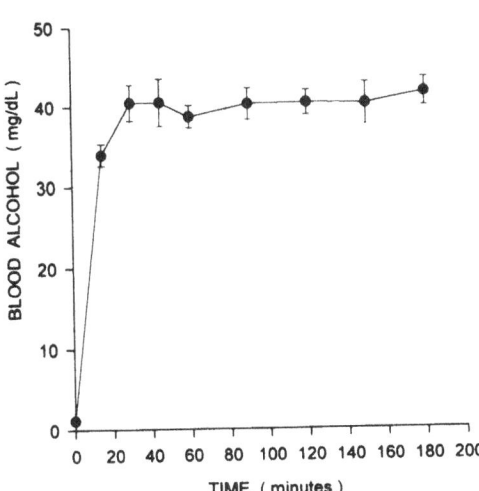

Figure 1. Effect of alcohol infusion on blood alcohol level in newborn piglets. Alcohol was infused as described in Material and Methods and measured in aliquots of blood taken at the indicated times. The results are means ± SEM for 6 experiments.

Table 1. Effect of hypoxia on the physiological parameters in newborn piglets

Conditions	MABP (Torr)	Blood pH	PaO_2(Torr)	$PaCO_2$(Torr)	Heart rate (beats/min)
Control	85.7 ± 3.2	7.43 ± 0.003	103.6 ± 3.7	35.3 ± 0.9	154 ± 6
Saline					
20 min	86.3 ± 5.1	7.41 ± 0.004	100.9 ± 5.1	40.0 ± 1.4	149 ± 7
40 min	81.9 ± 4.0	7.42 ± 0.002	107.3 ± 4.1	39.5 ± 0.8	151 ± 7
60 min	83.4 ± 3.7	7.40 ± 0.004	99.6 ± 3.3	41.4 ± 3.0	150 ± 8
30' Hypoxia	79.7 ± 6.4	7.35 ± 0.007	46.9 ± 4.0**	38.7 ± 1.7	203 ± 6*
10' Reoxygen	81.6 ± 4.1	7.28 ± 0.006*	100.4 ± 3.7	39.0 ± 1.3	207 ± 6*
20' Reoxygen	78.5 ± 3.6	7.29 ± 0.004*	102.9 ± 4.8	37.2 ± 0.9	215 ± 8*
40' Reoxygen	80.4 ± 4.6	7.34 ± 0.017	111.3 ± 3.5	38.2 ± 2.2	202 ± 5*
60' Reoxygen	77.6 ± 4.0	7.37 ± 0.009	109.3 ± 3.1	37.4 ± 3.0	177 ± 7

The measurements are presented as the means ± SEM for 8 experiments. * $p < 0.05$; ** $p < 0.01$ for difference from control as determined by one way analysis of variance.

rate of 50% ethanol caused blood alcohol levels to increase from control values (saline-infused animals) of < 1 mg/dL to approximately 40 mg/dL. They remained at this level during the maintenance infusion rate. In our experiments, animals infused with either saline or 50% ethanol at the above rates achieved comparable blood alcohol levels regardless of the experimental conditions of the study (i.e.normoxia vs. hypoxia).

Effects of Alcohol on the Physiological Parameters of Newborn Piglets

The effects of mild hypoxia and either saline or alcohol infusion on the physiological parameters of newborn piglets are shown in Table 1 and 2. Infusion of saline did not alter the measured parameters. Decrease in the FiO_2 to 14% resulted in a small, not statistically significant, decrease in blood pH and a statistically significant increase in heart rate. During the posthypoxic reoxygenation period the blood pH further decreased to 7.29 ± 0.004 (p<0.01) then increased again to values not significantly different from control. At the end of reoxygenation period, all of the physiological parameters were similar to control (Table 1).

Table 2. Effect of alcohol and hypoxia on the physiological parameters of newborn piglets

Conditions	MABP (Torr)	Blood pH	PaO_2(Torr)	$PaCO_2$(Torr)	Heart Rate (beats/min)
Control	88.2 ± 4.6	7.41 ± 0.004	110.2 ± 4.3	36.1 ± 0.6	164 ± 6.4
Alcohol					
20 min	80.1 ± 3.9	7.39 ± 0.007	105.3 ± 4.5	35.3 ± 0.8	200 ± 11
40 min	78.2 ± 3.7	7.40 ± 0.004	98.5 ± 2.6	37.8 ± 1.9	243 ± 16**
60 min	81.5 ± 5.2	7.38 ± 0.007	97.8 ± 4.9	37.5 ± 2.1	256 ± 16**
30' Hypoxia	51.3 ± 4.1**	7.26 ± 0.011*	47.2 ± 1.1**	36.5 ± 2.4	241 ± 12**
10' Reoxygen	50.7 ± 3.5**	7.13 ± 0.018*	105.3 ± 5.7	36.1 ± 1.1	235 ± 10**
20' Reoxygen	49.3 ± 1.9**	7.18 ± 0.024*	108.5 ± 7.2	36.2 ± 1.2	228 ± 12**
40' Reoxygen	47.1 ± 2.4**	7.24 ± 0.021*	108.7 ± 4.6	38.7 ± 1.9	233 ± 13**
60' Reoxygen	45.2 ± 2.2**	7.23 ± 0.016*	110.1 ± 2.2	38.1 ± 1.2	231 ± 16**

The results are presented as the means ± SEM for 8 experiments. * $p < 0.05$; ** $p < 0.01$ for differences from control as determined by one way analysis of variance.

Alcohol infusion alone caused a significant increase in heart rate from 164 ± 6 beats/min (control) to 256 ± 16 beats/min ($p < 0.05$). When the FiO_2 was decreased for the alcohol infused piglets (arterial hypoxia), the blood pressure decreased from 88.2 ± 4.6 Torr (control) to 51.3 ± 4.1 Torr ($p < 0.01$) and during the reoxygenation period further decreased to 45.2 ± 2.2 Torr (Table 2). Blood pH values decreased from control values of 7.41 ± 0.004 to 7.26 ± 0.011 ($p < 0.05$) with alcohol infusion alone and to 7.23 ± 0.016 ($p < 0.05$) during hypoxia and posthypoxic reoxygenation, respectively. Heart rate, which increased during alcohol infusion, stayed at this high level when the FiO_2 was decreased and through the period of reoxygenation.

The Effect of Alcohol Infusion on Cortical Oxygen Pressure

The effects of alcohol and alcohol plus arterial hypoxia on cortical oxygen pressure are shown in Figure 2. In control piglets, the cortical oxygen pressure was 57 ± 4 Torr and did not change during saline infusion. Infusion of alcohol caused progressive decrease in cortical oxygen pressure to 36 ± 5 Torr. Decrease in FiO_2 to 14% caused a decrease in cortical oxygen pressure to 39 ± 3 Torr in the saline-infused piglets and to 13 ± 4 Torr in the alcohol-infused piglets. In the saline perfused piglets, during reoxygenation the cortical oxygen pressures returned to levels comparable to those prior to hypoxia whereas in the alcohol treated group, cortical oxygen pressure stayed below the control level throughout the period of reoxygenation.

Effects of Alcohol and Alcohol Plus Arterial Hypoxia on the Extracellular Striatal Concentration of Dopamine and Its Metabolites in Newborn Piglets

Alterations in the extracellular level of dopamine in the striatum during the alcohol infusion following a period of arterial hypoxia and reoxygenation are shown in Figure 3.

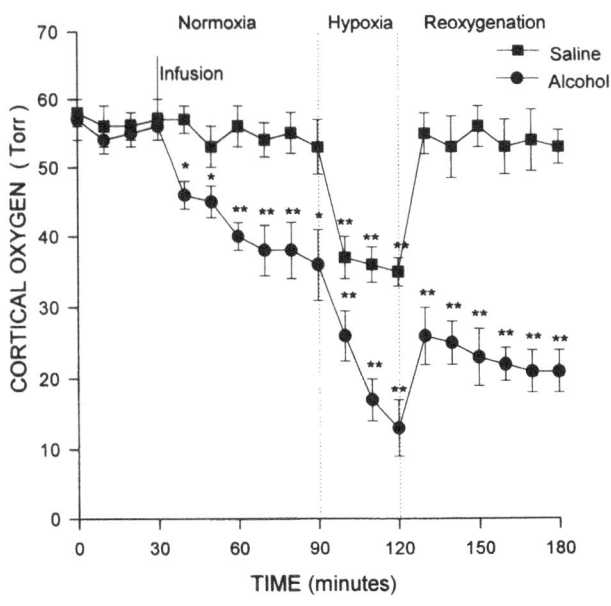

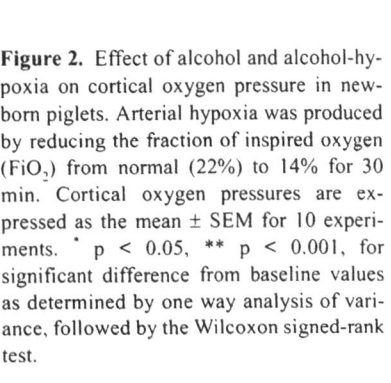

Figure 2. Effect of alcohol and alcohol-hypoxia on cortical oxygen pressure in newborn piglets. Arterial hypoxia was produced by reducing the fraction of inspired oxygen (FiO_2) from normal (22%) to 14% for 30 min. Cortical oxygen pressures are expressed as the mean $\pm$ SEM for 10 experiments. * $p < 0.05$, ** $p < 0.001$, for significant difference from baseline values as determined by one way analysis of variance, followed by the Wilcoxon signed-rank test.

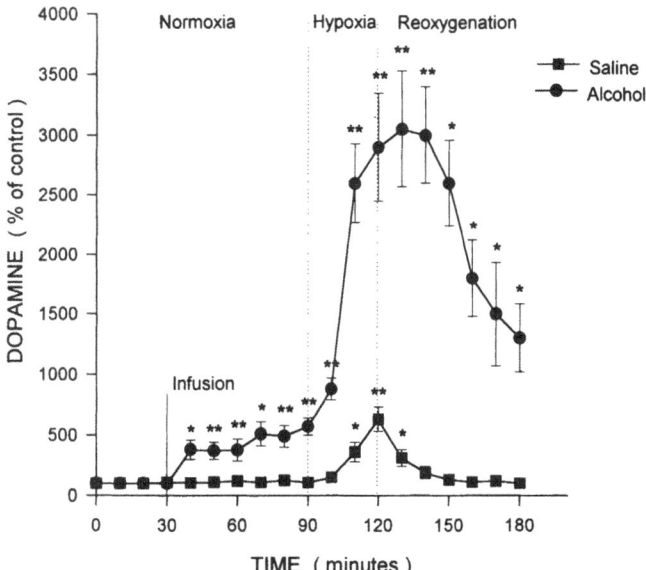

Figure 3. Changes in the extracellular concentration of dopamine in the striatum during alcohol infusion and alcohol plus arterial hypoxia treatment in newborn piglets. The microdialysis samples were obtained and analyzed as described in Materials and Methods. Samples were collected for 40 min prior to infusion of alcohol and the four measurements averaged and considered the baseline value (100%). The results are presented as the mean ± SEM for 8 experiments. * $p < 0.05$, ** $p < 0.001$ for significant difference from baseline values as determined by one way analysis of variance, followed by the Wilcoxon signed-rank test.

The control level of dopamine was 1.5 ± 0.8 pmole/ml. As can be seen, the extracellular level of dopamine significantly increased, to 580% of control, by 60 minutes of alcohol infusion ($p < 0.01$). When the FiO_2 was decreased for alcohol treated piglets, the dopamine levels rose to 3,000% of control by the end of the 30 minute period of arterial hypoxia. During the reoxygenation period, the dopamine level decreased again to 1,300% of control, but this value was still significantly greater than the control value. In control piglets, the level of dopamine did not change with saline infusion. Decrease in the FiO_2 caused an increase in extracellular dopamine to 610% of control and by 20 minutes after the start of reoxygenation, the dopamine level returned to control values.

Alcohol infusion and hypoxia also affected the extracellular levels of two major dopamine metabolites, DOPAC and HVA. Control levels of DOPAC and HVA were $2,070 \pm 410$ pmol/ml and $1,920 \pm 370$ pmol/ml, respectively. Infusion of alcohol caused a statistically significant increase in DOPAC to 134% of control ($p < 0.05$) without significantly altering the level of HVA. Decrease of FiO_2 during alcohol infusion caused decreases in the extracellular levels of DOPAC and HVA. During the reoxygenation period, there was a further decrease in the levels of both metabolites. At the end of the reoxygenation period, the level of DOPAC had decrease to $56 \pm 3\%$ of control ($p < 0.01$) while HVA decreased to $53 \pm 6\%$ of control ($p < 0.01$).

In saline infused piglets, reduction of the FiO_2 to 14% resulted in the extracellular levels of DOPAC and HVA decreasing to $69 \pm 4\%$ of control and $81 \pm 5\%$ of control, respectively. Both metabolites returned to control levels during the reoxygenation period.

DISCUSSION

The effect of alcohol on brain oxygenation and dopamine metabolism during normoxia and arterial hypoxia was investigated in newborn piglets. The oxygen pressures in the cortex, as measured by quenching of the phosphorescence of a phosphor in the blood, are representative of the mean oxygen pressures in the microvasculature of the tissue and therefore of the availability of oxygen to the tissue. Alcohol infusion resulted in a statistically significant decrease in cortical oxygen pressure. The level of alcohol used, 40 mg/dl blood (equivalent to 0.4g/kg), was below that generally considered to substantially impair mental capacity and it caused no significant changes in physiological parameters of the piglets, including mean arterial blood pressure and blood pH. It has been suggested that alcohol induces a transient increase in the permeability of the blood brain barrier.[21] Leakage of the phosphor from the blood into the extracellular space could lead to a decrease in measured oxygen pressure. It is highly unlikely, however, that such alterations, if they occur, result in permeability to molecules as large as the albumin phosphor complex.

A low dose of alcohol resulted in an increase in the extracellular dopamine level in striatum and this increase was in parallel to the decrease in cortical oxygen pressure. Because oxygen measurement in the striatum *in vivo* is not yet possible, we assume, with some reservation, that the oxygen pressure changes in different regions of the brain behave at least in a parallel manner. Thus, in the present study, the changes in extracellular dopamine and its metabolites in the striatum have been related to the oxygen pressure in the cortex. It is interesting that infusion of low doses of alcohol and mild arterial hypoxia, which resulted in similar decreases in cortical oxygen pressure, also caused similar increases in extracellular dopamine. It can be suggested that the primary mechanism responsible for the increase in extracellular dopamine during alcohol infusion is the decrease in tissue oxygen pressure. This is consistent with earlier data from our laboratory showing that, particularly in mild hypoxia, the extracellular level of dopamine depends primarily on the oxygen concentration in the tissue.[18,19] The brain of newborn piglets appears to have no "oxygen reserve" and the extracellular levels of dopamine in the striatum increase even with small decreases in oxygen pressure.

Our results also show that adminstration of mild arterial hypoxia during low dose alcohol infusion produced severe brain hypoxia. This is consistent with the report of Flamm et al.[8] that adminstration of alcohol to cats with spinal cord and cerebral cortex lesions resulted in significantly larger lesions in both groups. Similar results were presented by Albin and Bunegin[2] in an induced focal ischemia model in dogs, where the lesion volumes were significantly larger in alcohol treated animals. Franco et al.[9] reported significantly increased mortality in alcohol treated mice that received head injury. Kim et al.[14] and Zink et al.[28] reported that alcohol treated animals had significantly longer periods of post-injury apnea when compared to injured control animals. Studies on immature, halothane anesthetized, piglets also showed that traumatic brain injury increased the susceptibility of the brain to subsequent hypoxia, and prolonged apnea occurred in alcohol treated animals following brain injury.[27] These authors reported that combination of alcohol and trauma led to much greater respiratory depression and an impaired ventilatory response to increased PaCO$_2$ than did either alcohol or brain injury alone.

We suggest that a very important effect of alcohol in central nervous system is to decrease oxygenation, resulting in brain hypoxia, and that some of the toxicity of alcohol is mediated by the induced increase in extracellular dopamine. Increased extracellular dopamine has been proposed to play a major role in mediating neuronal damage in striatum in a variety of pathological conditions. Dopamine can exert deleterious effects by several different mechanisms, including direct neurotoxic effects, effects on the glutaminergic system, an increase in the production of free radicals, an increase in cerebral metabolism of glucose resulting from an uncoupling of blood flow and metabolism.

In summary, our data show that acute administration of a low dose of alcohol caused brain hypoxia which can be at least partly responsible for the observed increase in extracellular level of dopamine. It was also found that a combination of alcohol infusion with mild arterial hypoxia was additive and led to severe brain hypoxia. It is postulated that alcohol induces brain hypoxia which, when combined with hypoxia/ischemia or trauma, exacerbates the tissue hypoxia and subsequent injury to the brain.

ACKNOWLEDGMENT

This work was supported by NIH grant, NS-31465.

REFERENCES

1. Ahlenius S., Carlsson A., Engel J., Svensson T.H., and Sodersten P. (1973) Antagonism by alpha-methyltyrosine of the ethanol induced stimulation and euphoria in man. *Clin. Pharmacol. Ther.* 14, 586–591.
2. Albin M.S., and Bunegin L. (1986) An experimental study of craniocerebral trauma during ethanol intoxication. *Crit. Care Med.* 14, 841–846.
3. Barbaccia M.L., Bosio A., Spano P.F., and Trabucchi M. (1982) Ethanol metabolism and striatal dopamine turnover. *J. Neural Transm.* 53, 169–174.
4. Bustos G., Liberona J.L., and Gysling K. (1981) Regulation of transmitter synthesis and release in mesolimbic dopaminergic nerve terminals. Effect of ethanol. *Biochem. Pharmacol.* 30, 2157–2164.
5. Carlsson A., Engel J., Stromborn U., Svensson T.H., and Waldeck B. (1974) Supression by dopamine agonists of the ethanol induced stimulation of locomotor activity and brain dopamine synthesis. *Naunyn-Schmiedeberg's Arch. Pharmacol.* 283, 117–128.
6. Engel J., and Lilijequist S. (1983) The involvement of different central neurotransmitters in mediating stimulatory and sedative effects of ethanol, in *Stress and Alcohol Use* (Pohorecky L. and Brick J., eds) pp 153–169, Elsevier, New York.
7. Fadda F., Mosca E., Colombo G., and Gessa G.L. (1990) Alcohol preferring rats: genetic sensitivity to alcohol induced stimulation of dopamine metabolism. *Physiol. Behav.* 47, 727–734.
8. Flamm E.S., Demopoulos H.B., Seligman M.L., Tomasula J.J., De Crescito V., and Ransohoff J. (1977) Ethanol potentiation of central nervous system trauma. *J. Neurosurg.* 46, 328–335.
9. Franco C.D., Spillert C.R., Spillert K.R., and Lazaro E.J. (1988) Alcohol increase mortality in murine head injury. *J. Natl. Med. Assoc.* 80, 63–65.
10. Hunt W.A. (1981) Neurotransmitter function in the basal ganglia after acute and chronic ethanol treatment. *Fed. Proc.* 40, 2077–2081.
11. Imperato A., and DiChiara G. (1986) Preferential stimulation of dopamine release in the nucleus accumbens of freely moving rats. *J. Pharmacol. Exp. Ther.* 239, 219–225.
12. Khatib S.A., Murphy J.M., and McBride W.J. (1988) Biochemical evidence for activation of specific monoamine patways by ethanol. *Alcohol.* 5, 295–301.
13. Kim Y.H., Jones D.L., Natale A., and Klein G.J. (1993) Ethanol increases defibrillation threshold in pigs. *PACE.* 16, 19–25.
14. Kim H.J., Levasseur J.E., Patterson J.L., Jackson G.F., Madge G.E., Povlishock J.T., and Kontos H.A. (1989) Effect of indomethacine pretreatment on acute mortality in experimental brain injury. *J. Neurosurg.* 71, 565–572.
15. Leslie S.W., Woodward J.J., Wilcox R.W., and Farrar R.P. (1986) Chronic ethanol treatment uncouples striatal calcium entry and endogenous dopamine release. *Brain Res.* 368, 174–177.
16. Lucchi L., Lupini M., Govoni S., Covelli V., Spano P.F., and Trabucchi M. (1983) Ethanol and dopaminergic systems. *Pharmacol. Biochem. Behav.* 18, 379–385.
17. Lynch M.A., and Littleton J.M. (1983) Possible association of alcohol tolerance with increased synaptic Ca sensitivity. *Nature.* 303, 175–176.

18. Pastuszko A. (1994) Metabolic response of dopaminergic system during hypoxia-ischemia and reoxygenation in the immature brain. *Biochem. Med. Metab. Biol.* 51, 1–15.

19. Pastuszko A., Lajevardi S.N., Chen J., Tammela O., Wilson D.F., and Delivoria-Papadopoulos M. (1993) Effects of graded levels of tissue oxygen pressure on dopamine metabolism in striatum of newborn piglets. *J. Neurochem.* 60, 161–166.

20. Russell V.A., Lamm M.C.L., and Taljaard J.J.F. (1988) Effect of ethanol on [3H] dopamine release in rat nucleus accumbens and striatal slices. *Neurochem. Res.* 13, 487–494.

21. Thompsen, H., Kaatsch, H.J., and Asmus, R. (1994) Magnetic resonance imaging of the brain during alcohol absorption and elimination-a study of the "rising tide phenomenon". *Blutalkohol,* 31, 178–185.

22. Williams-Hemby L., and Porrino L.J. (1994) Low and moderate doses of ethanol produce distinct patterns of cerebral metabolic changes in rats. *Alcoholism: Clin. Exp. Res.* 18, 982–988.

23. Wilson D.F., Rumsey W.L., Green T.J., and Vanderkooi J.M. (1988) The oxygen dependence of mitochondrial oxidative phosphorylation measured by a new optical method for measuring oxygen concentration. *J. Biol. Chem.* 263, 2712–2718.

24. Wilson D.F., Pastuszko A., DiGiacomo J.E., Pawlowski M., Schneiderman R., and Delivoria-Papadopoulos M. (1991) Effect of hyperventilation on oxygenation of the brain cortex of newborn piglets. *J. App. Physiol.* 70, 2691–2696.

25. Woodward J.J. and Gonzales R.A. (1990) Ethanol inhibition of N-Methyl-D-Aspartate-stimulated endogenous dopamine release from rat striatal slices: reversal by glycine. *J. Neurochem.* 54, 712–715.

26. Wozniak K.M., Pert A., Mele A., and Linnoila M. (1991) Focal application of alcohols elevates extracellular dopamine in rat brain: a microdialysis study. *Brain Res.* 50, 31–37.

27. Zink B.J., and Feustel P.J. (1995) Effects of ethanol on respiratory function in traumatic brain injury. *J. Neurosurg.* 82, 822–828.

28. Zink B.J., Walsh R.F., and Feustel P.J. (1993) Effects of ethanol on traumatic brain injury. *J. Neurotrauma.* 10, 275–286.

29

INFLUENCE OF VARIOUS VENTILATORY PARAMETERS ON NIR REFLEXION SPECTROSCOPY BASED CEREBRAL OXYGEN MEASUREMENTS

An Experimental Animal Study

K. M. Scheufler,[1] Ch. Thees,[2] F. Steinberg,[3] and J. Zentner[1]

[1]Department of Neurosurgery
[2]Department of Anesthesiology
 University of Bonn, Germany
[3]Department of Medical Radiation Biology
 University of Essen
 Germany

INTRODUCTION

Sustained elevation of intracranial pressure (ICP) is commonly found as a sequel to severe head injury and has been demonstrated to adversely affect both cerebral metabolism (Bouma 91, Cruz 90), hemodynamics (Obrist 84) and outcome (Gopinath 94, Marshall 79) in this group of patients. Hyperventilation is in wide use as one of the immediate therapeutic measures in the treatment of head injury to control intracranial hypertension by reduction of cerebral blood volume (CBV) (Cruz 90, Greenberg 78, Owen-Reece 94, Pryds 90, Thoresen 79). However, questions still remain regarding the effects of carbon dioxide tension on cerebral oxydative metabolism in brain injured patients, especially with potential impairment of cerebrovascular reactivity and altered hemodynamics (Cruz 90, Harper 65, Langfitt 65, Muizelaar 91, Obrist 84, Sutton 90). The benefits of lowering raised ICP may sometimes be compromised by adverse effects on cerebral oxygenation (Cruz 90, Muizelaar 91, Sutton 90). The value of NIRS based monitoring of cerebral hemodynamics and oxygenation has already been demonstrated (Höper 94, McCormick 91, Owen-Reece 94, Steinberg 96, Wray 88). However, no conclusive data exist from a controlled experimental study addressing the effects of pCO_2 variation on cerebral tissue oxygenation during intracranial hypertension. Quantification of these effects is of importance, since clinical decision making would be greatly facilitated by definition of a therapeutic range, within which pCO_2 can be safely varied according to individual

Oxygen Transport to Tissue XIX, edited by Harrison and Delpy
Plenum Press, New York, 1997

requirements without causing cerebral desaturation. We therefore designed this experimental study to elucidate the impact of different ventilatory parameters on cerebral tissue oxygen saturation using a new reflectance spectroscopic system (Multiscan OS 10 (Steinberg 96)) under controlled baseline conditions and during ICP elevation.

ANIMALS AND METHODS

An animal model of intracranial hypertension has been described previously (Barzó 91, Marshall 75). The experiments were performed on 10 adult male New Zealand White rabbits according to the guidelines of the institutional committee for animal care. Anesthesia was induced with a sleeping dose of ketamin/xylazin followed by continuous administration of α-Chloralose to provide adequate anesthesia without influence on systemic circulatory responses to elevated ICP. Animals were intubated via tracheostomy and ventilated with an oxygen/air mixture to establish a baseline arterial pO_2 of 100–150 mmHg. Artificial ventilation was adjusted to three distinct levels by repeated measurements of arterial pO_2: Normoventilation (pCO_2: 35 mmHg), Hypoventilation (pCO_2: 50 mmHg) and Hyperventilation (pCO_2: 25mmHg). Arterial and central venous pressures were measured in the inferior aorta and vena cava, respectively. Electrocardiograms were recorded via needle electrodes placed in the lateral chest wall. The head of the animal was secured in a stereotactic frame and a trephination exposing the frontoparietal region bilaterally was performed. Two 22 G cannulas were introduced into the cisterna magna through the atlantooccipital ligament, one attached to a continuous infusion pump for infusion of mock CSF, the other one for continuous recording of ICP. The sensorhead of the reflection spectroscopy system (Multiscan OS 10, see Steinberg 96) was then placed on the intact dura using an arm holder. Special care was taken to achieve a bloodless field for continuous hemoglobin and oxygen measurements. Measurements of cerebral tissue oxygen saturation (range 100–0%) and regional oxyhemoglobin concentration as well as total hemoglobin concentration (both mg/ml tissue) were taken at baseline conditions (arterial pO_2: 100–150 mmHg, normoventilation), during controlled hyper-and hypoventilation and during hypoxic conditions (arterial pO_2=50 mmHg) for a minimum of 1 minute (1 sample/s, sampling frequency: 16 Hz). Subsequently, ICP was elevated to establish a „critical" $tiSO_2$ level of 30%, and the effects of hyperventilation and arterial hypoxia on the already compromised cerebral tissue oxygenation were recorded. Electrocortical activity was continuously monitored by an 8 channel EEG.

RESULTS

The effects of different pCO_2 levels, reflecting normoventilation (35 mmHg), hypoventilation (50 mmHg) and hyperventilation (25 mmHg) on cerebral tissue oxygen saturation ($tiSO_2$), tissue Hb concentration and HbO_2 concentration (average median, maximum and minimum values/ mean standard deviation (MSD) calculated from a minimum of 1000 single measurements) were recorded in 10 animals. Cerebral intracapillary hemoglobin oxygenation ($tiSO_2$) was significantly lowered during hyperventilation (median: 47%; MSD: 1.94) compared to normo-and hypoventilation (ANOVA, α: 0.05, P< 0.001; fig. 1). Median $tiSO_2$ levels of 59% (MSD: 1.85) were recorded during normoventilation (range 55–62%), whereas peak tissue oxygenation values of 71% (median: 69%; MSD: 1.42) were recorded during hypoventilation, reflecting significant vasodilatation

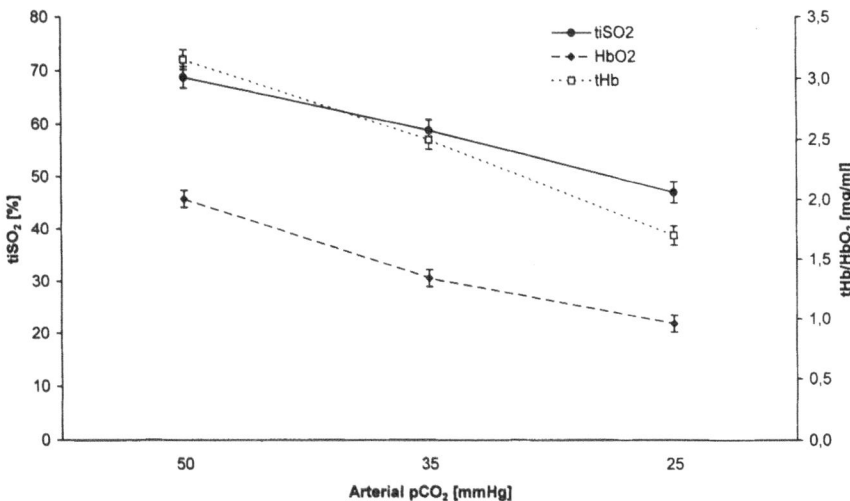

Figure 1. tiSO$_2$ - Local intracapillary hemoglobin oxygen saturation [%]. HbO$_2$ - Local tissue oxyhemoglobin concentration [mg/ml tissue]. tHb - Local tissue hemoglobin concentration [mg/ml tissue]. Median values from 10 animals. Bars indicate Mean Standard Deviation.

and increased tissue perfusion (fig. 1). Corresponding changes of cerebral tissue HbO$_2$ concentrations in response to pCO$_2$ variation are also shown in fig. 1: Cerebral oxyhemoglobin (HbO$_2$) concentrations drop significantly with decreasing carbon dioxide tension (ANOVA, α: 0.05, P< 0.001).

A significant decline of tissue hemoglobin concentration (average median, maximum and minimum values from 10 animals) in the rabbit cerebrum occurs with reduction of pCO$_2$ values (ANOVA, α: 0.05, P< 0.001; fig. 1). This finding, in unison with the results listed above, reflects cerebrovascular reactivity to arterial carbon dioxide tension (mediated by tissue pH).

The impact of hyperventilation compared to normoventilation on already compromised cerebral tissue oxygen saturation due to elevated ICP is illustrated in the left section of fig. 2: A mean ICP of 22 mmHg was necessary to establish constant tiSO$_2$ levels of 30% before hyperventilation was initiated; average median values from 6 animals are reported). A significant decrease of tiSO$_2$ down to a median value of 20% (range 25–17%) at arterial pCO$_2$ values of 20 mmHg was observed during hyperventilation (ANOVA, α: 0.05, P< 0.001). These saturation changes were accompanied by frequency reduction of electrocortical activity, and thus have to be ascribed to further impairment of oxygen availability rather than increasing neuronal oxygen consumption. In addition, tHb concentrations (as an estimate of cerebral blood volume) were not sufficiently lowered to reduce ICP effectively (fig. 2, left section).

The cerebrovascular effects of arterial pO$_2$ are reflected by the corresponding changes of cerebral tiSO$_2$, HbO$_2$ and tHb concentrations during induced arterial hypoxia (reduction of arterial pO$_2$ from 150 to 50 mmHg; right section of fig. 2). Whereas tiSO$_2$ remained unaltered at 69%, hypoxic vasodilatation was reflected by an increase in local tHb concentrations from 2.3 mg/ml to 3.4 mg/ml (fig. 2, right section; median values calculated from 6 animals). The increase in HbO$_2$ concentrations (1.14mg/ml to 1.63 mg/ml; median values calculated from 6 animals) is not equally pronounced due to an increase of O2 extraction during hypoxia to maintain constant tiSO$_2$ levels.

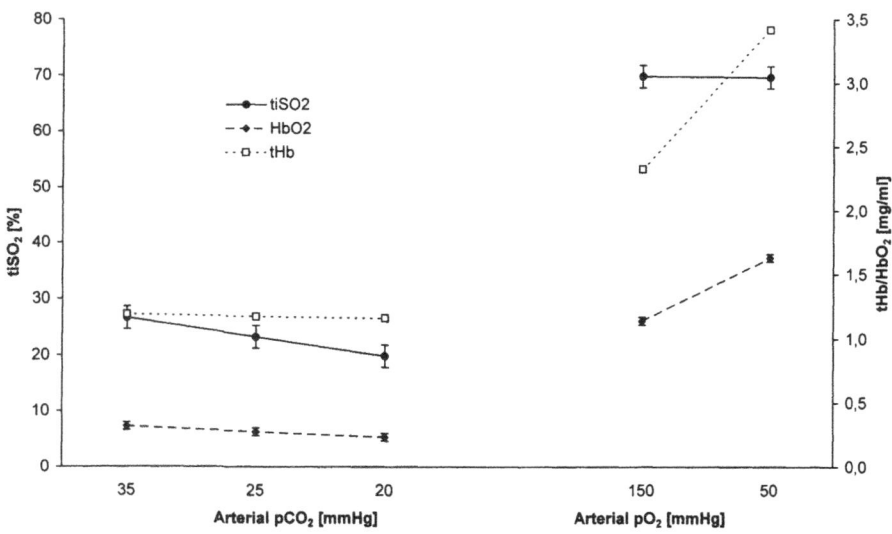

Figure 2. tiSO$_2$ - Local intracapillary hemoglobin oxygen saturation [%]. HbO$_2$ - Local tissue oxyhemoglobin concentration [mg/ml tissue]. tHb - Local tissue hemoglobin concentration [mg/ml tissue]. Median values from 10 animals. Bars indicate Mean Standard Deviation (MSD of tHb measurements: 0.02).

DISCUSSION

Control of increased ICP can be achieved by acting on three intracranial compartments: Cerebral tissue volume, cerebral blood volume and cerebrospinal fluid volume. On an emergent basis, besides administration of osmodiuretics, intracranial hypertension is usually treated by hyperventilation (Cruz 90, Muizelaar 91, Obrist 84) to reduce cerebral blood volume, a mechanism readily expressed by significant reduction of tHb in our study. However, induced hyperventilation may result in cerebral ischemia (Bouma 91, Cruz 90, Muizelaar 91, Sutton 90), reflected by a significant drop of cerebral tiSO$_2$ consistently found in all animals of our study. Hyperventilation has been shown to have a detrimental effect on outcome in one randomized trial (Muizelaar 91). Other studies have reported jugular venous oxygen desaturation periods related to hyperventilation in head injured patients (Bullock 93, Gopinath 94, Sutton 90). Such periods were clearly related to poor outcome (Gopinath 94). The physiologic rationale for routine hyperventilation has therefore been questioned, especially in view of the highly variable CO$_2$ reactivity found in head injured patients (Bouma 91, Muizelaar 91, Sutton 90). In addition, it has been demonstrated that normal coupling of CBF and cerebral metabolic rate for O$_2$ (CMRO$_2$) is retained in only 45% of comatose head injured patients (Sutton 90). However, those patients with coupled but reduced CBF and CMRO$_2$ may be vulnerable to excessive vasoconstriction during acute hyperventilation. Widening of cerebral arteriovenous oxygen difference (AVDO$_2$) during hyperventilation, indicating cerebral desaturation, has been consistently observed in experimental as well as clinical studies (Bouma 91, Obrist 84, Sutton 90). These findings are supported by the drop of tiSO$_2$ recorded during hyperventilation in our experimental series. On the other hand, the supranormal tiSO$_2$, HbO$_2$ and tHb values during hypoventilation indicate a certain degree of luxury perfusion. We observed a further decrease of tiSO$_2$ upon initiation of controlled hyperventilation in animals subjected to ICP elevation, which already had a compromised cerebral tiSO$_2$ of 30%. Consequently, the benefits of ICP reduction achieved by hyperventilation have to be carefully weighted against potentially detrimental effects on cerebral tissue oxygenation. Measurement of $AVDO_2$ carries the drawback of invasiveness and the inability to differentiate between relative ischemia and compensated hypoperfusion, in which decreased CBF is compensated for by increased oxygen extraction and may not reflect any significant alteration in brain metabolism (Sutton 90). Both of these conditions lead to an elevated $AVDO_2$. This problem can be overcome by NIR spectroscopy, since cellular oxydative metabolism (reflecting brain energy metabolism (Gilboe 84)) can be assessed directly (Steinberg 96, Wray 88).

CONCLUSION

Our study clearly underlines the importance of continuous monitoring of cerebral oxygenation to aid in the individual evaluation of ventilatory support in the brain injured patient. Continuous jugular venous oxygen saturation (SjO_2) measurements have been demonstrated to suffer considerably from technical problems encountered in routine clinical use (Bullock 93, Deardon 90). With the NIRS based technology, direct and continuous evaluation of cerebral oxygenation has become feasible (Höper 94, McCormick 91, Owen-Reece 94, Steinberg 96, Wray 88), eliminating some of the technical problems of SjO_2 or $AVDO_2$ measurements. Clinical studies will help to validate NIR spectroscopy as routine monitoring modality in neurosurgical intensive care.

REFERENCES

1. Barzó P, Dóczi T, Csete K, Buza Z, Bodosi M: Measurements of regional cerebral blood flow and blood flow velocity in experimental intracranial hypertension: infusion via the cisterna magna in rabbits. Neurosurgery 28: 821–825, 1991
2. Bouma GJ, Muizelaar JP, Choi SC, Newlon PG, Young HF: Cerebral circulation and metabolism after severe traumatic brain injury: The elusive role of ischemia. J Neurosurg 75: 685–693, 1991
3. Bullock R, Stewart L, Rafferty C, Teasdale GM: Continuous monitoring of jugular bulb oxygen saturation and the effect of drugs acting on cerebral metabolism. Acta Neurochir (Suppl) 59: 113–118, 1993
4. Cruz J, Miner ME, Allen SJ, Alves WM, Gennarelli TA: Continuous monitoring of cerebral oxygenation in acute brain injury: injection of mannitol during hyperventilation. J Neurosurg 73: 725–730, 1990
5. Dearden NM, Midgley S: Technical considerations in continuous jugular venous oxygen saturation measurement. Acta Neurochir (Suppl) 59: 91–97, 1993
6. Gilboe DD, Kintner D, Yanushka J: Cerebral oxygen utilization as a gauge of brain energy metabolism, in Bruley D, Bicher HI, Reneau D (eds): *Oxygen Transport to Tissue VI*. Plenum Press, New York, 1984, pp 169–177
7. Gopinath SP, Robertson CS, Contant CF, Hayes C, Feldman Z, Narayan RK, Grossman RG: Jugular venous desaturation and outcome after head injury. J Neurol Neurosurg Psychiatry 57: 717–723, 1994
8. Greenberg JH, Alavi A, Reivich M, Kuhl D, Uzzell B: Local cerebral blood volume response to carbon dioxide in man. Circ Res 43(2): 324–331, 1978
9. Harper AM, Glass HI: Effect of alterations in the arterial carbon dioxide tension on the blood flow through the cerebral cortex at normal and low arterial blood pressure. J Neurol Neurosurg Psychiatry 28: 449–452, 1965
10. Höper J, Gaab MR: Intraoperative monitoring of local Hb-oxygenation in human brain cortex, in Hogan MC (ed): *Oxygen Transport to Tissue XVI*. Plenum Press, New York, 1994
11. Langfitt TW, Weinstein JD, Kassell NF: Cerebral vasomotor paralysis produced by intracranial hypertension. Neurology 15: 622–641, 1965
12. Marshall LF, Smith RW, Shapiro HM: The outcome with aggressive treatment in severe head injuries. Part 1: The significance of intracranial pressure monitoring. J Neurosurg 50: 20–25, 1979

13. Marshall LF, Durity F, Lounsbury R, Graham DI, Welsh F, Langfitt TW: Experimental cerebral oligemia and ischemia produced by intracranial hypertension. Part 1: Pathophysiology, electroencephalography, cerebral blood flow, blood-brain barrier and neurological function. J Neurosurg 43: 308–317, 1975

14. McCormick PW, Stewart M, Goetting MG, Dujovny M, Lewis G, Ausman JI: Noninvasive cerebral optical spectroscopy for monitoring cerebral oxygen delivery and hemodynamics. Crit Care Med 19: 89–97, 1991

15. Muizelaar JP, Marmarou A, Ward JD, Kontos HA, Choi SC, Becker DP, Gruemer H, Young HF: Adverse effects of prolonged hyperventilation in patients with severe head injury: a randomized clinical trial. J Neurosurg 75: 731–739, 1991.

16. Obrist WD, Langfitt TW, Jaggi JL, Cruz J, Gennarelli TA: Cerebral blood flow and metabolism in comatose patients with acute head injury. Relationship to intracranial hypertension. J Neurosurg 61: 241–253, 1984

17. Owen-Reece H, Elwell CE, Goldstone J, Smith M, Delpy DT, Wyatt JS: Investigation of the effects of hypocapnia upon cerebral hemodynamics in normal volunteers and anesthetised subjects by near infrared spectroscopy, in Hogan (ed): *Oxygen Transport to Tissue XVI*. Plenum Press, New York, 1994, pp 475–482

18. Pryds O, Greisen G, Skov L, Fris-Hansen B: Carbon dioxide related changes in cerebral blood volume and cerebral blood flow in mechanically ventilated preterm neonates. Comparison of near infrared spectrometry and ^{133}Xenon clearance. Pediatr Res 27: 445–449, 1990

19. Steinberg F, Röhrborn HJ, Scheufler KM, Asgari S, Trost HA, Seifert V, Stolke D, Streffer C: NIR Reflection spectroscopy based oxygen measurements and therapy monitoring in brain tissue and intracranial neoplasms - correlation to MRI and angiographic data. Same Volume

20. Sutton LN, Mc Loughlin AC, Dante S, Kotapka M, Sinwell T, Mills E: Cerebral venous oxygen content as a measure of brain energy metabolism with increased intracranial pressure and hyperventilation. J Neurosurg 73: 927–932, 1990

21. Thoresen M, Walloe L: Changes in cerebral blood flow in humans during hyperventilation and CO_2 breathing. J Physiol 298: 53–54, 1979

22. Wray S, Cope M, Delpy DT, Wyatt JS, Reynolds OS: Characterization of the near infrared absorption spectra of cytochrome aa$_3$ and haemoglobin for the non-invasive monitoring of cerebral oxygenation. Biochimica et Biophysica Acta 933: 184–192, 1988

30

CONTINUOUS MEASUREMENT OF CEREBRAL OXYGENATION BY NIRS DURING INDUCTION OF ANAESTHESIA

A. T. Lovell,[1] H. Owen-Reece,[1] C. E. Elwell,[2] M. Smith,[1] and J. C. Goldstone[1]

[1]Division of Academic Anaesthesia
Department of Surgery
University College London Medical School
The Middlesex Hospital
Mortimer St., London, W1N 8AA, United Kingdom
[2]Department of Medical Physics and Bioengineering
University College London
Shropshire House, 11-20 Capper St., London, WC1E 6JA, United Kingdom

1. INTRODUCTION

Continuous intraoperative monitoring of cerebral oxygenation is not routine because existing techniques are either invasive, require a prolonged period of equilibration or involve the use of ionizing radiation. The potential of near infrared spectroscopy (NIRS) as a non invasive tissue oxygenation monitor was first outlined by Jöbsis (Jöbsis, 1977). NIRS enables the continuous measurement of oxyhaemoglobin (HbO_2) and deoxyhaemoglobin (Hb). To date most of the studies that have used NIRS during anaesthesia have considered either the effects on cerebral dynamics of extreme manoeuvres, such as clamping of a carotid artery and induction of ventricular fibrillation (Kirkpatrick et al., 1995; Mason et al., 1994; Williams et al., 1995; Levy et al., 1995), or have been confined to measuring cerebral blood flow (CBF) or cerebral blood volume (CBV) (Owen-Reece et al., 1994; Owen-Reece et al., 1996).

There are no reports on the effects of administration of drugs at clinically used doses, a manoeuvre likely to have far smaller effects than clamping major blood vessels supplying the brain or inducing circulatory arrest.

2. AIM

To determine whether NIRS measurements of cerebral oxygenation alter during induction of anaesthesia.

Oxygen Transport to Tissue XIX, edited by Harrison and Delpy
Plenum Press, New York, 1997

3. METHODS

Twelve healthy ASA I or II patients scheduled to undergo minor surgery were studied. All patients were unpremedicated.

$\Delta[HbO_2]$ and $\Delta[Hb]$ were recorded using a NIRO 500 spectrophotometer (Hamamatsu Photonics, Japan) with optodes located in the temporoparietal region in order to avoid interogating the frontal sinuses. An intraoptode spacing of 4 cm was used, and the optodes were shielded from ambient light. A sampling interval of 0.5 sec was used and the changes in $\Delta[HbO_2]$ and $\Delta[Hb]$ were calculated with a previously established algorithm (Wray et al., 1988; Essenpreis et al., 1993). A differential pathlength factor of 6.26 was used (Duncan et al., 1995). Beat to beat arterial saturation was recorded using a modified Novametrix 520A pulse oximeter attached to an ear lobe. End tidal carbon dioxide tension was monitored with a mainstream capnograph (Novametrix 7000A) directly attached to a tightly fitting facemask. Data were continuously recorded and sampled synchronously with the NIRS data. Blood pressure and heart rate were monitored non invasively with a Datex Cardiocap, sampling every 60 seconds.

3.1. Procedure

All measurements were made with the patients in a supine position. All patients breathed room air for the first four minutes of the study. Then propofol, an intravenous induction agent, was administered intravenously at a rate of 10 mg sec^{-1} until loss of verbal contact was achieved. At this stage ventilation was gently assisted with a circuit fed with 2.5 l min^{-1} O_2 and 7.5 l min^{-1} N_2O to maintain E_TCO_2 constant at its preinduction value. Data was recorded for a further 3 minutes before anaesthesia was deepened and surgery allowed to proceed.

3.2. Data Analysis

We averaged the data into 15 second epochs. All the data from the baseline epochs were averaged together. Statistical analysis was carried out using SPSS 4.0. Analysis was by a repeated measures ANOVA model with post hoc paired t-tests with Bonferroni's correction, a corrected p value < 0.05 was taken as significant. All results are expressed as mean ± SD.

4. RESULTS

The demographics of the patients are shown in Table 1.

There were no significant differences in SpO_2 or E_TCO_2 or heart rate at any stage of the study compared to baseline values of 98.6±1.28 %, 5.0±0.46 kPa and 79±10 min^{-1}, respectively. The fall in blood pressure following induction of anaesthesia is shown in Table 2. This shows that mean arterial pressure undergoes an immediate 11% reduction reaching a plateau 15% below baseline three minutes after induction of anaesthesia. The changes in $\Delta[HbO_2]$ and $\Delta[Hb]$ are shown in figure 1. The $\Delta[Hb]$ signal did not significantly change following induction. However there was a highly significant increase in $\Delta[HbO_2]$, p<0.000002. Further examination of the results showed that from 75 seconds following commencement of induction all values were significantly different from baseline. The changes in $\Delta[HbO_2]$ and $\Delta[HbTot]$, the sum of $\Delta[HbO_2]$ and $\Delta[Hb]$ are shown in

Table 1. Patient demographics

Age	$38 \pm 13(20 - 63)$ yr
Weight	69.8 ± 12.2 kg
Sex	7 Female 5 Male
Haemoglobin	13.2 ± 0.89 g dl-1
Propofol dose	215 ± 44 mg(3.09 ± 0.47 mg kg-1)
Time to loss of consciousness	52 ± 20 sec

figure 2. The changes in $\Delta[HbO_2]$ are mirrored by $\Delta[HbTot]$. Following induction of anaesthesia $\Delta[HbTot]$ also significantly increased, $p < 0.00002$. Further examination showed, that like $\Delta[HbO_2]$, $\Delta[HbTot]$ was significantly different from baseline from 75 seconds following commencement of induction until the end of the study.

5. DISCUSSION

NIRS was able to detect changes in $\Delta94Symbol"[HbO_2]$ and $\Delta94Symbol"[HbTot]$ associated with the induction of anaesthesia. The rise in both $\Delta94Symbol"[HbO_2]$ and $\Delta94Symbol"[HbTot]$ occurs very shortly after the loss of consciousness. These changes were at a time when no intravenous fluids were administered and so are unlikely to be related to changes in haematocrit. Since the haematocrit was constant, the $\Delta94Symbol"[HbTot]$ signal must represent an increase in the volume of blood in the field of observation. From scalp occlusion studies, NIRS signals do not originate solely from the scalp. The increase in $\Delta94Symbol"[HbTot]$ in this study may in part reflect an increase in cerebral blood volume (CBV). The changes in $\Delta94Symbol"[HbO_2]$ occurred at a time when the arterial saturation did not change. It seems likely that this increase in CBV was due to an increase in oxy-haemoglobin within the brain. The deoxy-haemoglobin signal did not significantly change. This data suggests an increase in the saturation of the brain. The NIRS signal has contributions from arterial, venous and capillary compartments and it is impossible with NIRS to determine whether the observed change occurred to an equal extent in all compartments.

In this study there may be many potential reasons for cerebral haemodynamics to change. Although mean arterial pressure fell during this study, it remained within the range of autoregulation and is less likely to contribute to the change in NIRS signals.

To explain our findings that CBV increased, we hypothesize that the brain either dilated or cerebral blood flow (CBF) increased. It is unlikely that CO_2 tension changes affected CBF in this study as it was held constant at each subject baseline.

Table 2. Changes in blood pressure following induction of anaesthesia

Time (min)	Systolic (mmHg)	Mean (mmHg)	Diastolic (mmHg)
Baseline	124 ± 22	90 ± 17	73 ± 15
1	$110 \pm 18*$	$80 \pm 15*$	$64 \pm 13*$
2	$110 \pm 15*$	$80 \pm 13*$	65 ± 13
3	$105 \pm 14*$	$76 \pm 12*$	$61 \pm 11*$

* $p < 0.05$ compared to baseline

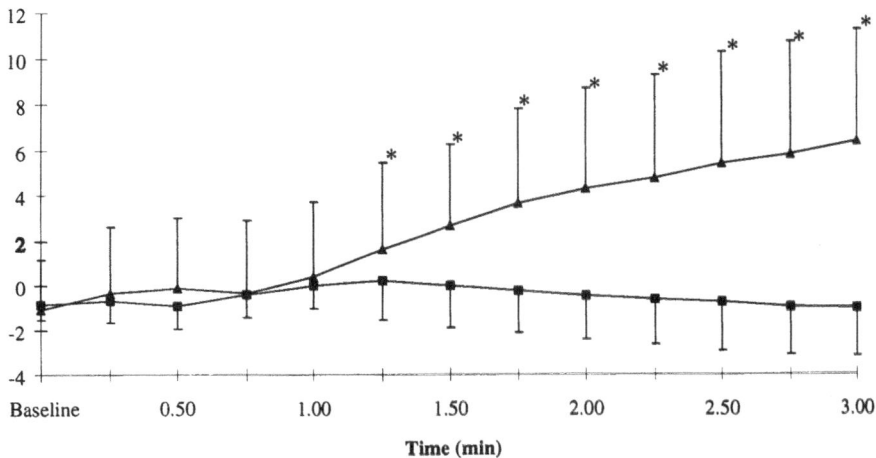

Figure 1. Changes in cerebral Δ[HbO₂] and Δ[Hb] following induction of anaesthesia with Propofol. s Δ[HbO₂] n Δ[Hb] * p < 0.05 compared to baseline.

Several investigators have demonstrated that N_2O vasodilates the cerebral circulation (Reinstrup et al., 1994), and in order to maintain anaesthesia during this study this drug was used.

The actions of intravenous anaesthetic agents on CBF are complex. Whilst cerebral vasoconstriction predominates clinically, direct application of thiopentone has demonstrated dilatation of pial arteries (Michenfelder, 1988). The net effect will be related to the balance between direct dilatation and vasoconstriction secondary to a fall in oxygen consumption ($CMRO_2$). It is known that propofol depresses $CMRO_2$ by 50% (Alkire et al., 1995).

We hypothesize that the increase in CBV that we observed following induction of anaesthesia was due to the direct vasodilating properties of propofol. It must take a period of time for the $CMRO_2$ to fall following propofol administration, and we suspect that only after this time would a vasoconstrictor response be seen. We did not observe the second phase of this response, but the timing of this study was limited by the desire to observe the

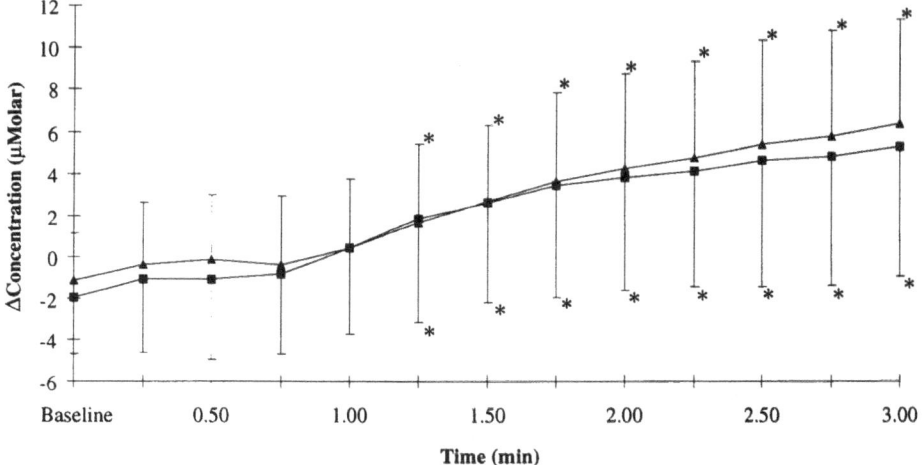

Figure 2. Changes in cerebral HbO₂ and HbTot following induction of anaesthesia with Propofol. s Δ[HbO₂] n Δ[HbTot] * p < 0.05 compared to baseline.

effects of a single administration of propofol. Single shot administration of propofol in unpremedicated patients only produces a brief duration of anaesthesia, thus limiting the duration of the study.

The relationship between the increase in CBV measured in 94Symbol"mMolar terms and that measured in ml 100g⁻¹ has previously been described (Wyatt et al., 1990). This technique requires a gradual change in arterial saturation, and our study was designed to keep arterial saturation constant.

6. CONCLUSIONS

Cerebral oxygenation and cerebral blood volume increases very rapidly following induction of anaesthesia with propofol. These changes are temporally associated with the onset of anaesthesia.

Although NIRS was able to detect changes in cerebral blood volume associated with the induction of anaesthesia, the clinical significance of this finding is unlikely to be important. Care must be taken in interpreting changes in NIRS data under anaesthesia to ensure that anaesthetic depth is held constant.

ACKNOWLEDGMENT

This work has been supported by a grant from the Wellcome Trust.

REFERENCES

Alkire, MT; Haier, RJ; Barker, SJ; Shah, NK; Wu, JC; Kao, J (1995): Cerebral Metabolism during Propofol Anesthesia in Humans Studied with Positron Emission Tomography. *Anesthesiology* 82, 393–403.

Duncan, A; Meek, JH; Clemence, M; Elwell, CE; Tyszczuk, L; Cope, M; Delpy, DT (1995): Optical pathlength measurements on the adult head, calf and forearm and the head of the newborn infant using phase resolved optical spectroscopy. *Phys Med Biol* 40, 295–304.

Essenpreis, M; Elwell, CE; Cope, M; van der Zee, P; Arridge, SR; Delpy, DT (1993): Spectral dependance of temporal point spread functions in human tissues. *Appl Opt* 32, 418–425.

Jöbsis, FF (1977): Noninvasive, infrared monitoring of cerebral and myocardial oxygen sufficiency and circulatory parameters. *Science* 198, 1264–1267.

Kirkpatrick, PJ; Smielewski, P; Whitfield, PC; Czosnyka, M; Menon, D; Pickard, JD (1995): An observational study of near-infrared spectroscopy during carotid endarterectomy. *J Neurosurg* 82, 756–763.

Levy, WJ; Levin, S; Chance, B (1995): Near-infrared measurement of cerebral oxygenation. *Anesthesiology* 83, 738–746.

Mason, PF; Dyson, EH; Sellars, V; Beard, JD (1994): The assessment of cerebral oxygenation during carotid endarterectomy utilising near infrared spectroscopy. *Eur J Vasc Surg* 8, 590–594.

Michenfelder, JD (1988): Anesthesia and the brain. Churchill Livingstone, New York.

Owen-Reece, H; Elwell, CE; Goldstone, J; Smith, M; Delpy, DT; Wyatt, JS (1994): Investigation of the effects of hypocapnia upon cerebral haemodynamics in normal volunteers and anaesthetised subjects by near infrared spectroscopy (NIRS). *Adv Exp Med Biol* 361, 475–482.

Owen-Reece, H; Elwell, CE; Harkness, W; Goldstone, J; Delpy, DT; Wyatt, JS; Smith, M (1996): Use of near infrared spectroscopy to estimate cerebral blood flow in conscious and anaesthetized adult subjects. *Br J Anaesth* 76, 43–48.

Reinstrup, P; Ryding, E; Algotsson, L; Berntman, L; Uski, T (1994): Effects of nitrous oxide on human regional cerebral blood flow and isolated pial arteries. *Anesthesiology* 81, 396–402.

Williams, IM; Mead, G; Picton, AJ; Farrell, A; Mortimer, AJ; McCollum, CN (1995): The influence of contralateral carotid stenosis and occlusion on cerebral oxygen saturation during carotid artery surgery. *Eur J Endovasc Surg* 10, 198–206.

Wray, S; Cope, M; Delpy, DT; Wyatt, JS; Reynolds, EOR (1988): Characterization of the near infrared absorption spectra of cytochrome aa$_3$ and haemoglobin for the non-invasive monitoring of cerebral oxygenation. *Biochim Biophys Acta* 933, 184–192.

Wyatt, JS; Cope, M; Delpy, DT; Richardson, CE; Edwards, AD; Wray, S; Reynolds, EOR (1990): Quantitation of cerebral blood volume in human infants by near-infrared spectroscopy. *J Appl Physiol* 68, 1086–1091.

31

MEASUREMENT OF ABSOLUTE CEREBRAL HAEMOGLOBIN CONCENTRATION IN ADULTS AND NEONATES

M. Wolf,[1] P. Evans,[2] H. -U. Bucher,[1] V. Dietz,[1] M. Keel,[1] R. Strebel,[1] and K. von Siebenthal[1]

[1]Clinic for Neonatology
University Hospital
8091 Zurich, Switzerland
[2]Johnson and Johnson Medical
European Development Centre
Gwent NP1 9UH, United Kingdom

1. INTRODUCTION

The concentration and saturation of the main oxygen carrier of the blood—the haemoglobin—may be an important clinical factor to judge the oxygenation of tissue. The brain is very sensitive to under-oxygenation. So far, the measurement of absolute cerebral haemoglobin concentration by near infrared spectrophotometry (NIRS) has not taken into account the contribution of skull and skin, which causes an underestimation by a factor of 3 [Owen-Reece, 1996] in adults.

2. METHODS

In this section a system is described, which uses near-infrared light to determine absolute concentrations of haemoglobin.

2.1. Dual Channels to Reduce Superficial Absorption of Scalp, Skull and Dura

The light reaching the detector 1 will pass from the emitter through several optically unpredictable layers of skin, bone, cerebro-spinal fluid and meninges (equation 1).

$$I_{t_1} = I_i * K_1 * 10^{-(\ell_1 k_1 c_1)} * 10^{-(C \alpha d_{11} B + G_1)} * 10^{-(\ell_1 k_1 c_1)} * K_2 \tag{1}$$

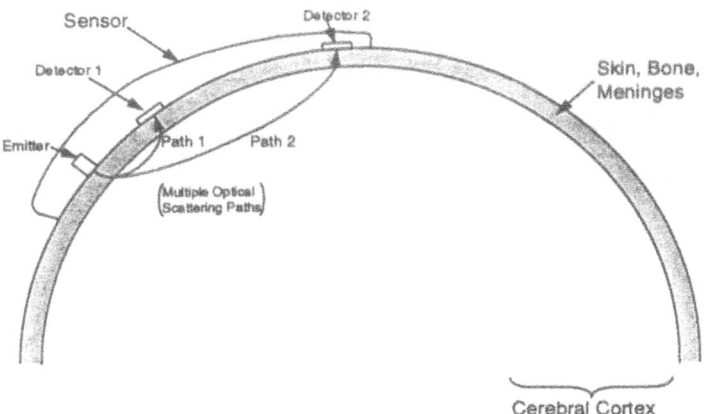

Figure 1. Schematic cross section of the frontal head with dual channel sensor.

where I_i = incident light intensity; I_{tx} = intensity measured at detector 1 (x=1) respectively 2 (x=2); K_1 = coupling efficiency of light source to skin surface; K_2 = coupling efficiency of the skin to the detectors; ℓ_x, k_x, c_x = collective coefficients for an absorbing layer consisting of skin, pigmentation, bone etc. beneath the emitter (x=1) and the detectors (x=2); C = unknown concentration of chromophore to be detected (e.g. haemoglobin); α = absorption coefficient of that chromophore; d_x = emitter-detector distance within head (x=1: detector 1; x=2 detector 2); d_{x1} = pathlength within head (x=1: detector 1; x=2 detector 2); B = differential pathlength factor [Delpy, 1988]; G_x = geometric factors [Delpy, 1988]

The same is true for the light received by detector 2 (equation 2):

$$I_{t_2} = I_i * K_1 * 10^{-(\ell_1 k_1 c_1)} * 10^{-(C\alpha d_{21}B + G_2)} * 10^{-(\ell_2 k_2 c_2)} * K_2 \tag{2}$$

By dividing the two intensities of the detectors we get (equation 3):

$$\frac{I_{t_2}}{I_{t_1}} = 10^{-C\alpha B(d_{21} - d_{11})} * 10^{(G_2 - G_1)} \tag{3}$$

where

$$C = \frac{-1}{B*(d_2 - d_1)} * \alpha^{-1}(\log_{10}(\frac{I_{t2}}{I_{t1}}) - (G_2 - G_1)) \tag{4}$$

To calculate the chromophore concentration out of the light intensities, equation 3 has to be inverted (equation 5).

$$d_{21} - d_{11} = d_2 - (\ell_1 + \ell_2) - (d_1 - (\ell_. + \ell_2)) = d_2 - d_1 \tag{5}$$

The parameters in equation 5 can be treated as vectors (C and $(\log_{10}(I_{t2}/I_{t1}) - (G_2 - G_1))$) and matrix ($\alpha$), if more than one wavelength and chromophore concentration are measured.

If more wavelengths are used than chromophores to be detected, then a least square fit (equation 6) can be applied [Draper, 1981].

$$[\alpha]^{-1} = \left([\alpha]^T * [\alpha]\right)^{-1} * [\alpha]^T \tag{6}$$

2.2. Beam Divergence

If a light source is emitting into a purely scattering spherical material, the photon flux per unit area will decrease with the squared distance to the light source even though no photons are absorbed (diffusion process). The correction for this effect is called divergence factor (equation 7).

$$\frac{I_{t_1}}{I_{t_2}} = \frac{d_1^2}{d_2^2} \tag{7}$$

2.3. Dynamic Compensation for Channel Gain

The coupling efficiency from skin to one of the detectors may vary depending on factors such as skin moisture, dirt on the detector and gain.

A light emitting diode (LED) is placed equidistant to the two detectors. The gains (LED_1, LED_2) on the two detection channels can be measured and corrected, because the intensity of the light originating from the LED should be the same for both detectors. The ratio q (equation 8) is used to calibrate the gains between the two channels.

$$q = \frac{I_{t_1}}{I_{t_2}} * \frac{LED_1}{LED_2} \tag{8}$$

2.4. Water Absorption and Wavelengths Dependent Scattering

Although the light absorption of water is low in the near infrared region, it constitutes a major component of tissue (85%) [Keele, 1971] and has to be taken into account.

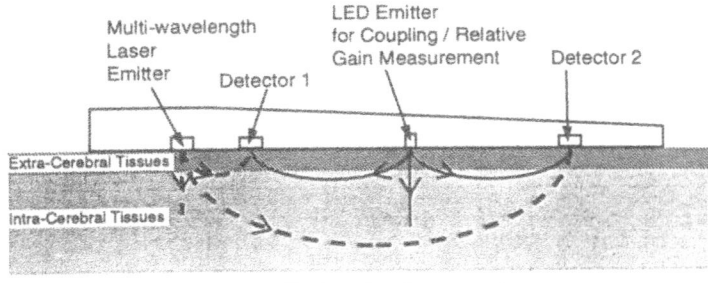

——————— Beam from LED Coupling Compensation source.

— — — — Beam from Multi-wavelength Laser sources.

Figure 2. Schematic diagram of the coupling compensation system.

Table 1. The coefficient matrix (mmol*cm/l)

	λ_1 776.5nm	λ_2 819.0nm	λ_3 871.4nm	λ_4 908.7nm
k_1 (HHb)	1.6436	-1.2510	-0.7075	0.6515
k_2 (O$_2$Hb)	-0.9384	-0.6945	0.5721	1.7344
k_3 (Cyt)	-01398	0.9945	0.1664	-0.9715

The absorption of water will cause an offset on the chromophore concentrations under consideration, which has to be subtracted.

The scattering is wavelength dependent and therefore B is wavelength dependent [Essenpreis, 1993 and Thorniley, 1994].

2.5. The Algorithm to Convert Light Attenuation into Chromophore Concentration

The Critikon Cerebral Redox Monitor 2020 (Johnson & Johnson Medical, UK) uses laser diodes at 4 wavelengths (776.5nm, 819.0nm, 871.4nm and 908.7nm). Three chromophores are detected: oxy-haemoglobin (O$_2$Hb in μmol/l), deoxy-haemoglobin (HHb in μmol/l) and cytochrome oxidase (Cyt in μmol/l). The total haemoglobin (tHb in μmol/l) corresponds to the sum of O$_2$Hb and HHb. Because there is one wavelength more than chromophore to be detected, the system of equations is over-determined. Hence a least square fit procedure (equation 6) is applied to improve the signal to noise ratio.

Taking into account the divergence and the offset by the water absorption the G_x of the previous equations can now be set to $G_x=0$.

The complete algorithm, which accounts for all the mentioned effects, is shown in equation 9.

$$
\begin{bmatrix} C_{HHb} \\ C_{O_2Hb} \\ \Delta C_{Cyt} \end{bmatrix} = \begin{bmatrix} k_{1\lambda_1} & k_{1\lambda_2} & k_{1\lambda_3} & k_{1\lambda_4} \\ k_{2\lambda_1} & k_{2\lambda_2} & k_{2\lambda_3} & k_{2\lambda_4} \\ k_{3\lambda_1} & k_{3\lambda_2} & k_{3\lambda_3} & k_{3\lambda_4} \end{bmatrix} * \begin{bmatrix} \frac{-1000}{B_{\lambda_1}*(d_2-d_1)}*\log_{10}((\frac{d_2}{d_1})^2 * \frac{I_{1,\lambda_1}}{I_{1,\lambda_1}} * \frac{LED_1}{LED_2}) \\ \frac{-1000}{B_{\lambda_2}*(d_2-d_1)}*\log_{10}((\frac{d_2}{d_1})^2 * \frac{I_{1,\lambda_2}}{I_{1,\lambda_2}} * \frac{LED_1}{LED_2}) \\ \frac{-1000}{B_{\lambda_3}*(d_2-d_1)}*\log_{10}((\frac{d_2}{d_1})^2 * \frac{I_{1,\lambda_3}}{I_{1,\lambda_3}} * \frac{LED_1}{LED_2}) \\ \frac{-1000}{B_{\lambda_4}*(d_2-d_1)}*\log_{10}((\frac{d_2}{d_1})^2 * \frac{I_{1,\lambda_4}}{I_{1,\lambda_4}} * \frac{LED_1}{LED_2}) \end{bmatrix} - \begin{bmatrix} O_{HHb} \\ O_{O_2Hb} \\ O_{Cyt} \end{bmatrix} \tag{9}
$$

The coefficient matrix is shown in Table 1.

The offsets (section 3.4) are shown in Table 2.

From these data the tissue saturation rSat can be calculated as the proportion of O$_2$Hb compared to the tHb.

Table 2. The fixed offsets, which are caused by the absorption of water

	HHb	O$_2$Hb	Cyt
Fixed offset	7.47μMol	28.75μMol	0

Table 3. Values obtained from 13 adult volunteers

Quantity	Mean (SD)
Total haemoglobin	93.2 (20.6) µmol/l
Oxy-haemoglobin	65.6 (14.9) µmol/l
Deoxy-haemoglobin	27.5 (6.6) µmol/l
Tissue saturation	70.3 (2.1) %

3. PATIENTS

The system was tested in 8 female and 5 male healthy adult volunteers (age median 31 (range 22 to 58 years) and 20 neonates (age: 33 4/7 (28 3/7 to 44) weeks gestation, actual weight 1890 (770 to 3740) g).

4. MEASUREMENT PROCEDURE

The sensor was attached to the forehead of the adult volunteers for at least one minute.

For the neonatal group the sensor was attached several times (range 3 to 9) by an elastic bandage for at least one minute. Thereby the pressure exerted on the sensor and its position on the head (mainly frontal) was deliberately varied.

The haematocrit of the neonates was determined by a blood sample. To determine one value for each period, a median was calculated.

Analysis of variance was used to separate the intra and inter subject components of variance. The square root of the intra and inter subject component of variance was divided by the global mean to calculate the reproducibility and inter subject variability, respectively.

5. RESULTS

Tables 3 and 4 give the values obtained in volunteers and neonates, respectively. All values refer to the tetrahaem molecule of haemoglobin.

For the 20 neonates, a linear regression reveals that there is a significant (p=0.015) correlation between the tHb and the haematocrit (figure 3).

The reproducibility of the measurements of tHb and rSat in neonates was 23.6% and 3.8%, respectively. The inter subject variability was 69.8% for the tHb and 13.5% for the rSat.

Table 4. Values obtained from 20 neonates

Quantity	Mean (SD)
Total haemoglobin	98.8 (35.0) µmol/l
Oxy-haemoglobin	65.7 (24.3) µmol/l
Deoxy-haemoglobin	32.0 (12.5) µmol/l
Tissue saturation	66.3 (4.4) %

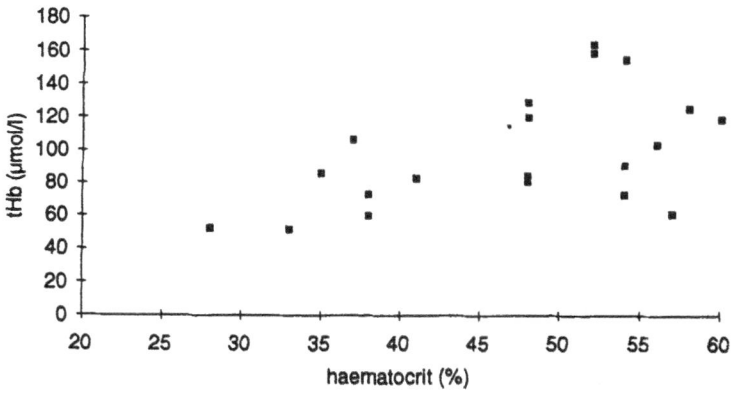

Figure 3. There is a significant linear correlation between haematocrit and total cerebral haemoglobin concentration (tHb).

6. DISCUSSION

The results of this new method can be compared to other methods, which measure cerebral blood volume (CBV in ml/100g) (table 5).

The CBV can be converted if the haemoglobin concentration in the blood is known (cHb in g/100ml). The normal cHb for adult females is mean 13.5 (95% confidence interval: 12.0–15.0) g/100ml and 15.1 (13.9–16.3) g/100ml for males [Ciba-Geigy, 1979]. The mean of these two cHb is 14.3 g/100ml. Taking the mean CBV from all authors, we get 4.6ml/100g. This corresponds to (equation 10):

$$4.6 \text{ ml}/100g = 14.3*0.69*4.6/100 \text{ g}/100g = 0.45g/100g \qquad (10)$$

The factor 0.69 [Lammertsma, 1988] accounts for the lower cHb in cerebral capillaries compared to the arterial cHb. The molecular weight of a tetrameric haemoglobin is

Table 5. Cerebral blood volume of healthy adults measured by different authors and methods

Author	Number of subjects	Method	CBV (ml/100g)	Reproducibility
Elwell (1994)	10	NIRS	2.9±1.0	±34%
Rempp (1994)	12	MRI	6.6±2.1	
Steiger (1993)	13	CT	5.5±1.1	±16%
Sakai (1985)	10	SPECT	4.8±0.4	
Phelps (1979)	5	PET	4.2±0.4	
Grubb (1978)	8	PET	4.3±0.4	±3%
Greenberg (1978)	10	PET	4.4±0.5	
Grubb (1973)	11	x-ray	3.2±0.9	±20%

NIRS=near infrared spectrophotometry, MRI=magnetic resonance imaging, CT=computer tomography, SPECT=single photon emission computed tomography, PET=positron emission tomography.

Table 6. Cerebral blood volume (CBV) measured by near infrare spectrophotometry using a method which takes oxy-haemoglobin as a tracer

Author	Number of subjects	Median gestational age (weeks)	CBV (ml/100g)	Reproducibility
Wyatt (1990)	12	29	2.2±0.4	±12%
Brun (1994)	8	30	3.7±1.1	±23%
Wolf (ISOTT 1995)	9	27 2/7	2.5±0.9	±17%

64500g. Hence 1g of haemoglobin corresponds to 15.5μmol. The density of brain corresponds to 1.05g/ml [Nelson, 1971].

$$0.45g/100g = 15.5*0.45*10.5 \ \mu mol/l = 73.8 \ \mu mol/l \qquad (11)$$

Hence the 4.6ml/100g correspond to approximately 73.8 μmol/l (equations 10 and 11). Our values for adults are slightly higher than expected.

We are not aware of any data obtained from neonates by methods other than NIRS. The mean CBV of three studies (table 6) corresponds to a tHb of 55.5μmol/l (equations 10 and 11) assuming a normal neonatal cHb of 19.7g/100ml [Ciba-Geigy, 1979]. The new tHb results are higher. However, the infants in the previous studies were considerably younger. A further explanation of this relatively high difference may be the reduction of the contribution of the outer tissue layers. These caused an underestimation of cerebral blood flow [Owen-Reece, 1996].

The significant linear correlation between haematocrit and tHb in the neonatal group was expected, because a high haematocrit indicates a high cHb, which should also be true for the brain. This evidence suggests that the inter subject variability encountered is physiological and that the instrument is working. However, it cannot be excluded, that the values are too high or low by factor, which is constant for all infants and therefore validation is still needed.

The reproducibility of 23.6% is in the same range as other NIRS studies [Elwell, 1994, Wolf, 1995]. The fact that the reproducibility is little (if at all) influenced by the variation of the position of and the pressure exerted on the sensor compared to the two studies mentioned, where the sensor remained on the same place, indicates that the two channel geometry substantially reduces such effects.

The inter subject variability is three times higher than the reproducibility, which indicates that most of the total variability is due to physiological factors.

The tissue oxygen saturation has been measured by several authors using optical methods (table 7).

Our values are in the same region as previously published data.

Table 7. Cerebral tissue oxygen saturation of healthy adults

Author	Number of subjects	Tissue saturation in %
Levy (1995)	10	56.5±3.8
De Blasi (1995)	12	60.3±3.8
Schindler (1995)	10	66.0±4.4
Elwell (1994)	10	67.6±13.8

We are not aware of any data on tissue oxygen saturation for neonates.

The inter subject variability is 3.6 times higher than the reproducibility, which indicates that also for the rSat the total variability is mainly due to physiological factors.

The advantages of this system compared to MRI, PET and CT are its continuous signal and its applicability at the bedside: furthermore, no tracer is required.

7. CONCLUSION

The values obtained from a two channel near infrared spectrophotometry system, which reduces the influence of skull and skin, yields values of absolute cerebral haemoglobin concentration and tissue saturation in a reasonable range.

ACKNOWLEDGMENTS

This research was funded in part by a Johnson & Johnson international academic research award to Prof. Hans Bucher. Section 3 of this paper is based on the theory presented in the Ph.D. submission by one of the authors (PE).

REFERENCES

Brun NC and Greisen G. Cerebrovascular responses to carbon dioxide as detected by near-infrared spectro-
photometry: comparison of three different measures. Pediatr Res, 1994, 36, 20–24.
Ciba-Geigy. Wissenschaftliche Tabellen: Physikalische Chemie-Blut-Humangenetik. Ciba-Geigy, Basel 1979.
De Blasi RA, Fantini S, Franceschini MA, Ferrari M and Gratton E. Cerebral and muscle oxygen saturation meas-
urement by frequency domain near-infrared spectrometer. Med Biol Eng Comput, 1995, 33, 228–230.
Delpy DT, Cope M, van der Zee P, Arridge S, Wray S and Wyatt J. Estimation of optical pathlength through tissue
from direct time of flight measurements. Phys Med Biol, 1988, 33, 1422–1433.
Draper NR and Smith H. Applied regression analysis. J. Wiley LTD New York, 1981.
Elwell CE, Cope M, Edwards AD, Wyatt JS, Delpy DT and Reynolds EOR. Quantification of adult cerebral
hemodynamics by near-infrared spectroscopy. J Appl Physiol, 1994, 77, 2753–2760.
Essenpreis M, Cope M, Elwell CE, Arridge SR, van der Zee P and Delpy DT. Wavelength dependence of the dif-
ferential pathlength factor and the log slope in time resolved tissue spectroscopy. In "Optical Imaging of
Brain Function and Metabolism", 1993, Plenum Press, New York.
Greenberg JH, Alavi A, Reivich M, Kuhl D and Uzzell B. Local cerebral blood volume response to carbon dioxide
in man. Circ Res, 1978, 43, 324–331.
Grubb RL, Phelps ME and Ter-Pogossian MM. Regional cerebral blood volume in humans. Arch Neurol, 1973,
28, 38–44.
Grubb RL, Jr., Raichle ME, Higgins CS and Eichling JO. Measurement of regional blood volume by emission to-
mography. Ann Neurol, 1978, 4, 322–328.
Keele CA and Neil E. Samson Wright's Applied Physiology. 12th edition 1971, University Press, London.
Lammertsma AA, Brooks DJ, Beaney RP et al. In vivo measurement of regional cerebral haematocrit using posi-
tron emission tomography. J Cereb Blood Flow Metab, 1988, 4, 317–322.
Levy WJ, Levin S and Chance B. Near-infrared measurement of cerebral oxygenation. Correlation with elec-
troencephalic ischaemia during ventricular fibrillation. Anesthesiology, 1995, 83, 738–746.
Nelson SR, Mantz ML, and Maxwell JA. Use of specific gravity in the measurement of cerebral edema. J Appl
Physiol, 1971, 30, 268–271.
Owen-Reece H, Elwell CE, Harkness W, Goldstone J, Delpy DT, Wyatt JS and Smith M. Use of near infrared
spectroscopy to estimate cerebral blood flow in conscious and anaesthetized adult subjects. Br J Anaesthesia, 1996, 76, 43–48.
Phelps ME, Huang SC, Hoffman EJ and Kuhl DE. Validation of tomographic measurement of cerebral blood vol-
ume with C-11-labled carboxyhemoglobin. J Nucl Med, 1979, 20, 328–334.

Rempp KA, Brix G, Wenz F, Becker CR, Guckel F and Lorenz WJ. Quantification of regional cerebral blood flow and volume with dynamic susceptibility contrast enhanced MR imaging. Radiology, 1994, 193, 637–641.

Sakai F, Nakzawa K, Tazaki Y, Ishii K, Hino H, Igarashi H and Kanda T. Regional cerebral blood volume and hematocrit measured in normal human volunteers by single-photon emission computed tomography. J Cereb Blood Flow Metab, 1985, 5, 207–13

Schindler E, Zickmann B, Muller M, Boldt J, Kroll J and Hempelmann G. Zerebrale Oxymetrie durch Infrarot Spektroskopie im Vergleich zur kontinuierlich gemessenen Sauerstoffsaettigung im Bulbus Vena Jugularis bei Eingriffen an der Arteria Carotis interna. Vasa, 1995, 24, 168–175.

Steiger HJ, Aaslid R and Stooss R. Dynamic computed tomographic imaging of regional cerebral blood flow and blood volume. A clinical pilot study. Stroke, 1993, 24, 591–597.

Thorniley MS, Lane NJ, Manek S and Green CJ. Non-invasive measurement of respiratory chain disfunction following hypothermic renal storage and transplantation. Kidney International, 1994, 45, 1489–1496.

Wolf M, Bucher H, Dietz V, Keel M, von Siebenthal K, Duc G. How to evaluate slow oxygenation changes to estimate absolute cerebral haemoglobin concentration by near infrared spectrophotometry in neonates. Adv Exp Med Biol, in press.

Wyatt JS, Cope M, Delpy DT, Richardson CE, Edwards AD, Wray S and Reynolds EO. Quantitation of cerebral blood volume in human infants by near-infrared spectroscopy. J Appl Physiol, 1990, 68,1086–1091.

NIR REFLEXION SPECTROSCOPY BASED OXYGEN MEASUREMENTS DURING INTRACRANIAL HYPERTENSION IN RABBITS

An Experimental Study

K. M. Scheufler,[1] Ch. Thees,[2] F. Steinberg,[3] and J. Zentner[1]

[1]Department of Neurosurgery
[2]Department of Anesthesiology
University of Bonn, Germany
[3]Department of Medical Radiation Biology
University of Essen, Germany

INTRODUCTION

Cerebral autoregulation is commonly defined as the sum of adaptive mechanisms of the brain in response to a variety of functional, i.e. physiological, and non-functional, i.e. pathological conditions. Its primary goal is to maintain adequate delivery of substrates and removal of waste products of cerebral metabolism. Cerebral autoregulation therefore involves adaptive mechanisms at subcellular, cellular and intercellular levels as well as regulation of local, regional and global cerebral circulation (Bruce 73, Grubb 75, Harper 66, Lowell 71, McGillicuddy 78, Weinstein 64). Autoregulation also includes systemic responses of metabolism, respiration and circulation. All of those cerebral autoregulatory mechanisms may be severely impaired in head injured patients (Langfitt 65, McGillicuddy 78, Obrist 84).

Recent strategies in neuro-intensive care focus on maintaining or restoring cellular energy metabolism using a wide array of diagnostic and therapeutic modalities (Cruz 90, Gilboe 84, McCormick 91, Obrist 84). However, difficulties remain in the definition of sufficiently sensitive, specific and reliable parameters of cerebral metabolism suitable for continuous functional monitoring in the management of brain injured patients, especially for the assessment of adequacy of therapeutic interventions (Bruce 73, Cruz 90, McGillicuddy 78, Obrist 84). Some of these difficulties can be attributed to measurement of indirect parameters of cellular function, such as cerebral perfusion pressure, (regional) cerebral blood flow, or tissue pO_2, reflecting cerebrovascular autoregulation, rather than monitoring cellular function itself (Hudetz 94, McGillicuddy 78). Others are due to tech-

Oxygen Transport to Tissue XIX, edited by Harrison and Delpy
Plenum Press, New York, 1997

nical limitations, such as insufficient temporal or spatial resolution, or lack of practical applicability within the numerous limitations of intensive care management (Bruce 73, Lammertsma 84). Another critical factor is the complex interaction between many of the autoregulatory systems, which become even more complex with pertubations of normal physiology (Grubb 75, Hayashi 83, Hudetz 94, Langfitt 65, Lowell 71, McGillicuddy 78). It is therefore important from both a scientific and clinical point of view to bypass the complexity of the multiple effector systems by definition of a simple reliable parameter, that will adequately reflect cellular integrity and function (Gilboe 84, Steinberg 96 Symon 77). With respect to these requirements, near infrared spectroscopic (NIRS) measurement of cerebral tissue oxygenation holds great promise (Elwell 92, Hazeki 87, McCormick 92, Steinberg 96, Wray 88). It potentially allows for real-time continuous monitoring of tissue oxygenation, cerebral blood volume and flow, and, most recently, direct monitoring of intracellular oxydative metabolism by measurement of cytochrome redox state (Elwell 92, Hazeki 87, Steinberg 96, Wray 88).

We have therefore employed a new NIR reflection spectroscopic system (NIOS Multiscan OS 10 (Steinberg 96) to characterize in detail the relationships between raised intracranial pressure, cerebral perfusion pressure, spontaneous cortical electrical activity and regional cerebral tissue oxygenation as well as hemoglobin parameters in vivo using a previously established animal model (Barzó 91, Marshall 75 I-III, Pearce 81).

ANIMALS AND METHODS

A model of intracranial hypertension in rabbits via infusion of mock CSF in the cisterna magna in rabbits has been described previously (Barzó 91, Marshall 75 I-III). The experiments were performed on 10 adult male New Zealand White rabbits weighing between 3.1 and 3.5 kg in accordance with the guidelines of the institutional committee for animal care. Anesthesia was induced with sleeping doses of ketamin (50mg/kg)/xylazin (7mg/kg) followed by continuous administration of α-Chloralose (20% solution, 5–7 ml/h) and repetitive doses of Suxamethonium-chloride: (1mg/kg) for relaxation to provide adequate anesthesia without influence on systemic circulatory responses to elevated ICP. Animals were intubated via tracheostomy and ventilated with an oxygen/air-mixture to maintain an arterial pO_2 in the range of 100–150 mmHg. Arterial blood gases were monitored on a regular basis to establish normoventilation (pCO_2: 35–45 mmHg) and an arterial oxygen saturation of 99% throughout the experiment. Mean arterial pressure (MAP) was measured in the inferior aorta through an indwelling arterial line introduced through the femoral artery. In analogy, central venous pressure was recorded in the inferior vena cava. Electrocardiograms were recorded via needle electrodes placed in the lateral chest wall. The head of the animal was secured in a stereotactic frame and a trephination exposing the frontoparietal region bilaterally was performed. Two 22 G cannulas were introduced into the cisterna magna through the atlantooccipital ligament, one attached to a continuous infusion pump for infusion of mock CSF, the other one for continuous recording of ICP. The sensorhead of the reflection spectroscopy system (Multiscan OS 10, see Steinberg 96) was placed on the intact dura. The CSF infusion rate was gradually increased from 1 to 90 ml/h during the experiment to achieve a steady reduction of CPP (MAP-ICP) at steps of 1 mmHg, until local cerebral tissue oxygen saturation reached zero. A constant CPP plateau was established for a minimum of 4 minutes following each stepwise increase of ICP to allow for adaptation of cerebral hemodynamics. Measurements of cerebral tissue oxygen saturation (range 100–0%) and regional oxy-

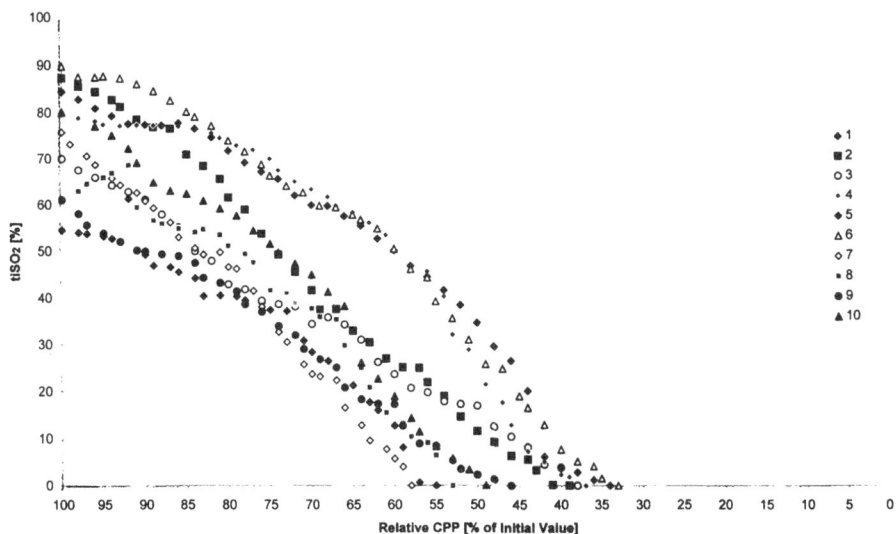

Figure 1. tiSO$_2$ - Local intracapillary hemoglobin oxygen saturation [%]. (median of tiSO$_2$ values calculated from approximately 1000 single measurements taken at each CPP level; MSD: 1.57).

hemoglobin concentration as well as total hemoglobin concentration (both mg/ml tissue) were taken at each CPP level for a minimum of 1 minute (1 sample/s, containing averaged data from 16 single measurements for each parameter taken at 16 Hz, i.e. $\approx$ 1000 sets of data per CPP level). Cortical electrical activity was monitored simultaneously on an 8-channel electroencephalograph. Data were recorded and stored on a PC system for analysis.

RESULTS

Figure 1 summarizes the effects of relative CPP (% of initial CPP before ICP elevation) on cerebral tissue oxygen saturation (median of tiSO$_2$ values calculated from approximately 1000 single measurements taken at each CPP level; mean standard deviation: 1.57). An almost linear correlation between the two parameters is apparent (r^2: 0.98, $P <$ 0.001). Absolute CPP values differ by up to 100% for a given tiSO$_2$ level interindividually, indicating that CPP is not predictive of tiSO$_2$.

Figure 2 demonstrates the effects of relative CPP (% of initial CPP before ICP elevation) on cerebral tissue HbO$_2$ concentration (median of HbO$_2$ values calculated from approximately 1000 single measurements taken at each CPP level; mean standard deviation: 0.04). As with tiSO$_2$, a linear relationship was found beween CPP and HbO$_2$ (r^2: 0.98, $P <$ 0.001). In analogy to the relationship between CPP and tiSO$_2$, CPP is not predictive of tissue HbO$_2$ concentration due to significant interindividual differences.

Figure 3 shows the relationship between relative CPP (% of initial CPP before ICP elevation) and total cerebral tissue hemoglobin concentration (tHb), reflecting cerebral blood volume (CBV) (median of tHb values calculated from approximately 1000 single measurements at each CPP level; mean standard deviation: 0.04). Lack of tight correlation ($r < 0.6$, $P > 0.1$) between these parameters was found in 5 of 10 animals.

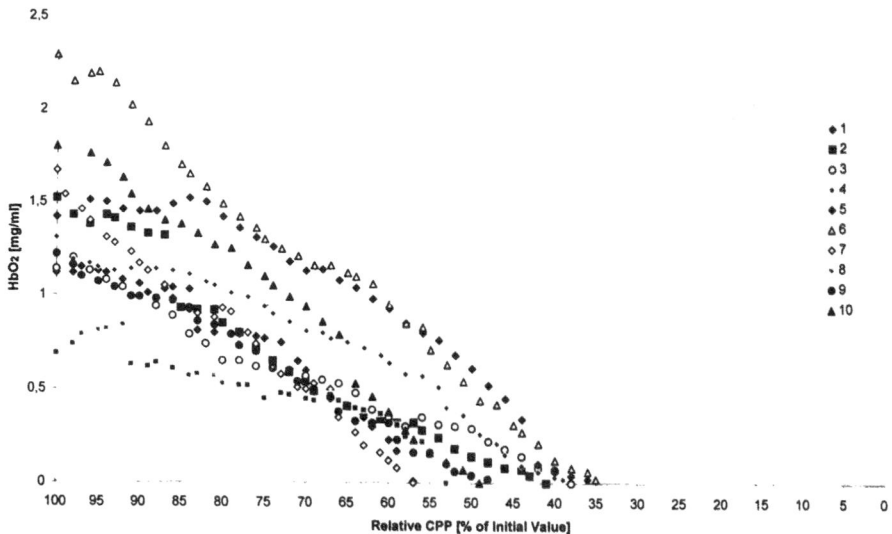

Figure 2. HbO₂ - Local tissue oxyhemoglobin concentration [mg/ml tissue]. (median of HbO₂ values calculated from approximately 1000 single measurements taken at each CPP level; MSD: 0.04).

DISCUSSION

Continuous monitoring of cerebral function and integrity is a mainstay of modern neuro-intensive care. It represents the basis for initiation and continuous evaluation of therapeutic measures. A multitude of experimental and clinical studies have led to conflicting results regarding the value of different hemodynamic or metabolic parameters as a predictor of neurologic function and outcome (Bruce 73, Cruz 90, Grubb 75, Marshall 75

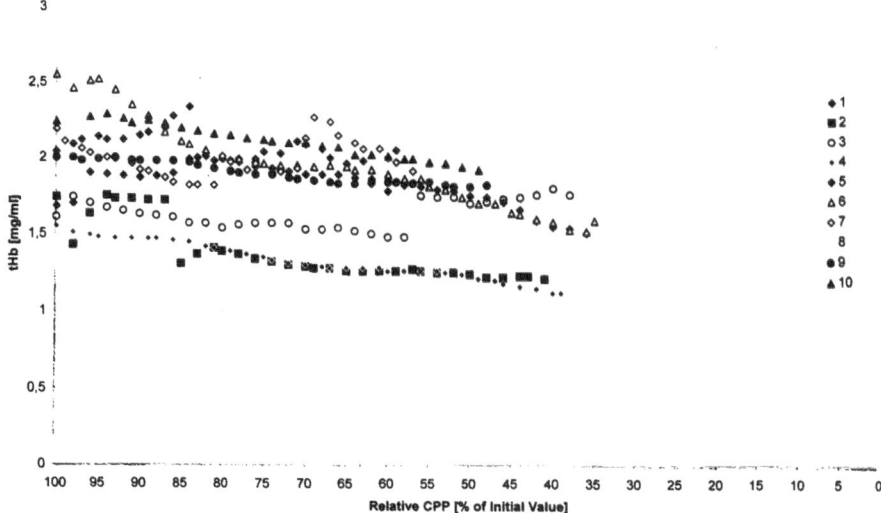

Figure 3. tHb - Local tissue hemoglobin concentration [mg/ml tissue]. (median of tHb values calculated from approximately 1000 single measurements at each CPP level; mean standard deviation: 0.04).

Table 1. Relationship between CPP and tiSO$_2$ levels at baseline conditions, beginning cerebral ischemia and during transition of the EEG towards electrocerebral silence (ECS) recorded by 8-channel EEG

	Baseline conditions		Decreasing EEG-frequency, ischemic arterial pressor response		Transition of spontaneous electrocortical activity towards ECS	
	CPP [mmHg]	tiSO$_2$ [%]	CPP [mmHg]	tiSO$_2$ [%]	CPP [mmHg]	tiSO$_2$ [%]
Max	70	88	50	40	42	5
Median	55	76	30	30	20	1
Min	45	51	20	21	9	0
MSD	7.57	12.14	9.1	5.5	9.8	1.63

Maximum, minimum and median values as well as mean standard deviation (MSD) are reported. The arterial vasopressor response to cerebral ischemia was recorded at tiSO$_2$ values ranging between 21% and 40% (median: 30%, MSD: 5.5%), and CPP values between 20 and 50 mmHg (median: 30 mmHg, MSD: 9.1 mmHg), respectively. ECS appears when tiSO$_2$ approaches 0% (mean: 1%) with low interindividual variability (MSD: 1.63). CPP values at ECS range from 9 to 42 mmHg (median: 20 mmHg), demonstrating considerable interindividual variability (MSD: 9.8)

I-III, Obrist 84). Definition of reliable threshold levels has turned out to be even more difficult (Bruce 73, Marshall 75 I, Marshall 75 III, McGillicuddy 78). NIRS based techniques have proven their usefulness in a variety of clinical settings, including management of severe head injury (Höper 94, McCormick 91, Steinberg 96). We are now able to provide data characterizing the relationship between cerebral tissue oxygenation and spontaneous electrocortical activity as indicators of cellular function during stepwise reduction of CPP. Cortical electrical activity was progressively suppressed at tiSO$_2$ levels around 30% and completely abolished when tiSO$_2$ approached 0%, demonstrating a remarkably homogenous pattern throughout the whole experimental series. Tissue oxygen saturation and tissue HbO$_2$ concentrations were closely correlated to CPP (r^2: 0.98). Overall tissue hemoglobin concentration (tHb) reflecting CBV was not correlated to CPP in 50% of our animals, while a previous study has reported increasing CBV with rising ICP[5]. Significant variability was noted with respect to individual CPP levels maintaining tiSO$_2$ at 30% (CPP range: 20–50 mmHg) and with tissue saturation levels near 0% (CPP range: 9–42 mHg). Our results indicate that CPP is not a good predictor of tissue oxygenation and therefore of cortical function or integrity. This corresponds favorably with the results from other experimental and clinical studies (Cruz 90, Marshall 75 I). It appears to be possible to define a safe range for cerebral tiSO$_2$ (above 40%), whereas a safe range for CPP cannot be generally determined due to large interindividual variability in CPP levels required for maintenance of adequate tissue oxygenation. However, it may be possible to define a safe range for CPP individually by determination of a minimum (critical) CPP based on measurements of cerebral tissue oxygen saturation.

CONCLUSION

Our results clearly demonstrate a close relationship between cerebral tissue oxygen saturation and spontaneous electrocortical activity as a determinant of cortical function. It therefore appears to be reasonable to assume that tiSO$_2$ is a valid and reliable parameter of functional cerebral metabolism. Measurement of tiSO$_2$ as single individual parameter can be performed independently from other diagnostic modalities and may provide a significant contribution towards a better understanding of information gained by measurements of ICP and CPP. Clinical studies will aid in further evaluation of the results obtained in this study with regard to application of this technique as routine monitoring in neuro-intensive care.

REFERENCES

1. Barzó P, Dóczi T, Csete K, Buza Z, Bodosi M: Measurements of regional cerebral blood flow and blood flow velocity in experimental intracranial hypertension: infusion via the cisterna magna in rabbits. *Neurosurgery* 28: 821–825, 1991

2. Bruce DA, Langfitt TW, Miller JD, Schutz H, Vapalahti MP, Stanek A, Goldberg HI: Regional cerebral blood flow, intracranial pressure, and brain metabolism in comatose patients. *J Neurosurg* 38: 131–144, 1973

3. Cruz J, Miner ME, Allen SJ, Alves WM, Gennarelli TA: Continuous monitoring of cerebral oxygenation in acute brain injury: injection of mannitol during hyperventilation. *J Neurosurg* 73: 725–730, 1990

4. Gilboe DD, Kintner D, Yanushka J: Cerebral oxygen utilization as a gauge of brain energy metabolism, in Bruley D, Bicher HI, Reneau D (eds): *Oxygen Transport to Tissue VI*. Plenum Press, New York, 1984, pp 169–177

5. Grubb RL Jr, Raichle ME, Phelps ME, Ratcheson RA: Effects of increased intracranial pressure on cerebral blood volume, blood flow and oxygen utilization in monkeys. *J Neurosurg* 43: 385–398, 1975

6. Elwell CE, Cope M, Edwards AD, Wyatt JS, Delpy DT, Reynolds EOR: Measurement of cerebral blood flow in adult humans using near infrared spectroscopy - methodology and possible errors. *Adv Exp Med Biol* 317: 235–245, 1992

7. Harper AM: Autoregulation of cerebral blood flow: influence of the arterial blood pressure on blood flow through the cerebral cortex. *J Neurol Neurosurg Psychiatry* 29: 398–403, 1966

8. Hazeki O, Seiyama A, Tamura M: Near infrared spectrophotometric monitoring of haemoglobin and cytochrome aa_3 in situ. *Adv Exp Med Biol* 215: 283–289, 1987

9. Hayashi N, Tsubokawa T, Moriyasu N: Cerebral venous circulation mechanism during intracranial hypertension, in Ishii S, Nagai H, Brock M (eds): *Intracranial Pressure V*. Springer, Berlin/Heidelberg, 1983, pp 348–351

10. Höper J, Gaab MR: Intraoperative monitoring of local Hb-oxygenation in human brain cortex, in Hogan MC (ed): *Oxygen Transport to Tissue XVI*. Plenum Press, New York, 1994

11. Hudetz AG, Feher G, Knuese DE, Kampine JP: Erythrocyte flow heterogeneity in the cerebrocortical capillary network, in Vaupel P (ed): *Oxygen Transport to Tissue XV*. Plenum Press, New York, 1994, pp 633–642

12. Lammertsma AA, Brooks DJ, Beaney RP, Turton DR, Kensett MJ, Heather JD, Marshall J, Jones T: In vivo measurement of regional cerebral hematocrit using positron emission tomography. *J Cereb Blood Flow Metab* 4: 317–322, 1984

13. Langfitt TW, Weinstein JD, Kassell NF: Cerebral vasomotor paralysis produced by intracranial hypertension. *Neurology* 15: 622–641, 1965

14. Lowell HM, Bloor BM: The effect of increased intracranial pressure on cerebrovascular hemodynamics. *J Neurosurg* 34: 760–769, 1971

15. Marshall LF, Durity F, Lounsbury R, Graham DI, Welsh F, Langfitt TW: Experimental cerebral oligemia and ischemia produced by intracranial hypertension. Part 1: Pathophysiology, electroencephalography, cerebral blood flow, blood-brain barrier and neurological function. *J Neurosurg* 43: 308–317, 1975 I

16. Marshall LF, Durity F, Lounsbury R, Graham DI, Welsh F, Langfitt TW: Experimental cerebral oligemia and ischemia produced by intracranial hypertension. Part 2: Brain morphology. *J Neurosurg* 43: 318–322, 1975 II

17. Marshall LF, Durity F, Lounsbury R, Graham DI, Welsh F, Langfitt TW: Experimental cerebral oligemia and ischemia produced by intracranial hypertension. Part 3: Brain energy metabolism. *J Neurosurg* 43: 323–328, 1975 III

18. McCormick PW, Stewart M, Goetting MG, Dujovny M, Lewis G, Ausman JI: Noninvasive cerebral optical spectroscopy for monitoring cerebral oxygen delivery and hemodynamics. *Crit Care Med* 19: 89–97, 1991

19. McCormick PW, Stewart M, Lewis G, Dujovny M, Ausman JI: Intracerebral penetration of infrared light. *J Neurosurg* 76: 315–318, 1992

20. McGillicuddy JE, Kindt GW, Miller CA, Raisis JE: The relation of cerebral ischemia, hypoxia, and hypercarbia on the Cushing response. *J Neurosurg* 48: 730–740, 1978

21. Obrist WD, Langfitt TW, Jaggi JL, Cruz J, Gennarelli TA: Cerebral blood flow and metabolism in coma-tose patients with acute head injury. Relationship to intracranial hypertension. *J Neurosurg* 61: 241–253, 1984

22. Pearce WJ, Scremin OU, Sonnenschein RR, Rubinstein EH: The electroencephalogram, blood flow, and oxygen uptake in rabbit cerebrum. *J Cereb Blood Flow Metab* 1: 419–428, 1981

23. Steinberg F, Röhrborn HJ, Scheufler KM, Asgari S, Trost HA, Seifert V, Stolke D, Streffer C: NIR Reflec-tion spectroscopy based oxygen measurements and therapy monitoring in brain tissue and intracranial neo-plasms - correlation to MRI and angiographic data. *Same Volume*

24. Symon L, Branston NM, Strong AJ, Hope TD: The concepts of thresholds of ischemia in relation to brain structure and function. *J Clin Pathol* (Suppl) 11: 149–154, 1977

25. Weinstein JD, Langfitt TW, Kassell NF: Vasopressor response to increased intracranial pressure. *Neurology* 14: 1118–1131, 1964

26. Wray S, Cope M, Delpy DT, Wyatt JS, Reynolds OS: Characterization of the near infrared absorption spec-tra of cytochrome aa$_3$ and haemoglobin for the non-invasive monitoring of cerebral oxygenation. *Bio-chimica et Biophysica Acta* 933: 184–192, 1988

33

HUMAN BRAIN FUNCTIONAL IMAGING WITH REFLECTANCE CWS

S. Nioka, Q. Luo, and B. Chance

Department of Biochem/Biophysics
University of Pennsylvania
Philadelphia, Pennsylvania 19104-6089

1. SUMMARY

Continuous Wave Spectroscopy (CWS) has been used in human subjects to detect brain function by observation of brain oxygenation and blood volume changes. In this research, we have developed a device for imaging brain oxygenation and blood volume with multiple sources and detectors in an area of 10 x 12 cm. Twelve, 1 watt tungsten bulbs were used as a light source, together with 8 detectors consisting of silicon-photodiodes with band-pass filters at 760 and 850 nms. The detectors were placed on a model system of the human head, providing 16 source-detector combinations with 4 cm separation. Eight to 16 seconds were required to image a brain function. Models using resin and a black object which have similar optical characteristics to a brain were studied. The model study showed that we can detect changes in absorption as deep as 2 cm from the surface. Human subjects with relatively thin hair, who volunteered for this study, were asked to tap fingers and rest, or to observe light flashing and darkness alternatively for 10 to 20 minutes while the images were being taken. In the finger tapping, the blood volume increased in a small, 1–2 cm area, while in the light stimulation, a much larger area was affected with a greater increased blood volume. Oxygenation occurred also in the area where blood volume increased. We concluded that the CWS can be used to image the location of brain activity by means of blood volume increase and oxygenation. In addition, brain disfunction and diseases, such as learning disability and brain hematoma may be detected with this imaging device.

2. INTRODUCTION

Imaging of the human brain function in vivo is relatively new and has been demonstrated with several technologies; PET, FMRI, EEG mapping, etc. The principle mechanisms for detecting brain function is through utilization and generation of metabolites,

Oxygen Transport to Tissue XIX, edited by Harrison and Delpy
Plenum Press, New York, 1997

therefore metabolites concentration such as glucose and oxygen can be used for imaging brain (1). Since optical probes using Near Infra Red (NIR) light can also measure oxygen concentration and blood volume, we aim to investigate the possibility of imaging brain function by NIR light. Maki has used a series of continuous wave reflectance probes to reconstruct images of the human brain cortex successfully (2). Here we used a simpler device; tungsten light bulbs and silicon photodiodes to image brain function. This simpler device has been used to measure oxygenation and blood volume changes (NIRS, Runman, NIM incorporated). This device is handy, safe and relatively cost efficient and therefore will provide flexibility in different avenues of human research in future.

3. METHODS

3.1. Optical Imaging Device

We have developed an optical imaging device, consisting of 9–12 small tungsten lamps (1 watt) and 8 silicon photodiodes as light sources and detectors, respectively. These were located over 4 x 9 cm area and mounted in an elastic, black plastic sheet. These lights and sources made up 16 combinations of light source-detector locations (Figure 1). The location of each light source-detector combination is designed to make matrix of 4 x 4 pixels. We assumed that the center of the location between each detector and source is considered to be a center of the volume (3), where the optical information came from; the center of the pixel. The distance from detector to source is 2.5 cm for human study, and depth of optical information is thought to be approximately half of the distance, 1.25 cm.

3.2. Brain Model Study

We have made a brain model to find the sensitivity and resolution of the imaging device. The model mimics optical characteristics of the brain (4). For this purpose, we used 1% intralipid with a black derlin block as a hematoma model. The 6 mm diameter block

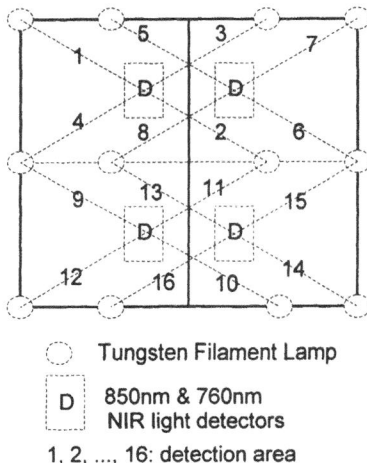

Figure 1. Design of a image probe which includes multiple light sources (open circle) and detectors (rectangle).

was localized 1 cm to 2 cm deep within the surface of the model, where the imaging device was set.

3.3. Human Brain Study

Ten human subjects volunteered for this study. The human subjects were tested for their brain activities. They sat on a chair and the imaging devices were located on their heads. The three areas of the brain, frontal, parietal, and occipital cortices were focussed for studying cognitive, motor nerve and visual activities with the probe located on the particular parts of the head. For the cognitive activity studies, French translation from English was questioned for 40 seconds followed by 2 minutes of resting. During the testing, brain activity was monitored with the optical imaging device on the forehead of the subject. For the motor nerve activity test, subjects were asked to tap fingers on either their left or right hand, while CWS imaging was taken on the parietal cortex. For the visual cortex activity, subjects were asked to watch a TV monitor, which had a 1 inch ball changing colors with 1 HZ frequency, while the occipital region of the head was monitored with the CWS imaging device. Each test had a testing period of 40 seconds followed by a 2 minute resting period.

4. RESULTS

4.1. Brain Model

With a black derlin (diameter 6 mm to 1 cm) located 1 to 4 cm from the surface, the CWS imaging device can detect the object as shown in figure 1. The sensitivity was assessed by signal to background noise ratios. With a device which has 4 cm distance between source and detector, the best S/N ratio was obtained at 2 cm depth, approximately half of the distance between source and detector. The location of the object was clearly detected by the device, and the resolution of the object was limited by the source-detector separation (Figure 2).

4.2. Human Brain Functional Imaging

4.2.1. Visual Cortex Stimulation. An example of the visual cortex activity is shown in Figure 3. In this case, a 54 year-old male subject was tested. He was asked to watch a monitor and showing a 5 cm diameter object, whose color was changing during the 40 second testing period. The imaging device was placed on the occipital region of the head

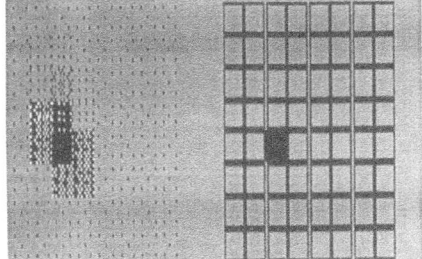

Figure 2. An image of a hematoma model (right). Left portion shows the location of a black derlin rod. The right picture shows an image of the black rod located with 1 cm depth from the surface.

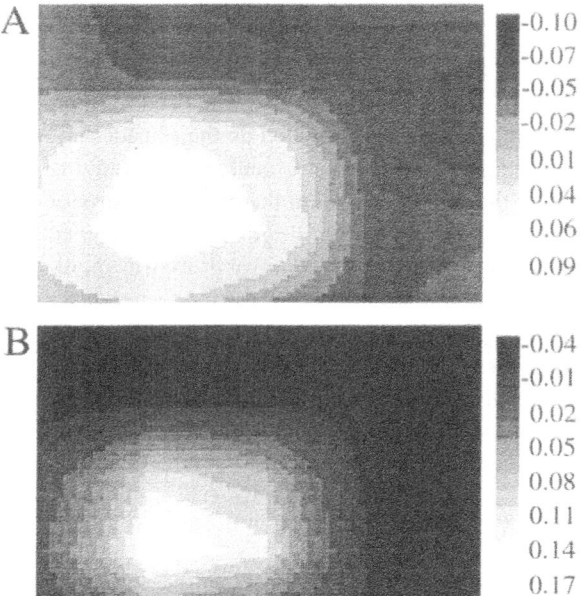

Figure 3. Blood volume images of a visual cortex of a man, viewing a monitor screen. Blood volume increase is depicted during the 1st test (A), and following 2nd test (B) after 2 min. of rest.

above the occipital processes. The change from the resting period before the test was analyzed as a difference of OD (DOD) in the Figure 3. The Figure shows that an area of 1 x 2 cm had an increased blood volume and desaturation of hemoglobin during the visual activation in this subject, and recovery phase shows that it reverses slowly during a 2 min period of rest.

4.2.2. Prefrontal Cortex Stimulation. A high school student was asked to translate words from English to French for 40 seconds with resting periods of 2 minutes in be-

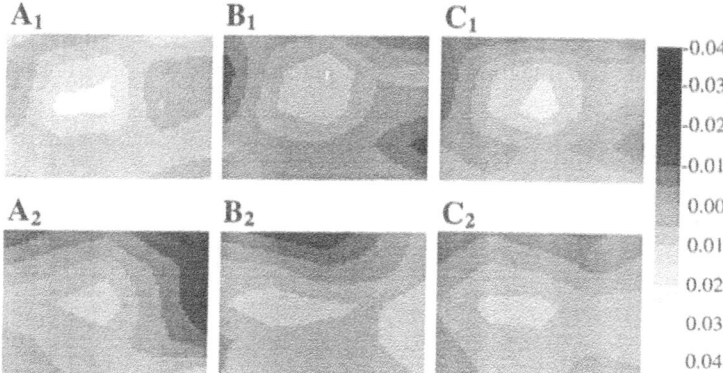

Figure 4. Images of the prefrontal cortex of a high school student who translated English to French. Oxygenation (A1,B1,C1) and blood volume increases (A2,B2,C2) are seen. This can be repeated as shown in order of time A to B to C.

tween. This sequence was repeated three times. The imaging probe was set on the subject's forehead. The results are presented in Figure 4. An increased oxygenation is associated with an increase in blood volume in an area of 0.5 x 2 cm of the prefrontal cortex. This characteristic was repeated and showed stable activation in a specific part of the prefrontal cortex. However, the first reaction of increase in blood volume and oxygenation was the greatest among the three sequential trials.

4.2.3. Motor Cortex Activation. A subject was asked to perform left finger tapping as fast as possible for 40 seconds and to rest for the following 2 minutes. The results showed an increase in blood volume and oxygenation in an area of 1.5 cm in the right parietal part of the brain. An increase in blood volume was high; an increase of 0.15 OD when the fingers were moving. In the recovery, a decrease in blood volume was observed. This change was associated with deoxygenation of the area.

DISCUSSION

As demonstrated in this study with the brain models and human functional brain studies, this CWS imaging device can localize the activated area of the human brain. The mechanism for detection is that human brain function is associated with blood volume increases and oxygen concentration changes. It is found that simple functions of the brain such as observing an object, moving a small part of the body and thinking, can activate as small an area as 0.5 to 1 cm of the brain cortex. The area of activation was expected to be seen in those cortices of interest.

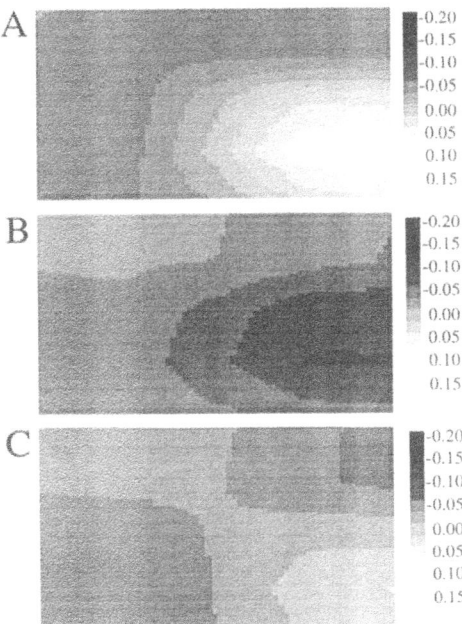

Figure 5. Blood volume images of the motor cortex of a man tapping his fingers. An increase in blood volume is seen during the finger tapping (A). After the tapping, recovery is seen at 30 seconds (B), and blood volume restored at 30–60 seconds (C).

With regard to the technical aspects of CWS imaging, the theoretical resolution and sensitive depth depends on the source-detector distance (half of the distance). We had 2.5 cm for source-detector distance that determined resolution of the image as 1.25 cm in the brain study. From the depth of 1.25 cm from the skin, we assume that most of the signal comes from blood vessels located in the gray matter and brain surface rather than inside of the brain; white matter. That should be sufficient for us to observe blood circulation and metabolism of the cortex, since major contributor is in gray matter. Even with 1.5 cm light source-detector separation, which light path is assumed to penetrate only 8 mm depth from the skin, human brain function has been well detected (5), suggesting that vessels located on the brain reflect metabolism of the brain cortex.

ACKNOWLEDGMENTS

This research was supported by NIH grant RO1 NS273460.

REFERENCES

1. Ogawa et al. Magnetic Resonance in Medicine. 29:205–210. 1993.
2. Maki et al. Med. Phys. 22:1997–2005. 1995.
3. Feng et al. SPIE. 2389:54–63. 1995.
4. Matcher et al. SPIE. 2389:486–495. 1995.
5. Hoshi et al. Neuroscience Letters. 150:5–8. 1993.

34

RELATIONSHIP BETWEEN REGIONAL MYOCARDIAL OXYGENATION AND FUNCTION DURING ACUTE ISCHEMIA SUPPORTED BY SELECTIVE SUCTION AND RETROINFUSION (SSR) IN PIGS

G. von Degenfeld,[1] W. Peter, H. Habazettl,[2] M. von Lüdinghausen,[3] K. Werdan,[4] and P. Boekstegers[1]

[1]Department of Internal Medicine I
Klinikum Großhadern
[2]Institute for Surgical Research
University of Munich, Germany
[3]Institute of Anatomy
University of Würzburg, Germany
[4]Department of Internal Medicine III
University of Halle-Wittenberg, Germany

INTRODUCTION

Retrograde infusion of arterial blood through coronary veins has cardioprotective effects during acute coronary ischemia. It has been shown to reduce infarct size and to preserve myocardial function, tissue oxygenation and regional blood flow in experimental[1,2,5,6,22] and clinical studies.[4,10,11,18] Nevertheless, the actual nutritional efficacy of retroinfusion in comparison with antegrade blood flow and the potential of retroinfusion to provide adequate oxygen and substrate supply and metabolite washout have not been fully established.

Reduction of antegrade coronary blood flow leads to well defined changes in myocardial oxygen tension, myocardial function and regional capillary blood flow.[3,7,8,16,21] Retrograde infusion of arterial blood, however, changes the relationship between these parameters and leads to reduced myocardial function compared with physiologic antegrade blood supply.[20] These findings, however, were made using an experimental nonsynchronized retroinfusion setup. They cannot be extrapolated to retroinfusion devices designed for clinical use, which operate in a synchronized mode (allowing draining of the coronary veins during systole) in order to avoid potentially harmful systolic peak pressures.[2,15] The aim of the present study was, therefore, to investigate the nutritional efficacy of Selective Suction and Retroinfusion (SSR), a percutaneous retroinfusion catheter device, in comparison with previously published data with antegrade perfusion[3] in a pig model of acute experimental coronary ischemia.

The nutritional efficacy of SSR was assessed by the same parameters as during antegrade perfusion: measurements of myocardial surface oxygen tension and regional myocardial function. Because polarographic oxygen probes located at different myocardial sites have shown partly conflicting results also during antegrade perfusion, intramyocardial oxygen electrodes were simultaneously used in this study.[3,12,13]

The relationship between regional myocardial surface pO_2, intramyocardial pO_2 and myocardial function could, thus, be analyzed in a pig model of acute coronary ischemia during SSR.

MATERIALS AND METHODS

Synchronized Suction and Retroinfusion of Coronary Veins

The system of selective synchronized suction and retroinfusion (SSR) has previously been described in detail.[1,2] Arterial blood is withdrawn from a femoral artery and selectively infused into the ischemic region during diastole through a retroinfusion catheter positioned in the coronary vein, an active suction device allowing rapid and complete draining of the vein during systole. Briefly, this system consists of a piston pump for retroinfusion (Prominent, FRG), a vacuum pump for suction (Medap, Switzerland), an ECG triggered electronic control device. The 4-lumen 7F SSR-catheter is connected to the infusion pump, the vacuum pump, the balloon pump and a pressure transducer. Infusion and suction are triggered by ECG in order to limit the infusion period to diastole. Retroinfusion is, thus, performed during every second diastole, whereas suction occurs during the remaining time of the SSR-cycle.

Multiwire Surface Electrode (MDO)

The clark-type multiwire surface pO_2 electrode (Mehrdraht-Dortmund-Oberflächenelektrode (MDO), Eschweiler, FRG) used for determination of myocardial surface oxygen tension has been described in detail elsewhere.[3,13,14] A two point calibration of the MDO was performed at 37°C in saline solution calibrated with pure N_2 (gas 1) and 10% O_2 (gas 2). The MDO was placed into an electrode support sutured to the epicardium. Correct isolation was checked by suffusing the electrode successively with pure O_2 and pure N_2. Keeping the position of the MDO (which features 8 cathodes) unchanged throughout the experiment, eight different points at the myocardial surface were continuously monitored. The results were corrected for drift (assumed to be linear) and tissue temperature. Myocardial surface pO_2-values were obtained by calculating the average of the 8 registrations of the MDO over 20 seconds (i.e. 1 measurement/second over 20 seconds with 8 cathodes): one myocardial surface pO_2 value represents, thus, the mean value of 160 single measurements.

pO_2-Microcatheter Probe

The clark-type polarographic pO_2-microcatheter was originally developed for intravascular oxygen monitoring and was adapted for intramyocardial measurements as described previously[1,2,19] (GMS, FRG). Two oxygen permeable silicone tubes were inserted 3–5mm subepicardially into the ischemic area (left anterior descending artery (LAD)) and the non-ischemic zone (circumflex artery (CX)) with an atraumatic needle. Following two point calibration in saline solution calibrated with N_2 (gas 1) and 10% O_2 (gas 2), 2 microcatheter probes were placed into the tubes and measurements were carried out as soon as stable baseline values were obtained. Measurements were corrected for drift, tissue temperature and 75% response time ($t_{75\%}$), the $t_{75\%}$ having been assessed in vitro for each microcatheter probe individually.

Regional Myocardial Function

Regional myocardial function was measured by means of sonomicrometry.[9] Two pairs of 5-MHz ultrasonic crystals (2mm diameter) were placed subendocardially into the ischemic (LAD) and non ischemic area (CX). The crystals were aligned perpendicular to the long heart axis 10 to 15mm from each other. Correct position was confirmed at postmortem. End diastolic (EDL) and end systolic (ESL) segment lengths were defined by dp/dt (EDL: before upstroke of dp/dt; ESL: 20ms before minimal dp/dt) for calculation of segment shortening (SS): SS [%] = (EDL−ESL) / EDL×100.[9] Segment shortening of 5 consecutive heart beats were analysed to obtain mean segment shortening at every time point.

Experimental Preparation

Following approval by the Bavarian Animal Care and Use Committee, this study was carried out on 20 German Farm pigs of either sex. The preparation was described in detail previously.[2] Briefly, following sedation and endotracheal intubation, anaesthesia was maintained by midazolam i.v., piritramide i.v. and enflurane. A microtip pressure transducer catheter (Altron, FRG) was introduced into the left ventricle and a thermodilution catheter into the pulmonary artery (Baxter, USA). Following thoracotomy, a snare was placed loosely around the left anterior descending artery (LAD) immediately distal to its first diagonal branch, and the SSR catheter was introduced into the anterior interventricular vein (AIV).

A first 60sec control occlusion of the LAD was performed with an atraumatic vessel clip (Frohnhäuser, FRG) to check correct position of the pO_2 electrodes and the ultrasonic crystals. Following at least 10 min reperfusion, baseline oxygen tension and regional myocardial function were measured. The LAD was then occluded for 10 min and SSR was performed in the SSR group (n=10), whereas the control group (n=10) was not treated. Myocardial oxygen tension and regional myocardial function were recorded 1, 2, 3, 5, and 7.5 min following LAD occlusion. Finally, circulatory standstill was induced by potassium chloride overdose.

Biopsies of the tissue surrounding the silicone tubes were taken for analysis of tissue injury caused by the intramyocardial pO_2 probes. Biopsies were processed as described previously.[2]

Data Acquisition and Statistical Analysis

Baseline measurements of tissue oxygen tension and regional myocardial function were obtained before the second LAD occlusion and all subsequent data were normalized to baseline values. During ischemia, the measurements were obtained as follows: since myocardial pO_2 and regional function changed rapidly during the first 2 minutes of ischemia, data obtained in this phase were discarded and only the time points baseline, 2, 3, 5 and 7.5 min after occlusion were included into the final analysis. A total of 5 measurements for each animal was, thus, obtained for every parameter.

Baseline data were analysed with the Kruskal-Wallis-test to assess similarity of both groups. Relationship between mean myocardial surface pO_2, intramyocardial pO_2 and regional myocardial function were tested by linear regression analysis and calculation of the Pearson's correlation coefficient (r).

RESULTS

Baseline haemodynamic variables (cardiac output, heart rate and mean arterial pressure), mean myocardial surface pO_2, intramyocardial pO, and regional myocardial function were comparable in both groups (table). Mean retrograde blood flow was 51 ±15ml/min and mean pressure in the AIV during retroinfusion 63 ±14mmHg.

Relationship between Myocardial Surface pO_2 and Intramyocardial pO_2

A significant linear relationship was found between baseline mean myocardial surface pO_2 and intramyocardial pO_2 under physiological, non-ischemic conditions (r=0.7; p<0.001). Mean surface oxygen pO_2 was, however, considerably higher than intramyocardial pO_2: 51.1 ±13.5 mmHg vs. 37.9 ±9.9 mmHg (fig. 1). Baseline values of the microcatheters showed similar values for different regions of the left ventricle (LAD: 37.9 ±9.9 mmHg vs. CX: 35.3 ±12.7 mmHg) and a correlation coefficient of r=0.71; p<0.005.

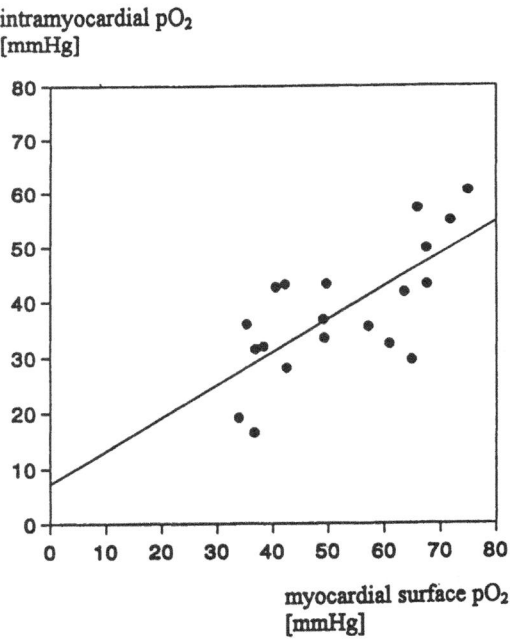

Figure 1. Baseline intramyocardial pO_2 and myocardial surface pO_2 in the LAD-region (r= 0.92; p<0.005; n=20) For calculation of myocardial surface pO_2: see text.

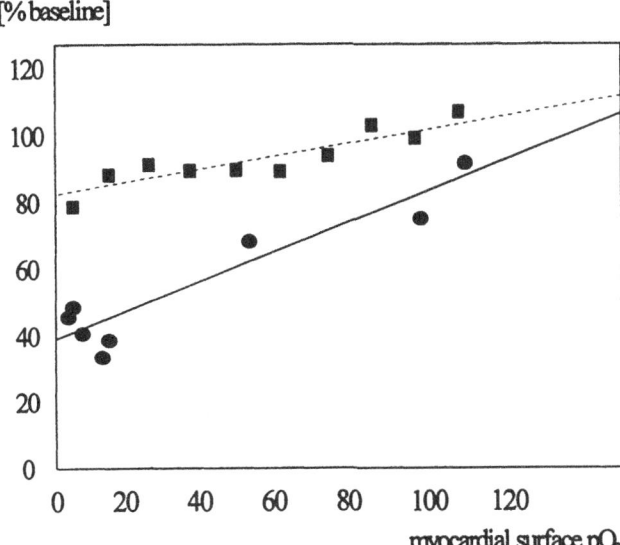

Figure 2. Myocardial surface pO$_2$ and regional myocardial function in the LAD-region during ischemia (see text for details): ● = during retroinfusion (r = 0.7; p < 0.001; n = 8) ■ = during antegrade perfusion (Conzen et al. *Cardiovasc Res* 91;25:207–216).

Following LAD occlusion, myocardial oxygen tension decreased during the first 2 minutes and remained stable thereafter. Since myocardial surface pO$_2$ decreased rapidly to 0mmHg during ischemia in the control group (in contrast to the SSR treated group), these data were excluded from further analysis. During ischemia supported by retroinfusion (SSR), there was a statistically significant correlation between mean myocardial surface pO$_2$ and regional myocardial function (r=0.92; p<0.001) (fig. 2). For statistical reasons the mean values of the time points 2, 3, 5 and 7.5 min of each animal was used for linear regression analysis. A similar correlation was found, however, if all pooled values of regional myocardial function and of myocardial surface oxygen tension were plotted. Intramyocardial oxygenation was preserved at a higher level than surface pO$_2$ during ischemia supported by SSR: 26.0 ±11.8 mmHg (intramyocardial pO$_2$) vs. 10.3 ±12.0 mmHg (surface pO$_2$). The correlation between mean intramyocardial pO$_2$ and myocardial surface oxygen was, however, not statistically significant during ischemia due to scattering of intramyocardial pO$_2$ data. However, a trend towards a relationship between these parameters was found (r=0.7; p=0.08).

Characteristics of the pO$_2$-Probes

Shorter t$_{75\%}$-response times were found for the MDO than for the microcatheter probes: 4 ±2 seconds (MDO) vs. 45 ±16 sec (microcatheters). Despite this comparatively slow response characteristic, however, the pO$_2$ microcatheter probes placed intramyocardially showed an initial pO$_2$ decrease within 5 seconds following LAD occlusion.

In contrast to the MDO, the pO$_2$ microcatheter probes showed a characteristic adjustment curve with an initial steep fall to values close to 10mmHg, followed by an asymptotic rise until steady state was reached after 30–45minutes.

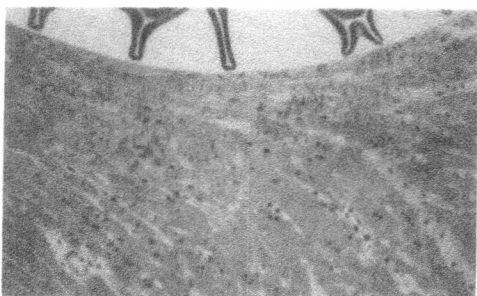

Figure 3. Cross-sectional view of the insertion channel of a pO₂ microcatheter probe, containing the silicone tube. Enlarged 100fold, Richardson staining.

Postmortem Studies

Correct placement of the pO₂ probes and the ultrasonic crystals in all experiments were confirmed post mortem. Using light microscopy, the cross sectional view of the myocardial biopsies including the silicone tube showed an irregular, 25–100 μm broad compression zone of the surrounding tissue (fig. 3). In addition, some necrotic cells were in the layers closer than 25 μm to the tube. Beyond this zone, no signs of tissue injury or compression were found, and all arterioles, venules and capillaries appeared open. Transmission electron microscopic inspection confirmed light microscopic findings and revealed a few segmented granulocytes near the silicone tube (fig. 4). No red blood cells were found in this area.

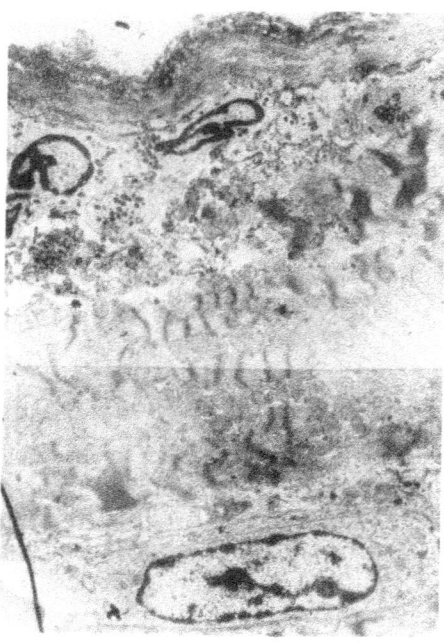

Figure 4. Cross-sectional view of the insertion channel of a pO₂ microcatheter probe, containing the silicone tube. Enlarged 100fold, Richardson staining.

No evidence was found for venous wall injury or tissue damage caused by retroinfusion.

DISCUSSION

Influences of Retroinfusion on Myocardial Oxygen Tension and Regional Myocardial Function

Retroinfusion therapy during acute coronary ischemia has cardioprotective effects in reducing infarct size and in preserving myocardial tissue oxygen tension, regional blood flow and myocardial function.[1,2,4,11,15,18,22] The present analysis addressed the issue of nutritional efficacy of selective suction and retroinfusion (SSR), a retroinfusion system designed for clinical use.

In a previous experimental study in pigs, the effects of graded reduction of antegrade blood flow on regional myocardial surface oxygen tension and regional myocardial function were investigated.[3] For comparison, the same parameters were measured during ischemia with retroinfusion by SSR in the present study. In both investigations, a significant correlation between tissue surface oxygenation and regional myocardial function was found (fig. 2). The regression line, however, showed a steeper slope during retrograde than antegrade blood supply, indicating a more pronounced loss of regional myocardial function at similar oxygen tensions. The intersections of the regression lines with the y-axis (corresponding to a calculated 0mmHg myocardial surface pO_2) showed 81% of baseline regional myocardial function with antegrade flow, in contrast to only 40% with SSR. Thus, at comparable levels of myocardial surface oxygenation, regional myocardial function was considerably less well preserved by retroinfusion with SSR. These results are consistent with a previous study, in which regional myocardial function during non synchronized retroinfusion was found to be reduced in comparison with continuous antegrade flow.[20]

Regional myocardial function was, thus, found to be impaired during retroinfusion despite considerable preservation of myocardial oxygen tension. Therefore, insufficient oxygen supply cannot explain the finding of reduced regional myocardial function. Transmission electron microscopic and light microscopic investigations carried out in the present study did not provide any evidence for impaired oxygen diffusion caused by alterations of the venous system or the surrounding tissue (e.g. edema or hemorrhage). Previous findings, however, indicate that retrograde diastolic filling of the coronary veins could lead to mechanical impairment of left ventricular function:[20] retrograde blood flow using an experimental non synchronized setup showed that gradual enhancement of retrograde blood flow led to higher coronary venous pressure and little additional functional gain. Especially in case of overperfusion, regional myocardial function was not further enhanced. Although extrapolation of these data to our analysis has to be interpreted with caution, mechanical impairment of myocardial function by retrograde blood flow could be a plausible explanation for the finding of less well preserved regional myocardial function at similar myocardial surface oxygen tension compared with antegrade blood flow.

Myocardial Surface pO_2 and Intramyocardial pO_2

Before ischemia, myocardial surface pO_2 was considerably higher than intramyocardial pO_2: 51.1 ±13.5 mmHg (myocardial surface) vs. 37.9 ±9.9 mmHg (intramyocardial). Although similar findings had been reported previously, the present investigation is the first to compare myocardial surface and intramyocardial pO_2 measurements directly by simultaneous measurement at both locations in the beating heart. During acute regional ischemia supported by retroinfusion (SSR), however, the relationship between mean myocardial surface and intramyocardial oxygen tension was reversed and the myocardial surface oxygenation was found to be lower than the intramyocardial pO_2: 10.3mmHg or 20.2%of baseline (myocardial surface) vs. 26.0mmHg or 68.6%of baseline (intramyocardial). Postmortem investigation carried out in the present study showed no evidence for tissue injury to cause artificially low intramyocardial oxygen levels[13,17]. Tissue compression was restricted to an irregular, 25–100μm broad zone around the insertion channel. In addition, the microcatheters were found to react to pO_2 changes following LAD occlusion within 5 seconds, suggesting that oxygen diffusion through the compression zone was not severely delayed by tissue compression. The main limiting factor, however, of the intramyocardial probes was the comparatively slow response time for complete assessment of a sudden change in tissue oxygen partial pressure. Although measurements of intramyocardial oxygen tension have been performed in numerous studies, the electrode sizes ranging between 1μm and 550μm, no clear relationship between measured intramyocardial pO_2 and the size of the electrode was found.[2,16,17,46] From our results we would infer, that pO_2 microcatheter probes are suitable for assessment of regional myocardial oxygen tension, representing the mean tissue pO_2 of this area, but not for pO_2 measurements with high spatial and temporal resolution.

REFERENCES

1. Boekstegers P, Diebold J, Weiss C: Selective ECG synchronised suction and retroinfusion of coronary veins: first results of studies in acute myocardial ischaemia in dogs. *Cardiovasc Res* 1990;24:456–464
2. Boekstegers P, Peter W, von Degenfeld G, Nienaber CA, Abend M, Rehders TC, Habazettl H, Kapsner T, von Lüdinghausen M, Werdan K: Preservation of regional myocardial function and myocardial oxygen tension during acute ischemia in pigs: comparison of selective synchronized suction and retroinfusion of coronary veins to synchronized coronary venous retroperfusion *J Am Coll Cardiol* 1994;23:459–469
3. Conzen PF, Habazettl H, Christ M, Baier H, Hobbhahn J, Vollmar B, Peter K: Left ventricular surface tissue oxygen pressures determined by oxygen sensitive multiwire electrodes in pigs. *Cardiovasc Res* 1991;25:207–216
4. Costantini C, Sampaolesi A, Serra CM, Pacheco G, Neuburger J, Conci E, Haendchen RV: Coronary venous retroperfusion support during high risk angioplasty in patients with unstable angina: preliminary experience. *J Am Coll Cardiol* 1991;18:283–292
5. Drury JK, Yamazaki S, Fishbein MC, Meerbaum S, Corday E: Synchronized diastolic coronary venous retroperfusion: results of a preclinical safety and efficacy study. *J Am Coll Cardiol* 1985;6:328–335
6. Farcot JC, Meerbaum S, Lang TW, Kaplan L, Corday E: Synchronized retroperfusion of coronary veins for circulatory support of jeopardized ischemic myocardium. *Am J Cardiol* 1978;41:1191–1201
7. Guth BD, Martin JF, Heusch G, Ross J: Regional Myocardial Blood Flow, Function and Metabolism Using Phosphorus-31 Nuclear Magnetic Resonance Spectroscopy During Ischemia and Reperfusion in Dogs. *JACC* 1987;10:673–681
8. Guth BD, Wisneski JA, Neese RA, White FC, Heusch G, Mazer CD, Gertz EW: Myocardial lactate release during ischemia in swine. Relation to regional blood flow. *Circulation* 1990;81:1948–1958
9. Heimisch W: Die Sonomikrometrie in der Herz- und Kreislaufforschung: Physikalisch-technische Grundlagen und Ergebnisse zur Myokard-Mechanik und Ventrikel-Geometrie. *Dissertation* Munich, FRG; 1989
10. Incorvati RL, Tauberg SG, Pecora MJ, Macherey RS, Krucoff MW, Dianzumba SB, Donohue BC: Clinical applications of coronary sinus retroperfusion during high risk percutaneous transluminal coronary angioplasty. *J Am Coll Cardiol* 1993;22:127–134
11. Kar S, Drury JK, Hajduczki I, Eigler N, Wakida Y, Litvack F, Buchbinder N, Marcus H, Nordlander R, Corday E: Synchronized coronary venous retroperfusion for support and salvage of ischemic myocardium during elective and failed angioplasty. *J Am Coll Cardiol* 1991;18:271–282

12. Kessler M, Harrison DK, Höper J: Tissue Oxygen Measurement Techniques, in Baker CH, Nastuk WL (eds): *Microcirculatory technology*. Orlando, Academic Press, 1986, pp 391–425

13. Kessler M, Hoper J, Krumme BA: Monitoring of tissue perfusion and cellular function. *Anesthesiology* 1976;45:184–197

14. Kessler M, Klövekorn WP, Höper J, Sebening F, Brunner M, Frank KH, Harrison DK, Kernbach C, Anderer W, Richter H, Ellermann R: Local oxygen supply and regional wall motion of the dogs heart during critical stenosis of the LAD, in Lübbers DW, Acker H. Leninger-Follert E (eds.): Adv Exp Med Biol Vol.169, Plenum Press, 1984;331–339

15. Meerbaum S, Lang TW, Osher JV, Hashimoto K, Lewis GW, Feldstein C, Corday E: Diastolic retroperfusion of acutely ischemic myocardium. *Am J Cardiol* 1976;37:588–598

16. Menke W; Schuchhardt S; Fritz H: Intramyocardial oxygen pressure and blood flow during experimental coronary stenosis. In: Lübbers DW, Acker H, Leninger-Follert E, Goldstick TK (eds), *Oxygen transport to tissue*, Vol.5, Plenum Press, New-York 1983 pp341–350.

17. Moss AJ: Intramyocardial oxygen tension. *Cardiovasc Res* 1968;2:314–318

18. Nanto S, Nishida K, Hirayama A, Mishima M, Komamura K, Masai M, Sakakibara T, Kodama K: Supported angioplasty with synchronized retroperfusion in high-risk patients with left main trunk or near left main trunk obstruction. *Am Heart J* 1993;125:301–309

19. Nollert G, Vetter HO, Martin K. Weinhold C, Kreuzer E, Schmidt W, Reichart B: Does cold blood cardioplegia offer adequate oxygen delivery to the myocardium during coronary artery bypass grafting?. *Adv Exp Med Biol Vol XV*, Plenum Press, Baltimore, 1994:283–290

20. Oh BH, Volpini M, Kambayashi M, Murata K, Rockman HA, Kassab GS, Ross J, Jr. Myocardial function and transmural blood flow during coronary venous retroperfusion in pigs. *Circulation* 1992;86:1265–1279

21. Vatner SF: Correlation between acute reductions in myocardial blood flow and function in conscious dogs. *Circ Res* 1980;47:201–207

22. Wakida Y, Nordlander R, Kobayashi S, Kar S, Haendchen R, Corday E: Short-term synchronized retroperfusion before reperfusion reduces infarct size after prolonged ischemia in dogs. *Circulation* 1993;88:2370–2380

35

POST-ISCHEMIC ^{31}P NMR DETERMINATION OF MYOCARDIAL INTRACELLULAR pH *IN VIVO* USING ATP PEAK

Ralph J. F. Houston,[1] Arend Heerschap,[2] Stefan H. Skotnicki.
Freek W. A. Verheugt,[4] and Berend Oeseburg[1]

[1]Physiology Department 237
[2]Radiodiagnostic Institute
[3]Department of Cardiothoracic Surgery
[4]Department of Cardiology
Faculty of Medical Sciences
University of Nijmegen
Box 9101, NL-6500 HB Nijmegen, The Netherlands

1. INTRODUCTION

During tissue hypoxia induced by arterial hypoxia as well as by ischemia, a fall in intracellular pH may indicate compromise, particularly in tissue with limited resistance to hypoxia or limited capacity for anaerobic metabolism, such as brain or heart. ^{31}P NMR is an attractive method to measure high energy phosphate levels and pH on line and non-invasively. The usual pH indicator used is the chemical shift, or frequency difference, between the NMR peaks of inorganic phosphate (P_i) and phosphocreatine (PCr). However, during reperfusion and subsequent recovery, when P_i is taken up to form PCr and ATP, in our model at least (an isolated, working rat heart perfused with erythrocyte suspension), the P_i concentration drops to such a low level that the peak often becomes undetectable. During ischemia, a phosphomonoester peak appears which overlaps the P_i peak, and is often larger than the P_i peak during reperfusion. Another difficulty with the use of the P_i peak is that it can be obscured by 2,3-diphosphoglycerol (2,3-DPG) peaks from blood. Since the positions of the three ATP peaks, particularly the γ peak, also depend on pH, this fact could be used to determine pH complementarily as P_i and ATP are unlikely to reach low levels simultaneously in tissue with the potential to recover.

2. METHOD AND MATERIALS

The animal procedures described were approved by the Faculty Ethical Committee on Animal Research. In each of a series of seven experiments, an adult male Wistar (WS) rat was anesthetized with diethyl ether, its heart was removed and placed in ice-cold buffer to cause arrest. The aorta was cannulated and Langendorff perfusion with buffer solution started within approximately 10 min of heart removal. The left atrium was also cannulated, and both sutures used to attach wires for pacing in synchronism with the NMR pulse sequence. After checking for leaks, the perfusion system was switched to working configuration with erythrocyte suspension. Fluid columns were used to maintain preload pressure at 2 kPa and afterload pressure at 13 kPa. The solutions were oxygenated with carbogen (95% O_2, 5% CO_2) using membrane oxygenators. Buffer composition was NaCl 118, $CaCl_2$ 3.0, KCl 6.1, $NaHCO_3$ 25, $MgSO_4$ 1.2, Na EDTA 0.5, glucose 11.1mm and bovine serum albumin (fraction V) to 15 g·L^{-1}, and the erythrocyte suspension was the same solution with washed bovine erythrocytes added to a hematocrit of 25%. The solutions were phosphate free, and bovine erythrocytes which lack 2,3-DPG were used, to avoid obscuring the P_i peak. A more detailed description of the perfusion setup has already been published (Olders *et al.*, 1990).

The heart was mounted in a custom-designed radio frequency probe and installed inside the vertical bore of a 4.3 T magnet attached to an NMR spectrometer (SMIS Ltd, Guildford, Surrey, UK). Phosphorus experiments were conducted at 72.918 MHz using a simple 90° pulse and acquire sequence with a repetition time of 2.5 s. 128 finite impulse decays (FIDs) averaged together gave an adequate signal-to-noise ratio for analysis, resulting in a new spectrum every 5⅓ min. Baseline readings were collected for 32 min, then the heart was made globally ischemic by clamping off the lines leading to and from the heart for a given time, after which it was reperfused in Langendorff mode until coronary flow reduced to a more normal level, and a further 32 min of recordings taken. The NMR spectra were analyzed using the NMR1 package (New Methods Research Inc, New York NY, USA), first converting a 10 Hz Lorentzian line to a 30 Hz Gaussian, deconvoluting the baseline, and integrating over 5 line widths. The area of each peak was taken to represent the quantity of the corresponding substance present. ATP was calculated from the average area of the β and γ peaks, as the α peak overlapped another peak, possibly that of nicotinamide adenine nucleotide (NAD).

Calculation of pH was done in two ways, first from the chemical shift between the P_i and PCr peaks (Moon and Richards, 1973) using the coefficients published by Jacobus *et al.* (1978), and secondly by using the pH value and chemical shift between the PCr and γ-ATP peaks during the stable, working periods as reference, and the change in this shift due to pH, derived graphically from the Moon and Richards paper.

3. RESULTS AND ANALYSIS

Fig. 1 shows selected spectra from a representative experiment, with duration of ischemia 32 min. Gaussian line broadening of 20 Hz was applied, and PCr taken as 0 ppm. Spectrum (a) was from the stable working heart; (b) was after 5⅓ min ischemia, when PCr was reducing and P_i increasing; (c) was after 16 min ischemia, when PCr reached a very low level; (d) was after 26⅔ min, with a PME peak visible; (e) was after 5⅓ min reperfusion, when P_i had become indeterminate; and finally spectrum (f) was obtained from the stable working heart near the end of the experiment, when P_i was again detectable.

Fig. 2 shows the results of an analysis of the data collected in the experiment described above. PCr and ATP were normalized to the average value over the first five data points; P_i and PME were normalized to the peak P_i value. One set of pH values was calculated from the PCr - P_i chemical shift as described above. No pH values could be calculated from the first two spectra obtained after reperfusion, as the P_i peak position was indeterminate.

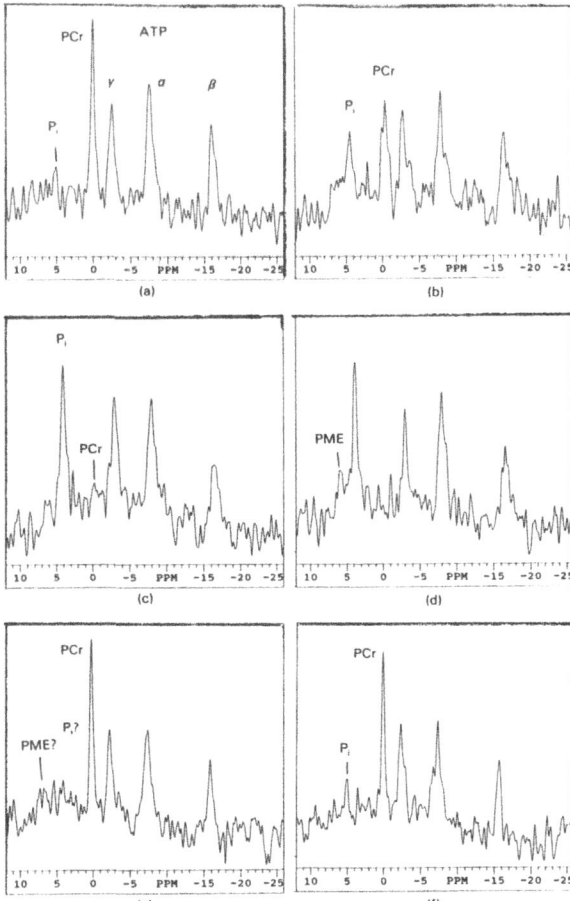

Figure 1. Selected spectra from a representative experiment, with ischemia lasting 32 min. Gaussian line broadening 20 Hz, PCr taken as 0 ppm. (a) stable working heart; (b) after 5⅓ min ischemia; (c) after 16 min ischemia; (d) after 26⅔ min, PME visible; (e) after 5⅓ min reperfusion, P_i indeterminate; (f) stable working near end of experiment.

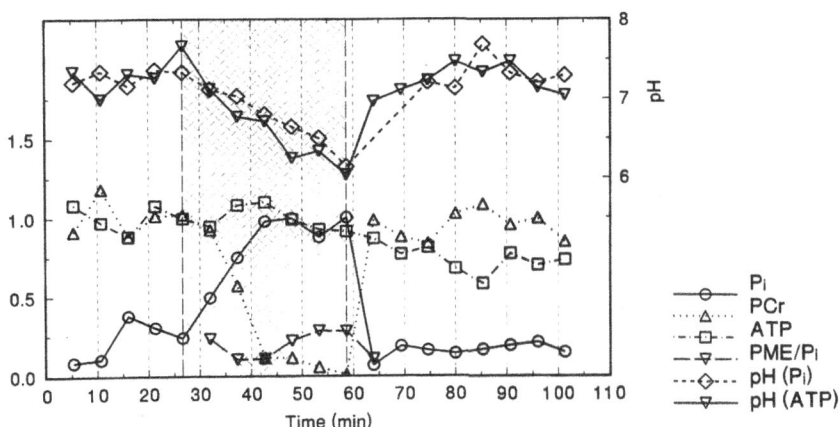

Figure 2. Results of analysis of spectra. PCr and ATP normalized to average value over first five data points; P_i and PME shown normalized to peak P_i value. pH values calculated from PCr - P_i and PCr - γ-ATP chemical shifts. Data points shown at end of averaging interval. Shaded area represents ischemic insult.

The average shift between the PCr and γ-ATP peaks for the first five and the last five data points from this experiment (stable working situation) was 2.393 ppm, and the average pH calculated from the P_i - PCr shift was 7.30. Referring to Moon and Richards (1973), a change from pH 7.3 to pH 8 reduces the shift between PCr and γ-ATP by 0.3 ppm, while a change from pH 7.3 to pH 6 increases it by 2.9 ppm. It is not justifiable to fit a curve more complex than a quadratic through these three points; such a formula derived from these data is

$$pH\ (ATP) = 0.59\ \delta^2 - 5.0\ \delta + 15.9$$

where δ is the shift between PCr and γ-ATP in ppm. This formula was used to produce the second pH trace. The average shift between γ-ATP and β-ATP for the stable working heart (the ten points referred to above) was 13.69 ppm.

Fig. 3 shows the correlation of the pH values calculated by the two methods described, excluding the first two data points after reperfusion in each experiment. The equation of the best fit straight line is

$$pH\ (ATP) = 1.036\ (SE\ 0.068)\ pH\ (P_i) - 0.244\ (SE\ 0.487)$$

which is very close to identity. The Spearman rank correlation coefficient was 0.63, two-sided p-value < 0.0001.

4. DISCUSSION

When using ^{31}P NMR, intracellular pH is usually calculated from the chemical shift between P_i and PCr (Moon and Richards, 1973; Jacobus *et al.*, 1978). This method becomes problematic if the signal-to-noise ratio is poor, if P_i drops to a very low level, or if its peak overlaps another. Moon and Richards also showed that the positions of the three ATP peaks also depend on pH, the γ shift being most sensitive, and the α shift least. It would be unusual

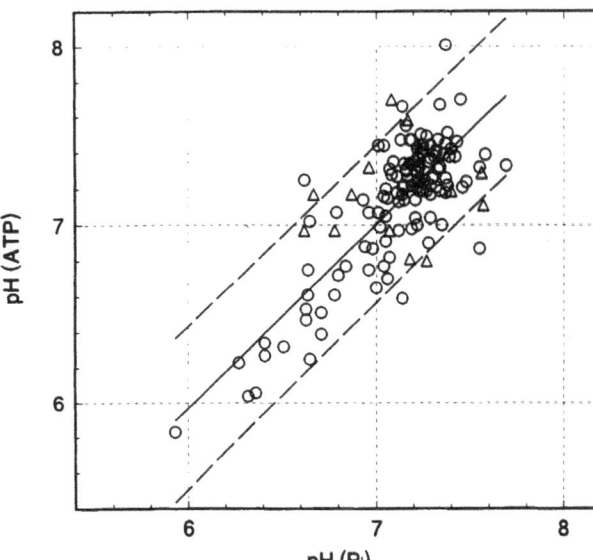

Figure 3. Correlation of pH values calculated by the two methods described. Solid line is best fit straight line, broken lines are 95% data confidence limits. Triangles represent first two data points after reperfusion in each experiment, and were excluded from correlation calculation. Excessive scatter of these points was due to difficulty in identifying a P_i peak.

for P_i and ATP to reach low levels together in tissue with the potential to recover, so pH determination from ATP shifts may be a complementary method. However, these shifts also depend on the degree of complexation with magnesium, which in turn depends on the intracellular free magnesium concentration, $[Mg^{2+}]_i$. Williams *et al.* (1993) published an elegant series of results allowing $[Mg^{2+}]_i$ and pH to be calculated from the relative chemical shifts among the three ATP peaks. Using this method to determine pH in skeletal muscle, Widmaier *et al.* (1996) found a good correlation with pH as calculated from the P_i shift. In our model, the Williams method is difficult to apply, as the α peak is distorted by another peak, possibly nicotinamide adenine nucleotide (NAD). Moon and Richards make no mention of magnesium, so we assume none was present; under this condition the shift of γ-ATP is most sensitive to pH around pH 6.5. Burt *et al.* (1976) included 12mm Mg^{2+} in their 3mm ATP solution, and demonstrated no noticeable change in the γ-ATP shift until pH fell below around 6.8, with maximum sensitivity nearer to pH 5.3. This is borne out by the data of Williams *et al.* which demonstrate that the shift between the β and γ peaks is essentially unaffected by pH values above 6.2 with an $[Mg^{2+}]:[ATP]$ ratio of 9, but is very sensitive to pH changes in the physiological range when $[Mg^{2+}]:[ATP]$ is 0. From their results (fig. 1), and given that the α shift is relatively insensitive to pH, for a given γ-α shift, an increasing $[Mg^{2+}]:[ATP]$ corresponds to a decreasing pH; in other words, if $[Mg^{2+}]$ is higher than expected, the pH reading will be higher than the actual pH.

Various researchers have investigated *in vivo* magnesium concentrations, and the effect of ischemia. Jung *et al.* (1996) reported a free $[Mg^{2+}]$ of 0.39 mm and 88% complexation of ATP to magnesium in human myocardium. These values were found from line splitting at a field of 1.5 T due to J coupling between the phosphate groups (which is not affected by pH); at the 4.3 T field strength we use, this splitting is not visible. Jelicks and Gupta (1991) found $[Mg^{2+}]$ of 0.76mm in Langendorff-perfused rat hearts; Barbour *et al.* (1991–2) used the α-β shift to measure an intracellular $[Mg^{2+}]$ of between 0.69 and 1.21 mm in isolated rat hearts under various conditions; and Altura *et al.*(1993) showed that reducing perfusate $[Mg^{2+}]$ from 1.2 to 0.3mm reduced intracellular $[Mg^{2+}]$ from 0.65 to 0.43 mm in isolated rat hearts. Rauch *et al.* (1994) found that $[Mg^{2+}]$ (as found from the γ-β shift) increased from 0.34 to 1.68 mm in guinea-pig hearts after 30 min global ischemia, while Schreur *et al.* (1993) showed an increase from 0.76 to 4.3 mm after 15 min ischemia in rat heart. However, Mosher *et al.* (1992) caution against the over-estimation of increased $[Mg^{2+}]$ due to the presence of Mg_2ATP. To summarize, it would appear that the chemical shifts of ATP ought to be quite sensitive to pH changes occurring in well-perfused tissue, but may become less sensitive during ischemia due to increasing magnesium concentration.

Referring back the results of Williams *et al.* (fig. 2B), our γ-β shift of 13.69 ppm at a pH of 7.3 in a stable, working heart corresponds to an $[Mg^{2+}]:[ATP]$ ratio of around 0.9, comparable to that found by Jung *et al.* As we have ignored the effects of changes in $[Mg^{2+}]$, then if $[Mg^{2+}]$ does increase during ischemia, we would expect to see a systematic error as pH decreases, with calculated pH being higher than actual pH. This would cause the line in fig. 3 to curve upwards with decreasing pH, which does not happen: the relationship between the pH values as calculated by the two different methods is close to identity. This discrepancy requires further investigation.

5. CONCLUSION

The ATP shifts depend on pH, and also on magnesium complexation which we have ignored; within the errors of the method, this appears to be a justifiable simplification. The good agreement between the results of the two methods, and the ability to determine pH during reperfusion, and also when the P_i peak overlaps other peaks, suggest that calculation of intracellular pH from the chemical shift of γ-ATP is a useful technique.

6. SUMMARY

^{31}P NMR allows non-invasive measurement of intracellular pH, which drops during tissue hypoxia or ischemia. Determination is usually based on the chemical shift between the inorganic phosphate (P_i) and phosphocreatine (PCr) peaks. During reperfusion, P_i is taken up to form PCr and ATP, and in our model at least (an isolated, working rat heart perfused with an erythrocyte suspension), the level of P_i reduces well below the pre-ischemic level, making pH determination difficult. The chemical shifts of the three ATP peaks also depend on pH, and the level of ATP remains high during reperfusion, so these might be used to determine pH. The results of one experiment are presented in detail, showing the time course of high energy phosphate levels before, during and after a 32 min ischemic insult, and close agreement between the pH determinations from the P_i and γ-ATP peaks can be seen. The formula used to calculate pH from the ATP peak was:

$$\text{pH (ATP)} = 0.59\ \delta^2 - 5.0\ \delta + 15.9$$

where δ is the shift in ppm between PCr and γ-ATP. All pH readings by both methods from a series of seven experiments were compared and a 1:1 agreement demonstrated (correlation coefficient 0.63, $p < 0.0001$). Although the ATP shifts also depend on magnesium complexation which we have ignored, this appears to be justifiable within the errors of the method; the good agreement between the results of the two methods, and the ability to determine pH during reperfusion suggest that calculation of intracellular pH from the chemical shift of γ-ATP is a useful technique.

ACKNOWLEDGMENTS

We acknowledge the technical contributions of J. Evers and Ir H.J. van den Boogert and the support of the Dutch Heart Foundation (investment subsidy 902–19–115) which made this work possible.

REFERENCES

Altura, B.M., Barbour, R.L., Dowd, T.L., Wu, F., Altura, B.T., and Gupta, R.K. (1993). Low extracellular magnesium induces intracellular free Mg deficits, ischemia, depletion of high-energy phosphates and cardiac failure in intact working rat hearts: a 31P-NMR study. Biochim. Biophys. Acta. 1182: 329–332

Barbour, R.L., Altura, B.M., Reiner, S.D., Dowd, T.L., Gupta, R.K., Wu, F., and Altura, B.T. (1991–92). Influence of Mg2+ on cardiac performance, intracellular free Mg2+ and pH in perfused hearts as assessed with 31P nuclear magnetic resonance spectroscopy. Magnes. Trace. Elem. 10: 99–116

Burt, C.T., Glonek, T., and Barany, M (1976). Analysis of phosphate metabolites, the intracellular pH, and the state of adenosine triphosphate in intact muscle by phosphorus nuclear magnetic resonance. J. Biol. Chem. 251: 2584–2591

Jacobus, W.E., Pores, I.H., Taylor, G.J., Nunnally, R.L., Hollis, D.P., and Weisfeldt, M.L. (1978). Tight coupling of intracellular pH and ventricular performance. J. Mol. Cell. Cardiol. 10 supp 1: 39

Jelicks, L.A. and Gupta, R.K. (1991). Intracellular free magnesium and high energy phosphates in the perfused normotensive and spontaneously hypertensive rat heart. A 31P NMR study. Am. J. Hypertens. 4: 131–136

Jung, W.I., Widmaier, S., Seeger, U., Bunse, M., Staubert, A., Sieverding, L., Straubinger, K., van Erckelens, F., Schick, F., Dietze, G., and Lutz, O. (1996). Phosphorus J coupling constants of ATP in human myocardium and calf muscle. J. Magn. Reson. B. 110: 39–46

Moon, R.B. and Richards, J.H. (1973). Determination of intracellular pH by 31P magnetic resonance. J. Biol. Chem. 248: 7276–7278

Mosher, T.J., Williams, G.D., Doumen, C., LaNoue, K.F., and Smith, M.B. (1992). Error in the calibration of the MgATP chemical-shift limit: effects on the determination of free magnesium by 31P NMR spectroscopy. Magn. Reson. Med. 24: 163–169

Olders, J., Boumans, T., Evers, J., and Turek, Z. (1990). An experimental set-up for the blood perfused working isolated rat heart. Adv. Exp. Med. Biol. 277: 151–160

Rauch, U., Schulze, K., Witzenbichler, B., and Schultheiss, H.P. (1994). Alteration of the cytosolic-mitochondrial distribution of high-energy phosphates during global myocardial ischemia may contribute to early contractile failure. Circ. Res. 75: 760–769

Schreur, J.H., de Beer, R., van Echteld, C.J., and Ruigrok, T.J. (1993). Post-ischemic contractile dysfunction does not correlate with an elevated intracellular free [Mg2+]: a 31P-NMR study on isolated rat and rabbit hearts. J. Mol. Cell. Cardiol. 25: 1015–1024

Widmaier, S., Jung, W.I., Bunse, M., van Erckelens, F., Dietze, G., and Lutz, O. (1996, in press). Change in Chemical Shift and Splitting of 31-P γ-ATP Signal in Human Skeletal Muscle During Exercise and Recovery. NMR Biomed.

Williams, G.D., Mosher, T.J., and Smith, M.B. (1993). Simultaneous determination of intracellular magnesium and pH from the three 31P NMR Chemical shifts of ATP. Anal. Biochem. 214: 458–467

36

INTRAMYOCARDIAL pO$_2$ MEASURED BY EPR

Oleg Y. Grinberg,[1] Bruce J. Friedman,[2] and Harold M. Swartz[1]

[1]Department of Radiology
[2]Section of Cardiology
Dartmouth-Hitchcock Medical Center
Hanover, New Hampshire 03755

1. INTRODUCTION

Oxygen is essential for normal cardiac function and plays an important role in cardiac regulation (Neely 1974, Vergroesen 1987). How pO$_2$ is controlled or even at what level this control occurs (Spaan, 1993), is still not clear. Myocardial pO$_2$ reflects the dynamic equilibrium between oxygen consumption (MVO$_2$) and oxygen delivery in the myocardium and direct measurement of myocardial pO$_2$ would help to understand this relationship.

Many methods have been developed to measure pO$_2$ in tissue (Vanderkooi, 1991), but their applicability to the functioning myocardium is limited. Electron paramagnetic resonance (EPR) oximetry appears to have some significant advantages for use in the beating heart (Zweier, 1991, Swartz, 1992). Recently it was shown (Friedman, 1995), that EPR oximetry with lithium phthalocyanine (LiPc) crystals as the O$_2$-sensitive material, provides accurate and dynamic evaluation of local myocardial pO$_2$ in the contracting isolated heart. Myocardial pO$_2$ and coronary vascular resistance were found to increase significantly after recovery from repetitive ischemia (Friedman, 1996). It also was demonstrated (Grinberg, 1995) that cardioactive drugs had characteristic effects on myocardial pO$_2$ and myocardial oxygen consumption. The findings were explained by relative changes of capillary density. Comparison of pO$_2$ versus flow in the constant-flow perfused rat heart demonstrated a threshold dependence; when a critical flow or pressure was reached myocardial pO$_2$ increased rapidly (Friedman, 1995). The pO$_2$ in two different regions measured simultaneously in the same heart also differed (Grinberg, 1995). Moreover, the pO$_2$ from the two sites responded differently to changes of end diastolic pressure: as end diastolic pressure was increased, the local pO$_2$ near the endocardial surface usually decreased while local pO$_2$ near the epicardial surface initially increased.

This paper considers the data obtained with EPR oximetry in 192 experiments with isolated rat hearts. Initial measurements were taken during the first 40 min. before any physiological or pharmacological interventions. Based on the previous findings and analysis of the initial measurements, we suggest a model to explain: the observed large variations in local pO_2 found in these isolated heart preparations; the occurrence of the threshold for effects of flow on myocardial pO_2; and the different changes in myocardial pO_2 observed with increased diastolic pressure.

2. EXPERIMENTAL

2.1. EPR Oximetry with LiPc Crystals

LiPc is a neutral p-radical crystalline substance that has a strong EPR signal whose line width is sensitive to oxygen (Turek, 1989) and therefore a convenient probe for EPR oximetry (Liu, 1993). The use of EPR oximetry with LiPc to measure pO_2 under constant flow in the isolated perfused rat heart has been described in detail (Friedman, 1995). Usually four to ten small LiPc crystals with total volume 0.5 x 0.2 x 0.2 mm^3 were implanted in the posterolateral wall of the left ventricle in the mid myocardium. EPR spectra, which can be obtained repeatedly, reflect pO_2 in the surrounding tissue with a resolution of $\leq 5\%$, and respond rapidly (30 sec) to changes in pO_2 in this volume. Hemodynamic parameters, including perfusion pressure, oxygen consumption, and left ventricular contractility were measured simultaneously. In six experiments the pO_2 at two different sites was studied simultaneously by use of a magnetic field gradient to separate the EPR spectra from two LiPc crystals implanted intra-myocardially 2–3 mm apart and at different depths.

2.2. Animal Preparation and Experimental Protocol

The animal protocol has been described in detail (Friedman, 1995). The study conformed to the guidelines of the animal research laboratory of Dartmouth Medical School which is accredited by the American Association of Laboratory Animal Care. During preparation and stabilization of the heart which required 20–40 min. the same procedures were followed and the same initial measurements were made before start of the various experimental protocols. During initial equilibration the rate of coronary flow was adjusted, with a variable speed peristaltic pump, using a protocol which specified a specific perfusion pressure (35–95 mm Hg). After obtaining the desired perfusion pressure, flow was kept constant. Measurements were taken in triplicate, five minutes apart.

Pressure and heart rate were monitored with a disposable transducer and a hemodynamic monitor (Hewlett Packard 78534c). The perfused heart with implanted crystals was placed just above the EPR resonator (external loop) and maintained at a temperature of $37°C$. Modified Krebs Henseleit solution in equilibrium with 5% CO_2 and 95% O_2 was used for perfusion.

After implantation of the LiPc crystals a latex balloon was inserted into the left ventricle and filled with a volume of distilled water sufficient to produce an end diastolic pressure of 8–12 mm Hg. It was used to determine left ventricular isovolumetric parameters including systolic and diastolic pressure, developed pressure (difference between peak systolic and end diastolic pressure), and left ventricular contractility (product of heart rate and developed pressure). Clark electrodes were used to monitor pO_2 in the influent and the effluent throughout the experiment. MVO_2 was calculated from the product of flow rate and the difference between influent and effluent pO_2, (with the assumption that the concentration of oxygen in media and pure water was the same, 210 micromol/l in air at 760 mm Hg, $37°C$) and normalized to the wet weight of each heart. The Clark electrode was calibrated with a gas mixture 5% CO_2 & 95% O_2 and verified with air.

2.3. Data Preparation

All initial measurements of local pO$_2$ were separated into various intervals and the number of experiments for each of these intervals was calculated. pO$_2$, flow rate, left ventricular contractility, and oxygen consumption were grouped by 5 mm Hg perfusion pressure intervals and the mean ± standard error calculated for each group and plotted versus perfusion pressure. An unpaired t-test was used to compare pO$_2$ for the different intervals and was considered to be significantly different at p<0.05.

3. RESULTS

The mean influent pO$_2^{in}$ for these 192 experiments is 539 ± 5 mmHg, the mean effluent pO$_2^{out}$ is 200 ± 4 mmHg and the mean myocardial pO$_2$ is 110 ± 8 mmHg.

4. DISCUSSION

The grouping of data by 5 mm Hg intervals facilitated interpretation of our results. Figure 3 shows that while flow rate (3B) increased proportionately over the range of perfusion pressures, oxygen consumption (3C) and left ventricular contractility (3D) initially increased but then reached a plateau. Figure 2 shows that myocardial pO$_2$ is maintained at a fairly constant level (about 100 mm Hg) for perfusion pressures of 35–70 mm Hg; however, as pressure and flow increase further, myocardial pO$_2$ increases sharply to a new plateau. One possible explanation may be the presence of an autoregulation system which maintains the baseline myocardial pO$_2$ but fails or is reset at higher flow rates or perfusion

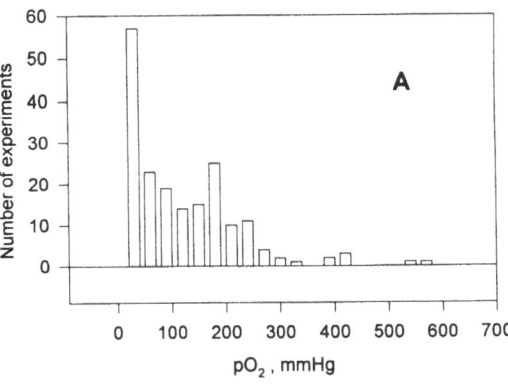

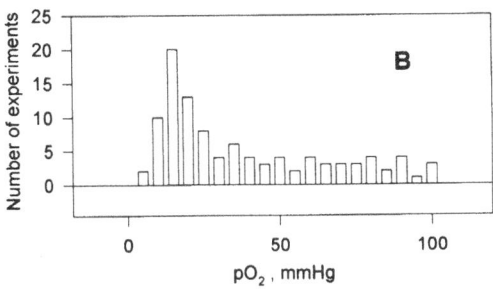

Figure 1. The measured myocardial pO$_2$ in 192 experiments in constant flow isolated rat hearts. A. Myocardial pO$_2$ was plotted as histogram with intervals of 30 mm Hg; B. Myocardial pO$_2$ with histogram intervals of 5 mm Hg for pO$_2$ from 0 to 100 mm Hg. Figure 1 shows a wide range of values with the intervals centered around 15 mm Hg and 170 mm Hg having the largest number of results.

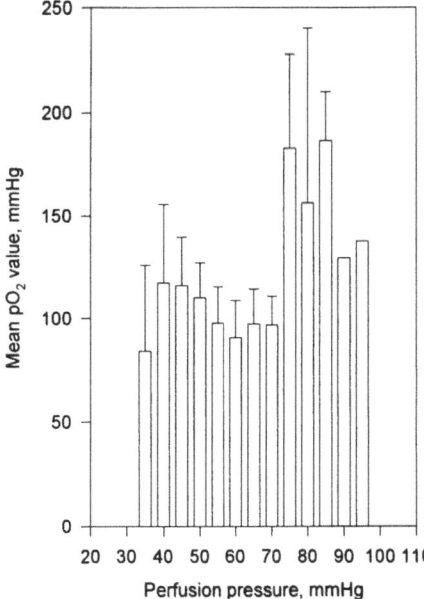

Figure 2. Mean pO_2 values versus perfusion pressure (combining data into 5 mm Hg intervals for perfusion pressure). The pO_2 values for the 5 mm Hg intervals of perfusion pressures from 35–70 mm Hg were not significantly different. The pO_2 values increased significantly at perfusion pressures greater than 70 mm Hg.

pressure. Another possible explanation is that myocardial oxygen consumption and work increase proportionately to flow only up to a point (70 mm Hg), after which oxygen delivery increases greater than consumption resulting in a net increase in myocardial pO_2.

In the early part of this century August Krogh (Krogh, 1919) proposed a model to describe tissue oxygen tension. The Krogh equation describes the relation between oxygen consumption, pO_2 in the capillary, pO_2 in tissue, and intercapillary distance 2R. According to this model the main circulatory unit is a cylinder with radius R around the capillary. Oxygen diffusion from the capillary in striated muscle produces an oxygen gradient, that can be characterized by the difference $G = P_c - P_R$:

$$G = MVO_2 F(R)/4D \qquad (1)$$

where: P_R is the tissue pO_2 at distance R, P_c is the capillary pO_2, $F(R)$ is a known function of R (Krogh , 1919) and D is the diffusion coefficient of oxygen in tissue . If capillary density does not change, equation (1) predicts the difference G increases with increasing MVO_2.

Local pO_2 (Friedman, 1995) and the Krogh equation (1) was used to estimate capillary density. For this calculation the Kety assumption (Kety, 1957) was used for P_c:

$$P_c = (pO_2^{in} + pO_2^{out})/2 \qquad (2)$$

Using this assumption the capillary pO_2 is $P_c = 366 \pm 5$ mmHg.

Application of the equations (1) and (2) to calculate capillary density using measured local pO_2 values gives a wide range of capillary density. Therefore the relative capillary density was used to explain myocardial pO_2 changes in our previous study (Friedman, 1995).

The Krogh cylinder is a simple model which provides an estimate of how much tissue can be supplied by oxygen from an isolated capillary. Other factors must also be taken

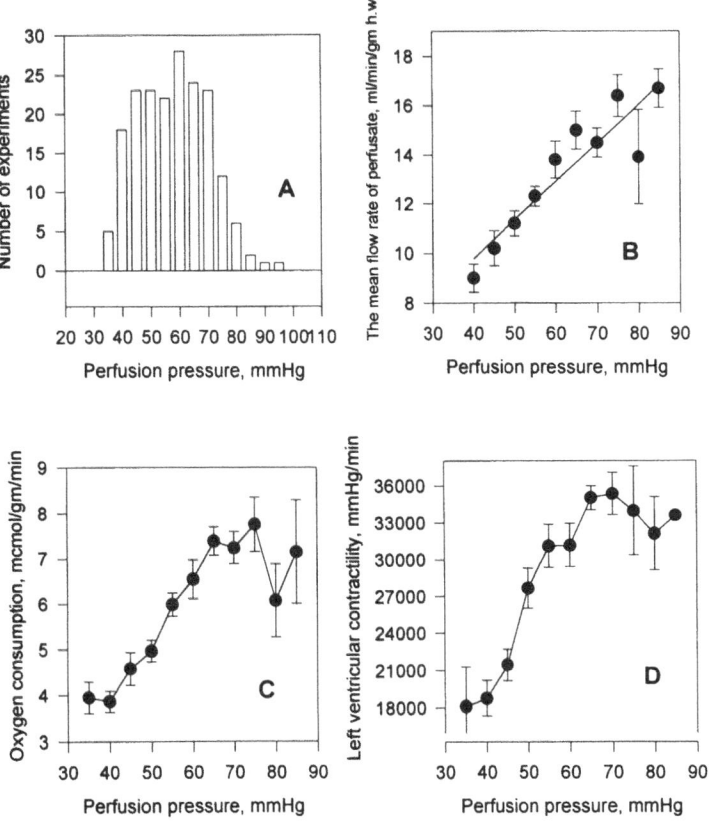

Figure 3. Relation of flow, oxygen consumption, and left ventricular contractility to perfusion pressure (Mean values versus perfusion pressure in intervals of 5 mm Hg). A - Number of experiments at various perfusion pressures ; B - Flow rate of perfusate vs. perfusion pressure ; C - Oxygen consumption vs. perfusion pressure; D - Left ventricular contractility vs. perfusion pressure. All three parameters increased with increasing perfusion pressure between 40 and 70 mm Hg.

into account, for instance oxygen lost from individual capillaries can be greatly affected by the presence of large micro vessels (Ellsworth, 1990). It appears that diffusive exchange between arterioles and capillaries can play an important part in the distribution of oxygen in tissues (Secomb, 1994). It was shown that the data obtained with a beating heart (Van der Ploeg, 1994) were better represented by a single mixed compartment than by a Krogh cylinder.

The Krogh model does not consider the heterogeneity of myocardial pO₂ which has been demonstrated by several investigators (Gayeski, 1988, Ashruf 1994, Rumsey 1994). Heterogeneity was visualized (Gayeski, 1988) as clusters of cells and capillaries. The measured radius of these clusters was 100 to 150 micrometers, which is approximately the same volume measured by LiPc crystals. Figure 1 shows there is a large range of values for local myocardial pO₂ . While this may be due to differences of perfusion pressure or myocardial clustering, the absence of significant myocardial pO₂ variation over the 35–70 mm Hg intervals of perfusion pressure (Figure 2), suggests the wide range of myocardial pO₂ is principally due to the presence of myocardial clusters.

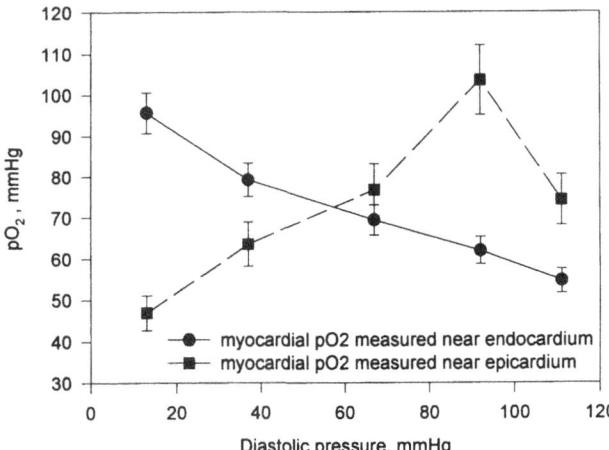

Figure 4. The local myocardial pO₂, derived from the line width of two crystals, versus end diastolic pressure. The figure summarizes the results of a typical experiment in which two sets of LiPc crystals were placed at different sites in the heart and used to measure the pO_2 simultaneously at both sites. The pO_2 from the two sites responded differently to changes of end diastolic pressure. As end diastolic pressure increased, local pO_2 near the endocardial surface decreased while local pO_2 near the epicardial surface initially increased.

Local flow rate can play an important role in myocardial clusters. The importance of flow rate in a single microvessel was shown spectrophotometrically (Maeda, 1994). The observed difference in pO_2 for two different regions in the same heart (Figure 4) and the discordant response to changes in end diastolic pressure are explained by differences in local flow. An increase of diastolic pressure causes a redistribution of myocardial flow, initially decreasing endocardial and increasing epicardial flow. Decreased local flow to the endocardium is reflected by decreased endocardial pO_2. The simultaneous increase of epicardial pO_2 reflects an increase of local flow in the epicardium.

Based on these findings a semi-empirical model is suggested to explain changes in myocardial pO_2 without changes in capillary density. Two principal assumptions of this model are: a) Myocardial pO_2 is determined by the local pO_2 of circulatory clusters (200–300 microns in size) which includes the precapillary arteriole, the capillaries, a venule, and the surrounding myocardium; b) The local pO_2 in the cluster is dependent on the effluent pO_2 in the precapillary arteriole, which in turn strongly depends on the local flow through this circulatory unit and pO_2 in the tissue surrounding the precapillary arteriole.

The effluent pO_2 in the precapillary arteriole can be different due to the distribution of transmural flow and/or variations in the vasculature of the heart. Assuming that r is the radius, l the length and d the wall thickness of the precapillary arteriole, the effluent pO_2 of this arteriole which is the capillary pO_2 for all capillaries in this cluster can be described by the expression:

$$P_c = P_t + (pO_2^{in} - P_t)\exp(-A/f) \tag{3}$$

where: pO_2^{in} is the influent pO_2 (for all these experiments it was 532 mm Hg), f is local flow through the cluster, P_t is the mean quasi steady state pO_2 in the tissue surrounding the precapillary arteriole, $A = 2Dlr\pi/d$, parameters P_t and A may be different for each cluster. By this formula (3) if arteriole flow is too low, P_c approaches P_t. A long transit time causes pO_2 in the precapillary arteriole to decrease to the level of pO_2 in the tissue surrounding the precapillary arteriole. In the absence of large micro vessels in the cluster or in the presence of low P_c, the difference G does not fit the Krogh model, formula (1). Figure 1 reflects these two possibilities: a) for the pO_2 clustered around 170 mm Hg presumably there is a larger vessel in close proximity to the cluster b) for the pO_2 around 15 mm Hg

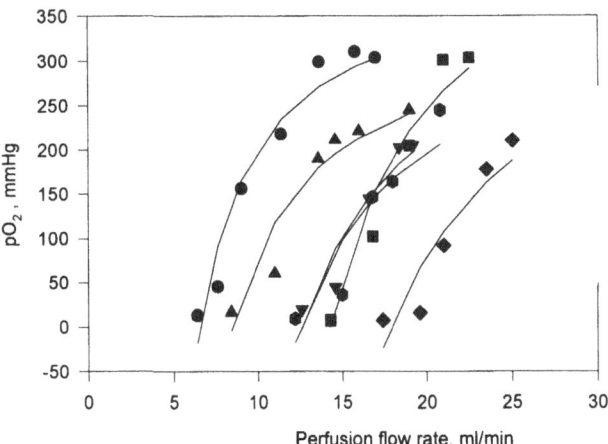

Figure 5. The local myocardial oxygen tension derived from lithium phthalocyanine line width versus flow rate for six hearts each represented by a different symbol. This figure demonstrates fitting of the changes of pO₂ based on formula 4, versus flow for six constant-flow perfused rat hearts. The separate lines represent a two parameter fitting for G' and the product of A and N where N is the number of the precapillary arterioles, assuming the local flow rate f = F/N where F is the flow rate. The range of the fitting parameter are 60–179 mm Hg for G' and 9–31 ml/min for A*N.

presumably the low level of oxygen in the capillary would result in anoxic clusters. According to formula (3), however, increasing local flow will increase P_c and avoid the occurrence of anoxic clusters. As flow in the precapillary arteriole is increased P_c increases and P_t', the mean quasi steady state myocardial pO₂ in the circulatory cluster, also increases:

$$P_t' = pO_2^{in} - G'exp(A/f) \qquad (4)$$

where: $G' = G - (P_t' - P_t)$, where G is defined by formula (1), $(P_t' - P_t)$ is the difference in the mean pO₂ of the tissue surrounding the precapillary arteriole and the circulatory cluster. This formula describes a threshold phenomenon for the effect of local flow rate on local myocardial pO₂.

6. CONCLUSIONS

EPR oximetry can provide accurate and sensitive measurement of local tissue pO₂ in the functioning myocardium. A model based on the presence of myocardial circulatory clusters is proposed to explain the experimental results, including the large variation of the initial local pO₂ values, the "threshold flow phenomenon" and the influence of diastolic pressure on local tissue pO₂. These macroscopic circulatory clusters (200–300 micrometers in size) consist of a precapillary arteriole, the capillaries, a venule, and the surrounding myocardium. The local tissue pO₂ measured by EPR oximetry using LiPc crystals reflects the effluent pO₂ from the precapillary arteriole and strongly depends on the local flow through this circulatory cluster and on pO₂ in the tissue surrounding the precapillary arteriole.

ACKNOWLEDGMENTS

This work was supported by AHA, New Hampshire Affiliate Grant in Aid 9407770S, and NIH grant GM 34250 and used the facilities of the EPR center at Dartmouth Medical School which was supported by NIH grant RR 01811.

REFERENCES

Ashruf JF, Ince C, Ischemic areas in hypertrophic Langendorf rat hearts visualized by NADH videofluorimetry. Adv Exp Med Biology, 1994; 345:259–62

Ellsworth ML, Pittman RN, Arterioles supply oxygen to capillaries by diffusion as well as by convection, Am J Physiol, 1990 Apr; 258 (4 Pt 2):H1240–3.

Friedman BJ, Grinberg OY, Isaacs KA, Walczak TM, Swartz HM, Myocardial oxygen tension and relative capillary density in isolated perfused rat hearts. J Molec Cell Cardiology, 1995;27:2551–58.

Friedman BJ, Grinberg OY, Isaacs KA, Ruuge EK, Swartz HM, Effect of repetitive ischemia on myocardial oxygen tension in isolated perfused and hypoperfused rat hearts, Mag Res Med, 1996;35:214–230.

Gayeski TE, Federspiel WJ, Honig CR, A graphical analysis of the influence of red cell transit time, carrier-free layer thickness, and intracellular pO_2 on blood-tissue O_2 transport., Adv Exp Med Biology, 1988;222:25–35.

Grinberg OY, Friedman BJ, Swartz HM, Myocardial oxygen tension and capillary density in isolated perfused rat heart, ISOTT XXV, Pittsburgh, Pensylvania, 1995;Abstracts:48. Grinberg OY, Grinberg SA, Friedman BJ, Swartz HM, Myocardial oxygen tension and capillary density in the isolated perfused rat heart during pharmacological intervention, Adv Exp Med Biology, 1995, in press.

Kety SS, Determination of tissue oxygen tension. Fed Proc 1957;16:666–670.

Krogh A. The number and distribution of capillaries in muscles with calculations of the oxygen pressure head necessary for supplying the tissue. J Physiol (Lond), 1919;52:409–415. Krogh A. The supply of oxygen to the tissues and the regulation of the capillary circulation. J Physiol (Lond), 1919;52:457–474.

Liu KJ, Gast P, Moussavi M, Norby S, Vahidi N, Walczak T, Wu M, Swartz HM. Lithium phthalocyanine: A probe for EPR oximetry in viable biological systems. Proc Natl Acad Sci USA, 1993;90:5438–5442.

Maeda N, Shiga T, Velocity of oxygen transfer and erythrocyte rheology, NIPS Volume 9/ Feb.1994:22–27.

Neely JR, Morgan HE, Relationship between carbohydrate and lipid metabolism and the energy balance of heart muscle. Ann Rev Physiol, 1974;36:413–459.

Rumsey WL, Pawlowski M, Lejavardi N, Wilson DF, Oxygen pressure distribution in the heart in vivo and evaluation of the ischemic "border zone", Am J Physiol 1994 Apr;266(4 Pt 2):H1676–80.

Secomb TW, Hsu R. Simulation of O_2 transport in skeleton muscle: diffusive exchange between arterioles and capillaries. Am J Physiol, 1994 Sep;(3 Pt 2):H121–21.

Spaan JA, Dankelman J. Theoretical analysis of coronary blood flow and tissue oxygen pressure-control. Adv Exp Med Biology, 1993;346:189–95

Swartz HM, Boyer S, Brown D, Chang K, Gast J, Glockner JF, Hu H, Liu J, Moussavi M, Nilges M, Norby SW, Smirnov A, Vahidi N, Walczak T, Wu MR, Clarkson RB. The use of EPR for measurement of concentration of oxygen in vivo in tissues under physiologically pertinent conditions and concentrations. Adv Exp Med Biology, 1992;317:221–228.

Turek P., Andre JJ, Moussavi M, Fillion G. Septet spin state in lithium phthalocyanine 1 radical compound. Mol Crys Liq Cryst, 1989;176:535–542.

Van der Ploeg CP, Dankelman J and Spaan JA, Classical Krogh model does not apply well to coronary oxygen exchange, Adv Exp Med Biology, 1994;345:299–304.

Vanderkooi JM, Erecinska M, Silver IA, Oxygen in mammalian tissue: methods of measurement and affinities of various reactions. Am J Physiol, 1991;260:C1131-C1150.

Vergroesen I, Noble MI, Wieringa PA, Spaan JA, Quantification of O_2 consumption and arterial pressure as independent determinants of coronary flow. Am J Physiol, 1987;252:H5435–553O.

Zweier JL, Thompson-Gorman S, Kuppusamy P, Measurement of oxygen concentrations in the intact beating heart using EPR spectroscopy: A technique for measuring oxygen concentrations in situ. J Bioener Biomem, 1991;23(6):855–871.

37

SIMULTANEOUS MEASUREMENT OF OXYGEN CONSUMPTION AND $^{13}C^{16}O_2$ PRODUCTION FROM ^{13}C-PYRUVATE IN DIABETIC RAT HEART MITOCHONDRIA

Nicolai M. Doliba, Ian R. Sweet, Andriy Babsky, Nataliya Doliba, Robert E. Forster, and Mary Osbakken

Biochemistry / Biophysics
University of Pennsylvania
Philadelphia, Pennsylvania 19144
Pharmacological Research Institute
Bristol-Myers Squibb
Princeton, New Jersey 08543

INTRODUCTION

Several investigators have shown defects in energy production in hearts from animal models of diabetes mellitus (DM). These defects generally involve changes in oxidative phosphorylation (ox-phos) or in Krebs cycle functions. Abnormalities in ox-phos have been observed in heart mitochondria isolated from streptozotocin-diabetic (Pierce and Dhalla, 1985) and alloxan-diabetic (Puckett and Reddy, 1979) models as well as in genetic diabetic models (Kuo et al., 1983). In DM mitochondria, State 3 is decreased when a variety of substrates are used: Pyr plus malate and palmitylcarnitine plus malate (Kuo et al., 1983), α-ketoglutarate (Taegtmeyer et al., 1987), glutamate (Mokhtar et al., 1993), 3-hydroxybutyrate and malate, and acetoacetate plus malate (Grinblat et al., 1986). The RCI also decreased in mitochondria from DM animals. The ADP/O ratio is depressed only in the early stages of chemically induced DM, returning to normal ranges later on, indicating that mitochondria from DM hearts have normal ability to transfer electrons from NADH to O_2 or to couple oxidation to phosphorylation. This suggests that the deficiency of energy production is related to a decrease in substrate oxidation through the Krebs cycle. Two important cellular functions which may contribute to abnormal ox-phos are alterations in mitochondrial transporter systems (Kazmi et al., 1985; Paulson et al., 1984) and decrease in substrate supply. One possible mechanism of impaired substrate oxidation in the DM heart is abnormal mitochondrial Ca^{2+} uptake (Baba and Kako, 1991). It is well known that Ca^{2+} plays an important role in regulation of several matrix dehydrogenases (including PDH, isocitrate dehydrogenase (IDH), α-ketoglutarate dehydrogenase (KGDH)) involved in oxidative ATP synthesis (McCormack et al., 1990). Increases in intramitochondrial Ca^{2+} concentration lead to an increase in the activity of dehydrogenases and turnover of the Krebs cycle (McCormack et al., 1990), whereas overload of Ca^{2+} leads to depolarization of mitochondrial membrane, and uncoupling of respiration and oxidative ATP synthesis (Minezaki et al., 1994). Few investigations have been done to evaluate the role of Ca^{2+} on abnormalities of ox-phos in DM heart, in contrast to the extensive investigation of energy metabolism. Because PDH is a key enzyme in oxidative function which can be regulated by ion concentration (in particular Ca^{2+}), abnormalities in PDH function may be important in the pathophysiology of DM. The purpose of the present study was to examine the function of PDH by measuring the oxidation of Pyr ([1–^{13}C]pyruvate) in cardiac mitochondria from DM rats using simultaneous recording of MVO_2 and $^{13}C^{16}O_2$.

MATERIALS AND METHODS

Animal Preparation

A chemical model of DM was made and used in this study (Friedman et al., 1985). Briefly, the rats were injected intraperitoneally with streptozotocin (60 mg/kg body wt dissolved in citrate buffer). Four-five days after injection of streptozotocin, the plasma glucose concentration was measured to verify the success of development of DM. Rats were maintained in the diabetic state for 4 weeks. All rats (DM and CON) received Purina laboratory chow and tap water ab libitum, and were housed individually and maintained on a 12 hour: 12 hour light-dark cycle at 22°C as per the University of Pennsylvania Animal Care Committee guidelines.

Preparation of Isolated Heart Mitochondria

Heart mitochondria was prepared using a method described previously (Kondrashova and Doliba, 1989). Mitochondria obtained via this method exist in suspension as aggregates corresponding to their native state in the intact cell. Briefly, hearts were quickly excised under nembutal anesthesia (50 mg/kg, injected intraperitoneally), exsanguinated by perfusion, chilled, weighed and homogenized in a glass Potter-Elvehjem homogenator with a motor-driven Teflon pestle for 25–35 sec. The composition of the medium used for perfusion and homogenization was: 0.3 M sucrose, 10 mM HEPES (pH=7.4), 1 mM Ethylene Diamine Tetracetic Acid (EDTA), and 0.25% BSA After initial centrifugation (at 480g for 7 min), the pellet, containing nuclei and other cellular debris, was discarded. The supernatant, which contains mitochondria, was centrifuged at 10000 g for 10 min. The mitochondria pellet was suspended in 0.15 ml of the homogenization solution. Mitochondrial function was studied either with routine polarographic methods or with mass spectroscopy. The incubation medium used for both methods contained: 250 mM sucrose; 50 mM KCl; 1 mM KH_2PO_4; 3 mM HEPES (pH = 7.4).

$^{13}C^{16}O_2$ Production and MVO_2 from Oxidation of Labeled [1-^{13}C] Pyruvate in CON and DM Heart Mitochondria: Mass Spectroscopy

The reaction took place in a 3 ml, temperature-regulated, magnetically stirred glass chamber, separated from the ion source of a Varian-Mat CH7 mass spectrometer by a

Table 1. Pyruvate and α-Ketoglutarate oxidation by heart mitochondria of control (CON) and diabetic (DM) rats

	State 3 ng-atoms O_2/min/mg protein	State 4 ng-atoms O_2/min/mg protein	State 3/State 4 = RCI	ADP/O ratio	ROP: nmoles ADP /s/mg protein
Pyruvate					
CON	192.50±16.09	29.36±4.68	6.70±1.20	2.79±0.18	9.10±1.57
DM	115.29±18.15*	26.44±5.01	4.40±0.54*	2.74±0.24	5.07±1.42*
αKetoglutarate					
CON	174.26±4.59	16.11±2.68	11.76±1.63	2.86±0.26	8.09±0.65
DM	156.81±3.45*	14.28±2.83	11.58±1.86	2.71±0.18	6.58±0.20*

Values are means ± SD; n=6. Mitochondria (1.5 mg of protein in mL) were prepared as described by Kondrashova and Doliba (1989); respiration in State 3 was measured in the presence of 0.3 mM-ADP, and respiration in State 4 was measured after ADP was completely phosphorylated. * P ≤ 0.02: CON vs DM.

0.0075 cm thick Teflon membrane supported by a sintered glass disc. Gas dissolved in the reaction mixture diffuses into the mass spectrometer ion source. The program scans 3 peaks in 51 seconds and the peak signals were recorded on a single channel strip chart recorder. Mass 32 peak corresponds to $^{16}O_2$, which allows monitoring of dissolved O_2 during the reaction and is used to calculate MVO_2. Mass 44 peak corresponds to the predominant isotopic form $^{12}C^{16}O_2$. Mass 45 peak corresponds to the isotopic form $^{13}C^{16}O_2$ and the result of [1–^{13}C] labeled pyruvate decarboxylation via PDH. It should be noted that $^{13}C^{16}O_2$ can also be produced in the IDH and KGDH steps after metabolism of [1–^{13}C] pyruvate via pyruvate carboxylase (PC); however, since heart has very low levels of PC, this contribution is negligible. There can also be an error associated with the presence of malic enzyme, which can result in an overestimation of PDH function.

RESULTS

Mitochondria Respiration Rate by Polarographic Techniques (Table 1)

State 3, RCI and the rate of ox-phos (ROP) of DM rat heart were depressed when using Pyr plus malate as substrates. State 3 and ROP were also decreased when α-ketoglutarate was used as substrate. The phosphorylation capacity, expressed as ADP/O ratio, appeared to be normal with both sets of substrates. The greatest decrease in substrate oxidation was observed with Pyr, suggesting that PDH activity is depressed in DM. It should be pointed out that in DM mitochondria, the decrease in State 3 was dependent on the concentration of Pyr (Figure 1); and that the K_m for Pyr was higher in DM (0.058±0.011 mM) compared to CON (0.0185±0.0014 mM), with no significant change in V_{max}. RCI was decreased approximately 35 % at all Pyr concentrations between 0.05 and 1 mM.

$^{13}C^{16}O_2$ Production from Oxidation of Labeled [1-^{13}C] Pyruvate in CON and DM Heart Mitochondria by Mass Spectroscopy

Figure 2A presents the time course of $^{13}C^{16}O_2$ production (1) and MVO_2 (2) during oxidation of [1–^{13}C]pyruvate by heart mitochondria from CON and DM rats. Both the $^{13}C^{16}O_2$ production and MVO_2 with and without ADP were much less in DM mitochondria

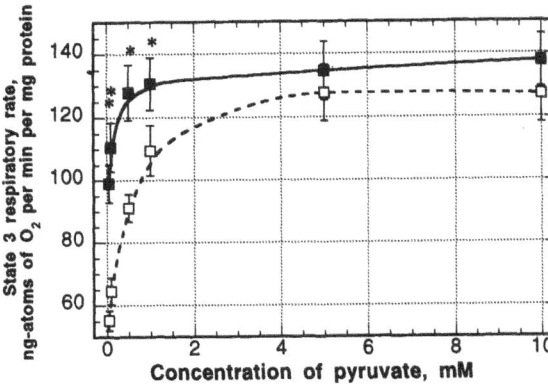

Figure 1. Pyruvate (Pyr) concentration dependency curve for state 3 respiration: CON ■; DM □* P < 0.05: CON vs DM.

compared to CON (with ADP, 35–50% less; without ADP, 20–30% less). In addition, during FCCP-induced uncoupling ox-phos, $^{13}C^{16}O_2$ production and MVO_2 were 40–50% less in DM than in CON, Figure 2B. To evaluate the effect of Ca^{2+} on respiratory function (and possible regulation of PDH), similar measurements were made after the addition of 1 μM Ca^{2+}. Free extramitochondrial Ca^{2+} concentration was controlled by the use of Ethylene Glycol-bis-(β-Amino Ethyl Ether) N′N′N′N′ Tetracetic Acid (EGTA)-Ca^{2+} buffer. Addition of Ca^{2+} caused minimal changes in $^{13}C^{16}O_2$ production in DM; whereas Ca^{2+} increased $^{13}C^{16}O_2$ production 33–40% in CON; Figure 3. Please note, no measurable $^{13}C^{16}O_2$ arises from metabolism of the second carbon of Pyr ([2–^{13}C]pyruvate).

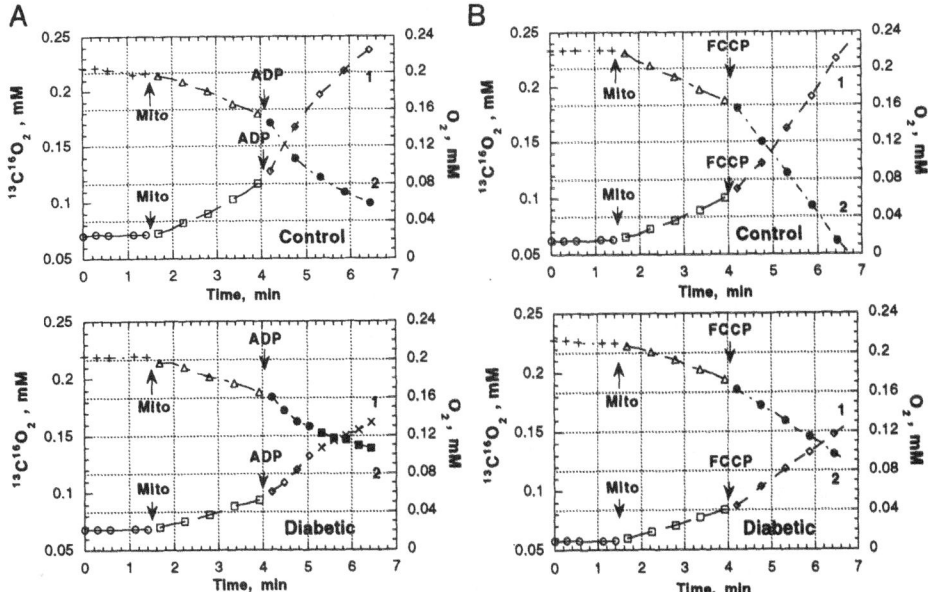

Figure 2. $^{13}C^{16}O_2$ production (1) and O_2 consumption (MVO_2) (2) during oxidation of (1–^{13}C) pyruvate by heart mitochondria from control and diabetic rats. A) $^{13}C^{16}O_2$ production (1) and O_2 consumption (MVO_2) (2) after addition of ADP. B) $^{13}C^{16}O_2$ production (1) and O_2 consumption (MVO_2) (2) after addition of FCCP to uncouple ox-phos.

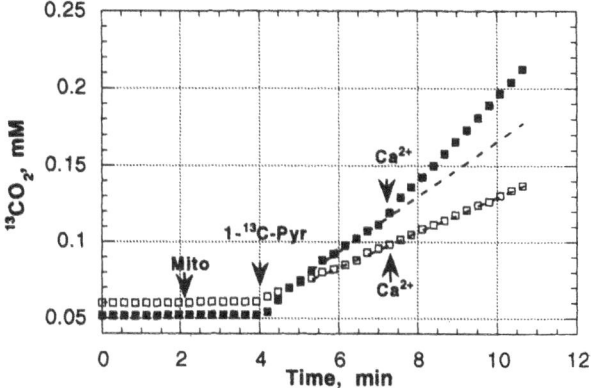

Figure 3. The effect of Ca^{2+} on CO_2 production in CON ■ and DM ❑ mitochondria.

Calcium Stimulated Respiration and Ca^{2+} Uptake

Heart mitochondria possess a uniporter that mediates Ca^{2+} uptake driven by the H^+-dependent mitochondrial membrane potential, and the egress of Ca^{2+} via the electrically neutral Na^+-Ca^{2+} antiport in the mitochondrial membrane (Crompton et al., 1976). In order to evaluate the changes in Ca^{2+} transport in diabetic heart, we have performed experiments where State 3 respiration was stimulated by $CaCl_2$ while Ca^{2+} uptake was estimated by using change in H^+ flux (Kondrashova et al., 1982). Figure 4 presents MVO_2 and Ca^{2+} up-

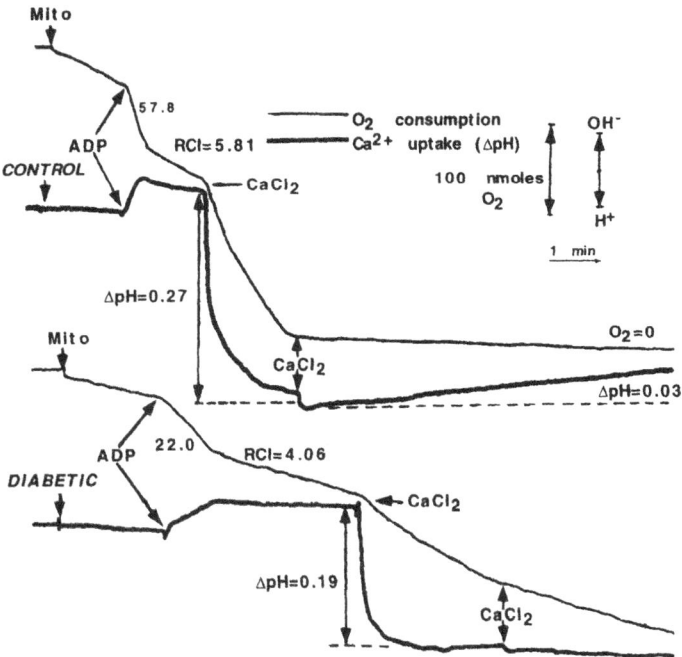

Figure 4. ADP and Ca^{2+} - stimulated respiration and Ca^{2+} uptake of heart mitochondria from control and diabetic rats.

take data in CON and DM heart mitochondria during Pyr + malate oxidation. State 3 was approximately 50% lower and RCI was 30% lower in DM mitochondria compared to CON. In order to measure Ca^{2+} capacity, we added 100 μM $CaCl_2$ to the incubation medium and followed Ca^{2+} uptake by changes in pH. CON mitochondria completely consumed the first addition of $CaCl_2$. After the second addition of $CaCl_2$, CON mitochondria consumed only a small amount of $CaCl_2$, followed by the release of $CaCl_2$ from organelles. In contrast, mitochondria from DM animals did not completely consume even the first addition of $CaCl_2$. These data suggest that the Ca^{2+} capacity in heart in DM rats is greatly decreased compared to CON.

DISCUSSION

Diabetic cardiomyopathy may be related to defective PDH function; this defect may be due to a deficiency in mitochondrial NAD and NADH or, alternatively, might be related to defects in the PDH complex (Kuo et al., 1983). In this regard, Kerbey et al. (1976) have shown that the regulation of PDH activity in rat heart may be affected by the availability of co-factors such as NAD and CoA. Mammalian PDH is a multienzyme complex that is regulated by phosphorylation and dephosphorylation. Kuo et al. (1985) demonstrated that the conversion of inactive (phosphorylated) enzyme into active (dephosphorylated) enzyme is inhibited in cardiac mitochondria from DM mice. These changes coincide with the onset of defective oxidative metabolism and can explain the depressed Pyr oxidation (Kuo et al., 1983,1985). The amount of active PDH in a tissue reflects the balance of the activity of the kinase which requires Mg^{2+}, and is inhibited by ADP and Pyr, and that of phosphatase, which requires a higher concentration of Mg^{2+} and is activated by Ca^{2+} (Wieland, 1983). The results of our study indicate that in DM, PDH has a decreased sensitivity to Pyr, with no change in the V_{max}. This explains the lack of change other investigators have observed when 5 mM pyruvate was used (Grinblat et al, 1986). Further, our results show that in CON heart mitochondria, Ca^{2+} stimulates Pyr oxidative function, whereas Ca^{2+} is ineffective in stimulating Pyr oxidation in DM mitochondria. The difference in Ca^{2+} response could be related to different baseline concentration of intramitochondrial Ca^{2+}. Others have shown that Ca^{2+} concentration in DM heart mitochondria is elevated (Zorov and Hansford, 1995). In support of this, we have demonstrated that total mitochondrial Ca^{2+} uptake (Ca^{2+} capacity) was less in DM rats, reflecting larger mitochondrial Ca^{2+} stores. Thus, the lack of stimulation of mitochondrial PDH by Ca^{2+} may be a causal factor in the development of DM cardiomyopathy and linked to the chronically high levels of Ca^{2+} that are found in mitochondria from DM rats.

REFERENCES

Baba A., Kako K. Calcium transport of sarcoplasmic reticulum and mitochondria *in situ* in heart cells of streptozocin-induced diabetic rats. *J. Appl. Cardiology, 1991, v.65, p.325–335.*

Crompton M., Capano M., Carafoli E. The sodium-induced efflux of calcium from heart mitochondria. - *Eur. J. Biochem., 1976, v.69, p.453–462.*

Friedman M., Ramirez I., Edens N., Granneman J. Food intake in diabetic rats: isolation of primary metabolic effects of fat feeding. *Am. J. Physiol., 1985, v.249, R44–51.*

Grinblat L., Pacheco Bolanos L., Stoppani A. Decreased rate of ketone body oxidation and decreased activity of D-3-hydroxybutyrate dehydrogenase and succinyl-CoA : 3-oxo-acid CoA-transferase in heart mitochondria of diabetic rats. *Biochem. J.,1986, v. 240, p.49–56.*

Kazmi S., Mayanil C., Baquer N. Malate-aspartate shuttle enzymes in rat brain regions, liver and heart during al-loxan diabetes and insulin replacement. *Enzyme, 1985, v.34, p.98.*

Kerbey A., Randle p., Cooper R., Whitehouse S., Pask H., Denton R. Regulation of pyruvate dehydrogenase in rat heart. Biochem. J., 1976, v.154, p.327–348.

Kondrashova M., Gogvadze V., Medvedev B., Babsky A. Succinic acid oxidationas only energy support of inten-sive Ca++ uptake by mitochondria. BBRC 1982. v.109, no2.,p.376–381

Kondrashova M., Doliba N. Polarographic observation of substrate-level phosphorylation and its stimulation by acetylcholine. *FEBS Letters, 1989, v.243, p.153–155.*

Kuo T., Moore K., Giacomelli F., Weiner J. Defective oxidative metabolism of heart mitochondria from geneti-cally diabetic mice. *Diabetes,1983, v. 32, p. 781.*

Kuo T., Giacomelli F., Wiener J. Oxidative metabolism of Polytron versus Magarse mitochondria in hearts of ge-netically diabetic mice. *Bioch. Bioph. Acta, 1985, v.806, p.9–15.*

Linn T., Pettit F.,Hucho F., Reed L. Ketoacid dehydrogenase complexes. Proc.Natl.Acad.Sci.USA 1969, v64, p.227–234.

McCormack J., Halestrap A., Denton R. Role of calcium ions in regulation of mammalian intramitochondrial me-tabolism. *Physiol. Rev., 1990, v.70, 391–425.*

Minezaki K.K., Suleiman M.-S., Chapman R.A. Changes in mitochondrial function induced in isolated guinea-pig ventricular myocytes by calcium overload. *J. Physiol., 1994, v.476, p.459–471.*

Mokhtar N., Lavoie J., Roussean-Migneron S., Nadean A. Physical training reverses defect in mitochondrial en-ergy production in heart of chronically diabetic rats. -*Diabetes, 1993, v.42, p.682–687.*

Paulson D., Schmidt M., Traxler J., Ramacci M., Shug A. Improvement of myocardial function in diabetic rats af-ter treatment with L-carnitine. *Metabolism, 1984, v.33, p.358.*

Pierce G., Dhalla N. Heart mitochondrial function in chronic experimental diabetes in rats. *Can.J.Cardiol.,1985, #1, p.48.*

Puckett S., Reddy W. A decrease in the malate-aspartate shuttle and glutamate translocase activity in heart mito-chondria from alloxan-diabetic rats. *J. Mol. Cell. Cardiol.,1979, #11, p.173.*

Taegtmeyer H., Zirafi C., Nguyen B. Function and metabolism of the heart in diabetes. A fresh look at an old prob-lem. *J. Appl. Cardiol., 1987, v.2, #1, p.37–48.*

Wieland O. The mammalian pyruvate dehydrogenase complex: structure and regulation. Rev.Physiol.Bio-chem.Pharmacol., 1983, v.96, p.123–170.

Zorov D.B., Hansford R.G. Altered dynamics of intramitochondrial calcium in cardiac myocytes from diabetic rats. *Biophys.J., 1995, v.68.#2, abstracts, W-Pos373.*

38

EXOGENOUS SURFACTANT AND NITRIC OXIDE HAVE A SYNERGISTIC EFFECT IN IMPROVING GAS EXCHANGE IN EXPERIMENTAL ARDS

A. Hartog, D. Gommers, A. van 't Veen, W. Erdmann, and B. Lachmann[*]

Department of Anesthesiology
Erasmus University Rotterdam
Postbox 1738, 3000 DR Rotterdam, The Netherlands

1. INTRODUCTION

In the acute respiratory distress syndrome (ARDS) instillation of exogenous surfactant has shown to improve blood gases and lung mechanics (Gommers and Lachmann, 1993; Hartog et al., 1995). Although these studies demonstrated that some patients did not respond to, or had an only transient improvement after a single dose of surfactant, better results were seen with higher or multiple surfactant doses (Gommers and Lachmann, 1993; Gregory et al., 1994). This implies that for treatment of ARDS high doses of surfactant are required. However, the non-availability of large amounts of surfactant and the high price restrict the use of exogenous surfactant therapy in adults.

Inhaled nitric oxide (NO) is a selective pulmonary vasodilator that has shown to improve ventilation/perfusion matching thereby increasing oxygenation in ARDS (Rossaint et al, 1993). However, in clinical studies evaluating the efficiacy of inhaled NO in the treatment of ARDS, a common finding is that a percentage of patients does not respond to NO inhalation (Rossaint et al., 1995). It has been suggested that strategies which optimize lung recruitment, such as exogenous surfactant therapy, are likely to enhance responsiveness to inhaled NO (Abman, 1996).

The aim of this study was to evaluate whether it is possible to save exogenous surfactant by combining administration of a low dose of exogenous surfactant with inhaled NO.

[*] Tel. 31-10-4087312, FAX 31-10-4367870.

Oxygen Transport to Tissue XIX, edited by Harrison and Delpy
Plenum Press, New York, 1997

2. MATERIALS AND METHODS

Adult New Zealand White rabbits (n=10; 2.5 ± 0.3 kg) were anesthetized, tracheotomized, intubated and cannulated with a carotid artery catheter. The animals were ventilated with 100% oxygen in a volume controlled mode using a Servo 300 ventilator equipped with a built-in NO administration module (Siemens, Solna, Sweden). All animals were surfactant depleted by repeated lung lavage, until PaO_2 was ≤ 75 mmHg at a positive end-expiratory pressure (PEEP) of 6 cmH$_2$O. Forty-five minutes after the last lavage, blood gases were measured and all animals were subsequently divided into two groups. One group received a low dose of surfactant (25 mg/kg) intratracheally followed by nitric oxide inhalation (surfactant group). Five different NO concentrations (4, 8, 10, 16 and 20 parts per million (ppm)) were used. Each dose was administered for 30 minutes, and was followed by a 30 minute wash out period. The other group served as control and received only inhaled NO in the same randomized order as the first group (non-surfactant group). In both groups, blood gases were measured every 30 min for 5 hour.

3. RESULTS

In the surfactant group, PaO_2 increased from 59 ± 10 to 272 ± 56 mmHg (mean ± SD, p<0.005) within 30 min after instillation of a dose of 25 mg/kg of surfactant. Nitric oxide inhalation resulted in an additional increase in PaO_2, and when NO inhalation was interrupted, PaO_2 decreased. In the non-surfactant group however, PaO_2 did not improve significantly after inhalation of NO.

For each NO concentration, there was a better improvement in PaO_2 in the surfactant group, than in the non-surfactant group (Figure 1).

4. DISCUSSION

Several studies have shown that both exogenous surfactant therapy and inhaled nitric oxide are able to improve arterial oxygenation in ARDS (Gommers and Lachmann,

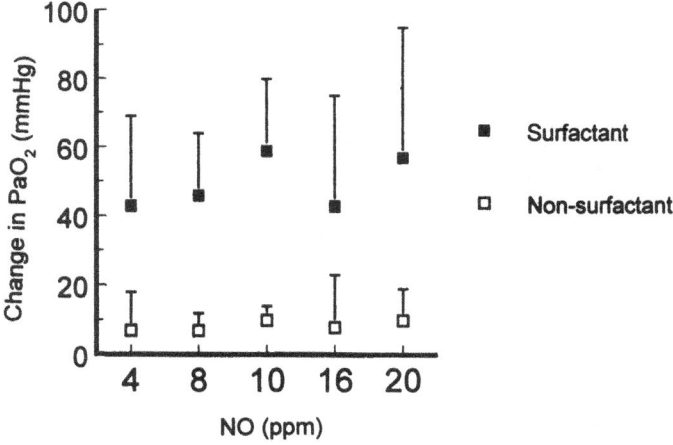

Figure 1. Change in PaO_2 (mean ± SD, in mmHg) after NO inhalation, at the doses indicated. The surfactant group received a low dose of surfactant (25 mg/kg) prior to NO inhalation, the non-surfactant group received only inhaled NO.

1993, Rossaint et al., 1993, Walmrath et al., 1996). Exogenous surfactant improves arterial oxygenation by re-areating atelectatic regions. Inhaled NO redirects pulmonary blood flow from lung units with low ventilation/perfusion ratios to units with normal ratios, thereby reducing ventilation/perfusion mismatch and thus improving oxygenation.

In the non-surfactant group, inhaled NO was not effective in increasing arterial oxygenation (figure). It is assumed that the inhalation technique delivers NO to the areated areas of the lung and that NO then diffuses passively to the smooth muscle of the pulmonary blood vessels to elicit their response (Gaston et al., 1994). When large areas of the lung are atelectatic, NO gas is unable to reach the target cells.

The results of this study show that inhaled NO leads to a further improvement of PaO_2 after pre-treatment with a low dose of exogenous surfactant. This is in accordance with the results of Karamanoukian et al. (1995), who reported that in congenital diaphragmatic hernia lambs NO inhalation alone was not able to improve oxygenation, whereas after pre-treatment with exogenous surfactant inhaled NO did improve oxygenation; in this study, a low dose of exogenous surfactant (50 mg/kg) was also used. From experimental and clinical observations it is known that a similar improvement in PaO_2, as seen after additional NO inhalation, can be obtained by increasing the amount of exogenous surfactant.

We did not find a dose-dependent NO effect, as also reported in other studies (Finer et al., 1994 Rossaint et al., 1993). In most animals, the largest increase in PaO_2 occurred at a different dose of NO. It appears that each lung has its own "optimal" NO dose, which is probably determined by the severity of the disease process, i.e. atelectasis, edema, etc.

We conclude that when the lungs are first re-areated by instillation of a low dose of exogenous surfactant, the efficacy of inhaled NO in increasing arterial oxygenation can be improved. Therefore, it may be possible to save exogenous surfactant by supplemental nitric oxide inhalation, but whether this combination leads to better results than treatment with a high dose of exogenous surfactant without NO needs to be determined by further clinical studies.

REFERENCES

Abman SH. Inhaled nitric oxide therapy in neonatal and pediatric cardiorespiratory disease. In:Tibboel D, van der Voort E (eds). Update in Intensive Care & Emerg Med 25. Springer 1996: 322–336.

Finer NN, Etches PC, Kamstra B, Tierney AJ, Peliowski A, Ryan CA. Inhaled nitric oxide in infants refered for extracorporal membrane oxygenation: dose response. J Pediatr 1994; 124: 302–308.

Gaston B, Drazen JM, Loscalzo J, Stamler JS. The biology of nitrogen oxides in the airways. Am J Resp Crit Care Med 1994; 149: 538–551.

Gommers D, Lachmann B: Surfactant therapy: does it have a role in adults? Clinical Intensive Care 1993; 4: 284–295.

Gregory TJ, Gadek JE, Hyers TM, Crim C, Hudson LD, Steinberg KP, Maunder RA, Spragg RG, Smith RM, Tierney DF, Gipe B, Longmore WJ, Moxley MA. Survanta supplementation in patients with acute respiratory distress syndrome (ARDS). Am J Resp Crit Care Med 1994; 149(Suppl): A567.

Hartog A, Gommers D, Lachmann B. Role of surfactant in the pathophysiology of the acute respiratory distress syndrome (ARDS). Monaldi Arch Chest Dis 1995; 50: 372–377.

Karamanoukian HL, Glick PL, Wilcox DT, Rossman JE, Holm BA, Morin FC. Pathophysiology of congenital diaphragmatic hernia VIII: inhaled nitric oxide requires exogenous surfactant therapy in the lamb model of congenital diaphragmatic hernia. J Pediatr Surg 1995; 30: 1–4.

Rossaint R, Falke KJ, López F, Slama K, Pison U, Zapol WM. Inhaled nitric oxide for the adult respiratory distress syndrome. N Eng J Med 1993; 328: 399–405.

Rossaint R, Gerlach H, Schmidt-Ruhnke H, Pappert D, Lewandowski K, Steudel W, Falke K. Efficacy of inhaled nitric oxide in patients with severe ARDS. Chest 1995; 107: 1107–1115.

Walmrath D, Günther A, Ardeschir H, Schermuly R, Schneider T, Griminger F, Seeger W. Bronchoscopic surfactant administration in patients with severe adult respiratory distress syndrome and sepsis. Am J Respir Crit Care Med 1996; 154: 57–62.

39

PARTIAL LIQUID VENTILATION AND INHALED NITRIC OXIDE HAVE A CUMULATIVE EFFECT IN IMPROVING ARTERIAL OXYGENATION IN EXPERIMENTAL ARDS

A. Hartog, R. J. M. Houmes, S. J. C. Verbrugge, W. Erdmann, and B. Lachmann[*]

Department of Anesthesiology
Erasmus University Rotterdam
Postbox 1738, 3000 DR Rotterdam, The Netherlands

1. INTRODUCTION

During the acute respiratory distress syndrome (ARDS), the decreased function of the pulmonary surfactant system leads to an increased surface tension at the alveolar air liquid interface. This increased surface tension is responsible for the end-expiratory alveolar collapse, atelectasis, right to left shunt and a decrease in PaO_2, finally resulting in hypoxemia (Hartog et al., 1995). A rational therapy to treat this condition is to eliminate the air-liquid interface by filling the lung up to the funtional residual capacity (FRC) level with perfluorocarbon (PFC) and ventilating these lungs with normal gas ventilation. This type of ventilation is called partial liquid ventilation, and has shown to improve gas exchange in several studies (Tütüncü et al., 1993, and Houmes et al., 1995).

Inhaled nitric oxide (NO) is a selective pulmonary vasodilator that has shown to increase oxygenation in patients with ARDS by decreasing pulmonary right-to-left shunt (Rossaint et al., 1993). However, in clinical studies evaluating the effectiveness of inhaled nitric oxide in the treatment of ARDS, a common finding is that a percentage of patients does not respond to NO inhalation (Rossaint et al., 1995).

It has been suggested that strategies which optimize lung recruitment, are likely to enhance responsiveness to inhaled NO (Abman, 1996). Since liquid ventilation with PFC has shown to enhance alveolar recruitment in atelectatic lungs (Tooley et al., 1996), we

---

[*] tel. 31-10-4087312, FAX 31-10-4367870.

Oxygen Transport to Tissue XIX, edited by Harrison and Delpy
Plenum Press, New York, 1997

hypothesized that PLV and inhaled NO may have a cumulative effect in increasing arterial oxygenation in the ARDS lung.

2. METHODS AND MATERIALS

Six female Yorkshire pigs (BW 7 ± 1 kg) were anesthetized, tracheotomized, intubated and cannulated with a carotid artery catheter. The animals were ventilated with 100% oxygen in a volume controlled mode with a Servo 300 ventilator equipped with a built-in NO administration module (Siemens, Solna, Sweden).

All animals were surfactant depleted by repeated lung lavage until PaO_2 was ≤ 100 mmHg at a positive end-expiratory pressure (PEEP) of 5 cmH_2O. Subsequently all animals were ventilated for 1 hour to obtain stable baseline values. Following this baseline period all animals received four intratracheal doses of 5 ml/kg perfluorocarbon. After each dose of perfluorocarbon, all animals received inhaled NO (10, 20, 30 and 40 parts per million (ppm) respectively, each dose for 10 min). Before administering a new dose of perfluorocarbon, NO was switched off for 10 min.

3. RESULTS

All animals survived the study period. Before and after lung lavage all animals were comparable for blood gases. No improvement in blood gases was observed during the one hour post-lavage period.

Administration of perfluorocarbon resulted in a significant dose-dependent increase in PaO_2 ($p<0.005$) (Figure 1). NO inhalation at any level of PFC resulted in an additional

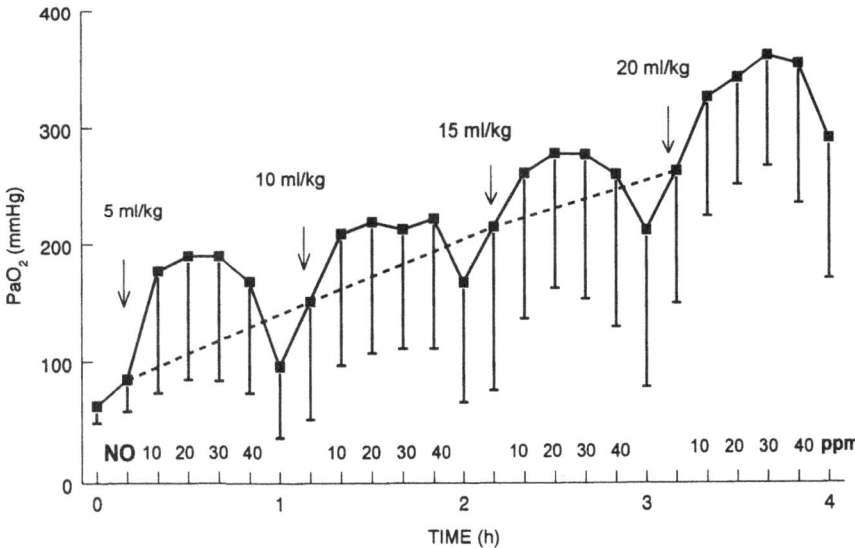

Figure 1. PaO_2 (Mean ± SD, mmHg). Arrows indicate PaO_2 after administration of a dose of perfluorocarbon to the total amount mentioned. The values 10, 20, 30 and 40 represent the doses of NO administered at each dose of perfluorocarbon (in parts per million (ppm)). The dashed line shows the dose-response effect of perfluorocarbon on arterial oxygenation.

increase in PaO_2 that was similar to the increase in PaO_2 that was obtained with the next increment of PFC alone ($p<0.05$) (Figure 1).

4. DISCUSSION

The results of the present study show that partial liquid ventilation is an effective therapy, and leads to a dose-dependent improvement in gas exchange in surfactant depleted animals. These findings confirm the results from previous reports on PLV (Tütüncü et al., 1993, and Houmes et al., 1995).

The beneficial effect of PLV on gas exchange is mediated through the physical presence of perfluorocarbon in the alveolus, preventing it from expiratory collapse. This results in an improvement of arterial oxygenation, since oxygenation of the blood can now continue during the expiratory phase.

Furthermore, the present study shows that NO inhalation after pre-treatment with perfluorocarbons results in a further improvement of gas exchange. This is in accordance with the results of Wilcox et al. (1995), who reported an increase in PaO_2 in a hypoplastic congenital diaphragmatic hernia lamb model after introducing PLV. In their study, inhaled NO was added for a short period of 10 minutes, which resulted in an additional increase in arterial oxygenation that was reversed when NO was switched off again.

In conclusion: The present study shows that the combination of perfluorocarbon and NO has a cumulative effect in improving arterial oxygenation. Whether the best results are obtained when combining lower doses PFC with inhaled NO, or with higher doses of PFC without inhaled NO, remains to be determined by future clinical studies.

REFERENCES

Abman SH. Inhaled nitric oxide therapy in neonatal and pediatric cardiorespiratory disease. In: Tibboel D, van der Voort E (eds). Update in Intensive Care & Emerg Med 25. Springer 1996: 322–336.

Hartog A, Gommers D, Lachmann B. Role of surfactant in the pathophysiology of the acute respiratory distress syndrome (ARDS). Monaldi Arch Chest Dis 1995: 50; 372–377.

Houmes RJ, Verbrugge SJC, Hendrik ER, Lachmann B. Hemodynamic effects of partial liquid ventilation with perfluorocarbon in acute lung injury. Int Care Med 1995; 21: 966–972.

Rossaint R, Falke KJ, López F, Slama K, Pison U, Zapol WM. Inhaled nitric oxide for the adult respiratory distress syndrome. N Eng J Med 1993: 328: 399–405.

Rossaint R, Gerlach H, Schmidt-Ruhnke H, Pappert D, Lewandowski K, Steudel W, Falke K. Efficacy of inhaled nitric oxide in patients with severe ARDS. Chest 1995: 107; 1107–1115.

Tooley R, Hirschl RB, Parent A, Bartlett RH. Total liquid ventilation with perfluorocarbons increases pulmonary end-expiratory volume and compliance in the setting of lung atelectasis. Crit Care Med 1996; 24: 268–273.

Tütüncü AS, Akpir K, Mulder P, Erdmann W, Lachmann B. Intratracheal perfluorocarbon administration as an aid in the ventilatory management of respiratory distress syndrome. Anesthesiology 1993: 79: 1083–1093.

Wilcox DT, Glick PL, Karamanoukian HL, Leach C, Morin II FC, Fuhrman BP. Perfluorocarbon-associated gas exchange improves pulmonary mechanics, oxygenation, ventilation, and allows nitric oxide delivery in the hypoplastic lung congenital diaphragmatic hernia lamb model. Crit Care Med 1995: 23: 1858-1863.

40

BLOOD OXYGEN DESATURATION HETEROGENEITY DURING MUSCLE CONTRACTION RECORDED BY NEAR INFRARED SPECTROSCOPY

P. G. Cooper, G. J. Wilson, D. T. A. Hardman, O. Kawaguchi, Y. -F. Huang,
A. Martinez-Coll, R. A. Carrington, E. Puchert, R. Crameri, C. Horamand,
and S. N. Hunyor

CRC for Cardiac Technology
Royal North Shore Hospital
St Leonards, NSW, 2065, Australia

INTRODUCTION

Non-invasive measurements of blood oxygen desaturation in electrically paced sheep latissimus dorsi muscle (LDM) were made using near infrared spectroscopy (NIRS). The theory of NIRS measurements of blood oxygen desaturation and blood volume is well established (Jobsis, 1977; Elwell et al., 1994, Chance et al., 1988). Our aim was to monitor changes in muscle vascular desaturation during electrical stimulation of the muscle using burst pacing. Establishing the properties of skeletal muscle during burst pacing is important because burst pacing is used to stimulate skeletal muscle, in particular the LDM, during cardiac assist.

NIRS is a useful technique for non-invasive monitoring of paced muscle because the characteristic desaturation occurring during induced contractions provides signals which can be unambiguously attributed to the vasculature of the contracting muscle. Moreover, it has become evident that NIRS desaturation signals can reveal interesting new metabolic phenomena not detectable using other techniques. For example, the NIRS signal collected trans-cutaneously from electrically paced muscle suggested an increase in blood oxygen saturation during contraction in 30% of our acquisitions. This was unexpected since muscle contraction usually depletes oxygen from the blood. The phenomenon was observed as a transient increase in saturation signal lasting from several seconds to one minute in regions being monitored. Subsequently, this study was supplemented by an invasive study which provided simultaneous measurements of arterial flow to the muscle, muscle shortening and NIRS monitoring during pacing.

Oxygen Transport to Tissue XIX, edited by Harrison and Delpy
Plenum Press, New York, 1997

Our findings are consistent with the reports of inhomogenous blood flow in muscle measured by microsphere distribution and xenon clearance (Cerretelli et al., 1984; Pendergast et al., 1985) and inert gas washout (Gronlund et al., 1989). We believe that we have observed using NIRS the same flow inhomogeny as measured by these authors.

Our results suggest that the blood flow heterogeneity in the contracting muscle is in part facilitated by transient relaxation of particular regions of the muscle, allowing "flushing" by the blood. We suggest that a possible reason for this transient "flushing" is a localised response to regional metabolite blood/tissue exchange requirements.

MATERIALS AND METHODS

Near Infrared Spectroscopy

We have designed and built two NIR-spectrometers. The first NIR-spectrometer used alternately pulsed laser diodes at 754 and 813 nm wave lengths mounted 30 mm from a photodiode used to detect scattered photons from tissue. The output voltage from the NIR-spectrometer was a 754 nm signal subtraction from the 813 nm signal, which in vitro calibration experiments, based on that of Chance et al. (1988), showed a linear signal output that corresponded to relative blood oxygen desaturation. The optode assembly containing the laser diodes and the photodiode were placed against the sheep's skin, from which the wool had been clipped, adjacent to the LDM. A second NIR-spectrometer that pulsed at three different wave lengths, 770, 790 and 850 nm, was subsequently constructed for an invasive study. This NIR-spectrometer measured relative blood volume (770 + 850 nm) as well as relative oxygen desaturation (770–850 nm). The optode assembly was designed to be sutured directly onto muscle.

Anatomy and Surgery of the Sheep LDM

The sheep LDM is a delta shaped flap of muscle, inserting proximally into the humerus and distally to the spine and ribs. It is some 15 mm thick at the narrow, proximal end, thinning out to 1–2 mm thickness in the broad, distal region. The mass with the fat trimmed is 90–290 g. The main supply is from the proximal, trifurcating neurovascular bundle containing the thoracodorsal artery (TDA) and thoracodorsal nerve (TDN). Anaesthesia was induced using 20 mg/kg of thiopentone (i.v.). The animals were intubated and anaesthesia was maintained using halothane (1.5–1.8% of inspired volume), N_2O (2L/min) and O_2 (1.5 L/min). Buprenorphine was injected for pain management (6 μg/Kg every 2 hours i.m.) during the procedure, and afterwards in sheep which recovered.

Non-Invasive NIRS Study of Muscle

Under anaesthesia, five sheep had burst pacers (Model 7220, Telectronics Pacing Systems Inc.) implanted subcutaneously. The muscle was stimulated using electrical field stimulation of the TDN via electrodes (Oscor Corp.) which were sutured close to the medial surface of the in situ LDM, 5 cm apart, at the origin and termination of the primary trifurcation of the TDN. After recovery from anaesthesia, the sheep were subjected to an electrical burst pacing protocol designed to mimic that used for cardiac assist (5 volts, 0.25 to 0.33 Hz burst rate, 33 Hz pulse rate, 1–4 pulses/burst). For each sheep, one minute of non-invasive NIRS acquisition was obtained from the contracting LDM three or more times a week. The traces were observed in real time on a computer and electronically stored for further analysis. The sheep were free to walk around their pen.

Invasive NIRS Study of Muscle

Under anaesthesia, an acute invasive study was conducted on one sheep to confirm heterogeneity of blood flow and blood oxygen desaturation. The sheep was lain on its right side. The left LDM and its main proximal supply, the TDA, were exposed. A Doppler flow probe cuff (Triton Inc., CA, USA) was applied around the TDA. Stainless steel stimulating electrodes were implanted adjacent to the TDN in a similar arrangement to that described above. A 70 mm length capacitive strain gauge (CSG) (AD Instruments, Castle Hill, NSW, Australia) was attached in parallel to the muscle fibres of the LDM. The CSG enabled recordings of muscle shortening velocity during electrical field stimulation of the TDN. During shortening, the left leg was free to move (unloaded). The optode assembly of the NIRS was positioned either directly on the costal surface of the exposed LDM opposite the CSG, or on the skin overlying the LDM. Data from the Doppler, CSG and NIRS sensors were simultaneously processed by an ADC (MacLab) (AD Instruments) at 200 samples/s, displayed on a Macintosh computer (Apple Computers) using Chart 3.4 data acquisition software (AD Instruments), stored and analysed off line. The LDM was stimulated via the implanted electrodes using burst pacing generated by a Grass stimulator (model S88, Grass Instruments, Mass, USA). The stimulation voltage was set at the minimum required to elicit the maximal shortening response (~5V, 100 µs pulse width, 32Hz pulse rate, 1–6 pulses/burst). The sheep was euthanased at the end of the procedure without recovery from anaesthesia by injecting KCl (0.1 g/ml, 30 ml i.v.).

RESULTS

A number of NIRS measurements gave complex responses to contractions, suggestive of photons received from both contracting and non-contracting regions. We were unable to pick up flow heterogeneity in resting muscle using NIRS. The muscle must be perturbed in order to show meaningful measurements in blood flow and oxygen saturation.

Non-Invasive Study

In the five sheep, 155 useable spectra of the LDM were acquired over a four month period of muscle stimulation. Of these acquisitions, 45 (29%) showed peaks suggestive of increased saturation in the paced muscle. Two of the five sheep had acquisitions taken from three regions of the muscle: Proximal (near the LDMs insertion into the humerus), mid and distal regions. For the total of 88 acquisitions in these two sheep, increased saturation peaks were acquired: proximal =30%, mid =33% and distal =28% of the time. No statistical significance was found for a regional preference for these saturation signals.

Non-invasive NIRS measurements were obtained from the LDM of conscious, standing sheep throughout a four to eight week period of LDM pacing. Figure 1a shows a four second segment of the NIRS acquisition in which induced contractions (4-pulses/burst) in the mid region of the LDM were observed. During each electrically induced contraction of the LDM, a characteristic peak, signifying blood oxygen desaturation, was obtained. This peak from the LDM could be distinguished from that reflected from the surrounding tissue because it was synchronised with pacing. The cyclic

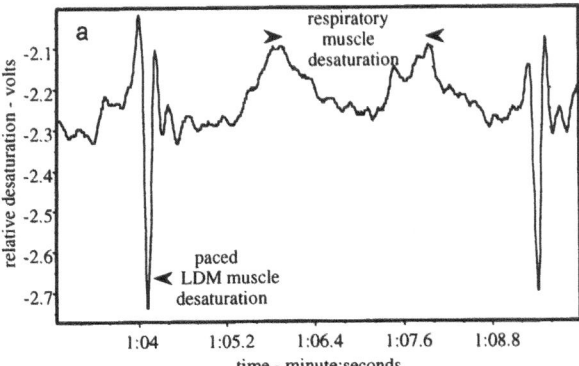

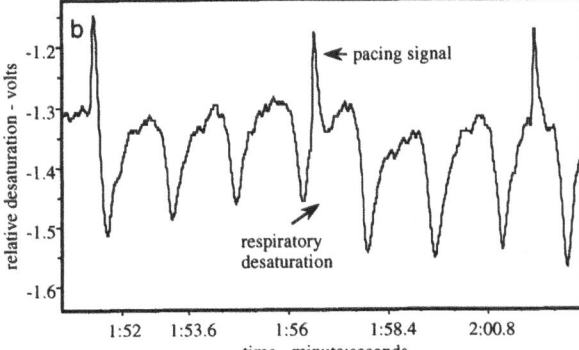

Figure 1. Non-invasive NIRS measurements of paced LDM (a) LDM blood oxygen desaturation distinct from background signal as sharp inverted peaks. Signal from underlying respiratory muscle also evident. (b) Signal acquired from the distal end of the LDM, approximately 2 mm thick, with the intercostal respiratory desaturation peaks dominating the signal. The pacing signal from the LDM is inverted.

desaturation peaks of the intercostal muscles, the respiratory muscles lying directly beneath the LDM, are also present in Figure 1. Smaller desaturation oscillations superimposed on the trace are due to pulsatile blood flow. The LDM pacing peaks result from an increase in regional deoxyhemoglobin content, corresponding to increased oxygen consumption during contraction. This is what would be expected from a contracting muscle. It is unlikely that desaturated myoglobin was contributing to the signal at this work load (Wang et al., 1990).

In some cases (Figure 1b), increased oxygen saturation was observed in contracting muscle. The desaturation peaks obtained during pacing indicate an increase in blood oxygen saturation rather than a decrease. The normal desaturation of the respiratory muscle in the same trace ruled out the possibility that the inverted pacing peaks were due to instrument artefact. In the distal region the signal from the respiratory muscle was greatly enhanced relative to the signal from the paced LDM overlying it, because the LDM is only about 1–2 mm thick in this region, whereas it is 15 mm thick in the proximal region. In all, some 30% of our acquisitions indicated an increase in saturation with muscle contraction.

Acquisition from the mid region of the LDM displays both saturation and desaturation in response to pacing (Figure 2). The first peak is a saturation peak. However, the signal reverts to normal desaturation troughs for about 18 seconds before reverting again to saturation peaks.

The average signal level in Figure 2 went from a high level of desaturation to a lower level. This is in response to a postural change by the sheep, resulting in altered tension on the LDM. At the start of the trace there is tension on the LDM. Over a 20 second period the sheep relaxes the muscle voluntarily, resulting in reduction of the desaturation level.

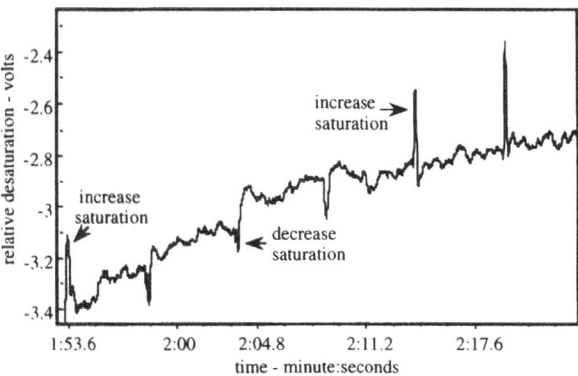

Figure 2. An acquisition segment in which the LDM paced signal changes between saturation to desaturation within an 18 second period. The decreasing desaturation background average signal was attributable to sheep postural change with decreased LDM tension.

Invasive Study

In an invasive study of one sheep, the animal was on its side with its left leg free from postural or other loading and thus free to move. The LDM therefore shortened upon stimulation. NIRS measurements during pacing of this sheep showed transient inversions in the oxygen saturation peaks obtained from contracting muscle similar to those observed in the non-invasive study above. The mid region of the LDM to which the NIRS optode assembly was attached characteristically elicited a signal like that shown in Figure 3a. Muscle shortening of around 9% of L_o was induced at 4-pulses/burst (3rd trace). Increases

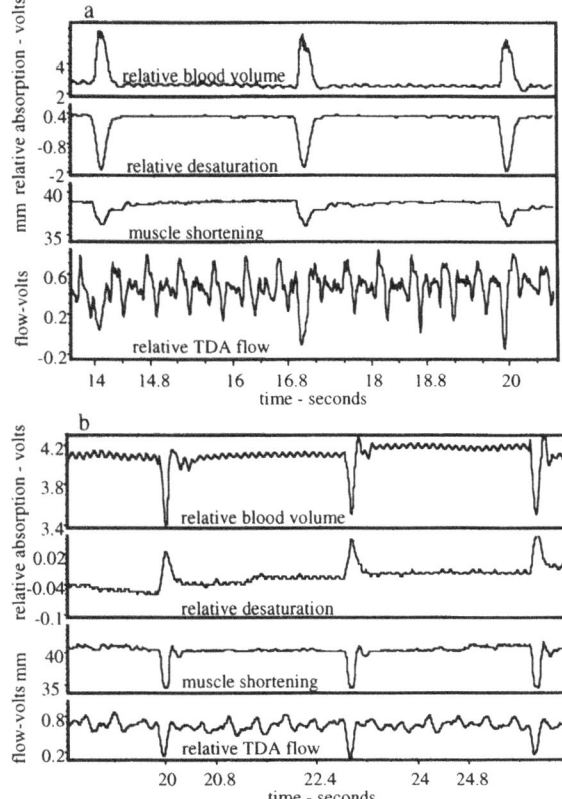

Figure 3. Invasive NIRS study with optode assembly attached to the costal side of the LDM. (a) Trace 1 - Relative blood volume (770 + 850 nm) decrease with muscle stimulation (4-pulses/burst). Trace 2 - With muscle contraction the blood vasculature desaturates (770–850 nm) giving a characteristic increase in relative absorption due to increased deoxyheamoglobin with contraction. Trace 3 - Muscle shortening, 6%, with each contraction. Trace 4 - Relative blood flow (phasic) in the main artery, TDA, supplying the LDM. Flow is restricted with each contraction, zero volts equals zero blood flow. (b) Muscle stimulated at 4-pulses/burst. Trace 1 - Blood volume increases (reduced attenuation) with each contraction, background oscillations are processing artefact. Trace 2 - At each contraction the vascularity has increased blood oxygen perfusion (decreased absorption). 30% of acquisitions showed this type of LDM increased saturation with contraction. Trace 3 - Muscle shortened during contraction, 7%, showing no global change associated increased saturation. Trace 4 - No significant change occurred in blood flow restriction during increased regional saturation.

in relative blood volume (top trace, 770 + 850 nm), relative desaturation (2nd trace, 770–850 nm) and mean arterial flow to the LDM (bottom trace) all suggest that blood volume decreased (decreased absorption) and blood oxygen desaturation increased, as expected, during muscle shortening. Decreases in phasic TDA blood flow (0 volts = zero flow) during muscle shortening probably accounts for the reduced blood volume during contraction.

In Figure 3b, however, traces are once again shown which indicate an increase in both blood volume and blood oxygen desaturation during contracttion. The muscle shortened by about 7% of L_o in response to stimulation at 4-pulses/burst (5.5 mm displacement for the 70 mm CSG) with each contraction, similarly to that in Figure 3a. Thus no global decrease in shortening occurred. Similarly, transient reduction in TDA flow is similar for both traces. This invasive study, where NIRS measurements were taken directly from the muscle surface, confirms the results obtained from the non-invasive, trans-cutaneous study on conscious animals.

DISCUSSION

Using NIRS, we have observed both decreases and increases in blood oxygen desaturation during pacing-induced contractions of the sheep LDM. Some 30% of the measurements indicated regional, transient increases in blood oxygen saturation, rather than the expected decreases which were more commonly seen. Similar results were obtained from both non-invasive studies where measurements were taken through the skin, and invasive studies where the NIR optodes were placed directly against the muscle. These transients lasted from several seconds to one minute.

Blood volume measurements performed simultaneously with desaturation measurements provided evidence that in the regions of increased oxygen saturation there were also transient reductions in muscle contraction. During normal contractions, as in Figure 3a, arterial flow is temporarily reduced during muscle stimulation. This would result in reduced venous pressure, given uninterrupted drainage (the LDM is above the heart). Hence a non-contracting region that shared a common venous drain with contracting muscle would have an increased pressure gradient across its vasculature. For instance, for a given decrease in venous pressure, there is an up to 5- to 10-fold increase in capillary hydrostatic pressure (Berne and Matthew, 1981). The haemodynamic response to decreased venous pressure would be to increase blood volume. For example in Figure 3b there is no increase in oxygen extraction (non-contracting) but an increase in blood volume and presumably also blood flow. This then resulted in a transient increase in blood oxygen saturation in this region.

No noticeable changes in Doppler blood flow or muscle shortening occurred that correlated with NIRS-monitoring of these non-contracting regions. Changes in blood flow and muscle shortening were evident when the number of pulses/burst in the pacing burst were altered (results not shown). This shows that the absence of an effect when saturation occurred was not due to instrument artefact. Hence we conclude that these regions, though transient, remained constant in number and size, so that global changes to arterial flow and muscle shortening remained constant.

These observations are consistent with reports of inhomogenous blood flow in muscle (Ellsworth and Pittman 1984; Pendergast et al., 1985; Piiper et al., 1990; Iversen and Nicolaysen, 1989). Microsphere entrapment was used to detect heterogeneity in muscle at rest and during treadmill exercise. Heterogeneity was shown to increase during exercise

(Pendergast et al., 1985). We suggest that when muscle contraction is increased during exercise, the arterio-venous pressure gradient would be even higher, assuming some restriction to flow in the dilated resistance vessels caused by the contraction. This would result in increased blood flow in non-contracting regions with a higher level of microsphere entrapment in such regions. Thus the frequency of occurrence of non-contracting transients may not increase in exercise, but rather the level of flow may increase at each occurrence.

The size of the non-contracting regions is not known. However, we can make some estimations based upon the anatomy of the LDM and the geometry of the NIR-spectrometer optode separation. The LDM is essentially a muscle flap. One can therefore approximate the volume of the transiently non-contracting regions. Considering that the photon path in the muscle approximates a two-dimensional geometry with a long axis of 3 cm (transmitting to receiving optode distance), an estimate of regional area from which photons were received would be approximately 6 cm^2. Therefore, for a muscle thickness of 0.5 cm, we can estimate that the optodes will detect changes from a mass of at least ~3 gm of muscle tissue. There is currently no way of knowing whether the transiently non-contracting regions were larger than this, although they could certainly be smaller in the thin, distal region of the LDM. The weight of the sheep LDM has been determined in our laboratory to be between 90 to 290 gm. Therefore the transient regions probably represent 1–3% of the total muscle mass.

What the NIRS measurements suggested, which could not be estimated by microsphere entrapment nor gas washout, was that the high flow regions in response to induced contraction resulted in reduced metabolism and not, as could otherwise be interpreted, in higher metabolism.

Given the frequency of occurrence of the non-contracting regions (~30% of NIRS acquisitions), it is probable that they would have a significant effect upon overall muscle power output. Interestingly, Piiper (1992) suggested that 30% of blood may be shunted in skeletal muscle, which supports our data. If approximately 30% of blood flow through skeletal muscle is shunted, there is a high likelihood that shunting serves a physiological function in skeletal muscle.

Pendergast et al. (1985) suggested two possible causes of flow heterogeneity: 1) a regional change in pattern of motor recruitment, or 2) a neuro-humoral mechanism for blood flow control of the muscle vasculature. A regional change in motor recruitment would be consistent with our results. One attribute of the increased flow during reduced metabolism would be an increased gradient for blood-to-tissue metabolite and gas exchange. As diffusion is driven by concentration gradients, periodic "flushing" may provide a mechanism to increase the exchange rate. As this occurs transiently and in a defined muscle volume at any one time, the overall work output of the muscle is not compromised. Experiments indicate that peripheral diffusion of oxygen is a limiting factor in maximal oxygen uptake (VO_2max) of working skeletal muscle (Hogan et al., 1988; Wagner et al., 1990). We suggest that periodic high flow rates as described here may provide the vascular-tissue gradient needed to optimise tissue oxygen saturation. However, as these heterogeneities are also present in resting muscle, other metabolites may also be involved. Clearance of muscle lactate build up (Connett et al., 1984) for example may also benefit from increased flow.

REFERENCES

Berne RM and Matthew MN (1981) Cardiovascular Physiology (4th ed.). CV Mosby Co, St.Louis. p115.

Cerretelli P, Marconi C, Pendergast D, Meyer M, Heisler N and Piiper J. (1984) Blood flow in exercising muscles by xenon clearance and by microsphere trapping. J. Appl. Physiol. 56(1): 24–30.

Chance B, Leigh JS, Miyake H, Smith DS, Nioka S, Greenfeld R, Finander M, Kaufmann K, Levy W, Young M, Cohen P, Yoshioka H, and Boretsky M (1988) Comparision of time resolved and -unresolved measurements of deoxyhemoglobin in brain. Proc. Natl. Acad USA 85: 4971–4975.

Connett RJ, Gayeski TEJ and Honig (1984) Lactate accumulation in fully aerobic, working dog gracilis muscle. Am. J. Physiol. 246: H120-H128.

Ellsworth ML and Pittman RN (1984) Heterogeneity of oxygen diffusion through hamster striated muscles. Am J Physiol. 246:H161-H167.

Elwell CE, Cope M, Edwards AD, Wyatt JS, Delpy L and Reynolds EOR (1994) Quantification of adult cerebral hemodynamics by near-infrared specxtroscopy. J Appl. Physiol. 77(6):2753–2760.

Gronland J, Maliru GM and Hlastala (1989) Estimation of blood flow distribution in skeletal muscle from inert gas washout. J. Appl. Physiol. 66(4): 1942–1955.

Hogan MC, Roca J, Wagner PD and West JB (1988) Limitation of maximal O_2 uptake and performance by acute hypoxia in dog muscle in situ. J. Appl. Physiol. 65(2): 815–821.

Iversen PO and Nicolaysen G (1989) Heterogeneous blood flow distribution within single skeletal muscles in the rabbit: role of vasomotion, sympathetic nerve activity and effect of vasodilation. Acta Physiol. Scand. 137:125–133.

Jobsis FF (1977) Non-invasive infrared monitoring of cerebral and myocardial oxygen sufficiency and circulatory parameters. Science 198: 1264–1267.

Pendergast DR, Krasney JA, Ellis A, McDonald B, Marconi C and Ceretelli P (1985) Cardiac output and muscle blood flow in exercising dogs. Respir. Physiol. 61:317–326.

Piiper J (1990) Unequal distribution of blood flow in exercising muscle of the dog. Res. Physiol. 80: 129–136.

Piiper J, (1992) Modeling of oxygen transport to skeletal muscle: blood flow distribution, shunt, and diffusion. In Oxygen Transport to tissue XIII. Ed. TK Goldstick et al., Plenum Press, NY.

Wagner PD, Roca J, Hogan MC, Poole DC, Bebout DC and Haab P (1990) Experimental support for the theory of diffusion limitation of maximum oxygen uptake. In Oxygen Transport to tissue XII. Ed. J Piiper et al., Plenum Press, NY.

Wang Z, Noyszewski EA and Leigh JS Jr (1990) In Vivo MRS measurement of deoxymyoglobin in human forearms. Mag. Res. Med. 14: 562–567.

41

DETERMINATION OF THE DIFFUSION COEFFICIENT OF MYOGLOBIN IN MUSCLE CELLS

Photo-Oxidation and Microinjection Method

Klaus D. Jürgens, Simon Papadopoulos, Thomas Peters, and Gerolf Gros

Zentrum Physiologie
Medizinische Hochschule
30623 Hannover, Germany

INTRODUCTION

Myoglobin (Mb) serves as an intracellular store for oxygen in red muscles, ensuring aerobic metabolic processes during temporary respiratory or circulatory deficits in oxygen supply. Myoglobin can also contribute to sarcoplasmic oxygen transport by loading oxygen near the capillaries, diffusing to sites where the PO_2 is low and releasing the oxygen there. To what extent this myoglobin-facilitated oxygen diffusion enhances the overall oxygen transport rate primarily depends on the cellular myoglobin concentration and on the mobility of this protein within the muscle cell. Myoglobin concentrations can easily be determined but until recently no direct measurements of the diffusion coefficient (D_{Mb}) of myoglobin in intact mammalian skeletal muscle cells have been performed. In many theoretical studies dealing with facilitated oxygen diffusion, therefore, estimates of the D_{Mb}-value representing results from self-diffusion measurements in highly concentrated myoglobin solutions have been applied instead.

Using a photo-oxidation method (Jürgens et al., 1994), we determined a D_{Mb}-value in intact mammalian muscle cells, which is considerably lower than the value obtained for concentrated myoglobin solution. This method has been criticised since it is not completely free of side effects of the UV-light, it requires an assumption about the kinetics of enzymatic metmyoglobin reduction and it measures axial instead of the possibly more important radial myoglobin diffusion (Groebe, 1995). In order to verify our result, we redetermined D_{Mb} with a completely different method, a microinjection technique (Papadopoulos et al., 1995). This method was also applied to measure axial intracellular diffusion coefficients (D) of other globular proteins of different sizes. Besides the verification of our previous D_{Mb}-value, the results allow us to characterize in more detail the properties of mammalian skeletal muscle cells

Oxygen Transport to Tissue XIX, edited by Harrison and Delpy
Plenum Press, New York, 1997

determining protein diffusion. They are also used to compare the data of axial protein diffusion with literature data on radial protein diffusion in skeletal muscle fibers. Based on our results the significance of myoglobin-facilitated oxygen diffusion is reexamined.

METHODS

1. Photo-Oxidation Method

The photo-oxidation method is based on the generation of intracellular gradients of different ligation states of myoglobin along the longitudinal cell axis (Jürgens et al., 1994). These concentration gradients are achieved by irradiating a small area of a rat diaphragm muscle with a UV-light pulse using a microscope-photometer. In this area the UV-light oxidizes about 10% of the native oxymyoglobin to metmyoglobin. The subsequent diffusion of metmyoglobin out of and of oxymyoglobin into this area is followed photometrically in the middle of the irradiated field at a wavelength of 420 nm. At this wavelength the absorbance coefficients of oxy- and metmyoglobin are markedly different. Non-myoglobin-specific UV-side effects influencing the absorbance change at 420 nm are eliminated by subtracting the simultaneously recorded absorbance changes occurring at 473 nm, which is an isosbestic wavelength for oxy- and metmyoglobin. The recovery of the myoglobin species in the irradiated field is described by a differential equation containing a diffusion term and a first order reaction term characterizing enzymatic metmyoglobin reduction. The values of D_{Mb} and the first order reaction constant k are calculated by fitting the measured curves to the numerical solution of the differential equation. Variation of the width of the irradiated field influences the diffusional part of the measured kinetics but not the reactional part and, therefore, is utilized to discriminate between D_{Mb} and k in a best-fit procedure.

2. Microinjection Method

A minute amount of metmyoglobin (Papadopoulos et al., 1995) or of another iron-containing protein (about 30 pl) is injected with a pneumatic piezo-pump under microscopic control into a single muscle cell of a fiber bundle of a rat leg muscle. The profiles of the generated protein concentration gradients along the longitudinal fiber axis are measured photometrically at a wavelength of about 410 nm by use of a scanning stage. The diffusion coefficients are calculated by fitting the measured curves to the numerical solution of the underlying differential equation.

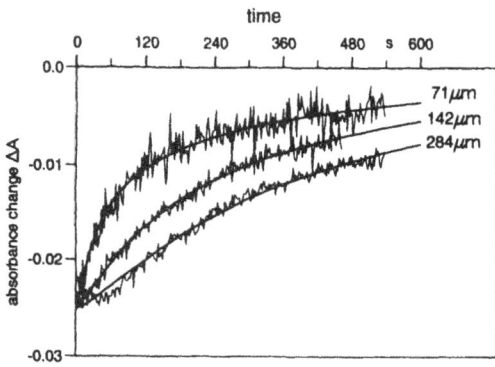

Figure 1. Absorbance changes in muscle tissue due to diffusion and metmyoglobin reduction following UV-photo-oxidation. Given are measured curves and the best fit for D_{Mb} for three different widths of the irradiated field.

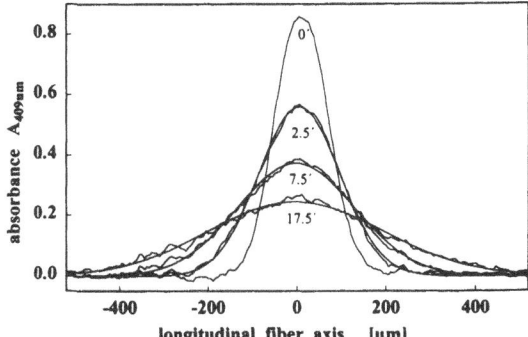

Figure 2. Concentration profiles of metmyoglobin in a soleus muscle fiber immediately after injection (0′) as well as 2.5, 7.5 and 17.5 minutes later (measured curves and best fit for D_{Mb}).

In order to see whether myoglobin behaves differently from other proteins inside skeletal muscle cells, we additionally injected five other globular proteins ranging in molecular diameter between 3 and 30 nm. Besides metmyoglobin (17 kD), we investigated cytochrome c (12.4 kD), human hemoglobin (64.5 kD), catalase (247.5 kD), ferritin (476 kD) and earthworm hemoglobin (3700 kD).

RESULTS

The diffusion coefficient (mean±SEM, n = sets of experiments) of myoglobin at T = 22°C obtained by the photo-oxidation method in rat diaphragm muscle amounts to $(1.17\pm0.08)\cdot10^{-7}$ cm^2/s (n = 6), the microinjection method led to $(1.25\pm0.13)\cdot10^{-7}$ cm^2/s (n = 12) in rat soleus muscle at this temperature. With the latter method we also measured at 37°C and found D_{Mb} to be $(2.2\pm0.12)\cdot10^{-7}$ cm^2/s (n = 24). From the temperature dependence of D_{Mb} a Q_{10} of 1.46 can be calculated.

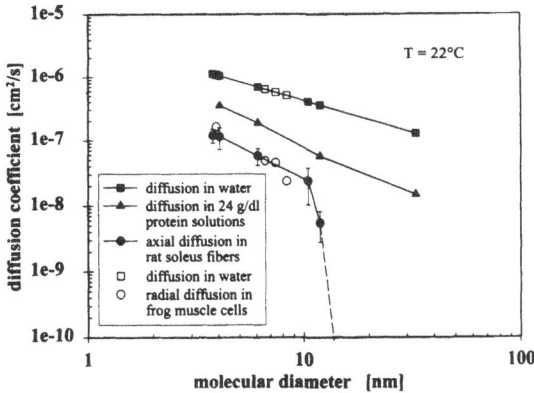

Figure 3. Axial diffusion coefficients of six globular proteins in rat soleus muscle (closed circles), in 24 g/dl concentrated protein solutions (triangles) and in dilute aqueous solutions (closed squares) as function of molecular diameter. The investigated proteins listed by increasing size are: cytochrome c, metmyoglobin, hemoglobin A, catalase, ferritin and earthworm hemoglobin. For comparison, literature data (Maughan and Lord, 1988) of the radial diffusion coefficients of the following proteins in frog skeletal muscle are shown (open circles): parvalbumin, triose-phospate isomerase, phosphoglyceromutase and glycerol-3-phosphate dehydrogenase. For these proteins the diffusion coefficients in water are also given (open squares).

The intracellular diffusion coefficients estimated for six injected proteins are shown on a double logarithmic scale as function of their molecular diameter in Fig. 3. They are compared with the results obtained from self-diffusion measurements in dilute and in highly concentrated (24g/dl, corresponding to the total protein concentration of muscle cells) aqueous protein solutions.

Compared with their diffusivities in dilute solution the diffusivities of the four proteins in the size range between 3 and 10 nm are all reduced to approximately 1/10 inside the muscle fiber. Larger molecules are hampered much more strongly, D of ferritin (12 nm diameter) is only 1/60 of the value in dilute solution, for earthworm hemoglobin (diameter 30 nm) no diffusion of the injected protein could be observed in the muscle cell. The intracellular diffusion coefficients of myoglobin and of hemoglobin amount to only 1/3 of their self diffusion coefficients in a 24 g/dl solution. A comparison of the range of axial diffusion coefficients determined by us with four radial diffusion coefficients of globular proteins of the same size range (Maughan and Lord, 1988) reveals no significant difference.

DISCUSSION

1. Experimentally Obtained D_{Mb} Values in Muscle Cells

Both measuring methods yielded identical myoglobin diffusion coefficients in diaphragm and soleus muscle of the rat, two muscles predominantly consisting of Mb-rich type I fibers. This result not only confirms the reliability of the photo-oxidation technique, it also shows that the mobility of myoglobin in the sarcoplasm is much lower than concluded from experiments with highly concentrated myoglobin solutions. The highly ordered cytoskeletal structure of the muscle cells seems to hamper protein diffusion even three times more strongly than expected from viscosity effects in a protein solution concentrated to 24 g/dl, the total protein concentration of a muscle cell. This has been proved for the 17 kD myoglobin as well as for the four times heavier hemoglobin. The intracellular hindrance of proteins larger than 250 kD (corresponding to a diameters > 10 nm) is much more pronounced. It may well be that these molecules are bigger than a critical pore-size given by the protein meshworks of the muscle cell.

It has been argued that axial protein diffusion coefficients may be different from radial protein diffusion coefficients. For small molecules like oxygen (Homer et al.,1984) and water (Cleveland et al., 1976) it has been shown that axial diffusivity is greater than radial diffusivity. Groebe and Thews ,1990, employed as radial D_{Mb} a value that is only 40% of the axial D_{Mb}. As can be seen from Fig. 3, anisotropic diffusion is not very likely for globular proteins of this size. For molecules with diameters between 3 and 10 nm the axial D-values measured by us and radial D-values from the literature do not differ significantly.

Our result for D_{Mb} is in good accordance with two other experimental results. Baylor and Pape, 1988, reported $1.6 \cdot 10^{-7}$ cm²/s (22°C) for the diffusivity of myoglobin in frog skeletal muscle, which is normally free from Mb. Moll, 1968, obtained a value of $1.5 \cdot 10^{-7}$ cm²/s at 20°C and $2.7 \cdot 10^{-7}$ cm²/s at 37°C from homogenates of rat skeletal muscles. All measured values, however, considerably differ from the myoglobin diffusivities used in theoretical studies of myoglobin-facilitated oxygen diffusion. Here a range of D_{Mb}-values between $5 \cdot 10^{-7}$ cm²/s and $23 \cdot 10^{-7}$ cm²/s (see e.g. Federspiel, 1986) is applied, which is 2 to 10 times the experimentally obtained value. Most frequently a D_{Mb} of $8 \cdot 10^{-7}$ cm²/s has been used, representing the result obtained from measurements of the self diffusion coeffi-

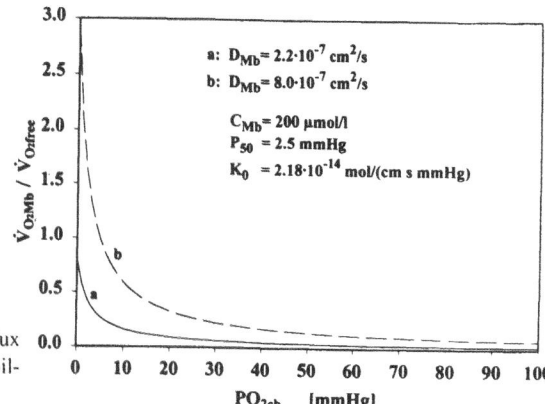

Figure 4. Ratio of facilitated and free oxygen flux as function of the oxygen pressure head at the capillary boundary of a Krogh-cylinder model.

cient in 18 g/dl myoglobin solution (Riveros-Moreno and Wittenberg, 1972). As a consequence the contribution of myoglobin facilitated oxygen diffusion has been overestimated in most theoretical studies.

2. Consequences for Intracellular Oxygen Transport

We, therefore, reexamined the role of myoglobin-facilitated oxygen diffusion for the intracellular oxygen transport rate. The ratio of facilitated oxygen flux (VO_{2Mb}) and free oxygen flux (VO_{2free}) is independent of geometrical factors and is given by

$$VO_{2Mb}/VO_{2free} = D_{Mb} \cdot C_{Mb}/(K_o \cdot (P_{50} + PO_{2cb})) \qquad (1)$$

where C_{Mb} is the myoglobin concentration, K_o Krogh's diffusion constant for oxygen, P_{50} the oxygen half-saturation pressure of myoglobin and PO_{2cb} the oxygen partial pressure head at the cell boundary (Jürgens et al.,1994). Fig. 4 shows this flux ratio as function of the PO_{2cb} for two different D_{Mb}-values, $8 \cdot 10^{-7}$ cm^2/s, which most often has been used in model studies, and our value of $2.2 \cdot 10^{-7}$ cm^2/s.

Applying $D_{Mb} = 2.2 \cdot 10^{-7}$ cm^2/s (37°C), we find that facilitated diffusion is smaller than free oxygen diffusion in the total PO_{2cb} range (curve a). The maximal flux ratio is 0.8, it is theoretically reached at $PO_{2cb} = 0$, i.e. anoxia. At a pressure head of 10 mmHg, which may be minimally required for a working muscle, facilitated oxygen diffusion makes up only 20% of free oxygen diffusion. The consequences of an overestimation of D_{Mb} for facilitated oxygen diffusion can be recognized from curve b.

From these results it can be concluded that under physiological conditions myoglobin-facilitated oxygen diffusion can play only a minor for intracellular oxygen transport. Only at severe hypoxia this mechanism can be of significance for protection of the integrity of the cell.

REFERENCES

Baylor, S.M. and Pape, P.C. (1988) Measurement of myoglobin diffusivity in the myoplasm of frog skeletal muscle fibers. J. Physiol. 406:247–275.

Cleveland, G.G., Chang, D.C., Hazlewood, C.F., and Rorschach, H.E. (1976) Nuclear magnetic resonance measurement of skeletal muscle anisotropy of the diffusion coefficient. Biophys. J. 16:1043–1053.

Federspiel, W.J. (1986) A model study of intracellular oxygen gradients in a myoglobin-containing skeletal muscle fiber. Biophys. J. 49:857–868.

Groebe, K. (1995) An easy-to-use model for O_2 supply to red muscle. Validity of assumptions, sensitivity to errors in data. Biophys. J. 68, 1246–1269.

Groebe, K. and Thews, G. (1990) Calculated intra- and extracellular PO_2 gradients in heavily working muscle. Am. J. Physiol. 259:H84-H92.

Homer, L.D., Shelton, J.B., Dorsey, C.H., and Williams, T.J. (1984) Anisotropic diffusion of oxygen in slices of rat muscle. Am. J. Physiol. 246:R107-R113.

Jürgens, K.D., Peters, T., and Gros, G. (1994) Diffusivity of myoglobin in intact skeletal muscle cells. Proc. Natl. Acad. Sci. USA 91:3829–3833.

Maughan, D. and Lord, C. (1988) Protein diffusivities in skinned frog skeletal fibers. Adv. Exp. Med. Biol. 226:75–84.

Moll,W. (1968) The diffusion coefficient of myoglobin in muscle homogenate. Pflügers Arch. 299:247–251.

Papadopoulos, S., Jürgens, K.D., and Gros, G. (1995) Diffusion of myoglobin in skeletal muscle cells -dependence on fibre type, contraction and temperature. Pflügers Arch. 430:519–525.

Riveros-Moreno, V. and Wittenberg, J.B. (1972) The self-diffusion coefficients of myoglobin and hemoglobin in concentrated solutions. J. Biol. Chem. 247:895–901.

42

IN VIVO ANALYSIS OF CAPILLARY LEUKOCYTE TRAFFICKING IN STRIATED MUSCLE ISCHEMIA/REPERFUSION

Martin Rücker, Brigitte Vollmar, and Michael D. Menger

Institute for Clinical and Experimental Surgery
University of Saarland
D-66421 Homburg/Saar, Germany

1. INTRODUCTION

During the last decade a considerable number of experimental studies has demonstrated that the activation of leukocytes, their adherence to the microvascular endothelium, and, consecutively, their transendothelial migration into tissue determine the manifestation of ischemia/reperfusion injury (Granger and Kubes, 1994; Menger and Vollmar, 1996; Carden et al., 1990; Vollmar et al., 1994; Vollmar et al., 1995). Based on the results of these studies, ischemia/reperfusion injury has to be considered as an inflammatory disease.

Within the scenario of postischemic inflammation, major attention has been drawn to elucidate the mechanisms, which mediate microvascular recruitment of the white blood cells. Intravital microscopic studies have demonstrated that in ischemia/reperfusion of striated muscle leukocytes do not plug nutritive capillaries, but adhere to the endothelial lining of postcapillary venules (Lehr et al., 1991, Menger et al., 1992a, Hansell et al., 1993) via specific adhesion receptors, including L-selectin and CD11b/CD18 on the surface of leukocytes and P-selectin and ICAM-1 on the surface of endothelial cells (Nolte et al., 1994; Nolte et al., 1995). The fact that the inhibition of venular leukocyte adherence by specific antibodies directed against these adhesion molecules abrogate the manifestation of postischemic reperfusion injury (Nolte et al., 1994; Nolte et al., 1995; Jerome et al., 1993; Jerome et al., 1994), indicates that the adherence of leukocytes in venules triggers the inflammatory response during postischemic reperfusion.

However, while characteristics of postischemic leukocyte flow behaviour in venules, including leukocyte rolling and firm adherence, have been studied extensively during the last years, little is yet known on postischemic leukocyte traffic through the nutritive capillary bed. To elucidate whether leukocytes use preferential pathways through the nutritive microcirculation during postischemic reperfusion, we studied individual capillary leukocyte flux in striated muscle tissue of Syrian golden hamsters using intravital fluorescence microscopy.

2. MATERIALS AND METHODS

2.1. Animals

Six to eight weeks old Syrian golden hamsters with a body weight of 60 to 80g were used for the microcirculatory analyses. The animals were housed one per cage and had free access to tap water and standard pellet food (Altromin, Lage, Germany) throughout the experiment.

2.2. Microcirculation Model

For intravital fluorescence microscopy we used the dorsal skinfold chamber preparation (Endrich et al., 1980), which contains striated muscle and skin tissue, and allows for the in vivo observation of the microcirculation in the *awake* animal over a prolonged period of time (Menger et al., 1987; Menger et al., 1994; Vajkoczy et al., 1995). Previous studies have demonstrated that the morphology and angioarchitecture of the skinfold striated muscle are similar when compared with that of true skeletal muscle tissue. The model permits quantification of microcirculatory parameters, including functional capillary density, heterogeneity of capillary perfusion, red blood cell (RBC)-velocity and diameters of individual microvascular segments (Menger et al., 1992b).

2.3. Surgical Technique

The chamber and its implantation procedure have been described previously in detail (Menger et al., 1992c). Briefly, under pentobarbital anesthesia (50 mg/kg body weight ip; Abbott, North Chicago, IL), the animals were fitted with two symmetrical titanium frames, positioned on the dorsal skinfold, sandwiching the extended double layer of skin. One layer of skin was completely removed in a circular area of about 15 mm in diameter, and the remaining layers (consisting of striated skin muscle and subcutaneous tissue) were covered with a removable cover slip, incorporated into one of the titanium frames. Fine polyethylene catheters (PE 10, inner diameter: 0.28mm; Portex LTD, Lythe, Kent, England) were inserted into the jugular vein, and allowed for the application of fluorescent markers with the aim of contrast enhancement for intravital fluorescence microscopy.

The chamber and catheters were implanted two days prior to the experiments in order to guarantee recovery of the tissue from surgery and anesthesia. The microcirculation of the striated muscle was then examined photomicroscopically; only skinfold preparations presenting without signs of inflammation and with normal microcirculation, as reported earlier (Endrich et al., 1980), were used for the experiments. The animals tolerated the chambers and catheters well and showed no signs of discomfort. In particular, no effect on sleeping and feeding habits were observed.

2.4. Intravital Fluorescence Microscopy

For in vivo microscopic observation, the awake animals were immobilized in a plexiglass tube, and the skinfold preparation was attached to the microscope stage. The stage was placed on a computer-controlled microscope desk, which enabled repeated scanning of identical segments of microvessels during the experiment. After intravenous injection of 0.2ml 5% fluorescein isothiocyanate (FITC)-labeled dextran (MW 150.000; Sigma Chemical Co., St. Louis, MO) and 0.1ml 0.2% Rhodamine-6G (MW 476; Sigma) in vivo microscopy was performed using a modified Leitz Orthoplan microscope with a 100W, HBO, mercury lamp, attached to a Ploemo-Pak illuminator with an I_2 blue and a N_2 green filter block (Leitz, Wetzlar, FRG) for epi-illumination. The observations were recorded by means of a CCD (charge-coupled device) video camera (FK 6990, Cohu; Prospective Measurements, San Diego, CA) and transferred to a video system for off-line evaluation.

2.5. Microcirculatory Analysis

Quantitative analysis of the striated muscle microcirculation included the determination of individual capillary diameter and leukocyte flux per minute within the individual capillaries. Perfused capillaries at baseline and only reperfused capillaries during postischemic reperfusion were considered for analysis of leukocyte flux. Using a water immersion objective with a magnification of x63 magnification at the 33cm video screen was ~2000x and resolution of the system was 0.2μm. The staining of plasma with FITC-dextran 150.000 allowed exact determination of inner vessel diameters, while the in vivo staining of white blood cells by Rhodamine 6G facilitated the analysis of capillary leukocyte flux. Quantitation was performed off-line by frame-to-frame analysis of the videotaped images (determination of capillary leukocyte flux) and by the use of a computer-assisted image analysis system (determination of capillary diameters; CapImage, Zeintl, Heidelberg, Germany).

2.6. Experimental Protocol

Baseline microcirculatory analysis included 3 regions of interest with a total of 10 to 15 capillaries per animal. Following baseline recording, 4h of tourniquet-(n=4) or pressure-induced ischemia (n=7) were applied by means of an O-shaped or circular silicon stamp (Menger et al., 1988; Menger et al., 1992b). Sequential microcirculatory analyses were performed 30min, 2h and 24h after onset of reperfusion. Using the computer-controlled stepping motor of the microscope desk, the technique allowed for repeated scanning of the identical capillaries, selected during baseline analysis of the experiment. Sham animals (n=4), undergoing microcirculatory analyses at the identical time points without induction of ischemia, served as controls.

2.7. Statistical Analyses

Results are expressed as means ± standard deviation (SD). The statistical procedure included analysis of variance and Student's t-test for comparison between the groups. Paired Student's t-test, including Bonferroni probabilities for repeated measurements, was performed for analyzing differences within each group (CSS, StatSoft, Inc.; Tulsa, OK). Differences were considered significant at a $P<0.05$ level.

3. RESULTS

There was a marked variation of the number of passing leukocytes per capillary already under baseline conditions, ranging between 0 and 22 cells/capillary x minute. Neither tourniquet ischemia nor pressure-induced ischemia did increase individual capillary leukocyte flux or variation of the number of passing leukocytes per capillary (Table 1).

Table 1. Capillary leukocyte flux in striated muscle ischemia/reperfusion

	Tourniquet-ischemia	Pressure-ischemia	Sham
Baseline	3.3 ± 2.0	4.4 ± 3.2	2.1 ± 1.5
30min reperfusion	2.3 ± 1.6	3.2 ± 2.3	1.5 ± 0.5
2h reperfusion	2.8 ± 1.2	3.1 ± 3.8	1.2 ± 0.4
24h reperfusion	3.1 ± 3.6	2.5 ± 0.9	3.3 ± 4.0

Capillary leukocyte flux (cells/capillary x min) in striated muscle before and 30min, 2h, and 24h after tourniquet- (n=4) or pressure-induced (n=7) ischemia. Sham animals with microcirculatory analyses at identical time points without ischemia served as controls (n=4).

In parallel, the fraction of capillaries with leukocyte traffic did not change significantly during postischemic reperfusion when compared with baseline conditions and sham controls (Table 2).

The analysis of the dependency of capillary leukocyte traffic on capillary diameters during baseline and sham conditions revealed a significant correlation ($p<0.001$), as indicated by increasing fractions of capillaries with leukocyte traffic dependent on increasing capillary diameters. This dependency of capillary leukocyte traffic on individual capillary diameter was found unchanged after both tourniquet and pressure-induced ischemia (Table 3).

4. DISCUSSION

The major finging of our study is that the inflammatory process of postischemic reperfusion in striated muscle, characterized by leukocyte accumulation and adherence in postcapillary venules (Granger et al., 1989; Menger et al., 1992a), is not associated with increased capillary leukocyte flux and does not reveal the development of preferential capillary pathways. Moreover, the dependency of individual capillary leukocyte flux on individual capillary diameter, which was observed under physiological (baseline/sham) conditions, was not found changed during reperfusion after both tourniquet- and pressure-induced ischemia.

Previous studies in rabbit tenuissimus muscle have clearly indicated that at individual arteriolar bifurcations, leukocytes are found to preferentially enter downstream branches, which present with a significantly higher flow rate (Ley et al., 1989). This preferential distribution of leukocyte flux according to the individual arteriolar flow condi-

Table 2. Fraction of capillaries with leukocyte traffic in striated muscle
ischemia/reperfusion

	Tourniquet-ischemia	Pressure-ischemia	Sham
Baseline	49.6 ± 15.5	53.1 ± 20.2	37.9 ± 20.7
30min reperfusion	45.8 ± 17.2	43.8 ± 14.9	34.8 ± 20.2
2h reperfusion	48.1 ± 10.9	32.9 ± 24.4	25.3 ± 6.7
24h reperfusion	37.8 ± 20.1	32.6 ± 15.9	34.9 ± 27.4

Fraction of striated muscle capillaries with leukocyte traffic (in percent of all capillaries analyzed) before and 30min, 2h, and 24h after tourniquet- (n=4) or pressure-induced (n=7) ischemia. Sham animals with microcirculatory analyses at identical time points without ischemia served as controls (n=4).

Table 3. Dependency of capillary leukocyte traffic on capillary diameter in striated muscle ischemia/reperfusion

Capillary diameter	Sham/baseline	Postischemic reperfusion
< 3.0 μm	8.3%	0.0%
3.0 to < 3.5 μm	25.6%	33.3%
3.5 to < 4.0 μm	33.3%	24.0%
4.0 to < 4.5 μm	39.5%	37.2%
4.5 to < 5.0 μm	50.0%	36.1%
5.0 to < 5.5 μm	56.5%	51.9%
> 5.5 μm	63.9%	61.3%

Fraction of capillaries with leukocyte traffic (in percent of all capillaries studied in the respective size range) in dependency of capillary diameter in striated muscle during reperfusion after tourniquet- and pressure-induced ischemia as well as during baseline conditions and in sham controls.

tions is similarly found at the capillary level. Using vital microscopic techniques, Blixt et al. (1987) have demonstrated volume flow dependent preferential pathways for leukocyte passage across the capillary network in the rat cremaster muscle. The results of the present study that capillary leukocyte flux strongly depend on the diameter of the individual capillaries support the observations of these previous studies. Moreover, in view that vessel diameter primarily determines vascular resistance, the knowledge that capillary leukocyte flux is dependent on capillary diameter with an increase of flux with increasing diameters further provides evidence that capillary leukocyte flux is directly determined by the individual vascular resistance.

Strikingly, reperfusion after both tourniquet- and pressure-induced ischemia did neither result in an increase of individual capillary leukocyte flux nor in changes of the dependency of capillary leukocyte flux on capillary diameter. These data indicate that the postischemic inflammatory response is confined solely to the postcapillary venules, and does not—at least primarily—represent a global inflammatory insult to the tissue. Since the leukocyte flux through the nutritive capillaries was not found enhanced during reperfusion, the postischemic recruitment of leukocytes in the postcapillary venules may be due to increased arterio-venous shunting of white blood cells, or, more probably, solely due to increased local leukocyte capture due to enhanced venular adhesion capacities in the absence of overall increased microvascular leukocyte flux conditions.

In summary, our study demonstrates that ischemia/reperfusion in striated muscle does not alter capillary leukcyte trafficking, and is not associated with an increased capillary leukocyte flux. Thus, we conclude that postischemic leukocyte accumulation in postcapillary venules is not due to an increased inflow of leukocytes into the nutritive microcirculation, but must rather be attributed to local mechanisms at the venular sites, such as the activation of individual adhesion molecules expressed on venular endothelial cells.

ACKNOWLEDGMENT

This study was supported by grants of the Deutsche Forschungsgemeinschaft (Me 900/1-1 and /1-2) and the European Community (BMH 4-CT95-0875).

REFERENCES

Blixt, A., Braide, M., Myrhage, R., and Bagge, U., 1987, Vital microscopic studies on the capillary distribution of leukocytes in the rat cremaster muscle, *Int. J. Microcirc.: Clin. Exp.* 6:273–286.

Carden, D.L., Smith, J.K., and Korthuis, R.J., 1990, Neutrophil-mediated microvascular dysfunction in postischemic canine skeletal muscle. Role of granulocyte adherence, *Circ. Res.* 66:1436–1444.

Granger, D.N., and Kubes, P., 1994, The microcirculation and inflammation: modulation of leukocyte-endothelial cell adhesion, *J. Leukoc. Biol.* 55:662–675.

Endrich, B., Asaishi, K., Goetz, A., and Messmer, K., 1980, Technical report-a new chamber technique for microvascular studies in unanesthetized hamsters, *Res. Exp. Med.* 177:125–134.

Hansell, P., Borgström, P., and Arfors, K.E., 1993, Pressure related capillary leukostasis following ischemia-reperfusion and hemorrhagic shock, *Am. J. Physiol.* 265:H381-H388.

Jerome, S.N., Smith, C.W., and Korthuis, R.J., 1993, CD18-dependent adherence reactions play an important role in the development of the no-reflow phenomenon, *Am. J. Physiol.* 264: H479-H483.

Jerome, S.N., Akimitsu, T., and Korthuis, R.J., 1994, Leukocyte adhesion, edema, and development of postischemic capillary no-reflow, *Am. J. Physiol.* 267:H1329-H1336.

Lehr, H.A., Guhlmann, A., Nolte, D., Keppler, D., and Messmer, K., 1991, Leukotrienes as mediators in ischemia-reperfusion injury in a microcirculation model in the hamster, *J. Clin. Invest.* 87:2036–2042.

Ley, K., Meyer, J.-U., Intaglietta, M., and Arfors, K.-E., 1989, Shunting of leukocytes in rabbit tenuissimus muscle, *Am. J. Physiol.* 256:H85-H93.

Menger, M.D., Hammersen, F., Barker, J.H., Feifel, G., and Messmer, K., 1987, Tissue pO2 and functional capillary density in chronically ischemic skeletal muscle, *Adv. Exp. Med. Biol.* 222:631–636.

Menger, M.D., Sack, F.U., Barker, J.H., Feifel, G., and Meßmer, K., 1988, Quantitative analysis of microcirculatory disorders after prolonged ischemia in skeletal muscle: Therapeutic effects of prophylactic isovolemic hemodilution, *Res. Exp. Med.* 188:151–165.

Menger, M.D., Pelikan, S., Steiner, D., and Messmer, K., 1992a, Microvascular ischemia-reperfusion injury in striated muscle: significance of *"reflow-paradox"*, *Am. J. Physiol.* 263:H1901-H1906.

Menger, M.D., Steiner, D., and Messmer, K., 1992b, Microvascular ischemia-reperfusion injury in striated muscle: significance of *"no-reflow"*, *Am. J. Physiol.* 263:H1892-H1900.

Menger, M.D., Vajkoczy, P., Leiderer, R., Jäger, S., and Messmer, K., 1992c, Influence of experimental hyperglycemia on microvascular blood perfusion of pancreatic islet isografts, *J. Clin. Invest.* 90:1361–1369.

Menger, M.D., Vajkoczy, P., Beger, C., and Messmer, K., 1994, Orientation of microvascular blood flow in pancreatic islet isografts, *J. Clin. Invest.* 93:2280–2285.

Menger, M.D., and Vollmar, B., 1996, Adhesion molecules as determinants of disease: from molecular biology to surgical research, *Br. J. Surg.* 83:588–601.

Nolte, D., Hecht, R., Schmid, P., Botzlar, A., Menger, M.D., Neumueller, C., Sinowatz, F., Vestweber, D., and Messmer, K., 1994, Role of Mac-1 and ICAM-1 in ischemia-reperfusion injury in a microcirculation model of Balb/C-mice, *Am. J. Physiol.* 267:H1320-H1328.

Nolte, D., Vestweber, D., Hecht, R., and Meßmer, K., 1995, Monoclonal antibodies to L- and P-selectin reduce postischemic reperfusion injury in striated muscle, *Langenbecks Arch. Chir. Forum* 95:399–403.

Vajkoczy, P., Olofsson, A.M., Lehr, H.A., Leiderer, R., Hammersen, F., Arfors, K.E., Menger, M.D., 1995, Histogenesis and ultrastructure of pancreatic islet graft microvasculature: Evidence for graft vascularization by endothelial cells of host origin, *Am. J. Pathol.* 146: 1397–1405.

Vollmar, B., Menger, M.D., Glasz, J., Leiderer, R., and Messmer, K., 1994, Impact of leukocyte-endothelial cell interaction in hepatic ischemia/reperfusion injury, *Am. J. Physiol.* 267: G786-G793.

Vollmar, B., Glasz, J., Menger, M.D., and Messmer, K., 1995, Leukocytes contribute to hepatic ischemia/reperfusion injury via intercellular adhesion molecule-1-mediated venular adherence, *Surgery* 117:195–200.

43

ASSOCIATION OF CAPILLARY DIAMETER RESPONSE AND NUTRITIVE PERFUSION FAILURE IN POSTISCHEMIC STRIATED MUSCLE

Martin Rücker, Brigitte Vollmar, and Michael D. Menger

Institute for Clinical and Experimental Surgery
University of Saarland
D-66421 Homburg/Saar, Germany

1. INTRODUCTION

Previous studies have underlined the pivotal role of nutritive capillary perfusion failure for the manifestation of postischemic reperfusion injury (Menger et al., 1993). This view is based on data derived from experiments, demonstrating a considerable fraction of capillaries with persistent stasis of blood perfusion, i.e "no-reflow", during reperfusion after prolonged periods of ischemia (Ames et al., 1968; Menger et al., 1992a; Vollmar et al., 1994).

Several mechanisms are discussed as the cause for postischemic capillary perfusion failure, including microvascular hemoconcentration (Fischer and Ames, 1972; Menger et al., 1988 and 1989a; Vollmar et al., 1994) and thrombosis (Quinones-Baldrich et al., 1991), endothelial cell swelling (Gidlöf et al., 1987; Hammersen et al., 1989; Menger et al., 1989b) as well as capillary plugging by leukocytes (Engler et al., 1986; Schmid-Schönbein, 1987). With the discovery of the endothelin - NO vasomotor control (see Moncada and Higgs, 1993; Lerman and Burnett, 1992), mechanisms, including vascular dysregulation and dysfunction, have gained primary interest to be studied as the cause of postischemic reperfusion failure (Loscalzo and Vita, 1994; Hagar, 1994; Amrani et al., 1995).

However, there is little information on microcirculatory vasomotor control in ischemia/ reperfusion. Therefore, we studied postischemic striated muscle capillary diameter response in the hamster, using the skinfold chamber model and intravital fluorescence microscopic techniques.

Oxygen Transport to Tissue XIX, edited by Harrison and Delpy
Plenum Press, New York, 1997

2. MATERIALS AND METHODS

2.1. Animal Model

For our study we used Syrian golden hamsters (6 - 8 weeks old and 60 - 80 g body weight), kept on standard laboratory chow (Altromin, Lage, Germany) and water ad libitum. For in vivo fluorescence microscopy a dorsal skinfold chamber was implanted, which contains striated muscle and skin, and allows for repeated analysis of the microcirculation in the awake animal over a prolonged period of time (Endrich et al., 1980). The model permits quantitative assessment of microhemodynamic parameters and cellular phenomena within the microcirculation of the striated muscle tissue in the awake animal. The chamber and its implantation procedure have been described previously in detail (Menger et al., 1992b). In brief: under pentobarbital anesthesia (50 mg/kg body weight i.p.; Abbott, North Chicago, IL), the animals were fitted with two symmetrical titanium frames, positioned on the dorsal skinfold, sandwiching the extended double layer of skin. One layer of skin was completely removed in a circular area of about 15 mm in diameter, and the remaining layers (consisting of striated muscle and subcutaneous tissue) were covered with a removable cover slip, incorporated into one of the titanium frames. In addition, fine polyethylene catheters (PE 10, inner diameter: 0.28 mm; Portex LTD, Lythe, Kent, England) were inserted into the jugular vein. A recovery period of 48 - 72 h between surgery and start of the experiments was allowed in order to eliminate the effects of anesthesia and surgical trauma on the microvasculature.

2.2. Intravital Fluorescence Microscopy

For in vivo microscopic observation, the awake animals were immobilized in a plexiglass tube, and the skinfold preparation was attached to the microscope stage. The stage was placed on a computer-controlled microscope desk, which allowed for repeated scanning of identical segments of microvessels during the experiment. Using a modified Leitz Orthoplan microscope with a 100W, HBO, mercury lamp, attached to a Ploemo-Pak illuminator with an I_2 blue filter block (Leitz, Wetzlar, FRG) for epi-illumination the microcirculation was recorded by means of a CCD (charge coupled device) video camera (FK 6990, Cohu; Prospective Measurements, San Diego, CA) and transferred to a video system for off-line evaluation. Microvascular dimensions were quantitatively analyzed after contrast enhancement by intravenous injection of the macromolecular tracer fluorescein isothicyanate (FITC)-labeled dextran (MW 150.000; 100 mg x kg^{-1}; Sigma Chemical Co., St. Louis, MO).

2.3. Microcirculatory Analysis

With the use of intravital fluorescence microscopy the diameters of a total of 211 capillaries were studied in conscious animals prior to and at different time points after a 4-h period of tourniquet- and pressure-induced ischemia. In vivo miroscopic observations (magnification: 1,900x on a 320mm diagonal video screen) were recorded on video-tape and were analyzed off-line, using a computer-assisted image analysis system (Zeintl et al., 1986) (CapImage, Zeintl, Heidelberg, Germany). Heterogeneity of capillary diameter distribution was calculated for each animal by the coefficient of variation (standard deviation devided by mean value). In addition, microcirculatory analyses included the determination of the functional capillary density (length of red blood cell perfused capillaries per area of observation) at each time point studied. Functional capillary density was determined by means of the computer-assisted image analysis system (CapImage), using pythagoreical techniques.

2.4. Experimental Protocol

Baseline recordings were followed by 4h of tourniquet-induced ischemia (n=4) or pressure-induced ischemia (n=7), respectively. Tourniquet ischemia was applied to the skinfold preparation by occlusion of all feeding and draining vessels at the margin of the observation window with an 'O'-shaped silicon stamp (Menger et al., 1992). Pressure-induced ischemia was performed using a circular silicon stamp, compressing the entire tissue under investigation with a pressure of ~50mmHg (Menger et al., 1988).

Postischemic analyses included measurements at 30 minutes, 2 hours and 24 hours after onset of reperfusion with 10 to 15 capillaries per animal at each time point. Sham animals without the induction of ischemia were studied at identical time points by intravital microscopy and served as controls.

2.5. Statistical Analyses

The statistical procedure included analysis of variance and Student's t-test for comparison between the groups. Paired Student's t-test, including Bonferroni probabilities for repeated measurements, was performed for analyzing differences within each group (CSS, StatSoft, Inc.; Tulsa, OK). All values are reported as mean ± standard error of the mean (SEM), and statistical significance was set at $P < 0.05$.

3. RESULTS

The early postischemic reperfusion period was characterized by a significant ($p < 0.05$) decrease of functional capillary density from 164.5 ± 8.3 cm^{-1} and 175.8 ± 7.8 cm^{-1} to 109.8 ± 20.4 cm^{-1} after tourniquet ischemia and 46.1 ± 12.5 cm^{-1} after pressure-induced ischemia. There was no complete recovery after the 24-h observation period.

Analysis of mean capillary diameters revealed no overall changes during postischemic reperfusion (Table 1). However, a slightly more heterogeneous distribution of diameter values was observed, as demonstrated by an increased coefficient of variation of diameter values during postischemic reperfusion (0.21) when compared with preischemic baseline conditions (0.16).

The differentiation between reperfused and non-reperfused ("no-reflow") capillaries revealed an up to 19% increase of mean capillary diameter of reperfused microvessels, while mean capillary diameter of those microvessels, which presented with postischemic perfusion failure, was found reduced, particularly at the end of the 24-h postischemic reperfusion period (Table 2). The increase of the mean diameter of the perfused capillaries resulted in a significantly ($p < 0.01$) increased fraction of microvessels with diameters greater than 5.5μm

Table 1. Overall capillary diameters in striated muscle ischemia/reperfusion

	Tourniquet-ischemia	Pressure-ischemia	Sham
Baseline	4.3 ± 0.3	4.9 ± 0.2	4.1 ± 0.1
30min reperfusion	4.6 ± 0.3	5.0 ± 0.5	4.1 ± 0.2
2h reperfusion	4.9 ± 0.6	4.7 ± 0.4	4.0 ± 0.2
24h reperfusion	4.1 ± 0.2	4.5 ± 0.3	3.7 ± 0.2

Overall mean capillary diameters (in μm) in striated muscle before and 30min, 2h, and 24h after tourniquet- or pressure-induced ischemia. Sham animals with microcirculatory analyses at identical time points without ischemia served as controls.

(29.5%) when compared with that during baseline conditions and sham controls (10.3%). Comparison of baseline diameters of postischemically reperfused versus non-reperfused capillaries did not show any significant difference ($4.53 \pm 0.12\mu m$ vs $4.54 \pm 0.07\mu m$).

4. DISCUSSION

The major finding of the present study is that in striated muscle reperfusion after both tourniquet- and pressure-induced ischemia does not change overall mean capillary diameters. However, differentiation between reperfused and non-reperfused capillaries revealed distinct responses, i.e. increased diameters of reperfused capillaries and narrowed diameters of those capillaries, which presented with *"no-reflow"*.

Previous studies have demonstrated that prolonged periods of ischemia followed by reperfusion and reoxygenation are characterized by the perfusion failure of individual capillaries, termed *"no-reflow"* (Ames et al., 1968; Menger et al., 1992a; Vollmar et al., 1994). According to these previous studies the present experiments confirm that a 4-h period of ischemia in the hamster dorsal skinfold striated muscle is associated with reperfusion failure of ~40% to 70% of the nutritive capillaries.

Several mechanisms have been suggested to promote the development of postischemic *"no-reflow"*, including i) thrombosis of microvessels (Quinones-Baldrich et al., 1991), ii) plugging of capillaries by leukocytes (Schmid-Schönbein, 1987), iii) swelling of capillary endothelial cells (Gidlöf et al., 1987; Hammersen et al., 1989), iv) impairment of microvascular blood fluidity with hemoconcentration (Menger et al., 1988; Menger et al., 1991; Vollmar et al., 1994), and v) increase of hydraulic resistance (Gidlöf et al., 1987; Mazzoni et al., 1989).

Although Quinones-Baldrich et al. (1991) proposed that capillary perfusion failure after ischemia and reperfusion is due to microvascular thrombus formation during ischemia, which may respond to fibrinolysis during early reperfusion, in vivo microscopic studies in tibialis anterior muscle revealed that thrombosis is not the cause for postischemic capillary perfusion deficits (Messina, 1990). In accordance with those in vivo findings, histological analyses of classical studies on ischemia/reperfusion were not able to demonstrate platelet or fibrin thrombi (Harman, 1948; Strock and Majno, 1969). The fact that neither heparin nor prostaglandin E_1 prevent capillary *"no-reflow"* (Strock and Majno, 1969; Rosolowsky and Weiss, 1987) further supports the view that blood coagulation and platelet aggregation are not primary causes for the development of postischemic capillary perfusion failure.

Table 2. Change of diameters of perfused and non-perfused capillaries in postischemic striated muscle reperfusion

	Tourniquet-ischemia		Pressure-ischemia	
	Perfused	Non-perfused	Perfused	Non-perfused
Baseline	4.3 ± 0.3	—	4.9 ± 0.2	—
30min reperfusion	4.7 ± 0.2	4.5 ± 0.5	5.8 ± 0.9	4.8 ± 0.5
2h reperfusion	5.0 ± 0.6	4.1 ± 0.1	5.0 ± 0.5	$4.4 \pm 0.4*$
24h reperfusion	4.4 ± 0.2	$3.6 \pm 0.1**$	4.7 ± 0.4	4.2 ± 0.4

Mean capillary diameters (in μm) of perfused and non-perfused capillaries in striated muscle before and 30min, 2h, and 24h after tourniquet- or pressure-induced ischemia. *p<0.05; **p<0.01 vs perfused capillaries

Electron microscopic analyses of postischemic myocardial tissue have suggested that plugging of capillaries by leukocytes may be the predominant mechanism of postischemic *"no-reflow"* (Engler et al., 1986; Schmid-Schönbein, 1987). However, this may vary between different organs. Although temporary capillary leukocyte plugging has also been observed in skeletal muscle, plug durations are in general less than one second (Harris and Skalak, 1993). Moreover, Hansell and coworkers (1993) have demonstrated that if stasis of leukocytes in skeletal muscle capillaries is observed after ischemia/reperfusion or hemorrhagic shock, this is solely related to decreased perfusion pressure conditions, but not due to adhesive interactions with the capillary endothelium or hindrance of lumenal passage (Hansell et al., 1993). This is in accordance with our intravital microscopic findings in postischemic striated muscle that plugging of capillaries by leukocytes is rarely observed as the cause for *"no-reflow"*.

There is major evidence that fluid shift-induced swelling of capillary endothelial cells due to the deficient metabolic energy reserve contributes to the pathogenesis of capillary *"no-reflow"* in ischemia/reperfusion (Gidlöf et al., 1987; Hammersen et al., 1989; Mazzoni et al., 1989). The narrowing of the capillary lumen by swollen endothelial cells elevates hydraulic resistance, thereby hindering restoration of capillary blood flow during resuscitation. Endothelial cell swelling has been demonstrated to occur during ischemia, however, was found potentiated during postischemic reperfusion (Gidlöf et al., 1987). In parallel, the shift of ions and water to the extravascular space also increases microvascular hematocrit, resulting in impairment of microvascular blood fluidity and intravascular hemoconcentration. This hypothesis is supported by experiments where postischemic capillary perfusion was found significantly improved when systemic hematocrit was lowered to a preischemic value of 30% by prophylactic isovolemic hemodilution (Menger et al., 1988). Thus, both endothelial cell swelling and intravascular hemoconcentration synergistically contribute to the manifestation of postischemic capillary perfusion failure.

Most of the mechanisms, potentially contributing to the manifestation of capillary *"no-reflow"*, are associated with an increase of vascular resistance. As a consequence vasodilation would be an effective mechanism to counteract capillary perfusion failure. Indeed, our present study demonstrates that postischemic capillary perfusion is maintained in those microvessels with increased diameters and, thus, lowered resistance, while there was lack of dilatation or even narrowing of those capillaries, which presented with perfusion failure. Narrowing or dilatation of capillaries may potentially be mediated by the contractile function of pericytes, involving both NO and endothelins, however, no proof is given yet for this hypothesis. Although the mechanisms of postischemic capillary dilatation are thus not yet known, we conclude from our study that the maintenance of postischemic nutritive perfusion is associated with capillary dilatation, and may, therefore, be included in novel therapeutic concepts which aim on the prevention of postischemic perfusion failure.

ACKNOWLEDGMENT

This study was supported by grants of the Deutsche Forschungsgemeinschaft (Me 900/1-1 and /1-2) and the European Community (BMH 4-CT95-0875).

REFERENCES

Ames, A., III, Wright, R.L., Kowada, M., Thurston, J.M., and Majno, G., 1968, Cerebral ischemia. II. The no-reflow phenomenon, *Am. J. Pathol.* 52:437–452.

Amrani, M., Chester, A.H., Jayakumar, J., Schyns, C.J., and Yacoub, M.H., 1995, L-arginine reverses low coronary reflow and enhances postischaemic recovery of cardiac mechanical function, *Cardiovasc. Res.* 30:200–204.

Endrich, B., Asaishi, K., Goetz, A., and Messmer, K., 1980, Technical report-a new chamber technique for microvascular studies in unanesthetized hamsters, *Res. Exp. Med.* 177:125–134.

Engler, R.L., Dahlgren, M.D., Petersen, M.S., Dobbs, A., and Schmid-Schönbein, G.W., 1986, Accumulation of polymorphonuclear leukocytes during 3-h experimental myocardial ischemia, *Am. J. Physiol.* 251:H93-H100.

Fischer, E.G., and Ames, A., 1972, Studies on mechanisms of impairment of cerebral circulation following ischemia: Effect of hemodilution and perfusion pressure, *Stroke* 3:538–542.

Gidlöf, A., Lewis, D.H., and Hammersen, F., 1987, The effect of prolonged total ischemia on the ultrastructure of human skeletal muscle capillaries. A morphometric analysis, *Int. J. Microcirc.: Clin. Exp.* 7:67–86.

Hagar, J.M., 1994, Endogenous endothelin-1 impairs endothelium-dependent relaxation after myocardial ischemia and reperfusion, *Am. J. Physiol.* 267:H1833–1841.

Hammersen F., Barker, J.H., Gidlöf, A., Menger, M.D., Hammersen, E., and Messmer, K., 1989, The ultrastructure of microvessels and their contents following ischemia and reperfusion, *Prog. Appl. Microcirc.* 13:1–26.

Hansell, P., Borgström, P., and Arfors, K.E., 1993, Pressure related capillary leukostasis following ischemia-reperfusion and hemorrhagic shock, *Am. J. Physiol.* 265:H381-H388.

Harman, J.W., 1948, The significance of local vascular phenomenona in production of ischemic necrosis in skeletal muscle, *Am. J. Pathol.* 24:625–641.

Harris, A.G., and Skalak, T.C., 1993, Effects of leukocyte activation on capillary hemodynamics in skeletal muscle, *Am. J. Physiol.* 264:H909–916.

Lerman, A., and Burnett, J.C.Jr., 1992, Intact and altered endothelium in regulation of vasomotion, *Circulation* 86:III-12-III-19.

Loscalzo, J., and Vita, J.A., 1994, Ischemia, hyperemia, exercise, and nitric oxide. Complex physiology and complex molecular adaptations, *Circulation* 90:2557–2559.

Mazzoni, M.C., Borgström, P., Intaglietta, M., and Arfors, K.-E., 1989, Lumenal narrowing and endothelial cell swelling in skeletal muscle capillaries during hemorrhagic shock, *Circ. Shock* 29:27–39.

Menger, M.D., Sack, F.U., Barker, J.H., Feifel, G., and Meßmer, K., 1988, Quantitative analysis of microcirculatory disorders after prolonged ischemia in skeletal muscle: Therapeutic effects of prophylactic isovolemic hemodilution, *Res. Exp. Med.* 188:151–165.

Menger, M.D., Sack, F.-U., Hammersen, F., and Messmer, K., 1989a, Tissue oxygenation after prolonged ischemia in skeletal muscle: Therapeutic effect of prophylactic isovolemic hemodilution, *Adv. Exp. Med. Biol.* 248:387–395.

Menger, M.D., Hammersen, F., Barker, J., Feifel, G., and Messmer, K., 1989b, Ischemia and reperfusion in skeletal muscle: Experiments with tourniquet ischemia in the awake Syrian golden hamster, *Prog. Appl. Microcirc.* 13:93–108.

Menger, M.D., Lehr, H.-A., and Messmer, K., 1991, Role of oxygen radicals in microcirculatory manifestations of postischemic injury, *Klin. Wochenschr.* 69:1050–1055.

Menger, M.D., Steiner, D., and Messmer, K., 1992a, Microvascular ischemia-reperfusion injury in striated muscle: significance of "no-reflow", *Am. J. Physiol.* 263:H1892-H1900.

Menger, M.D., Vajkoczy, P., Leiderer, R., Jäger, S., and Messmer, K., 1992b, Influence of experimental hyperglycemia on microvascular blood perfusion of pancreatic islet isografts, *J. Clin. Invest.* 90:1361–1369.

Menger, M.D., Vollmar, B., Glasz, J., Post, S., and Messmer, K., 1993, Microcirculatory manifestations of hepatic ischemia/reperfusion injury, *Prog. Appl. Microcirc.* 19:106–124.

Messina, L.M., 1990, In vivo assessment of acute microvascular injury after reperfusion of ischemic tibialis anterior muscle of the hamster, *J. Surg. Res.* 48:615–621.

Moncada, S., and Higgs, A., 1993, The L-arginine-nitric oxide pathway, *N. Engl. J. Med.* 329:2002–2012.

Quiñones-Baldrich, W.J., Chervu, A., Hernandez, J.J., Colburn, M., and Moore, W.S., 1991, Skeletal muscle function after ischemia: "No reflow" versus reperfusion injury, *J. Surg. Res.* 51:5–12.

Rosolowksy, M., and Weiss, H.R., 1987, Effect of blood coagulation and platelet aggregation on perfusable capillaries and arterioles in ischemic and nonischemic myocardium *Microvasc. Res.* 34:69–83.

Schmid-Schönbein, G.W., 1987, Capillary plugging by granulocytes and the no-reflow phenomenon in the microcirculation, *Fed. Proc.* 46:2397–2401.

Strock, P.E., and Majno, G., 1969, Microvascular changes in acutely ischemic rat muscle, *Surg. Gynecol. Obstet.* 129:1213–1224.

Vollmar, B., Glasz, J., Post, S., Leiderer, R., and Menger, M.D., 1994, Hepatic microcirculatory perfusion failure is a determinant for liver dysfunction in warm ischemia-reperfusion, *Am. J. Pathol.* 145:1421–1431.

Zeintl, H., Tompkins, W.R., Messmer, K., Intaglietta, M., 1986, Static and dynamic microcirculatory video image analysis applied to clinical investigations, *Prog. Appl. Microcirc.* 11:1–10.

44

INTRAMUSCULAR OXYGEN PARTIAL PRESSURE IN PATIENTS WITH CHRONIC EXERTIONAL COMPARTMENT SYNDROME (CECS)

B. Evers, V. Ödemis, and H. Gerngroß

Department of Surgery
Military Hospital of Ulm
89081 Ulm, Germany

INTRODUCTION

The incidence of chronic exertional compartment syndromes (CECS) in athletes has significantly increased during the past decade (Detmer, 1980; Puranen and Alvaikko, 1981; Rorabeck et al., 1988). Although making the exact diagnosis of CECS can be difficult due to its unspecific symptoms and limited diagnostic procedures, previous studies demonstrated that the diagnosis of a CECS can be established with high accuracy based on precise history, detailed physical examination and measurement of the intracompartmental pressure (ICP) during and after symptom-reproducing exertion (Detmer, 1980; Rorabeck et al., 1988). However, the exact etiology and pathophysiology of the CECS remain unclear. Therefore, the purpose of this study was to measure the oxygen partial pressure (pO_2) within the anterior tibial muscle in patients with a CECS to obtain information about the oxygen supply of the affected muscle group.

MATERIAL AND METHODS

Subjects

During a 15-months period 9 male athletes were seen with clinical symptoms of a CECS of the anterior tibial compartment. The median age was 23 (20–26) years; seven of them were long-distance runners, two were soccer players. Five of them showed bilateral symptoms, so that a total number of 14 cases was included. In four patients with unilateral CECS the intracompartmental pressure within the anterior tibial muscle of the asymptomatic leg was also investigated. Detailed history was taken. Physical examination was done at rest, during and after symptom-reproducing treadmill exertion.

METHODS

1. Measurement of Intracompartmental Pressure (ICP)

Measurement of the intracompartmental pressure (ICP) was based on a direct piezoresistive principle (Braun Dexon GmbH, Spangenberg, Germany) (Evers et al., 1995).

The system consists of a microprobe, which was inserted into the anterior tibial muscle via an iv needle, the microelectronic and the analyzer unit. After a short interval of automatic calibration the probe was inserted into the muscle and fixed with tape for monitoring of the ICP during exertion.

The pressure was determined at rest, during and after exertion.

2. Measurement of Intramuscular Oxygen Partial Pressure (pO$_2$)

The intramuscular oxygen partial pressure (pO$_2$) was polarographically measured with a pO$_2$-Histograph KIMOC (Eppendorf Netheler Hinze GmbH, Hamburg, Germany) using a sterilizable hypodermic needle (tip diameter: 350 μm) (Fleckenstein, 1984). The calibration of the probes was performed with room air and nitrogen in 0.9% saline solution. The probe was inserted in a well defined area of the anterior tibial muscle: 10 cm distal to the lateral knee joint space and 3 cm lateral to the anterior rim of the tibia. The probe was introduced through a 20 G Abbocath® at an angle of 45° to skin level, parallel to the muscle fibers and computer-assisted driven into the muscle. The needle probe was moved forward by 0.7 mm steps followed by a reverse movement of 0.2 mm each time (pilgrim's step principle). After each step the local pO$_2$ was measured within less than 1 sec. The device was inserted into the anterior tibial muscle for a total distance of 20 mm associated with the registration of 40 local pO$_2$ values within 80 sec. At the end of such a period the probe was automatically drawn back to the initial position, and its direction of insertion was slightly modified to ensure measurements in virgin tissue; 40 consecutive values of such a period were analyzed and displayed as median and range.

Measurements within the anterior tibial compartment were performed at rest and immediately after treadmill running.

3. Pain Score

For pain assessment a score ranging from 0 for no pain to 10 for maximal pain was applied before, during and after treadmill exertion. The maximal level of 10 was defined as severe pain forcing the athlete to quit treadmill running.

4. Statistical Analysis

Unless otherwise indicated all the data are expressed as median and range. The Mann-Whitney-U-test (for inter-group differences) and the Wilcoxon-test (for differences during the period of time) were used for statistical analysis.

RESULTS

All nine patients complained of diffuse pain within the anterior tibial compartment on exertion. Physical examination after treadmill exertion revealed tightness and tenderness in all cases. Transient foot extensor weakness was seen in three cases.

Table 1. Intracompartmental pressure (mm Hg) and pain score [0–10]
- median and range

	symptomatic (n =14)		asymptomatic (n = 4)		statistical significance	
	ICP	pain	ICP	pain	ICP	pain
pre exertion	33 (24-55)	0 [0]	18 (10-24)	0 [0]	p < 0.002	n.s.
immediately before pain-related stop	106 (80-130)	10 [10]	46 (26-52)	1 [1-2]	p < 0.002	p < 0.002
1 min. after exertion	85 (72 -110)	8 [7-9]	25 (18-38)	0 [0]	p < 0.002	p < 0.002
15 min. after exertion	54 (43-85)	3 [2-5]	17 (10-21)	0 [0]	p < 0.002	p < 0.002
25 min. after exertion	36 (26-48)	2 [1-3]	16 (10-19)	0 [0]	p < 0.002	p < 0.002

1. Measurement of Intracompartmental Pressure (ICP) (Table 1)

The preexertional ICP of the anterior tibial compartment in standing position was 33 (24–55) mm Hg (Table 1). During treadmill exertion the ICP increased significantly up to 106 (80–130) mm Hg immediately before the athlete had to quit running due to maximal pain ($p < 0.001$). One minute after exertion the ICP decreased significantly to 85 (72–110) mm Hg ($p < 0.001$). In eight symptomatic legs of five athletes persisting increased ICP values > 60 mm Hg were found for at least 15 min after exertion. The median ICP 15 min after exertion was 54 (43–85) mm Hg, representing a significant decrease ($p < 0.001$). Twenty-five minutes after exertion a return to baseline values occurred associated with no statistical differences in comparison with preexertional values; the median ICP was 36 (26–48) mm Hg.

Elevated ICP values immediately before the end of exertion and after exertion ($p < 0.001$) were related to significantly increased pain score levels ($p < 0.001$).

The group of the four asymptomatic legs showed significantly lower ICP levels during all time periods ($p < 0.002$), whereas significantly lower pain levels were observed during all intervals ($p < 0.002$) except the status before exertion ($p > 0.05$).

2. Measurement of Intramuscular Oxygen Partial Pressure (pO_2) (Table 2)

The preexertional pO_2 of the anterior tibial muscle in supine position was 19 (17–25) mm Hg (Table 2). Immediately after the end of exertion and during the first 15 min after exertion basically three different types of curves were seen:

1. In six cases an increased pO_2 of 35 (32–40) mm Hg was observed, followed by a slight decrease to 32 (26–35) mm Hg 5 min after exertion. Fifteen minutes after exertion a decrease to 22 (16–26) mm Hg was seen corresponding to preexertional values.

Table 2. Intramuscular oxygen partial pressure (mm Hg) and pain score [0–10]
median and range -

pre exertion	19 (17-25) mm Hg [0]		
post exertion	Group I n = 6	Group II n = 4	Group III n = 4
stat after exertion	35 (32-40) 9 [8-10]	14 (10-15) 9 [9-10]	4 (3-6) 9 [9]
5 min. after exertion	32 (26-35) 5 [3-6]	16 (13-24) 6 [5-7]	5 (3-8) 8 [8-9]
15 min. after exertion	22 (16-26) 2 [1-3]	25 (20-35) 2 [2-3]	26 (16-35) 4 [4-5]

Table 3. Intramuscular oxygen partial pressure (pO_2) and pain score
[pain] level of statistical significance

tested pair of groups	Group III vs. Group I		Group III vs. Group II		Group I vs. Group II	
tested parameters	pO_2	pain	pO_2	pain	pO_2	pain
pre exertion	n.s.	n.s.	n.s.	n.s.	n.s.	n.s.
stat after exertion	p = 0.01	n.s.	p < 0.05	n.s.	p = 0.01	n.s.
5 min. after exertion	p = 0.01	p = 0.01	p < 0.05	p < 0.05	p = 0.01	n.s.
15 min. after exertion	n.s.	p = 0.01	n.s.	p < 0.05	n.s.	n.s.

2. In the second group consisting of four cases, a mild decrease to 14 (10–15) mm Hg was found immediately after exertion, followed by a mild increase to 16 (13–24) mm Hg 5 min after exertion; 15 min after exertion a return to 25 (20–35) mm Hg occurred corresponding to preeexertional values.

3. In the third group consisting of four cases, a clearly decreased pO_2 of 4 (3–6) mm Hg was observed, persisting for at least 5 min after exertion with a median of 5 (3–8) mm Hg, followed by an increase to 26 (16–35) mm Hg 15 min after exertion, slightly exceeding the preexertional values.

Before exertion neither the pO_2 values nor the pain score results of the different groups showed any statistical differences (Table 3). However, immediately after exertion the pO_2 values in Group III were significantly lower than in Group I (p = 0.01) and II (p < 0.05), respectively. Additionally, the pO_2 values in Group II were significantly lower than in Group I (p = 0.01). On the other hand, immediately after exertion no statistical differences were found in terms of pain score. Five minutes after exertion the same statistical results occurred regarding the pO_2 levels, whereas the pain score in Group III was significantly higher in comparison with Group I (p = 0.01) and II (p < 0.05), respectively. While no significant differences in pO_2 were found between the different groups 15 min after exertion, the pain score in Group III was still significantly higher than in Group I (p = 0.01) and Group II (p < 0.05).

Based on the above mentioned three different types of pO_2 responses after exertion, these groups were also statistically analyzed for its corresponding ICP levels and pain score results (Table 4).

Table 4. Groups of different pO_2 response (I to III) and its corresponding ICP and pain score results - level of statistical significance 5 and 15 min after exertion, the Group III cases showed significantly higher pain score levels than the patients in Group I and II (p = 0.01, p < 0.05, resp.). This observation supports the hypothesis that the characterizing pain in patients with CECS may be of ischemic origin

tested pair of groups	Group III vs. Group I		Group III vs. Group II		Group I vs. Group II	
tested parameters	ICP	pain	ICP	pain	ICP	pain
pre exertion	n.s.	n.s.	n.s.	n.s.	n.s.	n.s.
immediately before pain-related stop	n.s.	n.s.	n.s.	n.s.	n.s.	n.s.
1 min. after exertion	n.s.	n.s.	p < 0.05	n.s.	p = 0.01	n.s.
15 min. after exertion	p = 0.01	p = 0.01	p < 0.05	p < 0.05	n.s.	n.s.
25 min. after exertion	n.s.	n.s.	n.s.	n.s.	n.s.	n.s.

Before exertion and immediately before pain-related stop of treadmill running, the three groups did not show any statistical differences in terms of ICP and pain score. However, 1 min after exertion the ICP levels in Group II were significantly lower than in Group I ($p = 0.01$) and III ($p < 0.05$), respectively, while no statistical differences were observed regarding the pain score levels. Fifteen minutes after exertion, both ICP and pain score results in Group III were significantly higher than in Group I ($p = 0.01$) and II ($p < 0.05$), respectively. Twenty-five minutes after exertion neither ICP nor pain score levels differed significantly between the three groups.

DISCUSSION

The chronic exertional compartment syndrome (CECS) is characterized by exercise-related pain, often forcing the athlete to either modify his training activities or to stop running. Precise history and clinical examination are important as screening factors. However, the only possibility to verify the diagnosis is to measure the ICP at rest, during and after symptom-reproducing exertion (Detmer, 1980; Rorabeck et al., 1988). In accordance with other studies we observed significantly increased ICP values before, during and after exertion in all symptomatic extremities ($p < 0.002$), associated with significantly increased pain score levels during and after exertion ($p < 0.002$) (Table 1) (Puranen and Alvaikko, 1981; Rorabeck et al., 1988).

However, the main purpose of the study was to obtain information about the oxygen partial pressure of the anterior tibial muscle in this group of patients.

Therefore, the Sigma-pO_2-Histograph KIMOC was used for the first time to determine the oxygen partial pressure within the anterior tibial muscle in patients with chronic exertional compartment syndrome (CECS).

The direct polarographic method proved to provide precise and reliable data of the intramuscular pO_2 (Fleckenstein et al., 1984). Computer-assisted introduction of the hypodermic needle and stepwise monitoring of pO_2 during the insertion procedure allow to measure a high number of different local pO_2 levels within one muscle compartment. The needle-related tissue trauma around the tip of the probe is reduced by the short reverse movement of the needle before each measurement (pilgrim's step procedure).

In the present series the direct polarographically measured preexertional median pO_2 level was 19 (17–25) mm Hg (Table 2). These findings correspond with those from the study of Heinrich et al. (1987) that revealed a mean pO_2 of 16.2 ± 3.5 mm Hg within the human anterior tibial muscle.

The intramuscular pO_2 after exertion showed three significantly different types of response (Tables 2, 3). While in six legs a mild increase of the intramuscular pO_2 was observed as sign of a reactive hyperemia (Group I), a moderate decrease to ischemic levels was found in four legs (Group II). Values between 3–8 mm Hg that were obtained in four symptomatic legs document a significant ischemic situation (Group III).

Furthermore, in Group III significantly higher corresponding ICP values were observed 1 and 15 min after exertion ($p = 0.01$, $p < 0.05$, resp.). So an increased ICP may lead to an impairment of oxygen supply to the muscle.

These findings support the hypothesis, that increased ICP may lead to decreased pO_2, causing increased pain as a result from compression and impaired oxygen supply.

Despite the limited number of investigated subjects the impaired muscular oxygen supply after exertion seems to play an important role in the pathogenesis of CECS. However, further studies are required for better understanding of the different behaviour patterns of the intramuscular pO_2 in patients with CECS, particularly after exertion. Whether the measurement of intramuscular pO_2 might become a prognostic factor in patients with CECS has also to be evaluated in future studies.

CONCLUSION

The present study showed that direct polarographic measurement of intramuscular pO_2 in patients with CECS reveals different pO_2 response types representing the different degree of muscle ischemia after exertion. Patients with severe ischemic levels after exertion had significantly higher ICP and pain score levels for 15 min after exertion. While the results of the study contribute to a better understanding of the relationship between ICP, intramuscular pO_2 and pain, further prospective studies are required to clarify the predisposing factors for the different behaviour concerning the pO_2 after exertion.

ACKNOWLEDGMENT

This research was supported by funds provided by the German Federal Army (InSan 03 K2-S-109193).

REFERENCES

Detmer DE: Chronic leg pain. Am J Sports Med 8: 141–144, 1980

Evers B, Becker HP, Rosenheimer M, Gerngroß H: A new piezoresistive system for measuring of intracompartmental pressure. J Jpn Orthop Assoc 69(2)(3): 755, 1995

Fleckenstein W, Heinrich R, Kersting T, Schomerus H, Weiss Ch: A new method for the bed-side recording of tissue pO_2-histograms. Verh Dtsch Ges Inn Med 90: 439, 1984

Heinrich R, Günderoth-Palmowski M, Grauer W, Machac N, Dette S, Egberts E: Gewebe-pO_2 M. tibialis ant. gesunder Probanden bei normovolämischer Hämodilution mit 10 %iger HES 200. VASA 16: 318–323, 1987

Puranen J, Alvaikko A: Intracompartmental pressure increase on exertion in patients with chronic compartment syndrome in the leg. J Bone Joint Surg 63A: 1304–1309, 1981

Rorabeck CH, Bourne RB, Fowler PJ, Finlay JB, Nott L: The role of tissue pressure measurement in diagnosing chronic anterior compartment syndrome. Am J Sports Med 16: 143–146, 1988

45

OXYGEN PARTIAL PRESSURE IN THE ANTERIOR TIBIAL MUSCLE DURING AND AFTER KNEE SURGERY WITH TOURNIQUET CONTROL

B. Evers, V. Ödemis, and H. Gerngroß

Department of Surgery
Military Hospital of Ulm
89081 Ulm, Germany

INTRODUCTION

Since first used by H. Cushing in 1904 for creating a bloodless field the pneumatic tourniquet has become an essential element in orthopaedic surgery, enabling the surgeon to perform meticulous dissections with a significant reduction of bleedings. These advantages are especially appreciated in micro-, hand and reconstructive surgery, arthroscopic surgery and fracture treatment.

The majority of surgical procedures on the knee is performed with tourniquet control (Sherman et al., 1986). However, despite the high clinical relevance of this instrument not only the exact skeletal muscle ischemic tolerance in humans, but also numerous important aspects of the pathophysiological background remain unclear.

A variety of studies on systemic and local biochemical and histological changes during ischemia and reperfusion of skeletal muscle has been published. Whereas systemic changes are primarily characterized by increased systemic potassium, lactate and pCO_2 concentrations and decreased pO_2, the most relevant pathophysiological processes during tourniquet-induced skeletal muscle ischemia and subsequent reperfusion concerning the cellular level proved to be the increase of the intracellular calcium concentration, the release of free radicals and the impairment of the arachidonic acid metabolism (Korthuis et al., 1985, 1988; McCord and Roy, 1982; Pedowitz, 1991). However, the use of protective agents (e.g. calcium antagonists, "scavengers" etc.) turned out to be less effective than for other tissues such as liver or heart (Korthuis et al., 1985, 1988; McCord and Roy, 1982).

Furthermore, numerous studies have reported on ischemia-related morphological changes of muscle and nerves distal to the tourniquet, as well as underneath the instrument (Pedowitz, 1991; Suval et al., 1987).

Oxygen Transport to Tissue XIX, edited by Harrison and Delpy
Plenum Press, New York, 1997

Whereas several articles were published on the *intravascular* pO_2 during and after muscle ischemia, the number of studies on *intramuscular* oxygen partial pressure remains very limited, mainly due to the lack of appropriate techniques. Since introduced by Fleckenstein (1984) the direct polarographic principle for measuring the intramuscular oxygen partial pressure was primarily used in healthy individuals, critically ill patients, in patients with occlusive vascular diseases and after vascular reconstructive procedures (Boekstegers et al., 1988; Heinrich et al., 1987). Since this technique proved to be exact and reliable, we decided to use it for the first time to determine the pre-, intra- and postoperative intramuscular oxygen partial pressure in patients undergoing knee operations with tourniquet control.

So the purpose of the study was: (1) to investigate the human pO_2 in the anterior tibial muscle before, during and after knee arthroscopy with pneumatic tourniquet control; and (2) to analyze the findings depending on the different types of ischemia: partial versus complete exsanguination.

MATERIAL AND METHODS

Subjects

According to the inclusion criteria only male patients between 18 and 45 years of age, who underwent a knee arthroscopy for cartilage surgery or meniscectomy, were included. Concerning the patient's past medical history neither any general diseases such as vascular, endocrinological, muscular or neural disorders, nor any previous trauma to the lower legs had to be reported. The study protocol was proven and agreed by the local ethical committee. Furthermore every patient had to sign an informed consent before entering the study.

The 40 male patients were randomly divided into two groups. The median age (range) in Group I and II was 24 (19–37) and 25 (20–42) years, respectively.

In Group I (n=20) all procedures were performed in a bloodless field using a pneumatic tourniquet after exsanguination of the leg with a rubber compression bandage (complete exsanguination). In Group II (n=20) the pneumatic tourniquet was applied after elevating the leg for one minute (partial exsanguination).

Perioperative Protocol

A pneumatic tourniquet with a straight cuff (110 x 800 mm) (Ulrich, Ulm, Germany) was applied at the middle of the patient's thigh in all cases. A tourniquet pressure of 400 mm Hg was used in all cases. All patients received spinal anaesthesia with 1 ml of 1% mepivicaine (Scandicain®) and 2 ml of 0.5% bupivicaine (Carbostesin® 0.5% hyperbaric). The average tourniquet time was 66 ± 13 minutes in Group I and 67 ± 14 minutes in Group II.

Method

The process of measurement as well as the analysis of pO_2 values and pO_2-histograms were realized with a Sigma-pO_2-Histograph KIMOC (Eppendorf-Netheler-Hinze GmbH, Hamburg, Germany). The intramuscular pO_2 was measured with a sterilizable hypodermic needle (tip diameter: 350 μm). The calibration of the probes was done with room air and nitrogen in 0.9% saline solution.

Experimental Protocol

Measurements were performed preoperatively, 5 minutes prior to reperfusion (PRP) until 12 minutes after reperfusion (ARP) with the patient in supine position.

The probe was inserted in a precisely defined area of the anterior tibial muscle: 10 cm distal to the lateral knee joint space and 3 cm lateral to the anterior edge of the tibia. The probe was introduced through a 20 G Abbocath® at an angle of 45° to skin level, parallel to the muscle fibers and computer-assisted driven into the muscle. The needle probe was moved forward by 0.7 mm steps followed by a reverse movement of 0.2 mm each time (pilgrim's step principle). After each step the local pO_2 was measured within less than 1 sec. The device was inserted for a total distance of 20 mm into the muscle associated with the registration of 40 local pO_2 values within 80 sec. At the end of such a period the probe was swiftly drawn back to the initial position, and its direction of insertion was slightly changed to measure in virgin tissue; 40 consecutive values of such a period were displayed as a histogram including the mean value. Pooled pO_2-histograms were calculated for the following intervals: preoperatively (PO), 5 minutes (PRP) before tourniquet deflation (TD), TD to 1 min 20 sec after reperfusion (ARP), 1 min 21 sec to 2 min 40 sec ARP, 2 min 41 sec to 4 min ARP, 4 min 01 sec to 5 min 20 sec ARP, 5 min 21 sec to 6 min 40 sec ARP, 6 min 41 sec to 8 min ARP, 8 min 01 sec to 9 min 20 sec ARP, 9 min 21 sec to 10 min 40 sec ARP, 10 min 41 sec to 12 min ARP. Unless otherwise indicated the data are expressed as mean ± standard deviation.

The Mann-Whitney-U-test was used for statistical analysis (alpha-error: 0.005).

RESULTS

1. Group I (Complete Exsanguination)

The preoperative pO_2 (PO) was 20.84 ± 8.6 mm Hg (Fig. 1A). The pO_2 before deflation of the tourniquet (PRP) decreased to 4.92 ± 5.3 mm Hg (Fig. 2A). One min 20 sec after reperfusion (ARP) of the limb almost no change of the pO_2 occurred; 2 min 40 sec after reperfusion the pO_2 value was still decreased (Fig. 3A); 4 min after reperfusion the pO_2 increased up to 10.94 ± 7.8 mm Hg (Fig. 4A); 5 min 40 sec after reperfusion the pO_2 reached its maximum level of 14.46 ± 7.6 mm Hg (Fig. 5A) and remained almost unchanged until 12 min after reperfusion (Fig. 6).

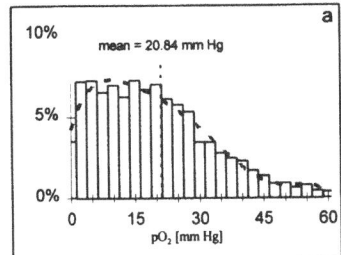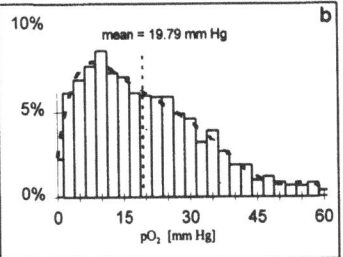

Figure 1. A: Group I (complete exsanguination): pO_2 [mm Hg] before tourniquet deflation (preoperatively). B: Group II (partial exsanguination): pO_2 [mm Hg] before tourniquet deflation (preoperatively).

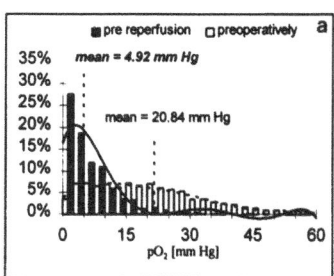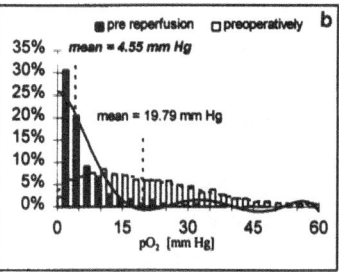

Figure 2. A: Group I (complete exsanguination): pO$_2$ [mm Hg] before tourniquet deflation (pre reperfusion) B: Group II (partial exsanguination):pO$_2$ [mm Hg] before tourniquet deflation(pre reperfusion).

2. Group II (Partial Exsanguination)

The preoperative pO$_2$ (PO) was 19.79 ± 6.9 mm Hg (Fig. 1B). The pO$_2$ before deflation of the tourniquet (PRP) decreased to 4.55 ± 4.1 mm Hg (Fig. 2B). One min 20 sec after reperfusion (ARP) of the limb almost no change of the pO$_2$ was observed; 2 min 40 sec after reperfusion the pO$_2$ increased up to 11.45 ± 6.5 mm Hg (Fig. 3B); 4 min after reperfusion the pO$_2$ climbed up to 16.80 ± 11.4 mm Hg (Fig. 4B); 5 min 40 sec after reperfusion the pO$_2$ reached its maximum level of 18.84 ± 12.8 mm Hg (Fig. 5B) and remained basically unchanged until 12 min after reperfusion (Fig. 6).

3. Comparison: Group I (Complete Exsanguination) vs. Group II (Partial Exsanguination)

The mean preoperative pO$_2$ value did not differ significantly between the two groups: 20.84 vs. 19.79 mm Hg (Figs. 1A, 1B, 6). The application of the tourniquet and the associated onset of muscle ischemia led to a remarkable decrease of the pO$_2$ in both groups: 4.92 vs. 4.55 mm Hg (Fig. 6). However, during the intervals 1 min 21 sec to 2 min 40 sec ARP and 2 min 41 sec to 4 min ARP the pO$_2$ value after partial exsanguination (Group II) was significantly (p < 0.005) higher than after complete exsanguination (Group I) (Fig. 6). In the subsequent intervals no significant differences in terms of pO$_2$ were observed between the two groups; 5 min 20 sec after reperfusion almost complete recovery was observed in both groups; at that time no significant differences were found in comparison with preoperative pO$_2$ levels.

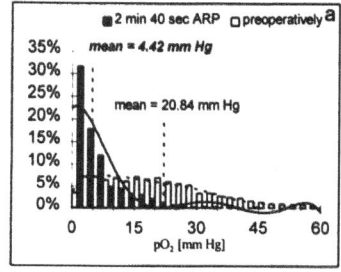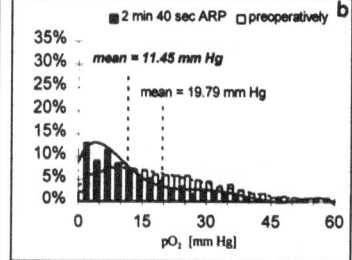

Figure 3. A: Group I (complete exsanguination): pO$_2$ [mm Hg]: 2 min 40 sec after reperfusion (ARP) B: Group II (partial exsanguination): pO$_2$ [mm Hg]: 2 min 40 sec after reperfusion (ARP).

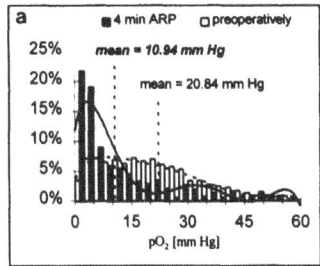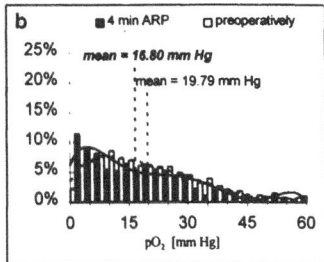

Figure 4. A: Group I (complete exsanguination): pO$_2$ [mm Hg]: 4 min after reperfusion (ARP). B: Group II (partial exsanguination): pO$_2$ [mm Hg]: 4 min after reperfusion (ARP).

DISCUSSION

In the present study the Sigma-pO$_2$-Histograph KIMOC was used for the first time to determine the oxygen partial pressure in the anterior tibial muscle in patients before, during and after knee arthroscopy after partial or complete exsanguination, respectively.

The direct polarographic method offers the advantage of precise and reliable measurement of intramuscular pO$_2$ (Fleckenstein et al., 1984). Furthermore, computer-assisted introduction of the hypodermic needle and stepwise monitoring of pO$_2$ during the insertion procedure allow to obtain a high number of different local pO$_2$ levels within a single muscle compartment. The comparatively high standard deviations that were seen in this study represent the great variation concerning local pO$_2$ concentrations within a single muscle compartment. The pilgrim's step procedure minimizes the needle-related tissue trauma around the tip of the probe.

In this series the direct polarographically measured preoperative mean pO$_2$ levels in Group I (complete exsanguination) and Group II (partial exsanguination) were 20.84 and 19.79 mm Hg, respectively (Fig. 6). These findings correspond with the results of other studies (Boekstegers et al., 1988). Heinrich et al. (1987) found a mean pO$_2$ of 16.2 ± 3.5 mm Hg within the human anterior tibial muscle; Santavirta et al. (1978a) reported a mean pO$_2$ of 22.6 ± 0.6 mm Hg within the anterior tibial muscle of rabbits.

The present study showed that the application of a pneumatic tourniquet for one hour during a knee arthroscopy leads to a remarkable decrease of intramuscular pO$_2$ distal to the tourniquet independent of the type of exsanguination. The tourniquet-ischemia asso-

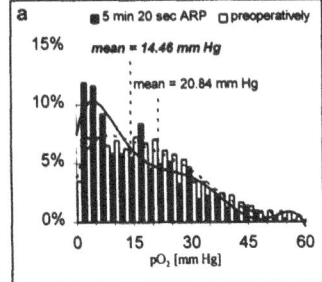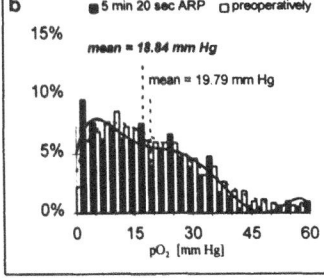

Figure 5. A: Group I (complete exsanguination): pO$_2$ [mm Hg]: 5 min 20 sec after reperfusion (ARP). B: Group II (partial exsanguination): pO$_2$ [mm Hg]: 5 min 20 sec after reperfusion (ARP).

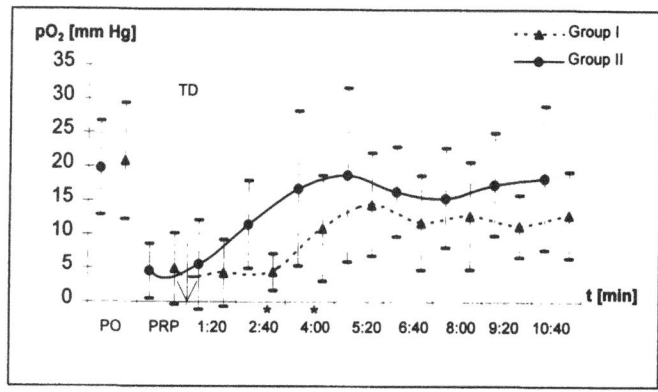

Figure 6. Time course diagram: PO = pre operation; PRP = pre reperfusion; TD = tourniquet deflation; * p<0.005.

ciated reduction of the intramuscular pO_2 to 4.92 vs. 4.55 mm Hg was observed in both groups (Fig. 6). However, even after complete exsanguination the pO_2 level did not decrease to 0 mm Hg. These findings correspond to the observation of other authors and are probably related to oxygen supply through bone vessels (Klenerman and Crawley, 1977; Santavirta et al., 1978a).

The same authors observed a decrease of the intramuscular pO_2 in rabbits and monkeys respectively within 20–30 minutes of ischemia to values between 3–10 mm Hg (Klenerman and Crawley, 1977; Santavirta et al., 1978a). Due to the design of the present study no measurements could be performed during the initial period of the surgical procedure. Therefore, the exact time course of the decrease of the intramuscular oxygen partial pressure was unable to be determined. The first data during ischemia were obtained 5 min before reperfusion, representing the status after about 60 min of tourniquet ischemia.

The tourniquet-induced skeletal muscle ischemia causes considerable alterations concerning both muscle morphology and biochemical metabolism. Due to the lack of oxygen, anaerobe glycolysis and consumption of phosphocreatinine and ATP are the most important pathways to provide energy. Subsequently, the concentration of metabolites like

Table 1. Comparison of the pO_2 [mm Hg] between Group I (complete exsanguination) and Group II (partial exsanguination). PRP = pre reperfusion; TD = tourniquet deflation; ARP = after reperfusion

Time	Group I (n=20)	Group II (n=20)	Level of significance
Preoperatively	20.84 ± 8.6	19.79 ± 6.9	n.s.
PRP–TD	4.92 ± 5.3	4.55 ± 4.1	n.s.
TD–1:20 min. ARP	4.28 ± 4.9	5.52 ± 6.6	n.s.
1:21–2:40 min. ARP	4.42 ± 2.7	11.46 ± 6.5	p< 0.005
2:41–4:00 min. ARP	10.94 ± 7.8	16.80 ± 11.4	p< 0.005
4:01–5:20 min. ARP	14.46 ± 7.6	18.84 ± 12.8	n.s.
5:21–6:40 min. ARP	11.79 ± 7.0	16.37 ± 6.6	n.s.
6:41–8:00 min. ARP	12.78 ± 7.9	15.44 ± 7.3	n.s.
8:01–9:20 min. ARP	11.29 ± 4.6	17.46 ± 7.6	n.s.
9:21–10:40 min. ARP	12.94 ± 6.4	18.36 ± 10.6	n.s.

lactate and creatinine increases associated with decreasing tissue pH (Haljamäe and Enger, 1975).

Characterized by a comparatively high ischemic tolerance, skeletal muscle tissue shows only limited and reversible damage after one or two hours of ischemia. Tourniquet-induced ischemia leads to a morphological picture of limited damage of muscle fibers, regional necrosis and pycnotic nuclei. Eckert and Schnackerz (1991) reported the critical ischemia time of human skeletal muscle to be 135 minutes under normothermic conditions.

In the present series the restoration of the intramuscular pO_2 after deflation of the tourniquet depended on the type of exsanguination. Whereas the pO_2 after partial exsanguination already increased during the interval between 1 min 21 sec and 2 min 40 sec after reperfusion, the restoration after complete exsanguination occurred with a delay of about 1 min 20 sec. Five min 20 sec after reperfusion after a one-hour period of tourniquet ischemia the preoperative levels were almost obtained in both groups. No further increase in pO_2 levels higher than before ischemia was observed (Fig. 6). Other authors reported similar results in animal experiments. After a one-hour period of tourniquet ischemia an increase of the intramuscular pO_2 up to pretourniquet levels was seen within 5 minutes in rabbits (Santavirta et al., 1978b) and within 5–10 minutes in dogs (Heppenstall et al., 1979; Nakahara, 1984). Several investigators found that an extension of the ischemia time leads to a prolongation of the restoration interval (Santavirta et al., 1978b; Heppenstall et al., 1979). So the normalization interval of the pO_2 in dogs after a two-hour period of ischemia was reported to be 10–15 minutes; after a three-hour period of ischemia an extension of the restoration interval up to 15–18 minutes was reported (Santavirta et al., 1978b).

Concerning the maximal increase of the intramuscular pO_2 during reperfusion of the extremity different results have been published. In the present study no increase of pO_2 to above preoperative levels occurred. Whereas some authors reported similar results (Santavirta et al., 1978b), others observed a threefold increase of pretourniquet intramuscular pO_2 levels during reperfusion (Heppenstall et al., 1979; Nakahara, 1984). These different observations seem to be related to the different animal species that were used; furthermore, varying methods were applied for the measurement of the intramuscular pO_2.

The reperfusion of skeletal muscle is characterized by the release of metabolites that were accumulated during ischemia (e.g. lactate) followed by normalization of the metabolism and restoration of ATP pools and phosphocreatinine levels. After an ischemia time of 2 hours the normalization of the majority of blood parameters, except the pH value, takes place within a few minutes after reperfusion. After the deflation of the tourniquet the muscle blood flow is significantly increasing. The status of reactive hyperemia persists for approximately 5 minutes (Santavirta et al., 1978b).

During reperfusion the release of free radicals leads to damage of the endothelial cells and permeability changes of the capillaries (McCord and Roy, 1982). The corresponding morphology is characterized by muscle cell edema. However, the release of free radicals seems to be more relevant after ischemia periods of more than 2 hours (Korthuis et al., 1985).

In the present study two different types of exsanguination before application of the pneumatic tourniquet were investigated: complete exsanguination with a rubber compression bandage and partial exsanguination by elevating the limb for one minute.

Concerning the pO_2 in the anterior tibial muscle during tourniquet ischemia no significant differences between both groups occurred. However, after partial exsanguination a significant faster restoration of the intramuscular pO_2 was seen after reperfusion. After complete exsanguination the pO_2 started to increase within the interval between 2 min 41 sec and 4 min, whereas the same increase after partial exsanguination occurred significantly earlier within the interval between 1 min 21 sec and 2 min 40 sec. Subsequently, the partial exsanguination before the application of the tourniquet was associated with an earlier recovery of preoperative pO_2 levels after reperfusion. After complete exsanguination the same process took about 80 sec longer.

Whereas after application of the rubber bandage the amount of persisting blood can be neglected, a rest amount of blood remains in the ischemic leg after elevation alone. This factor seems to play an essential role, since all other parameters were identical within the two groups.

One explanation for the significant difference between the two types of exsanguination concerning the intramuscular pO_2 after reperfusion could be a more intensive no-reflow phenomenon after complete exsanguination. This phenomenon occurs during and after reperfusion and is characterized by the absence of reflow within small vessels. Granulocytes interfere with the endothelial cells and cause the occlusion of the capillaries (Korthuis et al., 1985). After an ischemia of 2 hours 30% of the capillaries were not reperfused (Suval et al., 1987). The no-reflow phenomenon appears to have a protective effect on muscle cells. Some authors reported a diminished muscle cell damage in no-reflow areas compared with better perfused areas (Blebea et al., 1987; Suval et al., 1987).

The present investigations were performed before, during and after one-hour of tourniquet ischemia. As expected due to the comparatively short interval no irreversible changes occurred. Further studies are required to evaluate the intramuscular pO_2 during and after longer periods of ischemia. Future investigations may reveal a critical borderline intramuscular pO_2 representing the pO_2 level required to prevent irreversible muscle cell damage.

CONCLUSION

A one-hour period of tourniquet ischemia leads to a remarkable decrease of the pO_2 within the anterior tibial muscle distal to the tourniquet. Although the restoration of the intramuscular pO_2 is found to occur significantly faster after partial exsanguination than after complete exsanguination, the tourniquet-ischemia-associated alterations of the intramuscular pO_2 are reversible in both groups within 5 minutes after reperfusion.

ACKNOWLEDGMENT

This research was supported by funds provided by the German Federal Army (InSan 03 K2-S-109193).

REFERENCES

Blebea J, Kerr J, Shumko J, Feinberg R, Hobson R: Quantitative histochemical evaluation of skeletal muscle ischemia and reperfusion injury. J Surg Res 43: 311–321, 1987

Boekstegers P, Fleckenstein W, Rosport A, Ruschewsky W, Braun U: Überwachung der Sauerstoffversorgung des Skelettmuskels und der Gesamtsauerstoffaufnahme bei koronarchirurgischen Eingriffen. Anaesthesist 37: 287–296, 1988

Eckert P, Schnackerz K: Ischemic tolerance of human skeletal muscle. Ann Plast Surg 26: 77–84, 1991

Fleckenstein W, Heinrich R, Kersting T, Schomerus H, Weiss Ch: A new method for the bed-side recording of tissue pO_2-histograms. Verh Dtsch Ges Inn Med 90: 439, 1984

Haljamäe H, Enger E: Human skeletal muscle energy metabolism during and after complete tourniquet ischemia. Ann Surg 182: 9–14, 1975

Heinrich R, Günderoth-Palmowski M, Grauer W, Machac N, Dette S, Egberts E: Gewebe-pO_2 im M. tibialis ant. gesunder Probanden bei normovolämischer Hämodilution mit 10 %iger HES 200. VASA 16: 318–323, 1987

Heppenstall RB, Balderston R, Goodwin C: Pathophysiological effects distal to a tourniquet in the dog. J Trauma 19: 234–238, 1979

Klenerman L, Crawley J: Limb blood flow in the presence of a tourniquet. Acta Orthop Scand 48: 291–295, 1977

Korthuis RJ, Granger DN, Towensley M, Taylor A: The role of oxygen-derived free radicals in ischemia-induced increases in canine skeletal muscle vascular permeability. Circ Res 57: 599–609, 1985

Korthuis RJ, Grisham MB, Granger DN: Leucocyte depletion attenuates vascular injury in postischemic skeletal muscle. Am J Physiol 254: H823-H827, 1988

McCord JM, Roy RS: The pathophysiology of superoxide: Roles in inflammation and ischemia. Can J Physiol Pharmacol 60: 1346–1352, 1982

Nakahara M: Tourniquet effects on muscle oxygen tension in dog limbs. Acta Orthop Scand 55: 576–578, 1984

Pedowitz RA: Tourniquet-induced neuromuscular injury: A recent review of rabbit and clinical experiments. Acta Orthop Scand 62 (suppl 245): 1–33, 1991

Santavirta S, Höckerstedt K, Niinikoski J: Effect of pneumatic tourniquet on muscle oxygen tension. Acta Orthop Scand 49: 415–419, 1978a

Santavirta S, Höckerstedt K, Linden H: Pneumatic tourniquet and limb blood flow. Acta Orthop Scand 49: 565–570, 1978b

Sherman OH, Fox JM, Snyder SJ, Del Pizzo W, Friedman MJ, Ferkel RD, Lawley MJ: Arthroscopy -"No-problem surgery". J Bone Joint Surg 68A: 256–265, 1986

Suval W, Duran W, Boric M, Hobson R, Berendsen P, Ritter A: Microvascular transport and endothelial cell alterations preceding skeletal muscle damage in ischemia and reperfusion injury. Am J Surg 154: 211–218, 1987

46

DETECTION OF OXYGEN CONSUMPTION IN DIFFERENT FOREARM MUSCLES DURING HANDGRIP EXERCISE BY SPATIALLY RESOLVED NIR SPECTROSCOPY

Sachiko Homma and Atsuko Kagaya[*]

Research Institute of Physical Fitness
Japan Women's College of Physical Education
Setagaya-ku, Tokyo 157, Japan

INTRODUCTION

Muscle O_2 kinetics during exercise in humans have been estimated by measuring blood flow, and arterial and venous O_2 contents. To date, plethysmography and the thermodilution method have been used for blood flow measurement. The blood flow values estimated by these methods have been considered to represent the average value for several working muscles. To estimate the venous O_2 content, blood samples have been collected from the antecubital vein during handgrip exercise (Joyner *et al.* 1992, Hartling *et al.* 1989) and from the femoral vein during leg exercise (Costes *et al.* 1996, Richardson *et al.* 1995, Vollestad *et al.* 1990). Since these veins are relatively large, the venous O_2 content estimated from the blood collected from them has also been considered to represent the average for several working muscles. Thus, in humans, O_2 kinetics during exercise have been estimated as the average values for several working muscles, and differences in O_2 kinetics between individual working muscles which contribute to the exercise have not been detected. The recent development of near-infrared (NIR) spectroscopy makes it possible to estimate changes in O_2 levels in small blood vessels, capillaries, and intracellular O_2 uptake sites (Mancini *et al.* 1994). The estimation of changes in O_2 supply and consumption in muscle tissue has been achieved using the venous occlusion technique (Homma *et al.* 1996a), while tissue O_2 saturation has been demonstrated using spatially resolved NIR spectroscopy (Farell *et al.* 1992, Tsunazawa *et al.* 1996). The aim of this study was to investigate differences in the O_2 kinetics of two individual muscles, both working during handgrip exercise of various intensities, using spatially resolved NIR spectroscopy.

---

[*] With the technical assistance of Yoshio Tsunazawa, Technology Research Laboratory, Shimadzu Corp., Hadano city, Kanagawa 259-13, Japan.

MATERIALS AND METHODS

Subjects

Seven physically active women participated in this study after giving written informed consent. Their mean (±SD) age, height and body mass were 21.1±0.3 years, 162.4±3.9 cm and 55.8±5.5 kg, respectively.

Exercise

After 20 min of bed rest, pre-exercise control measurements were conducted over 5 min with the subjects in the supine position. A rhythmic handgrip exercise at a frequency of 60 repetitions·min^{-1} was then performed for 60 s, starting at a load equal to 10% of the each subject's predetermined maximum voluntary contraction (MVC). The load was increased by 10% MVC after a recovery period in the supine position. The exercise was stopped and did not proceed to the next intensity level if a subject could not continue at a given intensity for 60 s. The recovery times allowed between 10% and 20% MVC, 20% and 30% MVC, and 30% and 40% MVC exercise were 10, 15 and 20 min, respectively.

Near-Infrared Spectroscopy

We used two types of NIR spectrometers. One of them (OM100A, Shimadzu, Japan) adopted a single distance measurement, and the other was a multi-distance spectrometer for oxygen saturation (SO_2) measurement based on the spatially resolved method. The former detected diffusely reflected light at a position 30 mm from a light source of 780, 805 and 830 nm, and the amount of reflected light was converted to absorbance. From these changes in absorbance, relative 'changes' in oxygenated hemoglobin (HbO_2), deoxygenated hemoglobin (Hb) and total hemoglobin (HbT; HbO_2+Hb) content were calculated by solving the simultaneous equation. On the other hand, the multi-distance spectrometer measured the absorbances at two source and detector distances. The distances we used this time were 25 and 40 mm. The most important optical parameter, the effective attenuation coefficient μ_{eff} [$\mu_{eff} = (3\mu_a \mu_s')^{1/2}$] for the spatially resolved method, was calculated from the equation:

$$\mu_{eff} = (\text{unit: mm}^{-1}) = 1.54 \cdot (abs_{40mm} - abs_{25mm}) - 0.0627$$

In the above μ_a and μ_s' stand for the absorption coefficient and the reduced scattering coefficient, respectively. Then μ_a was easily obtained from μ_{eff} using the relation

$$\mu_a = \frac{\mu_{eff}^2}{3\mu_2'}$$

This μ_a is not a 'change' but an 'absolute value', except for the reciprocal of μ_s. (Note that μ_s' was considered approximately constant and was expected to be canceled out at the subsequent division stage of the SO_2 calculation.) We obtained μ_a values for three wavelengths, 727, 803 and 827 nm, and the μ_a values were substituted into a simultaneous equation similar to that used for the previous single-distance method. One difference was that we used 'μ_a' values in place of the 'changes' in absorbance. Therefore absolute HbO_2

and Hb values were obtained as the solution of this simultaneous equation, except for the factor $(1/\mu_s')$. Finally the SO_2 in tissue was calculated as the ratio of HbO_2 to HbT. The factor $(1/\mu_s')$ was canceled out in this division process, and we left it undetermined. Therefore values of HbO_2, Hb and HbT were expressed as an arbitrary units (AU).

Experimental Procedure and Physiological Measurements

Each subject was tested twice on two consecutive days using the same exercise protocol each time. Each light source and detector of the multi- and single- distance NIR spectrometers was applied to the brachio-radial (BR) and flexor carpi ulnaris (FCU) muscles. Subcutaneous fat thickness over the measured part of the BR and FCU was 3.3±0.8 mm (mean±SEM) and 3.3±0.7 mm, respectively. The distance between the light source on the BR and the nearest part of the FCU was over 35 mm. Similarly, the distance between the light source on the FCU and the nearest part of the BR was over 35 mm. The order of measurement of the BR and FCU muscles by each type of spectroscopy was randomized between subjects. SO_2, HbO_2, Hb and HbT were estimated using a multi-distance spectrometer. Also, HbO_2, Hb and HbT were observed by the single-distance photometer to estimate changes in O_2 supply and consumption. The changes in O_2 supply and consumption were evaluated from the rate of increase in HbT and Hb during pneumatic cuff inflation (60 mmHg) around the upper arm (Homma et al. 1996a). The cuff was inflated during pre-control conditions and immediately after each exercise load. O_2 supply and consumption were expressed as values relative to those under pre-exercise control conditions. Electromyograms (EMG) from the BR and FCU were also recorded by attaching electrodes close to the NIR optodes. The EMG signals were integrated for the last 20 s of exercise (iEMG). The iEMGs were then normalized relative to the values obtained during 10% MVC exercise in order to minimize differences between subjects, under the assumption that differences would be smallest during exercise at the 10% MVC level. The temperature and relative humidity of the laboratory was maintained at 24° and 60% , respectively, throughout the experiments.

Statistical Analysis

Differences between mean values were evaluated using two-way analysis of variance (ANOVA) with Student's paired t-test as a post hoc test. Differences with $P<0.05$ were considered significant.

RESULTS

All subjects were able to continue exercise at 10%, 20%, and 30% MVC for 60 s, but were able to continue for only 41.7±5.4 s at 40% MVC. The SO_2 at rest ranged from 63.4 to 73.2 (mean 68.8) % in the BR, and from 67.0 to 72.9 (70.5) % in the FCU. Typical changes in SO_2, HbT, HbO_2 and Hb estimated by spatially resolved NIR spectroscopy are shown for the BR (Fig.1.) and FCU (Fig. 2.) of a representative subject (Subject I). The BR (Fig.1) showed larger changes in SO_2, HbO_2 and Hb after the onset of exercise than the FCU (Fig. 2), and these changes became greater as the intensity increased from 10% to 30%. The same pattern of changes was also observed in the other 6 subjects. In the FCU (Fig.2), exercise at all intensities decreased SO_2, HbT and HbO_2, and increased Hb from

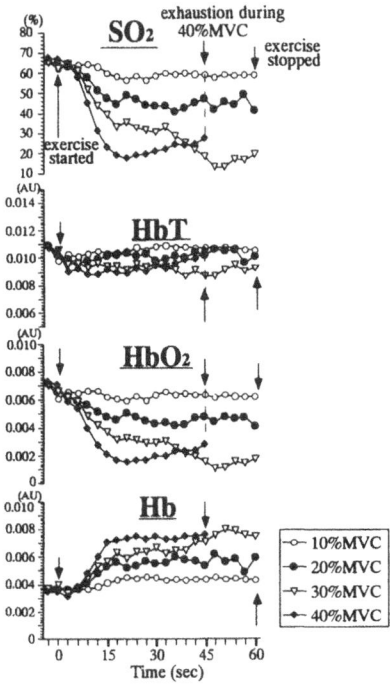

Figure 1. Changes in tissue oxygen saturation (SO$_2$), oxygenated hemoglobin content (HbO$_2$), deoxygenated hemoglobin content (Hb), and total hemoglobin content (HbT; HbO$_2$+Hb) in the brachio-radial muscle before and during exercise at each intensity for Subject I.

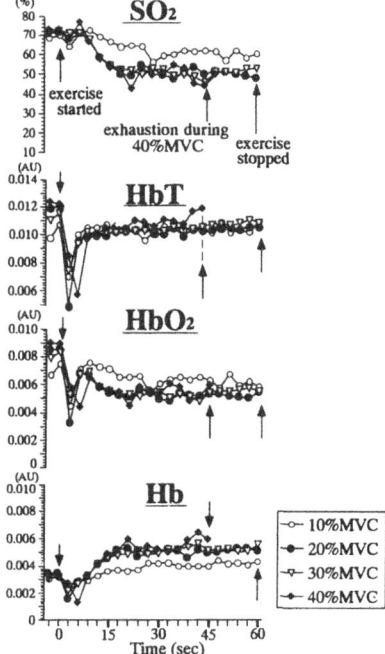

Figure 2. Changes in tissue oxygen saturation (SO$_2$), oxygenated hemoglobin content (HbO$_2$), deoxygenated hemoglobin content (Hb), and total hemoglobin content (HbT; HbO$_2$+Hb) in the flexor carpi ulnaris muscle before and during exercise at each intensity for Subject I.

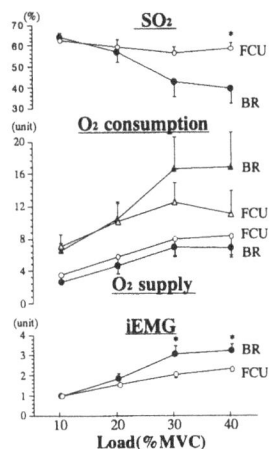

Figure 3. Changes in tissue oxygen saturation (SO$_2$), oxygen consumption, oxygen supply, and the integrated electromyogram (iEMG) in relation to exercise intensity. (Values are mean±SEM; *P<0.05, BR vs FCU.)

the pre-exercise level, however, no differences in these changes were found as the intensity increased from 20% to 40% MVC.

Fig.3 shows the mean (±SEM) changes in O$_2$ supply and consumption, SO$_2$ and iEMG for the BR and FCU at each exercise intensity in the 7 subjects. O$_2$ supply and O$_2$ consumption increased with exercise intensity up to 30% MVC, then levelled off at 40% MVC in both muscles. However, the degree of change in O$_2$ consumption tended to be higher in the BR than in the FCU at 30% and 40% MVC. SO$_2$ in the BR and FCU showed similar values at 10% and 20% MVC. However, SO$_2$ in the BR showed a greater decrease at 30% MVC than that in the FCU, and became significantly lower than in the FCU at 40% MVC. The iEMG also increased linearly as intensity rose from 10% to 30%, and was significantly higher in the BR than in the at 30% and 40% MVC.

DISCUSSION

At 40% MVC, the exercise time was less than 60 s for all subjects due to fatigue. Our previous study demonstrated that the exhaustion time for handgrip exercise at 30% MVC was 62.3±2.4 s for a similar group of subjects (Homma and Kagaya. 1996b). Therefore, the subjects appeared to be nearly exhausted after 60 s of exercise at 30% MVC in this study.

We applied a light source and detector to the BR and the FCU muscles in order to detect variations in O$_2$ kinetics in different muscles of the exercising forearm. These two muscles run parallel in the forearm, the BR being positioned on the radial side and the FCU on the ulnar side. According to Chance et al.(1992), the mean depth of penetration and the lateral spread of light into tissue is approximately equal to half the distance between the light source and the detector. The distance between the light source and the detector we used in this study was 25 and 40 mm for the multi-distance spectrometer and 30 mm for the single-distance spectrometer, and the distance between the light source on a muscle and the nearest part of the other muscle exceeded 35 mm. Therefore we think the detected NIR signal originated mainly from the BR or FCU when the light source and detector were attached to each muscle, respectively. The mutual interference of the signals was considered to be relatively small.

The most important finding of this study was that O$_2$ kinetics in different muscles were not equivalent during high intensity exercises. Although the changes in O$_2$ supply at each intensity were similar in both muscles, O_2 consumption tended to be higher in the BR than in the FCU during high intensity exercise. These differences were reflected in the SO_2, which was lower in the BR than in the FCU at high intensities. Since the iEMG value was higher in the BR than in the FCU at high intensity, it is considered that the contribution of the BR to the handgrip exercise became larger compared with that of the FCU as intensity increased. Therefore, it appears that energy demand was higher in the BR than in the FCU, and that O_2 utilization rose in the BR to meet the increased O_2 demand.

We have demonstrated that subcutaneous fat thickness influences the absorbance of NIR (Homma et al. 1996c). Matcher et al.(1996), using FEM simulation, also showed that the presence of a surface layer of different SO_2 has an effect of attenuating the magnitude of changes in SO_2 for a given change in internal SO_2. Since subcutaneous fat thicknesses over the measured parts of the BR and FCU were very similar in this study, the differences in SO_2 between the BR and FCU detected by spatially resolved NIR spectroscopy are considered to reflect the differences in SO_2 of muscle tissue.

In summary, we found that O_2 kinetics in the BR and FCU muscles were not equivalent during high-intensity exercise using spatially resolved NIR spectroscopy. These differences seem to be associated with changes in the relative contributions of these muscles to the exercise as its intensity increases.

REFERENCES

Chance, B., M. T. Dait, C. Zhang, T. Hamaoka, and F. Hagerman. Recovery from exercise-induced desaturation in the quadriceps muscles of elite competitive rowers. Am. J. Physiol., 262, C 766–775, 1992

Costes, F., J. C. Barthelemy, L. Feasson, T. Busso, A. Geyssant, and C. Denis. Comparison of muscle near-infrared spectroscopy and femoral blood gasses during steady-state exercise in humans. J. Appl. Physiol., 80, 1345–1350, 1996

Farrell, T. J., B. C. Wilson, and M. S. Patterson. The use of a neural network to determine tissue optical properties from spatially resolved diffuse reflectance measurements. Phys. Med. Biol. 37, 2281–2286,1992

Hartling, O. J., H.Kelbaek, T. Gjorup, B. Schibye, K. Klausen, and K. Trap-Jansen. Forearm oxygen uptake during maximal forearm handgrip exercise. Eur. J. Appl. Physiol. Occup. Physiol. 58, 466–470, 1989

Homma, S., H. Eda, S. Ogasawara, and A. Kagaya. Near-infrared estimation of O_2 supply and consumption in forearm muscles working at varying intensity. J. Appl. Physiol. 80, 1279–1284, 1996a

Homma, S., and A. Kagaya. Contribution of oxygen content in working muscle to prolonged exhaustive dynamic handgrip exercise. J. Exec. Sci. 6, 21–28, 1996b

Homma, S., T. Fukunaga, and A. Kagaya. Influence of adipose tissue thickness on near-infrared spectroscopic signal in the measurement of human muscle. J. Biomed. 1, 418–424, 1996c

Joyner, M. J., L. A. Nauss, M. A. Warner and D. O. Warner. Sympathetic modulation of blood flow and O_2 uptake in rhythmically contracting human forearm muscles. Am. J. Physiol., 263, H1078-H1083, 1992

Matcher, S. J., K. Nahid, M. Cope, D. T. Delpy. Absolute SO_2 measurements in layered media. OSA Trends in Optics and Photonics on Biomedical Optical Spectroscopy and Diagnostics, Eva Sevick-Muraca and Devid Benaron, eds. (Optical Society of America, Washington DC 1996), 83–90

Mancini, D. M., L. Bolinger, H. Lui, K. Kendrick, B. Chance, and J. R. Wilson. Validation of near-infrared spectroscopy in humans. J. Appl. Physiol. 77, 2740–2747, 1994

Richardson, R. S., D. R. Night, D. C. Pool, M. C. Hogan, B. Grassi, and P. D. Wagner. Determinants of maximal exercise VO_2 during single leg knee-extensor exercise in humans. Am. J. Physiol. 268, H1453-H1461, 1995

Tsunazawa,Y., Y. Okikawa, S. Iwamoto, H. Eda, and M. Takada. "Spatially resolved spectrosopy for extraction spectrum of hemoglobin diluted in a highly scattering medium" in Biomedical Optical Spectroscopy and Diagnostics, Technical Digest (Optical Society of America, Washington DC 1996), 119–121

Vollestad, N. K., J. Wesche, and I. Sejersted. Gradual increase in leg oxygen uptake during repeated submaximal contraction in humans. J. Appl. Physiol. 68, 1150–1156, 1990

47

SKELETAL MUSCLE FUNCTION, OXYGENATION AND BIOCHEMISTRY IN AN ENDOTOXEMIC MODEL OF SIRS

Ellen Dailor Iannoli and Thomas E. J. Gayeski

Departments of Anesthesiology and Pharmacology and Physiology
School of Medicine and Dentistry
University of Rochester
Rochester, New York

INTRODUCTION

Traditionally, "sepsis" refers to bacteremia leading to arteriolar vasodilatation and signs of generalized infection. Sepsis may progress to septic shock, often characterized by hypotension, classically explained as pooling of blood in the peripheral microcirculation.[1] A similar clinical picture in patients with no laboratory evidence of infection has been called the "septic syndrome."[9] Many of the characteristics of sepsis and the septic syndrome are mediated by host inflammatory products. The more comprehensive title systemic inflammatory response syndrome (SIRS) encompasses both states as well as states with similar clinical characteristics such as burn injuries.[16] The clinical signs of SIRS are fever, tachycardia, low systemic vascular resistance, tachypnea, and leukocytosis or leukopenia. Organ dysfunction may ensue despite adequate cardiac output and arterial oxygen tension. When more than 2 organs are dysfunctional, the patient is said to have multiple organ dysfunction syndrome (MODS). In patients the development of MODS leads to high mortality and great expense. Understanding the pathophysiology of SIRS may lead to improved therapy, lower mortality and reduced cost.[16]

The mechanism of organ dysfunction in SIRS is unknown. Two early hypotheses for explaining organ dysfunction were inadequate organ perfusion due to hypotension and a lack of organ oxygenation due to hypoxemia. However, neither hypotension nor hypoxemia is necessary for MODS to develop. Presently, the major hypotheses include: 1) a microcirculatory defect causing regional hypoperfusion and 2) biochemical defect(s) in which the tissue is adequately oxygenated but unable to use the available oxygen.

To make inferences about the mechanisms of pathophysiology in these syndromes, it is necessary to study the whole animal, individual organ and cellular levels. Observing the whole animal allows us to compare systemic changes in our model to those of the human syndromes. However, systemic observations do not predict individual dysfunction. At the indi-

Oxygen Transport to Tissue XIX, edited by Harrison and Delpy
Plenum Press, New York, 1997

vidual organ level, we can determine function, oxygen consumption, blood flow, and metabolic products in the blood and tissue. At the cellular level, we can determine cellular oxygenation and infer changes in the microcirculation and identify specific metabolic abnormalities.

Skeletal muscle is an organ with a wide range of oxygen consumption (100-fold range from rest to maximal exercise) and with extensive circulatory and biochemical reserves. By exercising the muscle during endotoxemia, we are able to determine if these reserves are affected. Examining recruitment of reserves in the early stages of SIRS before organ dysfunction exists at rest may allow us to determine which of the major hypotheses is likely to be the mechanism for subsequent organ dysfunction.

METHODS

Animal and Muscle Preparation

Hound-type mongrels weighing from 20 to 22 kg were anesthetized with intravenous sodium pentobarbital (30 mg/kg, followed by 6 mg/kg/hour infusion). Animals were mechanically ventilated with room air and intravenous normal saline was infused to maintain pH 7.25–7.45, PaO_2 greater than 90 and [Hb] 10–14 g/dL. They were placed on a heating pad and wrapped in plastic to maintain core temperature at 35–38° C. Systemic arterial pressure was monitored by an indwelling brachial artery catheter. A pulmonary artery catheter allowed central venous and pulmonary pressure monitoring as well as cardiac output measurement by the thermodilution method.

Both gracilis muscles were vascularly and anatomically isolated, wrapped with Saran® wrap to prevent O_2 exchange with the air, and maintained at 37° C. The entire blood supply of each muscle was preserved. A T-cannula in the femoral vein permitted blood sampling from the gracilis muscle only.

Resting muscles were left neurally intact. For exercising protocols, the obturator nerve was isolated and the distal tendon of each gracilis muscle was attached to a force transducer, with the resting tension set to 300 g. The cut obturator nerve was stimulated with a square-wave monophasic pulse 0.1 ms in duration and a voltage adequate to induce maximal isometric twitch contraction at the desired frequency (2 Hz or 4 Hz) for 3 minutes. The muscles were stimulated one at a time, with a one-hour recovery period before any further interventions.

Venous and arterial samples were simultaneously collected anaerobically. They were analyzed with an Instrumentation Laboratories 282 Co-Oximeter programmed for canine blood and an Instrumentation Laboratories 1304 Blood Gas Analyzer. Gracilis muscle flow was measured by timed collection of venous blood. VO_2 was calculated by the Fick principle.

Control State

The protocol was a paired design with the first muscle collected under control conditions and the contralateral muscle collected (frozen) under conditions of endotoxemia. Following muscle preparation, both muscles were left to recover for 1 hour, after which resting flow and VO_2 were measured for each muscle. In a resting protocol, the control muscle was collected shortly after resting values were determined. If it was an exercising protocol, each muscle was exercised and exercising flows, VO_2, and active tension development were recorded for each muscle. The control exercising muscle was frozen and collected at the end of three minutes of stimulation.

Endotoxemia

Endotoxin (0.015 mg/kg) was dissolved in approximately 80 mL normal saline (Sigma *E.Coli* LPS, 0128:B12) and was infused intravenously over one hour. At fifty-five minutes following the initiation of endotoxin infusion, whole animal hemodynamics were determined. At the end of this hour, the contralateral muscle was collected at rest or at the end of three minutes of exercise as appropriate. While the control muscle collection always preceded the endotoxin muscle as mandated by experimental constraints, right and left muscles were alternated as experimental or control in consecutive dogs to eliminate this potential systematic error.

Fast Freezing

The freezing method used for collecting the muscles has been described previously.[5] Briefly, muscles were frozen *in situ* with a liquid N_2-cooled copper block (5 x 5 x 5 cm) driven by a pneumatic piston. Descent of the piston turned off the nerve stimulator (if active) ~10 msec before contact to eliminate muscle contraction and prevent the muscle from pulling away from the copper block. The block was applied to the flat, distal portion of the muscle at 0.1 kg/cm². The entire assembly was then immersed in liquid N_2, and the muscle was separated from the block while the assembly remained immersed. Only the section of the muscle immediately subjacent to the copper block was stored in liquid N_2 and kept for analysis. Mathematical modeling indicates that in muscles frozen in this way, changes in Mb saturation due to O_2 consumption or O_2 unbinding during freezing would alter saturation by 0.1% at VO_{2max}.[4,5,12]

Myoglobin Saturation Estimation

The preparation of the muscle for cryomicrospectrophotometry is as described in Voter and Gayeski.[7] The catchment volume of the measurement is estimated to have a radius of 60 μm.[7] Stored specimens were fractured into tissue blocks 0.8 x 1.5 x 0.3 cm and a freshly cleaved cross-section prepared. The tissue block was transferred to the microscope cold stage, oriented with the surface in cross section placed perpendicular to the optical axis of the microscope. The cold stage was maintained at -120°C. Myoglobin saturation remains stable for > 5 hours at this temperature, and measurement across a block requires approximately 0.5 hours.

Cells were selected at a 500 um depth from the fascial surface, every 500 um across the block. The constant depth attempts to control for physiologic or freezing effects of depth on Mb saturation measurement (none in control muscles to 800 um[7]), while the 500 um steps take into account heterogeneity of O_2 delivery on the scale of a terminal arteriole.[8] This protocol was followed for 8 blocks (~100 cells); increasing the number of blocks from which myoglobin saturations were collected did not change the distribution in several control and endotoxemic muscles. The measuring diaphragm (20 um x 20 um square) was placed within the boundaries of a cell, and a spectrum collected at 1.8 nm steps from 541 nm to 590 nm (52 wavelengths). Saturations were calculated from this spectra using Lubber's non-linear multi-component analysis method, which calculates what percentage of each of a desaturated and highly saturated reference spectrum are required to recreate the unknown spectrum using a least mean squares fit.[24]

Sampling and Analysis of Metabolites

After completing spectroscopy, we made a cut approximately 1 mm along the longitudinal axis from the initial transverse surface. This 1 mm piece from each block had the fascial layer split off and the first 500 um of muscle tissue from the frozen surface isolated. This piece was minced using a scalpel blade and stored in liquid N_2. The kinetics of creatine kinase and adenylate kinase are such that the rate of progression of the freezing front is sufficient to stop reactions to a depth of at least 1000 um at VO_{2max}.[2,5] Several such pieces of the same muscle were pooled, weighed, and extracted in perchloric acid. Tissue metabolites (ATP, phosphocreatine, free inorganic phosphate, glucose-6-phosphate, free creatine, pyruvate, and lactate) were analyzed on neutralized aliquots of the perchloric acid extracts. All compounds except for free creatine were assayed using enzymatic reaction coupled to the production or consumption of reduced pyridine nucleotides as previously described.[3] Creatine concentration was determined using the method of Eggleton, et al., 1943.[11]

Statistics

P-values reflect outcome of 2-tailed paired t-test with values at 1 hour endotoxin infusion compared to control (pre-endotoxin) values. Changes are considered significant when $P < 0.05$.

RESULTS

Systemic Results

Systemic results are reported for all animals studied to date (N = 9). Leukocyte count decreased dramatically from $6–10 \times 10^3$/uL blood to approximately 2×10^3/uL blood at the one hour time point (Figure 1A). After beginning endotoxin infusion mean arterial blood pressure (MABP) remained unchanged for approximately 1 hour (Figure 1B). After one hour, MABP decreased rapidly due primarily to a decrease in systemic vascular resistance (SVR = MABP-RAP)/CO)(Figure 2A). There was an increase in the rate of pH decrease after endotoxin infusion although this pH change was not due to an increase in lactic acid in the blood.

Resting Skeletal Muscle

In contrast to systemic vascular resistance, resting muscle vascular resistance at one hour after the initiation of endotoxin infusion increased (range 3–250%) (Figure 2B). Consequently, muscle blood flow decreased by 42% (0–79%). Resting VO_2 was unchanged, thus the extraction ratio increased (mean 189%). Lactate production by resting muscles was unchanged.

Exercising Skeletal Muscle

Preliminary results from exercising muscles are shown (Figure 3). Active tension development (ATD) of one muscle stimulated at 2 Hz and one muscle stimulated at 4 Hz are shown in Figure 3. At 2 Hz stimulation, 1) VO_2 was also unchanged (Fig 3A) in all but one of the muscles studied (that muscle decreased its VO_2 and its ATD), 2) gracilis muscle

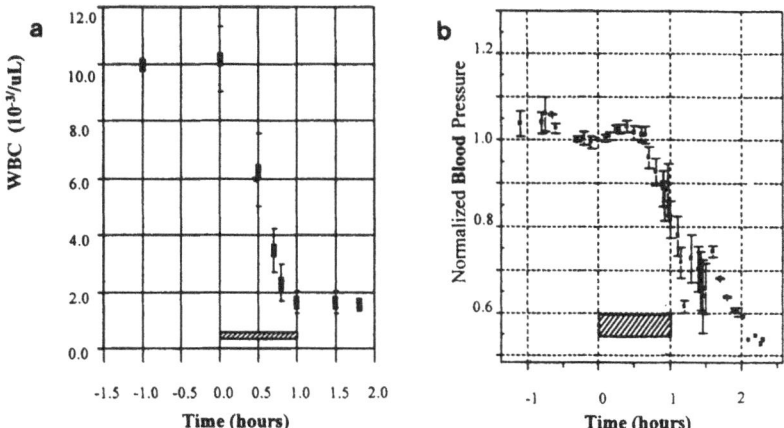

Figure 1. Time course of systemic parameters before and during endotoxin infusion (indicated by hatched bar). P-values reflect outcome of 2 tailed paired-t-test at the one hour collection time vs control. Data shown are mean +/- SE. **A)** White blood cell counts in thousands/microliter. WBC count is between 4.0 and 10 x 10^3 μL for control period, and below 4.0 x 10^3 μL by 1 hour for all animals in which hematology was studied. (P < 0.05, N = 5). **B)** Mean Arterial Blood Pressure normalized to control value. (P < 0.05, N=8)

peak ATD at 3 minutes was unchanged (Fig 3B), 3) muscles maintained their ability to autoregulate blood flow during exercise. However, at 4 Hz stimulation both ATD and VO$_2$ were reduced compared to controls (Fig 3A and 3B).

Myoglobin Saturation and Muscle Biochemistry

From preliminary data, Fig. 3C shows myoglobin saturation and Fig. 3D phosphocreatine/total creatine ratio for one resting, one 2 Hz exercise, and one 4 Hz exercising

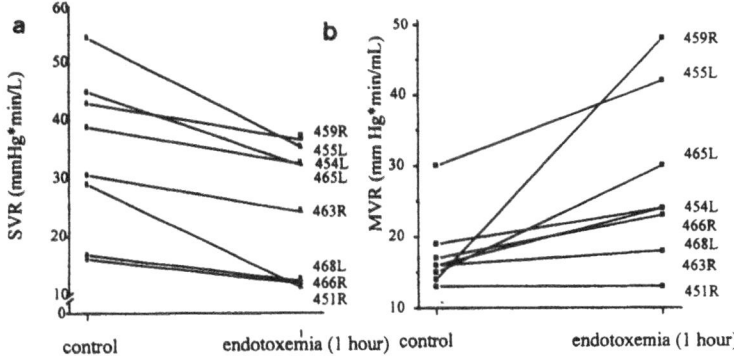

Figure 2. Labels indicate animal identification number. P-values reflect 2 tailed paired t-tests between measurements at control and 1 hour of endotoxemia. (N = 8). **A)** Systemic Vascular Resistance calculated by [MAP / CO]. The decrease in SVR with endotoxin infusion is consistent with a systemic inflammatory response. (P < 0.05). **B)** Muscle vascular resistance at rest calculated by [MAP / muscle blood flow]. In contrast to whole animal vascular resistance (SVR). MVR increases (P < 0.05), suggesting that skeletal muscle is still responsive to sympathetic stimulation and is acting to maintain systemic blood pressure at this point. Thus the decrease in SVR is not due to skeletal muscle vasculature.

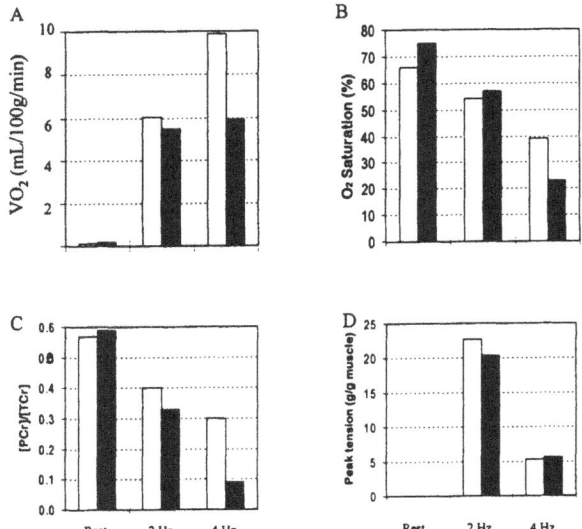

Figure 3. Physiologic variables for gracilis muscle at rest, 2 Hz exercise, and 4 Hz exercise. Gracilis nerve stimulated at indicated frequency maximally for 3 minutes. Open bars are control muscles, filled bars are endotoxin muscles. **A)** Exercising gracilis muscle oxygen consumption calculated by Fick equation. The muscle at 2 Hz might have a slight decrease in oxygen consumption , which would suggest that cellular energy reserves are challenged. This 4 Hz exercising muscle decreased its VO$_2$ after endotoxemia. **B)** For rest and 2 Hz, there is no change in median myoglobin saturation between control and endotoxemia. At 4 Hz, there is a decrease, suggesting a decrease in oxygen tension reserves which may be related to the decrease in VO$_2$. **C)** There is no change in [phosphocreatine]/[total creatine] between control and endotoxemia for rest and 2 Hz. Phosphocreatine represents biochemical reserves of high-energy phosphate. Normal [PCr]/[TCr] suggests that circulatory reserves are adequate to meet skeletal muscle energy requirements. At 4 Hz, [PCr]/[TCr] is decreased. **D)** There is no change in active tension development in these 2 Hz and 4 Hz muscles.

experiment (control and endotoxin). There is no difference in myoglobin saturation at rest or 2 Hz, however, there may be a decrease in myoglobin saturation in endotoxemia compared to control at 4 Hz. Energy charge is unchanged at rest, but may be decreased at 2 Hz and 4 Hz.

DISCUSSION

Model of SIRS

There is controversy regarding what constitutes an appropriate model of sepsis. Two comprehensive reviews, Wichterman, et al.[21] and Fink and Heard,[22] consider the merits of various animal models. The first major point of contention is whether or not endotoxemia is an appropriate model of sepsis. Wichterman cited reports in which bolus injections of LPS led to low CO and elevated SVR, rather than the preferred hyperdynamic model, but conceded, "...endotoxicosis in animals may be a reasonable paradigm for sepsis in humans, with the proviso that the animal model chosen must adequately replicate those features of the clinical syndrome that are the focus of the experiment." Fink and Heard concur that it is the clinical findings which are most important in evaluating the appropriateness of a model. They propose a working definition of the sepsis syndrome to apply to human and laboratory studies: "a constellation of clinical and laboratory findings indicative of a generalized inflammatory response that is otherwise unexplained and is accompanied by acute organ system dysfunction and is often, but not invariably, associated with the presence of a serious bacterial, fungal, or viral infection."

The animal model, the anesthetic technique, the LPS dose given, the mode of endotoxin infusion, the intravascular volume status of the animal, the time of observation, etc. could impact on whether or not a model fulfills criteria for SIRS. Vallet, et al.[19] and Fink and Heard[22] demonstrate hyperdynamic endotoxic states for endotoxemia. The advantages of the endotoxic model include that a known amount of endotoxin is given, so the model can be titrated to effect and be reproduced, and the experiment is acute and controlled.

We report a reproducible low-dose endotoxin model that induced SIRS as defined by the American College of Chest Surgeons.[16] That definition requires that 2 of the following criteria must be present: 1) temperature greater than $38^{\circ}C$ or less than $36^{\circ}C$; 2) heart rate greater than 90 beats per minute; 3) tachypnea (>20 respirations per minute or PCO_2 < 32 mm Hg); 4) white blood cell count greater than $12.0 \times 10^9/L$ or less than $4.0 \times 10^9/L$; or 5) more than 10% band forms (immature neutrophils). We maintain temperature by warming with external heat sources and anesthesia frequently blunts the hyperthermic response of infection. Tachycardia is present from the onset due to barbiturate anesthesia. Hence, criteria 1 and 2 cannot be used in this setting. Because the animals were anesthetized, respiratory responses were altered. However, it was necessary to increase ventilation to reduce PCO_2 to maintain pH. This pH change would have mandated a tachypnea in the unanesthetized animal as well. The white cell count fell well below $4.0 \times 10^9/L$ after endotoxin infusion.

According to Bone,[9] septic shock is defined as SIRS plus a documented infection (our model does not include infection), plus hypotension (decreased systolic systemic arterial pressure of more than 40 mm Hg) or signs of inadequate organ perfusion such as lactic acidosis. The blood pressure in our model decreases more precipitously in all animals after one hour. We observe an increase in the rate of hydrogen ion concentration shortly after beginning endotoxin infusion, although this pH drift is not due to an increase in blood lactate. Thus our animal model may mimic septic shock after one hour despite the absence of bacteremia.

The decrease in blood pressure following endotoxin infusion is due to a decrease in systemic vascular resistance with no compensatory increase in cardiac output. It has been demonstrated that inducible nitric oxide synthase (iNOS) is increased in response to endotoxin and TNF-α.[10,13–15] Nitric oxide activates guanylate cyclase, causing smooth muscle relaxation. The induction of iNOS would appear to be an unlikely candidate because physiologic effects are seen within 0.5 hr from the initiation of endotoxin infusion and the induction of iNOS requires at least 1 hr in some models.[13] Hence, the effect precedes the presence of the proposed mechanism.

"Inappropriate" nitric oxide production could also lead to decreased regulation of blood flow distribution between and within organs. This increased flow heterogeneity could lead to organ dysfunction due to either inadequate blood flow to an organ or maldistribution of flow within the organ despite apparent adequate whole organ flow. Our results do not suggest that either of these occur in skeletal muscle in early endotoxemia.

Venodilatation may lead to decreased ventricular preload and subsequent inadequate increase in cardiac output. There was not a decrease in cardiac filling pressure in our experiments as we maintained aggressive fluid infusion. We would not detect changes in ventricular compliance given the complexity of our protocol. In sepsis others have shown a decreased response to exogenously administered inotropes and vasopressors.[6] Perhaps decreased cardiac contractility is the cause of our decreased "compensatory" cardiac output.

Extremely high concentrations of endotoxin[25] are required to reduce mitochondrial function in vitro, thus our endotoxin concentrations should not have affected mitochondrial function. Since there is no evidence of altered mitochondrial function or cell biochemistry at rest or during 2 Hz stimulation, decreased mitochondrial function is unlikely to be the cause of decreased VO_2 and high energy phosphates in the 4 Hz muscle.

The decrease in white blood cell count has been observed in other models.[16,17,20] However, the mechanism for this change is unknown. Presumably leukocytes marginate causing inflammation. There is an association between adult respiratory distress syndrome and SIRS. However, pulmonary function as measured through the A-a gradient is not adversely affected in this early phase although pulmonary compliance subjectively decreases. Dog intestine has been shown to be severely affected in the setting of hemorrhagic shock,[23] suggesting leukocytes may marginate here. Gross observation of intestine does not reveal apparent edema or hyperemia.

Skeletal Muscle as an Organ Model

A priori, it seems likely that the etiology of dysfunction in MODS would be common amongst organs. Skeletal muscle is an organ with a wide range of VO2 as well as extensive circulatory and biochemical energy reserves. We propose that these reserves may be depleted during SIRS as compared to control states. Our preliminary experiments suggest that the decrement in circulatory reserves during early SIRS leads to apparent organ dysfunction only at very high work levels. In end-stage organ failure, all physiologic reserves are severely depleted so that an organ cannot continue minimal function. Discerning the order in which circulatory and biochemical reserves are depleted by studying skeletal muscle at different VO2 may yield knowledge of the mechanism of dysfunction of organs in MODS. In addition, it may also lead to an approach to therapy by identifying if increasing organ blood flow or MABP alters dysfunction. Thus the wide range of oxygen consumption of skeletal muscle permits the study of the relationship between organ function, physiologic and biochemical reserves during SIRS.

The data confirm that organ response cannot be inferred from whole animal data (e.g. SVR vs. MVR). Thus, combined studies of SIRS at the organ and cellular level are essential. Skeletal muscle dysfunction in this early stage of SIRS, cannot be detected at rest or at stimulation rates up to 2 Hz (35% of VO_2max). Our resting muscle results confirm those of Samsel, et al.[18] At 4 Hz there is a decrease in active tension development, oxygen consumption, high energy phosphates and tissue PO_2. We interpret these preliminary data as indicating that microcirculatory flow heterogeneity is the likely cause of organ dysfunction in this low dose, early state of endotoxemia.

SUMMARY

We have developed a reproducible low-dose endotoxin model which is useful for the investigation of early SIRS. The data confirm that organ function cannot be inferred from whole animal data (e.g. SVR vs. MVR). Thus, the study of SIRS at the organ and cellular level is essential. Decreased skeletal muscle oxygen consumption with 4 Hz exercise in early SIRS may be related to depletion of physiologic reserves, especially microcirculatory reserves, as suggested by decreased myoglobin saturation and decreased energy charge. Using this model, we will investigate whether organ dysfunction in SIRS is due to oxygen-limited cellular ATP production or impaired cellular metabolism.

REFERENCES

1. R.S. Cotran, V. Kumar, and S.L. Robbins, Robbins Pathologic Basis of Disease, W.B. Saunders Company, Philadelphia, (1989).

2. J. Olgin, R.J. Connett, and B. Chance, Mitochondrial redox changes during rest-work transition in dog gracilis muscle, *Advances In Experimental Medicine & Biology* 200:545–54 (1986).

3. R.J. Connett, T.E. Gayeski, and C.R. Honig, Lactate accumulation in fully aerobic, working, dog gracilis muscle, *American Journal of Physiology* 246:H120-H128 (1984).

4. T.E. Gayeski and C.R. Honig, O2 gradients from sarcolemma to cell interior in red muscle at maximal VO2, *American Journal of Physiology* 251:H789-H799 (1986).

5. T.E. Gayeski, R.J. Connett, and C.R. Honig, Minimum intracellular PO2 for maximum cytochrome turnover in red muscle in situ, *American Journal of Physiology* 252:H906–15 (1987).

6. W. Karzai, J.M. Reilly, W.D. Hoffman, R.E. Cunnion, R.L. Danner, S.M. Banks, J.E. Parrillo, and C. Natanson, Hemodynamic effects of dopamine, norepinephrine, and fluids in a dog model of sepsis, *American Journal of Physiology* 268:H692-H702 (1995).

7. W.A. Voter and T.E.J. Gayeski, Determination of Myoglobin Saturation of Frozen Specimens Using a Reflecting Cryospectrophotometer, *American Journal of Physiology* 269:H1328-H1341 (1995).

8. B.R. Berg and I.H. Sarelius, Functional capillary organization in striated muscle, *American Journal of Physiology* 268:H1215-H1222 (1995).

9. R.C. Bone, The pathogenesis of sepsis. [Review], *Annals of Internal Medicine* 115:457–469 (1991).

10. R.M. Palmer, The discovery of nitric oxide in the vessel wall. A unifying concept in the pathogenesis of sepsis. [Review], *Archives of Surgery* 128:396–401 (1993).

11. P. Eggleton, S.R. Elsden, and N. Gough, The estimation of creatine and diacetyl. *Biochem. J.* 37:526–529 (1943).

12. A. Clark, Jr. and P.A.A. Clark, Capture of spatially homogeneous chemical reactions in tissue by freezing, *Biophys J* 42:25–30 (1983).

13. N.K. Boughton-Smith, S.M. Evans, F. Laszlo, B.J. Whittle, and S. Moncada, The induction of nitric oxide synthase and intestinal vascular permeability by endotoxin in the rat, *British Journal of Pharmacology* 110:1189–1195 (1993).

14. C. Thiemermann, C.C. Wu, C. Szabo, M. Perretti, and J.R. Vane, Role of tumour necrosis factor in the induction of nitric oxide synthase in a rat model of endotoxin shock, *British Journal of Pharmacology* 110:177–182 (1993).

15. J.A. Lorente, L. Landin, E. Renes, R. De Pablo, P. Jorge, E. Rodena, and D. Liste, Role of nitric oxide in the hemodynamic changes of sepsis, *Critical Care Medicine* 21:759–767 (1993).

16. A.L. Beal and F.B. Cerra, Multiple organ failure syndrome in the 1990s. Systemic inflammatory response and organ dysfunction. [Review], *JAMA* 271:226–233 (1994).

17. M.S. Rangel-Frausto, D. Pittet, M. Costigan, T. Hwang, C.S. Davis, and R.P. Wenzel, The natural history of the systemic inflammatory response syndrome (SIRS). A prospective study [see comments], *JAMA* 273:117–123 (1995).

18. R.W. Samsel, D.P. Nelson, W.M. Sanders, L.D. Wood, and P.T. Schumacker, Effect of endotoxin on systemic and skeletal muscle O2 extraction, *Journal of Applied Physiology* 65:1377–1382 (1988).

19. B. Vallet, N. Lund, S.E. Curtis, D. Kelly, and S.M. Cain, Gut and muscle tissue PO2 in endotoxemic dogs during shock and resuscitation, *Journal of Applied Physiology* 76:793–800 (1994).

20. C. Lam, K. Tyml, C. Martin, and W. Sibbald, Microvascular perfusion is impaired in a rat model of normotensive sepsis, *Journal of Clinical Investigation* 94:2077–2083 (1994).

21. K.A. Wichterman, A.E. Baue, and I.H. Chaudry, Sepsis and septic shock—a review of laboratory models and a proposal. [Review], *Journal of Surgical Research* 29:189–201 (1980).

22. M.P. Fink and S.O. Heard, Laboratory models of sepsis and septic shock. [Review], *Journal of Surgical Research* 49:186–196 (1990).

23. A. Marston, The bowel in shock; the role of mesenteric arterial disease as a cause of death in the elderly, *Lancet* 2:365–369 (1962).

24. U. Heinrich, J. Hoffmann, and D.W. Lubbers, Quantitative evaluation of optical reflection spectra of blood-free perfused guinea pig brain using a nonlinear multicomponent analysis, *Pflugers Archiv - European Journal of Physiology* 409:152–7 (1987).

25. L. Mela, Direct and indirect effects of endotoxin on mitochondrial function, *Progress in Clinical & Biological Research* 62:15–21 (1981).

48

INCREASING OXYGEN CONSUMPTION AND SURVIVAL WITH FLUID RESUSCITATION THERAPY FOR HEMORRHAGIC SHOCK IN RATS

Matthew C. Graham and John L. Gainer

Department of Chemical Engineering
University of Virginia
Charlottesville, Virginia 22903-2442

1. BACKGROUND

Recovery from hemorrhagic shock has been reported to depend on the restoration of oxygen delivery to the tissues [Clowes, 1971; Crowell and Smith, 1964]. Oxygen consumption decreases during hemorrhage, due to reduced blood flow rates, and it has been noted that this decrease correlates with mortality [Wilson et al., 1972].

Factors usually considered to be of major importance in controlling oxygen delivery to tissues are blood flow rates, perfused capillary surface area, and blood oxygen concentration. It is frequently assumed that diffusion limitations are not important; however, it has been suggested that diffusion could be an important factor when a tissue is consuming oxygen at its maximum rate [Hogan et al., 1989, 1990a, 1990b, 1991a, 1991b; Gainer, 1994]. If this is also true for other "stressed" conditions, such as hypovolemia, then increasing oxygen diffusion might be beneficial for treating hemorrhagic shock.

It has been found that the carotenoid compound crocetin increases oxygen diffusivity through blood plasma (which has been suggested to offer the major resistance to oxygen transport [Holland et al., 1988; Huxley and Kutchai, 1981, 1983; Yamaguchi et al., 1985]). Crocetin has also been used to increase oxygen consumption and survival during moderate hemorrhagic shock (40% blood volume removed) in rats [Gainer et al., 1993]. Thus, it was decided to test an infusion fluid containing crocetin for the treatment of a severe shock condition (55-60% removal of blood volume) in rats. It has been noted [Tominaga and Waxman, 1992] that the major goal of fluid resuscitation should be the optimization of oxygen delivery and oxygen consumption. Thus, the combination of crocetin with fluid infusion might both replace lost fluid and increase oxygen consumption at the same time, resulting in an improved therapy for hemorrhagic shock.

Oxygen Transport to Tissue XIX, edited by Harrison and Delpy
Plenum Press, New York, 1997

2. EXPERIMENTAL METHODS

Rats were hemorrhaged at a rate of 1.5–2 ml/min. In one experiment, 55% of the estimated blood volume was removed and in the other one, 60% was removed. Both protocols utilized male Sprague-Dawley rats, which were fed ad libitum until the day of the experiment. An effort was made to ensure that the average weights of each group of rats were the same since it has been suggested that weight is an important consideration when comparing results from hemorrhagic shock experiments [Gainer, et al., 1995]. Fifteen animals were used in each experiment, divided into three equal groups as following: (A) one group was hemorrhaged but not treated; (B) another group was hemorrhaged and infused with isotonic saline at a volume of 3X the shed blood volume; and (C) a third group of was hemorrhaged and saline containing crocetin.

The rats were anesthetized with sodium pentobarbital, 50 mg/kg, i.p., and the right carotid artery exposed and cannulated with PE-50 polyethylene tubing. The cannula was then passed subdermally to the back of the neck and withdrawn through the skin. The incision was closed and 2% lidocaine applied to the wound. The rat was then placed in the oxygen consumption chamber.

Whole-body oxygen consumption rates were measured using an apparatus which is described in detail elsewhere [Arundel et al., 1984]. After the rat was placed in the chamber, the cannula was passed through the end by use of an air-tight septum and connected to a syringe pump.

Once the animal was awake, oxygen consumption rates were monitored for 15 to 30 minutes in order to establish the normal, non-hemorrhaged value. The animals were then hemorrhaged, with the estimated blood volume taken to be 60 ml of blood per kg of body weight [Altman and Dittmer, 1961; Bitterman et al., 1991; Wu et al., 1988].

For those animals which received fluid, the infusion began three minutes after the hemorrhage ended. Saline, or saline supplemented with crocetin, was infused at the rate of 1 ml/min using a syringe pump. The pH of all solutions infused was adjusted to 7.5, and the concentration of the crocetin was 0.04 units/ml. The number of crocetin units has been arbitrarily chosen to be the absorbance of a solution at 446 nm, and 1 unit corresponds to approximately 30 nmoles (or 11 mg). Thus, the crocetin dosage was around 40 mg/kg of body weight.

Oxygen consumption was continuously measured during the hemorrhage and infusion periods as well as for the next hour. If the rats survived that long, they were then removed from the chamber and placed in a cage with food and water, and observed for the next three hours. If still alive at this point (i. e., 4 to 5 hours after hemorrhage ended), they were considered to have survived.

3. RESULTS AND DISCUSSION

3.1. Survival

The results obtained when a 55% hemorrhage was used are shown in Table 1. All of the rats receiving crocetin in the infusion fluid survived, while only 40% of the saline-infused or untreated control rats survived. The condition of the rats at the arbitrarily-set survival time should also be noted. All rats receiving crocetin-supplemented saline were moving around and seemed to be alert. They responded to noise stimuli by jumping, moving, etc. In contrast, both of the saline-infused rats which survived were very listless and

Table 1. Survival after 55% hemorrhage

Group	Weight (grams)	Survivors	Condition of Survivors
Controls	324 ± 13	2/5	One was moving and responding to noise. The other one was listless and did not respond to noise.
Saline-Infused	323 ± 12	2/5	Both were listless and did not respond to noise.
Crocetin-Saline Infused	326 ± 12	5/5	All were moving and responded to noise

seemed to be "barely alive". They did not respond to noise. There were also two rats surviving in the untreated control group, and one of these was moving and alert while the other one was listless and did not respond to noise.

Thus, the addition of crocetin to a saline resuscitation fluid appears to enhance survival. Not only did more animals survive a 55% hemorrhage when treated with the crocetin-saline mixture, but they were also in much better shape as judged by their movements and alertness. In order to see if crocetin would be beneficial in more severe hemorrhage conditions, this same experiment was repeated except that 60% of the estimated blood volume was now removed. The survival results are shown in Table 2.

As might be expected, none of the non-infused controls survived a 60% hemorrhage. One of the five rats survived when infused with saline; however, all five of the rats treated with the crocetin-saline infusion survived. Again, the conditions of the survivors were noted. The ones receiving crocetin were not as lively as the crocetin-treated survivors which had been hemorrhaged 55% of their blood volumes. They were more listless, although some responded to noise. However, the lone survivor from the saline-infused group appeared to be "barely" alive, and seemed to be paralyzed.

3.2. Oxygen Consumption

It is interesting to compare the oxygen consumption values for the three groups that were hemorrhaged 60% of their blood volumes. Figure 1a shows the oxygen consumption values for the non-treated controls, none of which survived. It is obvious that their oxygen consumption rates declined continuously after they were hemorrhaged. The oxygen consumption for most (4/5) of the rats infused with saline alone (Figure 1b) also followed a pattern similar to that of the non-infused animals. One rat in that group survived, and its oxygen consumption increased as shown. All of the animals treated with crocetin (Figure 1c) showed the same initial decrease in oxygen consumption upon hemorrhage — but all of their rates then began to increase and approach the normal value. As noted before, all of these animals survived. Comparing all three groups, the data show that increasing oxygen consumption correlates with an increased rate of survival after hemorrhage.

Table 2. Survival after 60% hemorrhage

Group	Weight (grams)	Survivors	Condition of Survivors
Controls	351 ± 13	0/5	
Saline-Infused	347 ± 16	1/5	Barely alive, animal appeared to be paralyzed
Crocetin-Saline Infused	353 ± 9	5/5	All were listless and some responded to noise

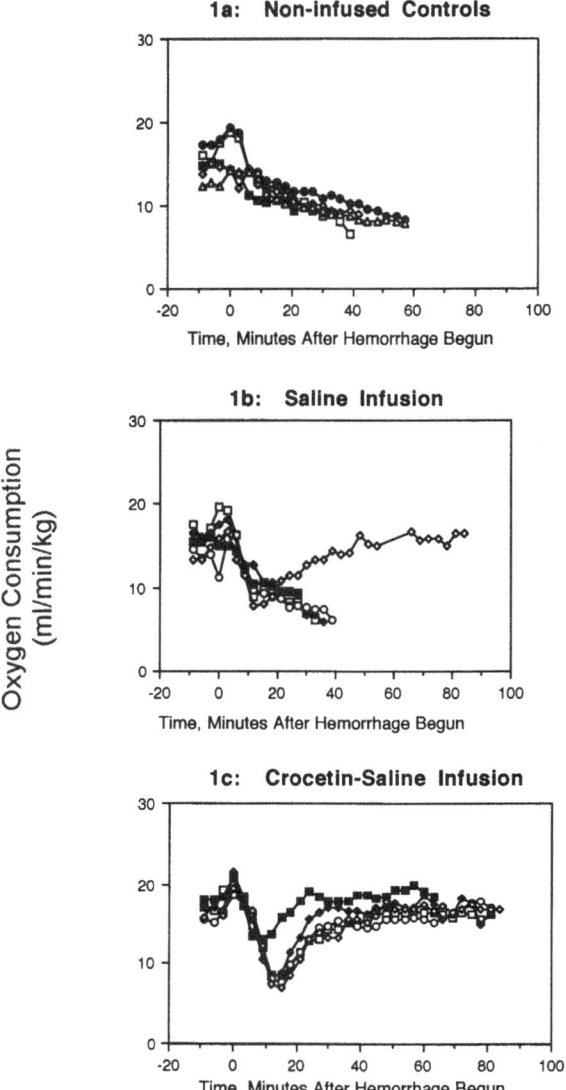

Figure 1. Oxygen consumption with 60% hemorrhage.

The oxygen consumption measurements corroborate the survival observations. When a severe hemorrhage is not treated, oxygen consumption declines continuously. It also decreases at first when rats are infused, whether or not crocetin is present. Oxygen consumption continued to decline in the saline-infused rats; however, when crocetin was used, the rate soon began to increase and returned to a normal, non-hemorrhaged value.

These results strongly suggest that crocetin increases the chance for survival after hemorrhagic shock. It is thought that the mechanism for this action is the ability of crocetin to increase the diffusivity of oxygen through blood plasma. Numerous experiments have been done in the past which show that crocetin has no effect on other factors frequently associated with changes in oxygen consumption such as blood flow, heart rate, ventilation rate, tissue perfusion, oxygen solubility, oxyhemoglobin saturation curve, or oxidative phosphorylation at the mitochondrial level.

Thus, it may be beneficial to add crocetin to resuscitation fluids for the treatment of hemorrhagic shock. Further experiments would be needed, though, in order to determine the optimum amount to be used, and if this compound can also be used with other commonly-used resuscitation fluids.

ACKNOWLEDGMENT

This research was sponsored by the Office of Naval Research, Arlington, Virginia, USA.

REFERENCES

Altman, P. L. and Dittmer, D. S., Federation of American Societies for Experimental Biology Handbooks, Washington, D. C., 1961, p. 5.

Arundel, P. A., Holloway, B. R., Mellor, P. M., J. Appl. Physiol., 57: 1591–1593 (1984).

Bitterman, H., Reissman, P., Bitterman, N., Melamed, Y., Cohen, L: Circ. Shock, 33: 183–191 (1991).

Clowes G. H., In "Septic Shock in Man." edited by S. Hershey, Boston: Little, Brown and Co., 1971, pps 85–106.

Crowell, J. W., Smith, E. E., Amer. J. Physiol., 206:313–316 (1964).

Gainer, J. L., J. Appl. Physiol., 76: 1826–1829 (1994).

Gainer, J. L., Rudolph, D. B. and Caraway, D. L., Circulatory Shock, 41: 1–7 (1993).

Gainer, J. L., Lipa, M. J. and Ficenec, M. C., Laboratory Animal Science, 45: 169–172 (1995).

Hogan, M. C., Roca, J., West, J. B. and Wagner, P. D. , J. Appl. Physiol., 66: 1219–1226 (1989).

Hogan, M. C., Bebout, D. E., Gray, A. T., Wagner, P. D., West, J. B., Haab, P. E., J. Appl. Physiol., 69: 830–836 (1990a).

Hogan, M. C., Bebout, D. E., Wagner, P. C., West, J. B., J. Appl. Physiol., 69: 570-576 (1990b).

Hogan, M. C. Bebout, D. E., Wagner, P. D., J. Appl. Physiol.,70: 1105–1112 (1991a).

Hogan, M. C., Bebout, D. E., Wagner, P. D., J. Appl. Physiol., 70: 2656–2662 (1991b).

Holland, R. A. B., Respir. Physiol. 59: 71–91 (1988).

Huxley, V. H. and Kutchai, H., J. Physiol., 316: 75–83 (1981).

Huxley, V. H. and Kutchai, H., Microvasc. Res., 26: 89–107 (1983).

Tominaga, G. T. and Waxman, K., In "Trauma 2000, Strategies for the New Millenium", edited by Gamelli, R. L. and Dries, D. J., R. G. Landes Co., Austin, Texas, 1992, pps. 20–29.

Wilson, R. F., Christensen, C., Leblanc, L. P., Ann. Surg., 176: 801–804 (1972).

Wu ,C. H., Bogusky, R. T., Holcroft, M. D., Kramer, G. C., J. Trauma, 28:757–764 (1988).

Yamaguchi, K., Nguyen-Phu, P., Scheid, P. and Piiper, J., J. Appl. Physiol., 58: 1215–1224 (1985).

49

INCREASED OXYGEN DISSOCIATION BY NITRIC OXIDE FROM RBC

Hiroaki Kosaka and Akitoshi Seiyama

1st Department of Physiology
Medical School, Osaka University
2-2 Yamadaoka, Suita, Osaka 565, Japan

INTRODUCTION

Endogenous carbon monoxide (CO), a product of heme oxygenase, is thought to share several biological actions with NO (Zhuo et al., 1993). Hemoglobin (Hb) tetramer in circulating erythrocytes traps the molecules so generated. CO increases the relative affinity of other heme sites for oxygen, thus shifting the remaining oxygen dissociation curve to the left, i.e., CO decreases oxygen (O_2) supply to tissues (Douglas et al. 1912). Exceedingly high concentration of NO also shifted the O_2 dissociation curve to the left (Kon et al., 1977). This may be because of the extremely high affinity of NO for Hb than CO. However, endogenous NO generated by stimuli (Green et al., 1981, Stuehr and Marletta, 1985, Hibbs et al., 1987) will not increase tissue hypoxia, because NO generated in vivo by cytokines or by nitrovasodilators (Kurz et al., 1993) is bound to the a subunit of erythrocyte Hb tetramer (Hb α-NO), composed of a and b subunits, and the Hb tetramer binding NO yields only less than 4% of total Hb tetramer in the circulating blood (Kosaka et al., 1994). Furthermore the present study shows that the small amount of NO decreases O_2 affinity of erythrocytes under low O_2 pressure (pO_2) and the effect is also observed in vivo.

MATERIALS AND METHODS

Rat care was in accordance with guidelines of the Animal Care Committee of Osaka University Medical School. After anesthesia with sodium pentobarbital (40 mg/kg), rats were administered nitroglycerin (20 mg/kg, i.p.). After 20 min, the heparinized arterial blood was deoxygenated under 95% N2 +5% CO_2 and aliquots of blood were examined until deoxygenation was complete. The pO_2 was analyzed by oxygen electrode. The oxygen saturation of the whole blood in a microslide (0.05 mm internal thickness) was determined by microspectroscopy (Seiyama et al., 1994). ESR spectra of the blood were recorded with a Bruker ESP 300E spectrometer at 110 K (incident microwave power, 5 mW; modulation frequency, 100 kHz; modulation amplitude, 5 G; time constant, 328 ms; sweep rate, 3 G s^{-1}). The concentrations of HbNO were determined by double integration of the ESR spectra, using HbNO as a standard, which was prepared with NO gas (via HbCO) anaerobically (Shiga et al., 1969).

Preparation of Animals

Each rat (Sprague-Dawley, male, 100–150 g body wt) was anesthetized with sodium pentobarbital (40 mg/kg body wt ip). An endotracheal tube was inserted into the trachea, and the rat was ventilated with a respirator. After transverse upper abdominal laparotomy, the liver was placed on a specially designed platform (made of transparent thin glass) on the stage of a microscope. A binocular microscope with an immersion objective lens was used. The liver was covered with a polychlorovinyliden film to prevent desiccation and possible gas exchange between the liver and atmosphere.

Microscopic Observation of Sinusoid in Liver

The apparatus was the same as that reported previously (Seiyama et al, 1994). A computerized microscope apparatus was used with: (i) charge-coupled device video camera, monitor, videotape recorder and image analyzer for measuring diameter (D) and length (L) between two spots on single sinusoids; (ii) two photomultipliers for measuring erythrocyte velocity; and (iii) a scanning spectrophotometer for measuring spectra of circulating blood. A single sinusoid and its neighboring hepatocytes at the edge of the liver were chosen for the spectral recordings by the following dual-spot microspectroscopy.

Dual-Spot Microspectroscopy

Transmitted lights from two spots (10 mm in diameter), upstream and downstream of single sinusoids, were simultaneously guided to a spectrophotometer, controlled with a computer and equipped with two photon counters after grating, through two separate quartz light guides inserted into two tiny deep holes of cylindrical block attached to the microscope instead of an eyepiece. To measure erythrocytes velocity according to the dual-spot cross-correlation method (Schmid et al., 1975), two photomultipliers were simultaneously connected to tiny deep holes of another cylindrical block through two separate quartz light guides.

Calculation of Rate of O_2 Release from Single Microvessels

Assuming a well-known concentric cylinder model of unit length composed of a single unbranched sinusoid and its surrounding tissue, rate of O_2 release per unit area of sinusoidal wall (RO_2) was calculated to be

$$RO_2 = \Delta SO_2 \cdot [Hb] \cdot Q/\pi DL \text{ or } RO_2 = D\Sigma O_2 \cdot [Hb] \cdot D \cdot v/(4L)$$

where ΔSO_2 is the difference in O_2 saturation of hemoglobin (SO_2, $0 \leq SO_2 \leq 1$) between the inflowing (upstream) and outflowing (downstream) erythrocytes in a single unbranched sinusoid, L is the length between upstream and downstream, D is the averaged diameter of the sinusoid between upstream and downstream, [Hb] is the averaged intrasinusoidal hemoglobin concentration obtained from five isosbestic points, and Q is the sinusoidal blood flow rate ($Q = (\pi D^2/4) \cdot v$). [Oxyhemoglobin (HbO_2)] inflow was calculated to be [HbO_2] inflow = $Q \cdot [Hb] \cdot SO_2$ (at upstream spot). The concentration of physically dissolved O_2 was neglected throughout, because of the low solubility of O_2 in plasma.

Nitrovasodilator Administration

For control studies, rats were given infusion of physiological saline containing heparin for 70 min with a syringe pump at a rate of 0.6 ml/h through a femoral vein. Before nitrovasodilator administration, rats were given infusion of the physiological saline for 20 min at the same rate. Then nitrovasodilator solution diluted with the physiological saline was infused for 50 min at the same rate.

RESULTS AND DISCUSSION

We found that NO is bound to the α subunits alone and the ESR spectra of α-NO heme species had a distinct three-line hyperfine structure in venous blood but not in arterial blood in rats treated with stimuli (Kosaka et al., 1994). ESR spectroscopy detects specifically HbNO alone and distinguish 5-coordinated Hb α-NO, where heme iron-proximal histidine bond in the Hb α-NO subunit is broken or stretched, from 6-coordinated Hb α-NO because the breaking or stretching of the iron-proximal histidine bond in the Hb α-NO induces a three-line hyperfine structure (Szabo and Perutz, 1976, Nagai et al., 1978, Hille et al., 1979).

The partially NO-bound Hb α in erythrocytes may enhance oxygen dissociation in other heme sites at tissues, because the distinct three-line hyperfine structure of HbNO appeared after circulation of organs despite that O_2 saturation of Hb in the venous blood is ~75%. Furthermore, the oxygen affinity was lowered in 5-coordinated α, but not β, heme model (Fujii et al., 1993). The heme iron-proximal histidine bonds in the α subunits play a role as the conduit of ligand-induced tertiary structural change between oxygenated and deoxygenated form (Perutz, 1970). Although Hb tetramer binding NO generated in vivo by cytokines or by nitrovasodilators yields only less than 4% of total Hb tetramer in the circulating blood, we assumed the possibility that these species of low O_2 affinity enhance facilitated diffusion of O_2 within erythrocytes.

To prove that, blood containing Hb α-NO alone should be prepared. We obtained the blood from rats treated with reagents to produce HbNO, because in vitro preparation of Hb α-NO accompanies simultaneous formation of Hb β-NO. Several hours after treatment with lipopolysaccharide (7.5 mg/kg) induced 26 mM of Hb α-NO and the oxygen dissociation curve was distinctly right-shifted. Furthermore, to exclude the effect on the blood pH caused by lipopolysaccharide, rats were infused intravenously with nitroglycerin (6 mg/kg/h). The oxygen dissociation curves of the blood containing 7 mM of Hb α-NO were also right-shifted. When nitroglycerin of 20 mg/kg was administered to rats intraperitoneally, Hb α-NO was produced up to 22 μM. Because Hb binds NO molecule in the α-subunit alone in vivo, there are 11~22 mM Hb tetramer binding NO in the blood containing ~2300 μM Hb tetramer. The oxygen saturation of the blood was the same as control at high pO_2, however, significantly lower with decrease in pO_2. The apparent Hill coefficient, n, increased from 2.4 to 3.6 (Table 1). The pO_2 at 50%-saturation of Hb of blood from a nitroglycerin-treated group (46 ± 1.0 (mean ± SD) mmHg) was significantly

Table 1. Effect of HbNO on pO_2 at 50% of oxygen saturation

HbNO (mM)	0	7	9	14	22
P50 (mmHg)	38 ± 0.8	44	45	46	48
n	2.6 ± 0.05	3.6	3.6	3.7	3.7

Effect of Hb α-NO concentration on oxygen dissociation curves. The P_{50} denotes $pO2$ at 50% of oxygen saturation of Hb in whole blood. These were obtained from the Hill plots, log $Y/(1-Y)$ versus log $pO2$, where Y is the fraction of oxygenated species.

higher than that from a control group (38 ± 0.4 mmHg) by Student's t-test (p < 0.001). Furthermore, the pO_2 at 50% of oxygen saturation of Hb in whole blood increased dependent on the HbNO concentration detected.

Taken together, these results show that the presence of Hb α-NO enhances the O_2 dissociation from erythrocytes, suggesting that the heme iron-proximal histidine bond in the NO-bound a subunit breaks or stretches during peripheral circulation; this change causes other heme sites to dissociate more oxygen, but oxygen binding capacity is restored in the lung by returning to the 6-coordinated form.

We then wanted to know whether it occurs in vivo. To prove that, because O_2 transport from flowing erythrocytes to tissues occurs in microvessels, especially in capillaries, quantitative measurement of O_2 release from single unbranched microvessels is preferable. We measured erythrocyte velocity, hemoglobin concentration, and difference of O_2 saturation of hemoglobin between upstream and downstream in single unbranched hepatic sinusoids and obtained O_2 release rate per units area of sinusoidal wall in male Sprague-Dawley rats. As a NO donor, we used nitroglycerin.

The rate of O_2 release per unit area of sinusoidal wall was constant before nitroglycerin administration. But administration of nitroglycerin significantly increased the rate of O_2 release (Fig 1). The control infused with saline at the same rate showed no significant change during the time course.

The rate of [oxyhemoglobin (HbO_2)] inflow in the sinusoid, which is responsible for O_2 inflow at upstream spot, was not significantly altered by administration of nitroglycerin. Despite the wide variation in the level of [HbO_2] inflow between individual rats, O_2 release was constant in the absence of nitroglycerin. Nitroglycerin infusion did not significantly alter the erythrocyte velocity and hemoglobin concentration ([Hb]) in the hepatic

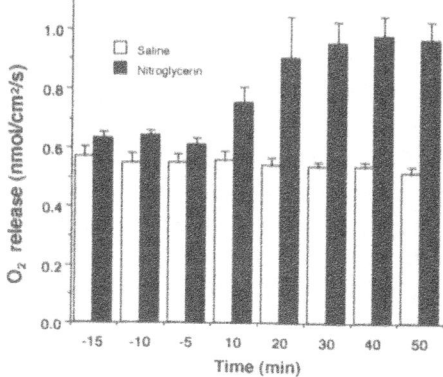

Figure 1. Effect of nitroglycerin on the rate of O_2 release and [oxyhemoglobin (HbO_2)] inflow in single unbranched hepatic sinusoids of rats. Values are mean ± SEM of four rats. For control studies (O), rats were given intravenous infusions of physiological saline containing heparin with a syringe pump (Harvard Apparatus, MA, USA) for 70 min at a rate of 0.6 ml/h. For nitroglycerin groups, rats were given intravenous infusions of the physiological saline for 20 min at the same rate of 0.6 ml/h. Then nitroglycerin solution diluted with the physiological saline was started (at time 0) to infuse for 50 min at the same rate of 0.6 ml/h. Infusion rate of nitroglycerin was 60 µg/kg/h. Friedman's test was used for statistical analysis. The rate of O_2 release increased significantly after administration of nitroglycerin (p < 0.05).

sinusoids, but significantly increased the difference in O_2 saturation per unit length of the sinusoid (ΔSO_2 /L). The increase in the O_2 release is therefore caused by the increased ΔSO_2/L, but independent of [HbO$_2$] inflow in the liver. Moreover, the diameters of the sinusoids were not changed after infusion of the NO donor.

The present study thus revealed another physiological role of NO as an emergency enhancer of oxygen supply to tissues. The present results may partly explain why cytokine-activated endothelium releases large amounts of NO (Radomski et al., 1990). Furthermore, endogenous NO can help to accelerate oxygen transfer from erythrocytes to tissues by this mechanism during such events as allograft rejection (Lancaster et al., 1992), and ischemic-reperfusion (Kumura et al., 1994). Such a beneficial interaction between erythrocyte Hb and endogenous NO generated from the vascular system may have been conserved.

CONCLUSIONS

We found that ESR spectra of Hb α-NO have a distinct three-line hyperfine structure in venous blood but not in arterial blood in rats treated with stimuli. Ex vivo study using whole blood revealed that oxygen affinity was significantly decreased in blood containing NO-bound Hb a. Effect of NO on oxygen release was then investigated by measuring oxygen transport parameters of flowing RBC between two points of single unbranched hepatic sinusoids with microspectroscopy. Oxygen release from RBC in the hepatic sinusoid was increased significantly by an NO donor, nitroglycerin, without increase in oxyhemoglobin inflow into sinusoids. The present study thus suggested a physiological ability of NO to increase oxygen release from flowing RBC to tissue.

ACKNOWLEDGMENTS

This research was supported in part by Grants-in-Aid from the Ministry of Education, Science and Culture of Japan.

REFERENCES

Douglas, C.G., Haldan, J.S., and Haldan, J.B.S. The laws of combination of haemoglobin with carbon monoxide and oxygen. J. Physiol., London 44: 275–304, 1912.

Green, L. C., S. R. Tannenbaum, and P. Goldman. Nitrate synthesis in the germfree and conventional rat. Science 212: 56–58, 1981.

Hibbs, Jb Jr, R. R. Taintor, and Z. Vavrin. Macrophage cytotoxicity: role for L-arginine deiminase and imino nitrogen oxidation to nitrite. Science 235: 473–476, 1987.

Hille, R., J. S. Olson, and G. Palmer. Spectral transitions of nitrosyl hemes during ligand binding to hemoglobin. J Biol Chem 254: 12110–12120, 1979.

Kon, K., N. Maeda, and T. Shiga. Effect of nitric oxide on the oxygen transport of human erythrocytes. J Toxicol Environ Health 2: 1109–1113, 1977.

Kosaka, H., S. Tanaka, T. Yoshii, E. Kumura, A. Seiyama, and T. Shiga. Direct proof of nitric oxide formation from a nitrovasodilator metabolised by erythrocytes. Biochem Biophys Res Commun 204: 1055–1060, 1994.

Kosaka, H., Y. Sawai, H. Sakaguchi, E. Kumura, N. Harada, M. Watanabe, and T. Shiga. ESR spectral transition by arteriovenous cycle in nitric oxide hemoglobin of cytokine-treated rats. Am J Physiol 266: C1400-C1405, 1994.

Kumura, E., H. Kosaka, T. Shiga, T. Yoshimine, and T. Hayakawa. Elevation of plasma nitric oxide end products during focal cerebral ischemia and reperfusion in the rat. J Cereb Blood Flow Metab 14: 487–491, 1994.

Kurz, M. A., T. D. Boyer, R. Whalen, T. E. Peterson, and D. G. Harrison. Nitroglycerin metabolism in vascular tissue: role of glutathione S-transferases and relationship between NO. and NO_2- formation. Biochem J 1993.

Lancaster, Jr Jr, J. M. Langrehr, H. A. Bergonia, N. Murase, R. L. Simmons, and R. A. Hoffman. EPR detection of heme and nonheme iron-containing protein nitrosylation by nitric oxide during rejection of rat heart allograft. J Biol Chem 267: 10994–10998, 1992.

Nagai, K., H. Hori, S. Yoshida, H. Sakamoto, and H. Morimoto. The effect of quaternary structure on the state of the alpha and beta subunits within nitrosyl haemoglobin. Low temperature photodissociation and the ESR spectra. Biochim Biophys Acta 532: 17–28, 1978.

Perutz, M. F. Stereochemistry of cooperative effects in haemoglobin. Nature 228: 726–739, 1970.

Radomski, M. W., R. M. Palmer, and S. Moncada. Glucocorticoids inhibit the expression of an inducible, but not the constitutive, nitric oxide synthase in vascular endothelial cells. Proc Natl Acad Sci U S A 87: 10043–10047, 1990.

Schmid, Schoenbein Gw, and B. W. Zweifach. RBC velocity profiles in arterioles and venules of the rabbit omentum. Microvasc Res 10: 153–164, 1975.

Seiyama, A., S. S. Chen, T. Imai, H. Kosaka, and T. Shiga. Assessment of rate of O_2 release from single hepatic sinusoids of rats. Am J Physiol 267: H944-H951, 1994.

Shiga, T., K. J. Hwang, and I. Tyuma. Electron paramagnetic resonance studies of nitric oxide hemoglobin derivatives. I. Human hemoglobin subunits. Biochemistry 8: 378–383, 1969.

Stuehr, D. J., and M. A. Marletta. Mammalian nitrate biosynthesis: mouse macrophages produce nitrite and nitrate in response to Escherichia coli lipopolysaccharide. Proc Natl Acad Sci U S A 82: 7738–7742, 1985.

Szabo, A., and M. F. Perutz. Equilibrium between six- and five-coordinated hemes in nitrosylhemoglobin: interpretation of electron spin resonance spectra. Biochemistry 15: 4427–4428, 1976.

Zhuo, M., S. A. Small, E. R. Kandel, and R. D. Hawkins. Nitric oxide and carbon monoxide produce activity-dependent long-term synaptic enhancement in hippocampus. Science 260: 1946–1950, 1993.

50

BEHAVIOR OF STIMULATED LEUKOCYTES IN THE PULMONARY MICROCIRCULATION OF PERFUSED RAT LUNGS

Takuya Aoki,[1] Yukio Suzuki,[2] Kazumi Nishio,[1] Kouichi Suzuki,[1] Atsusi Miyata,[1] Masaaki Mori,[1] Tomoaki Takasugi,[1] Hirofumi Fujita,[1] Harukuni Tsumura,[3] Yuzuru Ishimura,[4] Makoto Suematsu,[4] and Kazuhiro Yamaguchi[1]

[1]Cardiopulmonary Division
Department of Internal Medicine
[2]Department of Internal Medicine
Kitasato Institute Hospital
[3]Biomedical Department
Sankei Corporation
Tokyo, Japan
[4]Department of Biochemistry
School of Medicine
Keio University, Japan

1. INTRODUCTION

In a model of acute inflammation, neutrophil adhesion and transendothelial migration have been shown to be mediated through the leukocyte adhesion complex, CD11/CD18, and its endothelial ligand, ICAM-1 (7, 17). This extravasation is considered to occur from postcapillary venules. In the pulmonary microcirculation, however, more than 90% of activated neutrophils were observed to migrate from capillaries rather than postcapillary venules (8, 12), and neutrophil deformability (3–6, 21) and shear stress (2, 10) have been proposed to be major factors influencing neutrophil kinetics. Worthen et al. (22) reported that stimulated neutrophils remain in pulmonary capillaries as a result of decreased cellular deformability within the capillary. Vedder et al. (16) demonstrated that blocking of the CD11/CD18 glycoprotein adherence complex by the anti-CD18 monoclonal antibody (mAb) inhibited neutrophil accumulation in the gut but did not inhibit the pulmonary sequestration of neutrophils in ischemia and reperfusion injury. In addition, recent data obtained from studies utilizing P-selectin and ICAM-1 double mutant mice showed that neutrophil emigration into the peritoneum required both adhesion molecules, while emigration into the lung required neither (1). There have, however, been studies demonstrating that anti-CD18 mAb attenuates lung injury induced by tumor necrosis factor (9), gram-negative sepsis (19) and zymosan-activated plasma (11). These data suggest that the microcirculation of the lung differs from that of other systemic organs. Furthermore, little is known about the roles of adhesion molecules, especially the CD18-ICAM-1 pathway, in the pulmonary microcirculation. To investigate the dynamics of activated leukocytes and the roles of CD18-ICAM-1 pathway, we examined the effects of rat IL-8 and mAbs against CD18 and ICAM-1 on the behavior of leukocytes in microvessels of perfused rat lung under stable hemodynamic conditions.

2. METHODS

2.1. Reagents

Monoclonal antibodies against CD18 (WT-3) and against ICAM-1 (1A29) were generously provided by Dr. M. Miyasaka (Osaka University Medical School, Department of Bioregulation, Biomedical Research Center, Osaka, Japan). Fluorescein isothiocyanate (FITC), FITC-dextran (MW 145,000) were purchased from Sigma (St. Louis, MO), carboxyfluorescein diacetate succinimidyl ester (CFSE) from Molecular Probes (Eugene, Oregon). Rat cytokine-induced neutrophil chemoattractant (CINC/gro) is a peptide possessing biological activities analogous to those of the human IL-8 (15, 20) and was a generous gift from Dr. K. Watanabe (Institute of Cytosignal Research, Inc., Tokyo, Japan).

2.2 Experimental Groups

Four experimental groups; 1) control, 2) rat IL-8, 3) anti-CD18 and 4) anti-ICAM-1 were designed. 1) The control group (n = 6): CFSE-labeled leukocytes and perfused rat lungs had no pretreatment. 2) The rat IL-8 group (n = 6): Blood including CFSE-labeled leukocytes was incubated with 10 nM of rat IL-8 at 37°C for 10 minutes just before administration into the perfusion circuit. 3) The anti-CD18 group (n = 6): Blood including rat IL-8-activated CFSE-labeled leukocytes (same as the rat IL-8 group) was treated with WT-3 for 30 minutes at room temperature at a final concentration of 50 μg/ml. The blood was then administered into the perfusion circuit. 4) The anti-ICAM-1 group (n = 4): 1A29 was administered into the perfusion circuit at a perfusate concentration of 10 μg/ml. After a 10-minute perfusion period, blood including rat IL-8-activated CFSE-labeled leukocytes (same as the rat IL-8 group) was administered into the perfusion circuit.

2.3. Animal Preparation

Specific pathogen free male Sprague-Dawley rats (Sankyo Labo. Service, Tokyo, Japan), 8 weeks of age and weighing 250–300 g, were used. All of the following experimental protocols were approved by the animal committee of Keio University School of Medicine, Tokyo, Japan. Animals were anesthetized with pentobarbital sodium (50 mg/kg ip). The trachea was cannulated and connected to a ventilator, then ventilated at a tidal volume of 10 ml/kg and a respiratory rate of 60/min. Lungs were exposed by median sternotomy and blood was withdrawn from the heart. The trachea was ligated at the level of one tidal volume above functional residual capacity and fixed on a microscope stage in the supine position. The main pulmonary artery and left atrium were catheterized. Pulmonary arterial pressure was measured with a pressure transducer (SEN-6102M; Nihon Koden, Tokyo) connected to the pulmonary artery cannula and monitored continuously during the experiment (AcqKnowledge III;

BIOPAC Systems, CA). Krebs-Henseleit solution containing 3% albumin and auto-erythro-cytes was used as the perfusate (hematocrit = $6.5 \pm 0.5\%$). An extracorporeal membrane oxy-genator (Merasilox-S; Senkou-ikakougyou, Tokyo) was connected to the perfused lung circuit and equilibrated with 21% O_2–5% CO_2. The lungs were perfused at a rate of 10 ml/min with a peristaltic roller pump, and the perfusate from the left atrium was allowed to collect in the reservoir. The perfusate gas tension and pH were measured using a 1306 Blood Gas Analyzer (Instrumentation Laboratory, CA) at the beginning of and at intervals during each experiment. The pH was maintained between 7.38 and 7.42. The lung was humidified and the surface tem-perature was maintained at $37 \pm 0.5°C$. The left lingula of the lung was observed under an in vivo microscopic system (SO-MI, Sankei, Tokyo), employing normal and fluorescence objec-tive lenses (10x, 20x, 40x), connected to a closed-circuit video system. The in vivo micro-scopic system has three lights. The first is a normal lamp light (Techno Light KTS-150, Kenko, Tokyo), the second a xenon lamp light (Nikon, Tokyo) for fluorescence imaging, and the third a laser power supply (Omnichrome, Chino, CA) for confocal imaging. The image was displayed with a high sensitivity CCD camera (TEC-470, Optronics, CA) or a high sensi-tivity intensified imager (II) camera (EktaPro Intensified Imager, Kodak, San Diego, CA) and color video monitor (PVM-1444Q, Sony, Tokyo). The image was recorded on videotape with a tape recorder (SVQ-260, Sony, Tokyo). In order to determine the diameters of arterioles, capillaries and venules, and to analyze the high speed movement of leukocytes and erythro-cytes inside vessels, a confocal laser scanning microscope (Yokokawa, Tokyo) was used. Views of high speed movements of the cells were displayed by using the confocal laser scan-ning microscope with the II camera, and stored in a high speed video recorder system (Ek-taPro TR6,000 System, Kodak, San Diego, CA). Velocities of leukocytes and erythrocytes were recorded at a rate of 250 frames/sec using the high-speed video system. Leukocytes that remained in the same internal portions of capillaries were excluded from the speed analysis. Centerline blood flow velocity (Vc) was determined in arterioles and venules. The mean blood cell velocity (Vmean) was calculated from Vmean = Vc/1.6. The vessel diameters (Dv) of arterioles and venules were measured by processing the confocal video images with a com-puter-assisted digital image analyzing system (Apple Quadra 840-AV/Image 1.58). The vessel wall shear rates (g) were calculated based on the definition for a Newtonian fluid: $g = 8$ (Vmean/Dv) (S^{-1}), as described elsewhere (14).

2.4. Visualization of Vessel Networks, Erythrocytes, and Leukocytes

To visualize vessel networks, we administered FITC-dextran (MW 145,000), at a fi-nal concentration of 0.015%, into the perfusion circuit. Erythrocytes were labeled with FITC. Rat blood was diluted with phosphate buffered saline and centrifuged. The erythro-cyte pellet was then diluted with phosphate buffered saline. FITC was added at a final concentration of 0.1 mg/ml. After a 30-min incubation at 37°C, the solution was centri-fuged and diluted with 5 ml of phosphate buffered saline. Then, 1 ml of the dilute solution was administered into the perfusion circuit, when necessary. Leukocytes were labeled with CFSE according to our previously described method (14). Briefly, the CFSE solution (1.0 mg/kg) diluted with 1.5 ml of physiologic saline was injected at 0.3 ml/min into the femoral veins of the donor rats. After a 30-minute incubation in vivo, blood samples were collected by heart puncture. In order to avoid unnecessary activation of the naive cells, which can evoke shedding of L-selectin, no further cell separation was carried out. These procedures allowed us to obtain blood samples for perfusion in which $28.4 \pm 3.9\%$ of leu-kocytes were neutrophils. When used to study the pulmonary microcirculation, the blood samples containing CFSE-labeled leukocytes were directly injected into the perfusion circuit of the recipient rats to give a final leukocyte concentration of about 400–500 cells/μl.

2.5. Histological Examination

A catheter was inserted into the main bronchus after each experiment. The perfused rat lung was then fixed with formalin and embedded in paraffin. Six sections were cut at the same interval from the top to the bottom of the left lung, and stained with hematoxylin-eosin. In each section, 10 different microscopic fields selected at random were examined and the density of leukocytes was expressed as percentage compared with the control group (%). The leukocyte density of the control group is expressed as 100%.

2.6. Statistical Analysis

The results are presented as means ± SEM unless stated otherwise. All p values were determined using one-way analysis of variance (ANOVA) followed by the Scheffe-type multiple comparison test to detect differences between groups. A p value less than 0.05 was considered statistically significant.

3. RESULTS

3.1. Leukocyte Behavior in Pulmonary Microcirculation

Mean pulmonary arterial pressure was 12 ± 2 mmHg (mean ± SD), and did not differ significantly between groups. Mean pulmonary arterial pressure was constant throughout the experiments. The mean diameters of arterioles and venules used for measurements were approximately 15 μm and 18 μm, respectively, and exhibited no significant differences between the groups. The mean RBC velocities and wall shear rates in arterioles and venules of the control, rat IL-8, anti-CD 18 and anti-ICAM-1 groups are shown in Figure 1. There were no differences in mean arteriolar RBC velocities between the groups. Mean RBC velocities in venules were also comparable between groups. There are no statistically significant differences in wall shear rates in arterioles and venules between the groups. No rolling leukocytes were observed in either arterioles or venules of the pulmonary vasculature in any of the groups. Stationary adherent leukocytes were rarely observed in pulmonary arterioles and venules, and leukocytes remained in capillaries in all groups. Some leukocytes adhered along capillary walls, while other leukocytes stopped at the branching points and bifurcations of capillaries.

3.2. Representative Motion Patterns of Leukocytes

There were two distinct patterns of leukocyte behavior in capillaries of the control and the rat IL-8 groups. The majority of leukocytes in the control group were in "continuous" motion, with relatively little variation in velocity. When leukocytes were pretreated with Rat IL-8, many leukocytes stopped moving transiently (0.04 to 0.2 sec), and then some leukocytes resumed moving within the capillary. The latter "discontinuous" motion pattern, in which leukocytes stopped moving at least once for more than 0.04 sec during observation of a 200 μm x 200 μm area of the lung periphery, occurred at a frequency of 47.3 ± 8.2% of the total observed leukocyte number in the rat IL-8 group. The rat IL-8-in-

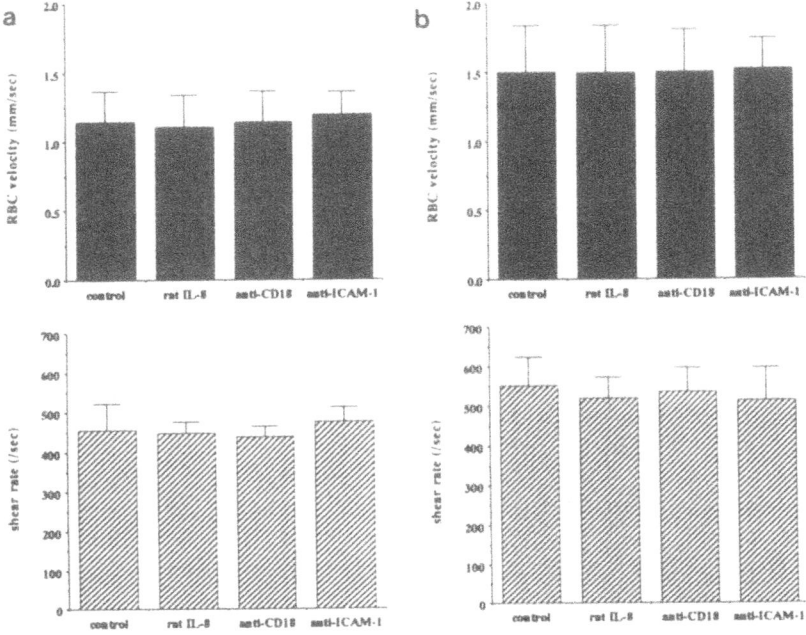

Figure 1. The mean RBC velocities (Vr) in arterioles and venules of all groups. There were no Vr differences in arterioles or venules between the groups. Values are means ± SEM. The control, rat IL-8 and anti-CD18 groups were n = 6, and the anti-ICAM-1 group was n = 4. There were no significant differences in A) arteriolar and B) venular wall shear rates between groups. Values are means ± SEM. The control, rat IL-8 and anti-CD18 groups were each n = 6, and the anti-ICAM-1 group was n = 4.

duced increase in the population of leukocytes showing the discontinuous motion pattern was attenuated when the lung and leukocytes were pretreated with mAbs against ICAM-1 and CD18, respectively.

3.3. Relative Leukocyte Velocities of Leukocytes

There were no differences in the relative leukocyte velocities (Vw/Vr) in arterioles between groups. The Vw/Vr values in capillaries of the control, rat IL-8, anti-CD18 and anti-ICAM-1 groups were 0.99 ± 0.15, 0.45 ± 0.11, 1.00 ± 0.18 and 1.00 ± 0.16, respectively. The Vw/Vr values in capillaries of the rat IL-8 group were decreased as compared with those of the control group. The capillary Vw/Vr values in the anti-CD18 and anti-ICAM-1 groups were restored to the control level. In postcapillary venules, the Vw/Vr values of the control, rat IL-8, anti-CD18 and anti-ICAM-1 groups were 1.00 ± 0.13, 0.65 ± 0.11, 0.85 ± 0.10 and 1.00 ± 0.15, respectively. The rat IL-8-induced reduction in Vw/Vr values was attenuated in part by anti-CD18 mAb and was abolished by anti-ICAM-1 mAb.

3.4. Histological Examination of Leukocyte Sequestration in Pulmonary Capillaries

The number of capillary leukocytes in the periphery of the perfused lung in the rat IL-8 group was increased (770 ± 140% control) as compared with that of the control group (100 ± 25%). The increased leukocyte numbers in the rat IL-8 group were partly attenuated by treatment with anti-CD18 (260 ± 48% control) and anti-ICAM-1 mAbs (340 ± 23% control), while those in the anti-CD18 and anti-ICAM-1 groups were significantly elevated as compared with that in the control group.

4. DISCUSSION

Rat IL-8 induced leukocyte velocity reductions and increased capillary neutrophil numbers, and these changes were attenuated by mAbs against ICAM-1 and CD18. These results indicate that the CD18-ICAM-1 interaction plays a key role in capillary leukocyte sequestration. Rat IL-8 increased leukocytes showing the discontinuous movement pattern, and this increase was also attenuated by mAbs against ICAM-1 or CD18. Therefore, interactions between CD18 and ICAM-1 appear to influence leukocyte kinetics in pulmonary capillaries, even with the existence of shear forces. Adhesion due to CD18 and ICAM-1 has been reported to occur in the static condition (13), and shear rates inhibit CD18-ICAM-1 dependent adhesion (18). The decreased leukocyte velocity and the transient cessation of movement in pulmonary capillaries may make it possible for CD18 and ICAM-1 to interact. Histological examination revealed that capillary neutrophil numbers in both anti-CD18 and anti-ICAM-1 groups were increased as compared with that of the control group, although mAbs against CD18 and ICAM-1 attenuated neutrophil elevation in the rat IL-8 group. Attenuated numbers of neutrophils are considered to be sequestered due to the interaction between CD18 and ICAM-1. On the other hand, the elevated neutrophils observed in both the anti-CD18 and the anti-ICAM-1 groups may have been sequestered owing to a mechanism other than the CD18-ICAM-1 interaction. Some leukocytes were observed to stop suddenly at the points at which capillaries branched-off from arteries as well as at capillary bifurcations. This leukocyte plugging may be attributable to mechanical forces rather than the CD18-ICAM-1 pathway. Neutrophil deformability has been proposed to be a major factor influencing neutrophil kinetics in pulmonary capillaries (3, 21), and activated neutrophils are reported to be stiff (22). Therefore, cellular deformability might influence the mechanism by which capillary neutrophils are increased in both the anti-CD18 and the anti-ICAM-1 groups. The existence of mechanisms other than capillary adhesion in pulmonary sequestration accounts for the finding that anti-CD18 mAb attenuates lung injury (9, 11, 19), while neutrophil emigration into the lung does not require ICAM-1 (1). In other words, inhibition of the CD18-ICAM-1 interaction may block CD18-ICAM-1-related capillary adhesion, while having no effect on the sequestration produced via mechanisms such as capillary plugging. Therefore, neutrophil emigration into the lung might occur even in ICAM-1 knockout mice.

5. SUMMARY

To investigate the dynamics of activated leukocytes and the roles of CD18-ICAM-1 pathway, we examined the effects of rat IL-8 and monoclonal antibodies (mAbs) against CD18 and ICAM-1 on the behavior of leukocytes in microvessels of perfused rat lungs. Specific pathogen free male Sprague-Dawley rats were used. Perfused rat lungs were prepared so as to obtain stable physiological shear rates. We used a confocal laser scanning microscope equipped with a high speed video analysis system to visualize pulmonary microcirculation. Rat leukocytes were activated with rat IL-8. No rolling leukocytes were observed in either pulmonary arterioles or venules, and leukocytes were sequestered in capillaries. The majority of unstimulated capillary leukocytes moved smoothly. About 50% of stimulated leukocytes, however, showed a transient cessation of movement in pulmonary capillaries. Rat IL-8 decreased the relative leukocyte velocities against mean blood velocities in capillaries (45%) and venules (65%), and increased intracapillary neutrophils. Anti-CD18 and anti-ICAM-1 mAbs attenuated these changes. These results suggest that unique features exist in the interaction between activated leukocytes and pulmonary microvessels, and that CD18-ICAM-1-dependent capillary sequestration is one of the major mechanisms by which activated leukocytes accumulate in lungs.

REFERENCES

1. Bullard, D. C., L. Qin, I. Lorenzo, W. M. Quinlin, N. A. Doyle, R. Bosse, D. Vestweber, C. M. Doerschuk, and A. L. Beaudet. 1995. P-selectin/ICAM-1 double mutant mice: acute emigration of neutrophils into the peritoneum is completely absent but is normal into pulmonary alveoli. *J. Clin. Invest.* 95: 1782–1788.
2. Doerschuk C. M., M. F. Allard, B. A. Martin, A. MacKenzie, A. P. Autor, and J. C. Hogg. 1987. Marginated pool of neutrophils in rabbit lungs. *J. Appl. Physiol.* 63 (5):1806–1815.
3. Doerschuk C. M., N. Beyers, H. O. Coxson, B. Wiggs, and J. C. Hogg. 1993. Comparison of neutrophil and capillary diameters and their relation to neutrophil sequestration in the lung. *J. Appl. Physiol.* 74 (6): 3040–3045.
4. Downey, G. P., D. E. Doherty, B. Schwab III, E. L. Elson, P. M. Henson, and G. S. Worthen. 1990. Retention of leukocytes in capillaries: role of cell size and deformability. *J. Appl. Physiol.* 69: 1767–1778.
5. Downey G. P., and G. S. Worthen. 1988. Neutrophil retention in model capillaries: deformability, geometry, and hydrodynamic forces. *J. Appl. Physiol.* 65 (4): 1861–1871.
6. Erzurum S. C., M. L. Kus, C. Bohse, E. L. Elson, and G. S. Worthen. 1991. Mechanical properties of HL60 cells: role of stimulation and differentiation in retention in capillary-sized pores. *Am. J. Respir. Cell Mol. Biol.* 5: 230–241.
7. Hernandez, L. A., M. B. Grisham, B. Twohig, K. E. Arfors, J. M. Harlan, and D. N. Granger. 1987. Role of neutrophils in ischemia-reperfusion-induced microvascular injury. *Am. J. Physiol.* 22: H699-H703.
8. Lien, D. C., P. M. Henson, R. L. Capen, J. E. Henson, W. L. Hanson, W. W. Wagner, Jr. and G. S. Worthen. 1991. Neutrophil kinetics in the pulmonary microcirculation during acute inflammation. *Lab. Invest.* 65 (2):145–159.
9. Lo, S. K., J. Everitt, J. Gu, and A. B. Malik. 1992. Tumor necrosis factor mediates experimental pulmonary edema by ICAM-1 and CD18-dependent mechanism. *J. Clin. Invest.* 89: 981–988.
10. Martin B. A., J. L. Wright, H. Thommasen, and J. C. Hogg. 1982. Effect of pulmonary blood flow on the exchange between the circulating and marginating pool of polymorphonuclear leukocytes in dog lungs. *J. Clin. Invest.* 69: 1277–1285.
11. Meyrick, B. O., and K. L. Brigham. 1984. The effect of a single infusion of zymosan-activated plasma on the pulmonary microcirculation of sheep. *Am. J. Pathol.* 114: 32–45.
12. Shaw, J. O. 1980. Leukocytes in chemotactic-fragment-induced lung inflammation. *Am. J. Pathol.* 101: 283–302.
13. Smith C. W., R. Rothlein, B. J. Hughes, M. M. Mariscalco, H. E. Rudloff, F. C. Schmalstieg, and D. C. Anderson. 1988. Recognition of an endothelial determinant for CD18-dependent human neutrophil adherence and transendothelial migration. *J. Clin. Invest.* 82: 1746–1756.
14. Suematsu, M., F. A. Delano, D. Poole, R. L. Engler, M. Miyasaka, B. W. Zweifach, and G. W. Schmid-Schönbein. 1994. Spatial and temporal correlation between leukocyte behavior and cell injury in postischemic rat skeletal muscle microcirculation. *Lab. Invest.* 70: 684–695.
15. Suzuki, H., M. Suematsu, M. Miura, Y. Y. Liu, K. Watanabe, M. Miyasaka, S. Tsurufuji, and M. Tsuchiya. 1994. Rat CINC/gro: a novel mediator for locomotive and secretagogue activation of neutrophils in vivo. *J. Leukoc. Biol.* 55: 652–657.
16. Vedder, N. B., R. K. Winn, C. L. Rice, E. Y. Chi, K.-E. Arfors, and J. M. Harlan. 1988. A monoclonal antibody to the adherence-promoting leukocyte glycoprotein, CD18, reduces organ injury and improves survival from hemorrhagic shock and resuscitation in rabbits. *J. Clin. Invest.* 81: 939–944.
17. Von Andrian, U. H., J. D. Chambers, L. M. McEvoy, R. F. Bargatze, K.-E. Arfors, and E. C. Butcher. 1991. Two-step model of leukocyte-endothelial cell interaction in inflammation: Distinct roles for LECAM-1 and the leukocyte B2 integrins in vivo. *Proc. Natl. Acad. Sci. USA.* 88: 7538–7542.

18. Von Andrian, U. H., P. Hansell, J. D. Chambers, E. M. Berger, I. T. Filho, E. C. Butcher, and K-E. Arfors. 1992. L-selectin function is required for B2-integrin-mediated neutrophil adhesion at physiological shear rates in vivo. *Am. J. Physiol.* 263 (Heart Circ. Physiol. 32):H1034-H1044.

19. Walsh, C. J., P. D. Carey, D. J. Cook, D. E. Bechard, A. A. Fowler, and H. J. Sugerman. 1991. Anti-CD18 antibody attenuates neutropenia and alveolar capillary-membrane injury during gram-negative sepsis. *Surgery.* 110: 205–212.

20. Watanabe, K., M. Suematsu, M. Iida, K. Takaishi, Y. Iizuka, H. Suzuki, M. Suzuki, M. Tsuchiya, and S. Tsurufuji. 1992. Effect of rat CINC/gro, a member of the interleukin-8 family, on leukocytes in microcirculation of the rat mesentery. *Exp. Mol. Pathol.* 56: 60–69.

21. Wiggs, B. R., D. English, W. M. Quinlan, N. A. Doyle, J. C. Hogg, and C. M. Doerschuk. 1994. Contributions of capillary pathway size and neutrophil deformability to neutrophil transit through rabbit lungs. *J. Appl. Physiol.* 77 (1): 463–470.

22. Worthen, G. S., B. Schwab III, E. L. Elson, and G. P. Downey. 1989. Mechanics of stimulated neutrophils: cell stiffening induces retention in capillaries. *Science.* 245: 183–186.

51

EFFECT OF BICARBONATE HAEMODIALYSIS ON PERIPHERAL OXYGEN AVAILABILITY IN MILD HYPERTENSIVE PATIENTS

G. Cicco,[1] A. J. van der Kleij,[3] G. Passavanti,[2] G. Baldassare,[2] A. Manicone,[1]
P. Vicenti,[1] M. Marra,[1] P. Izzo,[1] P. Coratelli,[2] and A. Pirelli[1]

[1]DIMO Internal Medicine and Hypertension
[2]Institute of Nephrology
University of Bari, Bari, Italy
[3]Department of Surgery/Hyperbaric Medicine
Academic Medical Center
University of Amsterdam
Amsterdam. The Netherlands

INTRODUCTION

Reduced peripheral oxygen availability is often observed in nephropatic hypertensive patients who require periodical haemodialysis and more often in patients with peripheral occlusive arterial disease (POAD). The aim of this study was to evaluate the effect of bicarbonate haemodialysis on peripheral oxygenation using transcutaneous pO_2 electrodes in a group of patients with mild hypertension (not treated with anti-hypertensive drugs) and POAD.

MATERIAL AND METHODS

Three times a week during five weeks twelve non-diabetic patients were submitted to a bicarbonate-haemodialysis using the same dialytic membrane (polysulphone) and the same dialytic protocol. The haemodialysis access was through an arterio-venous fistula on the forearm.

All the patients were non-smokers, non-dislipidemic, mild hypertensives with POAD stage I according the Leriche - Fontaine classification (Table I). Stage I implies patients with various undefined symptoms, paresthaesias, heaviness in legs and uncertain objectivity.

Before and after each session of three bicarbonate haemodialysis per week, transcutaneous PO_2 ($TcPO_2$) values in the subclavicular area and at the calve sites[1] were meas-

Table I. Patient characteristics (n=12) with
POAD Stage I (Leriche Fontaine classification)

M/f; mean age	7/5; 48±4 years
Mild hypertension	
Systolic blood pressure	150 ± 8 mmHg
Diastolic blood pressure	94 ± 6 mm Hg
Mean blood pressure	113 ± 3 mm Hg
Heart rate beats per minute	69 ± 9
Dialytic age	20 ± 5 months

ured and averaged for that week during four weeks of study. The Regional Perfusion Index (Calf $TcPO_2$ value/subclavicular $TcPO_2$ value = RPI) was calculated.[2-3] Three transcutaneous oxymeters were simultaneously used as previously described.[6] In the subclavicular region a Microgas 7640 MK_2 Kontron Instrument was applied with a Combi sensor to measure the transcutaneous PO_2 value and the PCO_2 value. At the right and left calve sites a Cutan PO_2 820 Monitor Kontron Instrument equipped with a Clark sensor was applied to measure peripheral transcutaneous PO_2 values. The temperature used for all sensors was 44°C. All measurements were performed in half supine position during standard environmental conditions.

In addition, the glucose 6-phosphate dehydrogenase (G6PDH) content in the serum and the 2.3 diphosphoglycerate (2.3DPG) content in the blood (Kit Boehringer Mannheim) were determined at the end of the first and fifth week. The level of G6PDH in the serum is an indication of haemolysis and is expressed in :/ml. The 2.3 DPG level in the blood indirectly indicates tissular oxygenation.[4]

RESULTS

At the end of each bicarbonate haemodialysis session mean blood pressure values were significantly decreased compared to the control values. The concomitantly increased heart rate may be associated with hypovolaemia (Table II). After the weekly three haemodialysis sessions significantly increased reference $TcPO_2$ values and peripheral

Table II. Haemodynamic parameters, central, and peripheral oxygenation parameters measured before and after bicarbonate haemodialysis (BD), three times a week and averaged at the end of week 1 and week 5. Values are expressed in mean ± sd. and % change

Variable	Week 1			Week 5		
	Before BD	After BD	% change	Before BD	After BD	% change
Systolic blood pressure mm Hg	150 ± 8	137 ± 10	-9*	148 ± 9	123 ± 9	-16**
Diastolic blood pressure mm Hg	94 ± 6	81 ± 4	-14**	95 ± 3	72 ± 4	-23**
Mean blood pressure mm Hg.	113 ± 3	98 ± 4	-12**	113 ± 2	89 ± 5	-21**
Heart rate per minute	69 ± 8	81 ± 10	+17**	71 ± 8	85 ± 10	+20*
$TcPO_2$ Subclavicular mm Hg	59 ± 6	65 ± 5	+10*	62 ± 6	68 ± 5	+10*
$TcPO_2$ R Calf mm Hg	49 ± 6	57 ± 5	+14**	50 ± 6	61 ± 5	+20*
$TcPO_2$ L Calf mm Hg	48 ± 6	55 ± 5	+12*	48 ± 6	59 ± 5	+22**
$TcPCO_2$ mm Hg	34 ± 1	32 ± 2	-4^{ns}	34 ± 1	32 ± 2	-5*
RPI	0.84 ± 0.04	0.87 ± 0.03	+4**	0.80 ± 0.02	0.86 ± 0.05	+8*

* $p < 0.01$ ** $p < 0.001$

Table III. Metabolic parameters, and Hct before and after bicarbonate haemodialysis (BD), three times a week, measured at the end of week one and week five. Values are expressed in mean ± s.d and % change

Variable	At the end of week 1			At the end of week 5		
	Before BD	After BD	% change	Before BD	After BD	% change
G6PDH (in serum)μ/ml	1.76 ± 0.73	2.44 ± 0.63	+48.93±34.94*	1.90 ± 0.68	2.41 ± 0.45	+35.58±31.76*
2.3-DPG (in blood) mmol/l	1.73 ± 0.24	2.23 ± 0.40	+29.05±17.62*	1.78 ± 0.24	2.45 ± 0.41	+38.03±19.22*
Hematocrit l/l	25.83 ± 2.55	28.25 ± 2.22	+9.56±2.75*	26.25 ± 2.73	28.42 ± 2.35	+8.50±3.36*

* $p < 0.01$

TcPO$_2$ were observed. Also the RPI revealed the same pattern. The increased G6PDH content may be associated with the physical effect of haemolysis. (Table III). The increased 2.3 DPG content in the blood after haemodialysis may be associated with an increased peripheral oxygen availability (Table III).

Linear Regression analysis (SPSS® for MS WINDOWS™ Release 6.1) revealed a significant correlation between the 2.3 DPG content in the blood and the TcPO$_2$ values after bicarbonate haemodialysis (Figure 1). After bicarbonate haemodialysis all hematocrit values were significantly increased compared with the values before bicarbonate haemodialysis. No

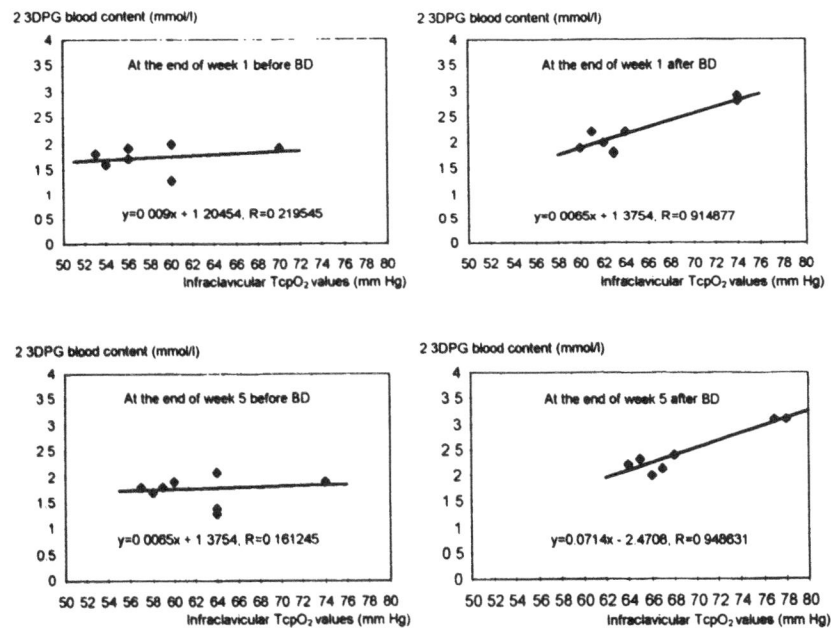

Figure 1. Regression analysis (dots represent inter-individual correlation) after bicarbonate haemodialysis between 2.3DPG blood content and infraclavicular TcPO$_2$ values at the end of week 1 before bicarbonate haemodialysis (left upper corner) and after bicarbonate haemodialysis (right upper corner) and at the end of week 5 before bicarbonate haemodialysis (left lower corner) and after bicarbonate haemodialysis (right lower corner).

significant correlation was found between 2.3 DPG and Hct before (y = -0,0437x + 2,8628/ R = 0,460217) as well after bicarbonate haemodialysis (y = -0,0343x + 3,1974/ R= 0,192873).

DISCUSSION

In an attempt to solve the problem of hypoxia during haemodialysis as reported by Johnson et al.[7] bicarbonate haemodialysis has been introduced.[8] However, a systemic study of the oxygen status[9] during 258 bicarbonate haemodialyses revealed a significant drop of the arterial pO_2 below 80 mmHg and these findings indicate that hypoxia occurs during bicarbonate haemodialysis. On the other hand this does not mean that after bicarbonate haemodialysis hypoxia (a decrease in arterial pO_2) itself influences the available amount of peripheral oxygen. In this study it was found that after a bicarbonate haemodialysis in nephropatic mild hypertensive patients with POAD, infraclavicular and peripheral $TcPO_2$ values were significantly increased compared to the initial values. At the same time an increased 2.3 DPG blood content was found.

It is known that a decreased affinity of haemoglobin for oxygen is expressed by a right shift of the Oxyhemoglobin Dissociation Curve (ODC) and may be caused by a lowered blood pH; an increased temperature, PCO_2 value, and 2.3 DPG blood content. The presence of high or low levels of 2.3 DPG in the blood indicates the tendency of the Oxyhaemoglobin Dissociation Curve (ODC) to shift to the right or left releasing more or less oxygen[5]. The increased central and peripheral oxygen availability at the end of the three weekly bicarbonate haemodialysis sessions may be associated with an increased 2.3 DPG blood content and consequently shifting the ODC to the right. Central oxygen availability was increased and it appeared that peripheral tissue also benefits as indicated by an significantly increased RPI.

It may be concluded from these data that hypertensive patients may benefit from bicarbonate haemodialysis in spite of reported hypoxia, especially when central and/or peripheral oxygen availability is attenuated in an impaired microcirculatory environment.

REFERENCES

1. Kleij van der. AJ., Bakker DJ. Oxymetry. In: "Handbook of Hyperbaric. Medicine."G.Oriani, A. Marroni, F. Wattel. (Eds) Springer - Verlag, Italia, Milano. 1996; pp.670–685.
2. Hauser, CJ., W Shoemaker, WC. Use of a Transcutaneous PO_2 Regional Perfusion Index to quantify tissue perfusion in peripheral vascular disease. Am. J Surg. 1983;197: 337–343.
3. Webster, MW., Steed, DL. $TcPO_2$ Theory and clinical applications in "Vascular Medicine" H.Boccalan. Editor. Elsevier. 1993;501.
4. Ganong, WF. Formazione e metabolismo del 2,3 DPG in: "Review of Medical Physiology" S.Francisco U.S.A. - Ed. Ital. Piccin (PD). 1973; 507:35–4.
5. F. Ganong WF. Curva di dissociazione dell'ossiemoglobina da Conneral et all. (The Lung - Clinical Physiology and pulmonary Function tests). In "Review of Medical Physiology" S. Francisco U.S.A. Ed. Ital. Piccin (PD). 1973;506: 35–2.
6. Cicco G., Dolce E.,Vicente P., Gigante P, Pirelli A. Peripheral oxygen release evaluation using 3 transcutaneous oxymeters contemporarily on patients with hypertension and lipoidoproteinosis. Clinical Hemorheology 1995;(15): 3: P 25:539.
7. Johnson NR, Bishel MD. Boylon CT. Hypoxia and hyperventilation in chronic haemodialysis. Clin Res 1974:19:145.
8. Nissenson AR. Prevention of dialysis-induced hypoxemia by bicarbonate dialysis. Trans Am Soc Artif Inter Organs 1980:26:339–341.
9. A.L. Nielsen, H.AE. Jensen, J. Hegbrant, H. Brinkenfeldt, P. Thuneborg and The Cord-Group-Oxygen status during haemodialysis Acte Aneasthesiologica Scand. 1995:39: Supplementum 107, 195–200.

52

SUBCUTANEOUS AND SPLANCHNIC TISSUE PERFUSION DURING HYPOVOLEMIA EVALUATED BY TISSUE GASES AND pH MEASUREMENTS

Michael Hartmann,[1] Åke Mellström,[1] and Kent Jönsson[2]

[1]Department of Anaesthesiology
[2]Department of Surgery and Department of Experimental Research
University Hospital MAS
Malmö, Sweden

1. BACKGROUND

Mortality during the last decades has not changed significantly despite advanced techniques for monitoring and treatment.[1] In patients subjected to major surgery a mortality of about 5 percent has been reported, whereas mortality increases to 15 percent if treatment in the intensive care unit is required. The major cause of death in these patients is due to multiple organ failure. In order to improve survival of these patients monitoring of cell function in the peripheral tissue and gastrointestinal tract could be of value.

Bowel hypoxia can produce irreversible damage which permits bacterial trans- location through ischemic but anatomically intact bowel.[2-4] An important reason for lack of progress in the development of equipment for peripheral and gastrointestinal perfusion monitoring might be the tendency to focus on central haemodynamic monitoring. Central haemodynamics and pulmonary function can be described by sophisticated measurements while assessments of the peripheral tissue perfusion are subjective and imprecise.[5] The concept of "compensated" shock is often used to describe a condition with apparently normal values of commonly used haemodynamic variables such as blood pressure, cardiac output, oxygen delivery, oxygen consumption and mixed venous oxygen tension.[6] This apparent normality may be a deception with persisting inadequate oxygenation in certain tissues. There is no absolute parallelity between central haemodynamics and the metabolic status in the peripheral tissues and the gastrointestinal tract. Several studies have demonstrated that the gastrointestinal tract develops ischemia during critical illness despite apparently normal blood flow supply. Therefore, compensated shock, not detectable with conventional monitoring equipment, is associated with tissue malperfusion that might lead to complication with further increase of mortality and morbidity.[7] Even small changes in circulating blood volume may have a dramatic effect on local perfusion long before any clinical signs of impaired circulation is available. Therefore information of adequate tissue oxygenation in relation to the metabolic demand is important if early signs of impaired peripheral tissue perfusion should be detected.[8] Subcutaneous and splanchnic tissues are among the first locations to be deprived of circulation during moderate hypovolemia.[9–12] Measurement of subcutaneous oxygen tension ($P_{SC}O_2$) provides a continuous measure of changes in the amount of oxygen delivered to the tissues. Reduced oxygen consumption as a result of reduced cardiac output implies that tissues are not adequately perfused and the oxygen delivery/demand ratio (DO_2/VO_2) is therefore from a physiological point of view of great interest, since it determines whether tissue oxygenation is sufficient or not.[8]

However, measurements of $P_{SC}O_2$ gives no indication of the ability of the tissues to extract and use oxygen in amounts that meet the metabolic demand of the tissues at the time. The only way to satisfy tissue demand when oxygen delivery is compromised is by increasing oxygen extraction. When oxygen availability of the tissue is limited an aerobic metabolism occurs, resulting in an increase in tissue carbon dioxide tension ($P_{SC}CO_2$) and decrease in tissue pH (pH_{SC}). Indirect measurement of intramucosal pH (pH_i) in the intestinal tract has shown to unveil defective tissue oxygenation in the splanchnic area.[12–17] Monitoring tissue perfusion and oxygenation in subcutis and the intestinal tract may be necessary in order to create an optimum physiologic balance of perfusion and cell function.

2. MEASUREMENTS OF TISSUE GASES AND PH

Evaluation of measurements of subcutaneous oxygen tension, subcutaneous carbon dioxide tension, measured pH_{SC} and calculated pH_{SCC} subcutaneous pH in comparison to intestinal pH_i, central and splanchnic haemodynamic parameters was performed. A shock model in pigs was studied where animals were subjected to bleeding in three steps of 10% of blood volume in each followed by resuscitation i.e. retransfusion of blood and in addition crystalloid solution. $P_{SC}O_2$, $P_{SC}CO_2$ and pH_{SC} were measured with a sensor (Paratrend 7^{TM}) placed subcutaneously in the groin. Intestinal intramucosal pH_i was measured with

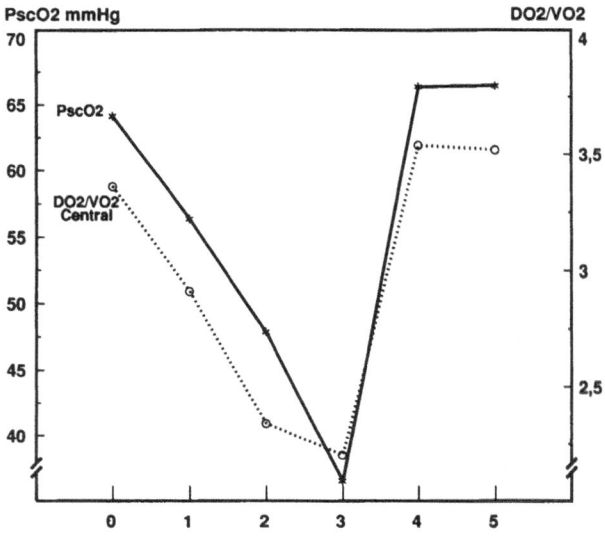

Figure 1. Subcutaneous oxygen tension (PscO2) and central oxygen delivery/demand ratio (DO2/VO2) before bleeding (= step 0), during bleeding (= step 1, 2 and 3) and after bleeding (= step 4 and 5).

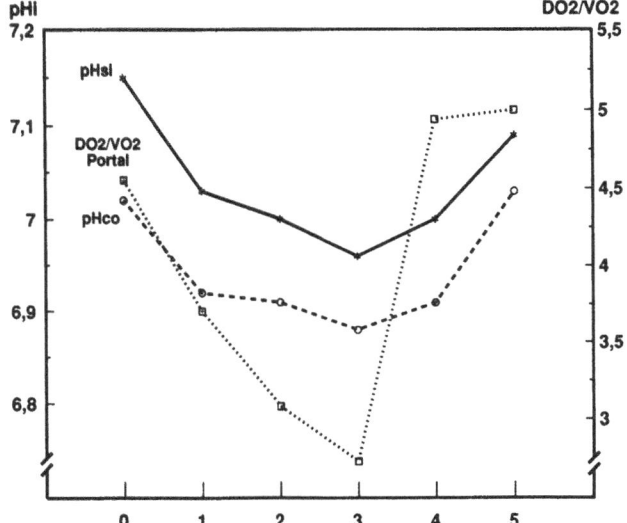

Figure 2. Intestinal pHi in sigmoid colon (pHco) and small intestine (pHsi) and oxygen delivery/demand ratio in the portal vein (DO2/VO2 portal) before bleeding (= step 0), during bleeding (= step 1, 2 and 3) and after bleeding (= step 4 and 5).

silicone tonometers placed intraluminally in the small intestine and sigmoid colon. Calculation of pH_{SCC} and pH_i was done according to the Henderson-Hasselbalch equation.[18] Intravascular catheters were used for blood gas and pH measurements. Blood flow in the portal vein was measured with an ultrasound probe. Cardiac output was determined by thermodilution technique and haemodynamic variables were calculated.

In the experimental setup shock was early indicated by a significant decrease in subcutaneous $P_{SC}O_2$ and intestinal pH_i, central and portal DO_2/VO_2 ($p<0.05$) (figs. 1-2). Calculated pH_{SCC} decreased significantly and subcutaneous $P_{SC}CO_2$ increased during the second step of bleeding ($p<0,05$) (fig 3). There was a significant correlation between subcutaneous $P_{SC}O_2$, central and portal DO_2/VO_2 and oxygen extraction ratio (ERO_2)($p<0.05$). Subcutaneous pH_{SC} and calculated subcutaneous pH_{SCC} were significantly correlated to each other and mixed venous and portal pH during all steps of bleeding ($p<0.05$). Subcutaneous pH_{SC} followed $P_{SC}O_2$ only after the first step of bleeding and during resuscitation ($p<0.05$) (fig 4).

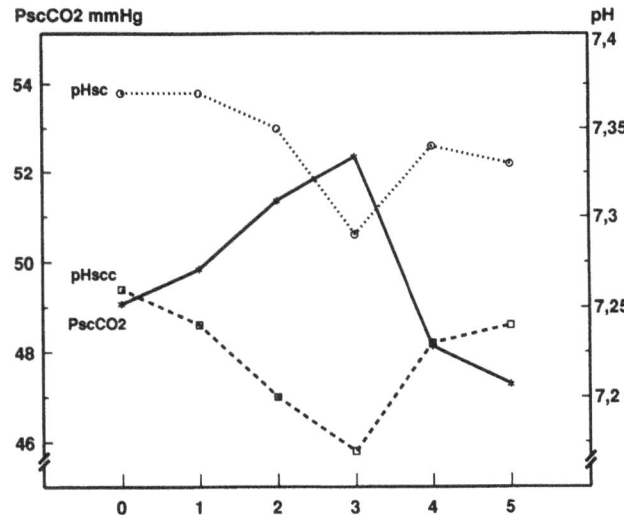

Figure 3. Subcutaneous carbon dioxide tension (PscCO2) measured (pHsc) and calculated (pHscc) subcutaneous pH before bleeding (= step 0), during bleeding (= step 1, 2 and 3) and after bleeding (= step 4 and 5).

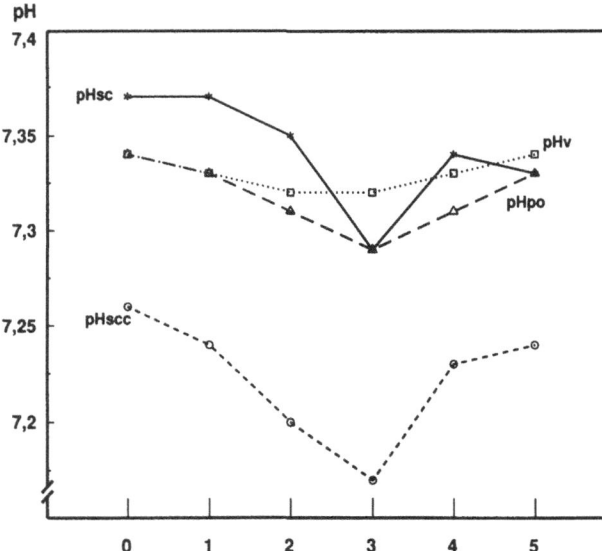

Figure 4. Subcutaneous oxygen tension (PscO2), measured subcutaneous pHsc and calculated subcutaneous pHscc before bleeding (= step 0), during bleeding (= step 1, 2 and 3) and after bleeding (= step 4 and 5).

3. DISCUSSION

Evaluation of the peripheral tissue perfusion relies on subjective clinical signs and symptoms such as weak pulse, clammy skin, low blood pressure, delayed capillary refill, cyanosis, altered mental status and thirst. Indirect parameters like puls, blood pressure, urine output, central venous pressure, and arterial blood gases are used, because they are readily available, not because they provide meaningful information of the peripheral or intestinal tissue perfusion. Measurement of oxygenation and perfusion in the peripheral tissue has revealed impaired response to oxygen administration in a face mask in suboptimally perfused patients.[10] Normal arterial oxygen tension does not guarantee optimum peripheral perfusion, but measurements of subcutaneous oxygen tension has proved to be a sensitive indicator of impaired tissue perfusion and oxygenation.[10,12] In this experimental model the subcutaneous oxygen tension again proved to be a sensitive indicator of impaired tissue perfusion. Subcutaneous tissue oxygen tension was well correlated to splanchnic tissue perfusion. This correlation between impaired perfusion and oxygenation in the intestinal tract and in the peripheral tissue might explain the findings of impaired healing in dehydrated rats. Healing of anastomoses in the small intestine and in subcutaneously implanted e-PTFE tubes were significantly correlated to each other and to dehydration in a study by Hartmann et al.[19]

Molecular oxygen is necessary for collagen accumulation.[20-21] It is therefore understandable that impaired tissue oxygenation may lead to impaired healing.

Subcutaneous $P_{sc}O_2$ and calculated tissue pH_{scc} decreased and subcutaneous $P_{sc}CO_2$ increased at the first step of bleeding. Measured tissue pH_{sc} decreased only between the first and the second step of bleeding. The sequence of changes clearly shows that changes in tissue oxygen tension and tissue carbon dioxide tension precede tissue pH changes. This indicates that the changes at the cellular level during haemorrhagic shock appears to be an oxygen deficit with subsequent metabolic acidosis, and that metabolic changes are counteracted before tissue acidosis develops and is possible to detect. A decrease in subcutaneous oxygen tension and the increase in subcutaneous carbon dioxide tension during bleeding demonstrates poor peripheral tissue perfusion with decreased oxygen supply and impaired removal of carbon dioxide. Measured values of tissue pH_{sc} were unexpectedly high and can not be considered to reflect the metabolic status of the tissue. The calculated tissue pH_{scc} values, according to the Henderson-Hasselbalch equation, were lower than the measured values. The principle of determining calculated pH is based on 1) measurements of tissue PCO_2 and on the assumption that 2) tissue HCO_3 equals that of arterial blood. During low flow states arterial HCO_3 probably overestimates true tissue HCO_3 because input is reduced and HCO_3 in fact may be consumed by local buffers during hypoxic conditions.[17] Thus calculated pH_{scc} in this experimental situation could be even lower and are probably a more true reflection of the actual metabolic status of the tissue than measured pH.

4. CONCLUSION

Measurements of subcutaneous oxygen tension can presumably be used as an indicator of splanchnic tissue perfusion. Measurements of subcutaneous pH_{sc} might be a valuable metabolic trendmarker providing additional information of disturbed metabolism, although measured subcutaneous $P_{sc}CO_2$ and calculated pH_{SCC} seem to reflect the absolute metabolic changes in the tissue to a greater extent.

REFERENCES

1. Ahnefeld FW, Heinrich H. Hat sich in der Routine des Anasthesisten die Berucksichtigung von Risikofaktoren bewahrt? Langenbeck's Arch Chir 1983; 361: 263.
2. Saydjari R, Beerthuizen GIJM, Townsend CM, Herndon DN, Thompson JC. Bacterial translocation and its relationship to visceral blood flow, gut mucosal ornithine decarboxylase activity, and DNA in pigs. J Trauma 1991; 31: 639.
3. Dietch EA, Bridges W, Baker J, Ma JW, Ma L, Grisham MB, Granger N, Specian RD, Berg RD. Hemorrhagic shock-induced bacterial translocation is reduced by xanthine oxidase inhibition or inactivation. Surg 1988; 104: 191.
4. Baker JW, Dietch EA, Li M, Berg RD, Specian RD. Hemorrhagic shock induces bacterial translocation from the gut. J Trauma 1988; 28: 896.
5. Shoemaker WC. Tissue perfusion and oxygenation: A primary problem in acute circulatory failure and shock states. Crit Care Med 1991; 19: 595.
6. Fiddian-Green RG. Associations between intramucosal acidosis in the gut and organ failure. Crit Care Med 1993; 21: S103.
7. Haglund U. Intramucosal pH. Intensive Care Med 1994; 20: 90.
8. Gottrup F. Tissue oxygen tension monitoring: Relation to hemodynamic and oxygen transport variables. In: Gutierres G & Vincent JL (eds). Update in intensive care and emergency medicine. Tissue oxygen utilization. Berlin, Heidelberg, New York, London, Paris, Tokyo, Honkong, Barcelona, Springer-Verlag 1991: 322.
9. Brantigan JW, Ziegler EC, Hynes KM, Miyazawa TY, Smith AM. Tissue gases during hypovolemic shock. J Appl Physiol 1974; 37: 117.
10. Gottrup F, Firmin R, Chang N, Goodson WH III, Hunt K. Continuous direct tissue oxygen tension measurement by a new method using an implanted silastic tonometer and oxygen polarography. Am J Surg 1983; 146: 399.
11. Jönsson K, Goodson WH III, West JM, Hunt TK. Assessment of perfusion in postoperative patients using tissue oxygen measurements. Br J Surg 1987; 74: 263.
12. Hartmann M, Montgomery A, Jönsson K, Haglund U. Tissue oxygenation in hemorrhagic shock measured as transcutaneous oxygen tension, subcutaneous oxygen tension, and gastrointestinal intramucosal pH in pigs. Crit Care Med 1991; 19: 205.
13. Fiddian-Green RG. A sensitive and specific diagnostic test for intestinal ischemia using silastic tonometers. Eur Surg Res 1984; 16: 32.
14. Grum CM, Fiddian-Green RG, Pittenger GL, Grant BJB, Rothman ED, Dantzker DR. Adequacy of tissue oxygenation in intact dog intestine. J Appl Physiol 1984; 56: 1065.

15. Fiddian-Green RG, Amelin PM, Herrmann JB, Arous E, Cutler BS, Schiedler M, Wheeler HB, Baker S. Prediction of the development of sigmoid ischemia on the day of aortic operations. Indirect measurements of intramural pH in colon. Arch Surg 1986; 121: 654.

16. Gys T, Hubens A, Neels H, Lauwers LF. Prognostic value of gastric intramural pH in surgical intensive care patients. Crit Care Med 1988; 16; 1222.

17. Antonsson JB, Boyle CC III, Kruithoff KL, Wang H, Sacristan E, Rothschild HR, Fink MP. Validation of tonometric measurement of gut intramural pH during endotoxemia and mesenteric occlusion in pigs. Am Physiol Soc 1990; 259: G519.

18. Fiddian-Green RG, Pittenger G, Whitehouse WM. Back diffusion of CO_2 and its influence on the intramural pH in the gastric mucosa. J Surg Res 1982; 33: 38.

19. Hartmann M, Jönsson K, Zederfeldt B. Importance of dehydration in anastomotic and subcutaneous wound healing: An experimental study in rats. Eur J Surg 1992; 158: 79.

20. Fujimoto D, Tamiya N. Incorporation of ^{18}O from air into hydroxyproline by chick embryo. Biochem J 1962; 84; 333.

21. Prockop DJ, Kaplan A, Udenfried S. Oxygen^{18} studies on the conversion of proline to collagen hydroxyproline. Arch Biochem Biophys 1963; 101: 499.

53

RELATIONSHIP BETWEEN HEMOGLOBIN OXYGENATION AND ANEMIA

Marcus Gliwitzki,[1,3] Rainer Gross,[1,2] Klaus Pietrzik,[3] Imelda Winoto,[4] and Soemilah Sastroamidjojo[1]

[1]SEAMEO-TROPMED Center Jakarta
University of Indonesia
PO Box 3852, Jakarta, Indonesia
[2]Deutsche Gesellschaft für Technische Zusammenarbeit (GTZ)
PO Box 5180, 6236 Eschborn, Germany
[3]Institute of Nutritional Sciences
Department of Pathophysiology of Human Nutrition
University of Bonn
Endenicher Allee 11–13, 53115 Bonn, Germany
[4]US Naval Medical Research Unit No.2 (US-NAMRU-2)
c/o American Embassy
Jalan Medan Merdeka Selatan V, Jakarta 10110, Indonesia

1. INTRODUCTION

The current definition of anemia is a state in which the quality or quantity of circulating red cells is reduced below a normal level (DeMaeyer et al, 1989). The quantity of circulating red cells is defined by the amount per volume blood. The quality of erythrocytes can be determined by both morphologic characteristics and the hemoglobin concentration (Hb). A World Health Organization (WHO) scientific group has suggested cut-off points of Hb below which anemia is likely to be present in individuals (WHO, 1968) (Table 1).

The main consequence of anemia is a reduced oxygen-carrying capacity of the blood. Oxygen dependent metabolism and organ functions are disturbed, and the patient suffers from tissue hypoxia. Muscular signs of tissue hypoxia are claudication and night cramps. Headache, lightheadedness, tinnitus, roaring in the ears, and faintness are cerebral signs (Erslev, 1991).

However, the body employs mechanisms to maintain oxygen supply to the tissues despite a hemoglobin concentration below the normal level. Increasing the cardiac output will decrease the fraction of oxygen which needs to be extracted during each circulation and thereby keep the oxygen pressure high. This is the most important mechanism in

Oxygen Transport to Tissue XIX, edited by Harrison and Delpy
Plenum Press, New York, 1997

Table 1. Hemoglobin values below which anemia is likely
to be present in populations living at sea level (WHO, 1968)

Age and Sex	Hemoglobin (g/L)
Children, 6 months - 6 years	110
Children, 6 - 14 years	120
Adult males	130
Adult females, non-pregnant	120
Adult females, pregnant	110

maintaining an adequate oxygen supply to the tissues (Varat et al, 1972). A decreased oxygen affinity of hemoglobin is one of the first mechanisms mobilized during anemia. This permits an increased oxygen extraction without jeopardizing the oxygen pressure (Erslev, 1991). Furthermore, pathological changes in oxygen supply may lead to centralization. This means an increased peripheral resistance of blood flow to the skin (Abramson et al., 1954), to the intestine, and to the kidneys (Bradley and Bradley, 1947). Thus, the central blood pressure will be stabilized and the oxygen supply to central organs such as brain and heart can be maintained (Erslev, 1991).

The objective of the present study was to investigate the effect of anemia on hemoglobin oxygenation in rabbit skin.

2. MATERIALS AND METHODS

Four 10 week old healthy New Zealand rabbits obtained from own laboratory breeding (US Naval Medical Research Unit No.2) were used in this experiment. They were housed individually in stainless steel cages and had access to food and water ad libitum. They were bled once weekly approximately 10 ml/ kg body weight for a period of 79 days via the lateral ear vein.

Both Hb and hemoglobin oxygenation (HbO_2) were determined once weekly. Hb was assessed according to the cyanomethemoglobin method (INACG 1986) and HbO_2 with the Erlangen micro-lightguide spectrophotometer (EMPHOII) (Frank et al, 1989). When measuring with the EMPHO II the lightguide tip was swept slowly across the depilated skin at the back of the rabbit. Five hundred spectra per measurement were recorded with the EMPHO 1.4 software and analysed with Offline 1.60 a.

Regression analysis has been performed with SPSS®for Windows™ 6.0.1 (Professional Statistics™, SPSS Inc, Chicago).

3. RESULTS

The Hb decreased about 54% from 146 g/L (± 10) to 67 g/L (± 4) and HbO_2 about 41% from 51% (± 3) to 21% (±10) (Table 2). HbO_2 started to decrease when Hb had dropped to 86 g/L and a further decrease was observed when the Hb dropped to 67 g/L. The correlation between Hb and HbO_2 was statistical significant (r^2=0.85, p<0.05).

Table 2. Changes in Hb and HbO$_2$ during the bleeding period ($\bar{x} \pm$ SD)

Parameter	Day 1	Day 38	Day 51	Day 60	Day 74	Day 79
Hb (g/L)	146 ± 10	106 ± 4	93 ± 10	86 ± 12	74 ± 9	67 ± 4
HbO$_2$ (%)	51 ± 3	47 ± 5	42 ± 11	31 ± 5	25 ± 11	21 ± 10

4. DISCUSSION

It is well known that oxygen supply to the tissue is jeopardized during anemia due to a decreased Hb. Therefore, HbO$_2$ may be a good parameter to detect anemia. Gross et al (1996) suggested that a decreased HbO$_2$ may be a better indicator for anemia than the Hb since it is referring to the function of hemoglobin and not to its quantity. The disadvantage of the Hb as an indicator for anemia is the fact that several mechanisms such as increased cardiac output, decreased oxygen affinity of hemoglobin compensate for a decreased Hb. It may be that despite a Hb below the normal level the oxygen supply to the tissue is still sufficient. Preliminary results of the present study support this hypothesis. Despite a decrease in Hb from 146 to 93 g/L no major changes in HbO$_2$ have been observed. Only after a further decrease to 67 g/L the HbO$_2$ started to decrease as well. This may be evidence that the above-mentioned mechanisms may be able to compensate for the decreased Hb to a certain level. When exceeding this level the HbO$_2$ eventually decreases as well. However, it may be that other variables such as changes in room temperature may also have caused the decrease in HbO$_2$. To get more accurate information the study needs to be repeated with both an experimental and a control group.

The correlation between Hb and HbO$_2$ was highly significant which is another argument for the applicability of HbO$_2$ for the detection of anemia. However, it is important to note that the rabbits were examined under resting conditions. During physical work the oxygen reserves of anemic subjects are exhausted faster than those of non anemic ones because the compensating mechanisms are already employed at rest (Sproule et al, 1960). Thus, it may be that the decrease of HbO$_2$ in anemic subjects starts even earlier during physical work.

The preliminary results of the present study suggest that HbO$_2$ determined by micro lightguide spectrophotometry may be a useful indicator to detect functional consequences of anemia. Further research is necessary to investigate both the sensitivity and specificity of micro lightguide spectrophotometry and the relation between anemia and HbO$_2$ in humans.

ACKNOWLEDGMENTS

The valuable recommendations of Dr Klaus Frank (BGT Überlingen, Germany) are highly appreciated.

REFERENCES

Abramson DJ, Fierst SM, and Flachs K. Resting peripheral blood flow in the anemic state. Am Heart J 1954;25:609–612.
Bradley SE and Bradley GP (1947). Renal function during chronic anemia in man. Blood 2:192–197.

DeMaeyer EM, Dallman P, Gurney JM, Hallberg L, Sood SK, Srikantia SG. Preventing and controlling iron deficiency anemia through primary health care. Geneva, World Health Organization, 1989.

Erslev AJ. Erythrocyte disorders: Classification and manifestations. In: Williams JW, Beutler E, Erslev AJ, Lichtman MA (eds.). Hematology. New York: Mc Graw-Hill publishing company, 1991;423–429.

Frank KH, Kessler M, Appelbaum K, Dümmler W. The Erlangen micro-lightguide spectrophotometer EMPHO I. Phys Med Bio. 1989;34(12):1883–1900.

Gross R, Gliwitzki M, Gross P, Frank K. Anaemia and haemoglobin status: A new concept and a new method of assessment. Food Nutrit Bull 1996;17(1):27–33.

INACG. Guidelines for the eradication of iron deficiency anemia. A Report of the International Nutritional Anemia Consultative Group (INACG). New York and Washington DC: The Nutrition Foundation, 1977.

Sproule BJ, Mitchell JH, Miller WF. Cardiopulmonary physiological response to heavy exercise in patients with anemia. J Clin Invest 1960;39:378–388

Norušis MJ. SPSS® for Windows™ 6.0.1, Professional Statistics™, Chicago: SPSS Inc, 1993.

Varat MA, Adolph JA, Fowler NO. Cardiovascular effects of anemia. Am Heart J 1972:83(3);415–426.

World Health Organization. Technical report series No.405; Nutritional anemias: a report of a WHO scientific group. Geneva: WHO, 1968.

54

PERIPHERAL OXYGEN SUPPLY IN CHILDREN DURING THERAPY WITH HUMAN GROWTH HORMONE (hGH)

Kirsten Höper,[1] H. G. Dörr,[1] and J. Höper[2]

[1]Universitätsklinik und Poliklinik für Kinder und Jugendliche
Loschgestr. 15
[2]Institut für Physiologie und Kardiologie
Waldstr. 6
Friedrich-Alexander-Universität Erlangen-Nürnberg
D-91054 Erlangen, Germany

INTRODUCTION

During normal growth, especially during the postpartum period, there is *de novo* synthesis of nailfold capillaries. Vital microscopy shows that in healthy human infants after birth only the subpapillary network is visible (Hoepfner 1929). During the next weeks, capillaries originating from this network are formed (Merlen 1969). At the age of 6 months short capillaries are found and during further maturation of the child these capillary loops become longer. The neonatal period is characterized by marked changes in intracapillary haemoglobin oxygenation (SO_2) followed by a slight continuous decrease in SO_2 until the age of approximately 15 years (Höper et al., submitted). This period is characterized by marked linear growth, which is normally regulated by growth hormone (GH), GH-induced release of IGF 1 and sexual steroids in puberty.

We measured noninvasively SO_2 and LASER doppler flux (LDF) in children treated with hGH (human growth hormone) in supraphysiologic doses as part of a multicenter study (for details see patients and methods). It is known from the literature, that hGH-treatment in adults is accompanied by an increase in basal metabolic rate of about 20% (Jorgensen et al. 1994). During GH-therapy energy- and thus oxygen-demand is increased. However, only if adequate compensatory mechanisms suffice the increased oxygen needs, disturbances in cellular function will be avoided. Even if energy production is normal, oxygen lack could become a limiting factor for vital metabolic pathways (e. g. metabolism of steroid hormones) and growth. To support this hypothesis, we studied the effect of hGH-therapy on peripheral oxygen supply.

PATIENTS AND METHODS

The investigations were performed in 6 healthy children aged 2.33 - 10 yrs. (3 females, 3 males; pubertal stage: Tanner 1) treated with hGH (dose: 1.4 IE/kg body weight * week, daily s.c. injections) as part of a multicenter-study (SGA-Studie II, Pharmacia; ethical consent: ethical committee of the universities of Munich and Erlangen, 1993 and 1994, resp.; written parental consent was obtained for each patient). All were born small for gestational age (birth weight: < mean - 2 SD), had low pretreatment heights (< HSDS (=height standard deviation score) - 2SD) and subnormal height velocities (< 50th percentile; all according to German growth charts; Reinken, v. Oost, 1992). GH-deficiency was excluded in all patients. One patient (AJ) has Russell-Silver syndrome.

As peripheral parameters, intracapillary oxygen saturation (SO_2) and laser doppler flux (LDF) were measured in the nailfold of the second finger. Before measurement the children were adapted to room temperature. SO_2 was measured using the Erlanger Microlightguide spectrophotometer EMPHO II (BGT, Überlingen; method see: Frank et al., 1989); within 45 secs. 3000 spectra were recorded in each patient moving the lightguide slowly over the area of interest. For measurement of LDF, a tool developed at our university was used (Wansch, 1995). The lightguide was fixed above the area of interest, measurements also lasted 45 secs., mean LDF in arbitrary units (arb. units) was calculated later. Results of a cross-sectional study in 29 healthy, age-matched healthy children (13 f, 16 m, ages: 2.4 - 12.8 yrs; Höper et al., submitted) served as normal values for SO_2

Body height was measured as mean of three measurements using a Harpenden Stadiometer. Body weight was measured with patients in underwear using a scales. German growth charts published by Reinken and v. Oost (1992) were used as normal values. Heart rate (bpm) and blood pressure (both upper arms using adequate blood pressure cuffs; Riva-Rocci-method) were measured.

Measurements were performed before and 4, 12, 26, 39, 52 weeks after start of hGH-therapy.

Growth velocity was calculated in cm/yr. To render correlation of growth to laboratory values, SO_2 and LDF, difference in body height between two appointments were also calculated as percent body height-difference per week (dBH %/wk; see Equ. 1 for 39th wk).

$$\Delta \text{ body height (dBH \%/wk)} = \frac{[BH_{(wk. 39)} - BH_{(wk. 26)}] * 100}{BH_{(wk. 26)} * (39 \text{ wks.} - 26 \text{ wks.})} \quad (1)$$

Haemoglobin (Hb) was measured with a standard laboratory method.

Table 1. Age at start of therapy, sex and GH-doses (mean ± SD; male (m), female (f))

Patient-initials	Age at start of therapy (yrs.)	Sex	Dose (IE/kg body weight * wk)
DK	10.0	f	1.41 ± 0.04
JM	9.05	m	1.41 ± 0.06
SJ	7.09	m	1.39 ± 0.05
SY	6.96	f	1.40 ± 0.03
MB	6.20	m	1.43 ± 0.03
AJ	2.33	f	1.37 ± 0.06

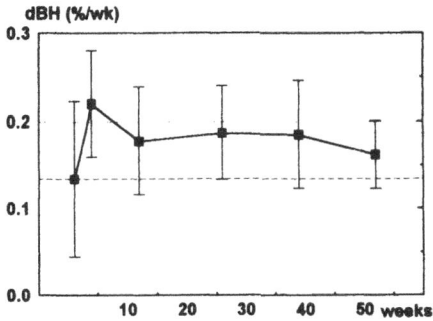

Figure 1. Percent body height difference/wk (dBH %/wk) during the first 52 weeks of hGH-therapy.

Results are given as mean value (MV) ± standard deviation (SD), unless stated otherwise. Statistical analyses were done using Student's t-test, Spearman-rank correlation and analysis of variance (WinStat 3.0, Kalmia und Co. Inc., G. Greulich Software, Germany).

RESULTS

There was a significant change in HSDS (height standard deviation score) during hGH-therapy (before: -3.98 ± 1.35; after 1 yr. of therapy: -2.75 ± 1.50; p < 0.005). Growth velocity was before therapy 5.81 ± 2.08 cm/yr. and increased in the first year of GH-therapy to 10.24 ± 2.97 (p < 0.005). As shown in figure 1, growth (calculated as dBH %/wk) was best within the first 4 weeks of therapy and remained high afterwards.

Changes in haemoglobin (Hb)-content were not significant during the therapy period (Fig. 2). There was a slight decrease within the first 12 weeks and a continuous rise afterwards. After 52 weeks pretreatment values were reached again.

Systolic blood pressure did not show significant changes. There was an initial rise (4 weeks) followed by a decrease until week 26 and another rise until week 52. Diastolic blood pressure dropped significantly within 26 weeks after start of therapy (p < 0.01) and rose significantly to values higher than before therapy at week 52 (p < 0.001). Blood pressure values were within the normal range throughout the study period (Fig. 3).

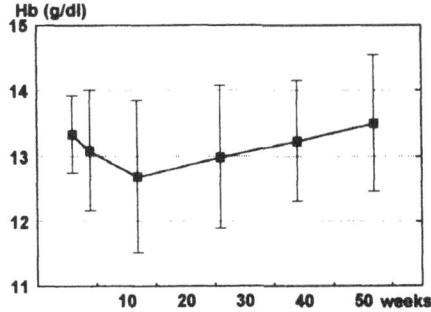

Figure 2. Hb-content during hGH-therapy.

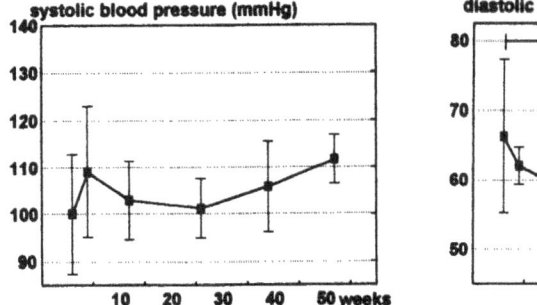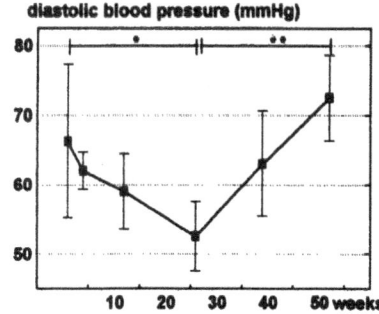

Figure 3. Systolic (left) and diastolic (right) blood pressure during 52 weeks of hGH-therapy. * p < 0.01, ** p < 0.001.

Heart rate of patients showed an initial rise (4 weeks) and fell to pretreatment values afterwards (n. s.) (Fig. 4).

LDF-values also showed an initial rise (4 wks), then returned to pretreatment values (12 wks) and decreased further to below pretreatment level (n. s.) (Fig. 5).

The intracapillary SO_2 was within the normal range (normal: 59.9 ± 6%) at start of therapy. After start, there was a dramatic decrease in SO_2 of almost 20% within 12 weeks (40.4 ± 6.9%; p < 0.05). From 12 weeks on there was a continuous increase until 52 weeks.

Figure 7 shows the correlation between percent body height difference per week and relative oxygen supply. The higher the relative growth velocity (dBH (%)) was, the higher oxygen supply rate (Hb * LDF) could be observed.

DISCUSSION

All children in our study showed an increased growth during the first year on hGH. Growth velocity increased from 5.8 to 10.2 cm/yr. During the first year, dramatic changes in local intracapillary haemoglobin oxygenation were observed in the nailfold of the second finger of the right hand. The lowest values were measured after 12 weeks of treatment. Age-matched healthy volunteers showed neither age- nor sex-dependent significant changes in SO_2 within one year (see shaded box in fig. 6). This leads to the conclusion, that the observed significant decrease in SO_2 was induced by hGH-therapy.

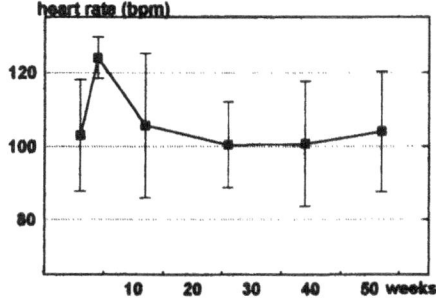

Figure 4. Heart rate during hGH-therapy.

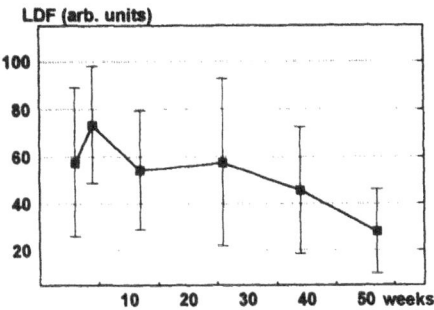

Figure 5. LDF-values during hGH-therapy.

Because local SO_2 is the result of oxygen supply rate and oxygen uptake rate, it can be assumed that during the first 12 weeks on high-dose hGH a disbalance between these parameters occurred. During this early treatment period, also a decrease in Hb was observed, which could cause a decrease in local oxygen supply rate. But at least during the first 4 weeks this is compensated by an increase in local flow (as measured by LDF). Between 4 and 12 weeks, both LDF and Hb show a decrease. Thus, an increased oxygen uptake rate as well as a decreased oxygen supply rate seem to be responsible for the low SO_2 values measured after 12 weeks.

Together with the changes in LDF and SO_2, alterations in central circulatory parameters were observed. The increase in systolic blood pressure can be divided into an early and a late phase. The early increase was observed within the first 4 weeks, the late one started in the 26th week. Early transient changes in blood pressure following hGH-therapy have been described before (Barton et al. 1992, 1993). The initial increase in systolic blood pressure occurs simultaneously with a decrease in diastolic pressure and an increase in heart rate. The latter has also been observed after application of IGF 1 (Vasconez et al. 1994).

Obviously, during the early phase a decrease in peripheral resistance occurs, which leads together with an increase in cardiac output to an increased peripheral flow. The latter is induced by IGF 1 (Copeland and Nair 1994; Franzeck et al. 1995). In the late phase, the increase in blood pressure was associated with a decrease in LDF, indicating an increase of peripheral vascular resistance.

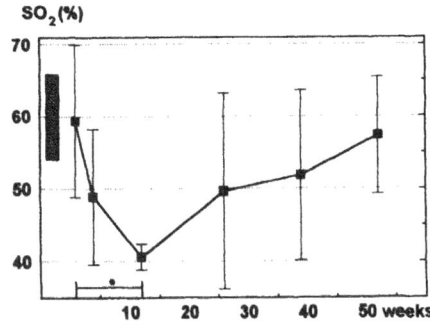

Figure 6. SO_2-values during hGH-therapy. Shaded box: age matched controls (MV ± SD). * p < 0.05.

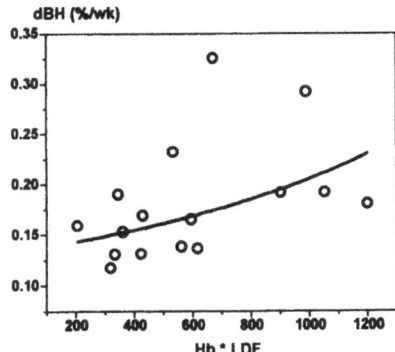

Figure 7. Percent body height difference per week (dBH (%/wk) in relation to relative oxygen supply (Hb * LDF; r = 0.51, p < 0.006).

The decrease in peripheral SO_2 may have consequences for local metabolic processes and might influence the success of hGH-treatment. Systemic hypoxaemia can cause a decrease in total body oxygen consumption (Frappell et al., 1992) or a decrease in brain glucose uptake (Hochachka et al., 1994). This phenomenon is called "hypoxic hypometabolism". A hypometabolic state could be the reason for the decrease in growth velocity. A hint for this assumption is the relation shown in Figure 7. This relation indicates that growth velocity depends on oxygen supply rate.

CONCLUSIONS

1. Within the first 3 months hGH therapy with supraphysiologic doses causes a disbalance between local oxygen supply and oxygen uptake. Therefore the intracapillary SO_2 declines to low values, at least in the nailfold.
2. The lowest values found were < 5% SO_2. This local hypoxia may cause alterations in cellular metabolism.
3. Furthermore it is known that hypoxia induces angiogenesis. This could account for the slow increase observed between 12 and 52 weeks of therapy.
4. From the present results of our preliminary study we can conclude that local oxygen may become a growth limiting factor during hGH therapy.

REFERENCES

Barton, JS, Cullen, S, Hindmarsh, PC, Brook, CG, Preece, MA (1992) Growth hormone treatment in idiopathic short stature: a preliminary analysis of cardiovascular effects. Acta Paediatr Suppl 383, 35–38.

Barton, JS, Hindmarsh, PC, Preece, MA, Brook, CG (1993) Blood pressure and the renin-angiotensin-aldosterone system in children receiving recombinant human growth hormone. Clin Endocrinol Oxf 38, 245–251.

Copeland, KC, Nair, KS (1994) Recombinant human insulin-like growth factor-I increases forearm blood flow. J Clin Endocrinol Metab 79, 230–232.

Frank, KH, Kessler, M, Appelbaum, K, Dümmler, W (1989) The Erlangen micro-lightguide spectrophotometer EMPHO I. Physics Med Biol 34, 1883–1900.

Frappell, P, Lanthier, C, Baudinette, RV, Mortola, JP (1992) Metabolism and ventilation in acute hypoxia: a comparative analysis in small mammalian species. Am J Physiol 262, R1040–1046.

Franzeck, UK, Dorffler-Melly, J, Hussain, MA, Wen, S, Froesch, ER, Bollinger, A (1995) Effects of subcutaneous insulin-like growth factor-I infusion on skin microcirculation. Int J Microcirc 15: 10–13.

Hochachka, PW, Clark, CM, Brown, WD, Stanley, C, Stone, CK, Nickles, RJ, Zhu, GG, Allen, PS, Holden, JE (1994) The brain at high altitude: hypometabolism as a defence against chronic hypoxia? J Cereb Blood Flow Metab 14, 671–679.

Hoepfner, Th (1929) Das Kropfproblem im Lichte der Kapillarmikroskopie und ihrer psychophysiologischen Beziehung. In W. Jaensch: Die Hautkapillarmikroskopie. Marhold C. Ed. Halle, 47–68.

Jørgensen, JO, Pedersen, SB, Borglum, J, Moller, N, Schmitz, O, Christiansen, JS, Richelsen, B (1994) Fuel metabolism, energy expenditure and thyroid function in growth hormone-treated obese women: a double-blind placebo-controlled study. Metabolism 43, 872–877.

Merlen, JF (1969) Essai de classification morphologique des boucles capillaires normales. Vie Méd. 40, 5169–5170.

Reinken L, van Oost G (1992) Longitudinale Körperentwicklung gesunder Kinder von 0 bis 18 Jahren - Körperlänge/-höhe, Körpergewicht und Wachstumsgeschwindigkeit. Klin Pädiatr 204, 129–133.

Vasconez, O, Martinez, V, Martinez, AL, Hidalgo, F, Diamond, FB, Rosenbloom, AL, Rosenfeld, RG, Guevera-Aguirre, J (1994) Heart rate increases in patients with growth hormone receptor deficiency treated with insulin-like growth factor I. Acta Paediatr Suppl 399, 137–139.

Wansch, R (1995) Konzeptionierung und Aufbau eines Sende- und Empfangssystems für einen Laser-Doppler-Velocimeter zur Untersuchung von Blutflußgeschwindigkeiten in kapillären Bereichen. Studienarbeit, Technische Fakultät der Universität Erlangen-Nürnberg.

55

OXYGENATION OF THE INTESTINAL MUCOSA IN ANAESTHETIZED DOGS IS ATTENUATED BY INTERMITTENT POSITIVE PRESSURE VENTILATION (IPPV) WITH POSITIVE END-EXPIRATORY PRESSURE (PEEP)

A. Fournell, T. W. L. Scheeren, and L. A. Schwarte

Zentrum für Anaesthesiologie
Heinrich-Heine-Universität
Düsseldorf, Germany

1. INTRODUCTION

Systemic inflammatory responses which alter the microcirculation leading to organ dysfunction and finally multiple organ failure (MOF) can be triggered by translocation of enteric bacterial endotoxin into the circulation following bowel ischaemia (Landow et al 1994). An adequate oxygen supply to satisfy O_2-demand in this critical region may therefore be essential for survival in critically ill patients, and therapeutic interventions to improve systemic oxygenation in MOF have to be safe for the splanchnic circulation, i.e., it is not acceptable for them to further compromise the oxygenation of the bowel.

ARDS (Adult Respiratory Distress Syndrome), usually a component of MOF, demands for mechanical ventilation of the patient. In this situation IPPV (Intermittent Positive Pressure Ventilation) with PEEP (Positive End-Expiratory Pressure) remains the mainstay of therapy (Barie 1995). The ability of PEEP during IPPV to improve systemic oxygenation has been proven and negative effects of PEEP on systemic circulation (i.e. reduced diastolic filling of the heart and diminished cardiac output) are well known; but there are only few data about its possible side effects on regional perfusion and, in particular, oxygenation due to the lack of clinically applicable methods. The development of tissue lightguide spectrophotometry and highly flexible probes which can be positioned using fiberendoscopes in the areas of interest allows the continuous and low-invasive measurement of regional oxygenation. The objective of our study was to evaluate the influence of PEEP during IPPV on the oxygen saturation (HbO_2) of the gastric mucosa.

Oxygen Transport to Tissue XIX, edited by Harrison and Delpy
Plenum Press, New York, 1997

2. METHODS AND MATERIALS

The experiments were conducted on six dogs of either sex, weighing 26 - 33 kg, which were repeatedly studied (10 experiments). The dogs were fasted with liberal access to water 12 hours prior to the experiment. The animals were trained to lie quietly on a laboratory table. Room temperature was kept at 24°C. Animals were anaesthetized with intravenous propofol (3 mg/kg). When anaesthesia was deep enough the trachea was intubated and the dogs were ventilated using a Starling pump (B. Braun Melsungen AG, Melsungen, Germany). Tidal volume was set to 10 ml/kg and rate of ventilation was adjusted to maintain endexpiratory CO_2 at 40 ± 2 mmHg. Anaesthesia was maintained with enflurane in air. The concentration of enflurane was monitored with a respiratory gas monitor (Capnomac Ultima SV™, Datex Instrumentation Corp., Helsinki, Finland) to maintain 2.5% end-tidal concentration.

The local intracapillary hemoglobin saturation (HbO_2) of the gastric mucosa was measured by tissue spectrophotometry using the Erlangen micro-lightguide Spectrophotometer EMPHO II (Frank et al 1989). The micro-lightguide was introduced through a flexible orogastric tube into the stomach of the dog after induction of anaesthesia. The correct position of the lightguide (facing the greater curvature of the stomach) was identified by means of gastroscopy. The flexible gastroscope was afterwards removed in order to avoid trauma to the gastric mucosa and to achieve a competent closure of the cardiac orifice of the stomach.

Systemic oxygen saturation was simultaneously measured by reflectance pulse oxymetry. The pulse oxymeter probe (Capnomac Ultima SV™, Datex Instrumentation Corp., Helsinki, Finland) was attached to the tongue of the dogs.

3. EXPERIMENTAL PROTOCOL

After obtaining control values during IPPV and ZEEP (0 cm H_2O endexpiratory pressure) PEEP was added in incremental steps from 5 to 10 and 15 cm H_2O. Each PEEP phase lasted 15 min (time needed to achieve steady state conditions).

4. DATA ANALYSIS

Data obtained in IPPV and ZEEP conditions and during IPPV and different PEEP levels were analyzed using Wilcoxon signed rank test. $p < 0.05$ was regarded as statistically significant.

5. RESULTS

Figures 1 and 2 show typical on-line registrations of gastric HbO_2 during IPPV with different levels of PEEP. In dog Charly (Fig. 1) each increment in PEEP (starting with 5 cm H_2O, going to 10 cm H_2O, and 15 cm H_2O, indicated per arrow) initiates an additional fall in the oxygen saturation of the gastric mucosa. The effects of all steps in increasing PEEP in this animal reduced HbO_2 to 58.2% of control value (obtained during IPPV and ZEEP) in spite of unaltered systemic oxygenation (98%, 98%, 99%, and 98% for ZEEP, 5, 10, and 15 cm H_2O PEEP).

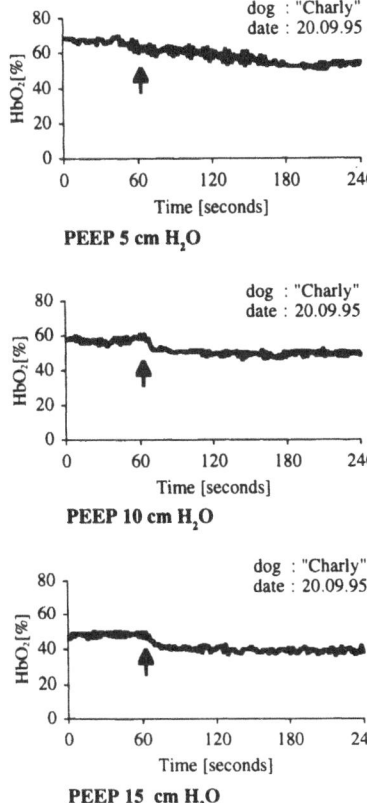

Figure 1. Mucosal oxygen saturation (HbO$_2$) in dog Charly exposed to increasing levels of PEEP during IPPV: each additional step produced a further attenuation in HbO$_2$.

In dog Cato (Fig. 2) HbO$_2$ is decreased due to 10 cm H$_2$O PEEP and is further attenuated when PEEP is increased to 15 cm H$_2$O PEEP (indicated by arrow, upper registration), but after returning to ZEEP (indicated by arrow, lower registration) HbO$_2$ starts to recover to the initial value registered during IPPV with ZEEP.

Similar results could be obtained in all dogs. The implementation of PEEP during IPPV reduced HbO$_2$ in each dog (Fig. 3). Despite unimpaired systemic oxygen saturation (98.1 ± 0.6%, 98.0 ± 0.6%, 97.8 ± 0.7%, and 97.3 ± 0.6% for ZEEP and PEEP-levels 5, 10, and 15 cm H$_2$O, values are mean ± SEM) HbO$_2$ of the gastric mucosa started to decline with the onset of PEEP and decreased further with increasing levels of PEEP in relation to control values (-4,9 ± 5.1%, -14.4 ± 5.9%, and -41.0 ± 6.1% for 5, 10, and 15 cm H$_2$O PEEP).

6. DISCUSSION

Ever since the first publication about the application of PEEP in patients with ARDS (Asbaugh et al, 1969) its effectiveness as a most useful tool to improve systemic oxygenation in ARDS has been recognized. Its mode of action, as already described in 1970 by Kumar et al, is most probably prevention of air-space collapse during expiration and increasing functional residual capacity (FRC). Increasing FRC from recruitment of terminal airspaces (alveoli or small airways) does not work in every patient (Falke 1980). In these patients the addition of PEEP during IPPV produces overdistension and increases shunt. Therefore, numerous approaches have been made to determine "optimal PEEP" or "best

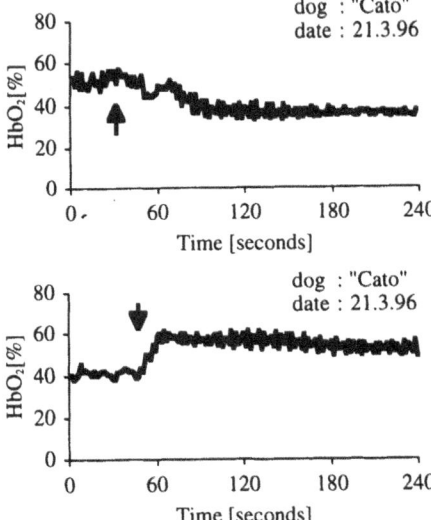

Figure 2. Depression in mucosal oxygen saturation (HbO$_2$) is reversible: when endexpiratory pressure is switched from 10 to 15 cm H$_2$O (upper registration) HbO$_2$ decreases, whereas it improves (lower registration) after PEEP has been turned off.

PEEP" (Suter et al, 1975), a level of PEEP with the least detrimental effects. Many physiologic variables have been used to determine the best level of PEEP (greatest values of PaO$_2$, static compliance, oxygen delivery, and the lowest values of intrapulmonary shunt), but all of these do not provide information about the oxygen status of the bowel, where ischaemia would enhance the development of MOF. This is due to the lack of methods in the past. We did not have methods to monitor the oxygenation in the splanchnic region, we had only methods to monitor hemodynamics. And monitoring hemodynamics in the bowel demonstrated negative effects of PEEP. Love et al in 1995 were able to show that PEEP caused a dose-related decrease in cardiac output (CO) and mesenteric blood flow (MBF), but whereas CO could be normalized by infusion of normal saline MBF remained diminished. But it is questionable if low blood flow is low tissue oxygen saturation in all circumstances. Hypoperfusion and/or increased oxygen extraction do not necessarily indicate tissue hypoxia. The question can only be solved by the direct measurement of oxygen tension or oxygen saturation in the tissue.

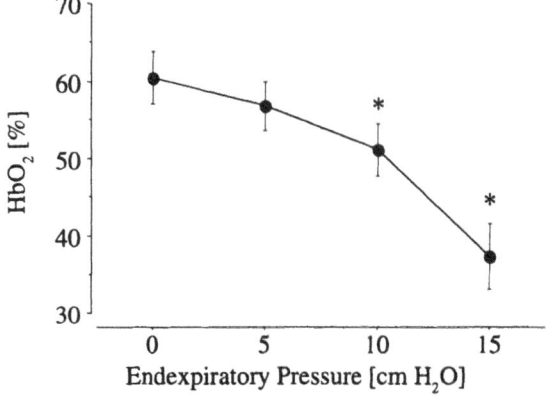

Figure 3. PEEP produced dose-related decreases in HbO$_2$, values are mean ± SEM, (*p < 0.05 by Wilcoxon signed rank test), 10 experiments in 6 dogs.

Splanchnic tonometry has been used in many studies as a metabolic measure of the adequacy of mucosal oxygenation. As Fiddian-Green pointed out in 1993 the measurement of intramucosal pH enables an inadequacy of tissue oxygenation in the gut to be detected within minutes of its onset. But gastric tonometry does not measure pO_2 or SO_2, it measures the pH of the mucosal lining of the stomach. It therefore represents an indirect measure of tissue oxygenation and it cannot be applied continuously.

To our knowledge this study is the first using lightguide spectrophotometry to assess oxygen saturation in the mucosa in the intact animal. The investigation in the intact animal without laparotomy and eventeration of parts of the small intestine is of great importance because these measures would have altered the pressures generated by IPPV and PEEP both in the thorax and in the abdomen with the consequence of unpredictable implications for HbO_2 in the bowel.

Our results show that all levels of PEEP during IPPV induce a reduction in HbO_2 of the gastric mucosa in dogs despite unimpaired systemic oxygenation. In MOF, including ARDS, this attenuation of local oxygen availability due to even moderate levels of PEEP (5 - 10 cm H_2O) during IPPV in order to augment systemic oxygen saturation may be deleterious because splanchnic ischaemia could be further increased and thus promote the development of MOF. If this finding in dogs can be verified in patients with MOF the definition of "optimal PEEP" has to be revised: it should no longer be only a level of PEEP where oxygen transport capacity is maximal and the parameters of lung function are not impaired but also a level where HbO_2 in the bowel is optimized or at least not further attenuated.

The model used in this study is healthy dogs with normal lungs and normal PaO_2. Lung mechanics are different in ARDS compared to this model: therefore the increase in arterial oxygenation from PEEP in a hypoxic model, i.e. an ARDS model, might overcome the decrease in local HbO_2 of the gastric mucosa due to PEEP. PEEP will not have only favorable effects in the hypoxic model, it might improve arterial oxygenation, but also side effects: mesenteric blood flow will be diminished. Since oxygen availability in the mesenteric region depends on regional blood flow and oxygen saturation and both variables are influenced by PEEP it remains questionable which of both effects predominates.

REFERENCES

Asbaugh D.G., Petty T.L., Bigelow D.B., Harris T.M. (1969) Continuous positive-pressure breathing (CPPB) in adult respiratory distress syndrome. *Thorac. Cardiovasc. Surg.* 57: 31

Barie P.S. (1995) Organ-specific support in multiple organ failure: pulmonary support. *World J. Surg.* 19: 581

Falke K.J. (1980) Do changes in lung compliance allow the determination of "Optimal PEEP"? *Anaesthesist* 29: 165

Fiddian-Green R.G. (1993) Associations between intramucosal acidosis in the gut and organ failure. *Crit. Care Med.* 21: 103

Frank K.H., Kessler M., Appelbaum K., Dümmler W. (1989) The Erlangen micro-lightguide spectrophotometer EMPHO I. *Phys. Med. Biol.* 34: 1883

Kumar A., Falke K.J., Geffin B., Aldredge C.F., Laver M.B., Lowenstein E., Pontoppidan H. (1970) Continuous positive-pressure ventilation in acute respiratory failure. *N. Engl. J. Med.* 283: 1430

Landow L., Andersen L.W. (1994) Splanchnic ischaemia and its role in multiple organ failure. *Acta Anaesthesiol. Scand.* 38: 626

Love R., Choe E., Lippton H., Flint L., Steinberg S. (1995) Positive end-expiratory pressure decreases mesenteric blood flow despite normalization of cardiac output. *J Trauma* 39: 95

Suter P.M., Fairley H.B., Isenberg M.D. (1975) Optimum end-expiratory airway pressure in patients with acute pulmonary failure. *N. Engl. J. Med.* 292: 284

56

MEASUREMENT OF CEREBRAL OXYGENATION AND HAEMODYNAMICS DURING HAEMORRHAGE/FLUID REPLACEMENT

M. S. Thorniley,[1] K. S. Khaw,[2] E. Balogun,[1] S. Simpkin,[1] C. Shurey,[1] I. A. Sammut,[1] and C. J. Green[1]

[1]Department of Surgical Research
Northwick Park Institute for Medical Research
Northwick Park Hospital
Harrow, Middlesex, HA1 3UJ, United Kingdom
[2]Department of Anaesthetics
Mount Vernon Hospital
Northwood, HA6 2RN, United Kingdom.

1. INTRODUCTION

Many major surgical procedures are accompanied by haemorrhage necessitating replacement with either fluid or blood. Despite apparently adequate replacement, tissue oxygen delivery may still be poor, resulting in cellular hypoxia and consequent metabolic dysfunction, which can ultimately lead to multi-organ failure. The overall aim of these investigations was to determine whether near infra-red spectroscopy (NIRS) can be used to measure changes in cerebral oxygenation and haemodynamics in response to controlled blood loss, with and without fluid replacement using a rat model of haemorrhage[1–6].

2. METHODS

2.1. Animal Preparation and Surgical Procedures

Male Lewis rats (400–500g) were used (n=10). All animal procedures were carried out according to the Animals (Scientific Procedures) Act, 1986. Surgical anaesthesia was maintained by artificial ventilation, using enflurane (1–2%) with O_2 delivered at 0.5 L min. Arterial oxygen saturation was maintained at between 95%–100% throughout the procedures (Ohmeda-Biox 3470). The temperature was maintained using a heating pad and monitored

Oxygen Transport to Tissue XIX, edited by Harrison and Delpy
Plenum Press, New York, 1997

throughout using a rectal probe. Continuous FiO_2, $EtCO_2$ measurements and intermittent blood pCO_2 and pO_2 determinations were made throughout the procedure, and the ventilation adjusted to maintain these within normal physiological parameters[7]. A Zeiss OPMI-6 operating microscope was used for all microsurgical procedures. The femoral artery was cannulated using a 20G portex catheter (flushed with heparinised saline 1000U/500ml), and used for direct blood pressure monitoring, removal of blood, and also for fluid administration.

2.2.1. NIRS Protocol. A CRITIKON™ Cerebral Redox Research Model 2001 (Johnson & Johnson Medical, UK), NIRS instrument was used with a sensor placed on the rat head[1-2]; with a 3.5 emitter-detector separation. NIRS measurements were made from induction of anaesthesia to termination. NIRS cerebral measurements were made during removal of blood (n=10) (approximately 10 min between each removal of 1 or 2 ml, to a maximum of 15 ml) with and without fluid replacement Haemaccel (Behring, Behringwerke AG, Marburg) in 1 ml aliquots to a maximum of 4 ml via a femoral artery cannula.

2.2.2. NIRS Measurements and Theory. NIRS is based on Beer Lambert's Law which relates optical absorption to the concentration of light absorbing chromophores present in a measured volume of tissue. Beer Lambert's Law strictly applies only to a non-scattering medium but the equation can be modified for measurements in scattering media such as biological tissue.

$$\text{Absorbance (A)} = a.c.l.B.$$

where a = molar extinction coefficient (mM^{-1} cm^{-1}), c = concentration (mM), l = emitter-detection separation in cm, B = pathlength factor.

It follows that by measuring changes in light absorption quantitated changes in chromophore concentration can be calculated, $\Delta c = \Delta A / a.l.B.$

For a volume of tissue containing several chromophores, changes in absorption due to each component can be calculated by using multiple interrogating wavelengths and applying the modified Beer Lambert's Law to each wavelength[1-2]. Quantitated concentration changes for each chromophore are then calculated by solving the resulting set of simultaneous equations. The calculation of chromophore concentration uses a linear summation of absorption terms containing specific multiplication factors at each wavelength.

The CRITIKON™ Cerebral Redox Research Monitor Model 2001 uses an algorithm based on published absorption spectra which have been obtained using isolated chromophores in non-scattering media. The haemoglobin and oxyhaemoglobin absorption spectra were obtained using haemolysed human blood[1-2]. The cytochrome difference absorption spectrum was obtained using purified cytochrome extracted from mitochondria[8]. The algorithm also incorporates wavelength dependent Differential Pathlength Factor data derived from "time of flight" studies.

Changes in the concentrations of O_2Hb, HHb and Caa_3 were calculated using the following table of multiplying factors:

Wavelength	776.5	819	871.4	908.7
HHb	+1.6436	-1.2510	-0.7075	+0.6515
O_2Hb	-0.9384	-0.6945	+0.5721	+1.7344
Caa_3	-0.1398	+0.9945	+0.1664	-0.9715

Units: mM.cm absorption^{-1}. The calculated concentration change is expressed in μM multiplied by pathlength in cm.

Summation of changes in the concentrations of O_2Hb and HHb provides a measure of changes in the total tissue haemoglobin (tHb), which reflect changes in tissue blood volume, hence giving an indication of perfusion. A further term, HbD or oxygenation index can be derived as: [HbD] = [O_2Hb]-[HHb]. This gives an indication of the net haemoglobin oxygenation status irrespective of any blood volume changes.

The instrument noise on the cytochrome and haemoglobin traces, measured with a sensor on a standard absorbing block and with the pathlength factor set to 1.0, is less than 0.05 and 0.5 μM respectively. As there is no recognised DPF which can be applied to measurements from the rat head we chose to use a value of 1 throughout the study. Measurements are represented in units of concentration multiplied by optical pathlength (μM.cm).As the emitter-detector separation is 3.5 cm and the DPF values published for various tissues range between 4 and 6.5 the concentration changes in μM units can be approximated by dividing measurements by a factor of between 14 and 22.75.

The mean changes in NIRS parameters at the end of the experimental period were determined and compared to baseline measurements. Data are presented as mean ± SEM. Statistical significance between groups was tested by unpaired Student's t-test and were considered significant when p<0.05.

3. RESULTS

Slow withdrawal of blood (n=10 rats) resulted in a progressive fall in blood volume (tHb) and oxy-haemoglobin (O_2Hb) with an increase in deoxygenated haemoglobin (HHb). Even at only 4% (1 ml) blood volume withdrawal the changes in the NIRS parameters were significant compared to baseline (p<0.05). At approximately 4% blood volume depletion the total haemoglobin concentration fell by -17.71 μM.cm; at 12 % (3 ml) -70.83μM.cm and 36% withdrawal -193.33±33.4 μM.cm.cm (mean±SEM). As the tHb decreased the O_2Hb level fell, by -24.57, -114.0 and a maximum drop of -330.77μM.cm at 4, 12 and 36 % blood volume depletion respectively. Addition of 1ml Haemaccel[R] (to a maximum of 4 ml, after 4 ml exsanguination) resulted in the level of HHb falling and a rise in the O_2Hb concentration.

In figure 1 is shown a representative example of the effect of progressive exsanguination on the cerebral NIRS parameters; a divergence in the concentration of oxy- and deoxy-haemoglobin is apparent as the exsanguination continues. There was a fall in the total haemoglobin concentration as 1 ml aliquots of blood were withdrawn from the femoral artery The fall in the *in vivo* tHb concentration (measured by NIRS) is comparable with the fall in the *in vitro* determination of the Hb concentration.

In figure 2 is shown the derived parameter, HbD ([HbD]=[O_2Hb]-[HHb], the oxygenation index) and mean arterial pressure. At the start of the exsanguination each withdrawal resulted in a temporary fall in the MAP followed by a slight recovery however there appears to be a critical point in the exsanguination after which the MAP fell with no recovery.

In the following 2 representative plots the fluid loss is replaced by Haemaccel. Figure 3 shows in oxy- and deoxy-haemoglobin as exsanguination and fluid replacement continues, it was observed that there was less divergence in the O_2Hb and HHb than that observed in figure 2. In figure 4 is shown the relationship between the HbD and the MAP in response to exsanguination and fluid replacement. A recovery of the MAP and HbD, the oxygenation index was observed with fluid replacement.

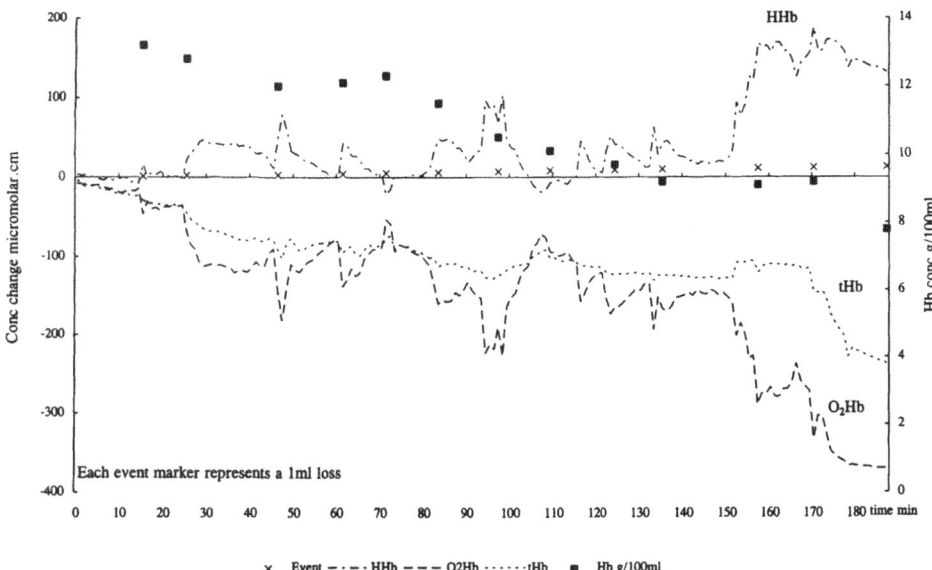

Figure 1. A plot of changes in the in vivo concentrations of oxy-, deoxy- and total haemoglobin and the in vitro determination of the haemoglobin concentration throughout exsanguination.

4. DISCUSSION

In this rat hypovolaemia model the results showed that changes in cerebral O_2Hb, HHb and tHb can be measured by NIRS even in response to only moderate blood loss. The haemoglobin concentration changes sampled intermittently correlated well with the tHb measured by NIRS. The oxygenation index, HbD, is a good indicator of oxygen deliv-

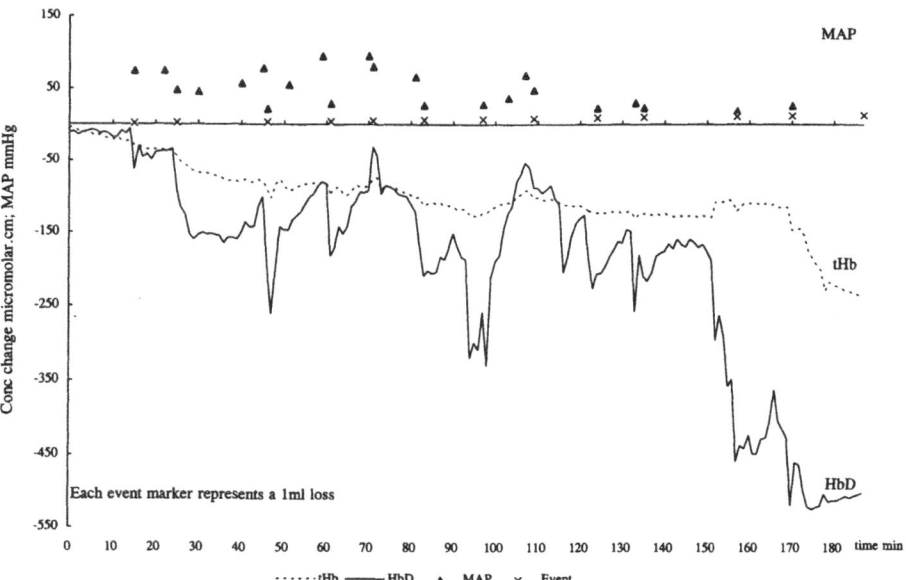

Figure 2. A plot of change in tHb, HbD and mean arterial pressure in response to progressive exsanguination.

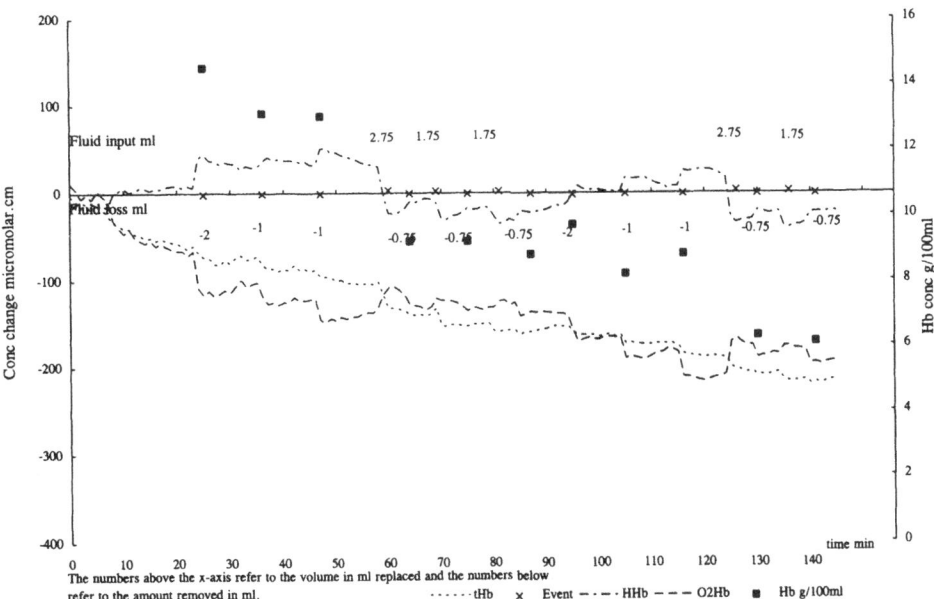

Figure 3. A plot of the change in oxy- and deoxy-haemoglobin as the procedure continues. The non-invasive NIRS total haemoglobin concentration is compared with the in vitro determination of haemoglobin concentration.

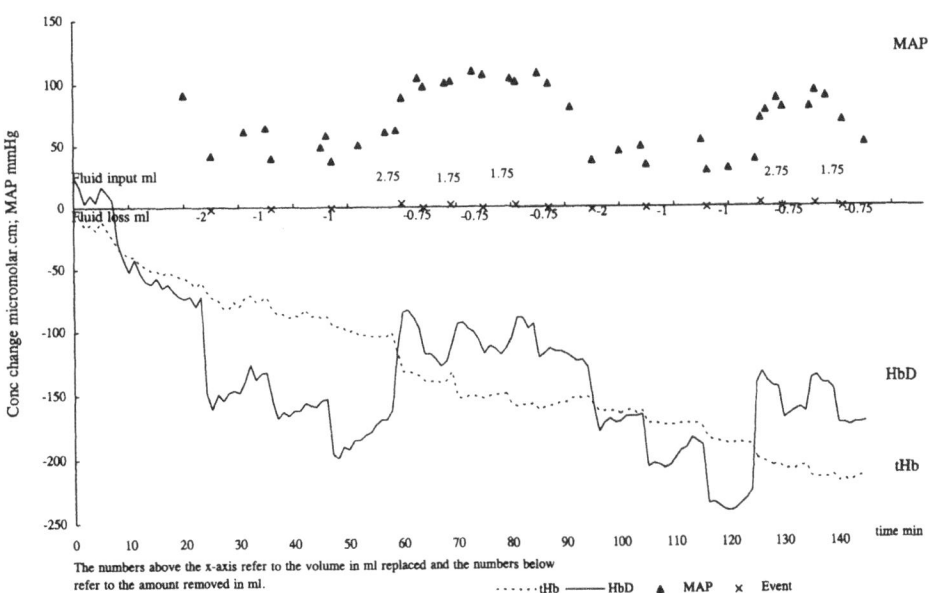

Figure 4. A plot of the change in cerebral tHb, HbD and MAP in response to exsanguination and fluid replacement with Haemaccel.

ery and, as can been seen, is affected by both a change in perfusion pressure as well as haemoglobin concentration.

These preliminary NIRS measurements suggest that fluid replacement restores cerebral oxygenation despite the haemodilution effect and the consequent decrease in haemoglobin oxygen carrying capacity.

The results indicate the potential use of NIRS intraoperatively or in the intensive care unit as a tool in assessing the benefits of fluid or blood replacement in meeting the challenges of various haemodynamic disturbance.

ACKNOWLEDGMENTS

The authors would like to acknowledge the support provided by Ms Kate Rider and Mrs Angela Kidd and Johnson & Johnson Medical especially Dr's Nick Barnett and Pete Evans. We would like to gratefully thank Dr Caroline Dore for her invaluable statistical guidance.

REFERENCES

1. Thorniley M.S., Simpkin S., Sammut I.A., and Green C.J. (1995) The use of a Critikon cerebral redox research monitor model 2001 for assessing cerebral oxygenation and haemodynamics following hepatic transplantation *Clin Sci* **89**: 31P
2. Thorniley M.S., Sammut I.A., Simpkin S., and Green C.J. (1995) An investigation into the effect of hepatic transplantation on cerebral oxygenation and haemodynamics. *Biochem Soc Trans* **23**: 525S.
3. Thorniley M.S., Lane N.J., Manek S. & Green C.J. (1994) Non-invasive measurement of respiratory chain dysfunction following hypothermic renal storage and transplantation. *Kidney Int* **45**: 1489–1496.
4. Lane N, Thorniley M.S., Manek S, Fuller B, Green C.J. (1996) Secondary ischaemia and tissue damage in transplanted stored kidneys *Transplantation* **61**: 689–696.
5. Irwin M.S., Thorniley M.S., Dore C., Green C.J. (1995) Near infra-red spectroscopy: a non-invasive monitor of perfusion and oxygenation within the microcirculation. *Br J Plastic Surg* **48**: 14–22.
6. Thorniley M.S., Lahiri A., Glenville B., Shurey C.B., Baker G., Ravel U., Crawley J., Green C.J. (1996) Non invasive measurement of cardiac oxygenation and haemodynamics during transient episodes of coronary artery occlusion and reperfusion. *Clin Sci* **91(1)**:51–58.
7. Thorniley M.S., Simpkin S., Fuller B., Jenabzadeh M.Z. & Green C.J. (1995) Monitoring of surface mitochondrial NADH levels as an indication of ischaemia during liver isograft transplantation *Hepatology* **21**: 1602–1609.
8. Brunori M., Antonini E. & Wilson M.T. (1981) Metal Ions in Biological Systems XXIII Siegel H (Ed), Marcel Dekker, New York, pp 187–228.

57

MONITORING OF MICROVASCULAR HEMOGLOBIN OXYGENATION IN LIVER AND SKELETAL MUSCLE TISSUE OF ENDOTOXIN-EXPOSED RATS USING REFLECTION SPECTROPHOTOMETRY

Brigitte Vollmar,[1] Dominik Rüttinger,[2] and Michael D. Menger[1]

[1]Institute for Clinical and Experimental Surgery
University of Saarland
66421 Homburg/Saar
[2]Institute for Surgical Research
University of Munich
81366 Munich, Germany

1. INTRODUCTION

Sepsis is considered a disorder of microvascular regulation and increased microcirculatory heterogeneity, producing a mismatch between cellular oxygen requirements and tissue perfusion (Menger et al., 1994; Reinhart et al., 1994). Tissue hypoxia results and may itself perpetuate the septic process (Reinhart et al., 1994). Thus, prevention of tissue hypoxia is a crucial part of the supportive treatment in patients with sepsis and septic shock. Consideration of the determinants of cellular oxygen supply on all levels would be ideal, however, interpretation is limited by the mostly global and indirect indicators of tissue oxygenation (Reinhart et al., 1994). Reflection spectrophotometry is a technique, which allows the determination of microvascular hemoglobin oxygenation non-invasively with high resolution (Frank et al., 1989). Thus, this technique may provide the basis for the study of sophisticated oxygen transport during sepsis or septic shock.

According to this, we studied microvascular hemoglobin oxygenation in liver and skeletal muscle tissue of rats after intravenous endotoxin/lipopolysaccharide (LPS)-exposure using reflection spectrophotometry.

Oxygen Transport to Tissue XIX, edited by Harrison and Delpy
Plenum Press, New York, 1997

2. MATERIALS AND METHODS

2.1. Surgical Procedure

Experiments were performed in accordance with the German legislation on protection of animals and the "Guide for the Care and Use of Laboratory Animals" [DHEW Publication No. (NIH) 85–23, Revised 1985, Office of Science and Health Reports, DRR/NIH, Bethesda, MD20205]. Male Sprague-Dawley rats with a body weight of 186–220g were housed in single cages in an environmentally controlled room with 12h light-dark cycle and had free access to water and standard pellet food. Under chloralhydrate (36 mg/100g body weight i.p.) anesthesia animals were tracheotomized to facilitate spontaneous respiration. The animals were placed in supine position on a heating pad for maintenance of body temperature between 36 and 37°C. Polyethylene catheters (PE 50, ID 0.58 mm, Fa. Portex, Lythe, UK) in the right carotid artery and jugular vein allowed for the assessment of systemic hemodynamics, blood sampling and injection of endotoxin. Following transverse laparotomy the left liver lobe and the left psoas muscle were exposed and covered with an oxygen impermeable saran wrap to prevent drying and interference with the ambient air.

With the use of reflection spectrophotometry (Frank et al., 1989; Mayevsky et al., 1992) (Erlangen microlightguide spectrophotometer EMPHO II, BGT Bodenseewerk Gerätetechnik, Überlingen, Germany), microvascular hemoglobin oxygenation was analyzed in both hepatic and skeletal muscle tissue. For quantitative evaluation of Hb/HbO_2 (desoxy-/oxyhemoglobin spectra) in diffuse reflection from tissues, the spectrophotometer uses a mathematical procedure, involving the backscattering properties and the absorption of hemoglobin (Frank et al., 1989; Kubelka and Munk, 1931). Values of microvascular hemoglobin oxygenation were calculated by the algorithm:

$$C_{Hbox}/(C_{Hbox} + C_{Hbdesox}) \tag{1}$$

where C_{Hbox} represents the concentration of oxygenated hemoglobin and $C_{Hbdesox}$ the concentra-tion of desoxygenated hemoglobin. The microlightguide technique allows analyses from tissue samples as small as 250µm in diameter, however, does not provide a resolution to estimate hemoglobin oxygenation in individual capillaries.

2.2. Experimental Protocol

Measurements (250 spectra from five randomly selected regions per animal and time point each) were performed in 6 animals before (pre-LPS) as well as 1h after LPS-exposure, respectively. Additional animals were analyzed either 6h (n=5) or 16h (n=5) after LPS-exposure. *E. coli* LPS (serotyp 0128:B12; 10 mg/kg; Sigma, Deisenhofen, FRG) was dissolved in sterile saline immediately before administration and applied in a volume of 5ml/kg as a single bolus injection through the penile vein.

2.3. Statistics

All values are given as means ± SEM. Inhomogeneity of microvascular hemoglobin oxygenation was analyzed from each animal by calculation of the coefficient of variation (relative dispersion) as standard deviation/mean of microvascular hemoglobin oxygenation. Correlations between data of mean arterial blood pressure, oxygen saturation and

Table 1. Arterial blood gas parameters and mean arterial blood pressure after LPS-exposure

	Pre-LPS	1h	6h	16h
art O₂sat (%)	96.4±0.7	98.0±0.5	98.7±0.3	98.8±0.1
art PO₂ (mmHg)	90±6	112±6*	130±7*	130±5*
MAP (mmHg)	84±2	67±5*	98±4	96±3

Abbreviations: art O₂sat, oxygen saturation in arterial blood; art PO₂, arterial oxygen partial pressure; MAP, mean arterial blood pressure. Values were obtained prior to and 1h after LPS-exposure (n=6)

PO_2 in arterial blood and microvascular hemoglobin oxygenation were tested using linear regression analysis. Differences between groups were tested by analysis of variance followed by Scheffé's test (appropriate multiple post hoc comparison). To test for time effects (pre-LPS vs. 1h after LPS-exposure) paired Student's t-test was applied. Differences were considered significant for $p<0.05$. Statistics were performed using the software package SigmaStat (Jandel Corporation, San Rafael, CA, USA).

3. RESULTS

In spontaneously breathing animals, LPS-exposure did not significantly influence arterial blood gas parameters, i.e. oxygen saturation, but increased ($p<0.05$) oxygen partial pressure, when compared to pre-LPS values (Table 1). LPS-exposure further resulted in an immediate decrease of mean arterial pressure from 84±2mmHg prior to LPS-exposure to 67±5mmHg at 1h after LPS-exposure (Table 1). At 6h and 16h after LPS-exposure mean arterial blood pressure was found in the range of pre-LPS-values (Table 1).

During pre-LPS baseline conditions, reflection spectrophotometry revealed an average hemoglobin oxygenation of 39.5±3.0% in liver and 46.9±1.2% in skeletal muscle tissue, respectively. Exposure of rats to LPS (1h) resulted in a significant ($p<0.01$) reduction of microvascular hemoglobin oxygenation in liver tissue with a mean value of 28.5±2.0%. Despite this reduction of microvascular hemoglobin oxygenation, heterogeneity of microvascular hemoglobin oxygenation did not markedly increase, as indicated by a coefficient of variation of 0.09±0.02, which was not significantly different from that at pre-LPS conditions (0.05±0.01).

At 6h and 16h after LPS-exposure, fraction of oxygenated hemoglobin in liver tissue was 43.5±2.7% and 45.2±3.8%, indicating complete restoration of hemoglobin oxygenation at later time points after LPS-exposure. Coefficients of variation of microvascular hemoglobin oxygenation in hepatic tissue at 6h and 16h after LPS-exposure were 0.06±0.01 and 0.09±0.03, and, thus, not significantly different from pre-LPS-values. Regression analysis revealed a significant ($p<0.05$) correlation between mean arterial blood pressure (r=0.6; Fig. 1), but not between oxygen saturation (r=0.3; n.s.) and PO_2 (r=0.4; n.s.) in arterial blood, and microvascular hemoglobin oxygenation in hepatic tissue.

In skeletal muscle tissue, reflection spectrophotometry revealed unchanged micro-vascular hemoglobin oxygenation throughout the observation period, demonstrating values of 46.9±1.2, 44.1±1.5, 42.9±3.7 and 47.3±5.4% at 1h, 6h and 16h following LPS-exposure, respectively. In addition, calculation of relative dispersion of microvascular hemoglobin oxygenation did not indicate an increase of heterogeneity upon LPS-exposure, since coefficients of variation did not significantly change with values of 0.07±0.01, 0.10±0.01, 0.09±0.01 and 0.12±0.06 prior to, as well as 1h, 6h and 16h after LPS-exposure, respectively.

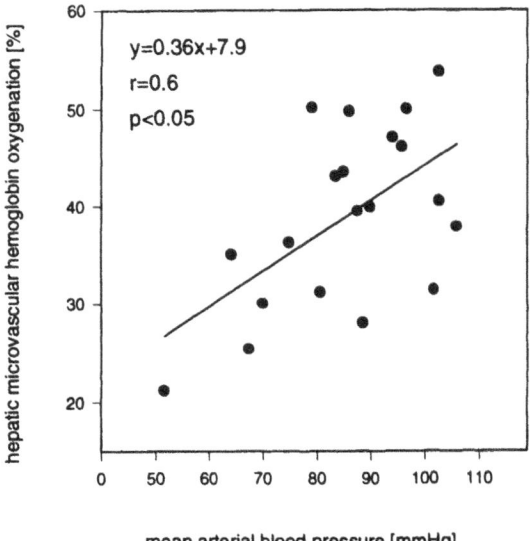

Figure 1. Regression analysis between hepatic microvascular hemoglobin oxygenation (reflection spectrophotometry) and mean arterial blood pressure, including data prior to and 1h (n=6), as well as 6h (n=5) and 16h (n=5) after LPS-exposure (10mg/kg i.v.). r, regression coefficient.

4. DISCUSSION

Beside cellular activation with subsequent cell-cell-interaction (Vollmar et al., 1993; Weingarten et al., 1993), transcellular signaling (Rehring et al., 1996) and release of toxic mediators (Decker, 1990; Lamy and Deby-Dupont, 1995; Thijs and Hack, 1995), endotoxins/LPS were demonstrated (i) to impair regional blood flow due to disturbed autoregulation and opening of arteriovenous shunts (Cunnion et al., 1986; Oettinger et al., 1986), (ii) to alter local vasomotor control (Baker et al., 1990; Cryer et al., 1988; McKenna, 1992) and (iii) to induce microcirculatory disorders by capillary engorgement, endothelial swelling and intravascular hemoconcentration (Balis et al., 1978). These microcirculatory disturbances obviously determine the local mismatch between oxygen supply and demand and the impairment of gas and substrate exchange observed during septic processes. Thus, circulatory disturbances rather than alterations of cellular metabolism are considered as major determinants of tissue oxygen uptake and cellular ischemia in sepsis/septic shock (Thijs et al., 1990).

However, there may be differential microvascular responses to sepsis among tissue, with supposedly skeletal muscle being less susceptible than the gut or the liver (Gutierrez, 1994). Gutierrez et al. (1991) examined the effect of endotoxin administration on rabbit hindlimb skeletal muscle oxygenation measured with surface PO_2 microelectrodes. They noted decreased tissue oxygenation, i.e. development of tissue hypoxia, with endotoxin exposure. However, the distribution of the tissue PO_2 histograms remained constant for 2h following the infusion of endotoxin, implying that skeletal muscle heterogeneity does not increase in early sepsis. This is in line with results of the present study inasmuch as microvascular hemoglobin oxygenation assessed by reflection spectrophotometry in skeletal muscle tissue did not reveal an increase of heterogeneity. The discrepancy between unchanged microvascular hemoglobin oxygenation (present study) and decreased tissue PO_2 (Gutierrez et al., 1991) in skeletal muscle tissue during early sepsis might be due to LPS-induced deterioration of transendothelial oxygen exchange/transport or due to LPS-induced increase of oxygen consumption.

In contrast to skeletal muscle tissue, microvascular hemoglobin oxygenation in hepatic tissue was found markedly reduced at 1h after LPS-exposure, but regained baseline values at 6h and 16h after LPS-exposure. *In vivo* microscopic studies of the liver (McCuskey 1993; Vollmar et al., 1993) revealed significant perfusion failure of individual sinusoids upon endotoxin exposure, probably due to swelling of endothelial cells and activation of Kupffer cells (Vollmar et al., 1996). As similarly proposed for hepatic postischemic reperfusion injury (Goto et al., 1992) impairment of nutritional blood flow following LPS-exposure might account for the reduction of microvascular oxygenation causing tissue hypoxia (Gutierrez et al., 1991), depletion of energy stores (Sugino et al., 1987) and, consequently, damage of parenchymal cells. However, significant correlation between microvascular hemoglobin oxygenation and mean arterial blood pressure both prior to and at 1h, as well as at 6h and 16h after LPS-exposure implies that systemic perfusion conditions are important determinants for oxygen supply to—at least—hepatic tissue.

In summary, we demonstrate that in liver, but not skeletal muscle tissue, LPS-induced transient drop of mean arterial blood pressure is associated with a decrease of microvascular hemoglobin oxygenation. This inadequacy of hepatic oxygen supply might be responsible for the early initiation of liver dysfunction, frequently observed in the clinical course of sepsis.

ACKNOWLEDGMENTS

This study was supported by a grant from the Wilhelm-Sander-Stiftung (Nr. 93.019.1). Empho II was kindly provided by Dr. K. Frank, BGT Bodenseewerk Gerätetechnik, Überlingen, Germany.

REFERENCES

Baker, C.H., Sutton, E.T., Zhou, Z., and Dietz, J.R., 1990, Microvascular vasopressin effects during endotoxin shock in the rat. *Circ. Shock* 30:81–95

Balis, J.U., Rappaport, E.S., Gerber, L., Fareed, J., Buddingh, F., and Messmore, H.L., 1978, A primate model for prolonged endotoxin shock. Blood-vascular reactions and effects of glucocorticoid treatment. *Lab. Invest.* 38:511–523

Cryer, H.G., Garrison, R.N., and Harris, P.D., 1988, Role of muscle microvasculature during hyperdynamic and hypodynamic phases of endotoxin shock in decerebrate rats. *J. Trauma* 28:312–318

Cunnion, R.E., Schaer, G.L., Parker, M.M., Natanson, C., and Parillo, J.E., 1986, The coronary circulation in human septic shock. *Circulation* 73:637–644

Decker, K., 1990, Biologically active products of stimulated liver macrophages (Kupffer cells). *Eur. J. Biochem.* 192:245–261

Frank, K.H., Kessler, M., Appelbaum, K., and Dümmler, W., 1989, The Erlangen micro-light guide spectrophotometer EMPHO I. *Phys. Med. Biol.* 34:1883–1900

Goto, M., Kawano, S., Yoshihara, H., Takei, Y., Hijioka, T., Fukui, H., Matsunaga, T., Oshita, M., Kashiwagi, T., Fusamoto, H., Kamada, T., and Sato, N., 1992, Hepatic tissue oxygenation as a predictive indicator of ischemia-reperfusion liver injury. *Hepatology* 15:432–437

Gutierrez, G., Lund, N., and Palizas, F., 1991, Rabbit skeletal muscle PO_2 during hypodynamic sepsis. *Chest* 99:224–229

Gutierrez, G., 1994, Sepsis and cellular metabolism. In: Reinhardt, K., Eyrich, K., Sprung, C. (eds). *Sepsis. Current Perspectives in Pathophysiology and Therapy.* Springer Verlag, Heidelberg, Germany, pp 181–190

Kubelka, P., and Munk, F., 1931, Ein Beitrag zur Optik der Farbanstriche. *Z. Tech. Phys.* 11a:76–77

Lamy, M., and Deby-Dupont, G., 1995, Is sepsis a mediator-inhibitor mismatch? *Intens. Care Med.* 21:S250-S257

Mayevsky, A., Frank, K., Muck, M., Nioka, S., Kessler, M., and Chance, B., 1992, Multi parametric evaluation of brain functions in the Mongolian gerbil in vivo. *J. Basic. Clin. Physiol. Pharmacol.* 3:323–342

McCuskey, R.S., 1993, Hepatic microvascular responses to endotoxinemia and sepsis. *Prog. Appl. Microcirc.* 19:76–84

McKenna, T.M., 1992, Recovery of vascular tissue contractile function during sustained endo toxin exposure. *Am. J. Physiol.* 263:H1628-H1631

Menger, M.D., Vollmar, B., and Messmer, K., 1994, Sepsis and nutritional blood flow. In: Reinhardt, K., Eyrich, K., Sprung, C. (eds). *Sepsis. Current Perspectives in Pathophysiology and Therapy.* Springer Verlag, Heidelberg, Germany, pp 163–173

Oettinger, W., Roscher, R., Jonescu, J., and Beger, H.G., 1986, Renal prostaglandin-release, blood flow, and function in porcine endotoxic shock. *Langenbecks Arch. Chir.* 187–190

Rehring, T. F., Brew, E.C., Friese, R.S., Banerjee, A., and Harken, A.H., 1996, Clinically accessible cell signaling: Second messengers in sepsis and trauma. *J. Surg. Res.* 60:270–277

Reinhart, K., Hannemann, L., Meier-Hellmann, A., and Specht, M., 1994, Monitoring of O_2 transport and tissue oxygenation in septic shock. In: Reinhardt, K., Eyrich, K., and Sprung, C. (eds). *Sepsis. Current Perspectives in Pathophysiology and Therapy.* Springer Verlag, Heidelberg, Germany, pp 193–213

Sugino, K., Dohi, K., Yamada, K., and Kawasaki, T., 1987, The role of lipid peroxidation in endotoxin-induced hepatic damage and the protective effect of antioxidants. *Surgery* 101:746–752

Thijs, L.G., and Hack, C.E., 1995, Time course of cytokine levels in sepsis. *Intens. Care Med.* 21:S258-S263

Thijs, L.G., Schneider, A.J., and Groeneveld, A.B.J., 1990, The haemodynamics of septic shock. *Intens. Care Med.* 16:S182–186

Vollmar, B., Glasz, J., Senkel, A., Menger, M.D., and Messmer, K., 1993, Role of leukocytes in the initial hepatic microvascular response to endotoxemia. *Zentralbl. Chir.* 118:691–696.

Vollmar, B., Rüttinger, D., Wanner, G.A., Leiderer, R., and Menger, M.D., 1996, Modulation of Kupffer cell activity by gadolinium chloride in endotoxemic rats. *Shock* 6:434–441

Weingarten, R., Sklar, L.A., Mathison, J.C., Omidi, S., Ainsworth, T., Simon, S., Ulevitch, R.J., and Tobias, P.S., 1993, Interactions of lipopolysaccharide with neutrophils in blood via CD14. *J. Leukoc. Biol.* 53:518–524

58

ASSESSMENT OF MICROVASCULAR OXYGEN SUPPLY AND TISSUE OXYGENATION IN HEPATIC ISCHEMIA/REPERFUSION

Brigitte Vollmar and Michael D. Menger

Institute for Clinical and Experimental Surgery
University of Saarland
66421 Homburg/Saar, Germany

1. INTRODUCTION

Reperfusion, following prolonged periods of ischemia, is characterized by nutritive perfusion failure (Menger and Meßmer, 1993; Menger et al., 1993; Suval et al., 1987). Moreover, reoxygenation of ischemic tissue may aggravate postischemic injury by increased oxygen radical production (Carden et al., 1990; Granger et al., 1981; Müller et al., 1996). The cytotoxic effect of reactive oxygen metabolites, in particular on endothelial cells, has been suggested to deteriorate the nutritive microcirculation, which, then, may lead to an alteration of microvascular oxygen supply, and, consequently, to a pronounced prolongation of tissue hypoxia during reperfusion (Menger et al., 1993).

To test this hypothesis, we studied hepatic microvascular oxygen supply by reflection spectrophotometry and hepatic tissue oxygenation by multiwire surface electrodes both during ischemia and postischemic reperfusion using a liver ischemia/reperfusion model in the rat.

2. MATERIALS AND METHODS

2.1. Animal Model and Preparation

Experiments were performed in accordance with the German legislation on protection of animals and the "Guide for the Care and Use of Laboratory Animals (NIH publication no. 86–23, revised 1985). After overnight fasting, but free access to tap water twelve male Sprague-Dawley rats (mean body wt 205±3g) were anesthetized by chloralhydrate (36 mg/100g body wt i.p.) and atropine (0.25 mg s.c). Tracheotomy was performed to facilitate spontaneous respiration (room air, flow rate 2 l/min). The animals were placed in supine position on a heating pad for maintenance of body temperature between 36°C and 37°C (Vollmar et al., 1994a,b).

Oxygen Transport to Tissue XIX, edited by Harrison and Delpy
Plenum Press, New York, 1997

The right carotid artery and jugular vein were cannulated with polyethylene catheters (PE 50, ID 0.58 mm, Fa. Portex, Lythe, UK) allowing for the assessment of arterial blood pressure and heart rate, as well as for withdrawal of arterial blood samples. Supplementary chloralhydrate (3.6 mg/100g body wt) was given i.p. if required.

Following transverse laparotomy the ligamentous attachments from the liver to the diaphragm and abdominal wall were dissected, and the liver lobes were covered with an oxygen impermeable saran wrap to prevent drying and interference with the ambient air (Vollmar et al., 1994a,b). Left hepatic lobar ischemia was induced by clamping the left hepatic artery and the portal branch. After an ischemic period of either 20min (n=7) or 60min (n=5), reperfusion for a total of 60min was initiated by removal of the vascular clamp. At the end of each experiment, arterial blood samples were withdrawn for spectrophotometric determination of serum aspartate (AST) and alanine (ALT) aminotransferase activities, serving as an indicator of hepatocellular damage.

2.2. Techniques

2.2.1. Platinum Multiwire Oxygen Sensitive Surface Electrode. The use of multiwire-surface electrodes for determination of surface tissue PO_2 has been described in detail elsewhere (Kessler et al., 1976). The measuring device consists of a Clark-type electrode incorporating eight thin platinum cathodes and one Ag/AgCl anode. At a constant polarization voltage of -700mV, a single platinum wire (diameter 15μm) registers a reduction current linearly dependent on partial pressure of oxygen (PO_2). The radius of the oxygen-sensitive surface area of the wires is 20–25μm with a membrane of 25μm thickness. Electrodes were calibrated in physiologic saline solutions equilibrated with pure nitrogen and 5% and 10% oxygen in nitrogen, before and after each measurement. Reduction current was found to be linear within this range. Reduction current signals from each of the eight channels were amplified, digitized and processed with a PDP 11/23 computer (Digital Equipment Corporation, Maynard, Mass).

At each experimental step, a total of 96–120 individual oxygen pressure measurements was obtained at 12–15 different electrode locations on the hepatic surface. The distribution of the PO_2 values reflects average tissue oxygenation, giving the net result of nutritive blood flow, microcirculatory blood flow distribution and oxygen consumption (Nylander et al., 1983). PO_2 histograms were constructed in the cumulative form, with the number of values in each new class (PO_2 class width: 5mmHg) added to the total of those in the preceding classes. From the point of view of clarity, this form of presentation is advantageous inasmuch as it allows to superimpose numerous curves in one diagram and is relatively easy to judge the homogeneity of the distribution of PO_2 values from the steepness of the curve.

Using the platinum multiwire oxygen sensitive surface electrode, hepatic tissue PO_2 was assessed during baseline, at the end of the ischemic period after either 20min or 60min and at 60min of reperfusion, respectively.

2.2.2. Reflection Spectrophotometry. With the use of reflection spectrophotometry (Frank et al., 1989; Mayevsky et al., 1992) (Erlangen microlightguide spectrophotometer EMPHO II, BGT Bodenseewerk Gerätetechnik, Überlingen, Germany), microvascular hemoglobin oxygenation was analyzed during baseline, during induction of ischemia and at the end of ischemia, as well as during onset of reperfusion and at the end of reperfusion. For quantitative evaluation of Hb/HbO$_2$ (desoxy-/oxyhemoglobin spectra) in diffuse reflection from tissues, the spectrophotometer uses a mathematical procedure, involving the backscattering properties and the absorption of hemoglobin (Frank et al., 1989; Kubelka and Munk, 1931). Values of microvascular hemoglobin oxygenation were calculated by the algorithm:

$$C_{Hbox}/(C_{Hbox} + C_{Hbdesox}) \qquad (1)$$

where C_{Hbox} represents the concentration of oxygenated hemoglobin and $C_{Hbdesox}$ the con-centra-tion of desoxygenated hemoglobin. The microlightguide technique allows analyses from tissue samples as small as 250µm in diameter, however, does not provide a resolu-tion to estimate hemoglobin oxygenation in individual capillaries.

2.3. Statistics

All values are given as means ± SEM. To test for time effects (baseline, ischemia, reperfusion) paired Student's t-test was applied. Differences between groups were tested by the unpaired Student's t-test and were considered significant for $p<0.05$. Statistics were performed using the software package SigmaStat (Jandel Corporation, San Rafael, CA, USA).

3. RESULTS

Twenty min and 60min left hepatic lobar ischemia followed by 60min of reperfusion did not significantly alter mean arterial blood pressure when compared to preischemic baseline conditions (see Table 1). In parallel, heart rate was not affected by either 20min or 60min of ischemia followed by reperfusion. In addition, analysis of systemic hematocrit revealed unchanged values throughout the experimental procedure (Table 1).

At baseline, distribution pattern of single PO_2 values showed physiologic tissue oxy-genation (20.5±1.5mmHg), indicated by the sigmoid shape of the cumulative histogram with less than 5% of PO_2 values ranging between 0–5mmHg (Fig.1). In contrast, 20min, and -in particular- 60min ischemia resulted in a left shift of the PO_2 histograms (mean: 3.5±0.7 and 1.2±0.5mmHg) with a left-skewed distribution pattern of single PO_2 values, indicated by the parabolic shape of the cumulative histograms. Reestablishment of hepatic

Table 1. Macrohemodynamic parameters and hematocrit in hepatic ischemia and reperfusion

		Baseline	I	R
MAP (mmHg)	20' ischemia	91±8	86±7	81±7
	60' ischemia	89±4	90±2	89±7
HR (min⁻¹)	20' ischemia	430±20	390±8	400±8
	60' ischemia	424±20	392±20	418±9
Hct (%)	20' ischemia	46±1	45±1	47±1
	60' ischemia	45±1	45±2	46±1

Abbreviations: MAP, mean arterial pressure; HR, heart rate; Hct, systemic he-matocrit. Values were obtained at baseline conditions (pre-ischemia), at the end of ischemia (I) and at the end of reperfusion in animals with 20min (n=7) and 60min (n=5) of ischemia of the left hepatic lobe. Mean±SEM. No statisti-cal significant differences between the two groups or within the groups at the different time points were observed.

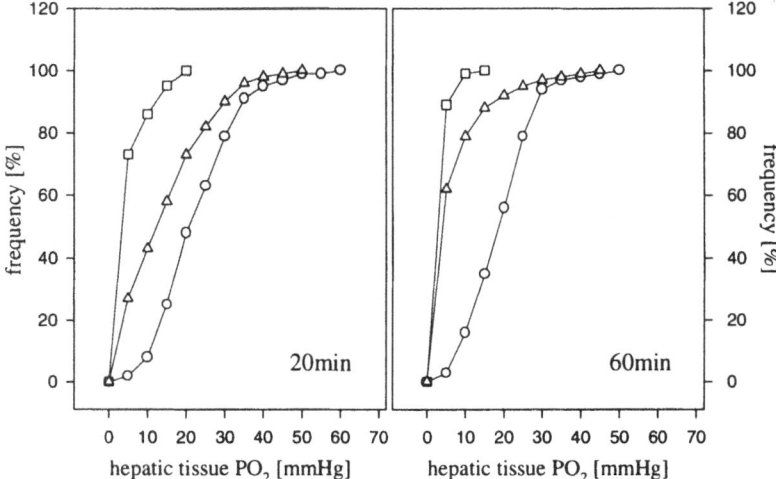

Figure 1. Cumulative histogram of hepatic surface tissue PO$_2$ (multiwire surface electrodes) at baseline (circle), at the end of ischemia (square), and after 60min of reperfusion (triangle) in animals with 20min (n=7; left panel) and 60min of left lobar ischemia (n=5; right panel).

arterial and portal venous lobar inflow after 20min and 60min ischemia did increase hepatic tissue PO$_2$ (15.0±3.0 and 6.0±0.9mmHg), but failed to restore preischemic values (Fig. 1).

Reflection spectrophotometry revealed a hepatic microvascular hemoglobin oxygenation of 43.5±1.6% at baseline conditions. After induction of ischemia, analysis of microvascular hemoglobin oxygenation verified a rapid desoxygenation within less than 10sec (9.8±0.9sec). Regardless the duration of ischemic time period, i.e. at the end of either 20min or 60min hepatic lobar ischemia, the fraction of oxygenated hemoglobin was 0.4±1.2%, indicating almost complete hemoglobin desoxygenation. In contrast, hepatic tissue PO$_2$ after 20min ischemia was still slightly higher when compared to values after 60min ischemia (3.5±0.7mmHg vs. 1.2±0.5mmHg). With reestablishment of hepatic blood flow after 20min ischemia, microvascular hemoglobin oxygenation quickly (7.9±2.4sec) increased, however without reaching baseline values. During reperfusion following 60min ischemia, impairment of hepatic microvascular hemoglobin oxygenation was more pronounced when compared with that following 20min ischemia, presenting with reduced values of 33.5±3.3%.

In animals subjected to 60min lobar ischemia, serum AST and ALT activities (1301±453 and 960±347 U/L) were found significantly ($p<0.01$) higher than those measured in sera of animals subjected to 20 min lobar ischemia (239±44 and 140±43 U/L).

4. DISCUSSION

Recent studies have demonstrated that reestablishment of blood flow after prolonged periods of ischemia aggravates ischemia-induced tissue damage, by causing additional injury and/or by unmasking injury sustained during the ischemic period. In accordance to the distinct key components in the pathogenesis of I/R-induced microvascular injury of striated muscle, i.e. no-reflow (Menger et al., 1992b) and reflow-paradox (Menger et al.,

1992a), hepatic lobar ischemia/reperfusion is associated with perfusion failure of sinusoids (Vollmar et al., 1994a), as well as microvascular accumulation of leukocytes, including their sinusoidal stasis and postsinusoidal venular adherence (Vollmar et al., 1994b, 1995, 1996). These microcirculatory disturbances are thought to alter hepatic tissue oxygenation during postischemic reperfusion thereby determining subsequent liver injury.

The present study reports on the useful application of reflection spectrophotometry and multiwire surface electrodes for assessment of either hepatic microvascular oxygen supply or hepatic tissue oxygenation in ischemia/reperfusion. With onset of lobar vessel occlusion, reflection spectrophotometry displays the desoxygenation of microvascular hemoglobin, which is a rapid process with—in case of complete stop of inflow— almost total desaturation of hemoglobin, independent to the subsequent ischemic time period. Therefore, values did not differ at the end of the 20-min- or 60-min period of ischemia. Since the assessment of tissue oxygenation by means of multiwire surface electrodes is thought to mirror the net result of oxygen supply to demand, hepatic tissue PO_2 values decrease with duration of ischemic time period, as given by the still higher mean value of 3.5 ± 0.7mmHg at the end of 20min of ischemia when compared with that of 1.2 ± 0.5mmHg at the end of 60 min of ischemia.

With reestablishment of hepatic lobar blood flow, intravascular oxygenation of hemoglobin measured by reflection spectrophotometry increased, however, without reaching baseline conditions. The decreased oxygen supply indicates decreased blood flow, and in fact, in this model, hepatic microvascular perfusion is significantly deteriorated, as given by $6.1\pm1.0\%$ and $12.6\pm2.9\%$ non-perfused sinusoids during reperfusion after 20min and 60min of lobar ischemia, respectively (Vollmar et al., 1994a). As similarily proposed by Goto and coworkers (1992), postischemic microvascular perfusion failure seems to be the dominant factor contributing to the decreased oxygen supply, i.e. reduced microvascular hemoglobin oxygenation, which in turn leads to prolongation of parenchymal tissue hypoxia despite onset of reperfusion and reoxygenation via the hepatic artery and portal vein. The fact that hepatic tissue PO_2 values at the end of reperfusion after 60min of lobar ischemia are found significantly lower than those after 20min of ischemia, verifies the dependency of tissue oxygenation to oxygen supply via sinusoidal blood flow.

We conclude that prolongation of tissue hypoxia via altered perfusion and, thus, insufficient energy supply significantly contributes to the manifestation of reperfusion injury, which is documented by the elevated liver enzyme activities in animals after 20min of lobar ischemia, but in particular after 60min of lobar ischemia followed by 60min of reperfusion.

ACKNOWLEDGMENTS

This study was supported by a grant from the Deutsche Forschungsgemeinschaft (Me900/1–2). Empho II was kindly provided by Dr. K. Frank, BGT Bodenseewerk Gerätetechnik, Überlingen, Germany.

REFERENCES

Carden, D.L., Smith, J.K., and Korthuis, R.J., 1990, Neutrophil-mediated microvascular dysfunction in postischemic canine skeletal muscle. *Circ. Res.* 66:1436–1444

Frank, K.H., Kessler, M., Appelbaum, K., and Dümmler, W., 1989, The Erlangen micro- lightguide spectro-photometer EMPHO I. *Phys. Med. Biol.* 34:1883–1900

Goto, M., Kawano, S., Yoshihara, H., Takei, Y., Hijioka, T., Fukui, H., Matsunaga, T., Oshita, M., Kashiwagi, T., Fusamoto, H., Kamada, T., and Sato, N., 1992, Hepatic tissue oxygenation as a predictive indicator of ischemia-reperfusion liver injury. *Hepatology* 15:432–437

Granger, D.N., Rutili, G., and McCord, J.M., 1981, Superoxide radicals in feline intestinal ischemia. *Gastroenterology* 81:22–29

Kessler, M., Hoeper, J., and Krumme, B.A., 1976, Monitoring of tissue perfusion and cellular function. *Anesthesiology* 45:184

Kubelka, P., and Munk, F., 1931, Ein Beitrag zur Optik der Farbanstriche. *Z. Tech. Phys.* 11a:76–77

Mayevsky, A., Frank, K., Muck, M., Nioka, S., Kessler, M., and Chance, B., 1992, Multi parametric evaluation of brain functions in the Mongolian gerbil in vivo. *J. Basic. Clin. Physiol. Pharmacol.* 3:323–342

Menger, M.D., Pelikan, S., Steiner, D., and Messmer, K., 1992a, Microvascular ischemia/re perfusion injury in striated muscle: Significance of "reflow-paradox". *Am. J. Physiol.* 263: H1901-H1906

Menger, M.D., Steiner, D., and Messmer, K., 1992b, Microvascular ischemia/reperfusion injury in striated muscle: Significance of "no-reflow". *Am. J. Physiol.* 263: H1892-H1900

Menger, M.D., and Meßmer, K., 1993, Die Mikrozirkulation des Skelettmuskels nach Ischämie und Reperfusion. *W.M.W.* 7/8:148–158

Menger, M.D., Vollmar, B., Glasz, J., Müller, M.J., Post, S., and Messmer, K., 1993, Hepatic ischemia/reperfusion. *Prog. Appl. Microcirc.* 19:106–124

Müller, M.J., Vollmar, B., Friedl, H.-P., and Menger, M.D., 1996, Xanthine oxidase and super oxide radicals in portal triad cross-clamping-induced microvascular reperfusion injury of the liver. *Free Rad. Biol. Med.* (in press)

Nylander, E., Lund, N. and Wranne, B., 1983, Effect of increased blood oxygen affinity on skeletal muscle surface oxygen pressure fields. *J. Appl. Physiol.* 54:99–104

Suval, W.D., Duran, W.N., Boric, M.P., Hobson, R.W., Berendsen, P.B., and Ritter, A.B., 1987, Microvascular transport and endothelial cell alterations preceding skeletal muscle damage in ischemia and reperfusion injury. *Am. J. Surg.* 154:211–218

Vollmar, B., Glasz, J., Leiderer, R., Post, S., and Menger, M.D., 1994a, Hepatic microcircula tory perfusion failure is a determinant of liver dysfunction in warm ischemia-reperfusion. *Am. J. Pathol.* 145:1421–1431

Vollmar, B., Menger, M.D., Glasz, J., Leiderer, R., and Messmer, K., 1994b, Impact of leukocyte-endothelial cell interaction in hepatic ischemia-reperfusion injury. *Am. J. Physiol.* 267:G786-G793

Vollmar, B., Glasz, J., Menger, M.D., and Messmer, K., 1995, Leukocytes contribute to hepatic ischemia-reperfusion injury via ICAM-1 mediated microvascular adherence. *Surgery* 117:195–200

Vollmar, B., Richter, S., and Menger, M.D., 1996, On leukocyte stasis in hepatic sinusoids. *Am. J. Physiol.* 270:G798-G803

59

OXYGEN SATURATION OF INTRACAPILLARY HAEMOGLOBIN IN PATIENTS WITH SYSTEMIC JCA (STILL'S DISEASE)

J. P. Haas,[1] K. Höper,[1] G. Leipold,[1] H. G. Dörr,[1] and J. Höper[2]

[1]Universitätsklinik mit Poliklinik für Kinder und Jugendliche
Loschgestr. 15
[2]Institut für Physiologie und Kardiologie
Waldstr. 6
Friedrich-Alexander-Universität Erlangen-Nürnberg
D-91054 Erlangen, Germany

1. INTRODUCTION

Juvenile chronic arthritis (JCA) is a group of diseases characterized by an arthritis with a pauciarticular (up to 5 joints involved) or a polyarticular type of onset, lasting for more than 6 weeks and an age of onset within the first 16 years of life [Wood P.N. 1977]. According to the clinical course, as well as to certain laboratory findings, different subgroups of JCA are defined. JCA is supposed to be an autoimmune disease of childhood and adolescence. The pathogenesis of the disease still remains unclear although numerous findings point to a disturbance in antigen-presentation between MHC (major histocompatibility complex) molecules and T-cells [Haas J.P. et al. 1991, Fink C.W. 1983]. The most dramatic clinical course of JCA is found in systemic JCA (S-JCA) characterized by pauci- or polyarticular course of arthritis accompanied by symptoms of severe general inflammation. High fever, exanthema and a general lymphadenopathy which are typical symptoms of S-JCA indicate that this disease is the simultaneous appearance of JCA and systemic vasculitis in childhood [Jakobs J.C. 1993].

We designed a study to gain precise data on the intracapillary haemoglobin oxygen saturation (SO_2-values) in different autoimmune diseases in children. The results were that SO_2-values in the skin of children suffering from JCA are not related to the clinical symptoms of hyperperfusion (rubor, hyperthermia) at the skin above affected joints and areas of exanthema [Haas J.P. et al. 1996]. The only exception was found in patients suffering from SLE and S-JCA. Here we report the results from measurements of SO_2-values in the skin of patients with S-JCA.

Oxygen Transport to Tissue XIX, edited by Harrison and Delpy
Plenum Press, New York, 1997

2. METHODS AND PATIENTS

2.1. Measurement of SO$_2$-Values in Skin's Capillaries

Oxygenation of intracapillary haemoglobin (SO$_2$) was measured with the micro lightguide spectrophotometer EMPHO (Erlanger Mikrolichtleiter Spektrophotometer, BGT Überlingen) [Frank K.H. 1989]. White light from a Xenon-lamp is parallelized, focused on to a micro-lightguide fibre and transmitted to the tissue's surface by a central micro-lightguide fibre closely surrounded by a hexagon of six detection fibres (diameter of each fibre 250 μm). The remitted light passes through a rotating interference bandpass disk (502–628nm). After photomultiplying the collected spectrum is analyzed. Collection of data lasts about 45 seconds for each area (measurement-areas approx. 1cm^2). The data render a calculation of the oxygenation of the intracapillary haemoglobin by a computer program according to Dümmler [Dümmler W. 1988]. The method is based on the Kubelka-Munk theory [1931]. The program allows on-line calculation of local SO$_2$ in intracutaneous vessels up to a depth of approx. 1 mm. More detailed informations about analysing the data are given by Höper J. et al. [1994] and Höper K. et al. [1996] elsewhere.

2.2. Statistical Analysis

Statistical evaluation used Spearman rank-correlation, Student's t-test, variation- and regression analysis considering $p < 0.05$ to be significant. The histograms show the frequency distributions of SO$_2$-values pooling the data measured at all different areas of our patients. From each area we analysed 3000 spectra with an average rate of 10. The curve over the histograms depicts the fitted normal distribution. Figure 1 shows the SO$_2$-values from 3 patients with S-JCA; 7 different areas were measured for each patient (total: 6044 averages each a. and b.). Measurements were performed in an active state of disease (a.) and as a follow up after treatment with glucocorticoids (b.). Figure 2 gives the values for one patient in 9 different areas (total: 2700 averages each a. and b.) in active disease (a.) and after treatment with acetylsalicylates (b). Areas of measurement were essentially identical before and after treatment in both groups.

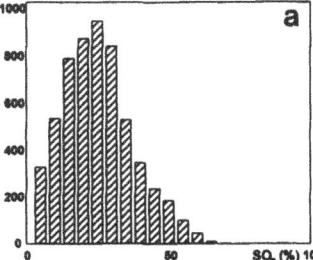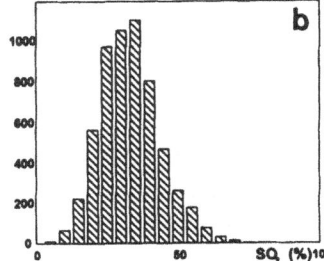

Figure 1. Patients T.D., Z.C., F.S. (a) Before treatment with glucocorticoids: High activity of S-JCA. (b) 2–6 weeks later after treatment with glucocorticoids (2mg/kg): Less activity of S-JCA.

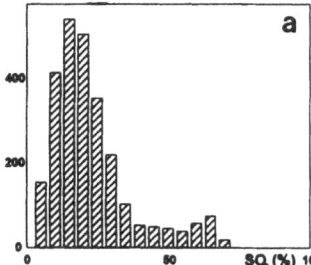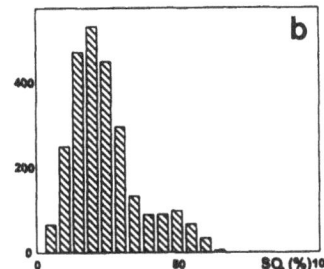

Figure 2. Patient Z.H. (a) Before treatment with salicylates: High activity of S-JCA. (b) 2 weeks later after treatment with salicylates (100mg/kg): Less activity of S-JCA.

2.3. Patients

Four patients suffering from S-JCA (age between 4–10 years) have been examined. They are under the care of the pediatric rheumatology division. None of them had lung disease, blood-gas values were in normal ranges. Patients were measured after adaptation to room-temperature in upright sitting position without being exposed to physical stress. Areas of measurement were marked in cases where comparison tests were planned. The patients were treated with different drugs commonly used in JCA [Jakobs J. C. 1993]. Between the two times of SO_2-measurement only glucocorticoid- or salicylate treatment was changed. As an example of laboratory data we present here the erythrocyte sedimentation rate (ESR) and the C-reactive protein (CRP) neither are specific for S-JCA but are good indicators for systemic inflammatory processes. Detailed information is given in table 1.

3. RESULTS

Figures 1a and 2a show a summary of the data pooling SO_2-values from all measured areas of the patients. In the active state of S-JCA all of the patients had significantly decreased SO_2-values at any area measured compared to those of healthy volunteers [Höper K. et al. 1996] and those of patients with others forms of autoimmune disease

Table 1. Clinical data of patients

Patient	AOM	n	Medication	Corticoids	Salicylates	ESR (mm), CRP (mg/dl) Before	After
T.D.	W, K, A	7	—	2mg/kg	50mg/kg	90/160, 14.0	52/98 4.2
Z.C.	W, K, A	7	MTX, AZA	2mg/kg	—	50/100, 7.4	23/56 2.6
F.S.	K, A, S	7	MTX, AZA	2mg/kg	—	100/152, 15.0	23/46 0.7
Z.H.	K, A, S	9	—	—	100mg/kg	132/145, 7.6	80/133 2.7

Areas of measurement (AOM): W = wrist, K = knee, A = ankle, S = skin (erythema). Medication: MTX = methotrexate (10mg/m^2 bodysurface area), AZA = azathioprin (2mg/kg).

Table 2. SO$_2$-values in S-JCA patients

SO$_2$-values	Glucocorticoid treatment (n=3)		Salicylate treatment (n=1)	
	Before	After	Before[#]	After
Median SD	22.9±12.4	30.8*±10.8	20.6±14.5	22.3*±12.3
SO$_2$-values < 10%	15.0	1.4*	21.6	12.2
SO$_2$-values < 20%	43.7	14.9*	61.2	51.0

SD = standard deviation
* significant with p < 10^{-5}
# significant with p < 0.001

[Haas J.P. et al. 1996]. As shown in table 2 the lowest category (SO$_2$ < 10%) was seen in 15% and 21.6%, respectively.

Significant correlation between SO$_2$-values and disease activity was found in all patients. In the group treated with glucocorticoids during the follow-up after therapy patients were found to have improved, with normal SO$_2$-values in all areas of investigation. SO$_2$-values below 10% as well as those below 20% were found with a significantly decreased frequency compared with the first measurements (p < 10^{-5}).

Figures 1b and 2b show the situation in follow-up measurements. All patients had improved according to their clinical state and their laboratory parameters. Nevertheless, the patient treated only with salicylates showed a general decrease with persisting subnormal SO$_2$-values (fig. 2b). Comparing the data of the glucocorticoid treated group with those of the patient treated with salicylates we found significant difference in SO$_2$-values (p < 0.001).

4. DISCUSSION

Although observed in a very limited number of patients the data indicate periods of severe intracapillary hypoxia in childhood systemic rheumatic diseases during acute states of disease. This hypoxia seems to be a local problem because systemic oxygen saturation shows no differences. As we have shown in our study on SO$_2$-values measured in other forms of juvenile autoimmune disease the decrease in SO$_2$-values seems not to be a consequence of subcutaneous edema, which one observes around the joints involved in any kind of arthritis [Haas J.P. et al. 1996]. The intracapillary hypoxia observed in S-JCA seems to correlate with disease activity and with the use of certain drugs. We propose it to be the consequence of a microangiopathy. It might be the basis as well as the result of immunologic interactions in the micro vessels. Only immunosuppression with glucocorticoids leads to normal SO$_2$-values in the skin of our patients. There seems to be less effect of other antiphlogistic and/or immunosuppressive drugs on SO$_2$-values in S-JCA.

The discussion of the data has to include the relations between intracapillar oxygenation and inflammation as well as the influence of glucocorticoids on small blood vessels. Glucocorticoids are both immunosuppresive and anti-inflammatory. They increase the synthesis of lipocortins (i.e. by macrophages) which inhibit the binding of phospholipase A$_2$ (PLA$_2$) to its substrate resulting in a decrease of prostaglandin and leukotrien synthesis [Schleimer et al. 1989]. Glucocorticoids also inhibit the upregulation of ICAM 1 (inter-cellular adhesion molecule) [Tessier P. et al. 1993] and the expression of E-selectin (ELAM 1) on endothelial cells [Cronstein B. 1993].

Hypoxia has been shown to be a potent stimulus in inflammation causing less efficient ATP sythesis and generation of reactive oxygen species leading to an activation of the arachidonic pathway via PLA_2 [Dennis E.A. 1994]. Reoxygenation after episodes of hypoxia corresponds with activation of neutrophils via platelet activating factor (PAF), interleukin 8 (IL8) and β-integrins (CD11/CD18) and ICAM 1 [Rainger G.E. et al. 1995]. Moreover hypoxia causes angiogenesis by VPF/VEGF (vascular permeability factor/vascular endothelial growth factor) expression [Dvorak H.F. et al. 1995]. The multiple interactions between tissue oxygenation, "hypoxic reperfusion" injury and endothelium have been shown to be relevant in the pathogenesis of arthritis [Edmonds S.E. et al. 1995].

With respect to the small number of cases measured so far, our data point out to the fact that in S-JCA acute inflammatory phases correspond with an intracapillary hypoxia which might be the consequence of immunological interactions at the endothelium of small vessels. We assume that local hypoxia is a relevant circumstance in the pathogenesis of S-JCA.

REFERENCES

1. Wood Ph.N.: Special Meeting on Nomenclature and Classification of Arthritis in Children. *EULAR Bull.*, 1977; No.4: 101.

2. Haas J.P., Andreas A., Rutkowski B., Brunner H., Keller E., Hoza J., Havelka St., Sierp G., Albert E.D. "A Model for the Role of HLA-DQ Molecules in the Pathogenesis of Juvenile Chronic Arthritis" *Rheumatol. Int.* 11: 191–197 (1991).

3. Fink C.W.: Clinical, Genetic and Therapeutic Aspects of Juvenile Arthritis. *Clinical Rheumatology in Practice*, 1983; 5/6: 100–115.

4. Jakobs J.C.: Pediatric rheumatology for the practitioner. 1993; J.C. Jakobs, 2nd ed. Springer-Verlag, New York, 241–272.

5. Haas J.P., Höper K., Leipold G., Dörr H.G., Höper J.: Oxygenation of intracapillary haemoglobin in children with autoimmune diseases. (manuscript submitted to *Int. J. of Microcirculation*, 1996).

6. Frank K.H., Kessler M., Appelbaum K., Dümmler W.: The Erlanger micro-lightguide spectrophotometer EMPHO I. *Physics Med. Biol.*, 1989; 34: 1883–1900.

7. Dümmler W.: Bestimmung von Hämoglobin-Oxigenierung und relativer Hämoglobinkonzentration in biologischen Systemen durch Auswertung von Remissionsspektren mit Hilfe der Kubelka-Munk Theorie. Dissertation, 1988. Friedrich-Alexander Universität, Erlangen-Nürnberg.

8. Kubelka-Munk : Ein Beitrag zur Optik der Farbanstriche. *Zeitsch. f. techn. Physik*, 1931, 12: 593.

9. Höper J., Gaab M.R.: Effect of arterial PCO_2 on local HbO_2 and relative Hb concentration in the human brain a study with the Erlangen micro-lightguide spectrophotometer (EMPHO). *Physiol. Meas.*, 1994, 15: 107–113.

10. Höper K., Dörr H.G., Höper J.: Age dependency of intracapillary haemoglobin oxygenation measured in nailfold of healthy volunteers. (Manuscript submitted *Physiol. Meas.*, 1996).

11. Schleimer R.P., Claman H.N., Oronsky A. (eds.): Anti-inflammatory Steroid action. Basic and clinical aspects. San Diego academic press, 1989.

12. Tessier P., Audette M., Cattaruzzi P., M Coll S.R.: Up-regulation by tumor necrosis factor of intracellular adhesion molecule 1 expression and function in synovial fibroblasts and its inhibition by glucocorticoids. *Arth. & Rheum.*, 1993; 36(11): 1528–1539.

13. Cronstein B.N.: The pharmacology of antiinflammatory agents. *The M. Sinai J. of. Med.*, 1993; 60(3): 209–217.

14. Dennis E.A.: Diversity of group types, regulation and function of phospholipase A_2. *J. Biol. Chem.*, 1994; 269: 13057–13060.

15. Rainger G.E., Fisher A., Shearmen C., Nash G.B.: Adhesion of flowing neutrophils to cultured endothelial cells after hypoxia and reoxygenation in vitro. *Am. J. Physiol.*, 1995; 269: H1398–1406.

16. Dvorak H.F., Detmar M., Claffey K.P., Nagy J.A., v.d. Water L., Senger D.R.: Vascular permeability factor/vascular endothelial growth factor: an important mediator of angiogenesis in malignancy and inflammation. *Int. Arch. Allergy Immunol.*, 1995; 107: 233–235.

17. Edmonds S.E., Ellis G., Gaffney K., Archer J., Blake D.R.: Hypoxia and the rheumatic joint: immunological and therapeutic implications. *Scand. J. Rheumatol.*, 1995; 24 (Suppl 101): 163–168.

60

OXYGEN DEPENDENT REGULATION OF CAPILLARY FLOW

Fact or Fiction?

Jens Höper,[1] Manfred Kessler,[1] Ronie Abi Raad,[1] and Richard Funk[2]

[1]Institut für Physiologie und Kardiologie
 Friedrich-Alexander-Universität Erlangen-Nürnberg, Germany
[2]Institut für Anatomie
 Universität Dresden, Germany

In order to carry out their normal functions, living organisms depend on a continuous supply of oxygen to all cells. Oxygen deprivation will inevitably lead to progressive alterations in biochemical processes and physiological disfunction. These disturbances will cause a breakdown of the homeostatic abilities of the organism.

Investigations of local oxygen supply have shown that under normal resting conditions the lowest pO_2 in tissue approaches 1–5 mm Hg, but no tissue anoxia is found (28, 29, 30). Measurements of local intracapillary oxygen saturation of haemoglobin (SO_2c) revealed similar results, under normal conditions no values below 20% SO_2c are found.

An adequate local oxygen supply can be maintained only by an optimal regulation of capillary flow. To be effective in the homeostasis of local oxygen supply the underlying mechanism should involve an oxygen dependent reaction with a K_mO_2 higher than that of cytochrome oxidase. This would be the sine qua non condition for the avoidance of local anoxia.

During recent years it has been shown that NO plays an important role in the regulation of vascular smooth muscle tone and thus in the regulation of peripheral resistance and capillary flow. Just recently Jia et al. (22) published clear evidence that haemoglobin is involved in the local control of NO and related compounds.

From results gained in different organs it must be concluded that this may not be the only mechanism controlling capillary flow.

INVESTIGATIONS IN BRAIN

It is well known that hypoxaemia results in an increase of global cerebral blood flow (CBF) accompanied by a decrease in cerebrovascular resistance. The increased flow in hypoxaemia is not evenly distributed throughout the brain but is related to the local normoxic flow rate (44) which in turn is proportional to the local metabolic consumption rate of glucose (45).

During hypoxaemia the most pronounced increase in local CBF is thus observed in structures having the highest metabolic rate during normoxia. Moreover, since overall oxygen uptake does not decline as long as the arterial PO_2 (P_aO_2) is above 25 mm Hg (2, 23) and in some structures even an increased O_2-uptake is observed (36, 49, 51) the local regulation of cerebral capillary blood flow must operate to cover these metabolic demands and to avoid local anoxia. Experiments performed by Kozniewska et al. (31) clearly showed that the changes in CMF found during hypoxia are closely related to the oxygen tension not in the arterial blood, but in the tissue. At similar P_aO_2 values (35.9 ± 6.1 mm Hg and 34.5 ± 1.7 mm Hg) the changes in CMF were dependent on local tissue pO_2. As long as only a limited number of pO_2 values were below 5 mm Hg only a redistribution of capillary flow was observed; i.e. , some CMF values increased and some decreased. The same had been described for the liver (28). With an increasing number of low pO_2 values not only a redistribution was observed but also an increase in mean CMF. From these experiments it was concluded that CMF changes during hypoxaemia are closely related to the prevailing local tissue pO_2. The "critical pO_2" at which an increase in capillary flow was always observed was approximately 5 mm Hg.

Experiments performed by Höper and Kozniewska (20) showed that MAO-B may participate in the regulation of capillary flow in cerebral cortex. Traystman et al. (50) also suggested from their experiments that an oxidase or oxygenase must be involved in the regulation of O_2-supply to the brain.

The administration of deprenyl, a specific blocker of MAO-B, during normoxia causes an increase of CBF and CMF by 29% and 28%, respectively. The increase of CBF during hypoxaemia ($P_aO_2 = 34.9 \pm 6.3$ mm Hg) following deprenyl administration , although diminished, was still present (101% versus 215% in control experiments). CMF decreased under these conditions by 30% on average. These results clearly suggest that MAO-B seems to be involved in the local regulation of cerebrocortical capillary blood flow during hypoxia.

INVESTIGATIONS IN RAT LIVER

Experiments performed in the isolated perfused rat liver have shown that hypoxia induces both an increase in total flow and a redistribution of sinusoidal flow. Similar changes have been observed after administration of monoamine oxidase (MAO) inhibitors.

In order to prove that MAO may be involved in the regulation of blood flow it was necessary to show that the enzyme is blocked when flow changes occur. Therefore experiments were performed on isolated perfused rat livers. After a control period 2µg norepinephrine were added to the perfusate, causing a flow decrease by 22%. This flow change did not induce local hypoxia (26). After 10 minutes pargyline, a blocker which predominantly blocks MAO-B (37, 41), was added. When the flow started to increase, the experiment was stopped. The organs were than homogenized and the mitochondria were isolated by differential centrifugation. The mitochondria from these and from control livers (perfused without pargyline) were incubated with 3H-pargyline. The mitochondria were then sonicated and gel electrophoresis was performed. After staining autoradiography was performed. Figure 1 shows the result. The right and the left track are molecular

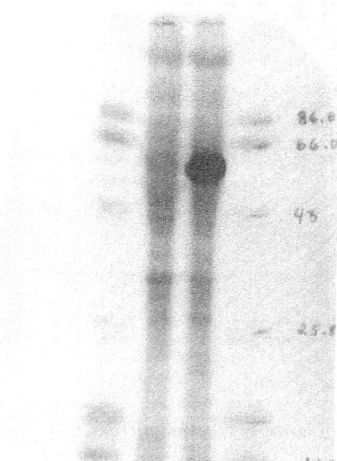

Figure 1. Polyacrylamide electrophoresis of [3H]pargyline bound to crude mitochondrial preparations. Track 1 and 4: molecular weight markers. Track 2: Proteins from the liver treated with pargyline during the experiment. Track 3: Proteins from the control liver perfused without pargyline. Note that only one labeled band was detected by autoradiography.

weight markers. As can be seen, only the control track (mitochondria from livers perfused without pargyline) shows the presence of radioactive labelled pargyline. This clearly indicates that in the livers perfused with pargyline the binding sites for this drug were already occupied by non-labelled pargyline. The molecular weight of the protein fraction binding tritiated pargyline is approximately 60.000 D. This is in agreement with results obtained by Costa and Breakefield (5) showing that pargyline binds to a protein fraction with a molecular weight between 57.000 and 60.000 D. The latter authors came to the conclusion that the 57.000 D protein labelled with pargyline containes sites for MAO activity.

Measurements with a liquid scintillation spectrometer showed that in the experimental livers (perfused with pargyline) 94.6% of MAO was blocked at the time when the flow increase occurred.

INVESTIGATIONS IN THE *Area vasculosa* OF THE EARLY CHICK EMBRYO

The extraembryonic vascular system of the early chick embryo develops during the first days of incubation. After approximately 4 days the secondary yolk sac circulation has developed. The vascularized part of the yolk sac is called area vasculosa. The blood vessels found are not innervated, have no smooth muscle cells and are formed merely by endothelial cells.

Figure 2 shows an electronmicrograph of a blood vessel of the *area vasculosa*.

In the extraembryonic vascular system of 4–5 days old eggs the vascular reactions after exposing the whole egg to various oxygen concentrations were studied. Jahn (21) first showed that yolk sac blood vessels may constrict on increasing oxygen tension and vice versa.

Figure 3 shows the oxygen saturation of intravascular haemoglobin after changing the environmental oxygen concentration to 100% oxygen and to 2% oxygen, respectively.

The simultaneously measured Laser Doppler signal (Figure 4) was seen to decrease by 11% when the eggs were subjected to an oxygen concentration of 2% oxygen. When combining Laser Doppler fluxmetry with video microscopy it is obvious that the decrease

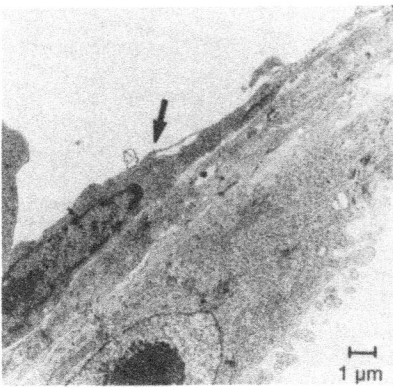

Figure 2. Electronmicrograph of an extraembryonic blood vessel wall. The bar corresponds to 1 μm. → = endothelial cell.

in the laser Doppler signal could be due to peripheral vasodilation occurring simultaneously resulting in a decreased RBC velocity.

With hyperoxia, the LDF signal (Figure 4) behaved differently. With an oxygen concentration of 100%, there was an increase in LDF. The early rapid increase in LDF occurred simultaneously with a decrease in vascular diameter, indicating that this decrease in vascular diameter is the cause for the increase in LDF.

The peripheral effects, i.e. the effects of oxygen on the extraembryonic blood vessels, are of special interest. According to Starck (46) the extraembryonal blood vessels develop from and are formed only by endothelial cells. This has been confirmed by Funk et al. (12). Nevertheless, these blood vessels show changes in vascular diameter.

There has been a long lasting debate on contractile properties of endothelial cells. Lübbers et al. (34) described specialized contractile endothelial cells in the frog mesen-

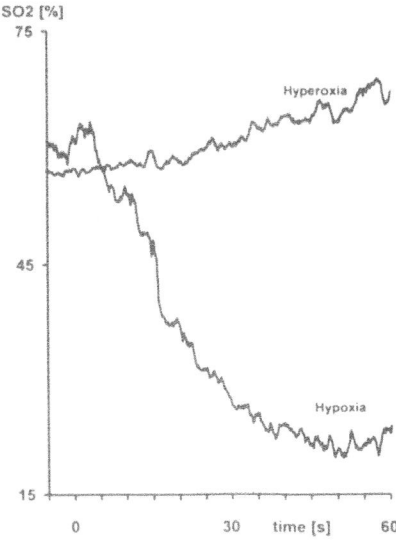

Figure 3. Saturation of intravascular haemoglobin (SO_2) measured during hypoxia (n=6) and hyperoxia (n=7). Only mean values are shown. Hypoxia or hyperoxia were induced at time 0s.

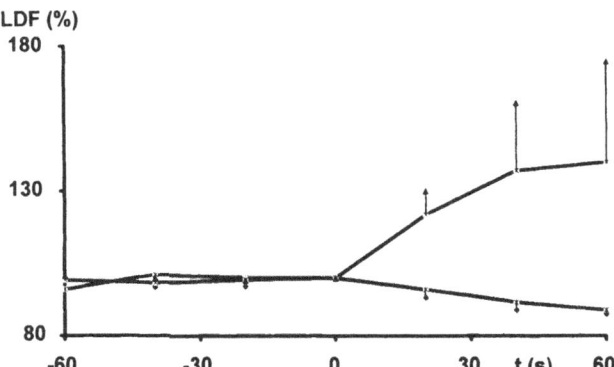

Figure 4. Mean LDF measured during normoxia, hypoxia and hyperoxia. The normoxic values measured at time 0 were taken as 100%. During hypoxia (n=6) a decrease in LDF was observed, whereas there was an increase during hyperoxia (n=7). Mean values ± SD.

tery. Drenckhahn (10) was able to demonstrate contractile elements in the endothelial cells in all vascular beds other than capillaries. McCuskey (35) presented results showing "sphincters" in liver sinusoids. It has also been shown that endothelial cells in the eye (25), spleen (40) and lung (39) are contractile. Ragan et al. (40) concluded that spontaneous capillary contractions were primarily due to endothelial contractility.

Hammersen et al. (13) came to the conclusion that the contractile elements in the endothelial cells play an important role in the formation of interendothelial gaps, but do not participate in the regulation of vascular diameter. Haraldson et al (14) showed that endothelial cell contractility is probably responsible for the histamine induced increase in capillary permeability.

The mechanism of activation and functional relevance of the contractility of endothelial cells has not yet been clarified. While Lübbers et al. (34) favoured the idea of a contractile response, Kessler et al. (27) proposed that changes in ion gradients across the endothelial cell membrane may cause a water shift and thus changes in cell volume. This latter idea is in accordance with results from Höper and Kessler (17, 19) showing that in the liver the norepinephrine induced decrease in flow is accompanied by a decrease in extracellular sodium activity. These authors later showed that a specific sodium channel is influenced by norepinephrine (18). McCuskey (35) has also pointed to the fact that endothelial cell swelling due to a change in intracellular osmolarity may cause cells to imbibe water.

The present results from the yolk sac indicate that in the area vasculosa all cells, including sparce pericytes (12), lining the blood stream may be involved in the diameter changes. This is in agreement with Das et al. (7) who showed that endothelial cells and pericytes may play a major role in the regulation of blood flow in the microcirculation.

Because the vascular system of the yolk sac membrane is not innervated (12) the observed changes must be induced by local and/or humoral mechanisms. There are no indications that NO, known to be a potent vasodilator, is involved (12). Kozniewska et al. (31) showed that changes in capillary blood flow in the brain cortex are closely related to the oxygen tension not in the arterial blood, but in the tissue. The same seems to be true for the yolk sac vasculature. Simultaneously and in parallel with a decrease in intravascular SO_2, a change in LDF occurred which at least partly was due to an increase in vascular diameter. Clearer information on a coupling between tissue oxygen supply and vascular diameter changes is given by the hyperoxic reaction. With almost no change in intravascular SO_2, a tremendous change in LDF and vascular diameter occurred. In this case, the vascular endothelial cells obviously do not react to an intravascular change, but to a change in tissue pO_2, which increases before intravascular SO_2 is altered. This indicates the existence of a signal transduction between the extravascular tissue and the endothelial cells. Dodge et al (9) suggested that endothelial cell-pericyte interactions may regulate at least in part microvascular contractility.

The question as to the underlying mechanism cannot be answered at the moment. As shown by Höper (15) lowering the tissue pO_2 causes a partial uncoupling of hepatocytes which may be accompanied by changes in local extracellular ion activities. Such changes in ion activities could also cause alterations in the membrane potential of endothelial cells and initiate the contractile mechanism.

The present results clearly show that, at least in a low pressure vascular system, endothelial cells can influence vascular diameter and thus alter the vascular resistance. Changes in the distribution of vascular resistance will change the pattern of capillary flow. Therefore this mechanism could be involved in the oxygen dependent regulation of capillary flow.

INTERCELLULAR COUPLING

Experiments performed in the isolated perfused rat liver revealed clear evidence that the intercellular coupling of hepatocytes can be influenced by the local oxygen supply. The experimental set up used for these experiments is shown in figure 5.

In figure 6 original registrations of V1 and V2 are shown. As can be seen, during hypoxia V1 increases whereas V2 decreases. The increase in V1 indicates that the resistance has increased. Thus the simultaneous decrease in V2 must be the result of a partial uncoupling of hepatocytes. The uncoupling starts within the first 20 seconds of hypoxia.

Further evidence for an uncoupling comes from direct measurements of the resistance between hepatocytes. As can be seen in figure 7 the change in resistance also starts within the first 20 seconds after onset of hypoxia.

These data indicate that the permeability of gap junctions decreases during hypoxia. Data from Flagg-Newton and Loewenstein (11) show that uncoupling occurs in a gradual manner, which means that first a "metabolic" uncoupling and only later an electrical uncoupling is observed.

Investigations on the permeability of gap junctions have shown that the intracellular calcium activity plays an important role in the regulation of intercellular coupling (6, 8, 32, 33). The permeability of gap junctions is high at low calcium activity ($<10^{-7}$ M/l) and decreases with increasing intracellular calcium activity. Above an intracellular calcium activity of approximately 5×10^{-5} M/l the gap junctions seem to be closed (38).

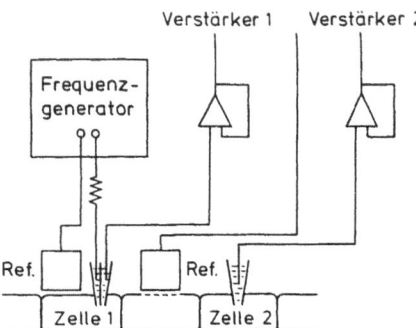

Figure 5. Setup for measuring intercellular coupling. A double-barrel microelectrode is inserted into cell 1, a second microelectrode is inserted into cell two, 500µm apart from cell 1. One channel of the double-barrel electrode is used for injecting current pulses (5×10^{-8} A). The potential change V1 induced by the current is measured with the other electrode channel. The second electrode measures the potential change V2 in cell 2.

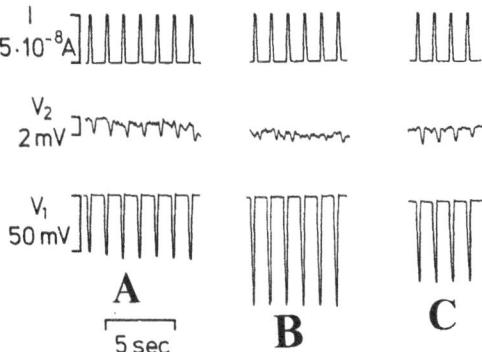

Figure 6. Original registration of injection current I, potential V1 and potential V2. A: Normoxia, B: Hypoxia, C: Normoxia.

The data from the literature as well as our own results indicate that during hypoxia an increase in intracellular calcium activity must occur. The resulting change in intercellular coupling may be involved in regulation of local capillary flow and may be part of a signal chain in tissue.

POSSIBLE ROLE OF MAO

Observations made by Cesarman (3) and Cossio (4) that MAO-I can lead to the disappearance of the complaints in angina pectoris may be indicative for the physiological significance. Furthermore, the results obtained by Höper and Kozniewska (20) indicate, that this mechanism is not only involved in the local regulation in liver and heart but also in the brain.

In order to avoid local anoxia the regulatory mechanism must have a $K_m pO_2$ higher than that of cytochrome oxidase, which is in the range of 0.1 mm Hg or even less. MAO fulfills these prerequisites because the $K_m pO_2$ is in the range of 7 mm Hg (24), thus being at least 70 times higher than that of cytochrome oxidase.

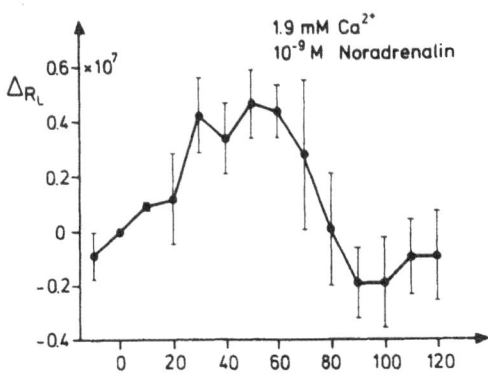

Figure 7. Change in longitudinal resistance during a short hypoxic period. Hypoxia was induced at time 0 s and lasted for 60 seconds.

Not only a decrease in oxygen supply rate but also an increased oxygen uptake can lead to an increase in capillary flow. Because there seems to be an interaction between respiratory chain and MAO (1, 43) this enzyme can be involved even in the coupling of oxygen uptake and blood flow.

According to Smith and Reid (43) an increasing mitochondrial oxygen uptake causes a decrease in MAO activity whereas low oxygen uptake seems to be accompanied by an increase in MAO activity. Thus the location of the enzyme in the outer mitochondrial membrane (42, 48) is of strategical importance for regulating blood flow.

Our data indicate that as long as the MAO activity is high an active metabolite is produced which causes a decrease in flow. Lowering oxygen supply or increasing oxygen uptake and thus lowering the local pO_2 will cause the level of the active metabolite to decrease and therefore the blood flow will increase. As shown by Tatyanenko et al. (47) the activity or efficiency of a calcium pump in the sarcoplasmic reticulum obviously can be influenced by the activity of MAO. This leads to the assumption that by the activity of MAO the intracellular calcium activity can be altered. This in turn can lead to a change in intercellular coupling as well as to changes in the contractile state of endothelial cells.

As long as the MAO activity is high, the active metabolite is produced, which keeps the flow at a level lower than the maximal flow possible. Increasing O_2-uptake by other oxidases or lowering O_2-supply will cause the level of the active metabolite to decrease and therefore the flow to increase.

The products of the breakdown of monoamines are aldehydes which are reactive chemical species. They can form Schiff bases and thus influence the conformation of proteins. Lowering the pO_2 or inhibiting MAO results in a decreased intracellular level of aldehydes. Because the formation of Schiff bases is a reversible process, the conformational change is reversed.

The data presented so far show that an oxidase may be involved in the O_2-dependent flow regulation. The effect of different MAO-I makes it probable that the regulatory enzyme is MAO-B.

Whether or not this mechanism is of physiological importance cannot be answered at this time. However, the observations made by Cesarman (3) and Cossio (4) that MAO-I can lead to the disappearance of the complaints in angina pectoris may be indicative for the physiological significance. The question of the localisation of this regulatory system needs further elucidation. It may be that only some specific cells (e.g. "sphincters") show this behaviour, but according to the result of extracellular ion measurements (19, 26) it is more likely that all cells, or at least many cells dispose this mechanism.

REFERENCES

1. Aebi, H, Stocker, F, Eberhardt, M (1963) Abhängigkeit der Monoaminoxydase-Aktivität und peroxydatischer Umsetzungen Umsetzungen von der Mitochondrienstruktur. Biochem. Z. 336, 526–544.

2. Artru, AA, Michenfelder, JD (1981) Canine cerebral metabolism and blood flow during hypoxemia and normoxic recovery from hypoxemia. J Cerebr Blood Flow Metabol 1, 277–283.

3. Cesarman, T.(1958) Marsilid in the treatment of Angina pectoris. J.clin.exp.Psychopath. 19,Suppl.1, 169

4. Cossio, P.(1958) The treatment of Angina pectoris and other muscle pain due to ischemia with iproniazid and isoniazid. Amer.Heart.J. 56, 113–118

5. Costa, MRC, Breakefield, X (1979) Electrophoretic characterization of monoamine oxidase by [3H] pargyline binding in rat hepatoma cells with A and B activity. Mol.Pharmacol. 16,242–249.

6. Dahl, G, Isenberg, G (1980) Decoupling of heart muscle cells: Correlation with increased cytoplasmic calcium activity and with changes of nexus ultrastructure. J Membrane Biol 53, 63–75

7. Das, A, Frank, RN, Weber, ML, Kennedy, A, Reidy, CA, Mancini, MA (1988) ATP causes retinal pericytes to contract in vitro. Exp Eye Res 46, 349–362.

8. DeMello, WC (1980) Influence of intracellular injection of H+ on the electrical coupling in cardiac purkinje fibres. Cell Biol Internat Rep 4, 51–58

9. Dodge, AB, Hechtman, HB, Shepro, D (1991) Microvascular endothelial-derived autacoids regulate pericyte contractility. Cell Motil Cytoskeleton 18, 180 - 188.

10. Drenckhahn, D (1983) Zellmotilität und zytoplasmatische Filamentsysteme in Gefäßendothelzellen. In: Struktur und Funktion endothelialer Zellen (K. Meßmer, F. Hammersen, eds.) Karger, Basel, pp 60–79.

11. Flagg-Newton, J, Loewenstein WR (1979) Experimental depression of junctional membrane permeability in mammalian cell culture. A study with tracer molecules in the 300 to 800 Dalton range. J Membrane Biol 50, 65–100

12. Funk, R, Höper, J, Abbi Raad, R (in preparation) Morphology of the area vasculosa blood vessels.

13. Hammersen, F, Hammersen, E, Osterkamp-Baust, U (1983) Bau und Funktion endothelialer Zellen. Eine Einführung. In: Struktur und Funktion endothelialer Zellen (K. Meßmer, F. Hammersen, eds.) Karger, Basel, pp 1–18.

14. Haraldsson, B, Zackrisson, U, Rippe, B (1986) Calcium dependence of histamine-induced increases in capillary permeability in isolated perfused hindquarters. Acta Physiol Scand 128, 247–258.

15. Höper, J (1984) Einfluß der Sauerstoffversorgung auf das Membranpotential und die intercelluläre elektrische Koppelung von Hepatocyten. Dissertation, Naturwissenschaftliche Fakultät, Erlangen.

16. Höper, J, Jahn, H (1985) Influence of environmental oxygen concentration on growth and vascular density of the area vasculosa in chick embryos. Int J Microcirc 15, 186–192.

17. Höper, J, Kessler, M (1981) pO_2 and sodium dependent mechanism regulating liver blood flow. In: Adv.Physiol.Sci. Vol 25. Oxygen Transport to Tissue (AGB Kovách, E Dóra, M Kessler, IA Silver, eds.) Akadémiai Kiadó, Budapest, pp163 - 164.

18. Höper, J, Kessler, M (1990) Sodium-dependence of the noradrenaline-induced flow change in the isolated perfused rat liver. Acta Physiol Scand 140, 447–448.

19. Höper, J, Kessler, M (1978) Norepinephrine induced changes in flow and extracellular activities of Na^+, K^+ and Ca^{++} in isolated perfused liver.In: Theory and application of ion-selective electrodes in physiology and medicine. Eds.:DW Lübbers, G Eisenman, M Kessler, W, Simon. Arzneim.-Forsch.(Drug Res.) 28,869–870.

20. Höper, J, Kozniewska, E (1992) Attenuation of hypoxic response in cerebral microcirculation following deprenyl. Int J Microcirc: Clin Exp: 11, 287 -295.

21. Jahn , H (1993) Der Einfluß der Luft-Sauerstoff-Konzentration auf die Entwicklung des Dottersack-Gefäßsystems beim Embryo des Haushuhns (Gallus gallus variatio domesticus) Thesis, Veterinary Medicine, München.

22. Jia, L, Bonaventura, C, Bonaventura, J, Stamler, JS (1996) S-nitrosohaemoglobin: a dynamic activity of blood involved in vascular control. Nature, 380, 221–226.

23. Johannson, H, Siesjö, BK (1975) Cerebral blood flow and oxygen consumption in the rat in hypoxic hypoxia. Acta Physiol Scand 93, 269–276.

24. Katz, IR, Wittenberg, JB, Wittenberg, BA (1984) Monoamine oxidase, an intracellular probe of oxygen pressure in isolated cardiac myocytes. J Biol Chem 259, 7504–7509.

25. Kelley, C, DÁmore, P, Hechtman, HB, Shepro, D (1987) Microvascular pericyte contractility in vitro: comparison with other cells of the vascular wall. J Cell Biol 104, 483–490

26. Kessler, M, Höper, J, Ji, S (1978) Action of norepinephrine on microcirculation and p02 distribution in the isolated perfused rat liver. Gerontology 24 (Suppl.l), 55–65.

27. Kessler, M, Höper, J, Krumme, B (1977) Influence of local ion activities in tissue on microcirculation. Bibl anat 15: 139–141.

28. Kessler, M, Höper, J, Krumme, BA (1976) Monitoring of tissue perfusion and cellular function. Anesthesiology 45, 184–197.

29. Kessler, M, Lang, H, Sinagowitz, E, Rink, R, Höper, J (1973) Homeostasis of oxygen supply in liver and kidney.In: Oxygen Transport to tissue Ed.:H.I.Bicher, D.F.Bruley, Plenum Publ.Corp. New York, Vol 37 A, 351–360.

30. Kessler, M, Lübbers, DW, Krumme, BA, Schönleben, K, Bünte, H (1977) Oxygen Tension in different tissues. Bibl. anat.16,146–149.

31. Kozniewska, E, Weller, L, Höper, J, Harrison, DK, Kessler, M (1987) Cerebrocortical microcirculation in different stages of hypoxic hypoxia. J Cerebr Blood Flow Metabol 7: 464–470.

32. Loewenstein, WR (1978) Cell-to-cell communication: Permeability, formation, genetics and functions of the cell-to-cell membrane channel. In: Physiology of membrane disorders. TE Andreotti, JF Hoffman, DD Fanestil (Eds) Plenum-New York, 335–356

33. Loewenstein, WR, Kanno, Y, Socolar, SJ (1978) The Cell-to-cell channel. Fed Proc 37, 2645–2650

34. Lübbers, DW, Hauck, G, Weigelt, H, Addicks, K (1979) Contractile properties of frog capillaries tested by electrical stimulation. Bibl anat 17: 3–10.

35. McCuskey, RS (1966) A dynamic and static study of hepatic arterioles and hepatic sphincters. Am J Anat 119, 455 - 478.

36. Miyaoke, M, Shinohara, M, Kennedy, C, Sokoloff, L (1980) Alterations in local cerebral glucose utilisation (LCGU) in rat brain during hypoxemia. Trans Ann Neurol Assoc 104, 1–4.

37. Murphy, DL (1976) In: Monoamine oxidase and its inhibition, Ciba Foundation Symposium, Elsevier,Excerpta Medica, North-Holland, 28–29.

38. Oliveira-Castro, GM, Loewenstein WR (1971) Junctional membrane permeability: Effects of divalent cations. J Membrane Bil 5, 51–77

39. Patton, WF, Alexander, JS, Dodge, AB, Patton, RJ, Hecht,man, HB, Shepro, D (1991) Mercury-arc photolysis: a method for examining second messenger regulation of endothelial cell monolayer integrity. Anal Biochem 196, 31–38.

40. Ragan, DM, Schmidt, EE, MacDonald, IC, Groom, AC (1988) Spontaneous cyclic contractions of the capillary wall in vivo, impeding red cell flow: a quantitative analysis. Evidence for endothelial contractility. Microvasc Res 36, 13 -30.

41. Sandler, M(1976) In: Monoamine oxidase and its inhibition, Ciba Foundation Symposium, Elsevier,Excerpta Medica, North-Holland, 29.

42. Schnaitmann, C, Erwin, VG, Greenawalt, JW (1967) Submitochondrial localization of monoamine oxidase. J.Cell.Biol.,34,719–735.

43. Smith, GS, Reid, RA (1978) The influence of respiratory state on monoamine activity in rat liver mitochondria. Biochem.J. 176, 1011–1014.

44. Smith, ML, Kagström, E, Siesjö, BK (1983) Local cerebral blood flow in the rat brain during hypercapnia and hypoxia. Acta Physiol Scand 118, 439–440.

45. Sokoloff, L (1981) Localization of functional activity in the central nervous system by measurement of glucose utilization with radioactive deoxyglucose. J Cerebr Blood Flow Metabol 1, 7–36.

46. Starck, D (1975) Embryologie.Thieme, Stuttgart.

47. Tatyanenko, LV, Raikhman, LM, Gorkin, VZ (1977) Type B monoamine oxidase and functions of Ca2+, Mg2+-dependent adenosinetriphosphatase in preparations from sarcoplasmic reticulum vesicles. Byull Éksp Biol Med 83, 283–284.

48. Tipton,KF, Houslay, MD, Mantle, TJ (1976) The nature and location of the multiple forms of monoamine oxidase. In: Monoamine oxidase and its inhibition, Ciba Foundation Symposium, Elsevier,Excerpta Medica, North-Holland, pp 5–16.

49. Traystman, RJ, Fitzgerald, RS (1981) Cerebrovascular response to hypoxia in baroreceptor- and chemoreceptor-denervated dogs. Am J Physiol 241, H724-H731.

50. Traystman, RJ, Gurtner, GH, Koehler, RC, Jones, MD, Rogers, MC (1983) Central chemoreceptor and oxygenase regulation of cerebral blood flow. J Cerebr Blood Flow Metabol 3 Suppl. 1, S180–181.

51. Weiss, HR, Buckweitz, E, Sinha, AK (1983) Effect of hypoxic-hypocapnia on cerebral regional oxygen consumption and supply. Microvasc Res 25, 194–204.

61

HETEROGENEITY OF BRAIN OXIDATIVE METABOLISM AND HYPOXIA RESPONSE

Mammalian Systems and Nature's Solutions

Jeff F. Dunn, Eric S. Rhodes, and Tomasz Panz

NMR Research Center
Dept. of Radiology
Dartmouth-Hitchcock Medical Center
7786, Hanover, New Hampshire 03755

1. INTRODUCTION TO HETEROGENEITY: THE PROBLEM

Brain ischemia (defined here as restricted flow) and/or hypoxia (defined here as a reduced oxygen supply) caused by stroke is a major cause of death and debilitation. In addition, approximately 50% of cardiac arrest survivors suffer from neurological deficits associated with a transient occurrence of hypoperfusion; neurosurgery is often associated with transient ischemia; and birth asphyxia, high altitude exposure or drowning may all cause brain damage due to lack of oxygen. Clearly the mammalian brain is sensitive to oxygen deprivation, making it a high priority to identify strategies to survive this damage.

Historically, the brain was considered as a single unit. Little change in metabolic rate was expected between regions and global indexes of perfusion and oxygen uptake were considered sufficient. Most work on brain metabolism used relatively global techniques such as freeze clamping for metabolite assay, or NMR with large regions of interest. The use of newer techniques with higher spatial resolution has changed our perception of brain metabolism, into one of a dynamic system with significant regional variability and function.

If brain oxidative metabolism is heterogeneous, then it is likely that different regions of brain will show variations in hypoxia sensitivity. This sensitivity can be defined as the capability of the cells to match metabolic requirements with oxidative energy production as oxygen supply is reduced. The goal of this paper is to prove that brain does have significant macroscopic variation in oxidative metabolism. Evidence to support this premise includes oxygen tension distributions, as well as studies of perfusion and metabolic rate. Comments are included about the underlying mechanisms for such variation. Data on perfusion and metabolic rate are also included from species with greatly differing hypoxia tolerance. Such data provides an opportunity to identify effective mechanisms for dealing with hypoxia/ischemia.

Oxygen Transport to Tissue XIX, edited by Harrison and Delpy
Plenum Press, New York, 1997

In summary, this paper investigates regional variations in oxygenation and perfusion in mammalian brain, and inferences are made about hypoxia tolerance from the growing body of information from more hypoxia tolerant species.

2. MICROELECTRODE MEASUREMENTS OF BRAIN pO_2

Models of tissue oxygenation indicate that pO_2 relates to perfusion and metabolic rate (Tenney, 1974). This is supported by surface electrode studies of rat cortex which show a linear correlation between pO_2 and metabolic rate in the normoxic brain (Gaab et al., 1990).

In this study, macroscopic variations in brain pO_2 were measured in male Sprague Dawley rats (250–350 g). Animals were anesthetized using isoflurane (2%) administered via a nose cone. Isoflurane was used as it has been previously shown to result in brain pO_2 values that are similar to those in non-anesthetized animals (Liu et al., 1995). The scalp was removed and a 2 mm hole was made 2 mm lateral to each side of the sagittal suture and 3 mm posterior to the bregma. Clark-style microelectrodes with 50–75 μm external tip diameters (Diamond General Corp., Ann Arbor, MI) were inserted using a micromanipulator and currents were recorded using a Keithley Model 485 picoammeter (Cleveland, OH).

A two-point *in vitro* calibration was performed by measuring pO_2 in 37°C physiological saline equilibrated with 100% helium gas or 21% O_2. The oxygen electrode was inserted sequentially to a depth of 5 mm. Readings were made in 0.5 mm intervals by inserting the electrode 0.7 mm and withdrawing it 0.2 mm to relieve tension at the tip.

Temperature was maintained at 37° C using a circulating water-heating pad and a heat lamp, and monitored with a rectal probe. Heart rate and blood pressure were monitored using a Biopac data acquisition system. Systolic blood pressure remained over 100 mm Hg throughout all experiments and no significant difference was observed between BP, heart rate, and temperature before and after drilling holes and insertion of the electrode. Electrode positions were determined by comparing gross fixed sections (4% neutral-buffered formalin) with a rat brain atlas (Paxinos et al., 1986).

3. HETEROGENEITY OF OXYGENATION

The pO_2 measurements ranged from < 2.0 to 76 mmHg. Figure 1 shows the spatial variation at different depths in the brain. The data from the left and right hemispheres were not significantly different and so these data were averaged. Depths of 0–2 mm correspond to neocortex, 2 mm to corpus callosum, 2.5–3.5 mm to hippocampus, and 4–5 mm to thalamus. The pO_2 measurements from the center of each depth are shown in Figure 2.

These results, showing that the brain has a large variability in tissue pO_2, support previous studies which indicate that brain pO_2 can range from 5–90 mmHg over sub-millimeter distances (Ivanov et al., 1982; Leniger-Follert et al., 1975; Lubbers, 1969; Seyde et al., 1986). These previous studies used electrodes with tip diameters of 15 μm or less. With 2 μm tip diameters, it was possible to show that large gradations in pO_2 occur with distance from capillaries (Ivanov et al., 1982).

Comparisons tested with ANOVA plus Tukey-B procedure for multiple tests, ($p <$ 0.05), indicated that the thalamus had a higher pO_2 than the corpus callosum (white matter) or hippocampus (32 ± 4 vs. 15 ± 3 and 20 ± 3, mm Hg mean ± SE, respectively), and if more data had been collected, the cortex (27 ± 6 mmHg) may also have had a higher

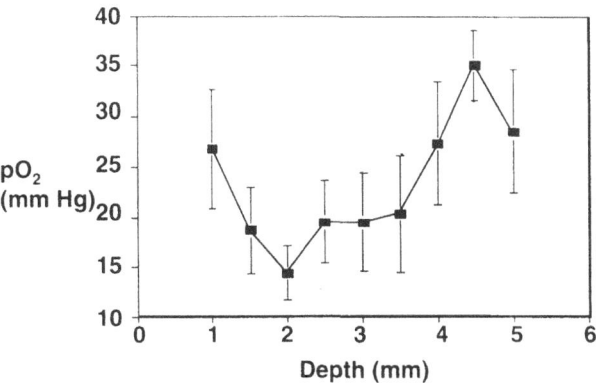

Figure 1. Microelectrode measurements with depth from cortex surface in a rat brain (mean±SE, n = 5).

pO_2 than corpus callosum or hippocampus (p = 0.06). The cortex values are in agreement with previous work in rats reporting 29 ± 5 mmHg using isoflurane (Seyde et al., 1986) and 31 ± 5 mmHg using urethane as the anesthetics (Metzger et al., 1980). It also agrees with previous work reporting significant macroscopic heterogeneity in pO_2. The rat hippocampus has been reported to have a lower pO_2 than the cortex (21 ± 6 vs. 31 ± 5)(Metzger et al., 1980), and pO_2 distribution histograms are left shifted (to a lower pO_2) in the hippocampus vs. the cortex (Nair et al., 1987). The values for white matter are not as consistent between studies. This paper's results indicate that white matter (corpus callosum) has one of the lowest values for mean pO_2 while Metzger (Metzger et al., 1980) reported corpus callosum values within the range of various grey matter regions.

These data confirm that there are significant variations in brain pO_2 between major structures, an observation which must indicate heterogeneity in the dynamic balance between oxygen delivery and utilization.

4. OXYGENATION, PERFUSION, AND METABOLIC RATE

Variations in metabolic rate and perfusion occur regionally across both macroscopic brain structures and locally within structures. This indicates that the brain has significant

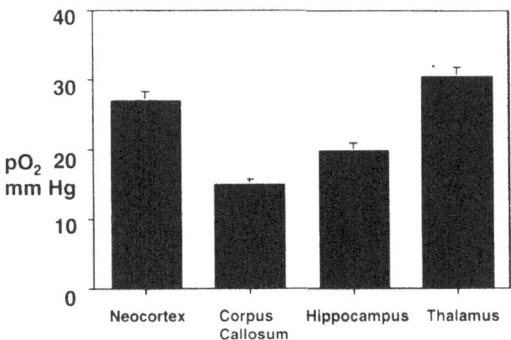

Figure 2. Average regional pO_2 values in a rat brain (mean±SE, n = 12).

oxidative heterogeneity. It is possible that there is a predictable relationship between oxygenation (pO_2), perfusion, and metabolic rate, and that these values may correlate with hypoxia sensitivity. This information might be useful for predicting regional sensitivity to hypoxia. Regions with high metabolic rates, for instance, may be more sensitive to reductions in arterial oxygen tension unless there is a large reserve capacity to increase flow or a large reserve in oxygen content at the end of the capillary, allowing for increased oxygen extraction. Local sites within macroscopic regions have been observed to have low pO_2 values and so may be at risk of becoming hypoxic (Leniger-Follert et al., 1975; Lubbers, 1969).

Adult human and rat brain cerebral metabolic rate for oxygen ($CMRO_2$) or glucose (CMRglu) has been shown to vary by a factor of 5 or more at any one time (Hatazawa et al., 1995; Kuschinsky et al., 1981; Lebrun-Grandie et al., 1983; Sokoloff et al., 1977). Perfusion shows similar variation, changing by a factor of 3–6 over regional, macroscopic structures in the whole brain and by 3 in gray matter structures (Bereczki et al., 1993; Hatazawa et al., 1995; Klein et al., 1986; Sakurada et al., 1978).

The relationship between perfusion and metabolic rate is remarkably consistent in brain between these regions. In both rats and humans, there is a linear relationship between metabolic rate and perfusion rate over the ranges measured. This linearity exists whether metabolic rate is assessed using CMRglu, or $CMRO_2$ (Klein et al., 1986; Kuschinsky et al., 1981; Lebrun-Grandie et al., 1983; Leenders et al., 1990).

Capillary density, an indirect assessment of perfusion, is also linear with CMRglu and perfusion rate (Gjedde et al., 1985; Klein et al., 1986). Glucose transport rate, an indirect correlate of metabolic rate, also varies regionally and changes in proportion to capillary density (Gjedde et al., 1985). This is interesting as it indicates that steady state metabolic rates in brain are predetermined and can be predicted by morphological indexes such as capillary length and density. Thus, in the brain there are variations in metabolic rate and perfusion, and there appears to be tight links between metabolic rate, perfusion rate and capillary density. It is not unreasonable to predict that such heterogeneity, coupled with the tight relationships between metabolism and supply, would be reflected in regional hypoxia sensitivity.

5. BLOOD FLOW, METABOLIC RATE AND HYPOXIA RESPONSE

Blood flow in the mammalian brain increases during hypoxia (Bereczki et al., 1993; Borgstrom et al., 1975; Kontos et al., 1987; Leniger-Follert et al. 1976). It is intriguing to suggest that the perfusion responses may reflect regional hypoxia tolerance. This relationship could take one of two forms, 1) a large response indicates good adaptation as the system is geared to maintain the brain while small responses may occur in regions which already are approaching maximum flow, or 2) large responses reflect an unstable system which is not capable of maintaining homeostasis. Certainly the % increase in regional blood flow induced by hypoxia (10% FiO_2), is extremely variable (10% to 70%), and does not correlate with initial blood flow (Bereczki et al., 1993).

Data from rat brain do not clarify these options although the CA1 region of the rat hippocampus, which is considered to be hypoxia sensitive, has an increase in hypoxia induced blood flow of 40%. This is a relatively average change in comparison to other regions (Bereczki et al., 1993).

The oxygen extraction fraction (OEF) may also reflect hypoxia tolerance as capillaries extracting most of the oxygen may not have significant capacity to tolerate a decrease in arterial pO_2. This parameter is also variable throughout the brain. One of the most striking examples of the extent of this variation in extraction was reported using a microspectrophotometric method for determining oxygen uptake in cats. On average, there were variations in OEF (based on venous hemoglobin saturation) between regions, but local variations in small vessels ranged from 0 to 96% (Buchweitz et al., 1980). Local areas of low pO_2 also occur (Leniger-Follert et al., 1975; Lubbers, 1969), and local sites within macroscopic regions show variable changes in both perfusion and pO_2 in response to reduced arterial oxygen tension (Leniger-Follert et al., 1976; Leniger-Follert et al., 1975). Could it be that some macroscopic regions have a higher proportion of capillaries with a high OEF, even though the average OEF is in a "normal" range. This may increase the proportion of local sites which are at risk for hypoxic damage, making that macroscopic region more hypoxia sensitive.

6. CAUSES OF REGIONAL HETEROGENEITY IN OXYGENATION

It is thought that basal or steady state metabolic rate in brain is determined by requirement for ionic flux associated with nerve function. Gray matter regions have higher metabolic rates and blood flow than white matter. However, there is still significant variation within gray matter itself. Even in these regions, morphological variation in macroscopic brain structures is still expected to play a greater determinant role than acute stimuli. This prediction is made on the basis of the relatively fixed values of capillary density and capillary length. Capillary density correlates linearly with both perfusion rates and metabolic rate in the brain (Gjedde et al., 1985; Klein et al., 1986). Also, although it is still a matter of some debate, it appears that capillary recruitment may not be occurring with hypoxia (Bereczki et al., 1993).

Variations in proportion of cell types would be expected to influence metabolic rate. Each macroscopic region in brain is composed of differing proportions of glia and neurons. This ratio, the glial index, can vary from 1.9 to 6 in the human brain (Blinkov et al., 1968). As a result, variability in hypoxia tolerance between these cell types could confer a variability in hypoxia tolerance which is in proportion to the glial index.

Subsets of glia as well as endothelial cells compose the blood brain barrier and so part of their metabolic requirement goes towards neuronal homeostasis. It has been shown that capillary endothelial cells have adapted to this increased energy demand by increasing oxidative capacity (Oldendorf et al., 1977). Mitochondrial volume from rat endothelial cells within the region of the blood-brain barrier is 8–11% of total cell volume, while endothelial cells from other tissues (non blood-brain barrier) only have mitochondrial volumes of 2–5% (Oldendorf et al., 1977). As a result of both the increased energy demand on brain endothelial cells, as well as their adaptation to increased oxidative capacity, it may be that they will also have an increased sensitivity to hypoxia. Neuronal function relies upon the function of the blood brain barrier and so metabolic failure in these cells would appear as a nervous system failure.

Even different neurons show varying hypoxia tolerance. It has long been known that the CA1 region of the hippocampus is one of the first regions to become electrically silent during hypoxia (Sugar et al., 1938). Also, a comparison between electrical activity in rat brain slice preparations of hypoglossal brain stem neurons and neurons from layer II and III of the neocortex during hypoxia, showed that the neocortical cells had a longer latency and were more likely to recover (O'Reilly et al., 1995).

7. NATURE'S SOLUTIONS TO HYPOXIA

A potentially useful approach to identifying the capacity for adaptation and intervention in human systems is to investigate the range of adaptive mechanisms that exist in other groups.

The freshwater turtles are masters of vertebrate hypoxia tolerance. Brain viability can be maintained even after months of anoxia (Musacchia, 1959; Ultsch et al., 1982). Perfusion in the turtle has a two phased response to anoxia. There is an initial increase in CBF followed by a return to control values in approximately 2 hrs (Hylland et al., 1994). The decline may be in response to the decline in metabolic rate induced by anoxic exposure in turtles. Cortical slice preparations show a 30% decline in ATP utilization with anoxic exposure (Doll et al., 1994) and whole brain studies indicate that brain metabolic rate can decline by over 80% with anoxic exposure (Kelly et al., 1988; Lutz et al., 1984). Clearly, the capacity to reduce metabolic rate is a major advantage. Also, perfusion changes appear to track changes in metabolic rate and so do not remain elevated.

Studies of high altitude adapted Quechua indians show that humans also have the capability of adjusting brain oxidative requirements. PET studies indicate that glucose uptake rates are greatly reduced in brain in Quechua indians compared to lowlanders (Hochachka et al., 1994). It is likely that this group also does not show a marked increase in perfusion, although this is not known.

Pekin ducks are capable of maintaining oxygen uptake in muscle at low arterial oxygen tensions. Their strategy is to have an excess of oxygen delivery at normal oxygen tensions, and so have a low OEF (20%) in normoxia. This translates as a high blood flow rate, high capillary density and high venous pO_2. When exposed to declining FiO_2, a resting oxygen uptake was maintained even at arterial oxygen contents of 5ml/100ml by increasing oxygen extraction up to 70%. Again, blood flow did not increase, even with such severe hypoxic exposure (Grubb, 1981).

Increases in OEF have also been observed in brain from hypoxia acclimated mammals. When sheep are acclimated to low oxygen (PaO_2 < 60mmHg for 72 hrs), they develop "cerebral hyperperfusion" and a decline in OEF (Curran-Everett et al., 1992). Rats exposed to an FiO_2 of 0.10 for 3 weeks respond with an increase in cerebral microvessel density ranging from 26% to 76%, but no increase in blood flow (LaManna et al., 1992). This may also indicate that OEF has declined in normoxic conditions.

A summary of these adaptations indicates that reduced OEF under normoxic conditions correlates with hypoxia tolerance. Increased blood flow is also minimized. These factors may relate back to the initial data on pO_2 heterogeneity. It is possible that areas of low mean OEF also have high pO_2 values relative to arterial oxygen tensions.

8. CONCLUSIONS

The brain is a very heterogeneous organ with respect to oxidative metabolism, based on observations of regional perfusion, metabolic rate, and pO_2. Most of this heterogeneity is fixed during development, as indicated by the relationship between capillary density and metabolic rate. Such variation is a strong indication that there will also be heterogeneity of hypoxia tolerance in brain on a macroscopic scale. Acute hypoxia causes a different % increase in perfusion rate between regions indicating that there is also heterogeneity in the brain's capacity to respond to hypoxia. Chronic hypoxia will change the capillary density/length and decrease the oxygen extraction fraction, indicating that regions with lower

OEF may be more hypoxia tolerant. Comparisons with relatively hypoxia tolerant groups indicate that there is significant potential for adaptation with respect to brain metabolic rate, perfusion and hypoxia survival. They also indicate that hypoxia adaptation correlates with 1) a reduced OEF during normoxia, and 2) minimal increases in perfusion rates upon chronic exposure to hypoxia. It may be that these adaptations will correlate with high regional pO_2 during normoxia.

ACKNOWLEDGMENTS

Thanks to Ellis Rolett and Hal Swartz for many helpful discussions, and to Ellis for his invaluable assistance with setting up the animal protocols and techniques.

REFERENCES

Bereczki, D., L. Wei, T. Otsuka, V. Acuff, K. Pettigrew, C. Patlak and J. Fenstermacher. *J Cereb Blood Flow Metab*, 1993; 13:475–86.

Blinkov, S.M. and I.I. Glezer. *The human brain in figures and tables* 1–482 (Plenum Press, N.Y., 1968).

Borgstrom, L., H. Johannsson and B. Siesjo. *Acta Physiol Scand*, 1975; 93:423–432.

Buchweitz, E., A.K. Sinha and H.R. Weiss. *Science*, 1980; 209:499–501.

Curran-Everett, D.C., M.P. Meredith and J.A. Krasney. *Resp Physiology*, 1992; 88:365–371.

Doll, C.J. and P.W. Hochachka. *J Exp Biology*, 1994; 191:141–153.

Gaab, M.R., B. Poch and V. Heller. *Adv Neurol*, 1990; 52:247–256.

Gjedde, A. and N. Diemer. *J Cereb Blood Flow Metab*, 1985; 5:282–289.

Grubb, B.R. *J Appl Physiol*, 1981; 50:450–455.

Hatazawa, J., H. Fuijta, I. Kanno, T. Satoh, H. Iida, S. Miura, M. Murakami, T. Okudera, A. Inugami, T. Ogawa, E. Shimosegawa, K. Noguchi, Y. Shohji and K. Uemura. *Ann. Nuclear Med.*, 1995; 9:15–21.

Hochachka, P., C. Clark, W. Brown, C. Stanley, C. Stone, R. Nickles, G. Zhu, P. Allen and J. Holden. *J Cereb Blood Flow Metab*, 1994; 14:671–679.

Hylland, P., G.E. Nilsson and P.L. Lutz. *J Cereb Blood Flow Metab*, 1994; 14:877–81.

Ivanov, K.P., A.N. Derry, E.P. Vovenko, M.O. Samoilov and D.G. Semionov. *Pflugers Arch*, 1982; 393:118–120.

Kelly, D.A. and K.B. Storey. *Am J Physiol*, 1988; 255:R774–R779.

Klein, B., W. Kuschinsky, H. Schrock and F. Vetterlein. *Am J Physiol*, 1986; 251:H1333–H1340.

Kontos, H.A., E.P. Wei, A.J. Raper, W.I. Rosenblum, R.M. Navari and J.L. Patterson. *Am J Physiol*, 1987; 234:H582–H591.

Kuschinsky, W., S. Suda and L. Sokoloff. *Am J Physiol*, 1981; 241:H772–H777.

LaManna, J.C., L.M. Vendel and R.M. Farrell. *J Appl Physiol*, 1992; 72:2238–2243.

Lebrun-Grandie, P., J.-C. Baron, F. Soussaline, C. Loch'h, J. Sastre and M.-G. Bousser. *Arch Neurol*, 1983; 40:230–236.

Leenders, K.L., D. Perani, A.A. Lammertsma, J.D. Heather, P. Buckingham, M.J.R. Healy, J.M. Gibbs, R.J.S. Wise, J. Hatazawa, S. Herold, R.P. Beaney, D.J. Brooks, T. Spinks, C. Rhodes, R.S.J. Frackowiak and T. Jones. *Brain*, 1990; 113:27–47.

Leniger-Follert, E., D.W. Lubbers and W. Wrabetz. *Pflugers Arch*, 1975; 359:81–95.

Leniger-Follert, E., W. Wrabetz and D.W. Lubbers. *Advances in Experimental Medicine and Biology*, 1976; 75:361–367.

Liu, K.J., G. Bacic, P.J. Hoopes, J. Jiang, H. Du, L.C. Ou, J.F. Dunn and H.M. Swartz. *Brain Res*, 1995; 685:91–98.

Lubbers, D.W. *Prog in Resp Res*, 1969; 3:112–123.

Lutz, P.L., P.M. McMahon, M. Rosenthal and T.J. Sick. *Am J Physiol*, 1984; 247:R740–R744.

Metzger, H., S. Heuber-Metzger, A. Steinacker and J. Struber. *Pflugers Arch*, 1980; 388:21–27.

Musacchia, X.J. *Physiol Zool*, 1959; 32:47–50.

Nair, R.K., D.G. Buerk and J.H. Halsey. *Stroke*, 1987; 18:616–622.

O'Reilly, J.P., C. Jiang and G.G. Haddad. *Brain Res.*, 1995; 683:179–86.

Oldendorf, W.H., M.E. Cornford and W.J. Brown. *Ann Neurol*, 1977; 1:409–417.

Paxinos, G.P. and C. Watson. *The rat brain in stereotaxic coordinates* (Academic Press, London, 1986).

Sakurada, O., C. Kennedy, J. Jehle, J.D. Brown, G.L. Carbin and L. Sokoloff. *Am J Physiol*, 1978; 3:H59-H66.

Seyde, W.C. and D.E. Longnecker. *Anesthesiology*, 1986; 64:480–485.

Sokoloff, L., M. Reivich, C. Kennedy, M.H. Des Rosiers, C.S. Patlak, K.D. Pettigrew, O. Sakurada and M. Shinohara. *J Neurochem*, 1977; 28:897–916.

Sugar, O. and R. Gerard. *J Neurophysiol*, 1938; 1:558.

Tenney, S.M. *Respir Physiol*, 1974; 20:283–296.

Ultsch, G.R. and D.C. Jackson. *J Exp Biol*, 1982; 96:11–28.

62

EFFECTIVENESS OF THE PERIPHERAL CHEMOREFLEX CONTROL SYSTEM IN THE ADJUSTMENT OF ARTERIAL O_2 PRESSURE AND O_2-Hb SATURATION

H. Kiwull-Schöne and P. Kiwull

Department of Physiology
Ruhr-University
D-44780 Bochum, Germany

INTRODUCTION

From the view of control engineering, the respiratory chemoreflexes can be described as negative-feedback circuits (Cunningham, 1974; Loeschcke 1973a), whereby most empirical data refer to the sensitivity of the controller in terms of pulmonary ventilation as a function of arterial blood gas partial pressures. Although considered in theoretical models, there are only few experimental reports to describe the effectiveness of the controller, e.g. by Loeschcke, 1973b, for the central respiratory chemoreflex system in restoring the arterial PCO_2 via brainstem extracellular fluid pH. Even less attention has been paid to the effectiveness of the peripheral chemoreflex during hypoxia.

To analyse the effectiveness of the peripheral chemoreflex for systemic oxygen supply in terms of arterial O_2 pressure (PaO_2) as a function of ventilatory drive, the present study compares values of PaO_2 resulting at different inspiratory O_2 fractions (F_IO_2) with intact and eliminated carotid chemoreflexes, i.e. under controlled and uncontrolled conditions. Besides PaO_2, the O_2-Hb saturation was considered as a target variable, assuming growing importance of arterial O_2 transporting properties under conditions of restricted inspiratory O_2 availability. Since considerable influences of both arterial CO_2 pressure and pulmonary vagal afferent input on the hypoxic response of ventilation have been demonstrated previously (Kiwull-Schöne and Kiwull, 1979; Kalhoff et al. 1994), the present experiments were carried out with either intact or cut vagus nerves and at different levels of $PaCO_2$. Based on the results, prerequisite properties of the adequate stimulus for the sensing element are discussed.

Oxygen Transport to Tissue XIX, edited by Harrison and Delpy
Plenum Press, New York, 1997

METHODS

The experiments were performed in 18 spontaneously breathing rabbits (average weight ± SEM: 2.6 ±0.1 kg), anaesthetized by an initial dose of 41 ±1.2 mg/kg pentobarbital sodium, followed by continuous infusion of 5.7 ±0.2 mg/kg/h. Tidal volume (V_T), inspiratory and expiratory durations (T_I, T_E) were measured by pneumotachography to calculate pulmonary minute ventilation ($\dot{V}$). To estimate dead space ventilation ($\dot{V}_D$), a combined anatomical and equipmental dead space volume was used, valid for tracheotomized rabbits of comparable size on Fleisch tube No. 0, as reported to be 2.3 ml/kg before and 3.2 ml/kg after vagotomy (Colinet-Lagneaux et al., 1969). Additionally, the non-perfused portion of the alveolar space was approached by considering the arterio-endtidal (a-et) PCO_2 difference in relation to the arterial CO_2 partial pressure ($PaCO_2$).

Endtidal CO_2 pressure ($PetCO_2$) was recorded by infrared absorption (Hartmann & Braun). From arterial blood samples, $PaCO_2$ and concentrations of actual and standard bicarbonate (HCO_3^-a, $HCOO_3^-t$) were indirectly determined by pH measurement (Radiometer) with the equilibration method (Astrup and Schrøder, 1956). For correction of the Haldane effect on the PCO_2/pH relationship during hypoxia, the desaturation-induced pH-shift depending on concentrations of Hb and $HCOO_3^-a$ in rabbit blood was considered (Kiwull-Schöne et al., 1992).

The concentration of hemoglobin (Hb) was measured photometrically as hemiglobin cyanide (Merckotest). Arterial O_2 partial pressure (PaO_2) was measured polarographically by an electrode (Beckman). O_2-Hb saturation (SaO_2) was calculated for any given PaO_2 and acid-base value by means of the standard O_2-Hb dissociation curve and Bohr-effect data previously determined for rabbits (Kiwull-Schöne et al., 1987).

Two groups of N=9 animals with either intact or cut vagus nerves were studied separately. They inhaled different inspiratory gas mixtures for up to 10 minutes, alternatively pure oxygen, air and a moderately (F_IO_2=0.12) or severe (F_IO_2=0.07) hypoxic mixture. The $PaCO_2$ was either kept at the hyperoxic control level by adding CO_2 to the inspiratory line or was allowed to decrease consequent to hypoxic hyperventilation. Each animal was investigated first with intact and finally with cut carotid sinus nerves (CSN).

Under the different experimental conditions, group mean values (±SEM) of the determined variables were calculated. Non-linear and linear regression analysis was performed to describe relationships between ventilation and PaO_2 or SaO_2, respectively. Differences between group means or between slopes of linear regression lines were tested by unpaired t-tests for significance at the level of $P_D < 0.05$.

RESULTS

Under hyperoxic conditions, the average pulmonary ventilation ($\dot{V}$ ±SEM) in the groups with intact and cut vagus nerves was 348 ±19.8 and 365 ±11.8 ml/min/kg, respectively, being not significantly different. In both groups, cutting the carotid sinus nerves (CSN) during hyperoxia had no significant influence on $\dot{V}$ (-3.5 ±2.8%, P>0.20).

With falling oxygenation, $\dot{V}$ with intact (cut) vagus nerves progressively increased, by a maximum at 0.07 F_IO_2 of 181 ±34.5% (125 ±19.8%) above the hyperoxic level, when $PaCO_2$ was kept constant. When $PaCO_2$ was allowed to fall in response to hyperventilation, the maximum hypoxic ventilatory drive was only 56 ±15.0% in the group with intact, but significantly higher in the group with cut vagus nerves (103 ±17.6%, P_D = 0.05).

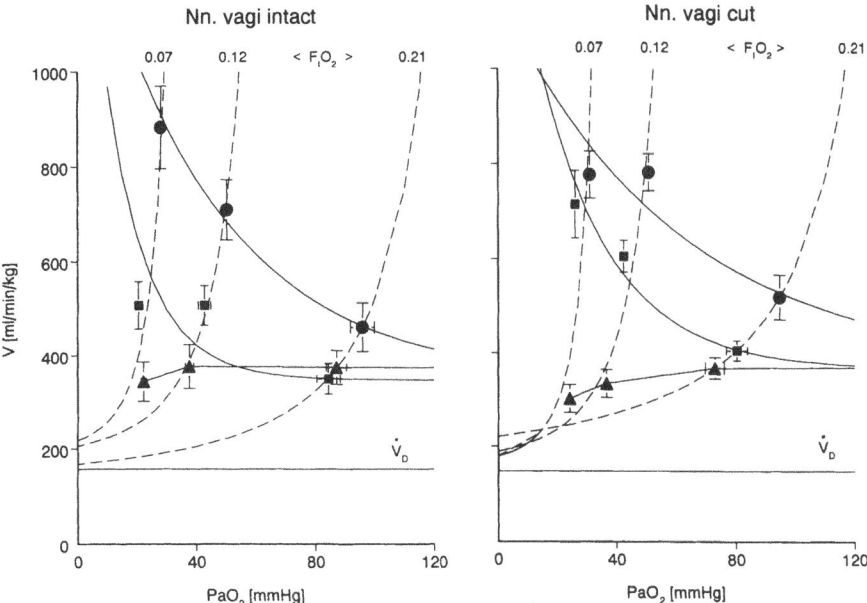

Figure 1. Hypoxic response curves of ventilation (solid lines) and restoring curves (broken lines) at different levels of inspiratory oxygen supply. Non-linear fit of mean values ±SEM (N=9) for two groups of rabbits with either intact or cut vagus nerves. Carotid sinus nerves (CSN) cut (). With intact CSN, the $PaCO_2$ was kept either isocapnic at the hyperoxic level (●) or was allowed to fall to hypocapnic values due to hypoxic hyperventilation (■). An average constant deadspace ventilation ($\dot{V}_D$) was assumed as horizontal asymptote.

The hypoxic response curves of ventilation as functions of PaO_2 are shown by Fig. 1, the solid lines representing non-linear exponential approaches to the measured mean values of $\dot{V}$ and PaO_2. Additionally, the effectiveness of the ventilatory chemoreflex drive to restore PaO_2 at different levels of F_IO_2 is shown, comparing the uncontrolled condition (CSN cut) with two levels (isocapnic and hypocapnic) of chemoreflex drive. The broken lines are restoring curves representing power-functions to approach the measured mean values, whereby limitations by inspiratory O_2 supply and dead space ventilation ($\dot{V}_D$) are considered. It can be seen that the gain in PaO_2 related to the same chemoreflex rise in $\dot{V}$ becomes smaller with lowered F_IO_2. The striking similarity of the restoring curves with either intact or cut vagus nerves suggests small vagal influence on the effectiveness of the peripheral chemoreflex. Likewise, average values for $\dot{V}_D$ (Fig. 1) are not significantly different between these groups and those for (a-et)PCO_2 are nearly identical (4.1 ±0.3 and 4.0 ± 0.3 mmHg).

For better quantitation of the chemoreflex effectiveness to restore PaO_2, the single values underlying the means of Fig. 1 are plotted in Fig. 2 as PaO_2 versus $\dot{V}$ and, for comparison, as SaO_2 versus V.

Linear regression analysis of the data shows different slopes at different levels of inspiratory O_2 supply. The numerical values of the slopes (±SD) of these regression lines, representing the gain in PaO_2 or SaO_2 for a given rise in pulmonary ventilation are shown by Table 1. There is a significant decrease in respiratory chemoreflex effectiveness to restore PaO_2 on going from air-breathing to severe hypoxia, although the arterio-endtidal PCO_2 difference, i.e. pulmonary gas exchange remained fairly constant.

In contrast, there is a considerable and significant increase in the effectiveness of the respiratory chemoreflex drive to restore SaO_2. This growing importance of the O_2-Hb satu-

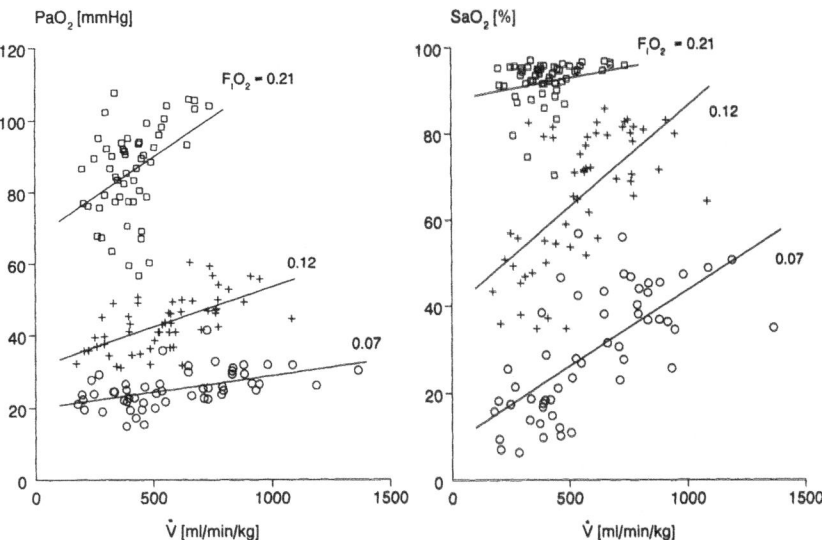

Figure 2. PaO$_2$ and SaO$_2$ in relation to pulmonary ventilation at different levels of inspiratory oxygen supply. The effectiveness of the controlling system, as judged from the slopes of the regression lines, becomes progressively smaller during hypoxia with respect to the adjustment of PaO$_2$ (left-hand), but increases to a maximum during moderate hypoxia with respect to the adjustment of SaO$_2$ (right-hand). F$_1$O$_2$: 0.21 (□), 0.12 (+) and 0.07 (○). For numerical values of slopes see Table 1.

Table 1. The effectiveness of respiratory chemoreflex drive to adjust the arterial O$_2$ pressure and O$_2$-Hb saturation at different levels of inspiratory O$_2$ supply

Experimental group				Air breathing F$_1$O$_2$=0.21	Moderate hypoxia F$_1$O$_2$=0.12	Severe hypoxia F$_1$O$_2$=0.07
Total	(n=54)	ΔPaO$_2$	[mmHg]	4.4 ±1.3	2.2 ±0.4	0.9 ±0.2*
		ΔSaO$_2$	[%]	1.0 ±0.6	4.7 ±0.8*	3.5 ±0.5*
		(a-et)PCO$_2$	[mmHg]	3.8 ±0.4	4.3 ±0.5	4.1 ±0.3
Isocapnia	(n=36)	ΔPaO$_2$	[mmHg]	5.2 ±1.5	2.4 ±0.4	1.0 ±0.2*
		ΔSaO$_2$	[%]	1.5 ±0.7	5.0 ±0.8*	3.5 ±0.5*
		(a-et)PCO$_2$	[mmHg]	3.9 ±0.4	4.2 ±0.6	3.9 ±0.4
Hypocapnia	(n=36)	ΔPaO$_2$	[mmHg]	1.3 ±2.3	1.4 ±0.6	0.4 ±0.3
		ΔSaO$_2$	[%]	0.3 ±1.2	6.3 ±1.3*	4.3 ±0.8*
		(a-et)PCO$_2$	[mmHg]	4.3 ±0.5	5.2 ±0.6*	4.8 ±0.5
Nn. vagi intact	(n=27)	ΔPaO$_2$	[mmHg]	3.4 ±1.7	2.1 ±0.6	1.1 ±0.3
		ΔSaO$_2$	[%]	0.7 ±0.5	4.1 ±1.2*	3.1 ±0.7*
		(a-et)PCO$_2$	[mmHg]	3.5 ±0.5	4.4 ±0.7	4.6 ±0.6*
Nn. vagi cut	(n=27)	ΔPaO$_2$	[mmHg]	6.8 ±1.8	2.4 ±0.5*	0.7 ±0.3*
		ΔSaO$_2$	[%]	1.8 ±1.1	5.3 ±1.0*	3.9 ±0.7
		(a-et)PCO$_2$	[mmHg]	4.2 ±0.5	4.2 ±0.6	3.6 ±0.4

Changes in PaO$_2$ and SaO$_2$ are related to a rise in of 100 ml/min/kg. Their mean values (±SD) are derived from the slopes of regression lines calculated either for all data or for subgroups of the data in Fig. 1. The arterio-endtidal PCO$_2$ difference remains mainly unchanged during hypoxia. There is neither significant effect of PCO$_2$ nor of pulmonary vagal afferents on the effectiveness of the respiratory chemoreflex, P$_D$ >0.05 by unpaired t-tests of subgroup differences.
* Significant difference compared to air-breathing, with P$_D$ <0.05 by unpaired t-tests.

ration over PaO_2 during hypoxia may depend on the Bohr effect and hence on the concomitant level of $PaCO_2$. Therefore, two subgroups (n=36) of the data in Fig. 2 were analysed, including either the hypocapnic or the isocapnic chemoreflex responses of ventilation. Derived from linear regression analysis of these subgroups, the Table shows that there is no significant difference between hypocapnia or isocapnia with respect to the gain in PaO_2 or SaO_2 upon 100 ml/min/kg rise in $\dot{V}$ at any level of F_iO_2. However, with concomitant hypocapnia the slightly smaller gain in PaO_2 is accompanied by a greater gain in SaO_2 during moderate hypoxia.

The influence of pulmonary vagal reflexes on the effectiveness to restore PaO_2 appears to be negligible (Fig. 1). This is demonstrated quantitatively by Table 1, showing the results of linear regression analysis for another two subgroups (n=27) of the data in Fig. 2, with either intact or cut vagus nerves. At no level of F_iO_2 is there a significant difference between these groups with respect to the gain in PaO_2 or SaO_2 at 100 ml/min/kg rise in $\dot{V}$.

Likewise, these groups do not differ significantly with respect to (a-et)PCO_2 at any level of F_iO_2 (Table 1).

DISCUSSION

The effectiveness of the peripheral chemoreflex loop in the respiratory control system can be judged from the rise in PaO_2 or SaO_2 due to a given rise in pulmonary ventilation. It could be shown in this study that the effectiveness to adjust PaO_2 progressively decreased when the inspiratory oxygen supply was lowered from normoxia to severe hypoxia. Even the over-proportionally augmented sensitivity of the chemoreflex controller during hypoxia, indicated by the hypoxic response curves in Fig. 1, did not compensate for the progressive insufficiency to restore the arterial PO_2. Likewise, despite the repeatedly demonstrated modulatory effect of pulmonary vagal reflexes on the sensitivity of the hypoxic ventilatory response (Kiwull-Schöne and Kiwull, 1979, and Fig. 1 of this study), their influence on the effectiveness to restore PaO_2 remained negligible (Fig. 1, Table 1). In much the same way, distinctly different respiratory chemoreflex drives as during concomitant hypocapnia or isocapnia, did not gain any importance for the overall chemoreflex effectiveness to adjust PaO_2.

These results demonstrate that sensitivity and effectiveness are completely different properties to characterize one and the same control system. Under the given experimental conditions, the main limiting factor for the effectiveness to adjust PaO_2 appears to be the inspiratory O_2 partial pressure (P_iO_2). This can be visualized by the restoring curves in Fig. 1, being progressively restricted during hypoxia by vertical asymptotes in terms of P_iO_2. The empirical restoring curves to fit our experimental data can be regarded as analogues of metabolic hyperbolae (Cunningham, 1974), their slopes depending on O_2 consumption and pulmonary gas exchange. Our data exclude the possibility that the loss in effectiveness to restore PaO_2 during hypoxia was additionally caused by impaired pulmonary gas exchange, since the arterio-endtidal PCO_2 difference remained unchanged (Table 1).

Most theoretical models of chemical respiratory control besides total ventilation refer to alveolar ventilation, which variable is directly determined by pulmonary gas exchange. In this study, dead space ventilation was only roughly estimated and found to be in the range between 20 and 45% of total ventilation. The complex interrelations of pulmonary and alveolar ventilation as well as of alveolar and arterial O_2 and CO_2 partial pressures are difficult to discern both theoretically and empirically. Local and chemoreflex actions of O_2 (and CO_2) do influence pulmonary circulation and bronchial tone in a varying manner and are additionally modified by vagal pulmonary reflexes.

Nevertheless, disregarding these multiple factors determining pulmonary gas exchange in detail, simply comparing the final result in terms of arterial O_2 partial pressure with either intact or cut carotid sinus nerves offers a useful overall estimate for the effectiveness of the peripheral respiratory chemoreflex under the condition of reduced inspiratory O_2 supply, as it normally may occur during short-term high altitude exposure.

In contrast, the effectiveness to adjust O_2-Hb saturation increased under the same conditions, reaching a maximum during moderate hypoxia. Additionally, there was a tendency to improve the gain in SaO_2 for a given rise in PaO_2 by hypocapnia. Thus, the O_2 binding properties of the blood as well as the Bohr effect contribute to improve the O_2 supply under conditions of restricted inspiratory O_2 availability. Since generally the limiting factor for SaO_2 is full saturation, the gain in SaO_2 will not be restricted by inspiratory oxygen supply. Therefore, O_2 transport properties of the blood were shown to gain growing importance over chemoreflex control of breathing for the oxygen supply of the tissues under conditions of restricted inspiratory O_2 availability.

On the other hand, from the view of this study, the less well restored PaO_2 appears as a more suitable adequate stimulus for the chemoreceptor than SaO_2. This view is directly supported by early microelectrode studies of Acker and Lübbers, 1976, and by recent phosphorescence quenching experiments of Lahiri et al., 1994. Both reports emphasize the leading role for the tissue O_2 partial pressure over O_2 content as the first step of the signal transduction cascade in the carotid body.

ACKNOWLEDGMENT

The expert technical assistance and help with the drawings by Ms. Claudia Bräuer and Ms. Sabine Adler is gratefully acknowledged.

REFERENCES

Acker, H., and Lübbers, D.W., 1976, Oxygen transport capacity of the capillary blood within the carotid body, Pflügers Arch. 366:241

Astrup, P., and Schrøder, S., 1956, A simple electrometric technique for the determination of carbon dioxide tension in blood and plasma, total content of carbon dioxide in plasma, and bicarbonate content in "separated" plasma at a fixed carbon dioxide tension (40 mmHg), Scand. J. Clin. Lab. Invest., 8:33

Cunningham, D.J.C., 1974, The control system regulating breathing in man, Quart. Rev. Biophys. 6:4

Colinet-Lagneaux, D., Troquet, J., Lambert, P., 1969, Echanges gazeux et méchanique respiratoire chez le lapin anesthésié. Influence de la vagotomie, Arch. Int. Physiol. Biochem. 77:88

Kalhoff, H., Kiwull-Schöne, H. and Kiwull, P., 1994, Pulmonary vagal afferents versus central chemosensitivity in the ventilatory response to hypoxia and lactic acid, Adv. Exp. Med. Biol. 345:121

Kiwull-Schöne, H. and Kiwull, P., 1979, The role of the vagus nerves in the ventilatory response to lowered PaO_2 with intact and eliminated carotid chemoreflexes, Pflügers Arch. 381:1

Kiwull-Schöne, H., Gärtner, B. and Kiwull, P., 1987, The effects of CO_2 and fixed acid on the O_2-Hb affinity of rabbit and cat blood, Pflügers Arch. 408:451

Kiwull-Schöne, H., Werkmeister, F., and Kiwull, P., 1992, The Haldane effect of rabbit blood under different acid-base conditions, Adv. Exp. Med. Biol. 316:11

Lahiri, S., Wilson, D.F., Iturriaga, R., and Rumsey, W.L., 1994, Microvascular PO_2 regulation and chemoreception in the cat carotid body, Adv. Exp. Med. Biol. 345:129

Loeschcke, H.H., 1973a, The respiratory control system: Analysis of steady state solutions for metabolic and respiratory acidosis-alkalosis and increased metabolism, Pflügers Arch. 341:23

Loeschcke, H.H., 1973b, The effectiveness of the control of pH in the extracellular fluid of the brain by the respiratory control system, Pflügers Arch. 341:43

63

EFFECT OF β-INTERFERON ON VASCULAR DENSITY, MITOCHONDRIAL METABOLISM AND ALKALINE PHOSPHATASE IN NORMOXIA AND HYPOXIA

Resit Demir and Jens Höper

Institut für Physiologie und Kardiologie
Friedrich-Alexander-Universität
Waldstraße 6, D-91054 Erlangen, Germany

INTRODUCTION

During the development of tissues, different genes must be up-regulated in a coordinated way in order to ensure a normal growth. Hypoxia is known to induce an up-regulation of different genes encoding for growth factors and a concomitant increase in these mitogens. As shown by Höper and Jahn (1995) a decrease in environmental oxygen concentration to 10% induces both an enlargement of the vascularized area and an increase in vascular density in the *area vasculosa* of the early chick embryo. An increase in oxygen uptake rate induced by radiation can also induce an increase in vascular density (Plaßwilm et al. 1996).

Interferons, on the other hand, can inhibit proliferation. Antiviral, immune modulating and antiproliferative effects have been described. Therefore, interferons are used in tumor therapy, but with varying success. While there is a good response in e. g. myoproliferative disease (Gisslinger et al. 1991, Seewann et al. 1991, Tichelli et al. 1989, Yataganas et al. 1991, Giles et al. 1988), renal cell carcinoma showed a response only in 16% (de Riese et al. 1991). This may be due to different local oxygenation and thus different up-regulation of mitogens. Thews and Vaupel (1996) reported that tumor oxygenation is an important factor influencing proliferation kinetics of tumors.

Antiangiogenetic activity of interferons has only recently been described (Stout et al. 1993, Angiolillo et al.1995). However, Cozzolino et al. (1993) found an angiogenic response in the rabbit cornea.

These obviously paradoxic results led to the hypothesis that there might be a competition between the proliferation-stimulating effect of hypoxia and the inhibiting effect of interferons.

Therefore, we investigated the effect of β-interferon on vascular density, local metabolism and local activity of alkaline phosphatase in the *area vasculosa* under different environmental oxygen concentrations.

Oxygen Transport to Tissue XIX, edited by Harrison and Delpy
Plenum Press, New York, 1997

MATERIALS AND METHODS

Eggs

For this study fertilized crossbred 'White-Plymouth-Rocks x Sussex' eggs (n = 51, Lehr- und Versuchsanstalt für Kleintierzucht, Kitzingen, Germany) were incubated in an upright position in a commercial incubator at 37.8 ± 0.1°C and 60 - 65% relative humidity. The daily weight loss of the eggs was 0.3 ± 0.05 g per day which was considered to be within the normal range.

From the first day the eggs were incubated either under conditions of atmospheric oxygen (20.9%, n = 22, group 1), in a mixture of 90% nitrogen and 10% oxygen (hypoxia, n = 17, group 2) or under severe hyperoxic conditions (5% oxygen, 95% nitrogen, n = 12, group 3).

After 48±1 hrs the egg shell was carefully opened using a dental drill (Eclo-S, Dentalwerk Buermoos, Salzburg, Austria). The inner shell membrane was not disrupted. Under the breeding conditions used the chick embryos were in stage 6 corresponding to Hamburger and Hamilton (1951).

Then 0.05 ml solution containing 30 000 IU recombinant β-interferon (IFN; Betaferon 1, Rentschler, Germany) were injected next to the germ disc. The maximal distance to the sinus terminalis was 2 mm. 0.05 ml Ringer's solution were injected in the control eggs.

The eggs were incubated for another 48±1 hrs. At this time the development of the embryo corresponded to stage 16 - 17 (Hamburger and Hamilton, 1951).

All examinations described in the following were performed on the fourth day of incubation under normoxic conditions.

Vascular Density

On day 4 of incubation, 48 hrs after application of β-interferon, the eggs were opened and the inner shell membrane was removed. The almost circular, vascularized part of the yolk sac membrane (*area vasculosa*) was photographed *in vivo*. Prints of known enlargement were evaluated for vascular density.

Furthermore, videomicroscopy was performed. For this purpose a CCD camera (VC3010, Euromex, the Netherlands) was adapted to a microscope (Leitz, Wetzlar, Germany). Vascular density was either determined from the photographs or from the video tape.

The vascular density was determined according to Höper and Jahn (1995). Results of vascular density are given as vascular intersections per squaremillimeter (VIS/mm^2). From all eggs at least two photographs were taken; one from the IFN- or Ringer-treated area. Vessel tortuosity and diameter have been shown to play only a minor role in the small areas investigated.

Erlangen Micro-Lightguide Spectrophotometer (EMPHO)

The EMPHO (BGT Überlingen, Germany) consists of four modules: the light source, the micro-lightguide, the detection device, and the computing system. Details have been published elsewhere (Frank et al. 1989). Light from a xenon arc lamp (Osram) is parallellised and focused onto the entrance plane of the illuminating micro-lightguide fibre. The light is transmitted to the tissue surface by a central micro-lightguide fibre closely surrounded by a hexagon of six detecting fibres. Each of the fibres has a diameter of 250 μm. Both the illuminating fibre and the detecting micro-lightguides are encased in a flexible rubber tube. At the measuring end they are inserted into a stainless steel cannula (outer diameter 1.6 mm). At the other end they are attached to the optical instrument by means of a special plug. The light transmitted by the detection micro-lightguide passes through a rotating interference bandpass filter disk (502–628 nm, Anders, Nabburg, Germany) and is measured by a photomultiplier. The output is connected to a computer used for signal storage and processing. Before each measurement a response spectrum for the apparatus is recorded.

Mitochondrial Metabolism

For investigations of the local mitochondrial metabolism of the *area vasculosa*, a micro-lightguide spectrophotometer (EMPHO II, BGT, Überlingen, Germany) using modified light guides was used (Höper 1996) to measure the cleavage of a tetrazolium salt (WST-1, Boehringer-Mannheim, Germany).

In recent years different tetrazolium salts have been described which can be used for the *in vitro* measurement of cell proliferation and viability (Vistica et al. 1991, Scudiero et al. 1988).

WST-1 is the tetrazolium salt (4-[3-(4-Iodophenyl)-2-(4-nitrophenyl)-2H-5-tetrazolio]-1,3-benzene disulfonate) which is cleaved to formazan by the "succinate-tetrazolium reductase" system (EC 1.3.99.1). This system belongs to the respiratory chain of mitochondria and is active only in viable cells (Slater, 1963). While WST-1 is slightly red, formazan is dark red. The maximal absorbance for the reaction product formazan is found between 420 and 480 nm. At 502 nm approx. 70% of the maximal absorption is found. Only the signal at 502 nm was evaluated from the spectra recorded. A detailed description of the method will be published (Höper 1996).

For the *in vivo* investigation of the cleavage of the tetrazolium salt a special light guide was developed. At the tip of the light guide a microcuvette was fixed. This cuvette was filled with 50 μl of the reagent. At the bottom the cuvette has an opening, which is placed at the surface of the *area vasculosa*. The WST-1 diffuses into the tissue and is cleaved to formazan. This is accompanied by a change in the absorption, which is measured.

Alkaline Phosphatase (AP)

The test also is based on a calorimetric principle. We used the precipitating BM purple alkaline phosphatase substrate (Boehringer Mannheim, Germany). As a result of the reaction a permanent dark blue precipitating product is formed, the change in absorption is measured.

For both tests, WST1 and alkaline phosphatase (EC 3.1.3.1), the absorption was determined after a reaction time of 15 minutes.

RESULTS

Figure 1a shows the vascular density (VIS/mm^2) determined in normoxia, mild hypoxia (10% oxygen) and severe hypoxia (5% oxygen). While mild hypoxia results in an increase in vascular density, during severe hypoxia a lower vascular density was observed.

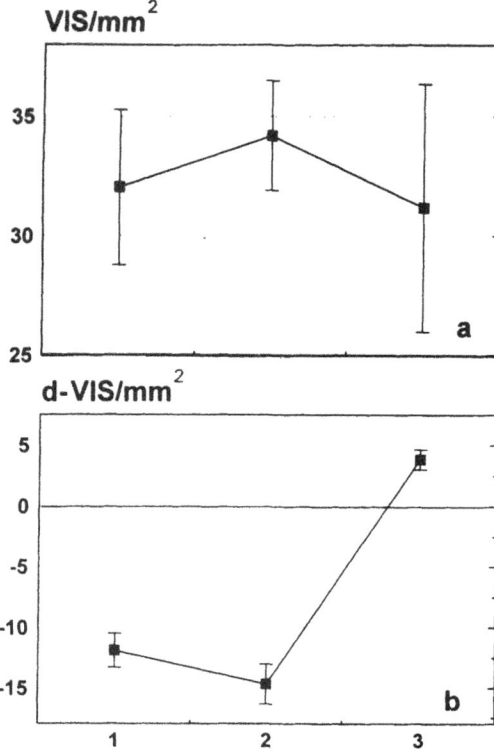

Figure 1. (a)Vascular density determined under normoxic conditions (1, n = 22), and during mild (2, n = 17) and severe hypoxia (3, n = 12). (b)Change in vascular density (d-VIS/mm^2 = VIS-IFN - VIS-control) after local administration of 30 000 IU β-IFN.

Under normoxic conditions as well as during mild hypoxia β-interferon causes a significantly lower vascular density (Fig. 1b, p 0.01 in both groups). In contrast, in severe hypoxia the number of vascular intersections/mm^2 was higher in the interferon-treated *area vasculosa* than in the controls.

Figures 2a and b depict the results of the WST-1 test. In the Ringer-treated eggs the smallest change in absorption was observed after incubation in 10% oxygen (mild hypoxia). After treatment with interferon, in the eggs incubated under normoxic and severe hypoxic conditions a depression of the WST reaction was observed, while in the 10% O$_2$ eggs a marked increase in absorption was observed. There was a significant difference between group 2 and group 3 (p 0.03). The changes in WST absorption in the control and interferon treated eggs were opposite to the changes in vascular density.

In contrast, in Ringer-treated eggs the alkaline phosphatase activity (Figures 3a and b) changes in parallel with the vascular density (Figure 1a and b). The largest absorption change and thus the highest enzyme activity was found during mild hypoxia. Alkaline phosphatase activity was significantly inhibited by IFN in mild and severe hypoxia (groups 1 vs. 2: p 0.01, groups 1 vs. 3: p 0.03).

DISCUSSION

The results of the present work show that under normoxic and mild hypoxic breeding conditions local application of β-interferon induces a significant decrease in vascular density of the *area vasculosa* of the early chick embryo. This indicates that β-IFN has an

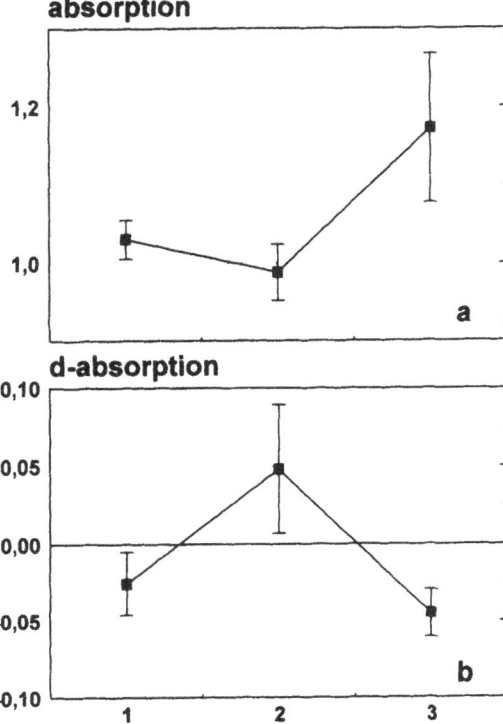

Figure 2. (a) Absorption measured at 502 nm due to the formation of formazan from the tetrazolium salt 4-[3-(4-Iodophenyl)-2-(4-nitrophenyl)-2H-5-tetrazolio]-1,3-benzene disulfonate (WST-1). 1 = Normoxia, n = 9; 2 = Hypoxia (10% oxygen), n = 6; 3 = Severe hypoxia (5% oxygen), n = 4. (b) Absorption difference between the controls and the β-IFN treated eggs (d-absorption (abs.) = abs.-IFN - abs.-control). It is obvious that in the mild hypoxic eggs after treatment with β-IFN more WST-cleavage took place, whereas in the normoxic and severe hypoxic eggs an inhi- bition was observed.

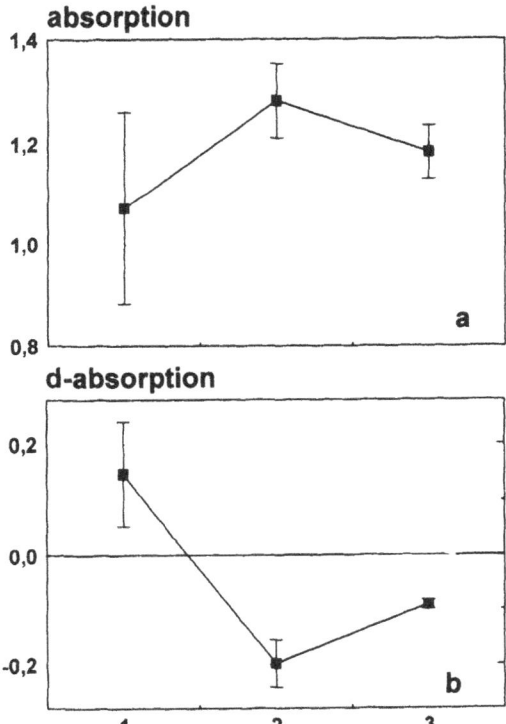

Figure 3. (a) Absorption induced by a permanent dark blue precipitating product formed by AP from the BM purple alkaline phosphatase substrate. 1 = Normoxia, n = 9; 2 = Hypoxia (10% oxygen), n = 5; 3 = Severe hypoxia (5% oxygen), n = 4. (b) Absorption difference between the controls and the β-IFN treated eggs eggs (d-absorption (abs.) = abs.-IFN - abs.-control). It is obvious that in the mild and the severe hypoxic eggs after treatment with β-IFN the formation of the dark blue precipitating product was smaller than in the non treated controls, whereas in the normoxic eggs a more pronounced change was observed.

antiangiogenic action. Further lowering of the environmental oxygen content to 5% abolishes this antiangiogenic effect.

In order to understand the different action of interferon on vascular density under different conditions it is first of all necessary to measure local parameters which may influence the efficacy of IFN and to study local physiologic reactions under *in vivo* conditions. A sine qua non condition is the availability of an adequate model. In the past we have shown that the *area vasculosa* of the early chick embryo allows investigations of vascular density under different conditions (Höper and Jahn 1995, Plaßwilm et al. 1996). Because vascular density in this tissue depends on both vasculogenesis and angiogenesis, conclusions on these processes may be drawn. In order to determine the local reactions the development of techniques allowing *in vivo* measurements of metabolic parameters was necessary.

In recent years, different tetrazolium salts (MTT, XTT and MTS) have been described which can be used for the measurement of cell proliferation and viability. All compounds are cleaved to formazan by the "succinate-tetrazolium reductase" system (EC 1.3.99.1) belonging to the respiratory chain of the mitochondria. According to Slater et al. (1963) this system is active only in viable cells. An increasing number of viable cells results in an increasing number of mitochondria and thus in an increase in the overall activity of mitochondrial dehydrogenases. These tests have been developed for *in vitro* investigations allowing the determination of an increasing number of cells in a sample. In contrast to MTT (Mosmann, 1983, Carmichael et al. 1987, Vistica et al. 1991) which is cleaved to water-insoluble formazan crystals, WST-1 yields water-soluble cleavage products like XTT (Scudiero et al. 1988, Weislow et al. 1989, Roehm et al. 1991).

First results with the *in vivo* WST-1 test were obtained in the *area vasculosa* in different developmental stages (Höper, submitted), showing that during the first 5 days there is an increasing reduction of the tetrazolium salt. Investigations performed with cultured fibroblasts and human glioblastoma cells showed that there is a linear relation between number of cells and absorption change (Frenzel et al., in preparation). Thus we may conclude that the increased formazan formation after incubation in 5% oxygen observed in this study primarily depends on number and/or activity of mitochondria.

Although we have not performed *in vitro* measurements with the AP test, we may conclude that we measure the extracellular activity of this enzyme. During angiogenesis the orientation, branching and three dimensional growth follows preformed extracellular matrix structures. Nees et al. (1996) showed that pericytes may participate in myocardial angiogenesis. The pericytes seem to be involved in the synthesis of the extracellular matrix. Beside other factors the pericytes synthesize alkaline phosphatase. As shown in figure 4, AP activity is found along the blood vessels of the *area vasculosa*. This indicates that with an increasing vascularity the AP activity should increase and vice versa. This is shown by our results.

Angiogenesis is under the control of different growth factors. Wilting et al. (1993) were able to show that vascular endothelial growth factor (VEGF) induces angiogenesis in the regions of precapillary vessels, capillaries and venules, respectively. VEGF is an important mitogen also for vasculogenesis (Flamme et al. 1995). Stewart et al. (1989) showed that epidermal growth factor (EGF) promotes extraembryonic angiogenesis in the *area vasculosa* of 3 day old chicks.

Vasoproliferative effects of hypoxia have already been described for the chorioallantoic membrane (CAM) of chick embryos (Strick et al. 1991). It has been shown that during hypoxia an up-regulation of the VEGF gene occurs (Shweiki et al. 1992). As shown by Höper and Jahn (1996), hypoxia not only induces an increase in vascular density but also an enlargement of the *area vasculosa*.

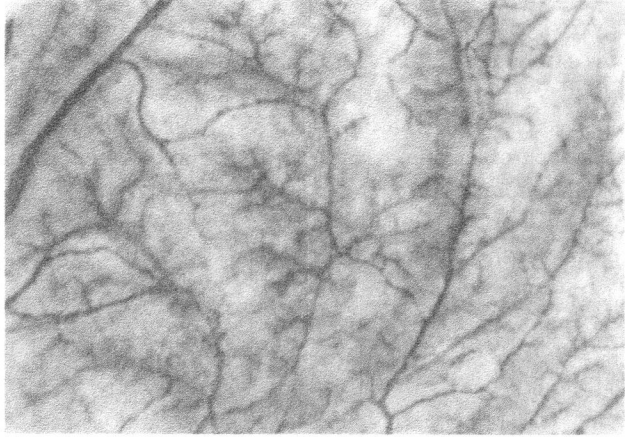

Figure 4. Part of the *area vasculosa* after staining for AP. It is evident that the AP activity (dark precipitation) is primarily found along the vascular structures.

A relation between differentiation, proliferation and AP activity was found in human ovarian adenocarcinoma cells (Brooks et al. 1991), in mitogen-stimulated B-lymphocytes (Kasyapa and Ramanadham 1992) and murine osteoblast-like cells (Noda et al. 1990). Chida (1990) came to the conclusion that the AP activity is one marker to justify the degree of differentiation of hepatocytes during the state of proliferation. This points to the fact that phosphorylation-dephosphorylation processes may be involved in the regulation of cell proliferation and differentiation.

To our knowledge no data are available on AP in the *area vasculosa*. Our data indicate that a high activity of AP may be associated with differentiation or proliferation of endothelial cells and thus with angiogenensis.

The significantly smaller vascular density found after IFN treatment in groups 1 and 2 must be the result of an inhibited vasculogenesis and/or angiogenesis. An antiangiogenic effect of interferon has been described (Stout et al. 1993, Angiolillo et al.1995). The basic mechanisms are still subject to discussion. Norioka et al. (1994), Angiolillo et al. (1995), and Strieter et al. (1995) reported that IFN inhibits bFGF induced angiogenesis. Matsubara et al. established already in 1986 that β-IFN increases superoxide anion release from human endothelial cells in a dose and time dependent manner. They suggested from their results that the superoxide anion toxicity is responsible for the antiangiogenic activity of IFN. These results could also explain why in severe hypoxia (group 3) the antiangiogenic effect of IFN is abolished. Noff et al. (1989) described that bFGF treatment increases alkaline phosphatase activity in rat bone marrow cells. On the other hand Kortsaris et al (1988) found that in breast carcinoma cells IFN in low dose (up to 100 IU/ml) induces an increase in AP activity whereas high doses cause a decrease. Thus we may conclude that the lower AP activity and the lower vascular density are linked.

Summarizing our results it can be concluded that the increase in vascular density induced by mild hypoxia is accompanied by an increase in AP activity whereas the local metabolic rate is not influenced. Severe hypoxia, on the other hand causes no increase in vascular density and AP activity but the local metabolic activity is increased.

In normoxia and mild hypoxia IFN treatment is followed by a decreased vascular density. Under severe hypoxic conditions IFN does not cause a decrease in vascular density. Obviously the stimulating effect of hypoxia abolishes the antiangiogenic effect of IFN. These results may also be important for tumor biology, where angiogenesis (Folkman et al. 1974) and hypoxia play an important role. Our results indicate that the IFN induced production of superoxide anions in tumor cells and non-tumor cells may be reduced during hypoxia. Furthermore, the effect of cytokines released from non-tumor cells during severe hypoxia may be abolished. Finally, hypoxia may lead tumor and non-tumor cells to induce angiogenesis in a synergistic way.

REFERENCES

Angiolillo, Al; Sgadari, C, Taub, DD; Liao, F; Farber, JM; Maheshwari, S; Kleinman, HK; Reaman, GH; Tosato, G. Human interferon-inducible protein 10 is potent inhibitor of angiogenesis in vivo. J Exp Med 182, 155–162, (1995)

Brooks, SE, Timmerman, J, Lau, CC, Tsao, SW, Knapp, RC, Sheets, EE. Effect of differentiation agents on the expression of CA 125, alkaline phosphatase, and cytokeratins in human ovarian adenocarcinoma cells (OVCA 433). Gynecol Oncol 42, 265–272 (1991).

Carmichael, J, DeGraff, WG, Gazdar, AF, Minna, JD, Mitchell, JB Evaluation of a tetrazolium-based semiautomated colorimetric assay: Assessment of chemosensitivity testing. Cancer Res. 47, 936 - 942, (1987)

Chida, K, Histochemical and immunohistochemical study on the expression of alkaline phosphatase, gamma-glutamyl transpeptidase and alpha-fetoprotein during the process of rat hepatocyte proliferation. Kitasato Arch. Exp. Med. 63, 91–98 (1990)

Cozzolino, F, Torcia,M, Lucibello,M, Morbidelli,L, Ziche,M, Platt,J, Fabiani,S; Brett,J; Stern,D. Interferon-alpha and interleukin 2 synergistically enhance basic fibroblast growth factor synthesis and induce release, promoting endothelial cell growth. J Clin Invest. 91, 2504–2512, (1993)

de Riese, W, Allhoff, E, Schuth, J. Das metastasierende Nierenzellkarzinom, Springer Verlag Berlin Heidelberg (1991)

Flamme, I, Breier, G, Risau, W. Vascular endothelial growth factor (VEGF) and VEGF receptor (flk-1) are expressed during vasculogenesis and vasular differentiation in the quail embryo. Dev. Biol. 169, 699–712 (1995)

Folkman, J. Tumor angiogenesis. In: Advances in Cancer Research. Eds. G. Klein and S. Weinhouse. New York, Academic Press (1974).

Frank, KH, Kessler, M, Appelbaum, K, Dümmler, W. The Erlangen micro-lightguide spectrophotometer EMPHO I. Phys. Med. Biol. 34, 1883–1900 (1989).

Giaccia, AJ, Chen, EY, Yeh, P, Waleh N, Laderoute, KR, Mazure, NM. Regulation of vascular endothelial growth factor by hypoxia. Proc. 10th Int. Congress of Radiation Research (abstract P11–11) p 218, Wuerzburg, Germany (1995)

Giles, FJ, Grey, AG, Brovic, M. Alpha interferon therapy for essential thrombocythemia. Lancet , II: 70–72, (1988)

Gisslinger, H, Chott, AM, Huber, H. Interferon in essential Thrombocytopenia. Br J Haematol , 79,suppl.4: 42–47,(1991)

Goldberg, MA, Schneider, TJ. Similarities between the oxygen sensing mechanisms regulating the expression of vascular endothelial growth factor and erythropoietin. J. Biol. Chem. 269, 4355–4359 (1994).

Hamburger, V, Hamilton, HL. A series of normal stages in the development of the chick embryo. J Morphol 88, 49–92, (1951)

Hlatky, L, Tsionou, C, Hahnfeldt, P,Coleman, N. VEGF expression in human mammary tumor cells: Upregulation with hypoxia. Proc. 10th Int. Congress of Radiation Research (abstract P11–12) p 218, Wuerzburg, Germany (1995).

Höper, J, Jahn, H. Influence of environmental oxygen concentration on growth and vascular density of the area vasculosa in chick embryos. Int J Microcirc 15, 186–192 (1995).

Höper, J. In vivo determination of local mitochondrial metabolism by use of a tetrazolium salt. Physiol. Measure. (submitted)

Kasyapa, CS, Ramanadham, M. Alkaline phosphatase activity is expressed only in B lymphocytes committed to proliferation. Immunol Lett 32, 111–116 (1992).

Kortsaris, A, Kyriakidis, DA. Ornithine decarboxylase and phosphatase activity can be stimulated by low concentrations of interferon in human breast cancer cell lines. Microbiologica 11, 347–353, (1988).

Matsubara, T, Ziff, M. Increased superoxide anion release from endothelial cells in response to cytokines, J Immunol 137, 3295–3298, (1986)

Mosmann, T. (1983) Rapid colorimetric assay for cellular growth and survival: application to proliferation and cytotoxicity assays. J. Immunol. Methods 65, 55–63

Nees, S, Fischlein, T, Juchem, G. Pericytes (P) participate essentially in myocardial angiogenesis. Pflügers Arch-Eur J Physiol 431(6) (Suppl) R129 (1996).

Noda, M, Vogel, RL, Hasson, DM, Rodan, GA. Leukemia inhibitory factor suppresses proliferation, alkaline phosphatase activity, and type I collagen messenger ribonucleic acid level and enhances osteopontin mRNA level in murine osteoblast-like (MC3T3E1) cells. Endocrinology, 127, 185–190 (1990).

Noff, D, Pitaru, S, Savion, N. Basic fibroblast growth factor enhance the capacity of bone marrow cells to form bone like nodules in vitro. FEBS Lett. ,250, 619–621, (1989)

Norioka, K, Mitaka, T, Mochizuki, Y ,Hara, M, Kawagoe, M, Nakamura, H. Interaction interleukin-1 and interferon-gamma on fibroblast growth factor induced angiogenesis. Jap J Canc Res 85, 522–529, (1994)

Plaßwilm, L, Höper, J, Sauer, R. Radiobiologische In vivo Untersuchung zur Angiogenese am Dottersackgefäßsystem des Hühnerembryos. Strahlenther. Onkol. 172, 260–264 (1996).

Poole, TJ, Coffein, JD. Vasculogenesis and angiogenesis: two distinct morphogenetic mechanisms establish embryonic vascular pattern. J Exp Zool 251, 224–231, (1989) .

Rodan, SB, Weselowski, G, Yoon, K, Rodan, Ga. Opposing effects of fibroblast growth factor and pertussis toxin on alkaline phosphastase, osteopontin, osteocalcine, and type I collagen mRNA levels in ROS 17/2.8 cells. J Biol Chem 264,1934–1941, (1989)

Roehm, NW, Rodgers, GH, Hatfield, SM, Glasebrook, AL An improved colorimetric assay for cell proliferation and viability utilising the tetrazolium salt XTT. J. Immunol. Methods. 142, 257 - 265,(1991)

Scudiero, DA, Shoemaker, RH, Paull, KD, Monks, A, Tierney, S, Nofziger, TH, Currens, MJ, Seniff, D, Boyd, MR. Evaluation of a soluble tetrazolium/formazan assay for cell growth and drug sensitivity in culture using human and other tumor cell lines. Cancer Res. 48, 4827–4833 (1988)

Seewann, HL, Zikulnig, R Gallhofer, G.Treatment of thrombocytosis in chronic myeloproliferative disorders with interferon alfa-2b. Eur J Canc , 27 suppl.4, 58–63, (1991)

Shweiki, D, Itin, A, Soffer, D, Keshet, E. Vascular endothelial growth factor induced by hypoxia may mediate hypoxia-initiated angiogenesis. Nature 39, 843–845 (1992).

Slater, TF, Sawyer, B, Sträuli, Ú. Studies on succinate-tetrazolium reductase systems. III. Points of coupling of four different tetrazolium salts.Biochem.Biophys. Acta 77, 383–393 (1963)

Stewart, R, Nelson, J, Wison, DJ. Epidermal growth factor promotes chick embryonic angiogenesis. Cell Biol. Int. Rep. 13, 957–965 (1989)

Stout, AJ, Gresser, I, Thomson, WD Inhibition of wound healing in mice by local interferon alpha/beta injection. Int J Exp Pathol , 74, 79–85, (1993)

Strick, DM, Waycaster, RL, Montani, JP, Gay, WJ, Adair, TH.Morphometric measurements of chorioallantoic membrane vascularity: effects of hypoxia and hyperoxia. Am. J. Physiol. 260, H1385-H1389 (1991).

Strieter, RM, Kunkel, SL, Arenberg, DA, Burdick, MD, Polverini, PJ. Interferon gamma-inducible protein 10 (IP-10), a member of the C-X-C chemokine family, is an inhibitor of angiogenesis. Biochem. Biophys. Res. Commun. 210, 51–57, (1995)

Thews, O, Vaupel, P, Relevant parameters for describing the oxygenation status of solid tumors. Strahlenther. Onkol. 172, 239–243 (1996).

Tichelli,A, Gratwohl,A, Berger,C. Treatment of thrombocytosis in myeloproliferative disorders with interferon alfa-2a. Blut 58, 15–19, (1989)

Vistica, DT, Skehan, P, Scudiero, D, Monks, A, Pittman, A, Boyd, MR. Tetrazolium-based assays for cellular viability: A critical examination of selected parameters affecting formazan production. Cancer Res. 51, 2515–2520 (1981)

Weislow, OS, Kiser, R, Fine, DL, Bader, J, Shoemaker, RH, Boyd, MR (1989) New soluble-formazan assay for HIV-1 cytopathic effects: Application to high-flux screening of synthetic and natural products for AIDS-antiviral activity. J. Natl. Cancer. Inst. 81, 577 - 586

Wilting, J, Christ, B, Bokeloh, M, Weich, HA. In vivo effects of vascular endothelial growth factor on the chicken chorioallantoic membrane. Cell Tissue Res. 274, 163–172 (1993)

Yataganas, Y Meletis, J Plata, E. Alpha interferon treatment of essential thrombocythemia and other myeloproliferative disorders excessive thrombocytosis, Eur J Cancer, 27 supp.4: 69–71, (1991)

64

THE CYTOCHROME OXIDASE REDOX STATE *IN VIVO*

Chris Cooper,[1] Martyn Sharpe,[1] Clare Elwell,[2] Roger Springett,[2] Juliet Penrice,[3] Lidia Tyszczuk,[3] Philip Amess,[3] John Wyatt,[3] Valentina Quaresima,[4] and David Delpy[2]

[1]Department of Biological and Chemical Sciences
Central Campus
University of Essex
Colchester CO4 3SQ, United Kingdom
[2]Department of Medical Physics and Bioengineering
University College London
Shropshire House, 11-20 Capper Street
London WC1E 6JA, United Kingdom
[3]Department of Paediatrics
The Rayne Institute
University College London
5 University Street, London WC1E 6JJ, United Kingdom
[4]Dip. Scienze e Tecnologie Biomediche
Universita L'Aquila
67100 L'Aquila, Italy

1. INTRODUCTION

The Cu_A redox centre of cytochrome oxidase has a distinct redox-sensitive absorbance band in the near infrared that enables it to be detected non-invasively (Jöbsis 1977). We have been recently developing novel *in vitro* (Matcher et al. 1995) and *in vivo* (Cooper et al. 1996a) tests for the validity of algorithms claiming to be able to measure these redox changes, successfully deconvoluted from the greater spectroscopic perturbations induced by haemoglobin oxygenation and concentration changes. However, even if this problem is solved completely there still remains the question as to what the redox state of this metal centre is actually telling us about mitochondrial function *in vivo*. We are therefore performing parallel *in vitro* studies of the effect of energization, electron flow and oxygen tension on the redox state of Cu_A, similar to those originally outlined by Chance as aids to understanding NADH surface fluorescence measurements (Chance et al. 1973). This paper studies the effect of varying the rate of electron entry to cytochrome oxidase on

Oxygen Transport to Tissue XIX, edited by Harrison and Delpy
Plenum Press, New York, 1997

the Cu_A redox state *in vitro* and *in vivo* and therefore relates to the value of the normal, physiological oxidation state of Cu_A.

2. METHODS

2.1. *in Vitro* Studies - Details

Bovine cytochrome *c* oxidase was purified according to the method of Kuboyama et al (1972), and had a maximum turnover (pH 7.4) of approximately 380 $e^- aa_3^{-1} s^{-1}$. 10 μM cytochrome oxidase was incubated at 30 °C. with 10 μM horse heart cytochrome c and 100 nM bovine liver catalase, in 1 ml of 25 mM K^+-HEPES, 0.1% lauryl maltoside, pH 7.8. 40 mM sodium ascorbate was added to the cuvette, to initiate turnover of the enzyme, and the solution allowed to go anaerobic. Following anaerobiosis hydrogen peroxide was added to the cuvette at a final concentration of between 0.5 and 2 mM. This was rapidly decomposed into oxygen by the catalase in the incubation medium, thus initiating enzyme turnover. Following the second anaerobiosis, the cuvette was re-oxygenated with hydrogen peroxide and a second steady state was produced. No difference was observed between the two steady states.

The absorbance changes from 520 to 800 nm were monitored using a Beckman DU 7400 diode array spectrophotometer. The redox changes of cytochrome c were monitored at 550–540 nm. The redox changes of cytochrome a were monitored at 605–630 nm, assuming a 14% contribution at this wave pair of cytochrome a_3 in the reduced enzyme and no contribution of cytochrome a_3 in the steady state (Nicholls 1993). The absorbance changes in the near infrared were recorded using a Aminco DW2 spectrophotometer under identical assay conditions and the Cu_A redox state was measured using the 780–830 nm wavepair. 100% oxidation of all chromophores was measured in the absence of added reductants and 100% reduction was measured upon anaerobiosis. Repeating the experiment at increasing concentrations of TMPD resulted in an increase in the cytochrome c redox state and a consequent increase in enzyme turnover. The latter was measured by using the time taken for the sample to go anaerobic (measurements with an oxygen electrode demonstrated that under these conditions the rate of oxygen consumption is essentially linear with respect to the complete time of the assay, due to the low Km of the enzyme for oxygen).

2.2. *in Vivo* Studies

Large white piglets (1.3–1.9 kg) were studied on the first or second day of life. After sedation with intramuscular midazolam (0.2 mg kg^{-1}) anaesthesia was induced with 5% isoflurane and then maintained with nitrous oxide, oxygen and < 1.5% isoflurane. An endotracheal tube was passed and a pressure limited ventilator set at a respiratory frequency of 10–20 breaths min^{-1}, an inspiratory time of 0.6–1.0s, a peak inspiratory pressure of 1.0–1.5 kPa and a positive end-expiratory pressure of 0.1–0.3 kPa. Arterial blood gases were routinely taken and these parameters and the FiO_2 were adjusted to provide a PaO_2 of 8.0–13.0 kPa and a $PaCO_2$ of 4.5–6.5 kPa during the surgery. Samples were drawn from the umbilical vein and artery which were cannulated with 3.5 mm outer diameter polyvinyl catheters. An infusion of 0.9% saline containing 1U of heparin was provided at 1–2mL.h^{-1}, through the arterial catheter, which was attached to a strain-gauge blood pressure transducer. A 10% glucose solution that also provided 0.4–0.6 mg.kg^{-1} morphine sulphate was infused at 3–6 mL.h^{-1} through the venous catheter (the glucose was omitted in studies involving insulin addition). Heart rate was monitored via ECG or from the pressure wave of the arterial transducer. Arterial saturation was monitored via a pulse oximeter and was >95% throughout the study protocol.

A four wavelength Hamamatsu NIRO500 (779, 821, 855, 908nm) was used to study changes in the haemoglobin oxygenation and cytochrome oxidase redox state. Optodes were placed directly on the animals head in a similar manner to that used in our neonatal human studies (Edwards et al. 1992). There was an approximately 90° angle between the source and detector and an interoptode spacing of 3.8–4.3 cm. An optical pathlength is necessary to convert the optical density changes into μM concentration changes (Wyatt et al. 1990). This was determined by multiplying the interoptode spacing by a differential pathlength factor (4.5) determined by measurements of optical pathlength on neonatal piglets using a multiwavelength CCD spectrometer, in a similar manner as has been applied to neonatal human babies (Cooper et al. 1996b).

Prior to the NIRS study, the inspired gases were changed to FiO$_2$ (99%) and isoflurane (1%). Compounds were injected i.v. into the umbilical catheter.

3. RESULTS

3.1. *in Vitro* Studies

There have been few systematic studies of the factors affecting the NIR-detectable Cu$_A$ redox centre in cytochrome oxidase (Cooper et al. 1994). We generated the steady state levels of the redox centres by first allowing purified cytochrome oxidase to go anaerobic in the presence of reductants - starting with the reduced enzyme ensures that the enzyme is in a homogenous "fast" form throughout the assay (Moody et al. 1991). Turnover was then initiated by adding oxygen to the cuvette. A rapid rate of oxygen consumption ensues due to the high concentrations of enzyme required for adequate signal:noise when measuring the NIR spectrum of cytochrome oxidase. Therefore up to 1mM oxygen was rapidly added to the solution by using a hydrogen peroxide/catalase oxygen generating system (Morgan et al. 1991). Enzyme turnover was modulated by changing the driving force of the reaction. This was accomplished by increasing the concentration of the redox mediator N, N, N', N' tetramethyl-*p*- phenylenediamine (TMPD) which increases the cytochrome *c* redox state and therefore enzyme turnover (Morgan et al. 1991; Nicholls 1993).

Figure 1 shows that as enzyme turnover increases there is a linear increase in the Cu$_A$ redox state. However, at all times the Cu$_A$ remains highly oxidised—even at maximal flux in this system when cytochrome *c* is 80% reduced, less than 30% of Cu$_A$ is reduced. In contrast a much higher level of reduction of haem *a* is observed. Under these conditions haem *a* is at least as reduced as cytochrome *c*. This high level of haem *a* reduction is consistent with previous studies of the intact enzyme (Nicholls et al. 1990) and myocytes (Brunori et al. 1992), although see (Morgan et al. 1991) for a conflicting view from studies on isolated mitochondria.

3.2. *in Vivo* Studies

The cytochrome oxidase concentration of the adult rat brain is 5.5 μM (Brown et al. 1991). Cerebral oxygen consumption is about 10 ml O$_2$ 100g^{-1} min^{-1} (Schlichtig et al.

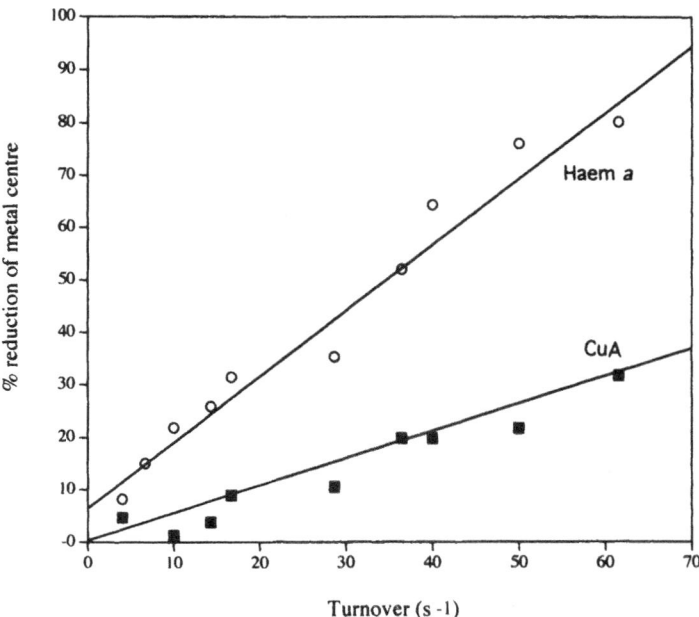

Figure 1. Variation of Cu_A and haem *a* redox states with varying flux. Turnover (measured in µM e- µM haem aa_3 s⁻¹) was varied by increasing the rate of electron entry to the enzyme as described in Methods.

1992). This suggests a turnover number through cytochrome oxidase in the rat brain *in vivo* of approximately 70 s⁻¹. Calculations for the neonatal brain are harder to make; in the human neonate the cerebral oxygen consumption is about 0.75 ml O_2 100g⁻¹ min⁻¹ (Altman et al. 1993). Using the method of Brown et al. (1991) we have measured the cytochrome oxidase concentration in the neonatal pig brain to be 3.2 ± 0.4 µM (n=5), which would imply a much lower turnover through neonatal cytochrome oxidase (10 s⁻¹) if consumption was as low as that observed in the neonatal human.

The data from the *in vitro* studies (Figure 1) therefore suggest that cytochrome oxidase Cu_A is largely, but perhaps not completely, oxidised at the electron fluxes observed *in vivo*. How can this be tested *in vivo*? *In vitro* studies of the redox state of electron carriers in the mitochondrial respiratory chain require that it be possible to obtain conditions where the centres are fully oxidised or fully reduced. In the purified cytochrome oxidase this is a trivial procedure for Cu_A as the presence of oxygen and absence of substrate allow full oxidation at the start of the study and the presence of reductants and lack of oxygen force full reduction at its end. In brain slices (Fujii 1991) or tissue homogenates (Brown et al. 1991) the rate of electron donation to cytochrome oxidase in the presence of oxygen cannot be assumed to be zero. In this case upstream inhibitors are employed to prevent electron flow. Examples used previously include the NADH dehydrogenase inhibitor, amytal (Chance et al. 1963; Fujii 1991) and the bc_1 complex inhibitor antimycin (Brown et al. 1991).

We therefore attempted to limit electron entry to cytochrome oxidase *in vivo* to determine if this would alter the Cu_A redox state. We used the neonatal pig as a model as due to its thin skull the neonate has been extensively studied by NIRS (the assumption being that a far larger proportion of the NIR light in this case enters the brain, as compared to the adult). Two separate approaches were adopted—limitation of substrate supply and in-

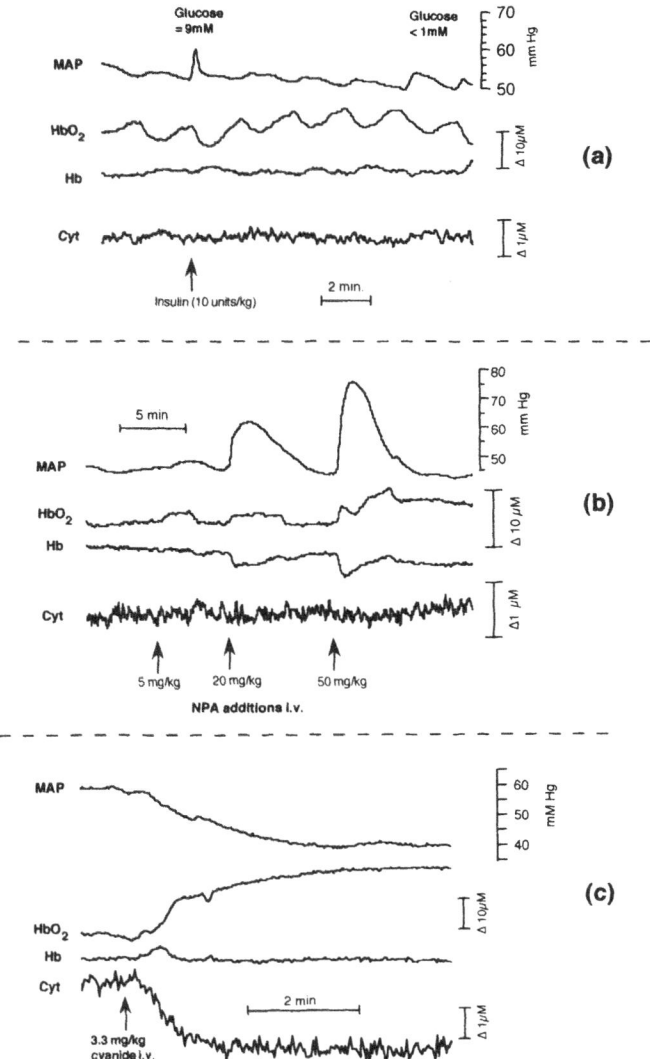

Figure 2. Near infrared spectroscopy of the neonatal pig brain. Changes in mean arterial pressure (MAP) and the redox state of cytochrome oxidase Cu_A (Cyt), deoxyhaemoglobin (Hb) and oxyhaemoglobin (HbO$_2$) were measured as described in Methods. Note the 10x expanded scale for Cyt. Reagents (insulin, 3-nitropropionic acid, sodium cyanide) were added i.v. at the indicated times. An upward trend indicates a rise in haemoglobin concentration or an oxidation of Cu_A.

hibitors of mitochondrial respiration acting upstream of cytochrome oxidase. All reagents were added i.v. The addition of insulin reduced blood glucose from 9mm to < 1mM with no detectable oxidation of cytochrome oxidase (Figure 2a). However, it might be expected that if this drop in blood glucose was having the desired effect of dropping the electron transfer rate to cytochrome oxidase that haemoglobin oxygenation in the brain would increase. This was not observed. We therefore added an inhibitor of complex II (succinate dehydrogenase), 3-nitropropionic acid (NPA), that is known to cross the blood-brain barrier (Simpson et al. 1993). In this case an increase in oxyhaemoglobin was observed, but still no oxidation of cytochrome oxidase (Figure 2b). We have observed a similar lack of Cu_A oxidation when the NADH dehydrogenase inhibitor, amytal is used (results not shown).

These results suggested that cytochrome oxidase Cu_A is predominantly oxidised in the steady state. We would therefore expect that an inhibitor that acts downstream of Cu_A

would cause a significant reduction of the enzyme. Sodium cyanide (Piantadosi et al. 1984; Tamura 1992) is such an inhibitor (CN binds to haem a_3). Following i.v. injection of NaCN we see the expected reduction of Cu_A (Figure 2c) and a large increase in cerebral oxyhaemoglobin and total cerebral haemoglobin concentration (as oxygen consumption is inhibited and vasodilatation occurs). At these levels of cyanide the cytochrome oxidase appears essentially completely reduced. Reducing the FiO_2 to zero subsequent to cyanide addition does not further reduce the Cu_A (Cooper et al. 1996a). The order of events is also not relevant i.e. reduction of cerebral Cu_A by anoxia is insensitive to subsequent cyanide addition. We therefore conclude that in the anaesthetised piglet brain the Cu_A redox state is primarily oxidised.

4. DISCUSSION

Our results show good agreement between *in vitro* and *in vivo* studies, suggesting that the redox state of Cu_A is *predominantly* oxidised *in vivo* .We do not believe that we have convincingly demonstrated that it is *completely* oxidised physiologically for the following reasons.

Firstly (given the low concentration of cytochrome oxidase in the neonatal brain) the signal:noise in these studies would find it difficult to detect a <10% change in the Cu_A signal.

Secondly without direct measurements of cerebral oxygen consumption we cannot be sure that the reagents used have completely prevented electron leakage to Cu_A. There is a limit as to how high a concentration of inhibitors can be used, because if the heart becomes compromised then oxygen delivery to the brain will fall, having the opposite of the intended effect (decreasing the rate of electron exit from Cu_A, rather than the rate of electron entry). We have seen this clearly when using excessive concentrations of the barbiturate, amytal, where haemoglobin oxygenation in the brain falls and cytochrome oxidase becomes reduced. These problems are not completely alleviated even when injecting directly into the carotid artery (results not shown). In contrast cyanide is a very strong, rapid inhibitor of cytochrome oxidase and it is possible to titrate cyanide into the animal (intravenously or in the carotid artery) and inhibit cerebral cytochrome oxidase without significantly compromising the heart in the short term. This may be partly due to the low metabolism of cyanide in the brain (McMahon et al. 1990), but the large glycolytic reserves in the neonatal heart may also play a role.

Thirdly if there is a fraction of cytochrome oxidase that is oxygen-limited at the start of the experiment, the upstream inhibitors used may not be strong enough to oxidise Cu_A. Although we do not see any changes in the Cu_A redox state when increasing the FiO_2 from 20% to 100%, we have previously reported an oxidation of cytochrome oxidase in the neonatal human brain following an increase in $PaCO_2$ (Edwards et al. 1991). If such an oxidation were confirmed it would not necessarily imply, as is generally stated, that a region of the brain is permanently oxygen-limited. Mitochondria are not static entities in cells and it is possible that in their movements (up and down the neurone for example) they encounter oxygen-depleted regions such that a small fraction of the mitochondria at any one time inside a cell are oxygen-limited (with a consequent reduction of cytochrome oxidase), but that the cell itself is not oxygen-limited in the time average.

Finally cerebral oxygen consumption is already reduced by 50% in the anaesthetised animal and it is theoretically possible that faster electron transfer rates in the unanaesthetised animal might lead to an increased reduction of Cu_A.

One important difference between the *in vitro* and *in vivo* results is that *in vivo* there is a proton electrochemical gradient across the inner mitochondrial membrane that inhibits oxygen consumption in non-phosphorylating mitochondria. The Cu_A site, in contrast to haem *a* sits on the outside of this membrane and electron exit from this centre to haem *a* (which sits in the membrane dielectric) is inhibited by the membrane potential - or more correctly the apparent midpoint potential of Cu_A will rise relative to that of haem *a* (Cooper et al. 1994).Therefore collapsing the membrane potential would be expected to oxidise Cu_A and reduce haem *a* - there is experimental support for this from studies in whole mitochondria (Rich et al. 1988) and reconstituted cytochrome oxidase proteoliposomes (Gregory et al. 1989; Nicholls 1993). We might expect the *in vitro* data (in the absence of a membrane potential) to underestimate the reduction of Cu_A and overestimate that of haem *a*.

The level of the membrane potential, and hence the rate of respiration, is dependant on a number of factors. One of the most significant of these is the ADP concentration; a rise in ADP will decrease the membrane potential and, at least in theory, partially oxidise Cu_A. These effects are observed in isolated mitochondria (Morgan et al. 1991), but whether they are large enough to see *in vivo* is not clear.

One final point should be made clear for those attempting to interpret the *in vivo* redox state behaviour of cytochrome oxidase. Surface reflectance spectroscopy measures the (visible) haem *a* centre of the enzyme (Kariman et al. 1985) whereas near infrared spectroscopy reflects the Cu_A redox centre (Cooper et al. 1994; Cooper et al. 1996a). The difference spectra peak at 605nm for haem *a* reduction and trough at 830nm for Cu_A reduction. Although both these centres will respond identically to reduced oxygen supply (by increasing their redox state) their starting redox state will be different; haem *a* will always be more reduced, making oxidations at this site easier to observe (LaManna et al. 1987). They also may respond differently to the energization state of the mitochondria (see above discussion). Interpreting data collected by these two different modalities is not therefore a trivial process and requires bioenergetic, as well as spectroscopic, expertise.

5. CONCLUSION

In vitro and *in vivo* studies of the NIRS-detectable Cu_A centre of mitochondrial cytochrome oxidase suggest that it is primarily oxidised in the anaesthetised brain *in vivo*.

ACKNOWLEDGMENTS

We would like to thank the MRC, Wellcome Trust, Royal Society, NATO and Hamamatsu Photonics for financial support. We also thank Dr. Huw Owen-Reece and Dr. Vincent Kirkbride for assistance with the animal preparations.

REFERENCES

Altman, D.I., Perlman, J.M., Volpe, J.J. and Powers, W.J. (1993). Cerebral Oxygen Metabolism in Newborns. Pediatrics 92: 99–104.

Brown, G.C., Crompton, M. and Wray, S. (1991). Cytochrome oxidase content of rat brain during development. Biochim. Biophys. Acta 1057: 273–275.

Brunori, M., Antonini, G., Malatesta, F., Sarti, P. and Wilson, M.T. (1992). The oxygen reactive species of cytochrome-c-oxidase: an alternative view. FEBS Lett. 314: 191–4.

Chance, B. and Hollunger, G. (1963). Inhibition of electron and energy transfer in mitochondria, 1; Effects of amytal, thiopental, rotenone, progesterone and methylone glycol. J. Biol. Chem. 278: 418–431.

Chance, B., Oshino, N., Sugano, T. and Mayevsky, A. (1973). Basic principles of tissue oxygen determination from mitochondrial signals. in Oxygen Transport to Tissue. (ed. H. Bicher and D. F. Bruley). New York, Plenum: 277–292.

Cooper, C.E., Cope, M., Quaresima, V., Ferrari, M., Nemoto, E., Springett, R., Matcher, S., Amess, P., Penrice, J., Tyszczuk, L., Wyatt, J. and Delpy, D.T. (1996a). Measurement of cytochrome oxidase redox state by near infrared spectroscopy. in Photon Migration and Optical Spectroscopy. (ed. A. Villringer and U. Dirnagl). Berlin, Springer. In Press.

Cooper, C.E., Elwell, C.E., Meek, J.H., Matcher, S.J., Wyatt, J.S., Cope, M. and Delpy, D.T. (1996b). The non-invasive measurement of absolute cerebral deoxyhaemoglobin concentration and mean optical pathlength in the neonatal brain by second derivative near infrared spectroscopy. Pediatr. Res. 39: 32–38.

Cooper, C.E., Matcher, S.J., Wyatt, J.S., Cope, M., Brown, G.C., Nemoto, E.M. and Delpy, D.T. (1994). Near infrared spectroscopy of the brain: relevance to cytochrome oxidase bioenergetics. Biochem. Soc. Trans. 22: 974–980.

Edwards, A.D., Brown, G.C., Cope, M., Wyatt, J.S., McCormick, D.C., Roth, S.C., Delpy, D.T. and Reynolds, E.O.R. (1991). Quantification of changes in the concentration of cerebral oxidised cytochrome oxidase. J. Appl. Physiol. 71: 1907–1913.

Edwards, A.D., McCormick, D.C., Roth, S.C., Elwell, C.E., Peebles, D.M., Cope, M., Wyatt, J.S., Delpy, D.T. and Reynolds, E.O. (1992). Cerebral hemodynamic effects of treatment with modified natural surfactant investigated by near infrared spectroscopy. Pediatr. Res 32: 532–6.

Fujii, T. (1991). Profiles of percent reduction of cytochromes in guinea pig hippocampal brain slices in vitro. Brain Res. 540: 224–8.

Gregory, L. and Ferguson-Miller, S. (1989). Independent control of respiration in cytochrome *c* oxidase vesicles by pH and electrical gradients. Biochemistry 28: 2655–2662.

Jöbsis, F.F. (1977). Non-invasive infrared monitoring of cerebral and myocardial oxygen sufficiency and circulatory parameters. Science 198: 1264–1267.

Kariman, K. and Burkhart, D.S. (1985). Non-invasive In Vivo Spectrophotometric Monitoring of Brain Cytochrome aa_3 revisited. Brain Res. 360: 203–213.

Kuboyama, M., Yong, F.C. and King, T.E. (1972). Studies on cytochrome oxidase VIII. Preparation and some properties of cardiac cytochrome oxidase. J. Biol. Chem. 247: 6375–6383.

LaManna, J.C., Sick, T.J., Pikarsky, S.M. and Rosenthal, M. (1987). Detection of an oxidizable fraction of cytochrome oxidase in intact rat brain. Am. J. Physiol.: C477-C483.

Matcher, S.J., Elwell, C.E., Cooper, C.E., Cope, M. and Delpy, D.T. (1995). Performance Comparison of Several Published Tissue Near-Infrared Spectroscopy Algorithms. Anal. Biochem. 227: 54–68.

McMahon, T.F. and Birnbaum, L.S. (1990). Age-related changes in toxicity and biotransformation of potassium cyanide in male C57BL/6N mice. Toxicol. Appl. Pharmacol. 105: 305–314.

Moody, A.J., Cooper, C.E. and Rich, P.R. (1991). Characterisation of 'fast' and 'slow' forms of bovine heart cytochrome *c* oxidase. Biochim. Biophys. Acta 1059: 189–207.

Morgan, J.E. and Wikström, M. (1991). Steady-State Redox Behavior of Cytochrome c, Cytochrome a, and CuA of Cytochrome c Oxidase in Intact Rat Liver Mitochondria. Biochemistry 30: 948–958.

Nicholls, P. (1993). The steady state behaviour of cytochrome c oxidase in proteoliposomes. FEBS Lett. 327: 194–8.

Nicholls, P., Cooper, C.E. and Wrigglesworth, J.M. (1990). Control of proteoliposomal cytochrome c oxidase: the overall reaction. Biochem Cell Biol 68: 1128–34.

Piantadosi, C.A. and Sylvia, A.L. (1984). Cerebral cytochrome aa_3 inhibition by cyanide in bloodless rats. Toxicology 33: 67–79.

Rich, P.R., West, I.C. and Mitchell, P. (1988). The location of Cu_A in mammalian cytochrome *c* oxidase. FEBS Lett. 233: 25–30.

Schlichtig, R., Klions, H.A., Kramer, D.J. and Nemoto, E.M. (1992). Hepatic dysoxia commences during O_2 supply dependence. J. Appl. Physiol. 72: 1499–1505.

Simpson, J.R. and Isacson, O. (1993). Mitochondrial impairment reduces the threshold for in vivo NMDA-mediated neuronal death in the striatum. Exp. Neurol. 121: 57–64.

Tamura, M. (1992). Protective effects of a PG12 analogue OP-2507 on hemorrhagic shock in rats. Jpn. Circ. J. 56: 366–375.

Wyatt, J.S., Cope, M., Delpy, D.T., van der Zee, P., Arridge, S.R., Edwards, A.D. and Reynolds, E.O.R. (1990). Measurement of optical pathlength for cerebral near infrared spectroscopy in newborn infants. Dev. Neurosci. 12: 140–144.

65

HYPOXIA/ISCHEMIA TRIGGERS A LIGHT SCATTERING EVENT IN RAT BRAIN

Britton Chance,[1] Avraham Mayevsky,[2] Bin Guan,[1] and Yutao Zhang[1]

[1]Johnson Foundation
Department of Biochemistry and Biophysics
University of Pennsylvania
Philadelphia, Pennsylvania 19104-6089
[2]Bar-Ilan University
Department of Life Sciences
Ramat-Gan, Israel

INTRODUCTION

Physiologists have shown that light scattering occurs during each stimulation of brain "slabs". Larger signals are obtained through depolarization of neurons caused by bioenergetic deficits in hypoxia/ischemia and by terminal inhibitors of cytochrome oxidase (8). In these studies we have used prolonged hypoxia and in some cases, cardiac arrest.

Detection of bioenergetic deficiency by minimally invasive methods is available from a number of sources, the principal one being phosphorus nuclear magnetic resonance (^{31}P NMR). This technique shows clearly, by the decline of the ratio of PCr/P_i or ATP/P_i, the deterioration of the bioenergetic state. The ^{31}P NMR method can be regarded to be the paragon of bioenergetic studies of tissues. The method is outstanding, and even though the time for data acquisition may be somewhat longer than desired, it can give an unlocalized spectrum in less than a minute from human cortical tissue (9). Due, however, to its expense and lack of portability, it can not serve adequately for the emergency room, where its measure of bioenergetic efficiency will be most needed.

While the ^{1}H NMR measure of deoxy hemoglobin (BOLD) (10) is most important in evaluating deoxyhemoglobin, both oxy and deoxy states are available in the capillary bed for brain function. The capillary tissue gradient is large and variable, and cannot be relied upon to indicate tissue and cellular hypoxia and bioenergetic deficiency. Cytochrome oxidase, due to its intracellular localization in the tissue mitochondria appears attractive for measuring bioenergetic deficiencies. However, there are two principle objections: the first is based upon the studies of Oshino et al. (11), who have indicated that the profile of reduction of NADH, not cytochrome oxidase, is closely linked to bioenergetic capability, ATP production, membrane potential, calcium transport, etc. However, studies of the ef-

Oxygen Transport to Tissue XIX, edited by Harrison and Delpy
Plenum Press, New York, 1997

458

fect of oxygen on respiratory activity clearly show that the redox state of NADH is insensitive to large changes of the redox state of cytochrome, particularly cytochrome oxidase. In other words, changes of redox state in cytochrome oxidase may have little effect upon the redox state of NADH until complete reduction of the cytochrome has occurred. Only then will the redox state of NADH be affected by hypoxia.

Rationale for Seeking a New Method for Detecting Brain Hypoxia

The effects of hypoxia upon bioenergetic function is illustrated by the equation for the reaction of cytochrome oxidase with oxygen, which has been studied in detail by Schindler and ourselves (12,13).

$$O_2 + aa_3^{2+} \xrightarrow{k_1} aa_a^{3+} \cdot O_2^- \tag{1}$$

$$-\frac{dO_2}{dt} = k_1 [O_2][aa_3^{2+}] \tag{2}$$

The ADP/O_2 ratio is taken to be 6,* and k_1 is ~10^8. Usual values of the dO_2/dt are 50 electrons/sec. Since each O_2 molecule accepts 4 electrons, we divide by 4: 50÷O_2 molecules/sec. In order to convert to ATP, 5 per sec, we multiply by 6: 6/4 x 50 ATP/sec = 75 ATP/sec. Using these values:

$$75 = 10^8 [O_2][aa_3^{2+}]$$
$$[O_2][aa_3^{2+}] = 7.5 \times 10^{-7} \tag{3}$$

Each requires a particular value to maintain ATP production; this is a self regulating system between zero $[aa_3]$ and its maximal concentration (~ 10 μM Cu_A^{1+}) at which the recalculated $[O_2]$=0.075 μM. Only at this low $[O_2]$ does the ATP flux diminish due to extreme hypoxia. Since the detectability limit for Cu_A^{2+} is ~ 1 μM, and the flux of ATP synthesis is regulated by aa_3 redox changes down to one-tenth of this value (0.1 μM), no optical signal of tissue hypoxia can be obtained by measuring Cu_A^{2+} at these low levels.

While NADH redox state is well correlated with mitochondrial dysfunction, direct measurement of NADH redox level non-invasively is frustrated by the fact that UV wavelengths are required for the measurement of its absorption or for the excitation of its fluorescence, and thus the penetration depth for the exposed tissue is small [1 mm] and the tissue cannot be reached with wavelengths appropriate for the study of NADH through the skin and skull.

Thus, it appears that in light scattering events measured in the NIR region, we have the advantage of non-invasive NIR studies through the skin and skull, and a signal directly related to bioenergetic deficiency is obtained from the increment of light scattering. This change would ultimately have to be correlated with the fraction of the neural material in

---

* Our determination in 1955 of 2.6±0.13 for bhydroxybutyrate, respectively, appears to have stood the test of time (13,14) as indicated by Lee, C.P. et al P/O = 2.78±0.05 (15).

the optical field that has been depolarized and rendered non-functional by a particular stress, for example, hypoxia/ischemia.

The Light Scattering Signal

The development of methods to separate light scattering from (17,18) absorption has stimulated us to reopen the use of light scattering as a physiological signal. For example, time and frequency domain photon migration devices readily separate absorption from scattering and continuous light methods with certain modifications do the same, although they rely upon a standard reference material. Thus, the application of these techniques to animal and human subjects, requires only the connection of source and detector to the tissues separated by an appropriate distance, either by direct coupling of source/detector or by light guide coupling. Tissue penetration is also linked to the input-output separation, [5–7 cm] in the case of the adult human head. Generally, the tissue penetration is, on the average, half the source detector separation.

Response time can be slow; a recent article (19) claims millisecond response time, but generally signal-to-noise ratio considerations lengthen this to one to two seconds for non-integrated detection. In the case of iterated phenomena, i.e. functional responses of the brain, computer averaging over several seconds permits response times of 10–100 msec.

Light scattering changes can be detected in two ways, as mentioned above, in time and frequency domain devices, absorption and scattering are readily deconvoluted by the following equations. The underlying theories for the relationship between scattering due to refractive index and cell volume changes is indicated in the following relationship which gives the dependence of the scattering factor $\mu_s{}'$, of tissue on the cell radius, a, cell volume fraction, ϕ, and refractive index of intra-/extra-cellular fluid, n_{in}/n_{ex}:

$$\mu_s{}' = \frac{\sigma_{tr}}{v_{par}}\phi = \frac{2.46}{a}\phi(1-\phi)\left(\frac{2\pi a}{\lambda}\right)^{0.37}\left(\frac{n_{in}}{n_{ex}}-1\right)^{2.09} \tag{4}$$

In case of no cell swelling or shrinkage, Eq. (4) becomes:

$$\mu_s{}' = \frac{\sigma_{tr}}{v_{par}}\phi = K\left(\frac{n_{in}}{n_{ex}}-1\right)^{2.09} \tag{5}$$

Thus, addition of a solute into tissue leads to a decrease of $\mu_s{}'$ since presence of the solute reduces the ratio of $n_{in}/n_{ex.}$

In Case of Cell Swelling or Shrinkage

An introduction of a solute into tissue can cause a osmotic pressure between the intra- and extra-cellular fluid, leading to cell shrinkage. Eq. (5) indicates that the scattering factor can be increased in tissue due to solute-induced cell shrinkage.

Wavelength Dependence

The second feature of light scattering is its wavelength indepen-dence in the region of 700–900 nm as contrasted to the well-characterized absorption bands of blood, water,

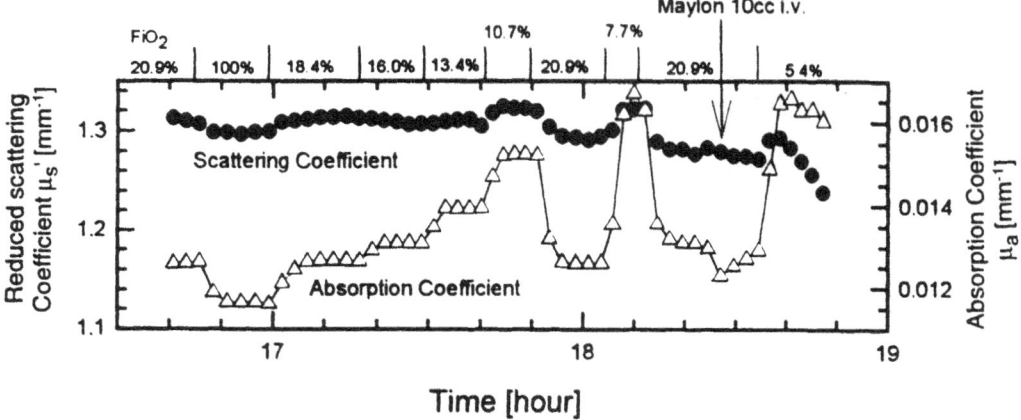

Figure 1. Reduced scattering and absorption scattering coefficients of piglet head obtained from the analysis of the time-response curves by the diffusion equation (reproduced with permission (21)).

and other biological pigments absorbing in this region. To this point, two and three wavelength phase modulation devices have been constructed that use wavelengths sensitive to the difference spectra of deoxygenated hemoglobin and to oxygenated hemoglobin and insensitive scattering effect as has been used in this study of light scattering.

PREVIOUS STUDIES

The simplest method employs continuous light and generates multiple photon migration patterns within the tissue. The mathematical analysis and the application to perfused liver is described elsewhere (20) and its application to the detection of the effect of osmotic stress in the perfused liver is illustrated by Table 5 of that paper (20).

Yamashita and his colleagues have employed time resolved spectroscopy to study light scattering changes in the pig brain following terminal hypoxia ([11]Maylon 10cc iv) (Fig. 1) (21). The μ_a and μ_s are readily deconvoluted from the photon kinetics with the TRS system (17). From their results (reproduced from (21)) it is seen that after the hemoglobin changes (hemoglobin deoxygenation) have subsided there is a decrease of $\mu_s{}'$ from 13 to 12 cm^{-1} in the last 10 minutes, a change of scattering.

EXPERIMENTAL METHODS

Phase Modulation Technique

Figure 2 illustrates a single side band phase modula-tion system for measuring scattering changes at a single wavelength, namely, 816 nm. A 1 KC local oscillator modulates the 50 MHz radio frequency signal which is time shared between the laser diodes illuminating the rat brain. An RF link between transmitter and receiver and a time shared phase lock loop serve to lock the phase of the transmitter andreceiver. A PMT detects photon propagation with an input/output separation of 1.5 cm and a photon migration pathlength of approximately 5 cm. The phase shift of the 1 KHz local oscillator through the RF pathway is measured with respect to the 1 KHz local oscillator phase by a zero crossing phase detector.

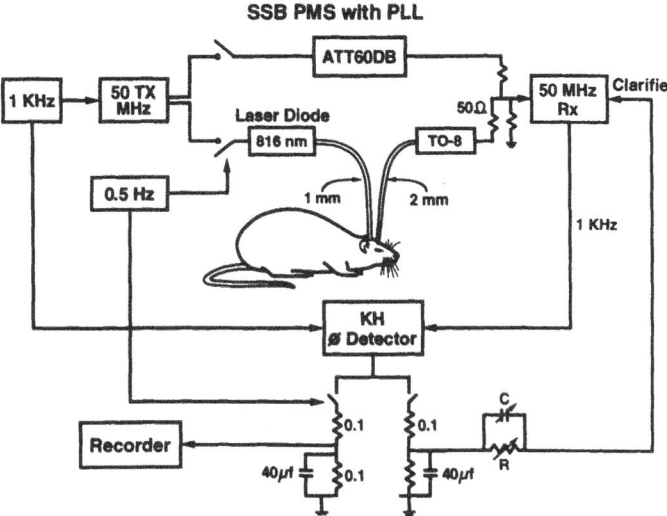

Figure 2. A simple single wavelength single side band phase modulation system for detecting light scattering changes in the rat brain. The system includes a time-shared phase locked loop (PLL) control system measurement at 50 MHz.

Figure 3 illustrates the increase of pathlength (decrease of scattering) by the tail vein injection of 1 gram per kilogram of sucrose. The light scattering decrease is thought to be due to the movement of sugar into the interstitial space followed by the uptake of sugar in the cortical neurons. The addition of sucrose is followed by an episode of hypoxia and recovery. At this wavelength (near the isosbestic point between oxy and deoxy hemoglobin), the traces show, initially, a small change (a pathlength increase) on initiating hypoxia, and a fast decrease on recovery therefrom. This change is followed by an anomalously large overshoot (a slow increase of optical pathlength) and a slow increase of pathlength with a recovery time relatively long compared to the recovery of blood dynamics. This change is comparable to the entry of sucrose from the vascular bed and is attributed to a light scattering change (see below).

Light Scattering in Human Brain

In order to validate that light scattering changes occur on solute addition in adult human brain we have been able to observe with the continuous light [Runman] changes in

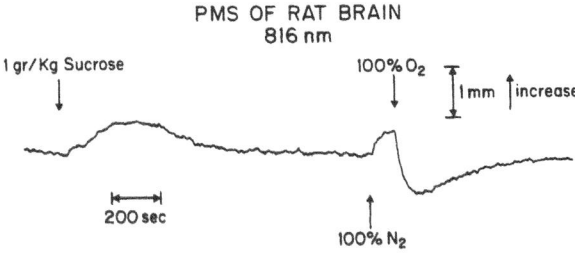

Figure 3. Illustrating a sucrose induced pathlength increase due to sucrose addition and a similar change following hypoxia and reflow.

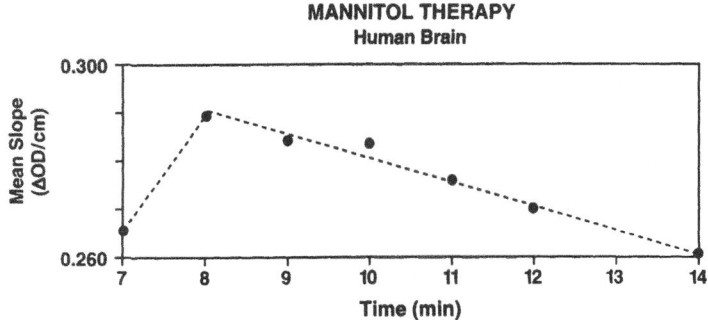

Figure 4. Illustrating the transferability of the NIR optical method from rat brain to adult human brain (IV). Delivery of one liter of mannitol therapeutic solution. Changes measured with variable source detector system.

the adult brain measured through the forehead of a human subject in the intensive care unit at Baylor Hospital following administration of the standard dose of mannitol [Fig. 4] (22,23). It is seen that the slope of the profile of light absorption against separation [see (20)], due to the administration of mannitol to the adult human corresponds almost exactly to the incremental slope observed with 40 millimolar mannitol for the perfused liver [Fig. 8 of (20)], taking into account that there is an offset of OD change. Therefore there appears to be agreement between the changes observed in the perfused liver and those in the brain on mannitol administration.

We have further studied light scattering events in cardiac arrest with a dual wave-length technique. In this experiment, a similar system was used (Fig. 5) except that the phase lock is obtained by rf coupling without the use of the PLL shown in Fig. 6. The coupling of the K+ electrode invasively through the skull is required. The fiber optics are attached to the rat brain with a separation of approximately 2 cm and the photon kinetics are recorded.

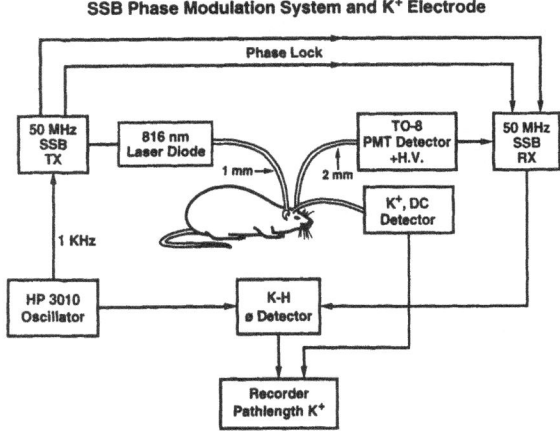

Figure 5. Experimental arrangement for correlating potassium release as measured by the potassium electrode with absorption/scattering changes as measured at 816 nm by a 50 MHz single side phase modulation system (no PLL).

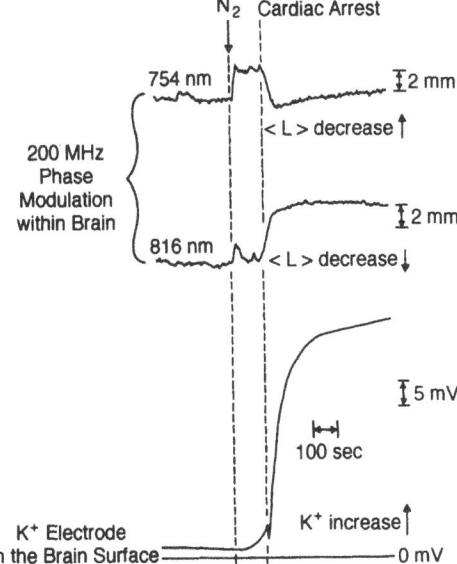

Figure 6. Experimental recording of potassium release on the pial-mater surface of the rat brain measured by the potassium electrode pathlength changes are measured by dual wave-length phase modulation system [200 MHz in this case]. The pathlength decreases are recorded in opposite directions on the two traces. The administration of nitrogen is indicated as is the moment of cardiac arrest, the latter causing pathlength increases at the two wavelengths.

The absorbance changes and the pathlength changes at 754 and 816 nm can be related to the spectrum of Hb/HbO$_2$. Thus, the pathlength changes at 754 nm are larger than those at 816 nm.

In this case, anoxia is followed by cardiac arrest after the deoxygenation signals are completed. Both traces respond with an equal increase of optical pathlength (albeit in opposite directions), consistent with a decreased scattering change (note that the pathlength calibrations for the two wavelengths are identical in magnitude, but reversed in sign). These pathlength changes at two wavelengths of 816 and 754 nm are not characteristic of a change due to blood volume in cardiac arrest; one would expect that there would be a decrease of oxyhemoglobin in the field of view of the brain in which the 754 nm change would exceed that at 816 nm, by a factor of > 3.

In order to further distinguish absorption and scattering and thus to verify the nature of the response, we have made two improvements. First, as Fig. 7 shows, we have elaborated the equipment of Fig. 2 to contain a third wavelength so that 754, 790, and 830 nm are available. Secondly, we have studied a rat model with bilateral carotid artery occlusion to create low flow which creates a low flow ischemic state as inspired oxygen is diminished by N$_2$ inhalation.

The performance of this apparatus is measuring the binding of Hb to oxygen and its deoxygenation is recorded in a blood lipid model as shown in Fig. 8, where in addition, the calibrations with intralipid verify the equality of the sensitivity of the phase traces to a light scattering change.

A new protocol has been employed which permits hypoxic/ischemic stress of the brain without compromising the cardiac function by bilateral carotid artery occlusion. In this case, it is possible to execute a complete cycle of hypoxia and reoxygenation in the initial baseline covering an interval of 10^5 sec (Fig. 9) giving pathlength changes closely related to those of Fig. 8. Several seconds after recovery from the hypoxic cycle, a scattering change with equal pathlength increases at the three wavelengths occurs. There-after, there is only partial recovery as evidenced by the displacement of the traces with respect to the baseline. The delayed release of K$^+$ from cortical neurons and partial reversibility

50 MHz Time-Shared 4λ 4 Channel PMS with PLL

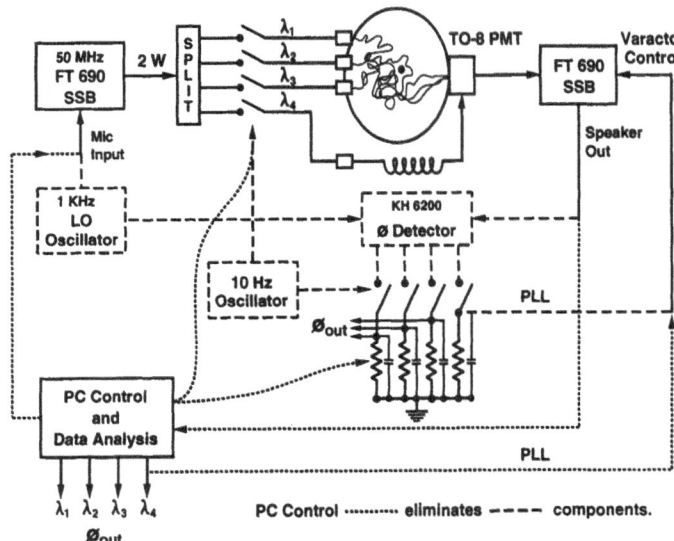

Figure 7. A 3-wavelength phase modulation system appropriate to the measurement of absorption and scattering changes in hypoxia (low flow ischemia) of rat brain with bilateral carotid artery occlusion. In this case a reference channel (λ_4) bypasses the rat brain to pro-vide phase lock loop (PLL) stabilization for the single side band 50 MHz spectrometer. The calibration with blood is shown in Fig. 8 and data from the rat brain are shown in Fig. 9.

after a severe hypoxic stress is proposed. Apparently, only a few small fraction of the neurons rapidly recover their bioenergetic state sufficiently to allow recovery of K^+ within the 5 min of the recording.

This protocol not only allows a separation of hypoxic stress in the brain with respect to the heart function but at the same time affords deoxygenation and reoxygenation as an enhanced stress on the brain neurons. Following the interval of hypoxia and recovery of 105 sec there is a delayed light scattering change common to all three optical channels

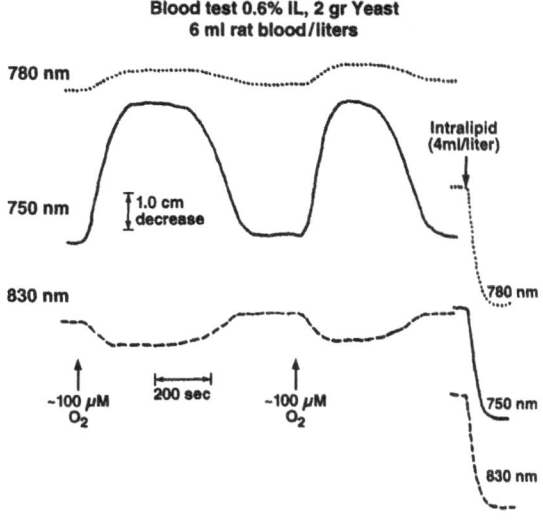

Figure 8. Illustrating the response of the 3-wave-length 50 MHz phase modulation system to oxygenation and deoxygenation of blood in a yeast/blood 0.5% intralipid brain model sample [1 liter, input/output separation 3 cm]. On the right is the calibration of the sensitivity to a scattering change by the addition of 4 ml per liter of interlipid. The system responds differentially to hemoglobin and similarly to interlipid at the 3 wavelengths.

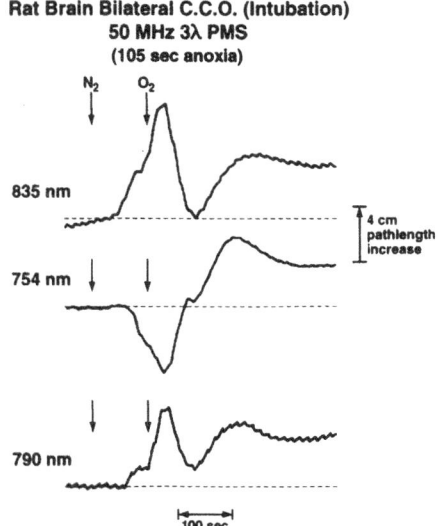

Rat Brain Bilateral C.C.O. (Intubation)
50 MHz 3λ PMS
(105 sec anoxia)

835 nm

754 nm

790 nm

4 cm
pathlength
increase

100 sec

Figure 9. Illustrating the response of rat brain with bilateral carotid artery occlusion to prolonged (10^5 sec) nitrogen hypoxia and reoxygenation with 100% O_2. The recovery of oxygenation of hemoglobin from the nitrogen/oxygen stress is complete prior to the pathlength increase observed at the three wavelengths. At these three wavelengths, the responses are very different from the hemoglobin changes of blood, supporting the idea of a post recovery light scattering change.

from which there is only partial recovery. Further cycles of hypoxia and recovery over a 30' interval reveal a smaller potassium release phenomenon consistent with the idea that an increasing population of neurons and astrocytes were depolarized and were incapable at this point of reaccumulating the effused potassium. Whether there is a slow recovery of K^+ over longer times is the topic of further study and highly significant to a better knowledge of recovery of the brain from hypoxic/ischemic stress.

The results may be compared to studies of dog brain by ^{31}P NMR where hypoxic stress caused a "split phosphate peak" in post-ischemic recovery, one portion of the tissue volume becomes highly acidic, the remainder neutral pH and capable of recovery. We propose to relate these two populations to the fraction of the light scattering change that is not recoverable to the initial base line (~ 80% in Fig. 9).

DISCUSSION

We have proceeded step by step in model systems, i.e., intralipid, in perfused organs (perfused liver), in rat brain without the potassium electrode, in rat brain with the potassium electrode, and finally a three wavelength system for consolidating that a scattering signal is involved. In all cases, the data are consistent with a scattering change attributable to an intracellular leakage of potassium following extreme hypoxia/ischemia.

We shall discuss the correlations separately:

1. Is there a light scattering event? The multisource single detector continuous light system, the phase modulation system and time resolved spectroscopy have been used in these studies, the former in intralipid model systems and in the perfused liver. The rat liver is of sufficient size to justify the use of separated sources and detectors with a continuous light system and to compute the absorption scattering changes on the basis of the appropriate algorithm (by Haida (24) and taken up in detail by Ferrall and Patterson (25), and by ourselves (26). In such cases, increase of pathlength was observed in the appropriate perturbations.

2. Correlation with the potassium electrode is superficially convincing, i.e., rapid decreases of light scattering are correlated with the measured rise of potassium on the surface of the brain. A flaw of the experiment is that the two measures are not at the same location, i.e., interstitial and on the pial mater. Thus, the correlation can be regarded as qualitative.

3. Finally, the use of the three wavelength technology, together with a bilateral carotid artery occlusion model enables a cycle of deoxygenation and oxygenation of hemoglobin to be completed before the light scattering change occurs as a consequence of hypoxia and reflow stresses, proposed here to be due to K^+ release and the partial recovery consistent with ^{31}P NMR studies.

Consequences on Brain Spectroscopy

The observations here that a light scattering change of magnitude about that half that of the largest oxy-deoxy signal (754 nm) suggest that extreme caution is recommended in the use a) of those instruments that do not indepen-dently measure absorption and scattering (most continuous wave instruments) and b) those time or frequency domain instruments that lack sufficient wavelengths or appropriate algorithms to distinguish large scattering and absorption changes.

Clinical Possibilities

The scattering change (detection of hypoxia/ischemia) is large and corresponds to a pathlength change of a magnitude of over half the oxy- and deoxy signal at 754 nm and is much greater than at 790 and 830 nm. Thus it is more accurate to distinguish from hemoglobin and easier to detect than cytochrome oxidase which is technically very difficult to detect in hypoxic tissues (see above). Also, the identification of a depolarization event associated with the cessation of function of cortical neurons appears to be a signal highly relevant to clinical studies and one demanding careful consideration in stroke and hematomas causing herniation (2).

The possibility that simple phase modulation systems or continuous light systems with variable input output separations could be used clinically is extremely attractive from the technical, and we believe, from the clinical standpoints.

ACKNOWLEDGMENTS

Supported in part by NIH grant NS 27346 and Human Frontier Science Program (HFSP).

REFERENCES

1. Cooper, C., this volume
2. Strong, A.J., Harland, S.P., Meldrum, B.S. and Whittington, D.J. (1996) The Use of In Vivo Fluorescence Image Sequences to Indicate the Occurrence and Propagation of Transient Focal Depolarizations in Cerebral Ischemia. Cereb. Blood Flow & Metab. 16:367–377.
3. Mayevsky, A. and Chance, B. (1982) Intracellular Oxidation Reduction State Measured *in situ* by a Multichannel Fiber-Optic-Surface Fluorometer. Science 217:537–540.

4. Bryant, S.H. and Tobias, J.M. (1952) Changes in Light Scattering Accompanying Activity in Nerves. J. Cell. Comp. Physiol 40:199–219.

5. Hill, D.K. and Keynes, R.D. (1949) Opacity Changes in Stimulated Nerve. J. Physiuol. 108:278–281.

6. Aubert, X., Chance, B. and Keynes, R.D. (1964) Optical studies of Biochemical events in the Electric Organ of Electrophorus. Proc. Royal Soc. B 160:211–245.

7. Salzberg, B.M., Okaid, A.M. and Gainer, H. (1985) Large Rapid Changes in Light Scattering Accompanying secretion by nerve terminals in the mammilian neuro-hypophysis. J. Gen. Physiol. 86:395–411.

8. Yoshioka, H., Miyake, H., Smith, D.S., Chance, B., Sawada, T. and , Nioka, S. (1995) The Effects of Hypercapnia on ECoG and Oxidative Metabolism in the Neonatal Dog Brain. J. Appl. Physiol. 78:2272–2278.

9. Chance, B. (1989) What are the Goals of Magnetic Resonance Research? NMR in Biomedicine, 2:179–187.

10. Kwong, K.K., Belliveau, J.W., Chesler, D.A., Goldberg, I.E., Weisskoff, R.M., Poncelet, B.P., Kennedy, D.N., Hoppel, B.E., Cohn, M.S., Turner, R., Chaneg, H.M., Brady, T.J. and Rosen, B.R. (1992) Dynamic Magnetic Resonance Imaging of Human Brain Activity during Primary Sensory Stimulation. Proc. Natl. Acad. Sci. USA 89:5675–5679.

11. Oshino, et al (1974) Mitochondrial Function under Hypoxic Conditions: The Steady States of Cytochrome aa3 and Their Relation to Mitochondrial Energy States. Biochim. Biophys. Acta 368:298–310.

12. Chance, B., Schoener, B. and Schindler, F. (1964) The Intracellular Oxidative-Reduc-tion State. In Oxygen in the Animal Organism (F.N. Dickens, ed.) Pergammon Press, London, pp. 367–388.

13. Chance, B. (1965) Reaction of Oxygen with the Respiratory Chain in Cells and Tissues. J. Gen. Physiol 49:163–188.

14. Chance, B. and Williams, G. Respiratory Enzymes in Oxidative Phosphorylation. I. Kinetics of Oxygen Utilization. J. Biol. Chem. 217:383–393. 1955.

15. Hinkle, P. Bioenergetics: A Practical Approach. Oxford University Press. 1994

16. Lee, C.P., Gu, Q., Xiong, Y., Mitchell, R.A. and Ernster, L. (196) P.O Ratios Reassessed: Mitochondrial P.O Ratios Consistently exceed 1.5 with succinate and 2.5 with NAD-Linked Substrates. FASEB J. 10:345–350.

17. Patterson, M.S., Chance, B. and Wilson, B.C. (1989) Time Resolved Reflectance and Transmittance for the Noninvasive Measurement of Tissue Optical Properties. J. Appl. Optics 28:2331–2336.

18. Sevick, E.M. and Chance, B. (1991) Quantitation of Time-and Frequency-Resolved Optical Spectra for the Determination of Tissue Oxygenation. Anal. Biochem. 195:330–351.

19. Yang, Y., Liu, H., Chance, B. and Zhang, Y. (1996) A Low Cost Phase Modulation System for Tissue Spectroscopy and Oximetry. Adv. Optical Imaging and Photon Migration. (March 18–20, Orlando FL) pp. 101–103.

20. Chance, B., Kitai, T., Liu, H. and Zhang, Y. (1995) Effects of Solutes on Optical Properties of Biological Materials: Models, Cells, and Tissues. Anal. Biochem. 227:351–362.

21. Yamashita, Y., Oda, M., Naruse, H. and Tamura, M. (1996) In Vivo Measurement of Reduced Scattering and Absorption Coefficients of Living Tissue Using Time-Resolved Spectroscopy. OSA TOPS on Advances in Optical Imaging and Photon Migration, Vol 2 (R.R. Alfano and James G. Fujiomoto (eds.) pp. 387–390.

22. Gopinath, S.P., Robertson, C.S., Contant, C.F., Narayan, R.K., Grossman, R.G. and Chance, B. (1995) Early Detection of Delayed Traumatic Intracranial Hematomas using Near Infared Spectroscopy. J. Neurosurgery, 83:438–444.

23. Robertson, C.S., Gopinath, S.P., and Chance, B. (1995) A New Application for Near-Infrared Spectroscopy: Detection of Delayed Intracranial Hematomas after Head Injury. J. Neurotrama (Suppl.) 12:591–600.

24. Haida, M., Miwa, M., Shiino, A. and Chance, B. (1993) A Method to Estimate the Ratio of Absorption Coefficients of Two Wave Lengths using Phase-Modulated Near Infrared Light Spectroscopy. Anal. Biochem. 208:348–351.

25. Farrell, T.J., Wilson, B.C. and Patterson, M.S. (1992) The Use of Neural Network to Determine Tissue Optical Properties from Spatially Resolved Diffuse Reflectance Measurements. Phys. Med. Biol. 37:2281–2286.

26. Beauvoit, B., Evans, S.M., Jenkins, T., Miller, E. and Chance, B. (1995) Correlation between the Light Scattering and the Mitochondrial Content of Normal Tissues and Transplantable Rodent Tumorks. Anal. Biochem.226:167–174.

66

SIMULATION OF THE OXYGEN DISTRIBUTION IN BRAIN TISSUE PERFUSED WITH FLUOROCARBONS

R. Stingele,[1] B. Wagner,[2] D. A. Wilson,[2] R. C. Koehler,[2] D. F. Hanley,[1]
R. J. Traystman[2]

[1]Department of Neurology
[2]Department of Anesthesiology/CCM
The Johns Hopkins Medical Institutions
Baltimore, Maryland 21287

1. INTRODUCTION

Perfluorocarbons (PFCs) are linear, branched or cyclic hydrocarbon chains that are substituted with fluorine atoms. This class of substances has a high solubility for respiratory gases such as oxygen and carbon dioxide. Due to their apolar structure, PFCs are immiscible in water, but stable emulsions of PFCs can be prepared by addition of surfactants.

PFC emulsions dissolve oxygen according with Henry's law, i.e. the amount of oxygen carried by a particular PFC emulsion is directly proportional to the partial pressure of oxygen and the concentration of the PFC in the emulsion. Oxygen transport by hemoglobin, on the other hand, is accomplished by chemical bonding. The cooperativity between the four oxygen binding sites of hemoglobin leads to non-linear oxygen binding behavior, represented by the sigmoid shape of the oxygen-hemoglobin dissociation curve. Hemoglobin, in contrast to PFCs, transports high amounts of oxygen at low partial pressures of oxygen and reaches close to full saturation at high partial pressure. The implications of the fundamentally different oxygen carrying mechanisms for oxygenation of tissue are unclear.

The purpose of the present study was to develop a theoretical model that allows modeling of oxygen delivery to tissue by blood, PFCs and mixtures thereof. The model takes into account the changes in flow of perfusate with PFC perfusion and the different oxygen carrying characteristics of blood and PFCs. In the model, oxygen is transported by convection with longitudinal flow of perfusate along the capillary and by diffusion into tissue following a radially directed gradient of PO_2 resulting from oxygen consumption in tissue.

To validate global oxygen extraction by the model, simulated values of venous outflow of oxygen are compared to measured mixed venous outflow from the superior sagittal sinus in an experimental series of nine cats that underwent complete FC-exchange followed by hypoxia. To validate the simulated distribution of oxygen, neuroelectrical function in the experimental series was compared to the predicted percentage of tissue below a PO_2 of 0.5 torr. Although imperfect, this method of validation covers both convectional and diffusional aspects of the model.

2. METHODS

2.1. Simulation

A modification of the model put forth by the Danish physiologist August Krogh in 1919[5–7] was used. Briefly, the classical Krogh model consists of a straight capillary (length L, radius R_c) perfused at constant flow. Oxygen enters a cylinder of tissue (radius R_t) concentric to the capillary by diffusion along a gradient perpendicular to the capillary. Inside the tissue cylinder oxygen is consumed at a constant rate (zero order). Neighbouring cylinders do not exchange oxygen and therefore the oxygen gradient is zero at the boundary of the cylinder. The solution of the cylindrical diffusion reaction equation with these conditions is the Krogh-Erlang equation.

This equation yields reasonable values of tissue PO_2 for high PO_2 values in the capillary. With hypoxia, on the other hand, the Krogh equation results in negative PO_2 values towards the boundary of the cylinder because of the constant rate of oxygen consumption in tissue. Since we intended to simulate the tissue distribution of oxygen in hypoxic situations we had to modify the Krogh model to allow for variable tissue oxygen consumption according to the local tissue PO_2 value. We used a Michaelis-Menten (MM) kinetic with a 'K_m' of 0.5 torr in accordance with in vitro studies on intact mitochondria,[1] allowing the rate of oxygen consumption to decrease towards zero for low PO_2 values in tissue. The cylindrical diffusion-reaction equation with MM kinetic cannot be solved analytically. We used a numerical method to solve the problem (shooting to a common fitting point,[8]), if the PO_2 in the capillary was less than 50 torr. We chose this cutoff value arbitrarily because of the observation that the difference between analytical and numerical solution was negligible above 50 torr in the present geometry (see Figure 1).

The numerical calculation of the distribution of PO_2 in the tissue cylinder was performed on a 50 by 50 grid. The capillary was divided into 50 segments of equal length (L/50) and the perpendicular slice of the tissue cylinder adjacent to each of these capillary segments was divided into 50 volume elements. The tissue volume elements therefore had the shape of an annulus with finite thickness $\Delta z = L/50$ that was concentric to the axis of the capillary. The volume of tissue elements in close proximity of the capillary was smaller than that of tissue elements farther away from the capillary.

Calculation of the radial drop of PO_2 in tissue can be broken down into five steps: First, the PO_2 in the capillary segment is calculated from the oxygen content at the arterial end of the capillary segment. Second, the capillary wall PO_2 is calculated in this segment, taking into account resistance to mass transfer inside the capillary. Third, the radial profile of PO_2 from the capillary wall to the outer boundary is obtained by either the Krogh equation (capillary PO_2 > 50 torr) or by numerical solution. The fourth step is to calculate the oxygen uptake of the slice based on the PO_2 values in the tissue voxels that make up the slice and the Michaelis-Menten equation. The fifth step is to decrease the flux of oxygen through the respective capillary segment by the amount of oxygen extracted by the tissue

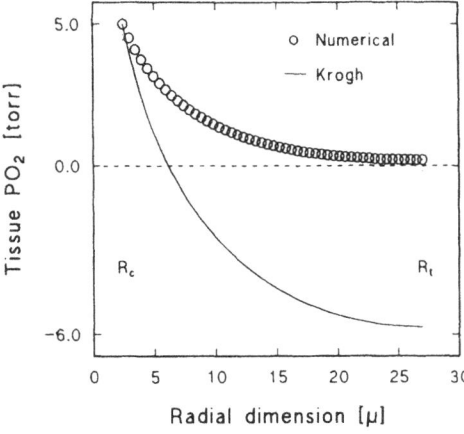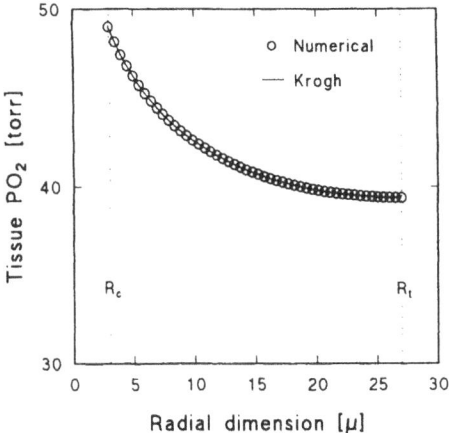

Figure 1. Comparison of analytical (Krogh equation) and numerical solution for the radial PO_2 gradient. For high capillary PO_2 the two solutions are identical (right) For low capillary PO_2 the analytical solution is negative towards the boundary of the tissue cylinder while the numerical solution remains positive. Both solutions respect the boundary condition of zero flux across the cylinder boundary ($dPO_2/dr = 0$ at R_t). R_t = cylinder radius, R_c = capillary radius.

slice. A comparison of numerical and analytical solution (Krogh equation) for high and low PO_2 values in capillary is given in Figure 1.

In order to assess the effect of random variability of geometric parameters (R_t, L) and of flow rate (Q) on the resulting tissue oxygenation, the simulation was recalculated 50 times with different values for R_t, L and Q. This was accomplished by multiplying each of these variables with an independent random factor before each of the 50 runs. The 50 triplets of random factors were generated such that the ratio of oxygen delivery to the cylinder volume (DO_2/Vol_{Cyl}) had a coefficient of variation of 20% around the average situation (for blood perfusion CBF = 76 ml/100g·min, R_t = 27μ, L = 400μ). The same set of random variables and random factors was applied for all treatments. This resulted in simulations of long and thick cylinders with high perfusion rates, short and thin cylinders with low perfusion rates and so forth for 50 combinations.

The average over the 50 recalculations of each of the tissue voxels in the 50 by 50 matrix was calculated and mapped into histograms with a bin width of 5, 10 and 15 torr for blood-, 10 cc/kg- and PFC-perfusion, respectively. The different bin widths were necessary to accommodate the wider range of occurring PO_2 values when PFC was present in the perfusate without changing the number of bins. Occurrences of P_tO_2 values were corrected for voxel volume in the histogram due to the radial geometry.

2.2. Experimental

Nine cats (2.7 - 4.0 kg) were anesthetized (Ketamine, 30 mg/kg i.m.; Acepromazine, 2 mg/kg i.m.), intubated and mechanically ventilated (70% N_2O, 30% O_2). A total of 7 lines were inserted: 1) Femoral artery: Reference withdrawal during injection of radiolabeled microspheres (15±1.5μ, reference sample method[3]), withdrawal of blood during PFC-exchange; 2) Left cardiac ventricle through contralateral femoral artery: Injection of radiolabeled microspheres for blood flow measurements; 3) Right brachial artery: Monitoring of blood pressure and arterial blood gases; 4) Femoral vein: Administration of PFC and drugs; 5) Inferior V. cava through contralateral femoral vein: Monitoring of central venous pressure during PFC-exchange. 6) Brachial vein: Continuous infusion of Fentanyl and epinephrine. 7) Superior sagittal sinus: Monitoring of blood gases.

A polarographic electrode (127μ, Au, polystyrene membrane, differential pulse voltammetry) was implanted 2 mm into cortex for measurement of tissue oxygen tension (P_tO_2). An Ag/AgCl reference electrode and a stainless steel screw auxiliary electrode were implanted through two 0.5 mm burr holes drilled over the left parietal brain. Epidural and rectal temperature was monitored. One channel evoked potentials (Spirit, Nicolet, Madison, WI) were recorded from a silver ball electrode located over the right somatosensory cortex.

After surgery was completed, ventilation was changed to 100% O_2 for adequate oxygenation during PFC-exchange. A continuous infusion of Fentanyl (2–5 μg kg^{-1} min^{-1}) was started for sufficient analgesia. After surgery, animals were allowed to stabilize for one hour.

Protocol: Monitoring of systemic blood pressure, central venous pressure and body temperature was continuous. All other measurements were taken at defined points during the protocol. One hour after surgery, a set of measurements was taken as baseline (control). Subsequently an intravenous bolus injection of 10cc FC-43 per kg body weight (Oxypherol, Green Cross, Japan) was administered over 10 minutes without removing blood from the animal. Ten minutes after the end of the infusion, a second set of measurements was taken (10cc/kg). Then, a complete PFC-exchange was performed. Exchange time was 90 minutes and the PFC-volume infused during this time period amounted to 1500 to 2000 ml FC-43. Volemia was adjusted according to central venous pressure, systemic blood pressure and end-diastolic left-ventricular pressure during the exchange transfusion. At the end of the PFC-exchange another set of measurements was taken (PFC-control). A graded hypoxia was induced by lowering the F_iO_2 to 0.8, 0.7 and 0.6, respectively. Each step of hypoxia was maintained for 10 minutes before a set of measurements was taken (F-80, F-70, F-60, respectively). During the hypoxia stages of the experiment, a continuous infusion of epinephrine was used to maintain perfusion pressure. Blood gas analysis was performed on samples from femoral artery and superior sagittal sinus (ABL3, Radiometer Copenhagen). Measurement of oxygen content was done with a coulometric oxygen analyzer (Lex-O_2-con, Lexington Instruments, MA) that allows measurement both in blood and FC-emulsions.

3. RESULTS

The measured experimental results are shown in table 1.

Exchange with FC decreased the arterial O_2 content to ~27% of control despite ventilation with 100% O_2. The reduction of C_aO_2 was partially offset by an increase of CBF (192% of control after FC-exchange) and OEF (157% of control after FC-exchange). Hypoxia in FC-exchanged cats led to a further increase of OEF and CBF. CMRO$_2$ was constant with FC-exchange and the initial steps of hypoxia and decreased only at $F_iO_2 = 0.6$.

The simulated distribution of oxygen is shown in Figure 2. The mean P_tO_2 was much lower with blood perfusion (61 torr) than with FC-perfusion (376 torr) at 100% O_2. P_tO_2 values were arranged closer around the mean value for blood perfusion than for FC-perfusion (25th and 75th percentiles: 55 to 67 torr for blood, 349 to 413 torr for FC-perfusion). Distribution of P_tO_2 was skewed towards low values with blood perfusion whereas perfusion with FC led to an even distribution of P_tO_2 values between maximal and minimal values.

Table 1. Measured experimental results

Treatment	Control	10cc/kg	PFC-con.	F-80	F-70	F-60
pH_a	7.33±0	7.34 ± 0.01	7.39[a] ± 0.02	7.36 ± 0.01	7.34[b] ± 0.02	7.31[b]±0.03
P_aO_2 [torr]	490±19	537[a] ± 12	633[a] ± 8	489[a,b] ± 10	416[a,b] ± 14	362[a,b]±11
P_aCO_2 [torr]	37.4±1.7	37.1 ± 1.2	36.4 ± 1.1	37.7 ± 1.5	35.6 ± 2.3	31.0[a,b]±2.2
C_aO_2 [ml/100ml]	16.9±0.7	14.9[a] ± 0.8	4.68[a] ± 0.2	4.13[a] ± 0.2	3.73[a] ± 0.1	3.13[a]±0.1
C_vO_2 [ml/100ml]	9.2±1	7.4 ± 0.9	1.3[a] ± 0.2	0.9[a] ± 0.1	0.8[a] ± 0.2	0.6[a,b]±0
MABP [mmHg]	132±10	132 ± 11	106[a] ± 6	94[a] ± 6	96[a] ± 3	92[a]±10
CBF [ml/100g/min]	76±9	85 ± 12	146[a] ± 11	155[a] ± 11	182[a,b] ± 15	153[a]±16
DO_2 [ml O_2/100g/min]	12.9±1.7	12.5 ± 1.7	6.9[a] ± 0.7	6.5[a] ± 0.6	6.9[a] ± 0.7	5.0[a]±0.8
$AVDO_2$ [ml/100ml]	7.9±0.8	7.5 ± 0.6	3.4[a] ± 0.2	3.2[a] ± 0.1	2.9[a] ± 0.2	2.6[a]±0.1
OEF	0.47±0.05	0.52 ± 0.04	0.74[a] ± 0.04	0.78[a] ± 0.03	0.79[a] ± 0.04	0.82[a]±0.01
$CMRO_2$ [ml/100g/min]	5.8±0.6	5.9 ± 0.7	5.0 ± 0.6	5.0 ± 0.5	5.3 ± 0.6	4.1[a]±0.7
P_tO_2 [torr]	26±14	42 ± 15	70[a] ± 23	47 ± 21	29[b] ± 14	18[b]±11
SEP-ampl. [μV]	55±14	54 ± 12	39 ± 12	40 ± 12	32 ± 11	17[a,b]±10

Values are reported as mean±SEM, [a]different from control (p<0.05), [b]different from PFC-control (p<0.05). Abbreviations: C_aO_2: arterial content of oxygen, CBF: global forebrain blood flow, DO_2: arterial oxygen delivery, $AVDO_2$: arterio-venous difference of oxygen content, OEF: oxygen extraction fraction, $CMRO_2$: cerebral metabolic rate of oxygen. Subscripts a and t refer to values in arterial perfusate and tissue.

The simulated results showed that with blood and FC-perfusion at high PO_2 (F_iO_2 = 1, 0.8 and 0.7), the entire tissue cylinder is at P_tO_2 values above 0.5 torr (P_tO_2 values below 0.5 torr were considered hypoxic; 0.5 torr = K_m used for MM-kinetic). With FC-perfusion at F_iO_2 = 0.6, 12% of tissue volume was below 0.5 torr although the average P_tO_2 in the tissue cylinder was 148 torr. In the experimental series, this corresponded to a significant drop of SEP amplitude at F_iO_2 = 0.6. Venous outflow of oxygen from sagittal sinus and the model was virtually identical (not shown).

Extraction of O_2 led to a curvilinear profile of PO_2 along the capillary with a rapid drop to low values at the arterial end (Figure 3). With FC perfusion the longitudinal profile of PO_2 was (almost) straight. For FC perfusion with F_iO_2 = 1, 0.8 and 0.7, the venous

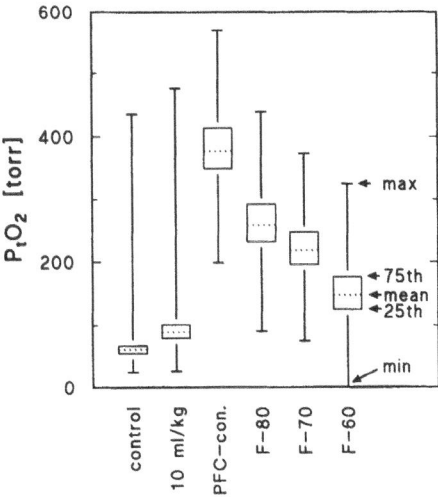

Figure 2. Mean, 25th and 75th percentile, maximum and minimum tissue PO_2 encountered for the six treatments (see text).

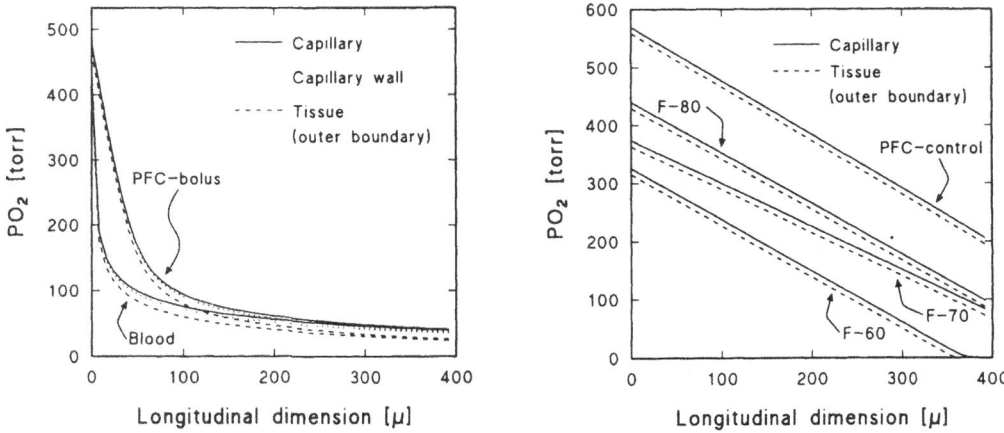

Figure 3. Simulated longitudinal PO₂ profile in capillary blood (solid) and at the outer cylinder boundary (dashed) for perfusion with blood and blood/FC mixture (left) and with FC at different F_iO_2 (right).

outflow PO₂ was above 0.5 torr but for $F_iO_2 = 0.6$ the profile approached zero torr asymptotically at the venous end. The infusion of a bolus of 10 cc/kg FC led to a PO₂ distribution very similar to perfusion with blood alone. The longitudinal profile of PO₂ in the capillary was curvilinear similar to the profile with blood perfusion. PO₂ values were considerably higher than with blood alone at the arterial pole of the capillary only (Figure 3). Towards the venous end of the capillary, PO₂ values for blood perfusion and for FC-bolus were almost identical (Figure 3). After infusion of the FC-bolus the amount of O₂ carried by the FC-fraction of the perfusate was 6% at the arterial extreme and this value dropped to below 1% at the venous end.

4. DISCUSSION

We here showed that the linear PO₂/O₂ content relationship of FC can result in a failure to evenly distribute O₂ in tissue. High P_iO_2 values measured in several types of tissue perfused with FC or blood/FC mixtures have been used as evidence for the ability of FC to adequately supply these tissues with O₂.[2] We showed that this can be misleading: Both our experimental and our simulated results suggest that FC-perfusion can result in hypoxic areas of brain tissue even though the average P_iO_2 is high. Simulated FC-perfusion at $F_iO_2 = 0.6$ resulted in 12 Vol% of hypoxic tissue (<0.5 torr), experimental FC-perfusion at $F_iO_2 = 0.6$ resulted in a significant drop of SEP amplitude. As a consequence of the inability of FC to carry large amounts of O₂ at low PO₂, most of the oxygen carried leaves the capillary in the arterial half of the vessel. This can result in a deficiency of O₂ in the venous part of the capillary with hypoxia in adjacent tissue. Figure 4 shows that the oxygen uptake in the arterial part of the tissue cylinder was higher with FC-perfusion than with blood perfusion even at an F_iO_2 of 0.6. FC-perfusion at 60% O₂ led to a zero O₂ uptake towards the venous part of the capillary. This shows that FC-perfusion might result in tissue hypoxia even though the global O₂-uptake is higher than with blood perfusion. It was shown that O₂ uptake was "improved" in working dog muscle by addition of perflubron to the perfusate.[4] Figure 4 shows that an increase in O₂ uptake is not necessarily an "improvement" of O₂ delivery to tissue. Adequate distribution of O₂ is an equally impor-

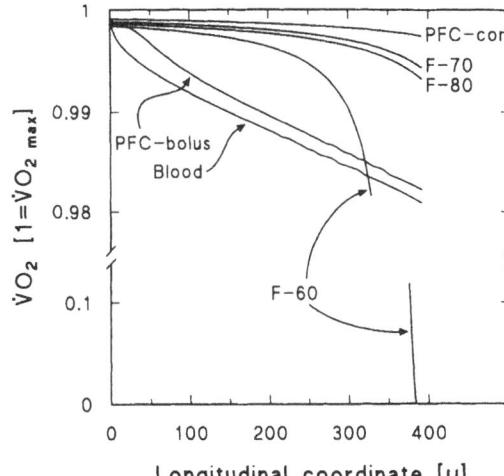

Figure 4. Simulated longitudinal drop of oxygen uptake for the six treatments. Note that at $F_iO_2 = 0.6$ (F-60) the O_2 uptake at the arterial end of the capillary is higher than with blood perfusion but it falls to zero before the end of the capillary is reached.

tant variable for integrity of tissue function. We therefore conclude that safety and efficacy of fluorochemicals should be assessed by measurement of functional parameters such as EEG and SEP rather than P_iO_2 or global O_2 delivery.

ACKNOWLEDGMENTS

We gratefully acknowledge the advice we received from Aleksander Popel and Tuhin K. Roy concerning the simulation throughout this study. This work was supported by a grant from the Deutsche Forschungsgemeinschaft (Sti 124/1).

REFERENCES

1. Chance, B. Cellular oxygen requirements. *Fed. Proc.* 16: 671–680, 1957.
2. Faithfull, N. S. Oxygen delivery from fluorocarbon emulsions—aspects of convective and diffusive transport. *Biomaterials, Artificial Cells, & Immobilization Biotechnology* 20: 797–804, 1992.
3. Heymann, M. A., B. D. Payne, J. I. Hoffman, and A. M. Rudolph. Blood flow measurements with radionuclide-labeled particles. *Prog. Cardiovasc. Dis.* 20: 55–79, 1977.
4. Hogan, M. C., D. C. Willford, P. E. Keipert, N. S. Faithfull, and P. D. Wagner. Increased plasma O2 solubility improves O2 uptake of in situ dog muscle working maximally. *J. Appl. Physiol.* 73: 2470–2475, 1992.
5. Krogh, A. The rate of diffusion of gases through animal tissues, with some remarks on the coefficient of invasion. *J. Physiol. (Lond).* 52: 391–408, 1919.
6. Krogh, A. The number and distribution of capillaries in muscles with calculations of the oxygen pressure head necessary for supplying the tissue. *J. Physiol. (Lond).* 52: 409–415, 1919.
7. Krogh, A. The supply of oxygen to the tissues and the regulation of the capillary circulation. *J. Physiol. (Lond).* 52: 457–474, 1919.
8. Press, W. H., B. P. Flannery, S. A. Teukolsky, and W. T. Vetterling. *Numerical Recipes in Cl.* Cambridge: Cambridge University Press, 1989.

67

SYNERGISTIC EFFECTS OF HAEMOGLOBIN AND *PLURONIC*® F-68 ON MITOTIC DIVISION OF CULTURED PLANT PROTOPLASTS

P. Anthony, K. C. Lowe, M. R. Davey, and J. B. Power

Department of Life Science
University of Nottingham
University Park, Nottingham NG7 2RD, United Kingdom

1. INTRODUCTION

The respiratory pigment, haemoglobin, has been evaluated for use as a vehicle for *in vivo* oxygen transport, for the perfusion of isolated organs and for facilitating gaseous supply to cultured animal cells (Goffe *et al.*, 1994; Zuck and Riess 1994; Tsuchida 1995; Winslow *et al.*, 1995). Haemoglobin solutions could also be valuable as media supplements in cultured plant cells and protoplasts to enhance division and to stimulate biomass production. However, this aspect has been relatively poorly studied.

This paper describes an assessment of the effects of the commercial bovine haemoglobin-based preparation, *Erythrogen*™, on the growth of cell suspension-derived protoplasts of *Petunia* and on leaf mesophyll-derived protoplasts of *Passiflora*. In some experiments, the culture medium was also supplemented with the non-ionic, polyoxyethylene (POE)-polyoxypropylene (POP) surfactant, *Pluronic*® F-68, which has been widely used as a cytoprotectant and growth-promoting additive to culture media for both animal and plant cells (Lowe *et al.*, 1993; Wu, 1995). One key objective was to determine whether *Erythrogen*™ and *Pluronic*® F-68 had synergistic effects on the division of cultured plant protoplasts.

2. MATERIALS AND METHODS

2.1. Protoplast Isolation and Culture

Protoplasts were isolated enzymatically from cell suspensions of an albino line of *Petunia hybrida* cv. Comanche (Power *et al.*, 1990) and from leaves of 6-week old glasshouse-grown seedlings of *Passiflora suberosa* (Otoni *et al.*, 1995). Protoplasts of *Petunia* were cultured for 9 days at a density of 2.0×10^5 ml^{-1} in liquid KM8P medium (Gilmour *et al.*, 1989) contained in 3.5 cm Petri dishes, while those of *Passiflora* were plated at $1.5 \times 10^5 \, \text{ml}^{-1}$ in liquid KPR medium (Thompson *et al.*, 1986) contained in 5.5 cm Petri dishes and incubated for a 15 day period. In some assessments with *Petunia* protoplasts, 0.01–1.0% (w/v) of *Pluronic*® F-68 (Sigma, Poole, UK) was also added to the medium. Protoplasts were maintained at $25 \pm 2°C$ in the dark.

2.2. Supplementation of Medium with Haemoglobin Alone or with Oxygen or Carbon Monoxide

A stabilised bovine haemoglobin solution ($103 \, \text{g} \, \text{l}^{-1}$, pH 7.42, *ca.* 3% methaemoglobin), (*Erythrogen*™; Biorelease Corporation, Salem, USA; obtained from TCS Biologicals, Botolph Claydon, UK), was added to KM8P medium to final concentrations of 1:50, 1:100 or 1:500 (v/v). Additional replicates were established, whereby at a dilution of 1:50 (v/v) the medium was gassed with 100% oxygen (10 mbar, 15 min) or 100% carbon monoxide (P/No: 850203; Phase Separations, Deeside, UK; 20 sec). Gassing was performed in 100 ml volume glass bottles containing 20 ml of medium. Aliquots (10 ml) were removed, avoiding the foamed layer at the meniscus, and added immediately to the protoplasts.

2.3. Assessment of Protoplast Viability, Growth and Plating Efficiency

The viability of freshly isolated protoplasts was assessed by the uptake and cleavage of fluorescein diacetate (Widholm, 1972), prior to counting of the isolated protoplasts and their plating in culture medium. The mitotic activity and growth of *Petunia* and *Passiflora* protoplasts was assessed after 9 days and 15 days of culture, respectively, by recording the initial plating efficiency (IPE). IPE was defined as the percentage of protoplasts originally plated that had regenerated a new cell wall and undergone at least one mitotic division. A minimum of 100 protoplasts per Petri dish were scored; each treatment was repeated 5 times.

2.4. Statistical Analyses

Means and standard errors (s.e.m.) were used throughout. Statistical significance between mean values was assessed using a parametric Moods median test and a Mann-Whitney U test (Snedecor and Cochran 1989). A probability of $P < 0.05$ was considered significant.

3. RESULTS

Enzymatic digestion of *Petunia* cell suspensions yielded $6.24 \pm 0.45 \times 10^5 \, \text{g.f.wt}^{-1}$ protoplasts with a mean viability of $94 \pm 1 \%$ (n = 3). *Erythrogen*™ at 1:50 (v/v) stimulated division of isolated protoplasts in KM8P medium, with an increase in mean IPE after 9 days of 64% ($P < 0.05$) over control (Figure 1). In contrast, the mean IPE values were not significantly different to control after 9 days in medium supplemented with 1:100 or 1:500 (v/v) of *Erythrogen*™ (Figure 1). Similarly, with 1:50 and 1:100 (v/v) of *Erythrogen*™, the mean IPE of *Passiflora* protoplasts was increased to 87% and 93%, respectively, above controls (Figure 2).

In cultures supplemented with 1:50 (v/v) *Erythrogen*™ and gassed with oxygen, the mean ($\pm$ s.e.m.) IPE (18.1 ± 1.5, n = 5) was significantly ($P < 0.05$) greater than the mean

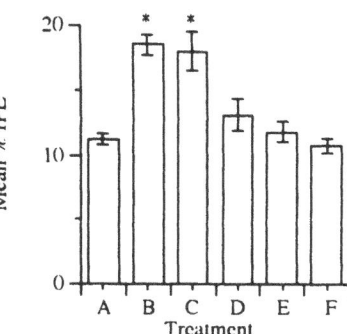

Figure 1. Mean IPEs of *Petunia hybrida* cv. Comanche protoplasts cultured A) without *Erythrogen*™ (control), and in the presence of B) 1:50 (v/v) *Erythrogen*™, C) 1:50 (v/v) *Erythrogen*™ gassed with oxygen, D) 1:50 (v/v) *Erythrogen*™ gassed with carbon monoxide, E) 1:100 (v/v) *Erythrogen*™ or F) 1:500 (v/v) *Erythrogen*™. n = 5 throughout; *P < 0.05.

control value (11.3 ± 0.4, n = 5). However, the IPE was not significantly different to the IPE of protoplasts cultured in the presence of the same, but ungassed, volume of *Erythrogen*™ (18.5 ± 0.8, n = 5). Similarly, for cultures containing 1:50 (v/v) *Erythrogen*™, but gassed with carbon monoxide, the mean IPE after 9 days (13.1 ±1.2, n = 5) was not significantly different to control (Figure 1).

The addition of 0.01% (w/v) *Pluronic*® F-68 to cultures containing 1:50 (v/v) of *Erythrogen*™ further increased the mean IPE (P < 0.05) after 9 days of culture by 38% (Figure 3). In contrast, there was no additional beneficial effect of supplementing culture medium with 0.1% or 1.0% (w/v) *Pluronic*® F-68, where the mean IPE after 9 days (19.1 ± 1.3% and 14.8 ± 0.8%, respectively, n = 5 throughout) was not significantly different to control (12.7 ± 2.5%, n = 5).

4. DISCUSSION

These experiments demonstrate a synergistic and beneficial effect of *Erythrogen*™, combined with *Pluronic*® F-68, in promoting mitotic division of isolated plant protoplasts in static liquid culture. This novel culture procedure, evaluated for protoplasts of the ornamental herbaceous plant *Petunia* and the woody plant *Passiflora*, should be applicable to isolated protoplasts and cells of other, biotechnologically-important plants. In addition, this approach may also be applicable to cells under agitated conditions and could be readily scaled-up to fermenters. In this respect, large-scale fermenter systems may be especially valuable for facilitating the *in vitro* biosynthesis of phytochemicals, particularly medicinally-important cell products (Toivonen, 1993). Related studies with animal cells have already demonstrated the stimulation of recombinant protein production by *Erythrogen*™ (Goffe *et al.*, 1994).

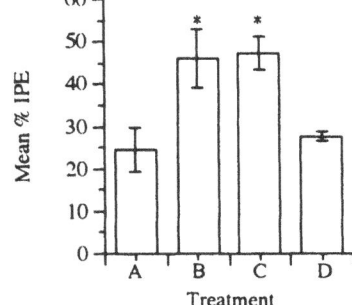

Figure 2. Mean IPEs of *Passiflora suberosa* protoplasts cultured A) without *Erythrogen*™ (control), and in the presence of B) 1:50 (v/v) *Erythrogen*™, C) 1:100 (v/v) *Erythrogen*™ or D) 1:500 (v/v) *Erythrogen*™. n = 5 throughout; *P < 0.05.

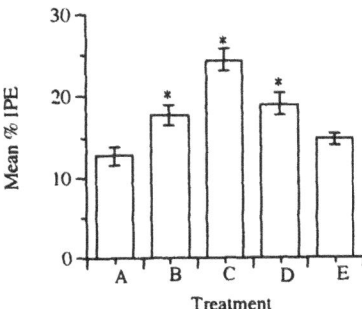

Figure 3. Mean IPEs of *Petunia hybrida* cv. Comanche protoplasts after 9 days of culture A) without *Pluronic*® F-68 and *Erythrogen*™ (control), or with 1:50 (v/v) *Erythrogen*™ alone B), in the presence of 0.01% (w/v) *Pluronic*® F-68 C), 0.1% (w/v) *Pluronic*® F-68 D), or 1.0% (w/v) *Pluronic*® F-68 E). n = 5 throughout; *P < 0.05.

The mechanism(s) by which *Erythrogen*™ and *Pluronic*® F-68 act on protoplast-derived cells is not clear. They may act by different mechanisms, especially as *Pluronic*® F-68 further enhanced the growth of *Petunia* protoplasts over that induced by *Erythrogen*™ alone. Conceivably, *Erythrogen*™ may adsorb oxygen from air/medium interfaces and facilitate gas delivery to cells, with a concomitant improvement in growth (Goffe *et al.*, 1994). We are aware that methaemoglobin, a constituent of *Erythrogen*™, would not function reversibly as an oxygen carrier. However, the relatively small amounts normally present (*ca.* 3% w/v) should not obscure the overall stimulation of mitotic division. Indeed, even if the methaemoglobin concentration increased progressively during the experiment, this would probably not lead to any significant alteration of division, since the initiation of mitotic competence and its sustainability in protoplasts is normally during the first few days of culture.

Pluronic® F-68 is known to be adsorbed onto cytoplasmic membranes of animal cells, increasing their resistance to mechanical damage (Goldblum *et al.*, 1990; Lowe *et al.*, 1994). This surfactant may have a similar effect on plant protoplasts. Additionally, it may also promote the uptake of nutrients, growth regulators and oxygen into protoplasts and protoplast-derived cells during culture. The concentrations of *Pluronic*® F-68 employed in the present study were selected on the basis of previous work, which demonstrated a growth-stimulating effect of *Pluronic*® F-68 during culture of protoplasts isolated from cell suspensions of *Solanum dulcamara* (Kumar *et al.*, 1992). Interestingly, the optimum concentration of 0.01% (w/v) of the surfactant, as determined in the present experiments, was identical to the concentration which acted synergistically with the oxygenated PFC liquid, *Flutec*® PP5 (BNFL Fluorochemicals Ltd., UK), in promoting mitosis in *P. hybrida* protoplasts in a static, two-phase culture system (Anthony *et al.*, 1994).

Future experiments will assess whether haemoglobin, combined with *Pluronic*® F-68 in aqueous culture medium overlaying oxygenated PFC, will stimulate protoplast growth over and above that already reported using a combination of physical and chemical parameters (Anthony *et al.*, 1995). Such studies should identify the most effective system for sustaining division of protoplast-derived cells prior to plant regeneration. This approach will be essential to maximise plant throughput, especially of material derived from experiments involving genetic manipulation.

ACKNOWLEDGMENTS

This work was supported by BNFL Fluorochemicals Ltd, UK. The authors acknowledge Biorelease Corporation, USA, for technical information.

REFERENCES

Anthony, P., Davey, M.R., Power, J.B., Washington, C. and Lowe K.C. Synergistic enhancement of protoplast growth by oxygenated perfluorocarbon and Pluronic F-68. *Plant Cell Rep 13*, 251–255 (1994).

Anthony, P., Lowe K.C., Davey, M.R. and Power, J.B. Strategies for promoting division of cultured plant protoplasts: beneficial effects of oxygenated perfluorocarbon. *Biotechnol. Techniques 9*, 777–782 (1995).

Gilmour, D.M., Golds, T.J. and Davey, M.R. *Medicago* protoplasts: fusion, culture and plant regeneration, in: *Biotechnology in Agriculture and Forestry: Plant Protoplasts and Genetic Engineering I*, Vol. 8 (Bajaj, Y.P.S. ed.), pp. 370–388, Heidelberg, Springer-Verlag (1989).

Goffe, R.A., Shi, J.Y. and Nguyen, A.K.C. Media additives for high performance cell culture in hollow fiber bioreactors. *Art. Cells, Blood Subs., and Immob. Biotechnol. 22*, A18 (1994).

Goldblum, S., Bae, Y-K., Hink, W.F. and Chalmers, J. Protective effect of methylcellulose and other polymers on insect cells subjected to laminar shear stress. *Biotechnol. Prog. 6*, 383–390 (1990).

Handa-Corrigan, A., Zhang, S. and Brydges, R. Surface-active agents in animal cell cultures, in: *Animal Cell Biotechnology* (R.E. Spier, and J.B. Griffiths, eds), Vol. 5, pp. 279–301, London, Academic Press (1992).

Kumar, V., Laouar, L., Davey, M.R., Mulligan, B.J. and Lowe, K.C. Pluronic F-68 stimulates growth of *Solanum dulcamara* in culture. *J. Exp. Bot. 43*, 487–493 (1992).

Lowe, K.C., Davey, M.R., Power, J.B. and Mulligan, B.J. Surfactant supplements in plant culture systems. *Agro-food-Ind. Hi-Tech. 4*, 9–13 (1993).

Lowe, K.C., Davey, M.R., Laouar, L., Khatun, A., Ribeiro, R.C.S., Power, J.B. and Mulligan, B.J. Surfactant stimulation of growth in cultured plant cells, tissues and organs, in: *Physiology, Growth and Development of Plants in Culture* (P.J. Lumsden, J.R. Nicholas, and W.J. Davies, eds.), pp. 234–244, Dordrecht, Kluwer (1994).

Otoni, W.C., Blackhall, N.W., d'Utra Vaz, F.B., Casali, V.W., Power, J.B. and Davey, M.R. Somatic hybridization of the *Passiflora* species, *P. edulis* f. *flavicarpa* Degener. and *P. incarnata* L. *J. Exp. Bot. 46*, 777–785 (1995).

Power, J.B., Davey, M.R., McLellan, M. and Wilson, D. Isolation, culture and fusion of protoplasts - 2. Fusion of protoplasts. *Biotechnol. Educ. 1*, 115–124 (1990).

Thompson, J.A., Abdullah, R. and Cocking, E.C. Protoplast culture of rice (*Oryza sativa* L.) using media solidified with agarose. *Plant Sci. 47*, 179–183 (1986).

Toivonen, L. Utilization of hairy root cultures for production of secondary metabolites. *Biotechnol. Prog. 9*, 12–20 (1993).

Tsuchida, E. *Artificial Red Cells - Materials, Performances and Clinical Study as Blood Substitutes*. Chichester, Wiley (1995).

Snedecor, G.W. and Cochran, W.G. *Statistical Methods*, 8th Edn. Iowa State University Press, Ames (1989).

Widholm, J. The use of FDA and phenosafranine for determining viability of cultured plant cells. *Stain Technol. 47*, 186–194 (1972).

Winslow, R.M., Vandegriff, K.D. and Intaglietta M. (eds.) *Blood Substitutes - Physiological Basis of Efficacy*. Boston, Birkhauser, (1995).

Wu, J. Mechanisms of animal cell damage associated with gas bubbles and cell protection by medium additives. *J. Biotechnol. 43*, 81–94 (1995).

Zuck T.F. and Riess, J.G. The current status of injectable oxygen carriers. *Crit. Rev. Clin. Lab. Sci. 31*, 295–324 (1994).

68

POST-THAW RECOVERY OF CRYOPRESERVED CELLS

Enhanced Viability by Oxygenated Perfluorocarbon or *Pluronic*® F-68

P. Anthony, K. C. Lowe, M. R. Davey and J. B. Power

Department of Life Science
University of Nottingham
University Park, Nottingham NG7 2RD, United Kingdom

1. INTRODUCTION

Cryopreservation is employed routinely to preserve a wide range of eukaryotic cells, with on-going research being directed primarily at enhancing and maximising post-thaw cell viability. The process consists of several steps, one of which involves controlled freezing. Overall, successful recovery of cells is dependent on the pre-freeze and post-freeze treatments. The transition of cells or tissues between low and physiologically normal temperatures with associated oxygen tensions causes respiratory imbalances which can, in turn, stimulate the production of toxic oxygen radicals (Fuller *et al.*, 1988). This serves to reduce physiological competence and mitotic potential, the combined effects of which reduce the overall efficacy of the cryopreservation process. Indeed, previous investigations of cryopreserved rice cells, have shown that cumulative respiratory impairment and dysfunction occur during the critical post-thaw recovery phase (Cella *et al.*, 1982).

One approach to overcome such imbalances involves the culture of cells in the presence of inert, oxygen-carrying perfluorochemical (PFC) liquids. The latter can dissolve substantial volumes of respiratory gases and have been shown to enhance, for example, oxygen supply to cultured cells maintained at temperatures *ca.* 25°C (Anthony *et al.*, 1994; Lowe *et al.*, 1995a,b, 1996). Furthermore, the non-ionic, polyoxyethylene (POE)-polyoxypropylene (POP) surfactant, *Pluronic*® F-68, which is a widely used cytoprotectant and growth-promoting media additive in cultured animal cell systems (Lowe *et al.*, 1993), is an ideal agent for potentially enhancing the recovery of plant cells following cryopreservation.

In the present investigation, the post-thaw viability of cultured plant cells has been studied (i) on medium overlaying oxygenated PFC, (ii) on media supplemented with

Pluronic® F-68, and (iii) with a combination of both treatments to evaluate possible synergistic interactions in the context of cryopreservation.

2. MATERIALS AND METHODS

2.1. Plant Materials and Maintenance of Cell Suspensions

Embryogenic cell suspensions of *Oryza sativa* L. cvs. Taipei 309 and Tarom (Japonica rices) were initiated from mature seed scutella (Finch *et al.*, 1991). Non-embryogenic suspensions of *Lolium multiflorum* were supplied by Dr E. Guiderdoni, IRAT/CIRAD, Montpellier, France. Rice suspensions of cv. Taipei 309 were maintained in AA2 liquid medium (Abdullah *et al.*, 1986) and those of cv. Tarom in R2 liquid medium (Ohira *et al.*, 1973). *Lolium* suspensions were cultured in N6 liquid medium (Chu *et al.*, 1975). All suspensions were maintained in the dark at $28 \pm 1°C$ and sub-cultured every 7 days by transferring 1 ml of settled cells with 7 ml of spent medium to 22 ml aliquots of fresh medium; *Lolium* suspensions were sub-cultured by transfer of settled cells (3 ml) with 7 ml of spent medium to 40 ml of fresh medium.

2.2. Cryopreservation of Cell Suspensions

Prior to cryopreservation, cells were cultured for 3–4 days in their respective culture medium supplemented with 60 g l^{-1} mannitol. Such preconditioned cells were harvested onto nylon mesh (45 µm pore size) and transferred to 2 cm³ polypropylene vials (Sarstedt, Leicester, UK) with approximately 0.2 g fresh weight of cells per vial. Approximately 0.75 ml of a cryoprotectant mixture consisting of 46.0 g l^{-1} glycerol, 39.0 g l^{-1} dimethyl sulphoxide (DMSO), 342.0 g l^{-1} sucrose and 5.0 g l^{-1} proline was added to each vial (Lynch *et al.*, 1994). Cells were cryoprotected for 1 h on iced water, prior to freezing at a controlled rate (-1°C min⁻¹) from 0°C to -35°C in a programmable freezer (Planer Cryo 10 Series, Planer Biomed, Sunbury-on-Thames, UK) and storage in liquid nitrogen at -196°C. Cells were stored at this temperature for 30 days (Tarom rice; *Lolium*) or 3 years (Taipei 309 rice).

2.3. Post-Thaw Recovery and Culture

Cells were thawed by plunging vials into sterile water at 45°C and excess cryoprotectants removed. Five ml aliquots of AA2 medium, semi-solidified with 0.4% (w/v) Sea Kem Le agarose (FMC Corporation, Rockland, USA) were overlaid onto 20.0 ml aliquots of oxygenated (10 mbar, 15 min) PFC (*Flutec*® PP6; BNFL Fluorochemicals Ltd, Preston, UK) contained in 100 ml screw-capped glass bottles. The cells (Taipei 309) from individual vials were placed onto two superimposed 2.5 cm diameter Whatman No. 1 filter paper discs overlaying the AA2 medium. Control treatments were the same, but lacking PFC. In some treatments, the medium was supplemented with 0.01% (w/v) of *Pluronic*® F-68 (Sigma, Poole, UK).

In a separate series of assessments, cells of the rice cvs. Taipei 309 and Tarom and those of *Lolium* were placed onto two superimposed 5.5 cm diameter Whatman No. 1 filter paper discs overlaying 20 ml aliquots of their appropriate culture medium contained in 9.0 cm Petri dishes. The medium was semi-solidified with 0.4% (w/v) Sea Kem Le agarose (FMC Corporation, Rockland, USA). In some treatments, the medium was supplemented with 0.01%, 0.1% or 0.2% (w/v) of *Pluronic*® F-68.

Cells of all treatments were cultured in the dark for 3 days at 28 ± 1°C, prior to transfer of the upper filter disk, supporting the cells, to the respective fresh media. Cells were cultured for a further 24 h prior to viability assessments, followed by a further 26 days under the same conditions, before biomass determinations. Each treatment was replicated 20 times.

2.4. Measurement of Post-Thaw Viability and Biomass

The post-thaw viability and metabolic capacity of cells was assessed by the reduction of triphenyl tetrazolium chloride and the absorbance measured at 490 nm (TTC; Steponkus and Lamphear, 1967). The fresh weight of thawed cells was recorded for biomass determinations (Lynch *et al.*, 1994).

2.5. Statistical Methods

Means and standard errors (s.e.m.) were used throughout. Statistical significance between mean values was assessed, as appropriate, using a conventional ANOVA and Student's *t*-test (Snedecor and Cochran, 1989). A probability of $P < 0.05$ was considered significant.

3. RESULTS

3.1. Cell Viability with Oxygenated PFC

The mean absorbance of cryopreserved rice cells of Taipei 309 following recovery in the presence of oxygenated PFC (0.63 ± 0.06; n = 20) was significantly ($P < 0.05$) greater than the corresponding mean control value (0.46 ± 0.06; n = 20).

3.2. Cell viability with *Pluronic*® F-68

The cell viability of cv. Tarom was optimally increased by 400% over control ($P < 0.001$) following supplementation of the recovery medium with 0.1% (w/v) *Pluronic*® F-68 (Table 1). Mean cell viability was also increased 3-fold ($P < 0.001$) and 2-fold ($P < 0.001$) with 0.2% (w/v) and 0.01% (w/v) *Pluronic*® F-68, respectively. A 2-fold increase ($P < 0.01$) in cell viability occurred with cells of Taipei 309 exposed to 0.01% (w/v) *Pluronic*® F-68. This concentration of surfactant also increased the mean viability of *Lolium* cells by 31% ($P < 0.01$). *Pluronic*® F-68 (0.01% w/v) also promoted biomass, in terms of fresh weight gain over control measured 30 days after thawing, by a maximum of 32% for cells of both Taipei 309 and *Lolium* (Table 2).

Table 1. Mean TTC absorbance of cells of rice and *Lolium* after 4 days of recovery from cryopreservation in the presence of *Pluronic*® F-68

Pluronic® F-68(% w/v)	*O. sativa* cv. Taipei 309	*O. sativa* cv. Tarom	*Lolium multiflorum*
0 (Control)	0.40 ± 0.06	0.19 ± 0.02	0.38 ± 0.03
0.01	0.98 ± 0.10	0.36 ± 0.06	0.50 ± 0.02
0.1	0.75 ± 0.13	0.76 ± 0.05	0.41 ± 0.03
0.2	0.59 ± 0.09	0.48 ± 0.05	0.36 ± 0.03

Table 2. Mean fresh weight (g) of cells of Taipei 309 and *Lolium* measured 30 days following recovery from cryopreservation in the presence of *Pluronic*® F-68

Pluronic® F-68(% w/v)	*O. sativa* cv. Taipei 309	*Lolium multiflorum*
0 (Control)	2.07 ± 0.28	3.49 ± 0.18
0.01	2.74 ± 0.13	4.50 ± 0.20
0.1	2.46 ± 0.11	3.29 ± 0.21
0.2	1.98 ± 0.16	3.32 ± 0.19

3.3. Cell Viability with Oxygenated PFC and *Pluronic*® F-68

In a separate series of assessments, cryopreserved rice cells were recovered in the presence of 0.01% (w/v) *Pluronic*® F-68, oxygenated PFC or a combination of these options. The mean absorbance of cells recovered in the presence of 0.01% (w/v) *Pluronic*® F-68 (1.27 ± 0.04; n = 20) or oxygenated PFC (1.13 ± 0.03; n = 20) were significantly greater (P < 0.01) than control treatments (0.93 ± 0.03, n = 20; Figure 1). A further synergistic increase (P < 0.01) in mean absorbance was observed when PFC and *Pluronic*® F-68 were used in combination (1.46 ± 0.08; n = 20), compared to all other experimental treatments. No significant differences were observed in mean absorbance between the oxygenated PFC and 0.01% (w/v) *Pluronic*® F-68 treatments.

4. DISCUSSION

The results of these experiments demonstrate that post-thaw viability and growth of cryopreserved rice and *Lolium* cells can be increased by supplementation of the recovery media with oxygenated PFC and/or *Pluronic*® F-68. These observations are consistent with previous studies in which the mitotic division of plant protoplasts was enhanced when the protoplasts were cultured at the interface between oxygenated PFC and aqueous culture medium (Anthony *et al.*, 1994). PFCs have been used in fermenter studies with micro-organisms (Junker *et al.*, 1990) and hybridoma cell cultures (Ju and Armiger 1992) to increase oxygen transfer rates. In addition to facilitating oxygen supply, PFCs are highly inert and can thus be recovered and re-cycled from aqueous culture media (Anthony *et al.*, 1994; Lowe *et al.*, 1995a,b, 1996). It is likely that culture of cells in the presence of oxy-

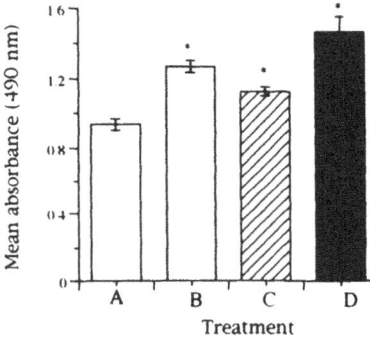

Figure 1. Mean absorbance (490 nm) of cells of Taipei 309 after 4 days following recovery from cryopreservation in the presence of (A) medium alone (control), (B) *Pluronic*® F-68 (0.01% w/v), (C) oxygenated PFC, and (D) a combination of treatments (B) and (C).

genated PFC overcomes the respiratory imbalances which exist in plant cells during the crucial post-thaw recovery period (Cella *et al.*, 1982).

Studies using animal cells have shown that *Pluronic*® F-68 adsorbs onto cytoplasmic membranes, conferring increased resistance to mechanical damage (Handa-Corrigan *et al.*, 1992; Lowe *et al.*, 1993, 1994). However, such effects would be less pronounced in cultured plant cells, because of the presence of cellulose-based cell walls. Nevertheless, adsorption of *Pluronic*® F-68 molecules onto the plasma membranes in post-thawed plant cells may also serve to reduce intracellular damage, which is known to occur during the rehydration phase when the DMSO cryoprotectant is removed progressively from the system (Benson and Withers, 1987).

Pluronic® F-68, at concentrations comparable to those used in the present investigation, can also stimulate both 2-deoxyglucose uptake and cellular amino acid incorporation into animal cells cultured under static conditions (Cawrse *et al.*, 1991). Similar increases in nutrient uptake or availability promoted by Pluronic may, in fact, explain the present findings. Indeed, such changes would be expected to positively alter metabolic flux, allowing core biochemical pathways to operate more efficiently and consistently, particularly under the high stress conditions of early post-thaw recovery. Nevertheless, further studies are required to determine the mechanism(s) by which oxygenated PFC and nonionic surfactants, including *Pluronic*® F-68, can facilitate post-thaw survival of plant cells. A focus of such work should be to investigate the effects of these agents on respiratory gas dynamics, since there is evidence that *Pluronic*® F-68 alters oxygen transport in agitated, sparged bioreactors (Murhammer and Pfalzgraf, 1992).

The present observations are consistent with previous, related studies, which demonstrated that PFC emulsified with *Pluronic*® F-68 was effective in prolonging the fertilisation capacity of turkey spermatozoa stored at reduced temperature (Thurston *et al.*, 1993). Thus, PFCs, used as oxygen-gassed liquids or as emulsions, coupled with *Pluronic*® F-68, offer potentially valuable alternative post-thaw handling strategies for both plant and animal cell systems, especially if these systems respond poorly to conventional recovery procedures.

ACKNOWLEDGMENTS

This work was supported by BNFL Fluorochemicals Ltd, UK.

REFERENCES

Abdullah, R., Thompson, J.A., and Cocking, E.C. Efficient plant regeneration from rice protoplasts through somatic embryogenesis. *Bio/Technol. 4*, 1087–1090 (1986).

Anthony, P., Davey, M.R., Power, J.B., Washington, C. and Lowe, K.C. Synergistic enhancement of protoplast growth by oxygenated perfluorocarbon and Pluronic F-68. *Plant Cell Rep. 13*, 251–255 (1994).

Benson, E.E., and Withers, L.A. Gas chromatographic analysis of volatile hydrocarbon production by cryopreserved plant tissue cultures: a non-destructive method for assessing stability. *Cryo-Letts. 8*, 35–46 (1987).

Cawrse, N., de Pomerai, D.I., and Lowe, K.C. Effects of Pluronic F-68 on 2-deoxyglucose uptake and amino acid incorporation into chick embryonic fibroblasts *in vitro*. *Biomed. Sci. 2*, 180–182 (1991).

Cella, R., Colombo, R., Galli, M.G., Nielsen, M.G., Rollo, F. and Sala, F. Freeze-preservation of rice cells: a physiological study of freeze/thawed cells. *Physiol. Plant. 55*, 279–284 (1982).

Chu, C.C., Wang, C.S., Sun, C.S., Hsu, C., Yin, K.C., Chu, C.Y., and Bi, F.Y. Establishment of an efficient medium for anther culture of rice through comparative experiments on the nitrogen sources. *Sci. Sinica 18*, 659–668 (1975).

Finch, R.P., Lynch, P.T., Jotham, J.P., and Cocking, E.C. Isolation, culture and fusion of rice protoplasts, in: *Biotechnology in Agriculture and Forestry*, (Y.P.S. Bajaj, ed.) Vol. 14, pp. 251–268, Heidelberg, Springer-Verlag (1991).

Fuller, B.J., Gower, J.D. and Green, C.J. Free radical damage and organ preservation: fact or fiction. *Cryobiology* 25, 377–393 (1988).

Handa-Corrigan, A., Zhang, S., and Brydges, R. Surface-active agents in animal cell cultures, in: *Animal Cell Biotechnology* (R.E. Spier, and J.B. Griffiths, eds.), Vol. 5, pp. 279–301, London, Academic Press (1992).

Ju, L-K. and Armiger, W.B. Use of perfluorocarbon emulsions in cell culture. *Biotechniques 12*, 258–263 (1992).

Junker, B.H., Hatton, T.A. and Wang, D.I.C. Oxygen transfer enhancement in aqueous/perfluorocarbon fermentation systems. I Experimental observations. *Biotechnol. Bioeng.* 35, 578–585 (1990).

Lowe, K.C., Davey, M.R., Power, J.B., and Mulligan, B.J. Surfactant supplements in plant culture systems. *Agro-food-Ind. Hi-Tech. 4*, 9–13 (1993).

Lowe, K.C., Davey, M.R., Laouar, L., Khatun, A., Ribeiro, R.C.S., Power, J.B., and Mulligan, B.J. Surfactant stimulation of growth in cultured plant cells, tissues and organs, in: *Physiology, Growth and Development of Plants in Culture* (P.J. Lumsden, J.R. Nicholas, and W.J. Davies, eds.), pp. 234–244. Dordrecht, Kluwer (1994).

Lowe K.C., Anthony, P., Davey M.R., Power J.B. and Washington, C. Enhanced protoplast growth at the interface between oxygenated fluorocarbon liquid and aqueous culture medium supplemented with Pluronic F-68. *Art. Cells, Blood Subs., and Immob. Biotech. 23*, 417–422 (1995a).

Lowe, K.C., Davey, M.R., and Power, J.B., Perfluorochemicals and plant biotechnology, in: *Fluorine in Agriculture* (R.E. Banks, ed.), pp. 157–164, Shrewsbury, Rapra (1995b).

Lowe, K.C., Anthony, P., Wardrop, J., Davey M.R., and Power J.B. Perfluorochemicals and cell biotechnology. *Art. Cells, Blood Subs., and Immob. Biotech.*, in press (1996).

Lynch, P.T., Benson, E.E., Jones, J., Cocking, E.C., Power, J.B., and Davey, M.R. Rice cell cryopreservation: the influence of culture methods and the embryogenic potential of cell suspensions on post-thaw recovery. *Plant Sci. 98*, 185–192 (1994).

Murhammer, D.W., and Pfalzgraf, E.C. Effects of Pluronic F-68 on oxygen transport in an agitated, sparged bioreactor. *Biotechnol. Tech. 6*, 199–202 (1992).

Ohira, K., Ojima, K., and Fujiwara, A. Studies on the nutrition of rice cell culture. I. A simple defined medium for rapid growth in suspension culture. *Plant Cell Physiol. 14,* 1013–1121 (1973).

Snedecor, G.W., and Cochran, W.G., in: *Statistical Methods*, 8th edn., Iowa State University Press, Ames (1989).

Steponkus, P.L., and Lamphear, F.O. Refinement of the triphenyl tetrazolium chloride method of determining cold injury. *Plant Physiol.* 42, 1423–1426 (1967).

Thurston, R.J., Rogoff, M.S., Scott, T.R., and Korn, N. Effects of perfluorochemical diluent additives on fertilizing capacity of turkey semen. *Poultry Sci.* 72, 598–602 (1993).

69

NOVEL FLUOROSURFACTANTS FOR PERFLUOROCHEMICAL EMULSIFICATION

Effects on Aggregation of Human Platelets *in Vitro* and Possible Applications as Anti-Thrombotic Agents

C. M. Edwards,[1] K. C. Lowe,[1] S. Heptinstall,[2] P. Lucas,[3] H. Trabelsi,[3] and A. Cambon[3]

[1]Department of Life Science
University of Nottingham
University Park, Nottingham NG7 2RD, United Kingdom
[2]Department of Cardiovascular Medicine
Queen's Medical Centre, Nottingham NG7 2UH, United Kingdom
[3]Laboratoire de Chimie Organique du Fluor
Université de Nice
B.P. 71, Parc Valrose, 06108 Nice, Cedex 2, France

1. INTRODUCTION

Liquid perfluorochemicals (PFC) are highly inert, can dissolve large volumes of respiratory gases and have been administered intravascularly as emulsions. One such "first-generation" commercial emulsion, *Fluosol*® (Alpha Therapeutic, U.K.), has been approved for clinical use as an adjunct to percutaneous transluminal coronary angioplasty (Lowe, 1994a,b). Current research is aimed at the development of more concentrated "second-" and "third-generation" formulations with improved stability and oxygen-transport properties. In this context, there is increasing interest in the use of tailor-made, "fluorophilic" surfactants, either as sole or co-emulsifiers, to enhance PFC emulsion stability (Kissa, 1993; Riess and Greiner, 1993). Ideally, such surfactants should consist of a hydrophobic moiety, which interacts with the PFC (disperse) phase, and a hydrophilic end, which interacts with the aqueous (continuous) phase. Such compounds reduce the interfacial tension between the two phases, thus facilitating emulsification and contributing to emulsion stability. In addition to their specific chemical features, fluoro-surfactants for PFC emulsification must have acceptable biocompatibility characteristics, particularly relating to their interactions with blood components.

The present investigation has evaluated the biocompatibility of a range of novel, non-ionic, glycosidic or polyol fluoro-surfactants in a human platelet aggregation bioassay

$$HO(CH_2CH_2O)_a(CH_2CHO)_b(CH_2CH_2O)_a - H$$
$$|$$
$$CH3$$

Polyoxyethylene Polyoxypropylene Polyoxyethylene

Figure 1. General chemical structure of *Pluronic®* F-68.

in vitro. A comparison has been made with the effects of the polyoxyethylene-polyoxypropylene block co-polymer surfactant, *Pluronic®* F-68 (poloxamer 188; Fig. 1; Edwards *et al.*, 1995, 1996), which has been widely used as a stabiliser in oxygen-carrying PFC emulsions prepared for intravascular applications (Lowe, 1994a,b). The results show that some of the fluoro-surfactants can inhibit spontaneous platelet aggregation, suggesting possible applications as anti-thrombotic agents, additional to their use in stabilising emulsified PFCs.

2. MATERIALS AND METHODS

2.1. Synthesis of Fluoro-Surfactants

A novel series of fluoro-surfactants, containing glycosidic ('S' series) or polyol ('P' series) derivatives, were synthesised via simple routes using highly fluorinated isocyanates with amino alcohols, polyethoxylated alcohols and partially protected sugars at anomeric carbon (Jouani *et al.*, 1994; Fig. 2). The yields of compounds were 88–95% and their purities were > 99%.

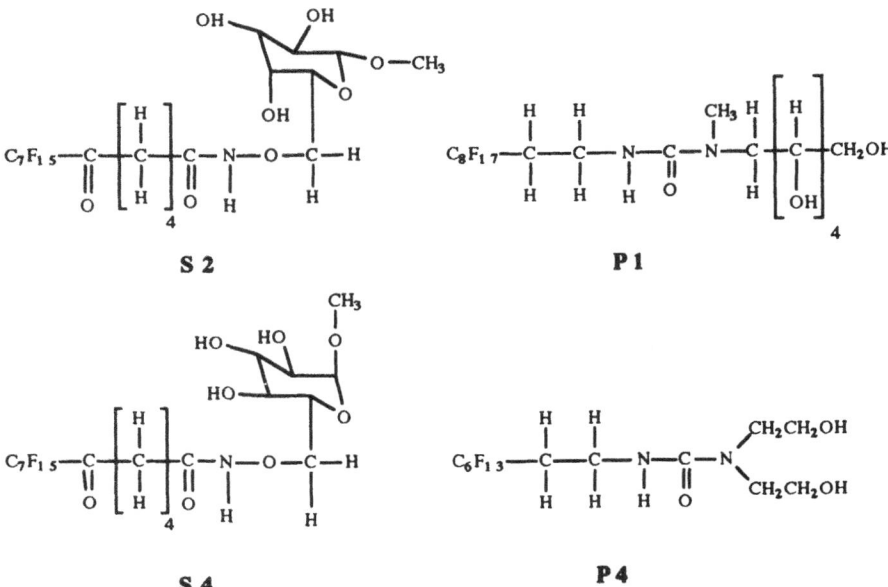

Figure 2. Chemical structures of fluoro-surfactants, S2 and S4, P1 and P4.

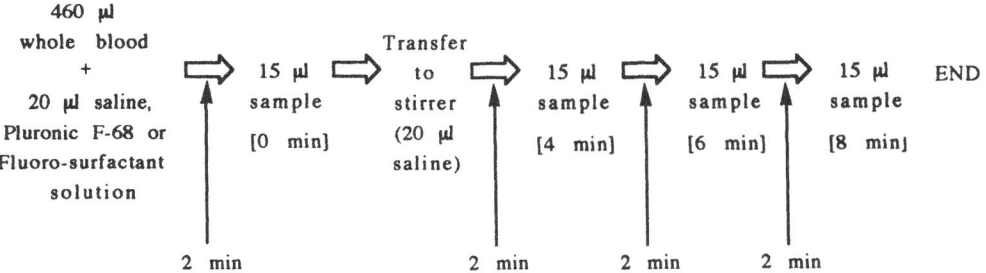

Figure 3. Flow diagram of the method for assessing surfactant effects on spontaneous platelet aggregation in human whole blood *in vitro*.

2.2. Blood Sampling and Platelet Aggregation Assay

Blood (25 ml) was obtained by sterile forearm venepuncture from healthy volunteers who denied taking aspirin or other non-steroidal anti-inflammatory drugs during the 7 day period prior to sampling. Nine ml aliquots were placed into tubes containing 50 μg ml^{-1} hirudin (*Revasc*™; Ciba, U.K.) and were incubated for 30 min at 37°C. Aliquots (460 μl) of blood were placed in plastic tubes in a water bath (37°C) and pre-incubated for 2 min with 20 μl of 0.001–10.0% (w/v) of fluoro-surfactant or *Pluronic*® F-68 (Fig. 3). Tubes were sampled (15 μl), 20 μl of saline added and stirred continuously (1000 rpm). Sampling (15 μl) was at 4, 6 and 8 min and samples were placed in 36 μl of formaldehyde fixing solution. Thirty six μl of fixed sample was mixed with 9.1 ml of saline and transferred to an Ultra-Flo 100 Automatic Platelet Counter (Coulter, U.K.) for counting. Platelet counts were calculated as percentage fall from time 0 and expressed as percentage aggregation.

2.3. Statistical Methods

Statistical analyses were performed according to Snedecor and Cochran (1989). Means and standard errors (s.e.m.) were used throughout and statistical comparisons were made by the paired Wilcoxon two-tailed test. A probability of $P < 0.05$ was considered significant.

3. RESULTS

The mean spontaneous platelet aggregation in normal whole blood after 8 min was $15 \pm 2\%$ (n = 15). The mean platelet aggregation after 8 min with compounds S2 or S4 at 0.32% (w/v) was $13 \pm 1\%$ and $12 \pm 2\%$ (n = 6), respectively (Fig. 4A). The polyol ('P') fluoro-surfactants significantly ($P < 0.05$) reduced spontaneous platelet aggregation in a dose-dependent manner, comparable to that caused by *Pluronic*® F-68. For example, the mean aggregation with 0.32% (w/v) of fluoro-surfactant P1 or P4 (8 min) was $8 \pm 1\%$ or $10 \pm 1\%$ (n = 6), respectively, compared to $6 \pm 1\%$ (n = 15) with 0.32% of *Pluronic*® F-68 (Fig. 4B). Overall, the inhibition of platelet aggregation by the 'S' compounds was much less pronounced than with the 'P' fluoro-surfactants or *Pluronic*® F-68.

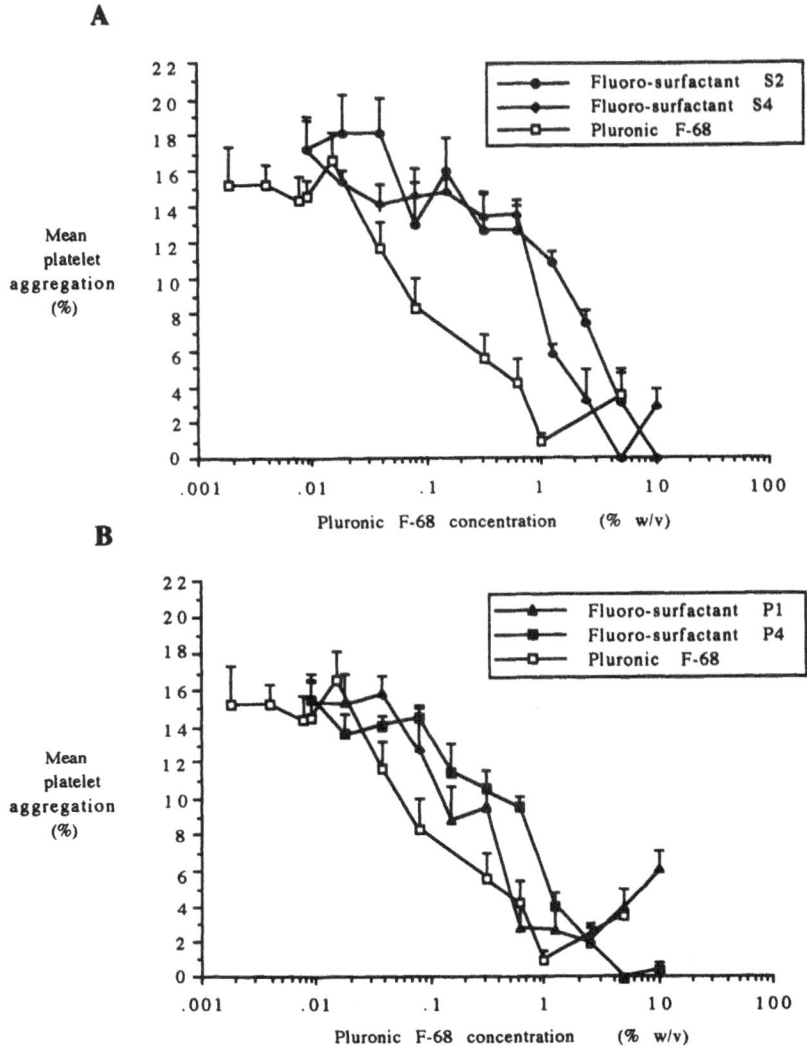

Figure 4. Mean spontaneous platelet aggregation (%) in human whole blood in response to **A**) fluoro-surfactants S2 and S4 or *Pluronic*® F-68 (% w/v), or **B**) fluoro-surfactants P1 and P4 or *Pluronic*® F-68 (% w/v). Values are mean of 6–15 values; vertical bars represent s.e.m.

4. DISCUSSION

This study has evaluated the effects of the novel fluoro-surfactants on spontaneous aggregation of human platelets *in vitro*. Previously, haemolysis assays, using human or animal blood, have been used to assess the biocompatibility of surfactants for intravascular applications (Krafft *et al.*, 1991; Riess and Greiner, 1993; Lowe *et al.*, 1995). The haemolytic activity of perfluoroalkylated surfactants generally decreases as the length of the hydrophobic chain increases (Riess and Greiner, 1993). Little attention has focused on the effects of fluoro-surfactants on platelets, despite increasing interest in their applications as stabilisers in PFC emulsions for intravascular uses (Riess and Greiner, 1993). We are aware that the *in vivo* behaviour of the fluoro-surfactants alone would be markedly different from that when administered as a constituent of a PFC emulsion. In the free form, the fluoro-compounds would be much more likely to interact with constituents of cell membranes, including those of platelets.

The polyol fluoro-surfactants, P1 and P4, inhibited platelet aggregation and mimicked the anti-thrombotic effects of *Pluronic®* F-68 more effectively than the glycosidic-derived compounds, S2 and S4. It has been suggested that the beneficial effects of *Pluronic®* F-68 in ischaemic injury (Schaer *et al.*, 1994) may involve inhibition of platelet aggregation in the microvasculature. Also, heart perfusion with PFC emulsions containing *Pluronic®* F-68 during coronary balloon angioplasty (Lowe, 1994a,b) may also involve direct effects of the surfactant on platelets. This would be beneficial since one complication of angioplasty is the re-occlusion of the blood vessel caused by platelet deposition in regions damaged by balloon inflation (Gershlick *et al.*, 1995). It is possible that some of the fluoro-surfactants could be beneficial in ischaemic tissues as anti-thrombotic agents.

ACKNOWLEDGMENTS

Supported by European Commission Brite-Euram Contract BRE2-CT94-0943.

REFERENCES

Edwards, C.M., Heptinstall, S., and Lowe, K.C. Inhibitory effects of the co-polymer surfactant, Pluronic F-68, on spontaneous platelet aggregation in human whole blood. *Br. J. Pharmacol. 116*, 159P (1995).

Edwards, C.M., May, J.A. Heptinstall, S., and Lowe, K.C. Effects of *Pluronic®* F-68 (Poloxamer 188) on platelet aggregation in human whole blood. *Thromb. Res. 81*, 511–512 (1996).

Gershlick, A.H., Spriggins, D., Davies, S.W., Syndercombe Court, Y.D., Timmins, J., Timmis, A.D., Rothman, M.T., Layton, C., and Balcon R. Failure of Epoprostanol (prostacyclin, PGI₂) to inhibit platelet-aggregation and to prevent restenosis after coronary angioplasty. Results of a randomised placebo-controlled trial. *Br. Heart J. 71*, 7–15 (1995).

Jouani, M.A., Szonyi, F., Trabelsi, H., and Cambon, A. Synthèse de tensioactifs non ioniques F-alkylés polyhydroxylés. *Bull. Soc. Chim. Fr. 131*, 173–176 (1994).

Kissa, E. *Fluorinated Surfactants. Synthesis, Properties, Applications*, Marcel Dekker, New York (1993).

Krafft, M.P., Vierling, P., and Riess, J.G. Synthesis and preliminary data on the biocompatibility and emulsifying properties of perfluoroalkylated phosphoramidates as injectable surfactants. *Eur. J. Med. Chem. 26*, 545–550 (1991).

Lowe, K.C. Perfluorochemicals in vascular medicine. *Vasc. Med. Rev. 5*, 15–32 (1994a).

Lowe, K.C. Properties and biomedical applications of perfluorochemicals and their emulsions, in: *Organofluorine Chemistry: Principles and Commercial Applications*, (R.E. Banks, B.E. Smart, and J.C. Tatlow, eds.), pp. 555–577, Plenum, New York (1994b).

Lowe, K.C., Furmidge, B.A., and Thomas, S. Haemolytic properties of Pluronic surfactants and effects of purification. *Art. Cells, Blood Subs., Immob. Biotechnol. 23*, 135–139 (1995).

Riess, J.G., and Greiner, J. Perfluoroalkylated sugar derivatives as surfactants and co-surfactants for biomedical uses, in: *Carbohydrates as Organic Raw Materials* (G. Descotes, ed.), pp. 209–259, VCH Verlagsgesellschaft, Weinheim (1993).

Schaer, G.L., Hursey, T.L., Abrahams, S.L., Buddemeier, K., Ennis, B., Rodriguez, R., Hubbell, J.P., Moy, J., and Parrillo, J.E. Reduction in reperfusion-induced myocardial injury in dogs by *RheothRx®* Injection (Poloxamer 188 N.F.), a hemorheological agent that alters neutrophil function. *Circulation 90*, 2964–2975 (1994).

Snedecor, G.W., and Cochran, W.G. *Statistical Methods*, 8th Edn., Iowa State College Press, Ames (1989).

70

PERFLUOROCHEMICAL EFFECTS ON NEUTROPHIL CHEMILUMINESCENCE

K. C. Lowe,[1] C. M. Edwards,[1] W. Röhlke,[2] P. Reuter,[2] U. Geister,[2] and H. Meinert[2]

[1]Department of Life Science
University of Nottingham
University Park, Nottingham NG7 2RD, United Kingdom
[2]Lehrstuhl Chemie
Universität Ulm
Parkstrasse 11, D-89073 Ulm, Germany

1. INTRODUCTION

Perfluorochemicals (PFC) are highly fluorinated, inert organic compounds which, in liquid form, can dissolve large volumes of respiratory gases. Emulsions of PFCs dispersed in physiologically-acceptable electrolyte solutions have been widely studied for intravascular applications as so-called "blood substitutes". One commercial, "first-generation" formulation, *Fluosol*® (Alpha Therapeutic, U.K.), has been approved for clinical use as an adjunct to percutaneous transluminal coronary angioplasty (Lowe, 1994a,b). In addition to enhancing oxygen delivery, emulsified PFCs may protect the heart and other tissues against inflammatory reperfusion damage through transient alterations in blood polymorphonuclear leucocyte (PMN; neutrophil) functions (Forman *et al.*, 1992a,b). At present, it is unclear whether such effects arise as a result of opsonisation of PFC emulsion droplets (typically *ca.* 0.2 µm diameter) by neutrophils, through direct effects of the essential surfactant component(s) on both PMNs and erythrocytes (Schaer *et al.*, 1994), or a combination of both.

The effects have been studied of a novel PFC emulsion and either the lecithin or *Pluronic*® (poloxamer) copolymer surfactants on human neutrophil function, as assessed by cellular chemiluminescence, *in vitro*. Measurement of chemiluminescence is a standard technique for evaluating the generation of the superoxide anion radical (O_2^-) by stimulated blood PMNs (Cheson *et al.*, 1976). It is therefore convenient for studying the effects of exposure of neutrophils to PFC-based respiratory gas carriers and their components.

Oxygen Transport to Tissue XIX, edited by Harrison and Delpy
Plenum Press, New York, 1997

2. MATERIALS AND METHODS

2.1. Blood Sampling and Preparation

Blood (25 ml) was obtained by sterile forearm venepuncture from healthy volunteers. Aliquots (4.5 ml) of blood were placed into polythene tubes containing 0.5 ml of 3.13% (w/v) trisodium citrate dihydrate as an anti-coagulant. Samples were incubated in a water bath (37°C) for 30 min before use. Whole blood was used to avoid pre-activation of neutrophils by the separation procedure, as reported previously (Neilson *et al.*, 1992).

2.2. Experimental Treatments

Fifty μl of blood was added to 10–40 μl of one of the following that had been pre-warmed to 37°C in a water bath: (i) a novel PFC emulsion containing 18.5% (w/v) perfluorodecalin and 1.5% (w/v) perfluorodimorpholine propane, prepared by high pressure homogenisation (Gaulin Micron Lab 40; 500 bar, 25°C) with 2.5% (w/v) lecithin (*Lipoid*® E100; Lipoid GmbH) (Table 1), (ii) lecithin (*Lipoid*® E100), (iii) commercial *Pluronic*® F-68 (poloxamer 188; ICI, UK), (iv) *Pluronic*® PE 6800 (BASF, Germany), or (v) isotonic saline [0.9% (w/v NaCl; Baxter, UK]. The emulsion was steam sterilised (121°C, 2 bar, 20 min) before use and had a mean droplet diameter of 210 nm (Malvern Autosizer IIc). Samples were re-incubated for a further 2 min at 37°C.

2.3. Measurement of Neutrophil Chemiluminescence

Fifty μl of sample was added to 1.0 ml of phosphate-buffered saline (PBS) containing *Luminol*® (5-amino-2,3-dihydro-1,4-phthalazine-dione; Sigma-Aldrich, UK; A8511; 100 μg ml^{-1}) to enhance the detection of the scintillation signal generated during the oxidative burst of the PMNs (Cheson *et al.*, 1976); samples were incubated for a further 5 min at 37°C. Twenty μl of phorbol myristate acetate (PMA; Sigma-Aldrich, UK) solution (100 μg ml^{-1}) was added to each sample (37°C) to stimulate the oxidative burst and the chemiluminescence recorded at 2 min intervals for up to 20 min using an LKB 2150 Luminometer (LKB, Sweden).

2.4. Statistical Methods

Statistical analyses were performed according to the methods of Snedecor and Cochran (1989). Means and standard errors (s.e.m.) were used throughout and statistical significance between mean values was assessed using a conventional Student's *t*-test. A probability of $P < 0.05$ was considered significant.

Table 1. Composition of the novel PFC emulsion

Constituent	Concentration (% w/v)
Perfluorodecalin	18.5
Perfluorodimorpholine propane	1.5
Lecithin	2.5
Isotonic saline [0.9% (w/v) NaCl]	to 100

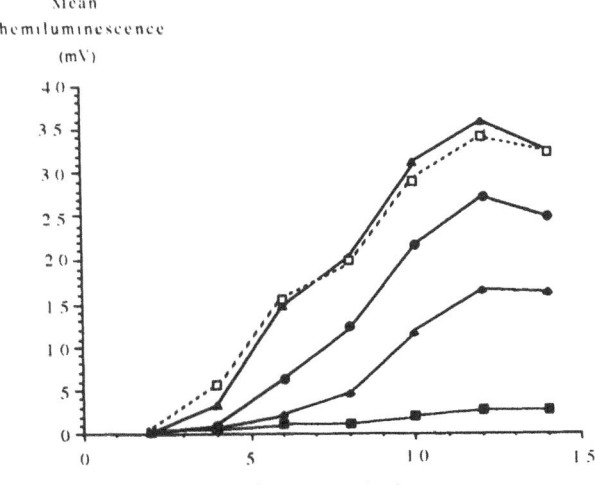

Figure 1. Effect of the novel PFC emulsion on the time course of changes in PMA-induced chemiluminescence of human neutrophils *in vitro*. Open square = control; triangle = 10 µl; circle = 20 µl; diamond = 30 µl; closed square = 40 µl.

3. RESULTS

The mean chemiluminescence, following stimulation of neutrophils with PMA, in controls, in a typical experiment, increased to a maximum of 34.5 mV within 10–15 min (Fig. 1). Dilution of samples with saline did not alter significantly the chemiluminescent response. A transient, dose-dependent, decrease in chemiluminescence, up to a maximum of 73% ($P < 0.05$), occurred when blood was pre-incubated with 10, 20, 30 or 40 µl of the PFC emulsion, compared to saline controls, an example of which is shown in Fig. 1. The mean ($\pm$ s.e.m., n = 6) peak chemiluminescence of neutrophils at 12 min after PMA stimulation and incubated with 10–40 µl of emulsion decreased progressively as the volume of emulsion added was increased (Table 2). Neutrophil chemiluminescence was unaffected by exposure to the PFC emulsion in the absence of PMA stimulation (data not shown). Incubation of blood with lecithin up to 16 mg ml^{-1} or either of the *Pluronics*® up to 65 mg ml^{-1} had no significant effects on chemiluminescence.

4. DISCUSSION

The results of these experiments show that incubation of human blood with the novel PFC emulsion attentuated the PMA-induced neutrophil chemiluminescence in a dose-dependent manner. Chemiluminescence decreased only when the whole emulsion was used and not when the lecithin or Pluronic surfactants were tested. Such transient changes in PMN function in response to phagocytosis of PFC emulsion droplets are con-

Table 2. Peak chemiluminescence of PMA-stimulated human neutrophils incubated with different volumes of the novel PFC emulsion *in vitro*

Volume of PFC emulsion (µl)	0	10	20	30	40
Peak chemiluminescence (mV)	26.8±4.7	25.9±6.9	19.3±4.8	11.1±3.3	7.3±2.4

Data are expressed as mean $\pm$ s.e.m. (n = 6).

sistent with previous studies using *Fluosol*® (Virmani *et al.*, 1984; Forman *et al.*, 1992a,b). However, the present finding that *Pluronic*® F-68 did not affect neutrophil chemilumines- cence, contrasted with an earlier report (Forman *et al.*, 1992a) that this surfactant was the principal component of *Fluosol*® responsible for O_2^- generation from human PMNs *in vi- tro*. Indeed, Ingram and others (Ingram *et al.*, 1992) claimed that the *Fluosol*® emulsion activated neutrophils, primarily through micelles formed by *Pluronic*® F-68.

Exposure of blood to the PFC emulsion in the absence of PMA stimulation did not induce a chemiluminescence response, suggesting that the inhibitory effects observed did not involve pre-activation of neutrophils. PMA is a diacylglyceol analogue which induces chemiluminescence by direct activation of protein kinase C (PKC; Castagna *et al.*, 1982). It is therefore unlikely that droplets of the novel emulsion inhibited PMA-induced chemi- luminescence by alteration of receptor-mediated activation pathways. However, other pos- sible mechanisms include coating of neutrophils with emulsion droplets, coupled to changes in membrane function prior to subsequent uptake. This, in turn, may lead to al- terations in PKC interaction with membrane-bound proteins, changes in calcium influx into the cells or direct interference of the nicotinamide adenine dinucleotide phosphate (NADPH)-oxidase enzyme which is key to the chemiluminescence response (Smith *et al.*, 1995). Blockade of PMA uptake is unlikely as exposure of neutrophils to the lecithin or *Pluronic*® surfactants did not affect the chemiluminescent response.

The present findings support the use of emulsified PFCs in the treatment of coronary angioplasty to facilitate improved oxygenation whilst concurrently reducing inflammatory damage caused by PMN-induced reperfusion injury. However, such potential anti-inflamma- tory effects of PFC emulsions must be balanced against possible risks of immunosuppression, especially on repeat, high dose administration. In this regard, it has been reported previously (Lowe, 1988) that injection of emulsified PFCs can alter immune system function *in vivo*, with the responses depending on dose and timing of administration relative to immunological challenge. Future studies should evaluate the extent to which the composition and physical characteristics of PFC emulsions can affect PMN functions and determine the time course and significance of any immunosuppressive responses in the recipient. Such investigations should take account of the preliminary report of Smith and Nesland (1994) in which proprietary PFC emulsions of different composition were shown to have contrasting effects on stimulated O_2^- production by human neutrophils *in vitro*.

ACKNOWLEDGMENTS

Supported by European Commission Brite-Euram Contract BRE2-CT94–0943.

REFERENCES

Castagna, M., Takai, Y., Kaibuchi, K., Sano, K., Kikkawa, U., and Nishizuka, Y. Direct activation of calcium acti-
    vated, phospholipid-dependent protein kinases by tumour-promoting phorbol esters. *J. Biol. Chem. 257,*
    7847–7851 (1982).
Cheson, B.D., Christiensen, R.L., Sperling, R., Kohler, B.E. and Babior, B.M. The origin of the chemilumines-
    cence of phagocytosing granulocytes. *J. Clin. Invest. 58,* 789–796 (1976).
Forman, M.B., Ingram, D.A., and Murray, J.J. Attenuation of myocardial reperfusion injury by perfluorochemical
    emulsions. *Clin. Hemorheol. 12,* 121–140 (1992a).

Forman, M.B., Pitarys, C.J., Vildibill, H.D., Lambert, T.L., Ingram, D.A., Virmani, R., and Murray, J.J. Pharmacologic perturbation of neutrophils by Fluosol results in a sustained reduction in infarct size in the canine model of reperfusion. *J. Am. Coll. Cardiol. 19*, 205–216 (1992b).

Ingram, D.A., Forman, M.B., and Murray, J.J. Phagocytic activation of human neutrophils by the detergent component of Fluosol. *Am. J. Pathol. 140*, 1081–1087 (1992).

Lowe, K.C. Emulsified perfluorochemicals for oxygen transport to tissues: effects on lymphoid system and immunological competence, in: *Oxygen Transport to Tissue*, Vol. 10 (M. Mochizuki, C.R. Honig, T. Koyama, T.K. Goldstick, and D.F. Bruley, eds.), pp. 655–663, Plenum, New York (1988).

Lowe, K.C. Perfluorochemicals in vascular medicine. *Vasc. Med. Rev. 5*, 15–32 (1994a).

Lowe, K.C. Properties and biomedical applications of perfluorochemicals and their emulsions, in: *Organofluorine Chemistry: Principles and Commercial Applications,* (R.E. Banks, B.E. Smart, and J.C. Tatlow, eds.), pp. 555–577, Plenum, New York (1994b).

Neilson, C.P., Crowley, J.J., Vestal, R.E., and Connolly, M.J. Impaired beta-adrenoceptor function, increased leukocyte respiratory burst, and bronchial hyperresponsiveness. *J. Allergy Clin. Immunol. 90*, 825–832 (1992).

Schaer, G.L., Hursey, T.L., Abrahams, S.L., Buddemeier, K., Ennis, B., Rodriguez, R., Hubbell, J.P., Moy, J., and Parrillo, J.E. Reduction in reperfusion-induced myocardial injury in dogs by *RheothRx*[8] Injection (Poloxamer 188 N.F.), a hemorheological agent that alters neutrophil function. *Circulation 90*, 2964–2975 (1994).

Smith, D., and Nesland, G. Influence of perfluorochemical (PFC) emulsions on superoxide production by polymorphonuclear leukocytes *in vitro*. *Faseb J. 8*, A409 (1994).

Smith, T.M., Steinhorn, D.M., Thusu, K., Fuhrman B.P., and Dandona, P. A liquid perfluorochemical decreases the *in vitro* production of reactive oxygen species by alveolar macrophages. *Crit. Care Med. 23*, 1533–1539 (1995).

Snedecor, G.W., and Cochran, W.G. *Statistical Methods*, 8th Edn., Iowa State College Press, Ames (1989).

Virmani, R., Fink, L.M., Gunter, K., and English, D. Effect of perfluorochemical blood substitutes on human neutrophil function. *Transfusion 24*, 343–347 (1984).

71

CELLULAR RESPONSES OF PLANT PROTOPLASTS TO CULTURE WITH OXYGENATED PERFLUOROCARBON

J. Wardrop, C. M. Edwards, K. C. Lowe, M. R. Davey, and J. B. Power

Department of Life Science
University of Nottingham
University Park, Nottingham NG7 2RD, United Kingdom

1. INTRODUCTION

Perfluorochemicals (PFCs) are inert compounds which can dissolve large volumes of respiratory gases, such as oxygen and carbon dioxide. They have been widely employed to regulate respiratory gas supply in eukaryotic and prokaryotic cell culture systems (King *et al.*, 1989; Junker *et al.*, 1990; Ju *et al.*, 1991; Lowe *et al.*, 1995) by growing cells at the interface between liquid culture medium and gassed PFC. Such use of PFC liquids as oxygen carriers during protoplast culture can stimulate mitotic division and differentiation of protoplast-derived cells and tissues of several species (Anthony *et al.*, 1994 a,b; Lowe *et al.*, 1995) including the major cereal rice (Wardrop *et al.*, 1996).

Whilst previous studies have investigated the effect of oxygenated PFC upon the growth of protoplast-derived cells, little information is available concerning the effect of gassed PFCs upon cell metabolism. This study has investigated the effect of PFC-facilitated oxygen enhancement using a commercially available perfluorodecalin liquid (*Flutec*® PP6; BNFL Fluorochemicals Ltd, UK), upon mitotic division of protoplast-derived cells. It focuses, in particular, on the rôle that the free radical scavenging enzyme, superoxide dismutase (SOD; superoxide oxidoreductase; EC 1.15.1.1), has during this process.

2. MATERIALS AND METHODS

2.1. Growth Assessment of Protoplast-Derived Cells

The initial plating efficiency of the culture system was defined as the number of viable protoplasts which had regenerated a cell wall and undergone mitotic division after a 14 day period (22 ± 2°C, dark). Mitotic division was assessed in protoplast-derived cells, the protoplasts being isolated enzymatically from cell suspensions of an albino line of *Salpiglossis sinuata*. Protoplasts were suspended at a density of 2 x 10^5 ml^{-1} in 3 ml aliquots of liquid KM8P medium (Gilmour *et al.*, 1989), either alone (control), overlaying 6 ml of untreated PFC (PP6) or overlaying 6 ml of oxygen-gassed (10 mbar; 15 min) PFC (PP6). The PFC liquid was sterilised by autoclaving prior to use. Viability was assessed by the uptake and enzymatic cleavage, by viable cells, of the dye, fluorescein diacetate (Widholm, 1972).

2.2. Measurement of Oxygen Tensions in PFC

Oxygen concentrations in gassed PFC were determined using a Jenway 9015 Dissolved Oxygen Meter with a POM102 probe (Scientific Laboratory Supplies Ltd., Nottingham, UK). Measurements of oxygen decline in initially fully saturated perfluorodecalin were made with time by linking the electrode to a chart recorder (J.J. Lloyd Instruments Ltd., Southampton, UK).

2.3. Superoxide Dismutase Assay

Superoxide dismutase was extracted from aliquots consisting of 6 x 10^5 cells and assayed spectrophotometrically (560 nm) following lysis of the protoplast-derived cells by freezing (at -20°C) and thawing the cells harvested after 1, 3, 7 and 14 days of culture. One unit (U) of enzyme was defined as that required to inhibit the reduction of nitro-blue tetrazolium by 50% (0.0125 absorbance U min^{-1}) through the xanthine-xanthine oxidase system (Beauchamp and Fridovich, 1971). The reaction mixture consisted of 50 mM K-phosphate buffer (pH 7.8), 0.1 µM EDTA, 3.9 x 10^{-3} M xanthine, 5.03 x 10^{-3} M nitro-blue tetrazolium and sufficient xanthine oxidase to induce a reaction rate of 0.025 absorbance U min^{-1}. The reaction was followed in a 3 ml capacity cuvette of 1.0-cm path length. The cuvette temperature was held at 25°C using a thermostated cuvette holder attached to a water bath. SOD concentrations were determined by a standard curve produced with commercially available SOD (Sigma, Poole, UK), using identical assay conditions as described earlier.

2.4. Protein Determinations

Protein content of the samples was determined spectrophotometrically (595 nm) using bovine serum albumin as a standard (Bradford 1976).

2.5. Statistical Methods

Statistical analyses were performed according to Campbell (1990). Mean and standard errors (s.e.m.) were used throughout. Statistical significance between mean values was assessed using ANOVA (incorporating Tukey's pairwise comparisons), or a conventional Student's *t*-test. A probability of $P < 0.05$ was considered significant.

3. RESULTS

3.1. Effect of Oxygenated PFC on Division of Protoplast-Derived Cells in Culture

After 14 days, control cultures (protoplast-derived cells in liquid medium alone) displayed a mean (± s.e.m) initial plating efficiency of 15.2 ± 2.6% (n = 5). Similar results

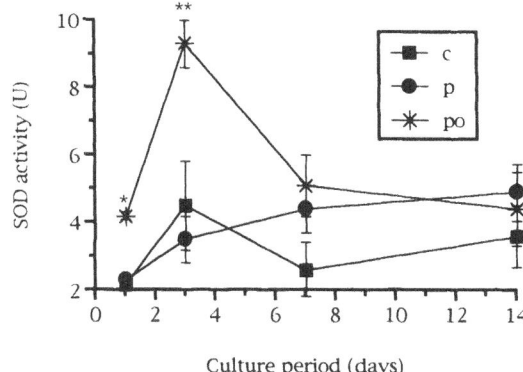

Figure 1. Changes in mean (± s.e.m.; n = 5) superoxide dismutase (SOD) activity in cultured protoplast-derived cells of *Salpiglossis sinuata.*

were obtained for protoplasts cultured overlaying non-oxygenated PFC, which had a mean division frequency of 13.5 ± 1.8% (n = 5) after 14 days. However, cells which had been cultured with oxygenated PFC (n = 5) showed a significant (P < 0.01) increase in their mean initial plating efficiency, which was 28.8 ± 3.1% at day 14.

3.2. Effect of Oxygenated PFC on Superoxide Dismutase Activity in Cultured Protoplast-Derived Cells

Control cultures displayed no significant changes in SOD activity with time. A similar result was obtained for protoplast-derived cells which had been cultured for 14 days in the presence of non-oxygenated PFC (Figure 1). In contrast, protoplast-derived cells cultured in the presence of oxygenated PP6 displayed significantly greater SOD activity (*P < 0.05; **P < 0.01) during the initial stages of culture (Figure 1; c - control, p - PFC, po - PFC and oxygen).

The decrease in SOD activity in protoplast-derived cells after 7 days of culture in the presence of oxygenated PFC, correlated closely with a progressive decline in the amount of oxygen (Figure 2) dissolved in the PFC phase (91 ± 1% to 19 ± 1%, n = 3) over the same period.

4. DISCUSSION

As demonstrated in this report, the use of oxygen-gassed PFC liquids as novel respiratory gas carriers during protoplast culture induces changes in cell biochemistry associ-

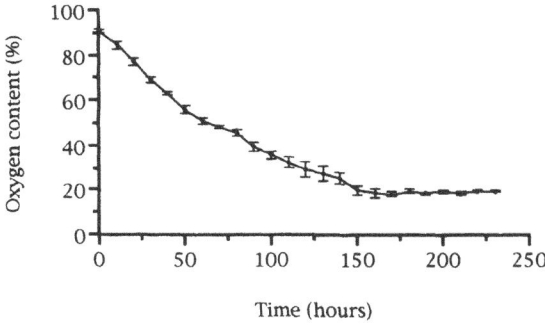

Figure 2. Changes in mean (± s.e.m.; n = 3) oxygen content of oxygen-saturated perfluorodecalin.

ated with exposure to elevated oxygen tensions. Enhancing the supply of molecular oxygen may induce the production of potentially deleterious free radicals, such as superoxide anions (O_2^-), hydrogen peroxide (H_2O_2) and the hydroxyl radical (OH^-). Cellular scavenging systems, including SOD enzymes, catalases and peroxidases, are responsible for the removal of these oxygen species. The results from the present experiments indicate that the level of SOD activity is correlated directly with the amount of available oxygen and that SOD production is induced by exposing the cells to high levels of molecular oxygen.

In this study, the SOD activity which was generated appears adequate to protect the dividing cells from oxidative damage associated with exposure to high levels of oxygen, whilst permitting a beneficial supply of oxygen to be maintained. The latter probably resulted in improved cellular respiration, accompanied by increased cell division and enhanced growth.

It has been noted previously that if a sustained supply of oxygen is available to cells in culture, there is still the possibility of inhibition of cellular metabolism through oxygen radical production and accumulation (Lutz *et al.*, 1992). Indeed, Lutz *et al.* (1992) showed that the delivery of radical-scavenging compounds (e.g. α-tocopherol), in conjunction with PFC-mediated oxygen delivery, prevented cell damage and such scavengers may work synergistically with endogenous SOD activity in culture systems. Whilst further studies are required, the present results suggest that supplementation of the culture medium with oxygen radical scavenging agents is unnecessary for plant cells, since the endogenous SOD activity is adequate to counteract any potentially damaging effects of oxygen radicals generated as a result of increased oxygen supplied by the PFC. In this respect, it is noteworthy that protoplasts have been cultured for up to 21 days in the presence of oxygenated PFC without any detrimental effects to the subsequent development and fertility of plants regenerated from protoplast-derived tissues (Wardrop *et al.*, 1996).

Future investigations into the activity of other oxygen radical scavenging enzymes, such as catalases and peroxidases, in the presence of oxygenated PFC may give an insight into other mechanisms which may be responsible, concomitantly, for the results found in this study.

ACKNOWLEDGMENTS

The authors thank BNFL Fluorochemicals Ltd. for gifts of *Flutec*® PP6 and a research studentship to JW.

REFERENCES

Anthony, P., Dave, M.R., Power, J.B., Washington, C. and Lowe, K.C. Synergistic enhancement of protoplast growth by oxygenated perfluorocarbon and Pluronic F-68. *Plant Cell. Rep. 13*, 251–255 (1994a).

Anthony, P., Davey, M.R., Power, J.B., Washington, C. and Lowe, K.C. Image analysis assessments of perfluorocarbon and surfactant-enhanced protoplast division. *Plant Cell Tiss. Org. Cult. 38*, 39–43 (1994b).

Beauchamp, C.O. and Fridovich, I. Superoxide dismutase: improved assays and an assay applicable to acrylamide gels. *Anal. Biochem. 44*, 276–287 (1971).

Bradford, M.M. A rapid and sensitive method for the quantitation of microgram quantities of protein utilising the principle of protein-dye binding. *Anal. Biochem. 72*, 248–254 (1976).

Campbell R.C. *Statistics for Biologists* (3rd ed.), Cambridge University Press (1990).

Gilmour, D.M., Golds, T.J. and Davey, M.R. *Medicago* protoplasts: fusion, culture and plant regeneration, in: *Biotechnology in Agriculture and Forestry: Plant Protoplasts and Genetic Engineering I, Vol. 8* (Bajaj, Y.P.S., ed.), pp. 370–388, Heidelberg, Springer-Verlag (1989).

Ju, L.K., Lee, J.F. and Armiger, W.B. Enhancing oxygen transfer in bioreactors by PFC emulsions. *Biotech. Prog.* 7, 323–329 (1991).

Junker, B.H., Hatton, T.A. and Wang, D.I.C. Oxygen transfer in aqueous - PFC fermentation systems: 1. Experimental observations. *Biotechnol. Bioeng. 35,* 578–585 (1990).

King, A.T., Mulligan, B.J. and Lowe, K.C. Perfluorochemicals and cell culture. *Bio/Technol. 7,* 1037–1042 (1989).

Lowe, K.C., Davey, M.R. and Power, J.B. Perfluorochemicals and plant biotechnology, in: *Fluorine in Agriculture* (Banks, R.E., ed.), pp. 157–164. Shrewsbury, Rapra (1995).

Lutz, J., Augustin, A.J., Purucker, E. and Milz, J. Combination of treatment with perfluorochemicals and free radical scavengers. *Biomat., Art. Cells & Immob. Biotech. 20,* 951–958 (1992).

Wardrop, J., Lowe, K.C., Power, J.B. and Davey, M.R. Perfluorochemicals and plant biotechnology: an improved protocol for protoplast culture and plant regeneration in rice (*Oryza sativa* L.). *J. Biotechnol.* (in press; 1996).

Widholm, J. The use of FDA and phenosafranine for determining viability of cultured plant cells. *Stain Technol. 47,* 186–194 (1972).

72

PERFLUOROCHEMICAL-FACILITATED CARBON DIOXIDE DELIVERY ENHANCES GROWTH OF SHOOTS *IN VITRO*

J. Wardrop, R. Marchant, K. C. Lowe, M. R. Davey, and J. B. Power

Department of Life Science
University of Nottingham
University Park, Nottingham, NG7 2RD, United Kingdom

1. INTRODUCTION

Perfluorochemicals (PFCs) have been used to regulate gas supply to prokaryotic and eukaryotic cells in culture (Ju *et al.,* 1991; Lowe *et al.,* 1995). One approach for delivery of respiratory gases involves the growth of cells at the interface between liquid medium and oxygenated PFC, thereby increasing the plating efficiency of cultured protoplasts (Anthony *et al.,* 1994 a,b). This technology has been extended to the recovery of fertile plants of rice (Wardrop *et al.,* 1996), confirming that exposure to PFCs has no adverse effects on cell totipotency.

Whilst most studies, using PFCs, have focused on oxygen regulation, PFCs provide an option for regulating gases, such as ethylene and carbon dioxide, in the context of plant growth. Rose was selected for studies with PFCs since it is cultivated as an ornamental, for cut flowers and as a source of oils for perfume manufacture. *In vitro* multiplication is widely used (Short and Roberts, 1991), with several factors influencing micropropagation. These include growth regulators (Barna and Wakhlu, 1995), the position from which axillary buds are excised from donor plants (Bressan *et al.,* 1982), the use of ethylene inhibitors (Mekers *et al.,* 1984) or ethylene *per se* (Kevers *et al.,* 1992), the culture medium (Horan *et al.,* 1995) and carbon dioxide enrichment *in vitro* (Woltering, 1990).

The present study has evaluated the potential of the PFC, perfluorodecalin (*cis* and *trans* $C_{10}F_{18}$; *Flutec*® PP6, BNFL Fluorochemicals Ltd, Preston, UK), for delivering carbon dioxide to improve micropropagation of cultured rose shoots. The use of PFCs in this way offers a novel approach to manipulation of the carbon dioxide environment *in vitro*.

Oxygen Transport to Tissue XIX, edited by Harrison and Delpy
Plenum Press, New York, 1997

2. MATERIALS AND METHODS

2.1. Tissue Culture of Rose

Shoot cultures of *Rosa chinensis* Jacq. cv. Baby Love and *R. hybrida* L. cv. Frensham were maintained on R7 medium [Murashige and Skoog (1962) salts and vitamins, 5.0 μM 6-benzylamino purine (BAP), 0.3 μM gibberellic acid (GA$_3$), 0.2 μM α-naphthaleneacetic acid (NAA), 40 g l^{-1} sucrose and semi-solidified with 0.8% (w/v) agar (Sigma, Poole, UK), pH 5.8]. Cultures were incubated at 23 ± 2°C with a 16 hour photoperiod (27 μE m^{-2} s^{-1}, Cool Light fluorescent tubes, Thorn EMI Ltd., Hayes, UK). At sub-culture (28 days), single shoots were transferred to either 80 ml aliquots of R7 medium alone, or to R7 overlaying 20 ml volumes of PFC saturated with CO$_2$ in Magenta vessels (Magenta Corp., Chicago, USA).

2.2. PFC-Mediated CO$_2$ Supply to Cultured Shoots

A sterile, 23 gauge needle (Becton Dickinson UK Ltd., Cowley, UK) was inserted through the side of the Magenta vessel to introduce CO$_2$ into the PFC layer. The needle was connected by sterile tubing, via a 2 μm pore Minisart filter (Sartorius AG, Göttingen, Germany), to a cylinder of 100% CO$_2$ (BOC, Guildford, UK). CO$_2$ was supplied at 2 mbar pressure (5 min). The aperture in the Magenta vessel was sealed with PVC tape following removal of the needle. Cultures were maintained for 42 days, during which time those treated with CO$_2$ were re-gassed every 7 days.

2.3. Morphological and Biochemical Analyses of Micropropagated Shoots and Roots

The fresh weight increase, number of shoots and leaves, days to rooting and number of roots per shoot were recorded. Chlorophyll and total protein of cultured shoots were measured as indicators of possible cellular metabolic effects induced by exposure to PFC gassed with CO$_2$. Chlorophyll was measured spectrophotometrically (647 nm and 664 nm), as was the total protein content of samples (595 nm), using bovine serum albumin as a standard (Bradford 1976).

For rooting, shoots were transferred to 80 ml aliquots of MS medium lacking growth regulators. Shoots cultured over CO$_2$-gassed PFC were rooted on the same medium overlaid on 20 ml of CO$_2$-saturated PFC. The phenolic content was estimated spectrophotometrically (765 nm) with a gallic acid standard (Slinkard and Singleton 1977). The intensity of colour, due to extracted phenolic compounds, was scored on a scale of 1 to 4 (0, no browning; 1, partial browning; 2, complete browning; 3, partial blackening; 4, complete blackening).

2.4. *Ex Vitro* Transfer of Plants

All plants studied were transferred to 7 cm pots containing a 1:1 (v:v) mixture of Levington M3 compost (Fisons plc, Ipswich, UK) and Perlite (Silvaperl, Gainsborough, UK). After watering, plants were covered with a polythene bag and maintained under natural daylight (23 ± 3°C). The bags were opened progressively for acclimation over a 21 day period.

2.5. Statistical Methods

Statistical analyses were performed according to Campbell (1990). Mean and standard errors (s.e.m.) were used throughout, and statistical significance between mean values was assessed using ANOVA (incorporating Tukey's pairwise comparisons) or a conventional Student's t-test. A probability of $P < 0.05$ was considered significant.

3. RESULTS

3.1. Micropropagation and Plant Acclimation

Culture of shoots for 42 days with CO_2-gassed PFC resulted in significant ($P < 0.001$) increases in both the mean number of shoots and leaves per explant (Table 1). This growth stimulation was reflected in significant increases in the mean fresh weight of shoots after both 42 days ($P < 0.001$) and 63 days ($P < 0.01$) of culture (Table 1).

The mean number of roots per shoot after 10 days on rooting medium was greater ($P < 0.01$) in response to CO_2-gassed PFC, compared to controls (Table 2). After a further 7 days of culture, a species-dependent response was apparent, with rooting in Frensham still showing significant improvement when compared to controls ($P < 0.01$); Baby Love exhibited no further promotion of rooting. Both fresh and dry weights of roots (63 days) were also significantly ($P < 0.01$) greater when CO_2-gassed PFC was employed during the rooting phase (Table 2), thereby confirming increased proliferation of these organs.

There were no differences between treatments upon transfer of plants to *ex vitro* conditions. All plants survived transfer to the glasshouse and, like the controls, were morphologically true-to-type.

3.2. Protein, Chlorophyll and Phenolic Compound Determinations

The mean total protein content of shoots cultured with CO_2-gassed PFC (183 ± 25 µg g^{-1}, n = 8) was significantly ($P < 0.01$) lower than in controls (489 ± 10 µg g^{-1}, n = 8). In contrast, the mean chlorophyll a concentration in shoots from CO_2-gassed PFC cultures (514 ± 37 µg g^{-1}) was not significantly different to control (486 ± 74 µg g^{-1}, n = 8). Similar results were also obtained for chlorophyll b concentrations in shoots from PFC and CO_2-supplemented cultures (252 ± 20 µg g^{-1}, n = 8), compared with those of controls (219 ± 36 µg g^{-1}).

An interesting observation was that those parts of roots which penetrated into the PFC layer did not exhibit the marked accumulation of phenolics characteristic of roots in contact with culture medium alone. This was reflected in significantly ($P < 0.01$) lower

Table 1. Effect of CO_2-gassed PFC on rose micropropagation

	R. chinensis Jacq. cv. Baby Love		*R. hybrida* L. cv. Frensham	
Parameters assessed	Control (n = 74)	CO_2-gassed PFC (n = 61)	Control (n = 32)	CO_2-gassed PFC (n = 32)
Mean shoot No. (42 d)	3.2 ± 0.2	4.7 ± 0.2[***]	3.0 ± 0.3	5.2 ± 0.3[***]
Mean leaf No. (42 d)	40.6 ± 2.3	65.2 ± 3.0[***]	40.2 ± 3.5	67.0 ± 4.6[***]
Mean % shoot fresh weight increase (42 d)	711.5 ± 38.9	1089.6 ± 46.3[***]	1270 ± 117	1613 ± 122[*]
Mean shoot fresh weight (mg) at 63 d	88.3 ± 11.0	136.3 ± 12.5[**]	55.5 ± 3.2	83.4 ± 6[***]

Values are mean ± s.e.m.; [*]$P < 0.05$; [**]$P < 0.01$; [***]$P < 0.001$.

Table 2. Effect of CO_2-gassed PFC on rooting of micropropagated shoots of rose

| | *R. chinensis* Jacq. cv. Baby Love | | *R. hybrida* L.cv. Frensham | |
Parameters assessed	Control (n = 74)	CO_2-gassed PFC (n = 61)	Control (n = 32)	CO_2-gassed PFC (n = 32)
Mean root No. (10 d)	2.3 ± 0.3	3.3 ± 0.2**	1.9 ± 0.3	5 ± 0.5***
Mean root No. (17 d)	4.3 ± 0.3	4.2 ± 0.2	3.7 ± 0.5	5.7 ± 0.6**
Mean root fresh weight (mg) at 63 d	11.3 ± 4.1	32.7 ± 3.8**	2 ± 0.2	7.3 ± 1.2***
Mean root dry weight (mg) at 63 d	2.1 ± 0.7	5.5 ± 0.7**	0.8 ± 0.9	3.5 ± 0.7***

Values are mean ± s.e.m.; *P < 0.05; **P < 0.01; ***P < 0.001.

browning index scores for roots cultured with CO_2-gassed PFC (0.5 ± 0.2, n = 4) compared to controls (4.0 ± 0), although the total phenolic compounds present in shoots and roots did not differ between controls (533 ± 111 µg g^{-1}, n = 4) and CO_2-gassed PFC-treated cultures (456.2 ± 25.9 µg g^{-1}).

4. DISCUSSION

This study shows, for the first time, that CO_2-gassed PFC promotes growth of rose shoots *in vitro*, defining a rôle for PFCs in the micropropagation of woody species. The present approach supports previous suggestions (King *et al.*, 1989; Lowe *et al.*, 1995) that PFC liquids are useful gas carriers in both plant and animal culture systems. Indeed, the beneficial effects of using CO_2-gassed PFC may also be allied to a potential reduction in the requirement for carbohydrate supplementation of media (Desjardins *et al.*, 1988; Kozai *et al.*, 1988), especially in cultures with some degree of inducible photo-autotrophy. Elevated CO_2 levels within culture vessels will also counteract the potentially deleterious effects of ethylene on the growth of rose shoots *in vitro* (Mekers *et al.*, 1984; Kevers *et al.*, 1992). Indeed, PFCs may act as scavengers of ethylene (King *et al.*, 1989; Lowe *et al.*, 1995), and such an effect may have contributed, indirectly, to the overall growth stimulation observed in the present investigation. The finding that total protein declined in response to culture with PFC and CO_2 can be explained by enhanced cell multiplication, leading to increased biomass, which would outstrip protein biosynthesis in the short term. The observation that roots of shoots cultured in the presence of carbon dioxide-gassed PFC lacked the marked browning of roots of shoots grown on culture medium alone, demonstrates that PFC-mediated CO_2 delivery suppresses phenolic oxidation. Whilst the assay for total phenolics showed no significant differences between treatments, the browning index confirmed the inhibition of phenolic oxidation.

Media supplements, such as polyvinylpyrrolidone (Preece and Compton, 1991), have been employed to suppress phenolic oxidation *in vitro*. The use of CO_2-gassed PFC provides a further option for inhibiting such biochemical changes, coupled with the capacity to promote plant growth. Delivery and regulation of CO_2 on a commercial scale to cultured plants is difficult and often costly. The use of PFC-supplemented systems for CO_2 delivery will avoid supply difficulties, as the gas-dissolving property of PFCs dictates that CO_2 would be released over a prolonged period (Anthony *et al.*, 1994a; Wardrop *et al.*, 1996). Additionally, PFC liquids are heat stable, readily sterilised and can be recycled from aqueous systems, making their use economically viable.

ACKNOWLEDGMENTS

The authors thank BNFL Fluorochemicals Ltd. for gifts of *Flutec*® PP6 and Dr N. Wright, Micropropagation Services (EM) Ltd. for provision of axenic shoot cultures of *R. chinensis* cv. Baby Love. JW was funded by BNFL Fluorochemicals Ltd. and RM by the Ministry of Agriculture, Fisheries and Food under Open Contract No. OC9312.

REFERENCES

Anthony, P., Davey, M.R., Power, J.B., Washington, C. and Lowe, K.C. Synergistic enhancement of protoplast growth by oxygenated perfluorocarbon and Pluronic F-68. *Plant Cell Rep. 13*, 251–255 (1994a).

Anthony, P., Davey, MR., Power, J.B., Washington, C. and Lowe, K.C. Image analysis assessments of perfluorocarbon and surfactant-enhanced protoplast division. *Plant Cell Tiss. Org. Cult. 38*, 39–43 (1994b).

Barna, K.S. and Wakhlu, A.K. Effects of thidiazuron on micropropagation of rose. *In Vitro Cell. Dev. Biol. 31P*, 44–46 (1995).

Bradford, M.M. A rapid and sensitive method for the quantitation of microgram quantities of protein utilising the principle of protein-dye binding. *Anal. Biochem. 72*, 248–254 (1976).

Bressan, P.H., Kim, Y-J., Hyndman, S.E., Hasegawa, P.M. and Bressan, R.A. Factors affecting *in vitro* propagation of rose. *J. Amer. Soc. Hort. Sci. 107*, 979–990 (1982).

Campbell, R.C. *Statistics for Biologists* (3rd ed.) Cambridge University Press, Cambridge (1990).

Desjardins, Y., Laforge, F., Lussier, C. and Gosselin, A. Effect of CO_2 enrichment and high photon flux on the development of autotrophy and growth of tissue cultured strawberry, raspberry and asparagus plants. *Acta. Hort. 230*, 45–53 (1988).

Horan, I., Walker, S., Roberts, A.V., Mottley, J. and Simpkins, I. Micropropagation of roses: The benefits of pruned mother-plantlets at Stage II and a greenhouse environment at Stage III. *J. Hort. Sci. 70*, 799–806 (1995).

Ju, L.K., Lee, J.F. and Armiger, W.B. Enhancing oxygen transfer in bioreactors by PFC emulsions. *Biotech. Prog. 7*, 323–329 (1991).

Kevers, C., Boyer, N., Courduroux, J-C. and Gaspar, T. The influence of ethylene on proliferation and growth of rose shoot cultures. *Plant Cell Tiss. Org. Cult. 28*, 175–181 (1992).

King, A.T., Mulligan, B.J. and Lowe, K.C. Perfluorochemicals and cell culture. *Bio/Technol. 7*, 1037–1042 (1989).

Kozai, T., Koyama, Y. and Watanabe, I. Multiplication of potato plantlets *in vitro* with sugar free medium under high photosynthetic photon flux. *Acta. Hort. 230*, 121–127 (1988).

Lowe, K.C., Davey, M.R. and Power, J.B. Perfluorochemicals and plant biotechnology in: *Fluorine in Agriculture* (R.E. Banks, ed.), pp 157–164, Shrewsbury, Rapra (1995).

Mekers, O., Meiresonne, L. and Meneve, I. Ethylene production inhibitors can improve *in vitro* propagation of roses. *Med. Fac. Landbouw. Rijksuniv. Gent. 49/3b*, 1139–1144 (1984).

Murashige, T. and Skoog, F. A revised medium for rapid growth and bioassays with tobacco tissue cultures. *Physiol. Plant. 15*, 473–497 (1962).

Preece, J.E. and Compton, M.E. Problems with plant exudation in micropropagation, in: *Biotechnology in Agriculture and Forestry, Vol 15, Medicinal and Aromatic Plants III* (Y.P.S. Bajaj, ed.), pp 168–189, Heidelberg, Springer-Verlag (1991).

Short, K.C. and Roberts, A.V. *Rosa* spp. (Roses). *In vitro* culture, micropropagation, and the production of secondary products, in: *Biotechnology in Agriculture and Forestry, Vol 15, Medicinal and Aromatic Plants III* (Y.P.S. Bajaj, ed.), pp 376–397, Heidelberg, Springer-Verlag (1991).

Slinkard, K. and Singleton, V.L. Total phenol analysis: automation and comparison with normal methods. *Am. J. Enol. Vitic. 28*, 49–55 (1977).

Wardrop, J., Lowe, K.C., Power, J.B. and Davey, M.R. Perfluorochemicals and plant biotechnology: an improved protocol for protoplast culture and plant regeneration in rice (*Oryza sativa* L.). *J. Biotechnol., 50*, 47–54 (1996).

Woltering, E.J. Beneficial effects of carbon dioxide on development of gerbera and rose plantlets grown *in vitro*. *Sci. Hort. 44*, 341–345 (1990).

73

REGULATION OF OXYGEN SUPPLY IN THE CEREBRAL CIRCULATION

Antal G. Hudetz

Departments of Anesthesiology and Physiology
Medical College of Wisconsin
Milwaukee, Wisconsin 53226

INTRODUCTION

Despite the ongoing intense research into the physiological mechanisms of cerebrovascular regulation, the precise degree to which the flow of blood cells and of plasma in the cerebral microcirculation is regulated and the tolerance of small areas of the brain to compromised flow and oxygen supply are poorly understood. Likewise, the physiological responses of cerebral blood flow (CBF) to changes in intravascular and extravascular pressures, blood oxygen and carbon dioxide tensions and systemic hematocrit have been well described; however, little attention has been paid to the question whether the perfusion of cerebral capillaries by blood cells follows these regional blood flow responses. Since cerebral oxygen supply is regulated mainly by a change in the rate of perfusion and much less by an alteration in diffusion distance for oxygen, an examination of the physiological and biophysical factors affecting the perfusion rate of cerebral capillaries by red blood cells is particularly important.

While the direct examination of cellular flow in cerebral capillaries has been a difficult task, recent advances in intravital video-microscopic and image analysis techniques now allow us to study the cerebal capillary circulation *in vivo* and have contributed to a better understanding of the regulation of cerebral capillary perfusion and oxygen supply. This paper focuses on the pattern and regulatory mechanisms of red blood cell (RBC) flow in cerebrocortical capillaries as derived from observations performed with these intravital microscopic techniques. The responses of capillary RBC flow to decreased perfusion pressure, hypoxia, hypercapnia and isovolemic hemodilution are described and the possible physiological regulatory mechanisms underlying these responses are discussed.

MATERIALS AND METHODS

The circulation of RBC in individual capillaries of the cerebral cortex was studied in adult Sprague-Dawley rats using fluorescence video-microscopy in a closed cranial win-

Oxygen Transport to Tissue XIX, edited by Harrison and Delpy
Plenum Press, New York, 1997

dow preparation as recently described (Hudetz *et al,* 1995). The cranial window was developed specifically to study the capillary circulation in the rat and consists of a rectangular chamber with a 3 mm diameter quartz window. Both intracranial pressure and perfusion rate are independently controlled in the subwindow space. The chamber is secured to the skull following craniotomy and opening of the dura. The flow in subsurface capillaries to about 50mm depth is visualized by the aid of fluorescently labeled RBC. The cells are obtained from a donor animal and injected intravenously after labeling in vitro. The velocity of RBC flow in individual capillaries is measured by tracking of the movement of labeled cells or by the dual-window cross-correlation technique. The latter technique allows the measurement of velocities up to 5 mm/s (Hudetz *et al,* 1992) which is necessary because of the high velocity of flow in cerebral capillaries. Fluorescent labeling also allows the automatic, on-line or off-line measurement of RBC flux by detecting fluctuations in the digitized optical signal from video windows superimposed on the microscopic image of the capillary network (Knuese *et al,* 1994). The technique measures labeled cell flux up to 12 cells/second and is able to follow time dependent variations in cell flux with periods of 3 seconds or longer. RBC flux, defined as the number of cells passing through a capillary per time, provides a qualitatively different measure of capillary flow from that provided by the linear velocity of cells. A divergence between changes in velocity and flow reflects a change in capillary hematocrit and/or in capillary diameter. Lineal cell density is calculated as RBC flux over RBC velocity, and can be taken as an index of capillary hematocrit if capillary diameter is assumed constant. Direct determination of path length and transit time of RBC is also possible and requires the reconstruction of capillary flow pathways *in vivo.* A technique has now been developed for this purpose and is based on computer-aided, three-dimensional tracing of capillaries from video-recorded optical sections of the microvascular network (Hudetz *et al,* 1993).

RESULTS

Spatial and Temporal Pattern of Capillary Perfusion

The most striking characteristics of the cerebrocortical capillary circulation are the irregular, tortuous course of capillaries and the remarkably high velocity of red blood cells. Viewing from the surface of the brain, capillaries appear to have no preferential orientation or characteristic length. The velocity of RBC in capillaries is typically around 1 mm/s and ranges from about 0.3 to 3.2 mm/s (Figure 1). 65% of all velocities fall in the range of 0.5 to 1.8 mm/s. The significance of this heterogeneity in flow velocity for the effectiveness of oxygen transport to tissue is not yet known. The direction of flow and the average rate of RBC perfusion are stable in most capillaries suggesting that at least part of the heterogeneity is due to the vascular architecture. The velocity in arteriolar capillaries is 2 to 6 times higher than in venular capillaries of similar size. However, most of the heterogeneity in flow velocity is present in capillaries of the same branching order.

Under normal physiological conditions, almost all capillaries appear to be perfused by both plasma and red blood cells. Despite the heterogeneity of flow among neighboring capillaries, some degree of RBC perfusion is present in every capillary segment. Stochastic fluctuations in RBC velocity are present in all capillaries. In addition, in a small fraction of capillaries RBC perfusion may stop for brief periods not longer than a few seconds. Under normal conditions, these flow cessations are rare and appear to occur at random. Periodic opening and closing (cycling) of capillaries is not found by intravital ob-

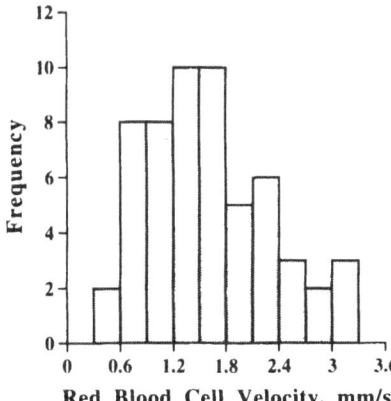

Figure 1. Frequency distribution of red blood cell velocity measured in cerebrocortical capillaries at normal perfusion pressure. Included in the histogram are data from 57 capillaries of 12 experiments. Reproduced with permission from Hudetz et al, 1996.

servations. Spontaneous periodic fluctuations in RBC velocity occur, however, when the perfusion pressure is decreased by systemic hypotension, cerebral artery occlusion or cerebral vasoconstriction elicited pharmacologically (Figure 2). Velocity oscillations in the frequency range of 4–12 cycles per minute are particularly interesting because these have no correlation with cardiac or respiratory cycles and are too slow to be caused by microrheological effects. In addition, the oscillations in RBC velocity are synchronous in neighboring capillaries of the cerebral cortex (Biswal and Hudetz, 1996).

Path length and transit time for RBC and plasma are important determinants of capillary gas and nutrient exchange. A longer transit time would allow more complete deoxygenation of blood, while a shorter transit time would enhance the diffusion gradient for oxygen from blood to tissue. Intravital microscopic observations yield RBC transit times between 100 and 300 ms and flow path lengths between 150 and 500 μm in the same capillary network (Hudetz et al, 1994). This short transit time of RBC may be in part due to the "network Fahraeus effect" which refers to the phenomenon that red cells preferentially follow the pathways of fast flow and leave more plasma to flow along the slower path-

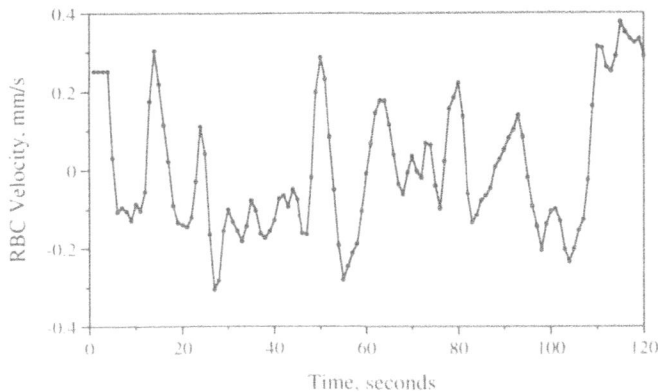

Figure 2. Spontaneous variations in RBC velocity obtained from a capillary after inhibition of nitric oxide synthase with L-NAME. The flow velocity was averaged over one-second intervals, normalized, and the mean of the data set was subtracted. The dominant frequency of velocity oscillation is approximately 4 cycles per minute. Modified with permisssion from Biswal and Hudetz, 1996.

ways. This is consistent with the finding of Wei et al (1993) that RBC transit time through the cerebral cortex is about 2.5-fold shorter than that of plasma.

Pressure Autoregulation

The general behavior of cerebrocortical capillary flow follows the autoregulation of CBF in face of changes in arterial pressure (Hudetz et al, 1995). When cerebral perfusion pressure is decreased by hemorrhagic hypotension or by an elevation of intracranial pressure, average red cell velocity is maintained as long as the perfusion pressure is above 60–70 mmHg, which coincides with the lower limit of CBF autoregulation. Autoregulatory maintenance of capillary flow velocity is equally effective during arterial hypotension and intracranial pressure elevation. RBC velocity is maintained in a slightly narrower range than RBC flux (the number of labeled cells traversing a capillary per time). This difference can be ascribed to either a slight collapse of capillaries or a decrease in hematocrit that accompanies hemorrhagic hypotension.

Although RBC velocity is "autoregulated" in most capillaries, a small fraction of capillaries demonstrates a different behavior. As shown in Figure 3, RBC velocity decreases with perfusion pressure in capillaries with resting flow velocity greater than 1 mm/s but not in those with resting velocity less than 1 mm/s. This difference indicates that the heterogeneity of RBC flow in cerebral capillaries varies with the perfusion pressure. Since the "autoregulating" capillaries are more numerous than the "non-autoregulating" capillaries, the overall response of RBC flow in the network is consistent with the normal autoregulatory pattern.

The capillaries with high resting flow and absent autoregulation are thought to represent functional thoroughfare channels (Hasegawa et al, 1967), while the capillaries with slow resting flow may represent the true "exchange" capillaries. The significance of this specialization is that the thoroughfare channels would provide a physiological reserve of flow which could be recruited when oxygen supply was challenged. Thus, the microcircu-

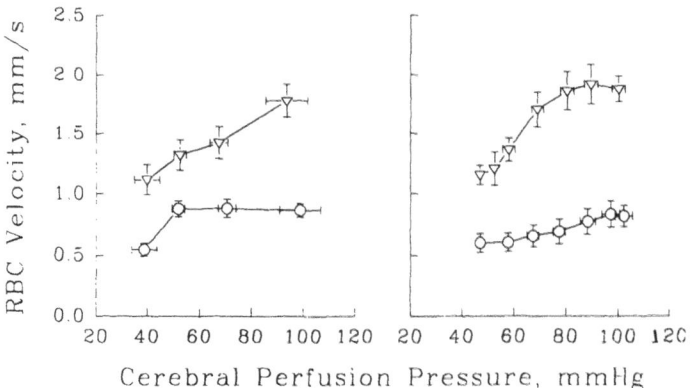

Figure 3. Dependence of red blood cell (RBC) flow velocity in cerebrocortical capillaries on cerebral perfusion pressure decreased by hemorrhagic hypotension (left) or intracranial pressure elevation (right). Cerebral perfusion pressure was calculated as mean arterial pressure minus intracranial pressure. Capillaries were sorted into two groups as a function of their resting velocity at normal perfusion pressure (triangles: fast flow capillaries, >1 mm/s; circles: slow flow capillaries, <1 mm/s). RBC velocity appears "autoregulated" in the slow flow (exchange) capillaries but not in the fast flow (thoroughfare) capillaries. Data shown are mean±SEM. Reproduced with permission from Hudetz et al, 1996.

latory response to a decrease in cerebral perfusion pressure would involve a redistribution of flow from the thoroughfare channels to the exchange capillaries. This may help to maintain a balanced perfusion and oxygen transport. The maintenance of flow velocity in the slower capillaries may also be important to protect the microcirculation from perfusion failure. This is because red cells fail to enter capillaries with slow flow (cell screening) and slow flow capillaries are more prone to plugging by leukocytes. A similar behavior of "functionally superperfused" and "adapted-flow" capillaries was suggested by Lübbers and Leniger-Follert (1978).

Hypoxia and Hypercapnia

Moderate hypoxia increases CBF and increases the velocity of flow in cerebral capillaries. We found that when the inspired O_2 is lowered to 15%, RBC velocity in individual capillaries increases by about 35%, which is consistent with the reported shortening of microvascular transit time, although at more severe hypoxia (Berecki et al, 1993). There was no visual evidence for opening of previously closed capillaries in response to hypoxia. Thus, the major mechanism by which regional CBF increases in hypoxia appears to be an increase in the velocity of flow in the capillaries.

The response of cerebral microcirculation to hypercapnia is also characterized by an increase in capillary flow velocity, but this appears to be more complex than the response to hypoxia. When the concentration of CO_2 in the inspired air is increased to 5%, RBC velocity increases by 31%. At an inspired CO_2 of 10%, the increase in RBC velocity from the normocapnic value becomes 41%. These changes in velocity are smaller than the corresponding changes in RBC perfusion as assessed by laser Doppler flowmetry at 69% and 128%, respectively. The relatively small increase in average RBC velocity during hypercapnia could indicate either shunting, capillary recruitment or an increase in capillary diameter. Intravital microscopic observations indicate no capillary recruitment during hypercapnia. However, significant increases in capillary diameter during hypercapnia have been detected using morphometric techniques (Duelli and Kuschinsky, 1993). In addition, we have preliminary evidence that the heterogeneity of capillary flow among neighboring capillaries is altered during severe hypercapnia such that RBC velocity increases in capillaries with low baseline velocity and decreases with high baseline velocity. These changes imply that RBC perfusion becomes more homogeneous during hypercapnia. A similar behavior of capillary plasma perfusion has recently been demonstrated (Abounader et al, 1995).

Hemodilution

Isovolemic hemodilution increases CBF dramatically, and this is thought to be due to either a direct effect of decreasing blood viscosity or to a vasodilator response to anemic hypoxia. Figure 4 shows that capillary RBC velocity increases greatly with a decrease in systemic hematocrit at maintained mean arterial pressure. Also shown there are the results from another series of experiments in which isovolemic hemodilution was preceded by bilateral occlusion of the carotid arteries in order to decrease the cerebrovascular vasodilator capacity. As shown there, the RBC velocity response to hemodilution was preserved after carotid occlusion, however, the critical hematocrit at which RBC velocity began to increase was shifted to the left, to lower hematocrit values. This finding suggests that the hyperemic response to moderate hemodilution was dependent on cerebral vasodilatation rather than a decrease in blood viscosity.

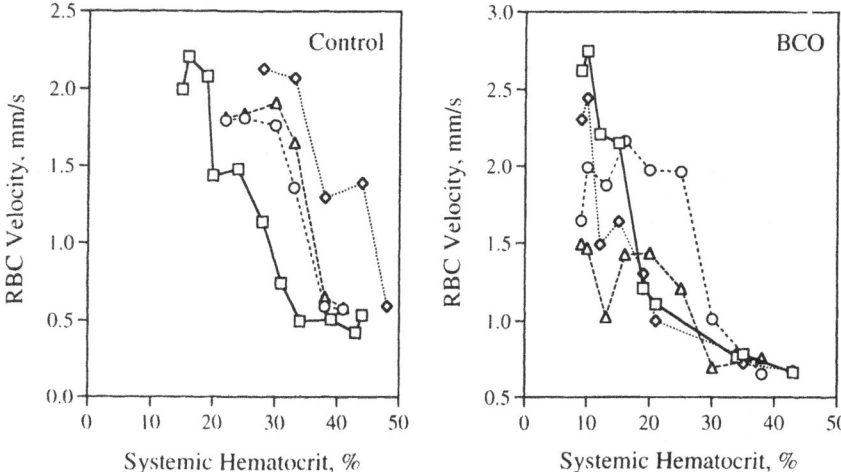

Figure 4. Effect of isovolemic hemodilution on red blood cell (RBC) velocity in cerebral capillaries in control (left) and after bilateral carotid occlusion (BCO, right). Note that after BCO, the curves are shifted to the left. Data shown are from measurements from individual capillaries from five experiments.

A similar increase is observed in capillary RBC flux (Figure 5) suggesting that a certain degree of compensation in convective oxygen transport in the face of hemodilution has occurred. Interestingly, the lineal density of RBC in capillaries (estimated as the ratio of fluorescent RBC flux and velocity) does not decrease at equal proportion to the decrease in systemic hematocrit. Thus, the direct effect of hematocrit on flow resistance in microvessels is probably small suggesting that a vasodilator mechanism rather than a hemorheological one may play the important role in the regulation of capillary flow and oxygen supply during anemic hypoxia.

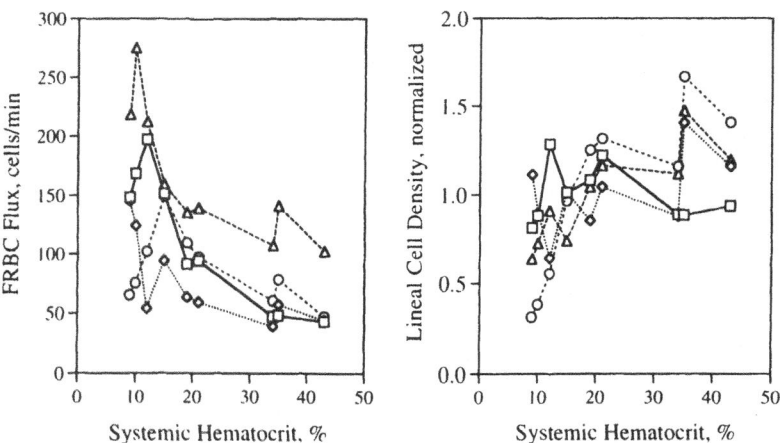

Figure 5. Effect of isovolemic hemodilution on fluorescent red blood cell (FRBC) flux (left) and lineal cell density (right) of cerebral capillaries. Note that cell density drops with systemic hematocrit to a moderate degree. Data are from several capillaries of one representative experiment.

Mechanisms of Capillary Flow Regulation

Our findings indicate that the regulation of CBF by cerebral arteries is not entirely complete and additional physiological mechanisms may participate in the regulation of capillary perfusion and oxygen supply in the brain. The differential behavior of RBC flow in the postulated thoroughfare and exchange capillaries is of particular interest. It is unlikely that rheological factors would play a major role in the capillary flow response because these factors would impede cell flow in capillaries with low velocity before affecting those with high velocity, just opposite to the changes observed in our autoregulation experiments. It is therefore postulated that cerebral capillary flow distribution is physiologically regulated by contractile cells of the microvascular bed. Smooth muscle cells with potential regulatory functions exist in precapillary areas of terminal arterioles and may regulate red cell entry into capillaries. Endothelial or pericyte contraction and relaxation may also be involved. Actin and myosin is present in brain capillaries, and endothelial cells can contract in response to mechanical, electrical and pharmacological stimulation. Retinal pericytes also contract or relax in response to a number of hormones and endothelial derived mediators such as angiotensin II, endothelin, serotonin, thromboxane A_2, prostacyclin and nitric oxide (NO) (Haefliger et al, 1994).

The next major challenge is to identify the sensing mechanism of capillary flow and/or tissue oxygen tension and the mediators of the postulated microvascular regulation. The cerebral microvascular response to hypoxia is brought about by a physiological mechanism that appears to depend on NO, the neuronal messenger and mediator thought to play an important role in neuro-vascular coupling mechanisms. We have preliminary evidence that specific inhibition of neuronal nitric oxide synthase (nNOS) abolishes the hypoxia-induced increase in RBC velocity in cerebral capillaries (Hudetz et al, 1996a). A similar result has been obtained after nNOS inhibition with respect to the capillary flow response to moderate hemodilution (Hudetz et al, 1996b).

SUMMARY

The dynamics and regulation of red blood cell flow in the cerebral microcirculation was studied by intravital fluorescence video-microscopy in a closed cranial window preparation in the rat. The studies revealed that capillary perfusion in the brain is essentially continuous but a stationary difference from capillary to capillary within the same microvascular network exists. The main mechanism of an increase in flow in cerebral capillaries is an increase in linear velocity with no or minor role for classical capillary recruitment. While cyclic opening and closing of capillaries is not evident, low frequency oscillations in capillary flow velocity are present when perfusion or oxygen supply to tissue is challenged. In hypoxic hypoxia and moderate hypercapnia, RBC velocity increases in all capillaries while in severe hypercapnia, redistribution of RBC velocity in the capillary network occurs. Both systemic hypotension and severe hypercapnia are accompanied by an increase in the homogeneity of capillary flow; this change involves the redistribution of RBC flow between thoroughfare channels and exchange capillaries. Thoroughfare channels may thus provide a recruitable flow reserve in the cerebral microcirculation. The capillary flow response to hypoxic and anemic hypoxia depends on the activity neuronal nitric oxide synthase. These findings suggest the presence of a physiological regulatory mechanism of cerebral capillary red blood cell flow and oxygen supply which may involve neuronal nitric oxide as a mediator.

ACKNOWLEDGMENTS

The author appreciates the contributions of Gabriella Fehér, Bharat Biswal, James D. Wood and Liezl Cueto to the experimental studies on which this review was based. This work was supported in part by grants from the American Heart Association GIA-95009340 and the National Science Foundation BES-9411631.

REFERENCES

Abounader, R., Vogel, J., Kuschinsky, W. (1995), Patterns of capillary plasma perfusion in brains of conscious rats during normocapnia and hypercapnia. Circ. Res. 76:120–126.

Berecki, D., Wei, L., Otsuka, T., Acuff, V., Pettigrew, K., Patlak, C., Fenstermacher, J. (1993), Hypoxia increases velocity of blood flow through parenchymal microvascular systems in rat brain. J. Cereb. Blood Flow Metabol. 13:475–486.

Biswal, B., Hudetz, A.G. (1996), Synchronous oscillations in cerebrocortical capillary red blood cell velocity after nitric oxide synthase inhibition. Microvasc. Res. 21:1–12.

Duelli, R., Kuschinsky, W. (1993), Changes in brain capillary diameter during hypocapnia and hypercapnia. J. Cereb. Blood Flow Metabol. 13:1025–1028.

Haefliger, I.O., Zschauer, A., Anderson, D.A. (1994), Relaxation of retinal pericyte contractile tone through the nitric-oxide-cyclic guanosine monophosphate pathway. Invest. Opthalamol. Vis. Sci. 35:991–997.

Hasegawa, T., Ravens, JR., Toole, JF. (1967), Precapillary arteriovenosus anastomoses. "Thoroughfare channel" in the brain. Arch Neurol. 16:217–224.

Hudetz, A.G., Biswal, B., Kampine, J.P. (1996a), 7-Nitroindazole abolishes cerebral capillary flow response to hypoxia. FASEB J. 10:3123.

Hudetz, A.G. Biswal, B., Wood, J.D., Kampine, J.P. (1996b), Neuronal nitric oxide synthase inhibitor abolishes cerebral capillary flow response to moderate isovolemic hemodilution. Anesth. Analg. (in press).

Hudetz, A.G., Fehér, G., Kampine, J.P. (1996c), Heterogeneous autoregulation of cerebrocortical capillary flow: Evidence for functional thoroughfare channels? Microvasc. Res. 51:131–136.

Hudetz, A.G., Fehér, G., Knuese, D.E., Kampine, J.P. (1994), Erythrocyte flow heterogeneity in the cerebrocortical capillary network. Adv. Exp. Med. Biol. 345:633–642.

Hudetz, A.G., Fehér, G., Weigle, C.G.M., Knuese, D.E., Kampine, J.P. (1995), Video microscopy of cerebrocortical capillary flow: response to hypotension and intracranial hypertension. Am. J. Physiol. 268 (Heart Circ. Physiol. 37): H2202-H2210.

Hudetz, A.G., Greene, A.S., Fehér, G., Knuese, D.E., Cowley, A.W. (1993), Imaging system for three dimensional mapping of cerebrocortical capillary networks in vivo. Microvasc. Res. 46: 293–309.

Hudetz, A.G., Weigle, C.G.M., Fenoy, F.J., and Roman, R. (1992). Use of fluorescently labeled erythrocytes and digital cross-correlation for the measurement of flow velocity in the cerebrocortical microcirculation. Microvasc. Res. 43: 334–341.

Knuese, D.E., Fehér, G., Hudetz, A.G. (1994), Automated measurement of fluorescently labeled erythrocyte flux in cerebrocortical capillaries. Microvasc. Res. 47: 392–400.

Lübbers, D.W., and Leniger-Follert, E. (1978), Capillary flow in the brain cortex during changes in oxygen supply and state of activation. In: "Cerebral Vascular Smooth Muscle and its Control", Ciba. Fnd. Symp. 56:21–47.

Wei, L., Otsuka, T., Acuff, V., Berecki, D., Pettigrew, K., Patlak, C., Fenstermacher, J. (1993), The velocities of red cell and plasma flows through parenchymal microvessels of rat brain are decreased by pentobarbital. J. Cereb. Blood Flow Metabol. 13:487–497.

74

OPTICAL IMAGING AND MEASURING OF LOCAL HEMOGLOBIN CONCENTRATION AND OXYGENATION CHANGES DURING SOMATOSENSORY STIMULATION IN RAT CEREBRAL CORTEX

Masahito Nemoto,[1] Yasutomo Nomura,[2] Mamoru Tamura,[2] Chie Sato,[2] Kiyohiro Houkin,[1] and Hiroshi Abe[1]

[1]Department of Neurosurgery
Hokkaido University School of Medicine
[2]Biophysics Division
Research Institute for Electronic Science
Hokkaido University
Sapporo 060, Japan

1. INTRODUCTION

The dynamic changes in local hemoglobin concentration and oxygenation in the cerebral cortex closely correlate with neuronal activity through the changes of cerebral metabolism and blood flow, but their mechanisms are not understood in detail. The purpose of this study is to reveal the changes in local hemoglobin concentration and oxygenation associated with neuronal activity using the techniques of charge-coupled device (CCD) imaging and microspectrophotometry, and to investigate the origin of the intensity changes of reflected light (intrinsic signals) in the activated cortex from the viewpoint of hemoglobin concentration and oxygenation changes.

2. METHODS

2.1. Animal Preparation and Somatosensory Electrical Stimulation

Male Wistar rats (180–250 g) were initially anesthetized with 2% halothane. The left femoral vein and tail artery were cannulated for intravenous drug administration, blood pressure monitoring and blood gas analyses. The animals were tracheotomized, immobilized with continuous infusion of pancuronium bromide (2 mg/kg/h iv) and artificially

Oxygen Transport to Tissue XIX, edited by Harrison and Delpy
Plenum Press, New York, 1997

ventilated with 70% N2 and 30% O2. Anesthesia was maintained with continuous infusion of α-chloralose (20 mg/kg/h iv). The rats were mounted in a stereotaxic frame and rectal temperature was maintained at 37.0±1.0°C with a heating pad.

The left frontoparietal bone (5×7 mm) was thinned over the left hind-limb cortex using a dental drill under continuous cooling with physiological saline. After hemostasis of the bleeding from the coronal suture, the thin skull window was filled with jelly made of carboxymethyl-polymer to make the bone more translucent and a cover glass (φ 15 mm) was placed over the jelly.

The contralateral posterior tibial nerve was stimulated with two subcutaneous needle electrodes along the paw. Stimulation parameters were a 100 μsec pulse, 4 mA intensity, 5 Hz frequency, 2 (or 5) sec duration and a 7 minute intertrial interval.

2.2. Imaging Setup and Analysis

Rats were placed on the stage of a Zeiss Axiotron microscope with a cooled CCD (C4880, Hamamatsu Photonics, Hamamatsu, Japan). The image data were acquired at 10- or 14-bit depth and were 500(W) × 508(H) pixels in size. An objective lens with a magnification of 2.5× or 10× and long working distances was used. This system provided a spatial resolution of 9 μm over a 4.6 mm square field or 2.4 μm over a 1.2 mm square field. The cortex was epiilluminated with white light from a stabilized 150W halogen bulb (KL1500 electronic, Schott Glaswerke, Wiesbaden, Germany). A sequence of digitized CCD images was acquired through a band-pass filter of 586 nm (Hw 2 nm), 577 nm (Hw 2 nm) or 603 nm (Hw 10 nm) at a rate of 500 msec synchronized to respiration. The images were analyzed by pixel-by-pixel subtraction and division (Æ reflectance image), or division and logarithm (Æ absorbance image) to detect intrinsic signals originating primarily from hemoglobin concentration and oxygenation changes.

$$\Delta \text{ reflectance} = \{ \text{Is}(\lambda) - \text{Ir}(\lambda) \} / \text{Ir}(\lambda) \qquad (\lambda = 586, 577, 603 \text{ nm}) \qquad (1)$$

$$\Delta \text{absorbance} = \ln \text{Ir}(\lambda) / \text{Is}(\lambda) \qquad (\lambda = 586 \text{ nm}) \qquad (2)$$

where Is(λ) represents reflected light intensity during stimulation or post-stimulation, and Ir(λ) reflected light intensity in the prestimulus state.

2.3. Microspectrophotometric Measurement

Reflected light from the activated cortex (capillary bed) in which the intrinsic signals had been detected by optical imaging was then guided to a spectrophotometer (MCPD 2000, Otsuka Electronics, Osaka, Japan) through a 20× objective lens and a quartz light guide (φ 2 mm) inserted into the eyepiece segment of the microscope. A sequence of visible absorption spectra (500–650 nm) from the spot (φ 100 μm) were obtained at a rate of 500 msec during and after the stimulation. The flattening of the absorption spectra due to light scattering in the brain tissue was corrected by the following equation:

$$\text{Corrected Abs}(\lambda) = 0.5 \times \{ exp\, (1.025 \times Abs\,) - 1.463 \times exp\, (-0.89 \times Abs\,) \} \qquad (3)$$

where corrected Abs(λ) and Abs(λ) are corrected absorption and actual absorption at wavelength λ, respectively. This equation was determined based on the relationship between the absorber concentration and actual absorption in the scattering materials measured by time-resolved spectroscopy. Corrected Abs(λ) was analyzed by a multicomponent least-square curve fitting for quantification of oxyhemoglobin and deoxyhemoglobin by the following equation:

$$Abs(\lambda) = L(\lambda)\{ \varepsilon1 (\lambda) [\text{oxyhemoglobin}] + \varepsilon2 (\lambda) [\text{deoxyhemoglobin}] \qquad (4)$$

$$+ \varepsilon3 (\lambda) [\text{oxidized cytochrome aa3}] + \varepsilon4 (\lambda) [\text{reduced cytochrome aa3}]\} \qquad (5)$$

where $L(\lambda)$ and $\varepsilon1-4 (\lambda)$ represent mean light path length and the molar extinction coefficients of each component at wavelength λ (500–650 nm), respectively. Oxygen saturation of hemoglobin (SO_2) and the total hemoglobin concentration (THB) were calculated as follows:

$$SO_2 = [\text{oxyhemoglobin}] / ([\text{oxyhemoglobin}] + [\text{deoxyhemoglobin}]) \times 100(\%) \quad (5)$$

$$THB = ([\text{oxyhemoglobin}] + [\text{deoxyhemoglobin}])/(\text{the sum of all components}) \quad (6)$$

Finally SO_2 and THB were expressed as changes relative to the control value.

3. RESULTS

3.1. Optical Imaging

Optical imaging of intrinsic signals was performed in 7 rats according to the above protocol. Intrinsic signals were detected centered 2–2.5 mm lateral and 1.5–2 mm caudal to the bregma. Intrinsic signal responses at each wavelength exhibited consistent characteristics in the time course, magnitude and distribution with the exception of the first time trial in which we frequently found an over-response.

3.1.1. Intrinsic Signal Responses at 586, 577 and 603 nm. Intrinsic signals at 586 nm were consistently detected as a reflectance decrease or absorbance increase. Figure 1-A shows sequential images of the reflectance decrease at 586 nm in the 2-sec stimulation. The signals appeared 1.0–1.5 sec after the stimulus onset, peaked at 3.0–4.0 sec, and disappeared by 10.0 sec. Large signals primarily originated from dilated pial arteries, that is, parietal branches of the middle cerebral artery and anterior cerebral artery.

Figure 1-B shows sequential images of the reflectance decrease at 577 nm in the 2-sec stimulation. The time course of reflectance signals at 577 nm was similar to that at 586 nm, but the amplitude at 577 nm was far larger than that at 586 nm, especially in the capillary bed; as a result, the signals of arterial components became inconspicuous. The spatial extent of the reflectance-decrease signals was about 2 mm in diameter at the peak time. The amplitude was maximal near the center and diminished toward the periphery but reflectance-increase signals were not detected in the exposed cortex.

Figure 1-C shows sequential images of the reflectance increase at 603 nm in the 2-sec stimulation. Large reflectance-increase signals came from parietal branches of the superior cerebral vein, and appeared 3–3.5 sec after the stimulus onset, peaked at 4.0–4.5 sec and gradually disappeared.

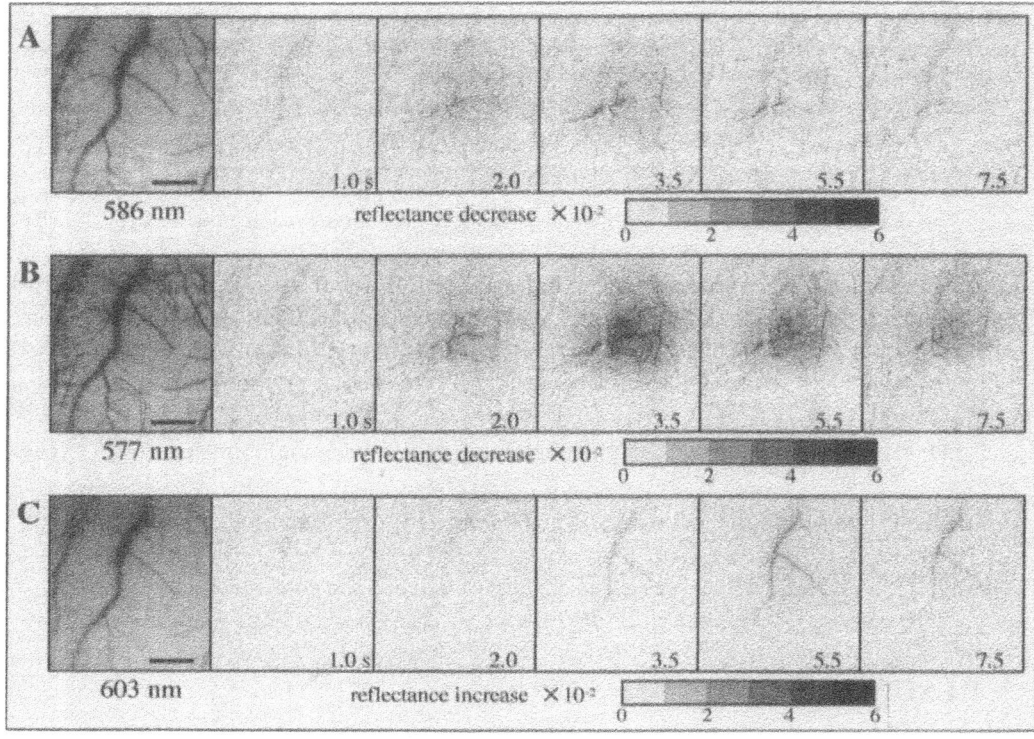

Figure 1. Typical time courses of intrinsic signal responses at wavelengths of 586, 577 and 603 nm in the 2-sec stimulation through a 2.5× objective lens (Scale bar = 1 mm).

3.1.2. High-Resolution Images of Intrinsic Signals. Figure 2 (A, B-left) shows the original images just before and after the 5-sec stimulation through a 10× objective lens. In 586 nm images the pial artery dilation from 30 μm to 45 μm can be recognized. In 603 nm images, it is difficult to discern pial arterioles but it is possible to observe the difference of light intensity at veins between the images before and after the stimulation (the cortical vein brightened after the stimulation). Figure 2-A shows sequential images of the absorbance increase at 586 nm in the 5-sec stimulation. The time course of absorbance-increase signals was similar to reflectance-decrease signals. The amplitude of signals in the 5-sec stimulation was larger than in the 2-sec stimulation and the time course was longer, but the spatial extent of signals in the 5-sec stimulation was almost the same as in the 2-sec stimulation. Figure 2-B shows sequential images of the reflectance increase and decrease at 603 nm. Large signals originating from the cortical vein appeared 3 sec after the stimulus onset, then became stronger during the stimulation. Large silhouettes in the reflectance-increase images arising from pial arteries were recognized after 2.0–3.0 sec and were detected as reflectance-decrease signals. Diffuse signals originating from the capillary bed showed a small reflectance decrease at 1.0–2.0 sec and a large reflectance increase after 3.0 sec.

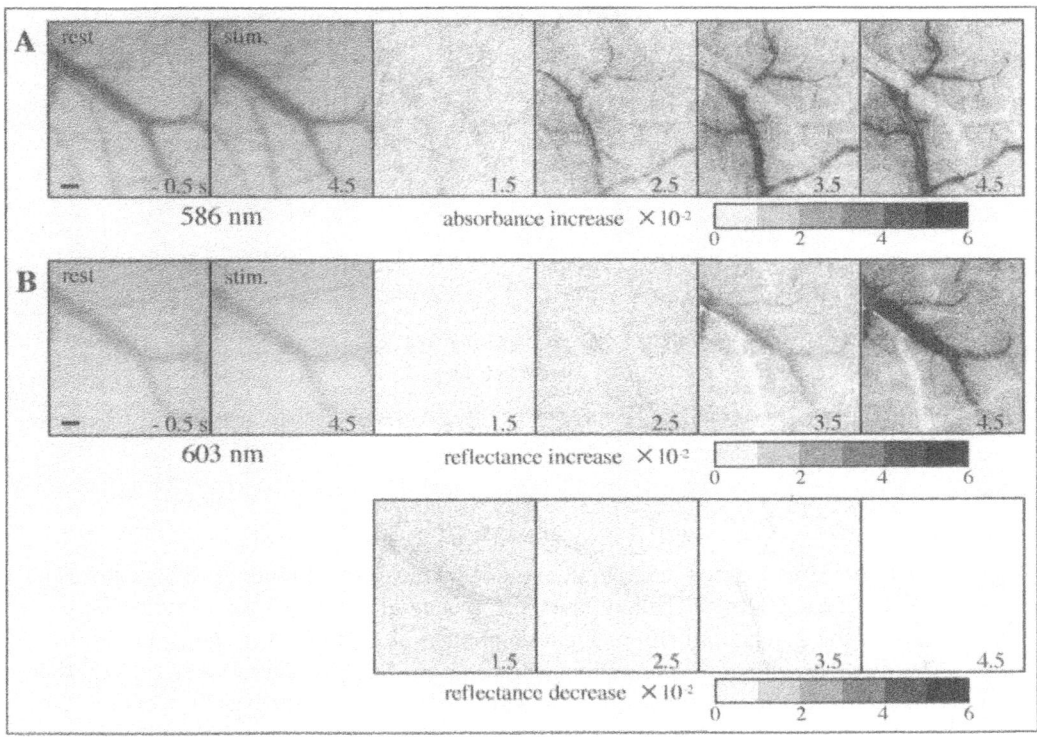

Figure 2. Typical time courses of intrinsic signal responses at wavelengths of 586 and 603 nm in the 5-sec stimulation through a 10× objective lens (Scale bar = 100 μm).

3.2. Microspectrophotometric Measurement

3.2.1. Activity-Related Changes in Absorption Spectra. The activity-related absorption spectra were obtained mainly from the capillary bed in which the intrinsic signals were detected by optical imaging. Figure 3 shows the absorption spectra for 10 sec in the 2-sec stimulation. The absorption in the wavelength range of 500–590 nm clearly increased after the stimulus onset. The absorption change at 590–650 nm was small. Figure 4 shows the time courses of absorption changes measured at three wavelengths (569, 577 and 586 nm). Error bars were calculated based on ten repeated measurements from the same capillary bed. The absorptions began to increase 1.5 sec after the stimulus onset, rose with a steep slope from 2.0 to 3.0 sec, peaked at 3.0–4.0 sec and returned to near the baseline level 8.0 sec after the stimulus onset.

Figure 5 shows the time courses of absorption changes at three other wavelengths (605, 610 and 630 nm). The absorption changes showed diphasic patterns: an increase in the early phase (with a peak at 2.0 sec) and a large undershoot in the late phase (with a negative peak at 4.0–5.0 sec) were observed. The amplitude of these absorption changes was about one-tenth as large as those at 569, 577 and 586 nm.

3.2.2. Hemoglobin Concentration and Oxygenation Changes. Each spectrum during and after the stimulation was corrected and the changes in hemoglobin concentration and

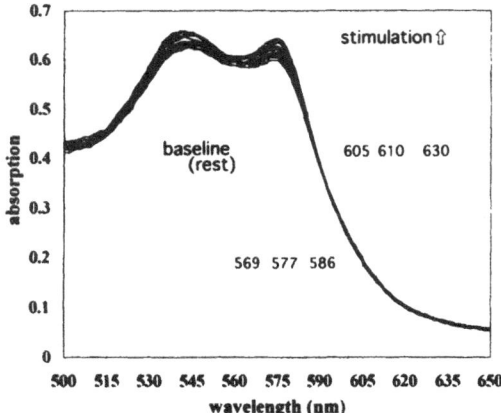

Figure 3. Absorption spectra of the capillary bed for 10 sec in the 2-sec stimulation.

oxygenation were calculated with multicomponent analysis. Figure 6 shows the time courses of the changes in oxyhemoglobin, deoxyhemoglobin and total hemoglobin concentrations in the 2-sec stimulation. Each component is expressed in the same arbitrary units. The oxyhemoglobin concentrations showed a tendency to decrease at 1.0 sec, then rapidly rose after 2.0 sec, peaked at 3.5 sec and returned to near the baseline level 8.0 sec after the stimulus onset. In contrast, deoxyhemoglobin slightly increased by 2.0 sec and decreased to a minimum value at 4.0 sec, then returned to near the original level 8.0 sec after the stimulus onset. The total hemoglobin, the sum of oxyhemoglobin and deoxy-hemoglobin, began to increase at 1.5 sec, peaked at 3.5 sec and returned to near the baseline level. Figure 7 shows the time course of oxygen saturation changes in the same stimulation. The saturation clearly increased after 2.0 sec, peaked at 3.5 sec and then returned to near the control value by 8.0 sec.

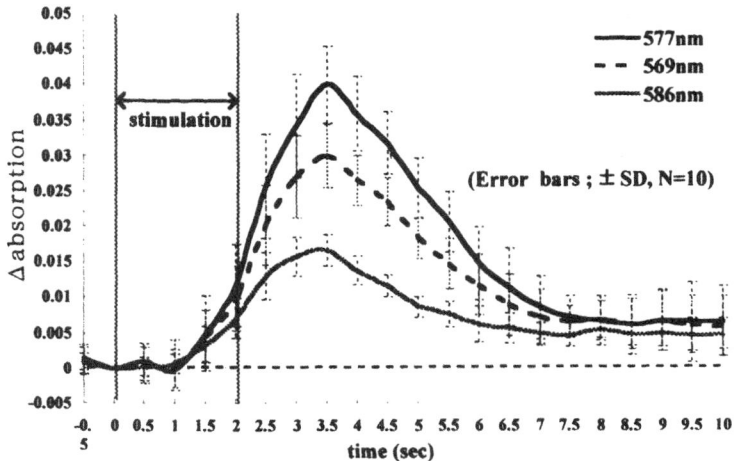

Figure 4. Typical time courses of absorption at wavelengths of 569, 577 and 586 nm in the 2-sec stimulation.

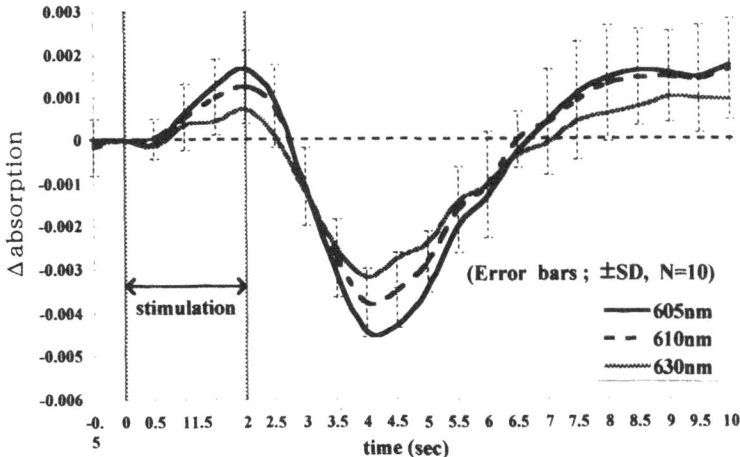

Figure 5. Typical time courses of absorption at wavelengths of 605, 610 and 630 nm in the 2-sec stimulation.

4. DISCUSSION

In this study we have shown the spatio-temporal responses of the activity-dependent intrinsic optical signals at wavelengths closely related to the hemoglobin concentration and oxygenation state, and the dynamic changes in hemoglobin concentration and oxygenation were measured by means of visible absorption spectra changes in the capillary bed. This study was designed to investigate the origin of intrinsic signals from the viewpoint of hemoglobin concentration and oxygenation changes, and to study local coupling between neuronal activity, metabolism and microcirculation. The higher resolution CCD images in this system (1 pixel represents a 2.4×2.4 µm region through a 10× objective) made it possible to determine whether the intrinsic signals originated from arterioles,

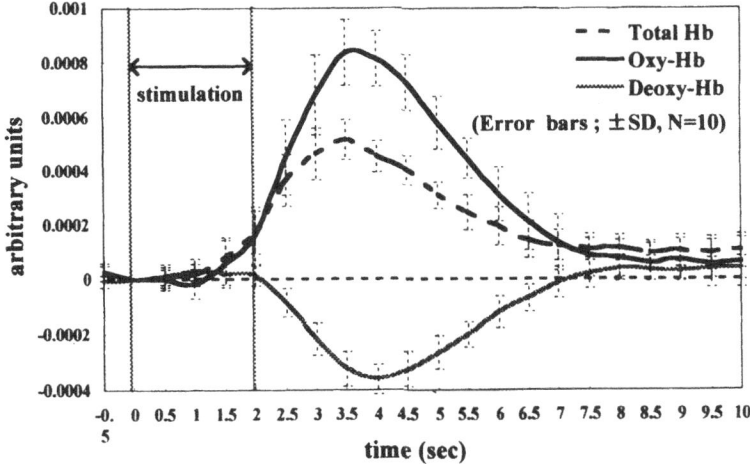

Figure 6. Typical time courses of the concentration changes in oxyhemoglobin, deoxyhemoglobin and total hemoglobin in the 2-sec stimulation.

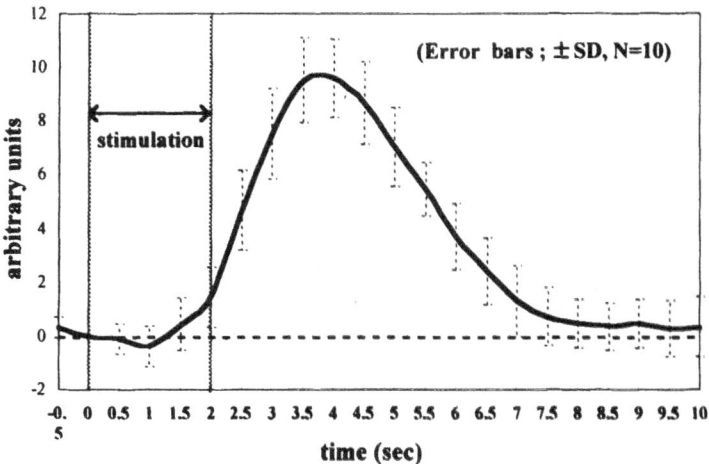

Figure 7. Typical time course of the hemoglobin oxygen saturation changes in the 2-sec stimulation.

venules or the capillary bed, and to measure the diameter changes of small vessels. The time courses of intrinsic signals in optical imaging corresponded to the findings of absorption changes with microspectrophotometry, by which quantitative analysis of the concentration and oxygenation state of hemoglobin and redox state of cytochrome oxidase is possible.

4.1. Optical Imaging

The three wavelengths of 586, 577, and 603 nm selected were sensitive to the concentration changes of total hemoglobin, oxyhemoglobin and deoxyhemoglobin, respectively. Optical imaging at 586 nm, an isobestic point of oxyhemoglobin and deoxyhemoglobin, reflects the concentration of total hemoglobin and the light scattering, not oxygenation. The large signals mainly originated from dilation of the arterioles, but it is unknown at present whether the small signals from the capillary bed indicate mean capillary recruitment. In these images at 586 nm, the macrovascular signals were frequently traced to the branches of not only the middle cerebral artery but also the anterior cerebral artery. This observation means that the primary sensory cortex of hind limb is anatomically located in the arterial border zone between the middle cerebral artery and anterior cerebral artery and is functionally supplied with blood from both of them. Thus, pial arteriolar responses near the activated cortex might be mediated in some part by diffusible substances.

Optical imaging at 577 nm, the peak wavelength of oxyhemoglobin, is greatly affected by the changes in the oxyhemoglobin concentration. Compared to Æ reflectance images at 586 nm, the images at 577 nm had high-amplitude signals originating from the capillary bed. This difference is explained by the changes in blood oxygenation in the capillaries: the regional blood flow increases over the oxygen consumption in the activated cortex, and the blood in the hyperoxygenation state flows in the capillary bed. The distribution of the reflectance decrease at 577 nm did not show any intracerebral blood steal phenomenon in this study.

Optical imaging at 603 nm is sensitive to changes in the deoxyhemoglobin concentration. The spatio-temporal aspects of Æ reflectance images at 603 nm were complicated, because the dilation of arterioles and capillary recruitment caused by the somatosensory stimulation reduced the reflectance at 603 nm, while the overcompensatory blood flow increase arising from the dilation of arterioles enlarged the reflectance in the capillary bed and draining veins because of the deoxyhemoglobin concentration decrease. Indeed, the reflectance of arterioles, the capillary bed and veins tended to decrease in the early phase. On the other hand, in the late phase, the reflectance of arterioles continued to decrease and that of the capillary bed and veins increased. Furthermore, the intrinsic signals of the superior cerebral vein were located in the lateral side of the vein. This observations might indicate the heterogeneity of blood oxygenation in the large vein.

4.2. Microspectrophotometric Measurement

The changes in absorption at wavelengths of 569 and 586 nm, isobestic points, reflect the changes in total hemoglobin concentration plus light scattering. The changes in absorption difference between the wavelengths of 569 and 586 nm predominantly correlated with the changes in total hemoglobin concentration. The time courses of these absorptions, and absorption differences revealed activity-related increases in the total hemoglobin concentration 1.5–8.0 sec after the onset of 2-sec electrical stimulation. The changes of absorption at 577 nm, as mentioned above, are sensitive to the concentration of oxyhemoglobin. The time course of absorption at 577 nm was similar to those at 569 and 586 nm, but the magnitude was larger than at 569 and 586 nm. These findings were compatible with the results of optical imaging in the capillary bed.

The changes in absorption at 605, 610 and 630 nm are greatly affected by the changes in the deoxyhemoglobin concentration. The time courses of these absorptions showed diphasic patterns, an increase in the early phase and a large undershoot in the late phase, which observations essentially coincided with previous reports (Grinvald et al 1986, 1991 and Narayan et al 1994, 1995). These changes were explained by changes in the hemoglobin oxygenation state or cerebral blood volume or the redox state of cytochrome oxidase (Frostig et al 1990, Turner et al 1994, Narayan et al 1995 and LaManna et al 1987). Recently an initial increase in deoxyhemoglobin concentration was corroborated in visual stimulation using imaging spectroscopy (Malonek et al 1996).

In this study we have investigated the origin of the intrinsic signals of the reflectance with spectrum correction and multicomponent analysis. The results suggested that (1) the time courses of the absorption decreases at 605, 610 and 630 nm in the late phase were similar to those of the deoxyhemoglobin concentration, and not the total hemoglobin concentration. These findings showed that the absorption decrease might represent overcompensatory hyper-oxygenation in the capillary bed but not cerebral blood volume reduction in our experiments. There is an obvious discrepancy between these results and Narayan's reports, which may possibly contribute to the difference in stimulation. In our parameters of electrical stimulation the dilator response of the resistance arteriole was so strong that local cerebral blood flow increased excessively and the blood oxygenation in the capillary bed developed significantly. (2) The time courses of the absorption increases at 605, 610 and 630 nm in the early phase were similar to those of the deoxyhemoglobin concentration by 1.0–1.5 sec after the onset of the stimulation, but might have been affected by the increase in the oxyhemoglobin concentration in addition to the changes in deoxyhemoglobin at about 2.0 sec. In other words, the absorption changes may be related to deoxygenation at about 1.0 sec, and related to both the oxygenation level and the total hemoglobin concentration at about 2.0 sec in our experiments. However, the amplitude of the early changes in absorption was so small and short that it was easily hidden by spontaneous oscillations and experimental artifacts. We have as yet insufficient data to shed light on the origin of the intrinsic signals in the early phase. Moreover, the exact time course of intrinsic signals is expected to be different according to the differences in animals, cortices activated, stimulus parameters and anesthetic state.

5. CONCLUSIONS

We have observed the time courses of activity-related intrinsic signals at visible wavelengths with a CCD imaging system and microspectrophotometry. The results confirmed that the hemoglobin concentration and oxygenation increased in the somatosensory hindlimb cortex during and after appropriate electrical stimulation of the posterior tibial nerve. However, the early changes in hemoglobin concentration and oxygenation require further investigation.

This microscopic system is simple and useful for studying local coupling between the neuronal activity, metabolism and microcirculation.

REFERENCES

1. Blood AJ, Narayan SM, Toga AW (1995) Stimulus parameters influence characteristics of optical intrinsic signal responses in somatosensory cortex. J Cereb Blood Flow Metab 15: 1109–1121
2. Cox SB, Woolsey TA, Rovainen CM (1993) Localized dynamic changes in cortical blood flow with whisker stimulation corresponds to matched vascular and neural architecture of rat barrels. J Cereb Blood Flow Metab 13: 899–913
3. Dirnagl U, Niwa K, Lindauer U, Villringer A (1994) Coupling of cerebral blood flow to neuronal activation: role of adenosine and nitric oxide. Am J Physiol 267: H296-H301
4. Fox PT, Raichle ME (1986) Focal physiological uncoupling of cerebral blood flow and oxidative metabolism during somatosensory stimulation in human subjects. Proc Natl Acad Sci USA 83: 1140–1144
5. Fox PT, Raichle ME, Mintun MA, Dence C (1988) Nonoxidative glucose consumption during focal physiologic neural activity. Science 241: 462–464
6. Frostig RD, Lieke EE, Ts'o DY, Grinvald A (1990) Cortical functional architecture and local coupling between neuronal activity and the microcirculation revealed by in vivo high-resolution optical imaging of intrinsic signals. Proc Natl Acad Sci USA 87: 6082–6086
8. Grinvald A, Lieke E, Frostig RD, Gilbert CD, Wiesel TN (1986) Functional architecture of cortex revealed by optical imaging of intrinsic signals. Nature 324: 361–364
9. Grinvald A, Frostig RD, Siegel RM, Bartfeld E (1991) High-resolution optical imaging of functional brain architecture in the awake monkey. Proc Natl Acad Sci USA 88: 11559–11563
10. Hoshi Y, Tamura M (1993) Detection of dynamic changes in cerebral oxygenation coupled to neuronal function during mental work in man. Neurosci Lett 150: 5–8
11. Kinuta Y, Kikuchi H, Ishikawa M, Hirai O, Imataka K, Kobayashi S (1987) Reflectance spectrophotometric measurement of in vivo local oxygen consumption in cerebral cortex. J Cereb Blood Flow Metab 7: 592–598
12. Kitai T, Tanaka A, Tokuka A, Tanaka K, Yamaoka Y, Ozawa K, Hirao K (1993) Quantitative detection of hemoglobin saturation in the liver with near-infrared spectroscopy. Hepatolgy 18: 926–936
13. LaManna JC, Sick TJ, Pikarsky SM, Rosenthal M (1987) Detection of an oxidizable fraction of cytochrome oxidase in intact rat brain. Am J Physiol 253: C477-C483
14. Lindauer U, Villringer A, Dirnagl U (1993) Characterization of CBF response to somatosensory stimulation: model and influence of anesthetics. Am J Physiol 264: H1223–1228
15. Malonek D, Grinvald A (1996) Interaction between electrical and cortical microcirculation revealed by imaging spectroscopy: implication for functional brain mapping. Science 272: 551–554
16. Masino SA, Kwon MC, Dory Y, Frostig RD (1993) Characterization of functional organization within rat barrel cortex using intrinsic signal optical imaging through a thinned skull. Proc Natl Acad Sci USA 90: 9998–10002

17. Menon RS, Ogawa S, Hu X, Strupp JP, Anderson P, Ugurbil K (1995) BOLD based functional MRI at 4 Tesla includes a capillary bed contribution: EPI imaging correlates with previous optical imaging using intrinsic signals. Magn Reson Med 33: 453–459

18. Narayan SM, Santori EM, Toga AW (1994a) Mapping functional activity in rodent cortex using optical intrinsic signals. Cereb Cortex 4(2): 195–204

19. Narayan SM, Esfahani P, Blood AJ, Sikkens L, Toga AW (1995) Functional increases in cerebral blood volume over somatosensory cortex. J Cereb Blood Flow Metab 15: 754–765

20. Ngai AC, Ko KR, Morii S, Winn HR (1988) Effect of sciatic nerve stimulation on pial arterioles in rats. Am J Physiol 254: H 133-H139

21. Ngai AC, Meno JR, Winn HR (1995) Simultaneous measurements of pial arteriolar diameter and Laser-Doppler flow during somatosensory stimulation. J Cereb Blood Flow Metab 15: 124–127

22. Sakatani K, Iizuka H, Young W (1990) Somatosensory evoked potentials in rat cerebral cortex before and after middle cerebral occlusion. Stroke 21:124–132

23. Seiyama A, Chen S, Imai T, Kosaka H, Shiga T (1994) Assessment of rate of O_2 release from single hepatic sinusoids of rat. Am J Physiol 267: H944-H951

24. Turner R, Grinvald A (1994) Direct visualization of patterns of deoxygenation in monkey cortical vasculature during functional brain activation. Proc Soc Magn Reson 1: 430

25. Watanabe M, Harada N, Kosaka H, Shiga T (1994) Intravital microreflectometry of individual pial vessels and capillary region of rat. J Cereb Blood Flow Metab 14: 75–84

75

INFLUENCE OF REPEATED ISCHAEMIA/REPERFUSION CYCLES (ISCHAEMIC PRECONDITIONING) ON HUMAN CALF ENERGY METABOLISM BY SIMULTANEOUS NEAR INFRARED SPECTROSCOPY, ^{31}P-NMR AND ^{23}Na-NMR MEASUREMENTS

T. Binzoni,[1] V. Quaresima,[2] E. Hiltbrand,[1] L. Gürke,[3] P. Cerretelli,[1] and M. Ferrari[2]

[1]Departments of Physiology and Radiology
University of Geneva
Switzerland
[2]Department of Biomedical Sciences and Technologies
University of L'Aquila
Italy
[3]Chirurgische Forschung
ZLF Kantonspital
Basel, Switzerland

1. INTRODUCTION

Ischaemic preconditioning (IP) (Lawson & Downey, 1993) with one or more cycles, consisting each of a short period of ischaemia followed by equal periods of reperfusion, improves myocardial tolerance to a subsequent ischaemic stress under experimental (Przyklenk et al., 1994) and clinical (Yellon et al., 1993) conditions. It has also been shown in rodents that IP improves ischaemic tolerance also in skeletal muscle (Gürke et al., 1996). So far, the effects of IP on human skeletal muscle are unknown.

The mechanism by which IP operates is still being investigated. Experimental studies evidenced the possible role of several substances like bradykinin, adenosine, nitric oxide, opioids and free radicals. The interrelation among these substances is unknown. However, it is well known that high energy phosphates decrease during ischaemia. This decrease could trigger the release of active substance and induce preconditioning. In addition, transient ischaemia induces tissue oedema with a consequent stretch of muscle fibers. It has been proposed that "[...] cellular stretch has the potential to provide a unified theory capable of linking and reconciling the different pathways that have been suggested to mediate ischaemic preconditioning" (Whittaker, 1996).

The aim of the present study was to investigate metabolic changes and the mechanism of development of oedema during preconditioning in human skeletal muscle. A quantitative assessment of oxidative and anaerobic (glycolytic) metabolism can be obtained combining near infrared (NIR) and ^{31}P-NMR spectroscopy. Tissue oedema may be investigated indirectly by ^{23}Na-NMR spectroscopy.

2. MATERIALS AND METHODS

Three periods of 5 min ischaemia of the thigh (with a counterpressure of 350 mmHg), followed by 5 min reperfusion, were induced in 8 healthy volunteers. The subject were laying supine in the magnet of a Picker modified (SMIS, Guilford, UK) NMR spectrometer (1.5 T). A 5 cm radio-frequency surface coil was positioned under the right calf. The two optodes of a NIRO500 (Hamamatsu Photonics, Japan) were positioned close to the coil. The length of the fibers was 6 m; the diameter of the active area of the optode was 3 mm; the interoptode distance was kept at 3 cm by a thermoplastic shell (Quaresima et al., 1995). ^{31}P-NMR spectra and NIR data were collected continuously with a resolution time of 96 s and 2 s, respectively. Creatine phosphate (PCr), ATP, pH and [Mg^{2+}] were obtained from ^{31}P-NMR spectra. Deoxyhaemoglobin (Hb) and oxyhaemoglobin (HbO$_2$) concentration changes (μM) were calculated using a differential pathlength factor of 5 cm (Duncan et al., 1995). Changes in total haemoglobin ([Hb$_{tot}$]=[HbO$_2$]+[Hb]) are strictly related to blood volume changes. Oxygen consumption ($\dot{O}_2$) was calculated during ischaemia from NIR data by a modified previously described method (De Blasi et al., 1993). O_2 values were expressed as percent of the value calculated during the first ischaemic period. The same protocol was performed on 3 subjects to acquire ^{23}Na-NMR spectra with a time resolution of 12.5 s.

3. RESULTS

In Figure 1 the typical time course of ATP, PCr and pH during the protocol is shown. No significant changes of PCr, ATP and [Mg^{2+}] were found during IP. A significant increase of pH (0.02 pH) was observed at the end of the experiment.

The typical time course of [Hb], [HbO$_2$] and [Hb$_{tot}$] changes during the protocol is described in Figure 2. As expected, a constant decrease of [HbO$_2$] was accompanied by an increase of [Hb]. The rapid [HbO$_2$] rise at the time of reperfusion is mainly due to the hyperhaemic response as evidenced by the [Hb$_{tot}$] tracing. [HbO$_2$] returned to the baseline 3–4 min after the cuff was released. A progressive rise of [Hb$_{tot}$] was found during each ischaemic period. A prompt rise of [Hb$_{tot}$] occurred when the cuff was released. A return to the baseline of [Hb$_{tot}$] was reached within 1–2 min when [HbO$_2$] was still recovering. The pattern of [Hb$_{tot}$] was variable amongst subjects. Therefore, it was necessary to adapt the $\dot{O}_2$ calculation method which requires a constancy of [Hb$_{tot}$]. Absolute values of [Hb] and [HbO$_2$] previously reported (Ferrari et al., 1995) were used to correct oxygenation changes following [Hb$_{tot}$] variations. No significant differences in $\dot{O}_2$ were found between the three ischaemic periods in all subjects. A typical time course of the total Na concentration during ischaemic preconditioning is shown in Figure 3. A continuous Na decrease was found

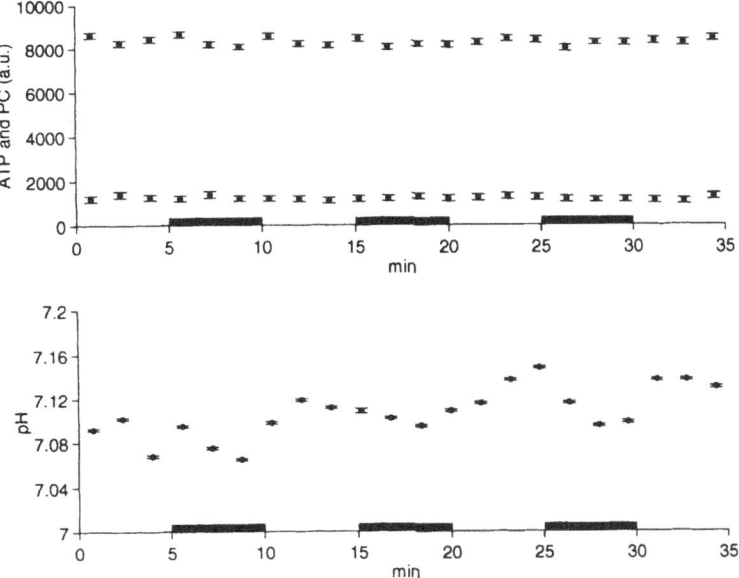

Figure 1. ^{31}P-NMR data during repeated 5 min human calf ischaemia followed by reperfusion of a sedentary subject. Top panel: PCr (upper tracing) and ATP concentrations (arbitrary units, a.u.). Bottom panel: pH values. Horizontal black bars indicate the ischaemic period. Vertical bars represent the calculated error of the single measurement.

in all subjects throughout the IP protocol interrupted by a rapid Na increase during each reperfusion phase. The time course of Na increase is similar to that of Hb_{tot} rise found after the release of the cuff.

4. DISCUSSION

In this study a commonly used IP protocol was applied to investigate the early metabolic and oxygenation changes of the human gastrocnemius.

No differences in high energy phosphates concentration were found in humans (Figure 1) confirming earlier results obtained in rat skeletal muscles (Gürke et al., 1996). By contrast, the same protocol performed on the heart of laboratory animals showed a consistent decrease of high energy phosphates. This difference may be explained by the very low metabolism at rest and the large oxygen stores of skeletal muscle.

The constancy of PCr and the lack of pH reduction suggest that anaerobic glycolysis (lactic and alactic) is not taking place during this protocol. Aerobic glycolysis is the main contributor to the energy metabolism as clearly shown by NIRS. In fact, as indicated in Figure 2, a constant desaturation occurred during the 3 ischaemic periods. The desaturation pattern of the calf is different from that found in the forearm during arterial occlusion (De Blasi et al., 1993). In the forearm [Hb] and [HbO_2] reached a plateau in about 4 min, suggesting an early occurrence of anaerobic metabolism. In the gastrocnemious $\dot{O}_2$ was constant during the 3 ischaemic cycles.

No [Hb_{tot}] changes were found during forearm ischaemia, whereas [Hb_{tot}] variations were found during calf occlusion. The high vascularization and the mass of the calf may

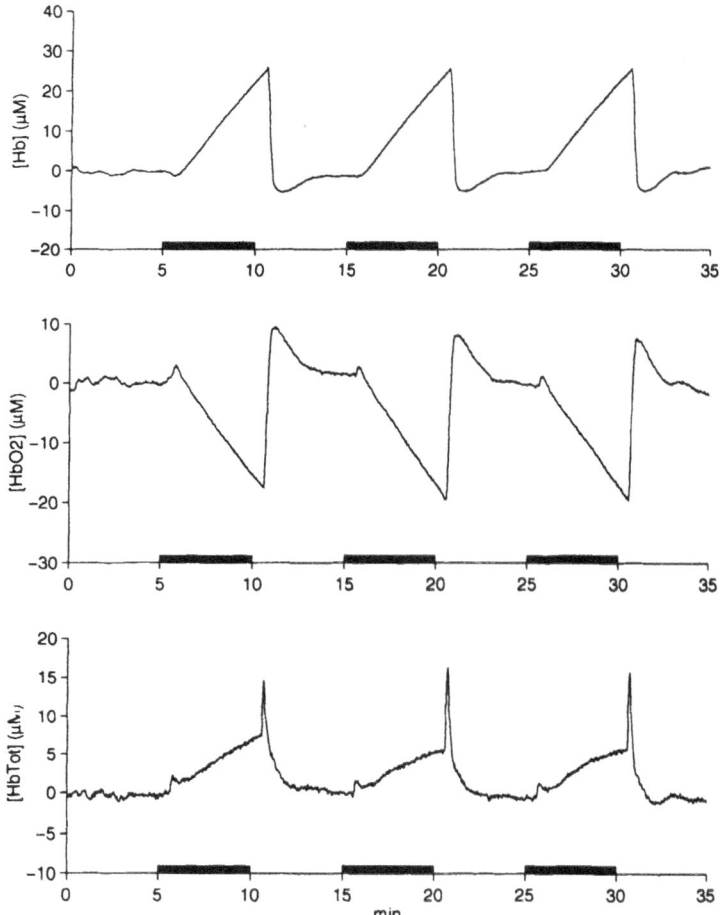

Figure 2. [Hb], [HbO$_2$] and [Hb$_{tot}$] changes during repeated 5 min human calf ischaemia followed by reperfusion of a sedentary subject. Horizontal black bars indicate the ischaemic period.

explain the blood volume shifts in the optical field. Hb$_{tot}$ recovered to the baseline values at the end of the protocol.

The slight rise of pH can be attributed to a small decrease of muscle temperature (~2 °C). On the other hand, the constancy of [Mg^{2+}], calculated from ^{31}P-NMR spectra, suggests that this ion is probably not involved in the changes of excitation-contraction properties observed after preconditioning.

The novel finding of this study is the continous Na decrease found in all subjects during IP. The passage of water and solutes from the intravascular compartment into the interstitial space likely induces a change of sodium transverse relaxation time. This change may be due to a shorter T$_2$ of the Na ion in the gel-like structure of the interstitial space. Considering that at the end of the protocol a) [Hb$_{tot}$] was unchanged and b) the Na signal represents the total amount of the Na ion in the muscle, the possible mechanism of Na disappearance is its trapping in the interstitial space occurring during oedema formation. The change of the Na ion environment increases the fraction of the Na "NMR invisible".

This study shows for the first time that, in the absence of high energy phosphates changes, short periods of ischaemia and reperfusion, as used during IP, lead to a continu-

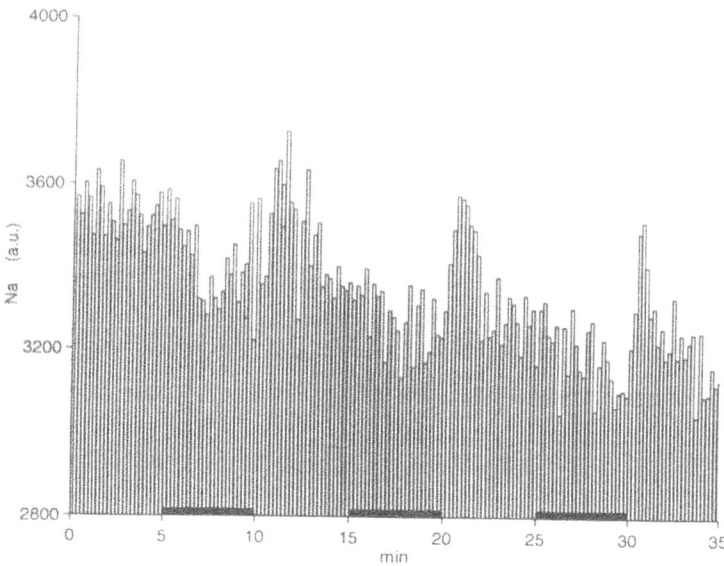

Figure 3. ^{23}Na-NMR data during repeated 5 min human calf ischaemia followed by reperfusion of a sedentary subject. Na concentration is expressed in arbitrary units (a.u.). Horizontal black bars indicate the ischaemic period.

ous reduction in "NMR visible" Na concentration. This reduction could be explained by interstitial oedema formation, leading to cellular stretch. The latter is known to improve ischaemic tolerance, as does IP. Consequently an oedema-induced cellular stretch may be at the basis of the beneficial effects of IP in skeletal muscle.

ACKNOWLEDGMENTS

We thank Dr C. Robino of Hamamatsu Italy for the loan of the NIRO500, the "Centro Interuniversitario Grandi Apparecchiature Biomediche nelle Neuroscienze" (University of Padova, Italy) and the Swiss National Foundation (#32-47075.96/1) for the financial support.

REFERENCES

De Blasi, R.A., Cope, M., Elwell, C., Safoue, F. & Ferrari, M. (1993). Non invasive measurement of human forearm oxygen consumption by near infrared spectroscopy. European Journal of Applied Physiology 67, 20–25.

Duncan, A., Whitlock, T.L., Cope, M. & Delpy, D.T. (1995). Measurement of changes in optical pathlength through human muscle during cuff occlusion on the arm. Optics and Laser Technology 27, 269–274.

Ferrari, M., De Blasi, R.A., Fantini, S., Franceschini, M.A., Barbieri, B., Quaresima, V. & Gratton, E. (1995). Cerebral and muscle oxygen saturation measurement by a frequency-domain near-infrared spectroscopic technique. Society of Photo-Optical Instrumentation Engineers Proceedings, SPIE, Bellingham, WA, USA. 2389, 868–874.

Gürke, L., Marx, A., Sutter, P.M., Frentzel, A., Harder, F., Seelig, J. & Heberer, M. (1996). Ischemic preconditioning - A new concept in orthopedic and reconstructive surgery. Journal of Surgical Research 61, 1–3.

Lawson, C. L. & Downey J. M. (1993). Preconditioning: state of the art myocardial protection. Cardiovascular Research 27, 542–550.

Przyklenk, K., Kloner, R.A. & Yellon, D.M. (1994). Ischemic preconditioning: The concept of endogenous cardioprotection. Kluwer Academic Publishers, Boston, Dordrecht, London.

Quaresima, V., De Blasi, R.A. & Ferrari, M. (1995). Customised optrode holder for clinical near-infra-red spectroscopy measurements. Medical & Biological Engineering & Computing 33, 627–628.

Whittaker, P. (1996). An alternative perspective on ischemic preconditioning derived from mathematical modeling. Basic Research in Cardiology 91, 47–49.

Yellon, D.M., Alkhulaifi, A.M. & Pugsley, W.B. (1993). Preconditioning the human myocardium. Lancet 342, 276–277.

76

NMR STUDIES OF HYPOXIC-ISCHAEMIC INJURY IN NEONATAL BRAIN USING IMAGING AND SPECTROSCOPY

R. Ordidge,[1] J. Thornton,[1] M. Clemence,[1] S. Punwani,[1] E. Cady,[1] J. Penrice,[2] P. Amess,[2] and J. Wyatt[2]

[1]Department of Medical Physics and Bioengineering
University College London, and
[2]Paediatric Department
University College London Hospitals
London, United Kingdom

1. INTRODUCTION

Brain injury arising in the perinatal period leads to death or permanent and severe impairment in more than 1000 infants per year in the UK (Alberman and Botting, 1991). The aim of this work is to use nuclear magnetic resonance (NMR) as a non-invasive measurement technique to investigate perinatal brain development in normal infants and those that have suffered hypoxic-ischaemic (H-I) brain injury.

Since 1982 the group has studied over 150 newborn infants with clinical evidence of acute perinatal asphyxia, using phosphorus (^{31}P) and proton (^{1}H) magnetic resonance spectroscopy (MRS). Marked increases in the cerebral concentrations of inorganic phosphate and lactate ([Pi] and [Lactate] respectively), and reductions in phosphocreatine concentration [PCr] and nucleotide triphosphate concentration [NTP] (mainly adenosine triphosphate, ATP) have been quantified within specified brain regions in asphyxiated infants, indicating failure of oxidative energy generation. In infants thought to have been subjected to an acute intrapartum insult there was usually a latent period of up to 15 hours before impairment of energy metabolism could be detected (Wyatt, 1994). This phenomenon was termed "secondary" or "delayed" energy failure. The delayed changes detected by MRS, especially the fall in [PCr]/[Pi] concentration ratio, were strongly predictive of death, or survival with reduced brain growth and severe neurodevelopmental disabilities, allowing the prognosis for long-term outcome to be assigned with reasonable precision, within a few days of birth (Azzopardi et al, 1989). The presence of a latent period before energy failure developed suggested that cerebroprotective interventions following resuscitation might have the potential to reduce or prevent delayed energy failure and the permanent brain injury with which it is associated.

Oxygen Transport to Tissue XIX, edited by Harrison and Delpy
Plenum Press, New York, 1997

Accumulating evidence points to a bi-phasic pattern of brain cell injury following a transient episode of severe hypoxia-ischaemia in the developing brain, with two distinct phases: early and delayed (Wyatt, 1994). The mechanisms underlying this pattern of injury remain unclear. Early neuronal injury is probably due largely to intracellular calcium toxicity, membrane failure and the effects of glutamate excitotoxicity (possibly mediated by the generation of nitric oxide). However, delayed cell injury appears to involve different mechanisms and there is some evidence to suggest that the process of programmed cell death (apoptosis) is involved (Rorke, 1982).

In order to test the efficacy of cerebroprotective strategies, a neonatal animal model has been developed which is capable of reproducing the development of delayed energy failure after resuscitation following an hypoxic-ischaemic episode. Using newborn piglets, the biphasic pattern of cerebral energy impairment has been reproduced following an acute reversed hypoxic-ischaemic insult (Lorek et al, 1994). The severity of delayed energy failure varies between individual animals, presumably due to biological variability in the metabolic and physiological responses to asphyxia. However, a strong relationship has been demonstrated between the severity of the acute energetic disturbance (derived from the time integral of NTP depletion) and the severity of delayed energy failure (particularly minimum [PCr]/[Pi]), indicating a clear dose-response effect (Lorek et al, 1994).

Diffusion-weighted MRI (DWI) has proven a valuable method of detecting acute cerebral ischaemia. Compared with MRS, DWI has the advantage that it provides high spatial resolution in the form of quantitative maps of the apparent diffusion coefficient (ADC) of brain water. The ADC value of brain water falls substantially during ischaemia (Knight et al, 1992). Although the mechanism for this change has not been unambiguously proven, it is thought to result from cell swelling. The redistribution of some water from the extracellular (high diffusion) environment to the intracellular (low diffusion) environment could change the average diffusion value, and might also reduce the diffusion coefficient in the reduced extracellular environment by increasing the barriers to water motion (the tortuosity of the water). Ionic homeostasis is normally preserved by the sodium/potassium pump and this function is responsible for a large proportion of cerebral energy consumption. During ischaemia, high energy phosphates are depleted leading to cell swelling through an osmotically driven influx of water in to cells associated with an increase in intracellular sodium concentration. Therefore, during ischaemia it is anticipated that changes in both ADC values and the concentrations of high energy phosphates, as measured by ^{31}P MRS, should occur almost simultaneously. However, such a relationship does not necessarily exist during the process of secondary energy failure that occurs in the re-oxygenated brain following H-I injury.

The purpose of this study was to investigate the temporal relationship between changes in relative high energy phosphate concentrations and ADC changes during secondary energy failure, with the goal of establishing DWI as a high resolution technique for monitoring cerebral hypoxic-ischaemic brain injury.

2. METHODS

Experiments were performed on healthy term Large white piglets, aged less than 24 hours. Anaesthesia was induced by inhalation of 5% isoflurane followed by tracheostomy and ventilation with a mixture of nitrous oxide, oxygen and isoflurane (<1.5%). Inflatable cuffs were positioned around both common carotid arteries. The anaesthetized animal was then positioned prone on a temperature-regulated water-filled mattress in a cylindrical perspex pod, designed to fit snugly into the bore of the magnet. The head was constrained using a stereotactic system and the pod inserted into the magnet. Heart rate, blood pressure, rectal and tympanic temperatures were monitored continuously.

Transient cerebral hypoxia-ischaemia was induced by reducing the inspired oxygen to 12% and occluding both common carotid arteries until severe depletion of high energy phosphates was detected by ^{31}P MRS. Animals were resuscitated by increasing the inspired oxygen to normalize arterial pO_2 and releasing the carotid occluders. The animals were maintained with full intensive care inside the magnet bore as observations continued throughout the next 2 days (Lorek et al, 1994).

MR measurements were performed using a 7 Tesla Bruker Biospec spectrometer system (Bruker, Karlsruhe, Germany). Data were acquired using a double-tuned (^{31}P and ^{1}H) 25mm diameter surface coil for radiofrequency (RF) transmission and reception. The coil was positioned on the intact scalp over the parietal lobes of the brain.

Fully relaxed ^{31}P pulse-acquire spectra were obtained with a relaxation delay of 10 seconds. Resonance peaks were measured by Lorentzian curve fitting. During the insult, energy status was monitored using 36 summed free induction decays (FIDs) and post-resuscitation 384 FIDs were summed to obtain a more accurate measurement.

ADC values were measured in a coronal plane using a diffusion-weighted stimulated echo acquisition (STEAM) sequence (Merboldt et al, 1991). The sequence timings were as follows: repetition time (TR) = 2 s, echo time (TE) = 34 ms, mixing time (TM) = 200 ms, acquisition time = 2.2 minutes per image. The slice thickness was 2.5 mm, the field of view was 3 cm and the image matrix of 128 x 64 was zero-filled to 128 x 128 to give an in-plane resolution of 0.23 mm.

Rectangular diffusion-sensitizing gradients of 10 ms duration were applied after the first and third RF pulses and gave "b-factors" of 0, 440, 900 and 1350 x 10^9 $m^2.s^{-1}$ for diffusion sensitization along the x and y directions and 0, 724, 1270 and 1780 x 10^9 $m^2.s^{-1}$ along the z axis after accounting for cross-terms arising from the slice selection (z) gradients. ADC maps were calculated using a log-linear fit algorithm. Average non-directional diffusion maps (ADC_{av}) were then obtained by averaging the ADC value from all three axes. This eliminates the confusing effects of diffusional anisotropy that is predominantly present in white matter. An average diffusion value for all brain tissue within the field of view of the surface coil was obtained by removing signal from cerebro-spinal fluid (CSF), scalp and low signal to noise ADC values, by use of an image segmentation algorithm. In this manner, an average non-directional ADC value was obtained that has been derived from approximately the same volume of brain tissue examined by ^{31}P MRS and allowed direct comparison between the two measurements.

3. RESULTS

Figure 1 shows a time series of ADC_{av} maps obtained in a single neonatal piglet following the H-I insult. The field of view of the surface coil extends to a depth of approximately 1cm into the piglet brain and is sufficient to encompass several gyri. The maps show a general pattern of ADC_{av} reduction (darkening of maps), with marked variations in time course between different anatomical regions. Cortical grey matter in the parasagittal regions show the earliest declines in ADC_{av} and are responsible for much of the differences seen in the map obtained 22 hours post-resuscitation. This early decline may reflect the varying sensitivity to damage of different brain regions and demonstrates the high regional sensitivity of DWI measurements. At 45 hours, ADC_{av} values from all brain regions were markedly reduced.

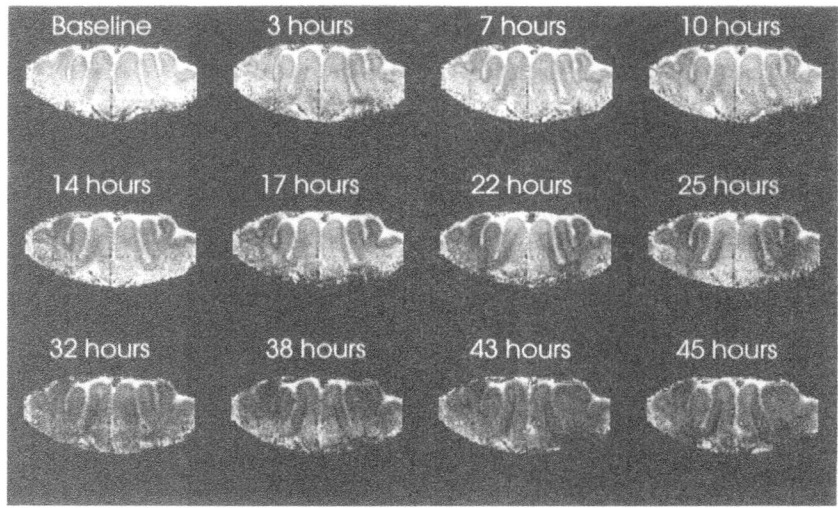

Figure 1. Time series of ADC$_{av}$ maps following H-I injury for a sinagle animal.

Figure 2 shows a comparison of the time course of [PCr]/[Pi] ratio and the ADC$_{av}$ value averaged over the whole brain section for a single animal. During the H-I insult, both values were markedly reduced, but recovered following resuscitation, which commenced at the origin of the X axis. The ADC$_{av}$ value then progressively declined during the subsequent secondary energy failure.

Figure 3 shows the ADC$_{av}$ values from 9 animals measured at different time points and plotted against the [PCr]/[Pi] index measured during the same experiments. A linear relationship between the two is suggested by a correlation coefficient of 0.876.

Figure 4 shows ADC$_{av}$ values averaged over the whole brain section at three time points for 6 animals. Control ADC$_{av}$ values were 0.94 ± 0.07 (s.d), compared with 0.88 ± 0.05 at 3 hours post-resuscitation, and 0.60 ± 0.02 at 46 hours post-resuscitation (x 10^{-9}

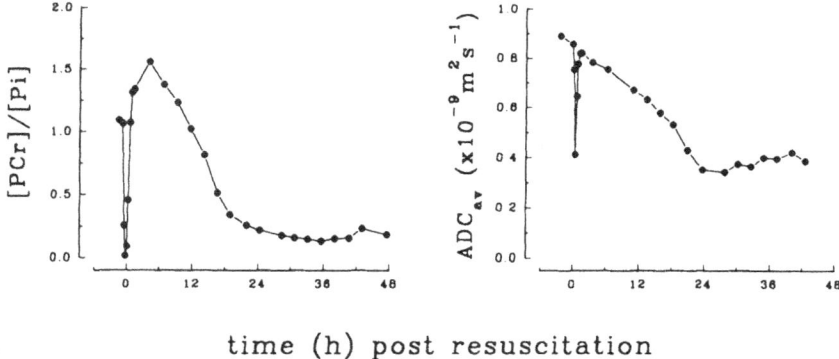

Figure 2. Whole brain slice ADC$_{av}$ and [PCr]/[Pi] ratios plotted versus time post-resuscitation following H-I injury for a single animal.

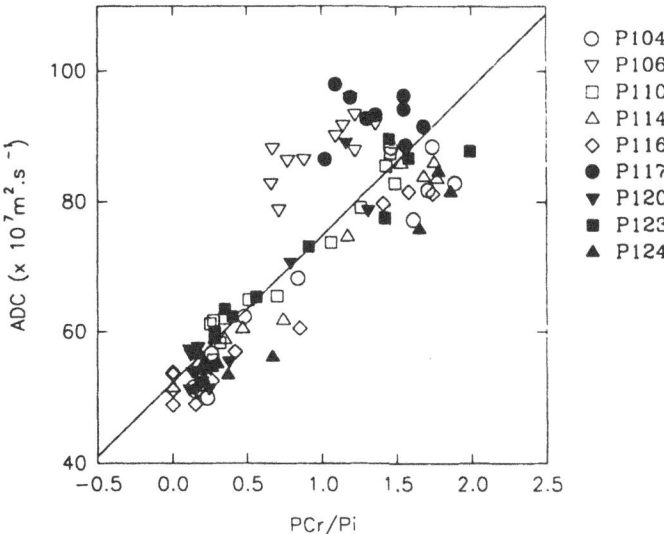

Figure 3. Whole brain slice ADC_{av} values plotted against [PCr]/[Pi] ratios (n=9).

$m^2.s^{-1}$). Although there appeared to be a slight difference between the control value and ADC_{av} measured at the 3 hour time point, this did not reach significance at the 5% level using Student's t-test. However, there were significant changes in ADC_{av} at both 24 hours (not shown) and 46 hours. The data suggests that there may be a slight decline in ADC_{av} in the acute period following H-I that could reflect the magnitude of early neuronal injury, but more data needs to acquired. This initial decrease, if present, could be of prognostic value in determining the eventual severity of H-I injury.

4. CONCLUSIONS

We have demonstrated a reduction in ADC values associated with a decline in energy metabolism following transient cerebral hypoxia-ischaemia in the newborn piglet.

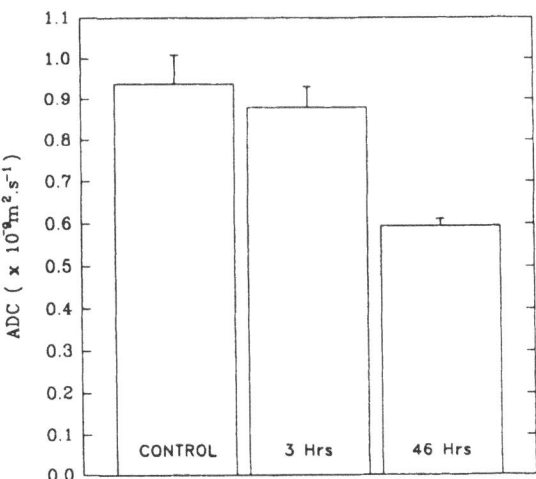

Figure 4. Whole brain slice ADC_{av} values for control brain and at 3 and 46 hours post-resuscitation following H-I injury (n=6).

ADC maps are capable of detecting variations in cerebral injury between brain regions at a spatial resolution that is not available using current ^{31}P MRS techniques. Changes in ADC values were found to be correlated with [PCr]/[Pi] ratio changes in this animal model, and may provide an alternative method of detecting delayed neuronal damage. ^{31}P MRS can only be performed in high field magnets (greater than 1.5 Tesla), and even at the field strength of 7 Tesla employed in the present study, necessitates a long examination time (approximately 40 minutes). The ADC value should be constant when measured at the different field strengths commonly employed in commercial MR scanners, which typically vary between 0.5 Tesla (or less) and 3 Tesla. However, both MRS and MRI procedures are non-invasive and allow repetitive measurements to be made, which may be necessary in determining the severity of H-I injury during the first 48 hours following birth. A potential problem with ADC measurements is the extreme sensitivity of the experiment to subject motion. This may be overcome by either using high-speed imaging techniques, such as echo-planar imaging, or by use of a motion-tracking and correction procedure (Ordidge et al, 1994). The ability to monitor delayed neuronal damage in this piglet model will be invaluable in assessing the efficacy of various proposed neuronal rescue strategies. Furthermore, the ability to perform the same measurements in infants should aid in patient selection and therapy monitoring when these procedures are applied to humans.

REFERENCES

Alberman ED, Botting B. Trends in prevalence and survival of very low birthweight infants, England and Wales: 1983–7. *Arch Dis Child* 1991;**66**:1304–1308.

Azzopardi D, Wyatt JS, Cady EB, Delpy DT, Baudin J, Stewart AL, Hope PL, Hamilton PA, Reynolds EOR. Prognosis of newborn infants with hypoxic-ischaemic brain injury assessed by phosphorus magnetic resonance spectroscopy. *Paediatr Res* 1989;**25**:445–451.

Knight RA, Ordidge RJ, Helpern JA, Chopp M, Rodolosi LC, Peck D. Temporal evaluation of ischaemic damage in rat brain measured by proton nuclear magnetic resonance imaging. *Stroke* 1992;**23**:576–582.

Lorek A, Takei Y, Cady EB, Wyatt JS, Penrice J, Edwards AD, Peebles DM, Wylezinska M, Owen-Reece H, Kirkbride V, Cooper CE, Aldridge RF, Roth SC, Brown G, Delpy DT, Reynolds EOR. Delayed cerebral energy failure following acute hypoxia-ischaemia in the newborn piglet: continuous 48-hour studies by phosphorus magnetic resonance spectroscopy. *Paediatr Res* 1994;**36**:699–706.

Merboldt KD, Hanicke W, Frahm J. Diffusion imaging using stimulated echoes. *Magn Reson Med* 1991;**19**:233–239.

Ordidge RJ, Helpern JA, Qing ZX, Knight RA, Nagesh V. Correction of motional artifacts in diffusion-weighted MR images using navigator echoes. *Mag Res Imag* 1994;**12**:455–460.

Rorke L. *Pathology of perinatal brain injury.* Raven Press, 1982.

Wyatt JS. Noninvasive assessment of cerebral oxidative metabolism in the human newborn. *J Roy Coll Phys Lond* 1994;**28**:126–132.

77

MAGNETIC RESONANCE SPECTROSCOPY IN DISORDERS OF O$_2$ DELIVERY AND UTILISATION

David G. Gadian[*]

Institute of Child Health
London, United Kingdom

1. INTRODUCTION

Magnetic resonance spectroscopy (MRS) provides a non-invasive method of investigating impaired oxidative metabolism, for example in ischaemia or in specific types of metabolic disease. This article provides a brief outline of the ways in which ^{31}P and ^{1}H MRS can be used to investigate impairments of oxidative metabolism in skeletal muscle and in brain.

2. ^{31}P MRS OF SKELETAL MUSCLE METABOLISM

^{31}P MRS studies of skeletal muscle featured prominently in the early demonstrations of NMR as a method of studying tissue metabolism, and in the progression from studies of isolated tissue to studies of man. There are several reasons for this. ^{31}P MRS provides a means of detecting ATP, phosphocreatine (PCr), and inorganic phosphate (P$_i$), all of which play key roles in muscle energetics; in addition to measuring the relative concentrations of these metabolites from the areas of their respective signals, it is also possible to measure the intracellular pH from the frequency of the P$_i$ signal, which is pH-sensitive in the physiological range. Series of spectra can be obtained over a period of rest, exercise and recovery, yielding a variety of biochemical parameters that can be used for the assessment of impaired energy metabolism. Also, for the majority of studies, adequate localization can be achieved by taking advantage of the localizing properties of surface coils, which were developed prior to more sophisticated methods of spectral localization. In addition, the first magnets that were available for clinical spectroscopy had a bore size which permitted studies of the human forearm, but they were not large enough (apart from neonatal studies) to accommodate the whole body.

* Address for correspondence: Professor David G. Gadian, Radiology and Physics Unit, Institute of Child Health, 30 Guilford Street, London WC1N 1EH, United Kingdom. Tel: 44 (0) 171 242 9789; Fax: 44 (0) 171 813 0399.

While a number of animal studies have been carried out, the advantages of spectroscopy are most apparent for investigations of exercise in man, where there are obvious benefits over biopsy procedures. These are illustrated by Figure 1, which shows four of a series of ^{31}P spectra obtained from forearm muscle at rest, during aerobic finger-flexing exercise, and during recovery. The spectra were obtained using a surface coil positioned over the finger flexor muscle. Figure 1(a) shows a typical resting spectrum, with signals from ATP, PCr and P$_i$. During aerobic exercise, PCr declines, P$_i$ increases (as seen in Figs. 1(b) and 1(c)) and there is a decline in intracellular pH from the resting value of about 7.0. The decline of PCr rather than of ATP reflects the known capacity of the creatine kinase reaction to maintain the ATP level during exercise. The recovery of PCr following exercise (Fig. 1(d)) is normally half completed in about a minute, but is slower when there is significant depletion of ATP during exercise (Taylor et al, 1983, 1986).

Studies of normal subjects, of the type illustrated in Figure 1, have provided fresh insights into several different aspects of muscle biochemistry and physiology (see Radda, 1990). They also give essential baseline information for studies of diseases involving impaired glycolytic or oxidative metabolism. For example, in patients with McArdle's syndrome (in which there is a glycogen phosphorylase deficiency), the muscle pH becomes alkaline on exercise rather than acid (Ross et al, 1981), which is totally consistent with the inability of these patients to generate lactic acid from glycogen. In other early studies, the muscle of a patient with phosphofructokinase deficiency not only became alkaline on exercise, but also accumulated sugar phosphate, as anticipated (Edwards et al, 1982). In two sisters with a mitochondrial disorder, recovery of phosphocreatine following exercise was very slow, demonstrating a reduced rate of oxidative metabolism (Radda et al, 1982). However, pH recovery was faster than in controls. Furthermore, in view of the high blood lactate levels that had been observed in one of the patients, the changes in intracellular pH on exercise were somewhat milder than might have been expected, suggesting that such patients might have an adaptive system for eliminating excess acid. Subsequent studies of the mitochondrial myopathies are consistent with these findings, commonly showing one or more of the following: (i) a reduced PCr/P$_i$ ratio at rest; (ii) a small decrease in pH dur-

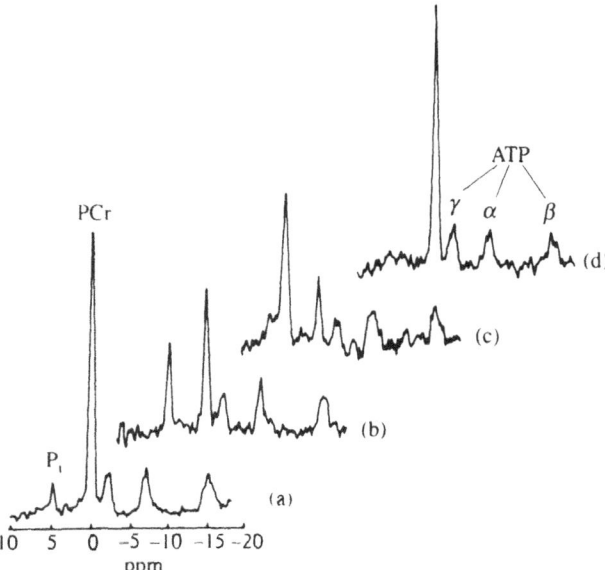

Figure 1. ^{31}P spectra obtained from forearm muscle before (a), during (b,c), and after (d) aerobic finger-flexing exercise. Signals are observed from ATP, phosphocreatine (PCr), and inorganic phosphate (P$_i$). (Reproduced with permission from *Magn. Reson. Med.* from Taylor et al, (1986) **3**, *Magn. Reson. Med.* 44–54.)

ing exercise; and (iii) slow PCr recovery (Matthews et al, 1991; Argov and Bank, 1991). Changes in any of these parameters can be used as an objective means of assessing response to therapy, the single-case study of Eleff et al (1984) providing one of the few examples of research in this area.

3. THE DEVELOPMENT OF ^{1}H MRS

MRS studies of skeletal muscle metabolism have largely exploited the ^{31}P nucleus, as did the initial studies of brain metabolism. However, the ^{1}H nucleus is now used more extensively than the ^{31}P nucleus for investigations of the brain. The development of ^{1}H NMR for metabolic studies lagged behind ^{31}P NMR for both technical and biochemical reasons. Technically, ^{1}H NMR is more complex than ^{31}P NMR because of the need to suppress the large signals from water and, in some cases, from fats, and because of the large number of metabolites that produce signals in a relatively narrow chemical shift range. However, techniques for solvent suppression and spectral 'editing' are now sufficiently well developed to permit the non-invasive monitoring of many metabolites of interest, provided that the field homogeneity is sufficently good. For the brain, excellent field homogeneity can often be obtained, so that even at the relatively low (by spectroscopists' standards) field strengths of 1.5–2.0T that are most commonly used for clinical spectroscopy, adequate spectral resolution can be achieved for many of the signals. In addition, the normal human brain (as opposed to the scalp and bone marrow) generates little, if any, signal from fats, and this facilitates the detection of the co-resonant signal from lactate. For other tissues, these problems are less easily overcome, which explains, at least in part, why most ^{1}H studies of tissue metabolism have focused on the brain.

From a biochemical viewpoint, ^{1}H NMR lagged behind ^{31}P studies because of the perceived strength of ^{31}P NMR in monitoring energy metabolism, as described above. However, it has become apparent that there are many disease states where ^{1}H spectroscopy might reveal abnormalities under circumstances where the ^{31}P spectra may appear normal.

A key feature of ^{1}H spectroscopy is the high sensitivity of the ^{1}H nucleus in comparison with other nuclei. In principle, this means that metabolites could be detected at relatively low concentrations. However, it is not necessarily straightforward to observe metabolites at low concentrations, as their signals may be masked by larger signals from other compounds that are present at higher concentrations. In practice, therefore, the higher sensitivity is generally exploited by trading signal-to-noise ratio with spatial resolution. The higher sensitivity of ^{1}H spectroscopy means that adequate signal-to-noise ratios can be obtained from smaller volume elements. For example the linear spatial resolution for clinical ^{1}H spectroscopy of the brain is typically 1–2 cm, which is about two-fold superior to the resolution achieved with ^{31}P spectroscopy.

Following earlier studies of metabolism in red blood cells (Brown et al, 1977), the first ^{1}H spectra of intact animals were reported by Behar et al, (1983). In their studies of anaesthetized rats, they were able to observe ^{1}H signals from a number of brain metabolites, including N-acetylaspartate (NAA), creatine + phosphocreatine (Cr), and choline-containing compounds (Cho), and they demonstrated an increase in the lactate signal on hypoxia, with a return to normal on subsequent oxygenation. Numerous technical developments followed, to the extent that many centres are routinely obtaining spectra of very high quality from well-defined localized regions of the human brain. In addition, an increasing number of groups are exploiting chemical shift imaging techniques to map the distribution of metabolites within selected planes of the brain, with spatial resolution of about 1 cm. The range of metabolite signals that have been unequivocally identified by ¹H MRS has widened considerably; one notable example is the detection of γ-aminobutyrate (GABA) both in normal subjects and, at elevated levels, in patients treated with the anticonvulsant vigabatrin, which is a specific inhibitor of GABA transaminase (Rothman et al, 1993). For the purposes of this article, however, we restrict our discussion primarily to the signals from N-acetylaspartate (at 2.01 ppm) and lactate (at 1.32 ppm). At its normal concentration of about 0.6 mmol/kg wet wt (Hanstock et al. 1988), lactate is barely, if at all, detectable in most ¹H spectra of the brain. However, several groups have shown the presence of large lactate signals associated with a variety of pathologies, including stroke and congenital lactic acidoses (see Figure 2).

The main interest in N-acetylaspartate centres around its potential use as a 'neuronal marker'. The function of this compound remains uncertain, but there is increasing evidence that, at least in the mature brain, it is present primarily within neurons (Moffett et al 1991; Urenjak et al 1993, and references therein), and so a loss of NAA is commonly interpreted in terms of neuronal loss or damage. In fact, a wide range of ¹H MRS studies have shown a reduction in signal intensity at 2.01 ppm that is entirely consistent with the neuronal/axonal loss or damage that might be anticipated on clinical grounds. This signal loss is commonly observed, for example, in stroke, gliomas, AIDS and epilepsy (see Gadian, 1995 for discussion). Some caution is necessary with this interpretation, however; for example the concentration of NAA increases in the developing brain (see Gadian, 1995), and this needs to be taken into account in the interpretation of spectra obtained from infants and young children. Nevertheless, it is reasonable to conclude that, at least in the mature brain, the observation of an abnormally low signal at 2.01 ppm will generally reflect a loss of NAA indicating neuronal (including axonal) loss or damage.

4. APPLICATIONS OF ¹H AND ³¹P MRS IN THE BRAIN

Birth asphyxia is the most common cause of impaired neurological function in full-term infants, and hypoxic-ischaemic injury is likely to be the main factor involved. ³¹P MRS of

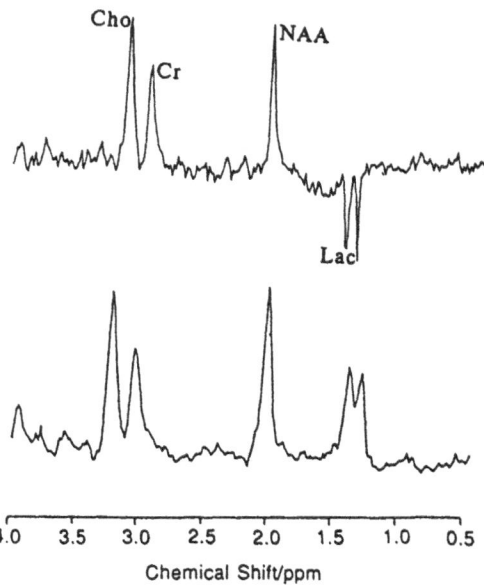

Figure 2. ¹H spectra obtained from two patients with disorders of oxidative metabolism, showing signals from N-acetylaspartate (NAA), creatine + phosphocreatine (Cr), choline-containing compounds (Cho), and lactate (Lac). Spectrum (a) was obtained with an echo time of 135 ms, under which conditions the lactate signal appears negative, while spectrum (b) was obtained with an echo time of 270 ms, under which conditions the lactate appears positive. (Spectrum (a) is reproduced with permission of Kluwer Academic Publishers, Lancaster, from Cross et al, (1993). *J. Inher. Metab. Dis.* **16**, 800–811, and spectrum (b) is reproduced with permission from Detre et al, (1991). *Ann. Neurol.* **29**, 218–221.)

neonates was started as soon as magnets of large enough bore size became available (Cady et al, 1983; Younkin et al, 1984), and its sensitivity to abnormalities of oxidative phosphorylation provides scope for investigating the mechanisms underlying hypoxic-ischaemic injury in newborns. In an extensive series of studies, it was shown that the PCr/P_i ratio provides a good prognostic indicator of the likely clinical outcome following birth asphyxia. PCr/P_i ratios within or close to the 95% confidence limits for normal controls were shown to be associated with normal outcome or with only minor impairment at 12 months, whereas low PCr/P_i ratios were associated with major neuromotor impairment (Azzopardi et al, 1989; Roth et al, 1992). The ratio of nucleoside triphosphates to the total NMR-visible phosphate pool was also related to outcome; values below the confidence limits for normal controls were almost always associated with fatal outcome. An important feature of sequential ^{31}P spectra obtained from severely affected infants is that an apparently normal spectrum (in terms of metabolite ratios) is commonly recorded soon after the birth asphyxia episode (8–17 h), but over the next few days the energy status declines. This raises the possibility that intervention during the intermediate 'normal' stage may be able to ameliorate the clinical outcome.

While the assessment of metabolite ratios, as in the above study, generally provides a sufficient basis for interpretation, it should be emphasized that this is not always the case. For example, the normality of metabolite ratios does not always imply that the high energy phosphate concentrations are normal throughout the area under investigation; it is possible that severely infarcted areas may produce no signal at all. Thus, decreases of up to 40% in the total ^{31}P MRS metabolite signals have been seen in chronic adult infarctions, with no accompanying changes in metabolite ratios or intracellular pH (Bottomley et al, 1986). This would be consistent with a substantial loss of viable brain cells, and serves to emphasize the increasing importance that is being placed on measuring absolute as well as relative concentrations.

An extensive series of ^{31}P studies have been carried out on patients with acute ischaemic stroke due to major cerebral vessel occlusion (Welch et al, 1988). All patients had focal neurological deficits, and they were serially studied at times ranging from 18 hours to 10–40 days after the onset of clinical deficit. Overall, there were distinct metabolic changes that were greatest during the acute or subacute (32–72 hours) stages. No significant abnormalities in high energy phosphates were observed beyond 7–9 days despite persistent neurological deficit and imaging evidence for infarction. It was suggested that, apart from partial recovery of some neurons, the return of high energy phosphates may also originate from glial cells or from macrophages that infiltrate the infarct. pH was initially acidotic, but there were subsequent alkalotic shifts which, it was suggested, may signify the end of active ischaemic cell metabolism and hence define the limit of a therapeutic window.

These ^{31}P studies of neonates and of adults suggest that ^{31}P MRS is of most value as a means of assessing acute damage, but that interpretation of later events is more complex, in part because gliosis and macrophage infiltration might lead to the reappearance of high energy phosphates in the spectra. It is clearly important to have a means of distinguishing the metabolic properties of different cell types, and in this respect ^{1}H MRS offers greater prospects. In particular, if, as discussed above, N-acetylaspartate is present in neurons but not in mature glial cells, then neuronal loss will result in a loss of the ^{1}H signal from NAA, and this signal would not recover with gliosis or macrophage infiltration. The initial ^{1}H MRS studies of stroke patients clearly showed a loss of NAA (Bruhn et al, 1989), and a range of subsequent studies have verified that infarction is indeed associated with loss of NAA, entirely consistent with expectation.

Of additional interest is the detection of lactate in ^{1}H spectra of stroke patients. One of the interesting observations has been the persistence of an elevated lactate signal for days or even months following stroke. In order to interpret this persistent elevation of lactate, it is important to determine its cellular localization, and also to establish whether it reflects continued lactate production or a pool of 'trapped' lactate. In one patient studied a month after a cortical stroke, the elevated lactate pool underwent complete metabolic turnover within 60 minutes, as indicated by the rate of incorporation of ^{13}C label from infused $1-^{13}C$-labelled glucose into lactate (Rothman et al, 1991). It would be surprising if this continued metabolism were associated with damaged brain cells so long after the onset of stroke. Indeed, a further study suggests that brain macrophages, which begin to appear three days after infarction and gradually disappear over several months, could be a major source of this persistently high lactate (Petroff et al, 1992). Taken together, all of these findings suggest that, while the *early* presence of lactate is presumably a reflection of abnormal brain cell metabolism, interpretation of its *continued* presence needs to take into account the changing cell populations associated with brain damage. More generally, these findings illustrate the point that interpretation of any spectral changes that are observed in diseased tissue requires an appreciation of the cell types that may contribute to such changes.

With the development of spectroscopic imaging techniques, it is now possible to generate images (albeit with relatively crude resolution when compared with conventional MRI) displaying the distribution of metabolites in selected planes within the brain (see Luyten et al, 1990; see also Nelson et al, 1991, for studies of skeletal muscle). In a recent study of 12 patients with acute middle cerebral artery stroke, Gillard et al (1996) showed that, in the early stages of stroke, they were able to detect elevated lactate in brain regions where no other spectroscopic or MR imaging abnormality was evident. It was concluded that such regions may represent an ischaemic zone at risk of infarction. These studies suggest a promising role for ^{1}H spectroscopic imaging in the selection and monitoring of therapeutic interventions for patients with acute stroke.

The detection of brain lactate is also of interest in investigations of the congenital lactic acidoses, which form a large group of disorders that are commonly associated with profound neurological dysfunction. Difficulties are frequently encountered in establishing a precise diagnosis, and the mechanisms underlying brain damage are poorly understood. As shown in Figure 2, ^{1}H MRS provides a means of monitoring elevated lactate in these disorders, and the technique can be used to examine regional variations in brain lactate (see, for example, Detre et al, 1991; Cross et al, 1993, 1994). Spectroscopic imaging investigations should be particularly useful in this respect. Such studies could help to elucidate the reasons why particular regions of the brain (e.g. the basal ganglia) are particularly susceptible to damage in disorders of lactate metabolism. MRS can also provide an objective means of assessing metabolic response to treatment. Recently, for example, the results of a short-term, double-blind placebo-controlled, crossover trial of sodium dichloroacetate (DCA) therapy were reported in 11 patients affected by various mitochondrial disorders (De Stefano et al, 1995). After one week of DCA treatment there were significant decreases in blood lactate, pyruvate, and alanine at rest and after bicycle exercise. ^{1}H spectra of the brain also showed significant changes in seven of the patients, with a decrease of the brain lactate/Cr ratio by approximately 40% during DCA treatment. However, ^{31}P spectra of skeletal muscle and self-assessed clinical disability remained unchanged. In spite of the latter finding, it is clear that MRS could play an important role in the development and evaluation of new forms of treatment.

5. CONCLUDING COMMENTS

These various spectroscopic observations complement the many ways in which MRI techniques can now be used for the investigation of disorders of O_2 delivery and utilisation. One MRI technique that is proving to be particularly important in the assessment of cerebral ischaemia is diffusion-weighted imaging, as discussed in the accompanying article by Ordidge. While the precise biophysical mechanisms underlying the changes in diffusion-weighted signal intensity remain uncertain, there is a strong consensus that the diffusion-weighted hyperintensity observed in acute stroke is associated with cell swelling. Thus diffusion-weighted MRI and MRS both provide a means of assessing early events in stroke; MRS provides a direct measure of impaired energy metabolism, while diffusion-weighted imaging probes the cytotoxic oedema associated with this impaired metabolism, and it does so with the fine spatial resolution of MRI, rather than the relatively crude resolution of MRS. It seems likely, therefore that the use of MRS in the assessment of cerebral ischaemia will be as part of a combined imaging/spectroscopy examination incorporating diffusion-weighted imaging as well as other MRI techniques such as magnetic resonance angiography. These techniques should together provide a powerful approach to defining the underlying pathophysiology of cerebral ischaemia, and to guiding and evaluating new forms of therapy.

REFERENCES

Argov, A. and Bank, W.J. (1991). Phosphorus magnetic resonance spectroscopy (^{31}P MRS) in neuromuscular disorders. *Ann. Neurol.* **30**, 90–97.

Azzopardi, D., Wyatt, J.S., Cady, E.B., Delpy, D.T., Baudin, J., Stewart, A.L., Hope, P.L., Hamilton, P.A. and Reynolds, E.O.R. (1989). Prognosis of newborn infants with hypoxic-ischaemic brain injury assessed by phosphorus magnetic resonance spectroscopy. *Pediatr. Res.* **25**, 445–451.

Behar, K.L., Den Hollander, J.A., Stromski, M.E., Ogino, T., Shulman, R.G., Petroff, O.A.C. and Prichard, J.W. (1983). High-resolution ^{1}H nuclear magnetic resonance study of cerebral hypoxia in vivo. *Proc. Natl. Acad. Sci. USA* **80**, 4945–4948.

Bottomley, P.A., Drayer, B.P., and Smith, L.S. (1986). Chronic adult cerebral infarction studied by phosphorus NMR spectroscopy. *Radiology* **160**, 763–766.

Brown, F.F., Campbell, I.D., Kuchel, P.W. and Rabenstein, D.C. (1977). Human erythrocyte metabolism studies by ^{1}H spin echo NMR. *FEBS Lett.* **82**, 12–16.

Bruhn H, Frahm J, Gyngell ML, Merboldt KD, Hanicke W and Sauter R. (1989). Cerebral metabolism in man after acute stroke: new observations using localised proton NMR spectroscopy. *Magn. Reson. Med.* **9**, 126–131.

Cady, E.B., Costello, A.M.deL., Dawson, M.J., Delpy, D.T., Hope, P.L., Reynolds, E.O.R., Tofts, P.S., and Wilkie, D.R. (1983). Non-invasive investigation of cerebral metabolism in newborn infants by phosphorus nuclear magnetic resonance spectroscopy. *Lancet* **i**, 1059–1062.

Cross, J.H., Gadian, D.G., Connelly, A. and Leonard, J.V. (1993). Proton magnetic resonance spectroscopy studies in lactic acidosis and mitochondrial disorders. *J. Inher. Metab. Dis.* **16**, 800–811.

Cross, J.H., Connelly, A., Gadian, D.G., Kendall, B.E., Brown, G.K., Brown, R.M. and Leonard, J.V. (1994). Clinical diversity of pyruvate dehydrogenase deficiency. *Ped. Neurol.* **10**, 276–283.

De Stefano, N., Matthews, P.M., Ford, B., Genge, A., Karpati, G., and Arnold, D.L. (1995). Short-term dichloroacetate treatment improves indices of cerebral metabolism in patients with mitochondrial disorders. *Neurology* **45**, 1193–1198.

Detre, J.A., Wang, Z., Bogdan, A.R., Gusnard, D.A., Bay, C.A., Bingham, P.M. and Zimmerman, R.A. (1991). Regional variation in brain lactate in Leigh syndrome by localized ^{1}H magnetic resonance spectrocopy. *Ann. Neurol.* **29**, 218–221.

Edwards, R.H.T., Dawson, M.J., Wilkie, D.R., Gordon, R.E., and Shaw, D. (1982). Clinical use of NMR in the investigation of myopathy. *Lancet* **i**, 725–731.

Eleff, S., Kennaway, N.G., Buist, N.R.M., Darley-Usmar, V.M., Capaldi, R.A., Bank, W.J., and Chance, B. (1984). ^{31}P NMR study of improvement in oxidative phosphorylation by vitamins K3 and C in a patient with a defect in electron transport at complex III in skeletal muscle. *Proc. Natl. Acad. Sci. USA* **81**, 3529–3533.

Gadian, D.G. (1995). NMR and its applications to living systems. 2nd Edition, Oxford University Press, Oxford.

Gillard, J.H., Barker, P.B., can Zijl, P.C.M., Bryan, R.N., and Oppenheimer, S.M. (1996). Proton MR spectroscopy in acute middle cerebral artery stroke. *Am. J. Neuroradiol.* **17**, 873–886.

Hanstock, C.C., Rothman, D.L., Prichard, J.W., Jue, T., and Shulman, R.G. (1988). Spatially localized ^{1}H NMR spectra of metabolites in the human brain. *Proc. Natl. Acad. Sci. USA* **85**, 1634–1636.

Luyten P.R., Marien, A.J.H., Heindel, W., van Gerwen, P.H.J., Herholz, K., den Hollander, J.A., Friedmann, G. and Heiss, W-D. (1990). Metabolic imaging of patients with intracranial tumours: H-1 MR spectroscopic imaging and PET. *Radiology* **176**, 791–799.

Matthews, P.M., Allaire, C., Shoubridge, E.A., Karpati, G., Carpenter, S. and Arnold, D.L. (1991). *In vivo* muscle magnetic resonance spectroscopy in the clinical investigation of mitochondrial disease. *Neurology* **41**, 114–120.

Moffett, J.R., Namboodiri, M.A., Cangro, C.B. and Neale, J.H. (1991). Immunohistochemical localization of N-acetylaspartate in rat brain. *NeuroReport* **2**, 131–134.

Nelson, S.J., Taylor, J.S., Vigneron, D.B., Murphy-Boesch, J. and Brown, T.R. (1991). Metabolite images of the human arm: changes in spatial and temporal distribution of high energy phosphates during exercise. *NMR Biomed.* **4**, 268–273.

Petroff, O.A.C., Graham, G.D., Blamire, A.M., Al-Rayess, M., Rothman, D.L., Fayad, P.B., Brass, L.M., Shulman, R.G., and Prichard, J.W. (1992). Spectroscopic imaging of stroke in humans: histopathological correlates of spectral changes. *Neurology* **42**, 1349–1354.

Radda, G.K. (1990). Some new insights into biology and medicine through NMR spectroscopy. *Phil. Trans. R. Soc. Lond. A.* **333**, 515–524.

Radda, G.K., Bore, P.J., Gadian, D.G., Ross, B.D., Styles, P., Taylor, D.J., and Morgan-Hughes, J. (1982). ^{31}P NMR examination of two patients with NADH-CoQ reductase deficiency. *Nature* **295**, 608–609.

Ross, B.D., Radda, G.K., Gadian, D.G., Rocker, G., Esiri, M., and Falconer-Smith, J. (1981). Examination of a case of suspected McArdle's syndrome by ^{31}P nuclear magnetic resonance. *N. Engl. J. Med.* **304**, 1338–1342.

Roth, S.C., Azzopardi, D., Edwards, A.D., Baudin, J., Cady, E.B., Townsend, J., Delpy, D.T., Stewart, A.L., Wyatt, J.S. and Reynolds, E.O.R. (1992). Relation between cerebral oxidative metabolism following birth asphyxia and neurodevelopmental outcome and brain growth at one year. *Dev. Med. Child Neurol.* **34**, 285–295.

Rothman, D.L., Howseman, A.M., Graham, G.D., Petroff, O.A.C., Lantos, G., Fayad, P.B., Brass, L.M., Shulman, G.I., Shulman, R.G. and Prichard, J.W. (1991). Localized proton NMR observation of [3–13C]lactate in stroke after [1–13C]glucose infusion. *Magn. Reson. Med.* **21**, 302–307.

Rothman, D.L., Petroff, O.A.C., Behar, K.L., and Mattson, R.H. (1993). Localized ^{1}H NMR measurements of γ-aminobutyric acid in human brain. *Proc. Natl. Acad. Sci. USA* **89**, 9603–9606.

Taylor, D.J., Bore, P.J., Styles, P., Gadian, D.G., and Radda, G.K. (1983). Bioenergetics of intact human muscle: A ^{31}P nuclear magnetic resonance study. *Mol. Biol. Med.* **1**, 77–94.

Taylor, D.J., Styles, P., Matthews, P.M., Arnold, D.A., Gadian, D.G., Bore, P., and Radda, G.K. (1986). Energetics of human muscle: exercise-induced ATP depletion. *Magn. Reson. Med.* **3**, 44–54.

Urenjak, J., Williams, S.R., Gadian, D.G. and Noble, M. (1993). Proton nuclear magnetic resonance spectroscopy unambiguously identifies different neural cell types. *J. Neurosci.* **13**, 981–989.

Welch, K.M.A., Gross, B., Licht, J., Levine, S.R., Glasberg, M., Smith, M.B., et al. (1988). Magnetic resonance spectroscopy of neurologic diseases. *Curr. Neurol.* **8**, 295–331.

Younkin, D.P., Delivora-Papadopoulos, M., Leonard, J.C., Subramanian, V.H., Eleff, S., Leigh, J.S. and Chance, B. (1984). Unique aspects of human newborn cerebral metabolism evaluated with phosphorus nuclear magnetic resonance spectroscopy. *Ann. Neurol.* **6**, 581–586.

78

NIR REFLECTION SPECTROSCOPY BASED OXYGEN MEASUREMENTS AND THERAPY MONITORING IN BRAIN TISSUE AND INTRACRANIAL NEOPLASMS

Correlation to MRI and Angiography

F. Steinberg,[1] H. -J. Röhrborn,[1] K. M. Scheufler,[2] S. Asgari,[3] H. A. Trost,[4*]
V. Seifert,[3] D. Stolke,[3] and C. Streffer[4]

[1]Institute of Medical Radiation Biology Essen
[2]Neurosurgery Bonn
[3]Neurosurgery Essen
[4]Neurosurgery München-Bogenhausen
45122 Essen, Hufelandstr. 55, Germany

INTRODUCTION

Cerebral autoregulation involves adaptive mechanisms at subcellular, cellular and intercellular levels as well as regulation of local, regional and global cerebral circulation (Bruce et al. 1973, Grubb et al. 1975, Harper et al. 1966, McGillicuddy et al. 1978). Autoregulation also includes systemic responses of metabolism, respiration and circulation. All of those cerebral autoregulatory mechanisms may be severly impaired in head injured patients (Langfitt et al. 1965, Höper et al. 1994, McGillicuddy et al. 1978) and in patients with brain tumors (DeChiro et al. 1986/1987, Kusske et al. 1974, Jeske et al. 1993).

Recent strategies in neuromonitoring focus on maintaining or restoring cellular energy metabolism using a wide array of diagnostic and therapeutic modalities (Cruz et al. 1990, Gilboe et al. 1984, DeChiro et al. 1986, Edwards et al. 1988, Elwell et al. 1992). Difficulties remain in the definition of sensitive, specific and reliable parameters (Scheufler et al. 1996a+b). Some of these difficulties can be attributed to measurement of indirect parameters of cellular function, others are due to technical limitations, such as insufficient temporal or spatial resolution or lack of practical applicability within the numerous limitations of intensive care management (Bruce et al. 1973, DeChiro et al. 1986, Delpy, 1996).

---

* FAX ++49 201 7235966.

Brain function and integrity is strongly dependent on oxygen supply. Brain damage often arises from reduced or interrupted oxygen supply. Measurement of absolute tissue hemoglobin content is essential for the following purposes: To declare data from different patients as pathologic or physiologic, the measurements must be comparable to each other. Therefore, all systems measuring relative changes are not useful for these kinds of measurements. Also, their spatial resolution is not good enough to determine the pathologically changed volume as precisely as it would be necessary (Edwards et al. 1988, Delpy, 1996). We have therefore employed a new NIR reflection spectroscopy system (MULTISCAN OS 10, NIOS GmbH, Germany) to characterize the brain oxygen status during routine dignostic procedures and intraoperative functional measurements of brain and tumor in neurosurgical patients. Correlation of NIRS data to MRI and angiography were performed. The therapeutic effect of tumor extirpation on the surrounding tissue was assessed.

MATERIAL AND METHODS

Patients

14 patients (Neurosurgery Essen) were treated because of solid SOL (angioma, glioblastoma, menigeoma).

Methods

HbO_2, Hb, tHb (in mg/ml tissue) and saturations values (in %) were measured simultaneously in real time, displayed online and stored on hard disk. Each location was measured for 15 s (one set of data/s) and mean values were calculated. Additionally, blood gas analysis was performed every 30 minutes. Tissue surfaces were kept free of hemoglobin during measurements. Measurement protocols were performed within less than 2 minutes. The measured tissue volume is defined by a cone (diameter of the sensor fiber = 1 mm and depth of 4 mm). Because of the varying diameter of tissue layers and the irregular shaped brain surface each individual layer was measured separately. Functional assessment of the primary sensory cortex and tumor infiltrated tissue within this region was performed via tibial nerve stimulation (somato sensory evoked potentials SSEP) (30mA, duration 10–15 s). The sensor was sterilized with GIGASEPT® The surface of the tissue must be kept blood free.

RESULTS

Figure 1 shows measurements of tHb, HbO_2 and $tiSO_2$ taken from the dura (above tumor and healthy cortex), occipital cortex (above tumor and unaltered white matter) and exposed tumor tissue (at 2 positions). No difference was noted at the dura level. Significant alterations of all three parameters are shown at the cortical level resulting from the tumor, which was seated $\approx$ 1 cm below the measurement site.

Figure 2 shows results from measurements performed in a 57-year-old male patient, who underwent surgery for removal of a sphenoideal ridge menigeoma. MRI (Fig 2a) shows a hypodense rim surrounding the tumor. NIRS data (Fig 2b) demonstrates high tHb values ($\approx$ blood volume) within the tumor and low values in the surrounding, edematous brain tissue.

Localisation	Saturation (%)	HbO2 (mg/ml)	Total-Hb (mg/ml)
DURA			
above healthy tissue	68	2.66	3.94
above tumour tissue	55	1.81	3.26
CORTEX			
above healthy tissue	47	1.09	2.31
above tumour tissue	28	0.27	0.98
TUMOUR			
Position 1	33	0.25	0.75
Position 2	19	0.22	1.09

Figure 1. Oxygenation in brain and tumor tissue of a 46-year-old male patient with glioblastoma tumor. tO_2, HbO_2 and tHb were measured in dura, cortex and tumor at 2 different positions: the measurement positions distinguish between locations above healthy and tumor tissue respectively.

Embolization of menigeomas with Ethibloc® is performed prior to surgery, if localization of tumor and vessel anatomy is suitable. Fig 3a shows MRI of a 56-year-old male patient with a parasagittal menigeoma in the precentral area. The appertaining angiograms before and after embolisation are demonstrated in Fig 3b + c. A complete occlusion can be seen. The NIR data (Fig 3d) demonstrate the strong effect from bilateral embolisation of the A. carotis ext. on tumor oxygenation in contrast to the unaltered values within brain tissue. Because of the occlusion, all three parameters were reduced significantly within the tumor.

Figure 2. 57-year-old male patient with a menigeoma. a) MRI shows hypodense area perifocal of the tumor. b) tO_2, HbO_2 and tHb were measured at 3 positions on the healthy tissue near the tumor and at 2 positions on the tumor.

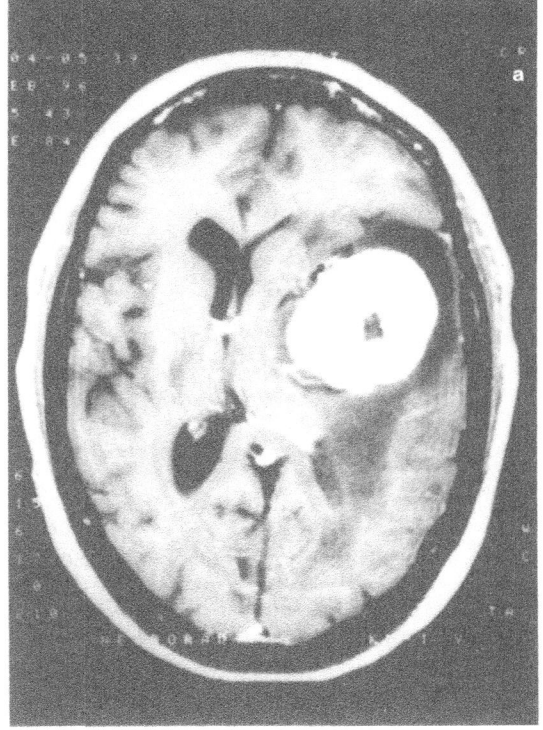

	Saturation (%)	HbO$_2$ (mg/ml)	Total-Hb (mg/ml)
Frontal 1	19	0.44	2.32
Temporal	25	0.64	2.55
Frontal 2	19	0.41	2.38
Tumor 1	25	1.82	7.20
Tumor 2	43	2.72	6.22

b

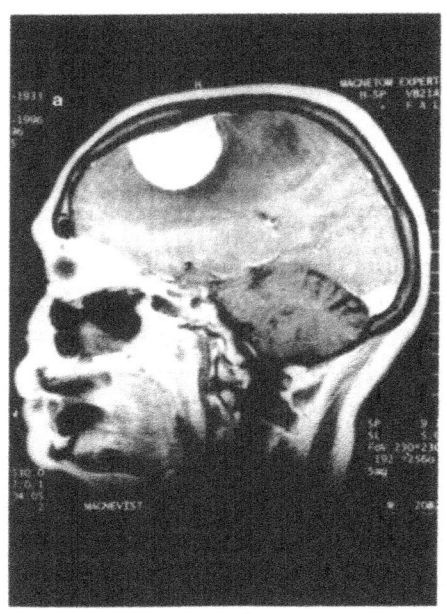

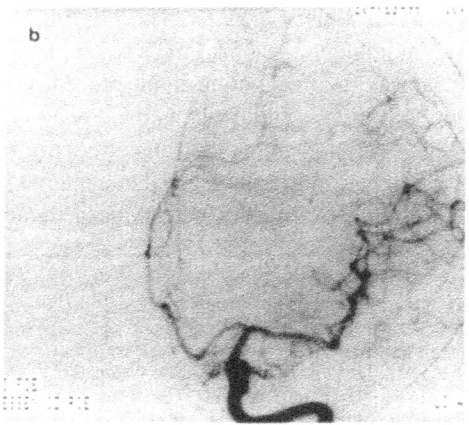

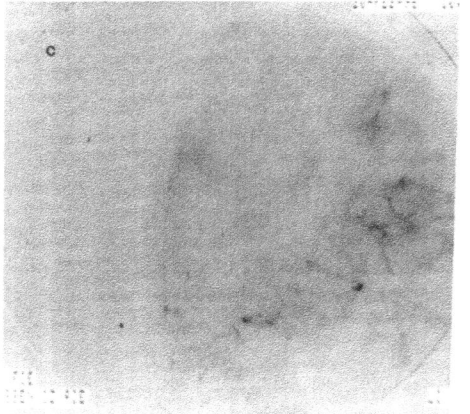

d	Saturation (%)	HbO2 (mg/ml)	Total-Hb (mg/ml)
Frontal 1	74	3.69	4.97
Frontal 2	50	2.04	4.08
Tumor 1	39	0.21	0.53
Tumor 2	38	0.25	0.66

Figure 3. 56-year-old male patient with a menigeoma parasagittal, precentral after embolisation of both tumor feedings of the A. carotis ext. Fig 3a) shows the MRI, b) + c) show the corresponding angiograms before and after embolisation, d) shows the NIR data. Both cortex values were significant higher compared to tumor values as an effect of the nearly complete occlusion of these tumor feedings.

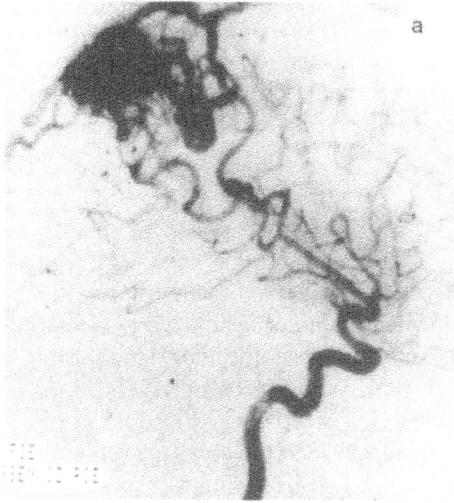

before Tumour Extirpation   **b**

Saturation (%)	HbO2 (mg/ml)	Total- Hb (mg/ml)	
47	1.80	3.45	Parietal-sagital
66	1.27	1.92	1 cm lateral
41	0.85	2.03	lateral1
30	0.67	2.22	lateral2
32	0.76	2.42	parietal-lateral
20	0.28	1.38	1 cm dorsal
49	1.73	3.42	1 cm medial
71	4.46	6.28	medial-parietal

after Tumor Extirpation

66	2.05	3.05	sagital
67	1.61	2.42	lateral
79	3.25	4.09	lateral
79	2.40	3.03	parietal-lateral
75	4.55	6.06	dorsal
75	4.92	6.56	medial
79	3.21	4.06	paramedial

Figure 4. 35-year-old female patient with AV angioma (localisation: cortex parieto-occipital. Fig 4a shows the diagnostic angiogram, b) shows the NIR data. The surrounding tissue was measured before (8 positions) and immediately after extirpation (7 positions). A significant increase in tO_2, HbO_2 and tHb was measurable.

Arteriovenous malformations (AVM) can exert significant cerebral effects by diversion of blood from the cerebral circulation and inducing a steal phenomenon. Fig 4 a shows a 35-year-old female patient with an AVM. The surrounding tissue was measured at 8 positions before extirpation of the AVM and immediately after (< 5 minutes)(Fig 4b). All localisations showed significant increase in saturation, HbO_2, and tHb (except in 2 positions).

A functional activation study (SSEP) was performed in a 31-year-old female patient with cerebral metastasis from malignant melanoma. The metatasis was located in the postcentral gyrus. $TiSO_2$, tHb and HbO_2 were measured before, during and after stimulation in healthy (= pre and post central gyrus) and tumor tissue (Fig 5). Data shown are mean values before, during and after stimulation. No effect on saturation values was

Saturation (%)	HbO2 mg/ml tissue	Total Hb mg/ml tissue	
32	0.36	1.10	**Healthy cortex, post central region, before stimulation**
38	0.82	2.11	**during stimulation**
34	0.56	1.60	**after stimulation**
29	0.38	1.22	**Healthy cortex, pre central region, before stimulation**
26	0.32	1.20	**during stimulation**
26	0.33	1.26	**after stimulation**
64	0.77	1.18	**Tumour, post central region, before stimulation**
48	0.42	0.88	**during stimulation**
50	0.46	0.93	**after stimulation**

Figure 5. 31-year-old female patient with cerebral metastasis of malignant melanoma. Before, during and after brain stimulation with SSEP (N. tibialis, 30 mA) in tO_2, HbO_2 and tHb were measured in the pre central region (= healthy tissue) and post central region (= healthy and malignant tissue).

noted. A 2-fold increase of HbO$_2$ and tHb was measured in the postcentral region. This can be understand as change in oxygen consumption. No changes were measured in the immediately adjacent precentral area, indicating highly selective, spatially confined changes of tissue perfusion resulting from functional neuronal activation. The changes were measurable within 2 - 3s following onset of stimulation.

DISCUSSION

Many of the lesions treated by neurosurgeons comprise relatively small tissue volumes (< 2 cm^3; e.g. tumors, aneurysms, hematoma) located within or near the grey matter. The grey matter has a thickness of 4 to 7 mm. Currently, no system reliably measuring such small tissue volumes transcranially and distinguishing between tissue layers (dura, grey and white matter) is available (Delpy, 1996). Thus, for practical reasons and in the situations described above, NIRS can only be performed intraoperatively with reflexion spectroscopy based systems measuring small and well defined tissue volumes.

In the past, conventional NIRS systems measured the whole brain tissue (e.g. fetal or neonatal brain or great volumes of the brain (Edwards et al. 1988, Elwell et al. 1992). All these measurements are based on the transmission mode. Reflection-mode based systems like MULTISCAN OS 10 provides two benefits: 1: It is possible to determine a small tissue volume (i.e., the neurosurgical target volume), 2: It is possible to determine absolute values. This is the condition sine qua non to interindividual comparison of NIRS measurements and to distinguish between normal and pathological values.

Claims to the technical system: The system must be tailored for the specific situation within the operation theatre. Long sensor cables (> 3 m), and resterilizeable sensors are required and the results must be monitored in real time on the screen, which is readable for the surgeon.

The results shown in Fig 1 clearly demonstrate a significant difference in oxygenation of different tissue layers. In a malignant process (= tumor), additional influencing factors are measurable. A strong interdependency of tumor and normal tissue was demonstrated. The tumor can influence the surrounding normal brain tissue in two ways: 1. Through increasing pressure caused by tumor growth or by peritumoral oedema blood flow is reduced. 2: AVM may exert cerebral ischemic effects by steal phenomena (Kusske et al. 1974).

Considerable heterogeneity could be demonstrated in tumors (see Fig 1: Glioblastoma, Fig 2: Menigeoma, Fig 5: Metastasis). This phenomenon has already been described by morphological methods (e.g. corrosion casts; Steinberg et al. 1990), quantification of tumor vessels (Steinberg et al. 1991), blood flow measurements (Kallinowsky et al. 1989), and in correlation to other biogical tumor parameters (Steinberg et al. 1994). Polarographic oxygen measurements (pO$_2$) have shown the same tendency in different tumor entities (Füller et al. 1994, Feldmann et al. 1995, Streffer et al. 1995).

A number of clinical symptoms leading patients to diagnostic evaluation can mainly be attributed to reduced (= pathological) tissue oxygen concentration. In these cases, therapeutic success depends on increasing the available tissue oxygen concentration. This therapeutic effect can be monitored (see Fig 4).

The functional stimulation experiments showed three important effects. 1: The vascular system responds immediately (within 2 to 3 seconds following functional neuronal activation) with an increase in blood volume. 2: The difference in oxygen consumption between non-activated and functionally stimulated tissue is significant. 3: The microcirculatory response to functional activation is precisely limited to a circumscribed small tissue volume.

CONCLUSION

MULTISCAN OS 10 can be used for intraoperative measurements and routine assessment of oxygen status in neurosurgical patients. It can be used for diagnostic purposes, therapy monitoring and cerebral function testing. Experimental studies can disclose elementary correlations of physiological parameters like various ventilatory parameter (see Scheufler et al. 1996b) or intracranial hypertension (see Scheufler et al. 1996a) on cerebral oxygen parameters. Further studies are required to prove reliability of the system under different pathologic conditions in patients.

ACKNOWLEDGMENT

The authors would like to express their thanks for the technical assistance of Mrs. Hildenhagen-Brüggemann, and for technical and financial support from NIOS Medizintechnik GmbH/Germany and Krebshilfe e.V. Herdecke.

REFERENCES

1. Bruce DA, Langfitt TW, Miller JD, Schutz H, Vapalahti MP, Stanek A, Goldberg HI, Regional cerebral blood flow, intracranial pressure, and brain metabolism in comatose patients. J Neurosurg 38, 131–144, 1973.
2. Cruz J, Miner ME, Allen SJ, Alves WM, Gennarelli IA, Continuous monitoring of cerebral oxygenation in acute brain injury: injection of mannitol during hyperventilation. J Neurosurg 73, 725–790, 1990.
3. Delpy DT, Instruments and methods for quantitative monitoring and imaging of brain function. Brain Science Conference, Hamamatsu, Japan, Febr. 28th - March 1st, 47–51, 1996.
4. DeChiro D.G. Positron emission tomography using ^{18}F-fluorodeoxyglucose in brain tumors. Invest.Radiol., 22, 360–371, 1986.
5. DeChiro D.G, Hatazawa J, Katz D, Rizzoli H, de Michele D. Glucose utilization by intracranial meningiomas as an index of tumor agressivity and probability of recurrence: a PET study. Radiology, 164, 521–526, 1987.
6. Edwards AD, Wyatt JS, Richardson C., Delpy DT, Cope M, Reynolds EOR., Cotside measurements of cerebral blood flow in ill preterm infants by near infrared spectroscopy. Lancet, 770–771, 1988.
7. Elwell CE, Cope M, Edwards AD, Wyatt JS, Delpy DT, Reynolds EOR: Measurement of cerebral blood flow in adult humans using near infrared spectroscopy - methodology and possible errors. Adv Exp.Med Biol. 317, 235–245, 1992.
8. Feldmann HJ, M Molls, T Auberger, G Stüben. Oxygenation and perfusion status of recurrent human tumors. In: Funktionsanalyse biologischer Systeme, 24.pp. 319–326. Editors: PVaupel, DK Kelleher and M Günderoth. Akademie der Wissenschaften und der Literatur, Mainz, 1995.
9. Füller J, Feldmann HJ, Molls M, Sack H, Untersuchungen zum Sauerstoffpartialdruck im Tumorgewebe unter Radio- und Thermoradiotherapie. Strahlenther Onkol 170, 453–460, 1994.
10. Gilboe DD, Kintner D, Yanushka J, Cerebral oxygen utilization as a gauge of brain energy metabolism. In: Bruley D, Bicher HI, Reneau D (eds.): Oxygen Transport to Tissue VI. Plenum Press, New York, 169–177, 1984.
11. Grubb RL, Raichle ME, Phelps ME, Ratcheson RA: Effects of increased intracranial pressure on cerebral blood volume, blood flow and oxygen utilization in monkeys. J Neurosurg 43, 385–398, 1975.
12. Harper AM, Autoregulation of cerebral blood flow: influence of the arterial blood pressure on blood flow through the cerebral cortex. J Neurol Neurosurg Psychiatry 29, 398–403, 1996.
13. Höper J, Gaab MR: Intraoperative monitoring of local Hb-oxygenation in human brain cortex. In: Hogan MC (ed):Oxygen Transport to Tissue XVI. Plenum Press, New York, 1994.

14. Jeske J, Tamulevicius P, Fuhrmann C, Steinberg F, Stolke D, Streffer C, Biochemical findings in human brain tumours. Europ J Cancer 29, 193, 1993.

15. Kallinowski F, Schlenger KH, Runkel S, Kloes M, Stohrer M, Okunieff P, Vaupel P, Blood flow, metabolism, cellular microenvironment, and growth rate of human tumor xenografts. Cancer Res 49, 3759–3764, 1989.

16. Kusske JA, Kelly WA, Embolization and reduction of the steal syndrome in cerebral arteriovenous malformations. J Neurosurg 40, 313–321, 1974.

17. Langfitt TW, Weinstein JD, Kassel NF, Cerebral vasomotor paralysis produced by intracranial hypertension. Neurology 15, 622–641, 1965.

18. McGillicuddy JE, Kindt GW, Miller CA, Raisis JE, The relation of cerebral ischemia, hypoxia and hypercarbia on the cushing response. J Neurosurg 48, 730–740, 1978.

19. Scheufler K.M, Thees CH, Steinberg F, Zentner J, NIR reflexion spectroscopy based oxygen measurements during intracranial hypertension in rabbits: an experimental study. This issue, 1996a.

20. Scheufler KM, Thees CH, Steinberg F, Zentner J, Influence of various ventilation parameters on NIR reflection spectroscopy based cerebral oxygen measurements: an experimental animal study. This issue, 1996b.

21. Steinberg F, Konerding MA, Sander A and Streffer C, Vascularisation, proliferation and necrosis in untreated human primary tumours and untreated human xenografts. Int J Rad Biol, 60, 161–168, 1991.

22. Steinberg F, Konerding MA, Streffer C. The vascular architecture of tumors: Histological, morphometrical and ultrastructural studies. J Canc Res Clin Oncol, 116: 517–524, 1990.

23. Steinberg F, M Leßmann, C Streffer; Non-invasive infrared-spectroscopic measurements of intravasal oxygen content in xenotransplanted tumours on nude mice - correlation to growth rate. Adv. Exp. Med. Biol. 345, 549–555, 1994.

24. Steinberg F, Röhrborn HJ, Otto T, Lubold K, Streffer C, NIR reflexion measurements of hemoglobin and cytochrome aa3 - correlations to oxygen consumption and energy metabolism - preclinical and clinical data. This issue, 1996.

25. Streffer C, Steinberg F, Tamulevicius P, Oxygen content and energy metabolism in Tumours - are they correlated? in: Funktionsanalyse biologischer Systeme (Eds. Vaupel, Kelleher, Günderoth), Gustav Fischer Verlag, Stuttgart, New York, 195–204, 1995.

79

AN ANALYTICAL METHOD FOR DETERMINING CEREBROVASCULAR TRANSIT TIME USING NEAR INFRARED SPECTROSCOPY

C. E. Elwell, M. Cope, and D. T. Delpy

Department of Medical Physics and Bioengineering
University College London
United Kingdom

1. INTRODUCTION

The measurement of cerebral blood flow (CBF) using near infrared spectroscopy (NIRS) is already well described (Edwards et al., 1988, Elwell et al., 1994, Owen-Reece et al., 1996). The technique requires the use of a small *change* in oxyhaemoglobin concentration (HbO_2) as an inert tracer. The peripheral concentration of the tracer is measured using pulse oximetry and the accumulation of tracer in the brain is measured using NIRS.

Since NIRS cannot inherently discriminate between the vascular compartments of the brain, an assumption of the cerebrovascular transit time is necessary to ensure that the measured increase in HbO_2 in the brain is purely the arterial accumulation of the tracer with no contamination arising from loss of tracer due to venous outflow. This paper describes a novel analysis method, using a stepwise regression model, which determines the initial linear portion of the measured CBF response curve and from this the first (or fastest) detected transit time, t_1. The model has been tested on simulated data sets with theoretical CBF responses. It has also been applied to CBF data sets collected from adult volunteers and ventilated neonates.

2. THEORY

As with most CBF measurement techniques, NIRS uses a modification of the Fick principle. This principle states that the amount of tracer taken up by an organ per unit time is equivalent to the difference between the rate of arrival of the tracer and rate of departure of the tracer from the organ. For cerebral measurements, the rate of arrival of the tracer in the brain is equivalent to the product of the blood flow through the brain (CBF)

Oxygen Transport to Tissue XIX, edited by Harrison and Delpy
Plenum Press, New York, 1997

and the arterial concentration of the tracer (Ca). The rate of departure of the tracer is therefore the product of CBF and the venous concentration of the tracer, (Cv). Hence:

$$\frac{dCt}{dt} = CBF(Ca - Cv)$$

where Ct is the concentration of tracer in tissue. Integrating and rearranging this expression:

$$CBF = \frac{Ct}{\int_0^t (Ca - Cv)dt}$$

If measurements are made within the cerebrovascular transit time, t_1, then the tracer will not have reached the venous circulation, $Cv = 0$, and CBF can be calculated from the following equation:

$$CBF = k \bullet \frac{\Delta[HbO_2]}{\int_0^t \Delta fSaO_2 dt}$$

where $\Delta[HbO_2]$ is the NIRS measured change in concentration of cerebral oxyhaemoglobin (measurement of tracer concentration in the brain, Ct) and $\Delta fSaO_2$ is the fractional change in arterial oxygen haemoglobin saturation (measurement of arterial concentration of tracer, Ca). k is a constant incorporating the molecular weight of haemoglobin, cerebral tissue density and the concentration of haemoglobin in the whole blood which are required in order to express CBF in units of ml.$100g^{-1}$.min^{-1} (Elwell, 1995). It is clear that knowledge of the cerebrovascular transit time, t_1, is essential to ensure that the assumption that $Cv = 0$ is valid within the measurement period, t.

To obtain data from which to calculate CBF, simultaneous arterial pulse oximetry (to measure $\Delta fSaO_2$) and NIRS measurements are made while the subject's inspired oxygen fraction (FiO_2) is first slowly reduced to provide a stable baseline and then rapidly increased to produce the required sudden rise in SaO_2. Figure 1 shows data collected from an adult volunteer during a CBF measurement.

Measurements are also possible, and indeed were pioneered, in the ventilated neonate (Edwards et al., 1988). In most protocols a cerebrovascular transit time is assumed and a measurement period, t, is defined during which the input function ($\Delta fSaO_2$) and the cerebral response function (ΔHbO_2) is calculated. Given the quoted differences between neonatal and adult cerebrovascular transit time, and the importance of this parameter for the accurate measurement of CBF, an analytical method for determining t_1 for each individual measurement on any subject is of obvious value.

3. ANALYTICAL MODEL

The methodology involved in the calculation of CBF has been described fully elsewhere (Elwell, 1995). Briefly, a response curve is formulated from the cerebral accumulation of tracer (ΔHbO_2) and the peripherally measured tracer input ($\int \Delta fSaO_2$). Theoretical response curves for single and dual flow compartments are shown in Figures 2(a) and 3(a).

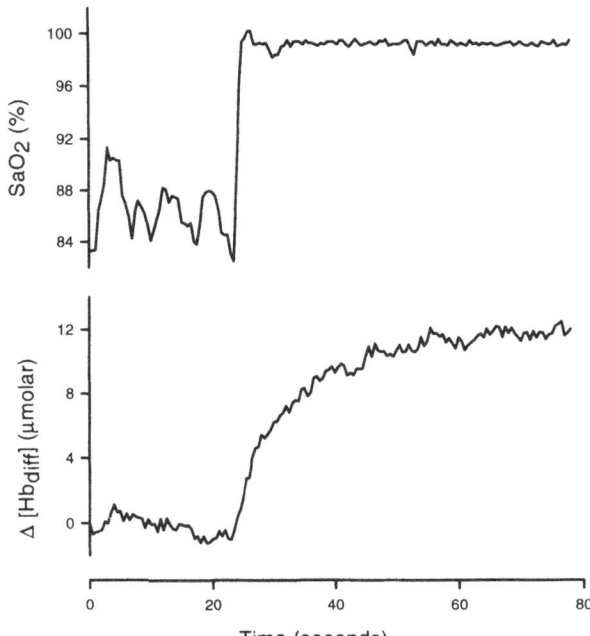

Figure 1. NIRS and pulse oximetry data collected from an adult volunteer during the measurement of CBF. FiO_2 is gradually lowered to produce the required fall in SaO_2, and then increased over the period of a single breath to provide the input bolus of ΔHbO_2.

The first (or fastest) cerebrovascular transit time is identified on these curves as the point where the response becomes non linear. At this point some of the tracer (ΔHbO_2) is beginning to leave the arterial compartment, signalling the start of venous outflow (i.e. where the Fick principle assumption that Cv=0 is no longer valid). For example in Figure 2(a) the response curve remains linear for the entire 20 second period - the transit time has not been reached. However in Figure 3(a) the response curve is linear only up until the

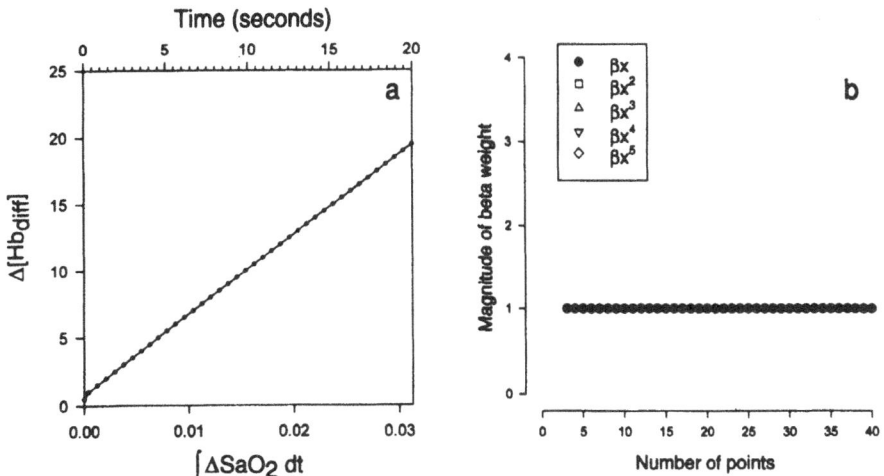

Figure 2. (a) The theoretical response from a single flow compartment with a transit time, $t_1 > 20$ seconds, and (b) the stepwise regression results graph for this data. Since the response contains only a linear term, the stepwise regression has accepted only the x term for each successive data point added to the analysis and $\beta x^2 = \beta x^3 = \beta x^4 = \beta x^5 = 0$.

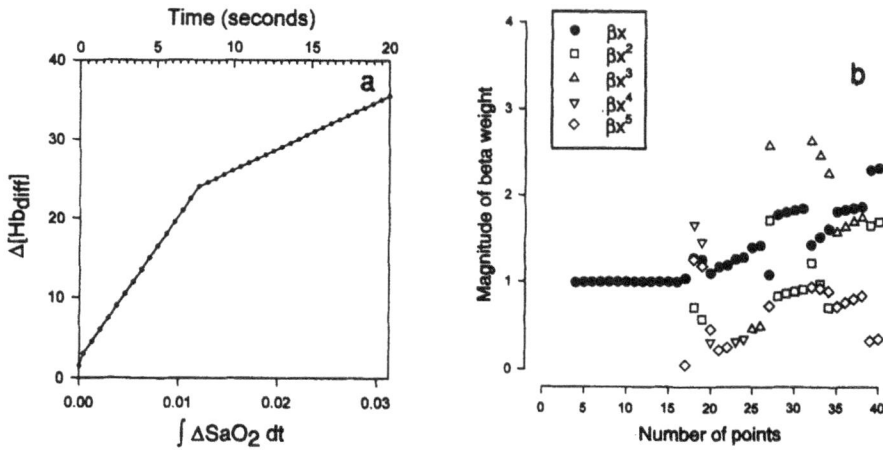

Figure 3. (a) The theoretical response from a dual flow compartment model with transit times of 8 seconds and > 20 seconds, respectively. The stepwise regression results graph (b) shows that the x^5 term is introduced when the 17th point is added to the analysis. This result indicates that CBF should be calculated using the first 8 seconds of the response.

16th point indicating a cerebrovascular transit time of 8 seconds (NIRS CBF data is typically collected every 0.5 seconds).

In these theoretical examples visual inspection is sufficient to identify the transit time, however with most experimental data some form of curve analysis is required.

3.1. Stepwise Regression Analysis

(Note: the terms and descriptions given here are those used by SPSS for Windows, SPSS Inc. Chicago, Ill.).

Stepwise regression analysis has been used to examine CBF response curves from experimental data. Initially it is assumed that the curve may be fully described by an nth order polynomial, where n, the number of independent variables, will be determined by the complexity of the response. For the purposes of this model a fifth order polynomial is chosen:

$$y = \beta_0 + \beta_1 x + \beta_2 x^2 + \beta_3 x^3 + \beta_4 x^4 + \beta_5 x^5$$

Stepwise regression analysis uses a combination of both forward and backward elimination techniques for determining "best fits" on given data sets. In the forward elimination method the independent variable with the strongest correlation with the dependent variable is first added to the fit. An F test is performed and from this an F value (or significance) is computed which will depend upon the R^2 (regression coefficient) of the fit, the degrees of freedom, the total number of independent variables and the number of independent variables already accepted. This value is then compared with pre established criteria, usually PIN (the probability of-F-to-enter), to decide whether the independent variable should be included in the best fit equation. Once tested, if this independent variable is accepted, the forward selection continues and the statistics for the variables not in the equation are used to determine which variable should be selected next. This process is repeated until there are no other independent variables which meet the entry criteria.

Backward elimination is the reverse process. All the independent variables are entered into the equation first and removal criteria, POUT (probability of-F-to-remove), are applied to assess whether they should be included in the fit.

As the name suggests, stepwise regression is a combination of both forward and backward elimination. Each independent variable is tested for inclusion and exclusion from the fit. The acceptance and rejection criteria are simultaneously set by PIN and POUT. Stepwise regression is therefore used to determine the minimum number of independent variables required to fully describe the response (y). Starting at the beginning of the data set describing the CBF response, where t=0, successive data points are added to the analysis and the appropriate independent variables (i.e. x, x^2, x^3, x^4, x^5) required to describe the function are added to the stepwise regression. In this way the analysis clearly identifies the point (e.g. 10th, 11th) at which the response becomes complex enough to require description by a term other than the linear x term.

For example in Figure 2(b) we see that for the entire data set (i.e. points 1 - 40) only the x term was required to fully describe the theoretical single flow response shown in Figure 2(a), whilst in Figure 3(b) the stepwise regression shows that the theoretical response shown in Figure 3(a) could be described by the x term alone only until the 16th point. Beyond this point other terms were required to describe the entire cumulative non linear function. Hence two parameters can be used to describe the initial linear portion of this curve, the first transit break point τ_{bp} and the first (or fastest) transit time, t_1. The first transit break point is the point in the response curve where a variable other than the linear x term is first introduced (i.e. point 17 in this case). The first transit time can then be very easily calculated from this using the following relationship: (Note that this equation assumes a sample time of 0.5 seconds)

$$t_1 = (\tau_{bp} - 1)/2$$

When $\tau_{bp} = 17$ the response is linear until the 16th point. Since each point represents a 0.5 second data collection, this is equivalent to a t_1 of 8 seconds. In this way stepwise regression may be used to identify the initial linear portion of a CBF response curve and hence the vascular transit time, t_1.

The pre established acceptance and rejection criteria, PIN and POUT, play an important role in determining the sensitivity of the model for detecting the best fit on any given set of data, and hence detecting the initial linear portion of a curve. For this reason this analysis routine was first tested on simulated model data. Theoretical CBF response curves of the type shown in Figures 2(a) and 3(a) were formulated for t_1 values between 2.5 and 10 seconds and τ_{bp} was successfully detected in all cases. Gaussian noise at three different levels was then applied to another data set where $t_1 = 8$ seconds. PIN and POUT were then varied to determine empirically the relationship between their appropriate values and the level of noise on the signal.

The analysis described above was applied to experimental CBF data collected from ten adult volunteers and ten ventilated neonates in order to determine the mean cerebrovascular transit time in each group. Appropriate values for PIN and POUT were chosen from a measurement of the noise on the measured signal.

4. RESULTS

Figure 4(a) shows the CBF response curve calculated from the adult volunteer data shown in Figure 1. Figure 4(b) shows the corresponding stepwise regression results graph.

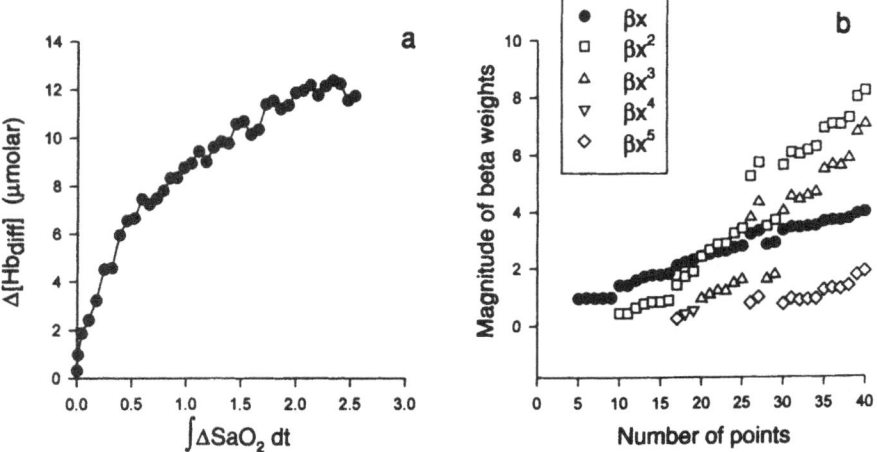

Figure 4. (a) The CBF response curve for data collected from an adult volunteer and (b) the corresponding stepwise regression results graph. In this subject $\tau_{bp} = 11$, equivalent to a first transit time, t_1, of 5 seconds.

In this subject $\tau_{bp} = 11$, equivalent to a first transit time, t_1, of 5 seconds. Figure 5(a) shows a CBF response curve for data collected from a ventilated neonate and (b) the corresponding stepwise regression results graph. In this infant $\tau_{bp} = 19$, equivalent to a first transit time, t_1, of 8 seconds. The results from the grouped adult and neonatal data are summarised in Table 1.

5. DISCUSSION

The analytical model described in this paper forms the preliminary stages of a more automated and objective approach to the manipulation of NIRS CBF data. Many other

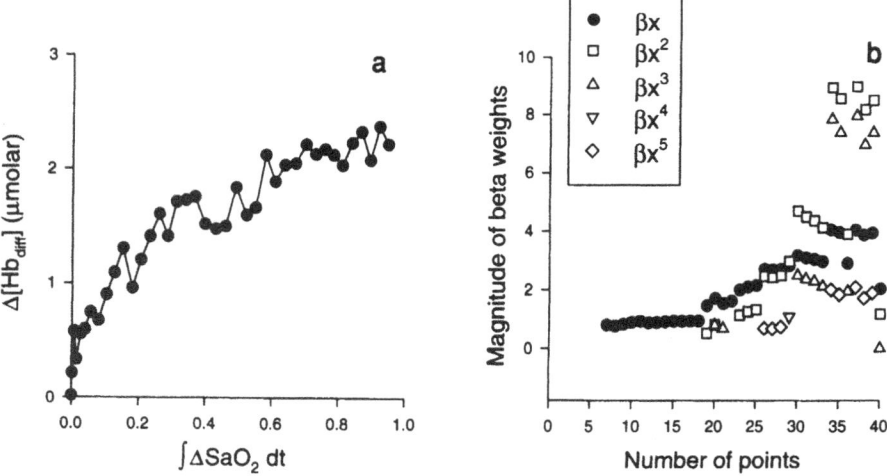

Figure 5. (a) The CBF response curve for data collected from a ventilated neonate and (b) the corresponding stepwise regression results graph. In this infant $\tau_{bp} = 19$, equivalent to a first transit time, t_1, of 8 seconds.

Table 1. Results from stepwise regression analysis of CBF data sets from ten adult and ten neonatal subjects. τ_{bp} is the point on the CBF response curve where a variable other than the x term is required to describe the function. t_1 is the calculated first (fastest) cerebrovascular transit time

	n	τ_{bp}	t_1 (seconds)
Adult	10	11.0 ± 2.1	5.0 ± 1.0
Neonatal	10	19.9 ± 5.9	9.4 ± 2.9

CBF methods use a tracer technique either in the form of wash-in or wash-out curves and in many cases a convolution of several effects must be considered.

The stepwise regression method is an attempt to separate out at least one important parameter from the CBF response curve, i.e. the definition of the period within the cerebrovascular transit time. Although the model has been tested on sets of simulated data, further work is required to determine the effect of, for example, the slurring of the input function ($\int \Delta fSaO_2$), variations in arrival time of the tracer and high levels of noise on clinical data (Elwell, 1995).

Measurement of cerebrovascular transit time may itself be a useful indicator of cerebral wellbeing. Previous studies using PET (Leenders et al., 1990) have demonstrated a mean transit time in adults of 5.5 seconds. The results of this analysis on ten adult volunteers (mean $t_1 = 5$ seconds) compares well with this estimate. Direct measurements of neonatal cerebrovascular transit time are less well reported in the literature although a study using labelled erthytrocytes by Arnot (Arnot et al., 1970) suggests a value of 8 seconds, again close to our derived mean value of 9.4 seconds.

The transit time data depends purely upon the time course of the CBF response rather than its exact magnitude and hence bypasses all the problems of extracerebral tissue contribution which have been shown to hamper the absolute quantification of CBF (Elwell, 1995). It is encouraging that these measurements of cerebrovascular transit time may provide a reliable and reproducible measurement of cerebral haemodynamics.

6. CONCLUSION

The first (fastest) vascular transit time can be estimated using stepwise regression analysis to determine the initial linear portion of a CBF response curve obtained with NIRS. Measurements of adult and neonatal vascular transit times using this method agree with those previously published. The ability to calculate cerebrovascular transit time for each data set should improve the accuracy of CBF calculations.

REFERENCES

Arnot, R.N., Glass, H.I., Clark, J.C. (1970) Methods of measurement of cerebral blood flow in the newborn infant using Cyclotron-produced isotopes. *Radioaktive Isotope in Klinik und Forschung* 9: 60–74

Edwards, A.D., Wyatt, J.S., Richardson, C.E., Delpy, D.T., Cope, M., Reynolds, E.O.R. (1988) Cotside measurement of cerebral blood flow in ill newborns by near infrared spectroscopy. *Lancet* 2: 770–771

Elwell, C.E. (1995) Measurement and data analysis techniques for the investigation of adult cerebral haemodynamics using near infrared spectroscopy. *PhD Thesis, University of London, UK.*

Elwell, C.E., Cope, M., Edwards, A.D., Wyatt, J.S., Delpy, D.T., Reynolds, E.O.R. (1994) Quantification of adult cerebral haemodynamics by near infrared spectroscopy. *J. Applied Physiol.* 77:2753–2760

Leenders, K.L., Perani, D., Lammertsma, A.A., Heathers, J.D., Buckingham, P., Healy, M.J.R., Gibbs, J.M., Wise, R.J.S., Hatazawa, J., Herold, S., Beaney, R.P., Brooks, D.J., Spinks, T., Rhodes, C., Frackowiak, R.S.J. (1990) Cerebral blood flow, blood volume and oxygen utilisation. Normal values and effect of age. *Brain* 113:27–47

Owen-Reece, H, Elwell, C.E., Harkness, W., Goldstone, J., Delpy, D.T., Wyatt, J.S., Smith, M. (1996) Use of near infrared spectroscopy to estimate cerebral blood flow in conscious and anaesthetised adult subjects. *Br. J. Anaesth.* 76:43–48

80

MEASUREMENT OF OXYGEN TENSIONS IN THE ABDOMINAL CAVITY AND IN THE SKELETAL MUSCLE USING 19F-MRI OF NEAT PFC DROPLETS

J. Lutz,[1] U. Nöth,[1,2] S. P. Morrissey,[3] H. Adolf,[3] R. Deichmann,[2] and A. Haase[2]

[1]Department of Physiology
University of Würzburg
Röntgenring 9, D-97070 Würzburg, Germany
[2]Department of Biophysics
University of Würzburg
Am Hubland, D-97074 Würzburg, Germany
[3]Department of Neurology
University of Regensburg
Universitätsstr. 84, D-93042 Regensburg, Germany

INTRODUCTION

Perfluorochemicals (PFCs) are not only of interest as artificial oxygen carriers but also as 19F-NMR (19F nuclear magnetic resonance) contrast agents for the determination of partial oxygen pressures *in vivo*. Since the longitudinal relaxation rate R_1 increases linearly with the partial oxygen pressure (pO_2), pO_2 values can be determined in regions of PFC accumulation. After the i.v. injection of a PFC emulsion, pO_2 measurements are possible in large vessels until most of the PFC particles are removed from the circulation by macrophages. Then intracellular pO_2 measurements can be performed in regions with high macrophage concentration, i.e. in the liver and spleen, and in pathological tissues with high macrophage activity, e.g. in tumors and in regions of inflammation. In many studies with PFC emulsions, negative apparent pO_2 values were reported under hypoxic conditions ($F_1O_2 < 0.21$) [1–7]. Purpose of this work was to measure extracellular pO_2 values in a depot of neat PFC in the abdominal cavity and in the skeletal muscle under hyperoxic, normoxic and hypoxic conditions, and further to determine whether neat PFC also yields negative apparent pO_2 values under hypoxic conditions.

MATERIALS AND METHODS

Perfluorochemical

The PFC used in this study was neat trans-1,2-bis(perfluorobutyl)ethylene, $CF_3(CF_2)_3HC=CH(CF_2)_3CF_3$, , abbreviated F-44E, with a molecular mass of 464 g/mol and a vapor pressure of 12.6 mmHg at 37 ° C [8]. The resonance line of the two chemically equivalent CF_3 groups (33% of the total fluorine signal) was set on resonance and used for imaging, while the three lines corresponding to the CF_2 groups were suppressed as described in [5] and [9].

Calibration

In vivo pO_2 values are obtained by solving the equation $R_1 = 1/T_1 = c + s \ pO_2$ for pO_2, using the calibration constants c (anoxic intercept) and s (slope) which were determined as follows: six vials with neat F-44E were equilibrated with gas mixtures of oxygen and nitrogen at atmospheric pressure, the oxygen content ranging from 0 % to 100 % (0 to 760 mmHg). T_1 values were measured with a spectroscopic inversion recovery experiment at (36.5 ± 0.5) °C and 7.05 Tesla. The linear fit of R_1 versus pO_2 yielded the calibration constants c = 0.342 /sec and s = 0.0158 /(sec %).

Animals and Anesthesia

All experiments were performed with Wistar rats of both sexes and 140 - 170 g body weight. Three rats received an intraperitoneal injection of neat F-44E at a dose of up to 7ml/100g body weight in order to determine the pO_2 value in the abdominal cavity. For the determination of the extracellular pO_2 value in the skeletal muscle F-44E droplets of 0.2 to 0.4 ml were injected with a 25 G needle in ca. 4 mm depth into the M. quadriceps femoris of 3 rats. Inhalation anesthesia was introduced with 5 % isoflurane and during the NMR experiments maintained at 1.5 % isoflurane in the breathing gas flowing at a rate of 1.5 l/min, while the rats were lying on a water heating pad to maintain a constant body temperature of 36.5 °C.

NMR Measurements

All NMR experiments were performed on a 7.05 T Bruker Biospec System with a 21 cm horizontal bore operating at a fluorine frequency of 282.543 MHz. *In vivo* pO_2 measurements were performed in the described regions while the animals were ventilated successively with three different oxygen concentrations (F_iO_2) in the breathing gas: pure oxygen, air and a mixture of 10 % oxygen and 90 % nitrogen corresponding to a F_iO_2 of 1.0, 0.21 and 0.10 respectively. Before 19F-MRI, animals inhaled the respective breathing gas during 30 min to achieve equilibration.

For quantitative 19F-MRI we used a frequency selective FLASH method which yields only the image of the on resonance line [5,9]. For each T_1 measurement, the magnetization was inverted and 8 T_1-weighted 19F-images were acquired successively without any delay inbetween. This experiment was averaged 16 times with a recovery delay of 15 sec between averages resulting in a total acquisition time of less than 6 min. From the T_1-weighted images the T_1 map is calculated pixelwise according to [10] and converted into a pO_2 map by using the calibration constants c and s. The pO_2 maps had a FOV of 8.5 cm x

8.5 cm for the abdominal cavity and 6.4 cm x 6.4 cm for the muscle. The matrix size was 128 x 128. Further experimental parameters were 50 kHz sweepwidth and 4.4 ms repetition time.1H-NMR images with the same resolution were acquired for anatomical orientation.

RESULTS AND DISCUSSION

The transaxial pO_2 maps and the 1H-NMR images had a slice thickness of 2 mm and an in-plane resolution of 0.66 mm x 0.66 mm for the abdominal cavity and of 0.5 mm x 0.5 mm for the skeletal muscle. A representative pO_2 map and the corresponding 1H-NMR image of the skeletal muscle is given in Figure 1.

In Table 1, pO_2 values are given in mmHg as mean values of 3 animals ± standard deviation. In both cases we obtained negative apparent pO_2 values under hypoxic conditions, showing that the problem with the negative apparent pO_2 values is not limited to PFC emulsions. Therefore we assume that the negative apparent pO_2 values result mainly from the fact that conditions in the *in vitro* calibration and in the *in vivo* experiment are not the same. Since the calibration constant c (anoxic intercept) is much more sensitive to changes in the experimental setup than the calibration constant s (slope) [11], we propose the following correction: Instead of c we use a constant c* which is the lowest R_1 value (highest T_1 value) found in the *in vivo* experiments under hypoxic conditions, while we

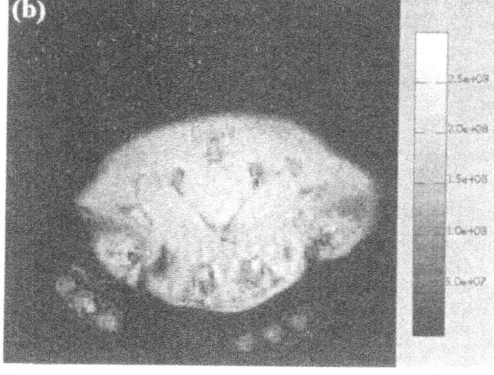

Figure 1. Pixelwise calculated pO_2 map (a) and 1H-NMR image (b) (for anatomical orientation) of a 2 mm transaxial slice through the hind legs of a rat breathing pure oxygen. An F-44E droplet of 0.2 ml was injected into the quatriceps femoris of the rat. The resolution is 0.5 mm x 0.5 mm. The pO_2 values are given in 10^{-1} %, the average pO_2 value in the F-44E depot was 53 mmHg.

Table 1. Oxygen tensions in the examined regions, given in mmHg ± SD

F_iO_2	100 %	21 %	10 %
pO_2 calculated with c			
Abdominal cavity	54±14	17±7	−15±7
Skeletal muscle	36±23	6±10	−27±6
pO_2 calculated with c*			
Abdominal cavity	88±14	52±7	19±7
Skeletal muscle	70±23	41±10	7±6

(c = 0.319/sec, c* = 0.270/s)

leave s unchanged. The value c* was 0.270/sec. The corrected pO_2 values (calculated with c*) are also given in Table 1.

This is the first approach to avoid negative apparent pO_2 values which are physiologically impossible. However, further experiments are necessary to investigate whether the constant c* is tissue-dependent and whether the constant s from the *in vitro* calibration is valid *in vivo*.

ACKNOWLEDGMENT

The authors thank Dr. J.L.Howell of DuPont de Nemours (Wilmington, DE) for supplying the neat F-44E. This work was supported by the Wilhelm Sander Stiftung (grant 93.001.1).

REFERENCES

1. Mason RP, Nunnally RL, Antich PP: Magn. Reson. Med. **18**, 71–79 (1991)
2. Mason RP, Jeffrey FMH, Malloy CR, Babcock EE, Antich PP: Magn. Reson. Med. 27, 310–317 (1992)
3. Hees PS, Sotak CH: Magn. Reson. Med. 29, 303–310 (1993)
4. Dardzinski BJ, Sotak CH: Magn. Reson. Med. 32, 88–97 (1994)
5. Nöth U, Morrissey SP, Deichmann R, Adolf H, Schwarzbauer C, Lutz J, Haase A: Magn. Reson. Med. 34, 738–745 (1995)
6. Shukla HP, Mason RP, Bansal N, Antich PP: Magn. Reson. Med. 35, 827–833 (1996)
7. Lutz J, Nöth U, Morrissey SP, Adolf H, Deichmann R, Haase A: Advances Exp. Med. Biol. 388, 53–57 (1996)
8. Riess JG, Le Blanc M: Pure & Appl. Chem. 54, 2383–2406 (1982)
9. Nöth U, Deichmann R, Adolf H, Schwarzbauer C, Haase A: J. Magn. Reson. B 105, 233–237 (1994)
10. Deichmann R, Haase A: Quantification of T_1 values by SNAPSHOT FLASH NMR imaging. J. Magn. Reson. 96, 608–612 (1992)
11. Mason RP, Shukla H, Antich PP: Magn. Reson. Med. 29, 296–302 (1993)

81

SKIN OXYGEN SATURATION IMAGER

D. J. Clark, T. J. H. Essex, and B. Cater

Regional Medical Physics Department
Newcastle General Hospital
Newcastle upon Tyne
NE4 6BE, United Kingdom

INTRODUCTION

The fraction of incident light which is returned from a turbid sample contains information about the pigments within the sample. Analysis of the spectra of this returned light allows detailed information of the relative quantity of these pigments to be determined. Termed remittance spectroscopy or reflectance spectrophotometry this technique has been widely employed; for example, Edwards and Duntley (1939) and Feather et al (1989) used it to investigate skin colour and cutaneous pigments; Malm and Tornquist (1988) and Bjerring and Andersen (1991) investigated port-wine stains and laser treated skin and Harrison et al (1992) looked at haemoglobin index and oxygen saturation. Acquiring a full remittance spectrum allows complex algorithms to be employed and detailed investigations of the various chromophores to be undertaken. However, this has the disadvantage of necessitating the use of expensive spectrophotometric equipment, time-consuming processing and the restriction of single point analysis. Whilst newer instruments employing rapidly rotating filters (Frank et al 1989) or photodiode arrays (Hagihara et al 1988) enable faster spectral acquisition and new super fast computers have and will continue to speed up processing, the restriction on single point investigations remains.

In this study we wanted to move away from point measurements to produce images of skin blood content and oxygen saturation. Acquiring full spectral information for every point within the desired patient image would obviously prove impracticable. However, the use of smaller wavelength ranges or even reducing the analysis to one or two wavelengths has been shown to produce clinically useful data. Pulse oximeters, for example, often employ only two wavelengths (Temper and Barker 1989). We therefore decided to use conventional video imaging techniques and restrict our investigations to a limited spectral region. Due to the restricted nature of the data acquired in the current system, exact measurements of oxygenation and blood content are not claimed, only relative levels across the image.

Oxygen Transport to Tissue XIX, edited by Harrison and Delpy
Plenum Press, New York, 1997

MATERIALS AND METHODS

The remittance spectrum of haemoglobin contains information about the oxygen status of the blood. The predominant double peak of the spectrum between 500 and 600 nm is a distinctive feature of oxyhaemoglobin whereas deoxyhaemoglobin has a modified spectral shape which, if fully deoxygenated, assumes a single peak. Several workers have described algorithms for determining haemoglobin content and oxygen saturation from this spectral information (Harrison et al 1992, Feather et al 1989, Turnbull 1977, Sato 1983). We assessed these methods and have adopted an approach based most closely on that of Harrison et al (1992) which requires the use of 6 discrete wavelengths and two equations:$A\lambda i$ is the absorbance at wavelength i.

A video camera operated in single frame mode is used to acquire the image of the patient surface. The wavelength selection is provided by use of interference filters. The filters used in this case have peak transmission wavelengths (λ_1-λ_6) of: 500, 530, 550, 570, 590 and 560nm and bandwidths of between 8 and 12nm. Six images are obtained in this way, each representing the remittance at a single wavelength (actually a range of wavelengths, determined by the bandwidth, centred on the peak wavelength of the filter). For any pixel within the image, a very crude remittance spectrum can be reproduced by plotting the camera intensity versus filter wavelength. It is, therefore, possible to use equations (1) and (2) to calculate the haemoglobin index and oxygen saturation index for each pixel in the frame and hence to produce images.

In its current form the instrument consists of several elements: a specialised optical radiation source which projects polarised light over the patient surface to produce uniform illumination; a CCD camera element and a set of interference filters to allow light from a specified wavelength range only to be detected; a computer to store and manipulate this information so as to calculate oxygen saturation and haemoglobin index; software to provide the means to control the hardware, process the information and display it in the form of an image - colour- or grey-scale coded so as to impart the information about the blood tissue status.

The optical radiation source consists of four 45W xenon flash lamps mounted on a rectangular frame (the scanner frame) of dimensions 400 x 500mm. The frame is covered with polarising film. Each flash tube requires 340 volts across it and a 6000 volt trigger pulse to fire it. The 340 volt supply is provided from a 1000μF capacitor which is repeatedly charged and discharged as the flash tube is fired. Charging time is approximately 1 second. As a safety precaution an automatic discharge circuit is incorporated.. The exposure time, and hence the light output, is varied by the on time of the flash lamp. This is controlled by varying the pulse width of the isolated trigger pulse from the computer. When the trigger pulse occurs across the isolator, a thyristor is triggered which generates a firing pulse to the flash tube via a trigger transformer. At the trailing edge of the trigger pulse the flash is extinguished and the capacitor charging circuit re-energised.

The camera used is a Plunix TM765E with manual gain control. The frame mode option is fitted to allow single frame data collection. The lens fitted is a standard 9mm f16 C-mount, focusing at about 0.2 metres. The camera lens is covered with crossed polarising film. The camera and filter slide assembly are constructed from 8mm aluminium plate and the base plate holds the assembly in position on the scanner frame. The 25.4mm diameter interference filters (Ealing Electro-Optics) are mounted in a slider mechanism on bearings and mounted between the camera lens and the CCD element. The slider is driven backwards and forwards by an ac motor and a toothed belt. An indexing bar is mounted on the filter carriage that allows the filters, by use of a photocell, to be placed in the correct alignment with the camera before an image is grabbed.

The computer system used in this system was a 486 DX-2 with 16Mb RAM. The graphics card is capable of 1024 x 768 @ 24bit colour. The software was developed in Delphi under Windows95. The software has been written so as to be user friendly with standard window interface format and performs the following tasks: controls the camera, filter carriage and flash lamp elements; acquires single frame images for each filter and stores them sequentially in a single file; displays the six separate grey scale images which allows the operator to check for over-exposure; uses a precision spot marker held in the background of the image to automatically re-align the six images; performs a normalisation procedure to account for the different peak transmission of the filters. This is achieved by use of two neutral density reflectors (one at approximately 0.2 reflectance and one at 0.8) also held in the background of the image. Finally, equations (1) and (2) are used to produce a haemoglobin and oxygen saturation map of the image which are displayed simultaneously (either image may be displayed in full screen format if desired). In the current system, only one image per filter is obtained and therefore only one value for each of the unknowns in the equations. It may be possible in the future to average many images to improve data quality.

$$Hb = \frac{(A_{\lambda_2} - A_{\lambda_1})}{\lambda_2 - \lambda_1} - \frac{(A_{\lambda_3} - A_{\lambda_2})}{\lambda_3 - \lambda_2} + \frac{(A_{\lambda_3} - A_{\lambda_4})}{\lambda_4 - \lambda_3} - \frac{(A_{\lambda_4} - A_{\lambda_5})}{\lambda_5 - \lambda_4} \; x100 \tag{1}$$

$$Ox = \frac{(A_{\lambda_4} - A_{\lambda_6})}{\lambda_6 - \lambda_4} - \frac{(A_{\lambda_6} - A_{\lambda_3})}{\lambda_6 - \lambda_3} \; x \frac{100}{Hb} \tag{2}$$

RESULTS

Haemoglobin and oxygen saturation images have been produced for a range of physiological conditions. The instrument in its current form is restricted to images of area up to 400x500mm with a resolution of 581 x 756 pixels. We present a set of illustrative images.

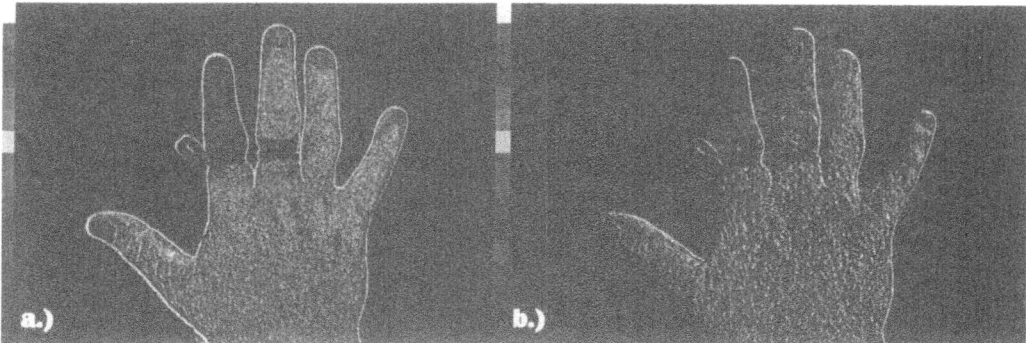

Figure 1. Grey-scale images of the hand showing a) haemoglobin content and b) oxygen saturation. An increase in haemoglobin or saturation is indicated by lighter grey pixel colour. The index finger has had the blood drained by using a "rolled-on" tourniquet and therefore has low oxygen saturation and low haemoglobin content. The middle finger has been occluded using a "snapped-on" tourniquet. The finger has increased haemoglobin due to partial engorgement caused during application of the tourniquet, but low oxygen saturation. The images show an apparent local heterogeneity which is probably due to local tissue scattering and is currently under investigation.

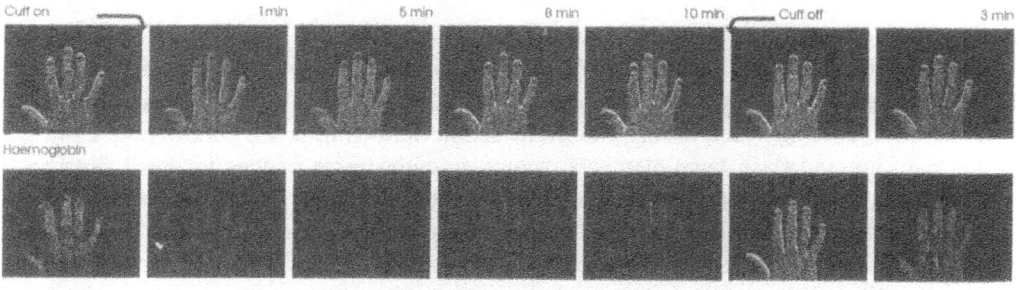

Figure 2. A series of sequential images of haemoglobin content and oxygen saturation taken during 10 minutes of arterial cuff occlusion. The cuff was inflated to 250mmHg and images were acquired every 60 seconds over a 14 minute period. (Only representative images are shown.)

The occlusion de-saturation and post-ischaemic reactive hyperaemic response are clearly evident from the oxygen saturation images (figure 2) and graph in figure 3. A gradual increase in skin blood content is also evident from figures 2 and 3 with a rapid increase immediately upon cuff release.

CONCLUSION

This technique permits the visualisation of haemoglobin content and oxygen saturation of skin of any area of the body surface and has many applications including: peripheral vascular disease; assessment of critical limb ischaemia; skin grafting; management of burns patients; port-wine stain therapy; allergy testing; psoriasis management; quantifica-

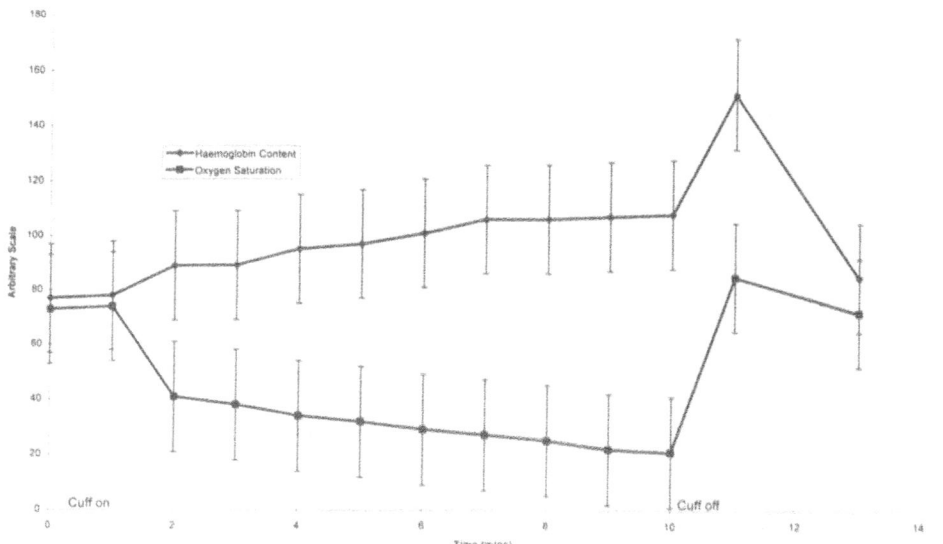

Figure 3. The average values taken from a 20x20 pixel area from the centre of the back of the hand in figure 2 plotted against time.

tion of erythemal responses. The validity of the technique has been demonstrated through the acquisition of images for a range of physiological conditions. The current system has limitations due to hardware and software as well as the physical principles of the technique. Further research and development in all these areas is ongoing.

ACKNOWLEDGMENTS

We wish to thank Mr P.O.Byrne for help with the power supply electronics and his general assistance and many suggestions; the staff of the departments of plastic surgery and vascular surgery at Newcastle General Hospital and the Royal Victoria Infirmary; and Professor K. Boddy for his continual support and encouragement.

REFERENCES

Bjerring P and Andersen PH 1991, Reflectance spectrophotometry - an objective method for quantification of skin blanching after laser treatment. Lasers Med.Sci. **6** 74.

Edwards EA and Duntley SQ 1939, The pigments and color of human skin. Am. J.Anat. **65** 1.

Feather JW, Hajizadeh-Saffer M, Leslie G and Dawson JB 1989, A portable scanning reflectance spectrophotometer using visible wavelengths for the rapid measurement of skin pigments. Phys. Med. Biol.**34** 807–820.

Frank KH, Kessler M, Appelbaum K and Dummler W 1989, The Erlangen micro-lightguide spectrophotometer EMPHO I. Phys.Med.Biol. **34** 1883–1900.

Hagihara B, Okutani N, Nishioka M, Negayama N, Ohtani W, Takamura S and Oka K 1988, Measurement of rat brain spectra. Adv.Exp.Med.Biol. **222** 351–357.

Harrison DK, Evans SD, Abbot NC, Swanson Beck J and McCollum PT 1992, Spectrophotometric measurements of haemoglobin made saturation and concentration in skin during tuberculin reaction in normal human subjects. Clin. Phys. Physiol. Meas. **13** 349–363.

Malm M and Tornquist G 1988, Telespectrophotometric reflectance measurements for evaluating results after argon laser treatment of port-wine stains compared with natural colour system notations. Ann.Plast.Surg. **20** 403–408.

Sato N, Hayashi N, Kawano S, Kamada T and Abe H 1983, Hepatic hemodynamics in patients with chronic hepatitis or cirrhosis as assessed by organ-reflectance spectrophotometry. Gastroenterology **84** 611–616.

Temper K and Barker SJ 1989, Pulse oximetry. Anesthesiology **70** 98–108.

Turnbull FW. 1977, The optical measurement of blood content and blood oxygen saturation in superficial tissue. PhD thesis. University of Strathclyde, UK.

82

AN IN-LINE OXYGEN GAS-FRACTION SENSOR FOR ANESTHESIA AND INTENSIVE CARE

P. E. M. Huygen,[1] A. Hartog,[1] C. Kolle,[2] E. Oosterbosch,[1] and B. Lachmann[1]

[1]Department of Anesthesia
Erasmus University Rotterdam
The Netherlands
[2]Joanneum Research
Graz, Austria

1. INTRODUCTION

Mechanically ventilated, critically ill patients are completely dependent on external care with respect to nutrition and oxygen supply. Adequate nutrition can have a strong influence on the outcome in critically ill patients, especially in COPD patients. Monitoring the uptake of oxygen provides the opportunity to control the enteral or parenteral nutrition[1] to monitor the effect of therapeutic measures to reduce energy expenditure,[2] and to monitor transient variations of the oxygen uptake caused by sudden changes of the right-to-left shunt e.g. when parts of the lungs are opening or closing.[3] However, measuring the metabolic rate is prohibitively complicated to be performed on a routine basis. Devices to accurately measure the rate of metabolic gas exchange have been described,[4,5,6] but they tend to be bulky and complicated, encorporating a mixing chamber for the expiratory gas. This is partly due to the fact that an affordable device to measure the oxygen fraction in the inspiratory and expiratory gas with short response time is lacking. As a result, measurement of metabolic rate can only be done occasionally, using a single device for all the patients in the ward, without facilities for on-line connection to the patient monitoring database.

The oxygen-induced quenching of fluorescence in porphyrines offers the possibility to measure the partial oxygen pressure in the inspired and expired gas. The porphyrine can be placed inside the tube through which the patient is ventilated. This has the advantage that the oxygen partial pressure is measured immediately, without delay due to transport of sample gas into a separate measuring chamber, as is often the case with common gas measuring devices. Especially in mechanical ventilation, transport of sample gas to a measuring chamber is unfortunate, because this causes a pressure-dependent delay whereas pressure is highly variable during mechanical ventilation.

In this paper we evaluate a prototype device (the oxymeter) to measure the oxygen partial pressure using fluorescence quenching.

Oxygen Transport to Tissue XIX, edited by Harrison and Delpy
Plenum Press, New York, 1997

1.1. Theory

Fluorescence means the immediate luminescence of materials (fluorophores), stimulated by radiation. Platinum(II)-octaethylporphyrin-keton is a fluorophore. When it is irradiated with light with a wavelength of 520 nm, molecules absorb photons, and emit light with a longer wavelength, 730 nm. The difference of wavelength between stimulation radiation and fluorescence radiation is called the Stokes shift. Oxygen molecules can release the energy of the excited porphyrin molecules, so that no photon is emitted, thus causing quenching of the luminescence. The relationship between the partial pressure of the oxygen (P_{O_2}) and the fluorescence intensity I can be described by the Stern-Volmer equation:

$$P_{O_2} = \frac{1}{K_{SV}} \left(\frac{I_0 - S}{I - S} - 1 \right)$$

where I_0 is the fluorescence intensity in the absence of oxygen, S is the fluorescence of an "unquenchable fraction" and K_{SV} is a constant, the Stern-Volmer constant.

Once the values of K_{SV}, I_0 and S have been determined by a calibration with three gases with different oxygen fractions, the partial oxygen pressure can be calculated from the measured intensity of the fluorescence.

2. DESCRIPTION OF THE DEVICE

Platinum(II)-octaethylporphyrin-keton is used as an oxygen indicator. It is immobilized in polyvinylchloride (PVC) and applied on a transparent polyester carrier foil (wet layer thickness: 20 µm). This foil is attached covering a hole in a tube through which a patient can be ventilated (figure 1). The foil is irradiated by a yellow Light Emitting Diode (LED) radiating through a suitable band-pass filter (colored glass). A silicon photodiode, placed behind a narrow-band filter that passes light with wavelength of 720 nm, measures the intensity of the fluorescence radiation. Another silicon photodiode measures the intensity of light from the LED that is reflected by the surface of the foil. This intensity is used as a reference to account for long-term deviations of the optical path and the LED intensity. The LED and phododiodes are heated to a constant temperature of 37°C.

To reduce ambient light interference and electrical drift, the LED generates a square wave signal and a synchronous demodulation system obtains the intensity of the lumines-

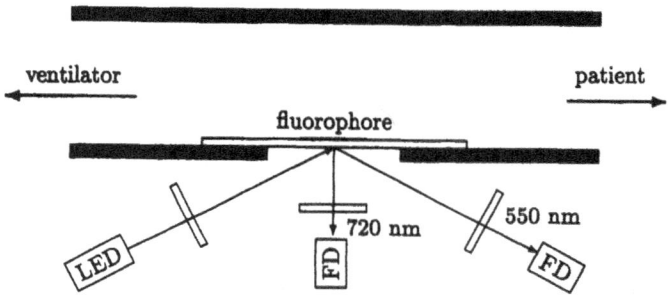

Figure 1. Scheme of a prototype to measure partial oxygen pressure in the ventilation tubing. See text for explanation. LED: Light emitting diode; FD: Photodiode.

cence and reference. Detection and demodulation are done by an integrated circuit involving a 20 bit ADC. An additional low-cost microprocessor provides a digital serial (RS232) output of signal and reference with a sample frequency of 25 Hz. The RS232 port is connected to a computer (PC) that performs calibration, displays the calibrated signal and stores the uncalibrated signals in a disk file.

3. MEASUREMENTS

3.1. Measurement of Temperature Dependence

In order to determine the sensitivity of the device to variations in temperature, the integrated thermostat/heater was disabled and the device was placed in an oven. The oven was heated to temperatures of 21°C, 30°C, 35°C and 40°C. When equilibrium was reached, the oxymeter was flushed with air, oxygen and nitrogen. After establishing equilibrium with the flushed gas the signal and reference intensity was measured. In order to account for temperature sensitivity of the opto-electronic equipment, measurements were also done with a piece of aluminium foil replacing the porphyrine-carrying foil.

Table 1 shows the measured ratios of signal to reference intensity. Figure 2 shows the net temperature effect of the foil, when the signal/reference ratio is divided by the signal/reference ratio found in the measurement with the aluminium foil at the same temperature. As an extra normalization the obtained results are divided by the result obtained at 35°C.

Table 1. Signal-Reference ratios measured at different temperatures while the system was filled with nitrogen, air or oxygen or when the fluorophore carrying foil was replaced by a sheet of aluminium foil

t(°C)	N_2	Air	O_2	Alu
21	2.12	1.64	0.95	0.0168
26	2.09	1.57	0.86	0.0146
30	2.06	1.50	0.82	0.0192
35	1.98	1.37	0.71	0.0109
40	2.00	1.28	0.60	0.0092

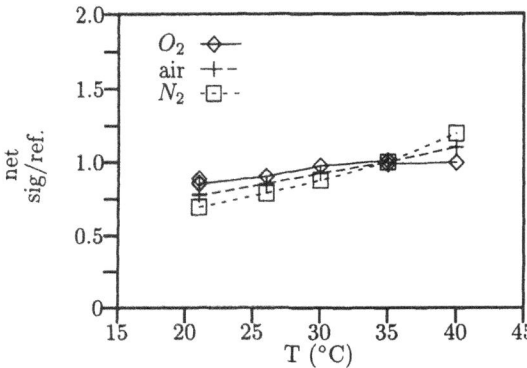

Figure 2. Temperature dependence of the aluminium foil. For each of the gases oxygen, nitrogen and air the quotient of the signal/reference and the signal/reference measured using aluminium foil is plotted, relative to the value found at 35 °C.

3.2. Measurement Accuracy and Dynamic Response

The noise level has been measured for several oxygen partial pressures. The standard deviation of the oxygen partial pressure signal was measured to be less than 0.1 kPa in air, less than 0.2 kPa in gas with 50% oxygen, and less than 0.5 kPa in pure oxygen.

The dynamic response has been measured by performing expirations followed by rapid inspirations through the oxymeter, and measuring the time needed for the oxymeter signal to follow the stepwise oxygen fraction change from 90% back to 10%. It was measured to be less than 150 ms.

3.3. An Example

Figure 3 shows the calibrated signal from the oxymeter while a healthy subject breathes through the device. The oxygen fraction has been measured simultaneously by a mass spectrometer (MGA2000, Case Scientific, UK). The obtained curves are similar, but the oxymeter signal underestimates the oxygen partial pressure during the expirations due to heating of the fluorophore by the warm expired air.

4. DISCUSSION

The potential of oxygen-induced fluorescence quenching as a new method to measure the oxygen fraction in the inspiratory and expiratory gas has been evaluated. The

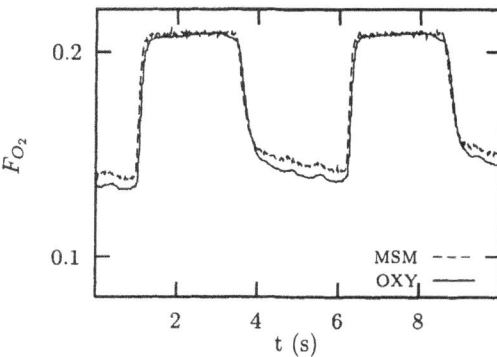

Figure 3. Oxygen signals measured with the prototype oxymeter (OXY) and with a mass spectrometer (MSM) while a subject breathes through the device.

10–90% rise-time of app. 150 ms that we found is sufficient for accurate determination of the end-tidal partial oxygen pressure,[7] but for accurate measurement of oxygen consumption it might be necessary to improve the rise time by modification of the immobilizer of the fluorophore and mathematical compensation for rise time.[8,9,10]

The main advantages of this method are that it is possible to build an affordable, small device, that can be incorporated in the tubing system of a mechanical ventilator, that the response on variations of the oxygen partial pressure is short, and that transport of sample gas to a separate measuring chamber, causing pressure dependent delay, is not needed. The main problem that has to be solved is the influence of the temperature sensitivity of the fluorescence. The fluorescence intensity varies 10 to 25% in the temperature range between 20 °C and 40 °C. If not corrected for, a large systematic error in the measurement of oxygen uptake will result. To account for the temperature dependence either the temperatures of the inspired and expired gas have to be made equal within at least 1%, or the temperature of the porphyrine will have to be determined during the measurement.

REFERENCES

1. JJB van Lanschot, BWA Feenstra, CG Vermey, HA Bruining. 1986. Calculation versus measurement of total energy expenditure. *Crit Care Med*; 14: 981–985.
2. DL Swinamer, PT Phang, RL Jones et al. 1988. Effect of routine administration of analgesia on energy expenditure in critically ill patients. *Chest*; 92: 4–10.
3. B Lachmann. 1992. Open up the lung and keep the lung open. *Intens Care Med*; 18; 19–32.
4. BWA Feenstra, WPJ Holland, JJB van Lanschot, HA Bruining. 1985. Design and validation of an automatic metabolic monitor. *Intens Care Med*; 11: 95–99.
5. H Stam, B van den Berg, JM Bogaard, A Versprille. 1987. A simple and accurate automated system for continuous long-term metabolic studies during artificial ventilation. *Clin Phys Physiol Meas*; 8: 261–269.
6. C Weissman, A Sardar, M Kemper. 1994. An in vitro evaluation of an instrument designed to measure oxygen consumption and carbon dioxide production during mechanical ventilation.. *Crit Care Med*; 22: 1995–2000.
7. JX Brunner, DR Westenskow. 1988. How the rise time of carbon dioxide analysers influences the accuracy of carbon dioxide measurements. *Br J Anaesth*; 61: 628–638.
8. A Arieli, HD van Liew. 1981. Corrections for the response time and delay of mass spectrometers.. *J Appl Physiol*: 51: 1417–1422.
9. JHT Bates, GK Prisk, TE Tanner, AE McKinnon. 1983. Correcting for the dynamic response of a respiratory mass spectrometer.. *J Appl Physiol*; 55: 1015–1022.
10. RR Mitchell. 1979. Incorporating the gas analyzer response time in gas exchange computations. *J Appl Physiol*; 47: 118–1122.

83

IMAGING HYPOXIA AND BLOOD FLOW IN NORMAL TISSUES

Cameron J. Koch,[1] Edith M. Lord,[2] Irving M. Shapiro,[3] Ronald I. Clyman,[4] and Sydney M. Evans[5]

[1]School of Medicine
[2]Cancer Center
 University of Rochester
 Rochester, New York
[3]School of Dentistry
[4]School of Medicine
 University of California
 San Francisco, California
[5]School of Veterinary Medicine
 University of Pennsylvania
 Philadelphia, Pennsylvania

INTRODUCTION

Oxygen deficiencies in normal tissue are assumed to be rare. This assumption is based upon detailed and time-tested mathematical and physiological models (for reviews see Oxygen Transport to Tissue Series). The input parameters to such models require detailed estimates of various circulatory and tissue properties: blood flow, distribution, oxygenation and pH; tissue cellularity, respiration rate and oxygen diffusion constant. Many of these parameters are only available in a global sense; thus there have been many experimental approaches to the assessment of tissue oxygenation. These include measurements by invasive devices such as microelectrodes. More recently, non-invasive techniques have been introduced, such as phosphorescence decay of blood-born dyes activated by short light pulses (Rumsey et al., 1988; Lo et al., 1996) and EPR of carbon particles (Vahidi et al., 1994). The former technique, introduced by Dr David Wilson, has the dual advantages of accurate temporal and spatial resolution (see reports this conference). One indirect method of tissue oxygen assessment utilizes the known oxygen-dependence of cellular radiation response - cells with oxygen tensions below about 20mm of Hg become progressively more radiation resistant. By measuring the radiation response of stem cells in the skin, investigators have concluded that some degree of hypoxia exists in this tissue (Douglas et al., 1975).

We have been interested in a technique which has the capability to measure hypoxia with very fine spatial resolution. This technique depends on the bioreductive metabolism of 2-nitroimidazoles (Varghese et al., 1976). 2-nitroimidazoles are thought to be reduced in one-electron steps at a relatively constant rate by cellular nitroreductases such as cytochrome P450 reductase. The reduced drugs (2 or 4 electron products, nitroso or hydroxylamine) bind to protein thiols forming covalent adducts which can be detected by various means (Raleigh and Koch, 1990). The oxygen dependence of the process occurs at the stage of the first electron transfer producing nitro-radical anion, which is back-oxidized at a rate which depends on the local oxygen concentration (Koch et al., 1984). Detection strategies include both non-invasive and invasive measurement of specific chemical groups on the drug sidechains. Examples of the former include magnetic resonance imaging or spectroscopy (MRI-MRS) of ^{19}F-fluorine (Raleigh et al., 1984), radioactive ^{18}F-fluorine using positron emission tomography (PET) or other isotopes monitored by single photon emission computed tomography (SPECT) cameras (Rasey et al., 1987; Parliament et al., 1992). Examples of invasive techniques are autoradiography of radioactive drug (Franko et al., 1987), electron energy loss spectroscopy (EELS) of electro-dense atoms (Aboagye et al., 1995) and immunohistochemical localization of the protein-drug adducts by polyclonal antibodies (Raleigh et al., 1987).

Our studies have focused on the 2-nitroimidazole EF5; [2-(2-nitro-1h-imidazol-1-yl)-n-(2,2,3,3,3-pentafluoro-propyl)acetamide]. Because of the sidechain fluorines, it is possible to use all presently-known invasive and non-invasive detection techniques with this drug. We have developed very sensitive and specific monoclonal antibodies against EF5 and its adducts, allowing immunohistochemical and flow cytometric analyses which have the potential to discriminate varying levels of hypoxia at the cell-cell level (Lord et al., 1993; Koch et al., 1995; Evans et al., 1995; Laughlin et al., 1996). This assay system has a signal which increases as the level of oxygen decreases; thus it is most suited for evaluation of pathological conditions of very low oxygen tension. To date, our studies on the use of EF5 have emphasized the investigation of the extent and degree of hypoxia in tumors, since hypoxia has a negative impact on tumor therapy (Evans et al., 1996). In the present investigations, we consider the application of EF5 binding to the detailed study of oxygen distribution in normal tissues, with the view towards investigating disease states which might involve low tissue oxygen - ischemia, tissue injury, reduced blood flow, infections, etc. Our results show that this technique may be suitable for such studies, particularly when the tissue involved is relatively homogeneous. In situations where the tissue is heterogeneous, variations in tissue cellularity and nitroreductase activity may cause difficulty in the interpretation of binding. We then demonstrate the use of EF5 to assess oxygenation in two normal tissues whose oxygenation state has been controversial; the growth plate of bone (Shapiro et al, 1996; Ye and Silverton, 1994), and the ductus arteriosus (Clyman et al, 1996). We find no evidence for severe hypoxia in growth plate, but profound hypoxia in closed ductus.

MATERIALS AND METHODS

EF5 was prepared as a 10mM solution in physiological saline. Drug was injected intravenously at a dose equivalent to 100µM whole-body. Detailed pharmacological studies confirm that this drug distributes evenly to all tissues of the body including brain - (Laughlin et al., 1996). After 3–24 hr, depending on serum half-life in the species being studied, tissues were removed from anesthetized animals and quickly frozen. Frozen sections (14µ) were prepared, fixed in 4% paraformaldehyde for 1hr at 0°, blocked with a solution of albumin, milk and mouse IgG, and stained with the monoclonal antibody ELK3–51, conjugated with the fluorescent dye Cy3 (Evans et al., 1995). Normal film or digital images were made using a Nikon or Zeiss fluorescent microscope, and the lamp source was calibrated with a standard dye solution (Laughlin et al., 1996). The presence of patent blood flow was imaged via the intravenous administration of the DNA-binding dye Hoechst 33342 just before animal sacrifice, at an equivalent whole body concentration of 20µM. This dye is removed from the circulation very quickly (Durand, 1982), and can be imaged by fluorescence microscopy on the same or adjacent frozen sections as used for immunohistochemistry of the anti-EF5 antibodies. The excitation and emission characteristics of the Hoechst dye (360nm→450nm) do not overlap with Cy3 (565nm→585nm).

Binding to tissue may be calibrated for the case of either high oxygen (minimal binding) or severe hypoxia (maximum binding) by incubating small samples of tissue in a stirred solution of medium containing EF5 (typically 100 µM) and equilibrated with gas containing 5% CO_2 and either 40% oxygen or <0.01% oxygen. After a suitable incubation time at 37°C, typically 3 hr, the samples were frozen on dry ice, and 14µ sections were collected; sectioning was performed as close to the surface of the tissue as possible and then at a point approximately half-way through the sample. The half-way point was determined by sectioning the entire sample and retaining every 7th section (approximately 100µ), then selecting the section closest to the middle of the sample. Sections were stained with antibodies to EF5 as described above. For samples maintained under anoxic conditions, maximal binding at the tissue surface would be expected with the possibility that there would be a decrease in binding as the distance increased between a region of interest and the closest surface (due to possible limitations in diffusion of drug or nutrients). In contrast, for the high oxygen studies, minimal binding at the tissue surface would be expected with the likelihood that there would be some increase as the distance increased between a region of interest and the closest surface (due to possible limitations in diffusion of oxygen - Franko et al., 1987).

RESULTS

EF5 binding in liver and growth plate of 3 week old chicks is illustrated in Figure 1. The upper two panels illustrate antibody staining of liver with (-6 hr) vs without administration of EF5 to the animal. The lower left panel illustrates binding of EF5 to growth plate after EF5 exposure *in vivo*, while the lower right illustrates binding to samples of similar tissue when exposed to EF5 *in vitro* under conditions of severe hypoxia. For the growth plate samples, the orientation (diagonally upper left to lower right) runs from the trebeculae, often lined with osteocytes, to the terminally differentiated hypertrophic chondrocytes, to the proliferative chondrocytes to the articular cartilage. No binding was observed in the relatively acellular articular cartilage of either section showing that this tissue cannot metabolize the drug. Each panel of the figure in this and other gray-scale photographs represents a field size of 3.8 x 3.0 mm^2. The camera exposure of the digital images was 5 sec for the liver, 15 sec for the growth plate *in vivo*, and 1.5 sec for growth plate *in vitro*.

Figure 2 illustrates binding of EF5 and Hoechst dye to mouse liver and heart. For both tissues, an identical section was photographed at the wavelengths appropriate for Cy3 vs Hoechst 33342. In liver (upper two panels) every nucleus was stained by the Hoechst dye, but there was a gradient in intensity from the arterial to venous vasculature (upper

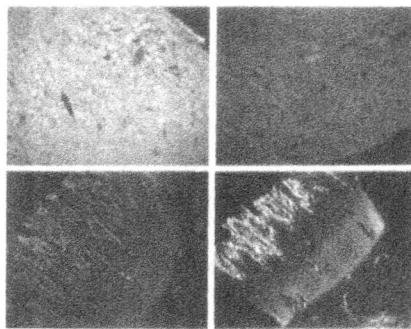

Figure 1. EF5 binding in liver and growth plate of 3 week old chicks. Upper panels: antibody staining of liver with (left) vs without (right) administration of EF5. Lower left panel: growth plate after EF5 exposure *in vivo*. Lower right panel: growth plate exposed to EF5 *in vitro* under conditions of severe hypoxia. Growth plate orientation (diagonally upper left to lower right): trebeculae, often lined with osteocytes → terminally differentiated hypertrophic chondrocytes → proliferative chondrocytes → articular cartilage. Each panel 3.8 x 3.0 mm^2. Camera exposures: liver, 5 sec; growth plate *in vivo*, 15 sec; growth plate *in vitro*, 1.5 sec.

right to lower left). EF5 binding was relatively uniform throughout the liver tissue. In normal heart tissue, every nucleus was equally stained by the Hoechst dye. The field sizes for these and other color panels is 1.2 x 0.8 mm^2. The camera exposure for the EF5 images was 150 seconds. (Use of color slide film and a conventional fluorescent microscope is many-fold less sensitive than the digital camera used for the gray-scale photographs, and for an approximate direct comparison, the exposure times should be divided by 20 for 'equivalent' brightness.

Binding of EF5 to several rat tissues is illustrated for liver, spleen, colon and esophagus (Figure 3). Camera exposure times for liver (upper left) and spleen (upper right) were 200 seconds, while those for colon (lower right) and esophagus (lower right) were 75 seconds. Binding to liver and spleen was relatively homogeneous, while that in colon and esophagus was extremely heterogeneous, with the brightest staining at the lumen surfaces (particularly prominent in esophagus).

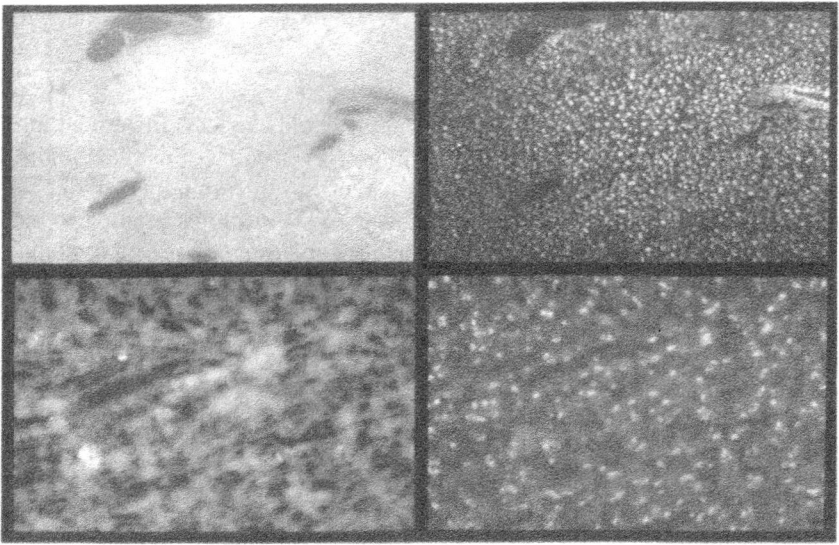

Figure 2. EF5 (left) and Hoechst (right) binding to mouse liver (upper) and heart (lower). Each panel 1.2 x 0.8 mm^2. Camera exposure for the EF5 images was 150 seconds.

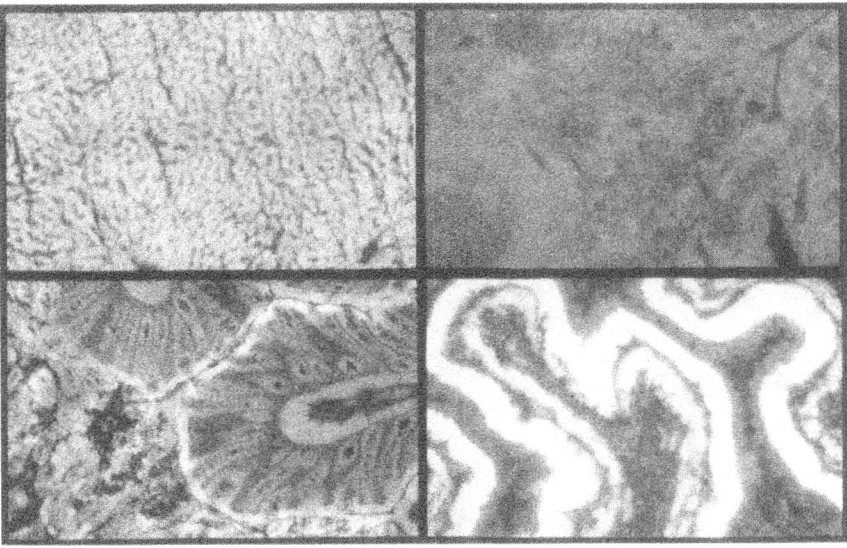

Figure 3. Binding of EF5 to rat tissues with camera exposures: liver (upper left - 200 sec), spleen (upper right - 200 sec), colon (lower left - 75 sec); esophagus (lower right - 75 sec).

Binding in newborn baboon tissue is illustrated for liver, kidney and ductus arteriosus in Figure 4. While the camera exposure time for liver (upper left) kidney (upper right) and patent ductus (lower right) was 5 seconds, that for closed ductus (lower right) was 0.5 seconds. The somewhat higher binding seen at the junction of the ductus and outer surface of the aorta (lower left of both lower panels) was a feature of this junction only, and was not seen in other regions of the aorta (data not shown).

Examples of the use of tumor (9L glioma) and liver tissue samples from a rat, incubated *in vitro* under conditions of severe hypoxia is illustrated in Figure 5. Each tissue sample had spatial dimensions of 2–4 mm. Surface sections from tumor (upper left) and liver (lower left) had relatively uniform binding rates which were similar for the two tissues. Sections from the center of the tissue samples showed that EF5 could diffuse and be metabolized for distances of about 1 mm in tumor tissue (upper right) but some aspect of drug diffusion or metabolism prevented significant binding in liver past 0.5 mm (lower right). Parts of these images were overexposed (camera exposure time 0.5 sec) to allow

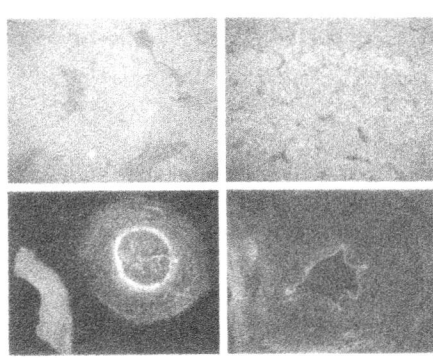

Figure 4. Binding of EF5 to baboon tissues with camera exposures: liver (upper left - 5 sec), kidney (upper right - 5 sec), closed ductus (lower left - 0.5 sec); patent ductus (lower right - 5 sec).

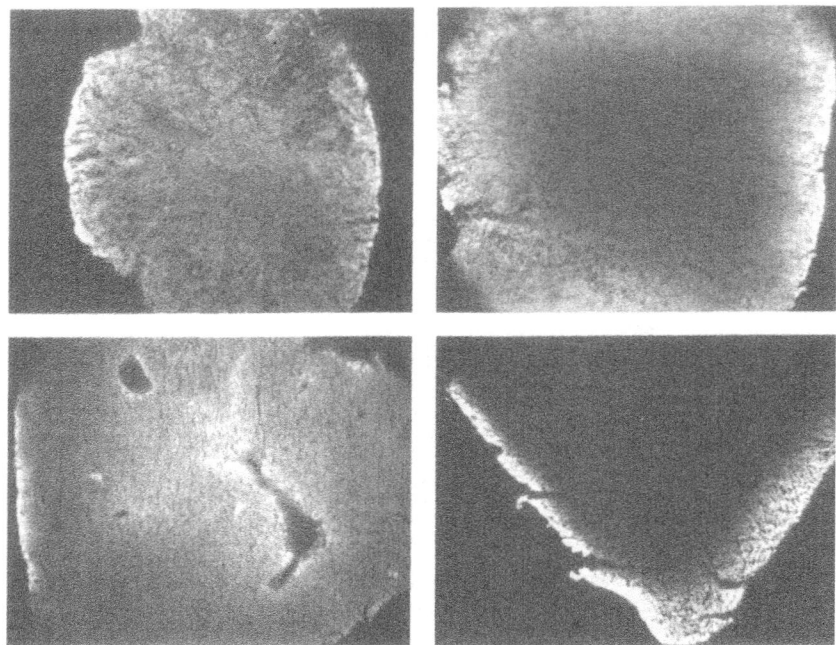

Figure 5. Binding of EF5 to rat tumor and liver tissue samples in vitro (100µM drug, 3 hr, extreme hypoxia). Upper panels: antibody staining of tumor from surface (left) and central (right) sections. Lower panels: antibody staining of liver from surface (left) and central (right) sections. Camera exposure 0.5 sec.

adequate visualization of the entire tissue. The rim of high binding is decreased along the left diagonal of the central section from liver (lower right panel) because of tissue compression artifacts often seen at some edges of the sections.

DISCUSSION

The serum half-life of EF5 increases substantially with animal size and longevity - mice, ~50 min; rats, ~150 min; cats, ~6 hr; baboons, ~8 hr (data not shown). Such half-life increases are a common feature of xenobiotic compounds. The consequence of increasing drug half-life is an effective increase in drug exposure (concentration x time) when tissues were collected several hr post-injection (typically 3–6 hr for rodent and chick tissue; 24–36 hr for cat and baboon). Despite this progressive increase in drug exposure, there is a progressive loss of EF5 binding in liver and many other soft tissues tissues as the animal size and lifespan increases. An exception to this trend was the chicken. Birds have a substantially higher metabolic rate than mammals of similar size; drug lifetime and liver binding in chick was similar on an absolute basis to that of mouse.

For rodent tumor cells in tissues culture, EF5 binding has a consistent, characteristic dependence on oxygen partial pressure, and this dependence can be calibrated (Koch et al., 1995). In addition, the antibody detection technique has been found to provide a quantitative measure of drug-adduct formation (i.e. absolute levels of antibody fluorescence correspond to absolute levels of EF5 uptake measured using radioactive drug - Koch et al.,

1995). Thus, even though there is an indirect relationship between binding and oxygen concentration, it is possible to use EF5 as an absolute oxygen sensor for cells in tissue culture.

In vivo, several factors could interfere with the use of antibody-based detection of EF5 binding as an absolute oxygen sensor: (a) potential limitations in drug access, especially to ischemic regions (b) variability in the levels or oxygen dependence of the appropriate cellular nitroreductase enzymes (c) tissue cellularity (d) uniformity of the presumed linear relationship between actual adduct concentration and the antibody staining process and (e) non-linearity in the fluorescence measurement process. Each of these potential limitations is under current investigation. A method which gives simultaneous information about drug access and binding capability has been termed the 'euthanized animal' control. In this method, drug is administered and the animal euthanized after a suitable time to provide adequate biodistribution (typically 0.5 hr). The animal is then maintained at 37°C for approximately 1 hr. It is assumed that all tissues will quickly become anoxic, at which point available drug will be completely metabolized (the time course of drug disappearance can be determined by homogenizing small tissue samples in trichloroacetic acid (TCA) and measuring non-metabolized drug by high performance liquid chromatography). The expectation is that relatively uniform binding should be observed in every tissue. For non-ischemic tissues this has been our finding (data not shown).

Information on the kinetics of adduct formation can be obtained using various techniques *in vitro*. In particular, the use of the stirred tissue samples incubated with EF5 under conditions of very high vs zero oxygen can provide information on the minimum and maximum binding rates, and on the diffusion properties of non-perfused tissue. An example of the tissue binding rates for rat 9L glioma tumor and rat liver was illustrated in figure 5. This experiment showed that the maximum binding rates, as assessed by immunofluorescence, are not different even though liver tissue is often considered much different than most other tissues from an enzyme complement standpoint. Minimum binding rates for sections from the surface of highly oxygenated tissue are almost undetectable using this technique, thus demonstrating that the dynamic range of the binding method is large and consistent for these tissues (data not shown). With the maximum and minimum binding rates calibrated in this manner, the level of EF5 binding for liver *in vivo* is about 20-fold less than the maximum rate for extreme hypoxia. Assuming that the oxygen dependence of binding is similar for liver and tumor cells, it is possible to estimate the tissue pO2 for normal liver to be in the range of 3% or 22 mm of Hg (torr) oxygen partial pressure. This seems quite reasonable based on the 70% portal blood flow in liver. Prior studies have addressed the question of oxygen levels in liver tissue using other 2-nitroimidazole compounds, and different techniques. The much lower estimates of liver pO2 estimated by the former studies are under investigation.

Experiments using tissue samples of gut and esophagus incubated *in vitro* have proven to be technically difficult, since the tissues contract and reorient under the culture conditions employed to date. Thus, for the tissues which express the most heterogeneity in binding *in vivo*, we still have inadequate data to determine whether these rates of binding reflect low tissue pO2, or anomalous nitroreductase characteristics. Again, the literature data is controversial; for example Cobb has suggested that high binding in esophagus is due to anomalous nitroreductase activity (Cobb et al., 1992), while Franko and colleagues have shown that the binding in esophagus is nevertheless still highly oxygen sensitive (Parliament et al., 1993).

Despite substantial variation in absolute binding rate and tissue cellularity, our data for growth plate in young chicks shows no evidence for severe hypoxia *in vivo*, since binding is about 10-fold less than for tissue specimens incubated under severe hypoxia *in vitro* (Fig 1). Although the mechanical characteristics of this tissue allowed the *in vitro* calibration procedure, it is not possible to accurately assess the precise oxygen level in this tissue.

In contrast, the evidence for extreme hypoxia in closed ductus arteriosus is quite compelling, particularly in view of the dramatic differentiation between patent *vs* closed examples (Fig 4). Although space does not allow a detailed description, we have observed a continuum of time-dependent and degree of closure-dependent binding in newborn baboons (Clyman *et al*, in preparation). The time factor (time after birth) is important because this tissue undergoes substantial remodelling during and after closure, in contrast to other large arteries such as the aorta which, in addition to retaining their complement of muscle cells, do not bind EF5 to any significant extent.

Comparisons of liver binding in the various species investigated (chicken, mouse, rat, cat - data not shown, and baboon) illustrate the consistently moderate binding rate by this tissue. In mice, both absolute uptake of radioactive drug and binding as determined by antibody have very consistent values which substantially exceed the binding rate in many tissues at risk for ischemic injury (e.g. heart, brain - see (Laughlin et al., 1996). Binding in normal brain is almost undetectable, while that in heart is similar to the very low levels found in spleen (see Figs. 1–4 and paper by Evans, this conference). Binding in kidney is similar to that of liver. This is likely caused by a high drug exposure, associated with the dominant urinary excretion route for this drug, rather than a low tissue pO2 as in liver, and similar considerations may apply to other excretory tissues such as gut and colon.

In summary, we believe that binding of the 2-nitroimidazole EF5 may provide a new method to monitor normal tissue pO2 at a very fine spatial resolution. While the heterogeneity of several tissues may prevent a consistent interpretation of the relationship between binding and pO2, it is likely that all pathological conditions resulting in subnormal tissue oxygenation will be amenable to study with this technique.

ACKNOWLEDGMENT

Work supported by grants CA56679 (CJK,SME) CA28332 (EML) DE10875 (IMS) and HL56061 (RIC) from the NIH.

REFERENCES

Aboagye, E.O., Lewis, A.D., Johnson, A., Workman, P., Tracy, M. and Huxham, I.M. (1995). The novel fluorinated 2-nitroimidazole hypoxia probe SR-4554: reductive metabolism and semiquantitative localisation in human ovarian cancer multicellular spheroids as measured by electron energy loss spectroscopic analysis. *Br. J. Cancer,* **72**, 312.

Clyman, R.I., Goetzman, B.W., Chen, Y.Q., Mauvay, F., Kramer, R., Pytela, R., and Schnapp, L.M. (1996) Changes in endothelial call and smooth muscle cell integrin expression during closure of the ductus arteriosus. *PED. RES.* **40**, 198.

Cobb, L.M., Nolan, J. & Hacker, T. (1992). Retention of misonidazole in normal and malignant tissues: interplay of hypoxia and reductases. *Int. J. Rad. Onc. Biol. Phys.,* **22**, 655.

Douglas, B.G., Fowler, J.F., Denekamp, J., Harris, S.R., Ayres, S.E., Fairman, S., Hill, S.A., Sheldon, P.W. & Stewart, F.A. (1975). The effect of multiple small fractions of x-rays on skin reactions in the mouse. *In: Cell Survival After Low Doses of Radiation: Theoretical and Clinical Implications. (T. Alper Ed) Wiley, Chichester,* 351.

Durand, R.E. (1982). Use of Hoechst 33342 for cell selection from multicell systems. *J. Histochem. Cytochem.,* **30**, 117.

Evans, S.M., Koch, C.J., Joiner, B., Jenkins, W.T., Laughlin, K.M. and Lord, E.M. (1995). Identification of hypoxia in cells and tissues of epigastric 9L rat glioma using EF5 [2-(2-nitro-1H-imidazol-1-yl)-N-(2,2,3,3,3-pentafluoropropoly)acetamide]. *Br. J. Cancer*, **72**, 875.

Evans, S.M., Jenkins, W.T., Joiner, B., Lord, E.M. & Koch, C.J. (1996). 2-nitroimidazole (EF5) binding predicts radiation sensitivity in individual 9L subcutaneous tumors. *Cancer Res.*, **56**, 405.

Franko, A.J., Koch, C.J., Garrecht, B.M., Sharplin, J. & Howorko, J. (1987). Oxygen dependence of binding of misonidazole to rodent and human tumors in vitro. *Cancer Res.*, **47**, 5367.

Koch, C.J., Stobbe, C.C. & Baer, K.A. (1984). Metabolism induced binding of 14C-misonidazole to hypoxic cells: kinetic dependence on oxygen concentration and misonidazole concentration. *Int. J. Radiat. Oncol. Biol. Phys.*, **10**, 1327.

Koch, C.J., Evans, S.M. & Lord, E.M. (1995). Oxygen dependence of cellular uptake of EF5 [2-(2-nitro-1H-imidazol-1-yl)-N-(2,2,3,3,3-pentafluoropropoly)acetamide] : analysis of drug adducts by fluorescent antibodies vs bound radioactivity. *Br. J. Cancer*, **72**, 869.

Laughlin, K.M., Evans, S.M., Jenkins, W.T., Tracy, M., Chan, C.Y., Lord, E.M. and Koch, C.J. (1996). Biodistribution of the nitroimidazole EF5 [2-(2-nitro-iH-imidazole-1yl)-N-(2,2,3,3,3-pentafluoropropyl)-acetamide] in mice bearing subcutaneous EMT6 tumors. *J. Pharmacol. Exptl. Therapeut.*, **277**,

Lo, L.-W., Koch, C.J. and Wilson, D.F. (1996). Absolute calibration of oxygen dependent quenching of the phosphorescence of Pd-meso-tetra (4-carboxyphenyl) porphine: a phosphor with general application for measuring oxygen concentration in biological systems. *Analyt. Biochem.*, **In Press**,

Lord, E.M., Harwell, L.W. and Koch, C.J. (1993). Detection of hypoxic cells by monoclonal antibody recognizing 2-nitroimidazole adducts. *Cancer Res.*, **53**, 5271.

Parliament, M.B., Chapman, J.D., Urtasun, R.C., Mcewan, A.J., Golberg, L., Mercer, J.R., Mannan, R.H. and Wiebe, L.I. (1992). Non-invasive assessment of human tumour hypoxia with I-iodoazomycin arabinoside: preliminary report of a clinical study. *Br. J. Cancer*, **65**, 90.

Parliament, M.B., Wiebe, L.I. and Franko, A.J. (1993). Nitroimidazole adducts as markers for tissue hypoxia: mechanistic studies in aerobic normal tissues and tumour cells. *Brit. J. Cancer*,

Raleigh, J.A., Franko, A.J., Treiber, E.O., Lunt, J.A. and Allen, P.S. (1984). Covalent binding of a fluorinated 2-nitroimidazole to EMT-6 tumours in Balb/c mice : Detection by F-19 nuclear magnetic resonance at 2.35 T. *Int. J. Radiat. Oncol. Biol. Phys.*, **10**, 1337.

Raleigh, J.A., Miller, G.G., Franko, A.J., Koch, C.J., Fuciarelli, A.F. and Kelley, D.A. (1987). Fluorescence immunohistochemical detection of hypoxic cells in spheroids and tumours. *Br. J. Cancer*, **56**, 395.

Raleigh, J.A. and Koch, C.J. (1990). The importance of thiols in the reductive binding of 2-nitroimidazoles to macromolecules. *Biochem. Pharmacol.*, **40**, 2457.

Rasey, J.S., Grunbaum, Z., Magee, S., Nelson, N.J., Olive, P.L., Durand, R.E. and Krohn, K.A. (1987b. Characterization of radiolabelled fluoromisonidazole as a probe for hypoxic cells. *Radiat. Res.*, **111**, 292.

Rumsey, W.L., Vanderkooi, J.M. and Wilson, D.F. (1988). Imaging of phosphorescence: a novel method for measuring oxygen distribution in perfused tissue. *Science*, **241**, 1649.

Shapiro, I.M., Mansfield, K.D., Evans, S.M., Lord, E.M. and Koch, C.J. Chondrocytes in the endochondral growth cartilage are not hypoxic. (1996) *Am. J. Physiol.* Submitted.

Vahidi, N., Clarkson, R.B., Liu, K.J., Norby, S.W., Wu, M. and Swartz, H.M. (1994). In vivo and in vitro EPR oximetry with fusinite: a new coal-derived, particulate EPR probe. *Mag. Res. in Med.*, **31**, 139.

Varghese, A.J., Gulyas, S. and Mohindra, J.K. (1976). Hypoxia-dependent reduction of 1-(2-nitro-1-imidazolyl)-3-methoxy-2-propanol by Chinese hamster ovary cells and KHT tumor cells in vitro and in vivo. *Cancer Res.*, **36**, 3761.

Ye, G.F. and Silverton, S.F. Computer modeling of oxygen supply to cartilage: addition of a compartmental model. (1994) *Adv. Exptl Med and Biol.* **361**, 31.

84

IMAGING HYPOXIA IN DISEASED TISSUES

S. M. Evans,[1] M. Bergeron,[2] D. M. Ferriero,[2] F. R. Sharp,[2] H. Hermeking,[3]
R. N. Kitsis,[4] D. L. Geenen,[4] S. Bialik,[4] E. M. Lord,[5] and C. J. Koch[6]

[1]Department of Clinical Studies
School of Veterinary Medicine
University of Pennsylvania
[2]Department of Neurology
University of California
San Francisco, California
[3]Molecular Genetics Laboratory
Johns Hopkins University
Baltimore, Maryland
[4]Department of Cardiology
Albert Einstein College of Medicine
Bronx, New York
[5]Cancer Center
University of Rochester
Rochester, New York
[6]Department of Radiation Oncology
School of Medicine
Philadelphia, Pennsylvania

INTRODUCTION

Hypoxia is a physiologic condition which is a component of many disease processes. Hypoxia can be mediated by low inspired oxygen, lung disease or organ ischemia. Although, in general, the delivery of oxygen to tissue has sufficient excess capacity to compensate for a wide range of demand (luxury perfusion), a marked deficiency of intracellular oxygen results in immediate tissue dysfunction. Prolonged hypoxia is the prevalent cause of irreversible injury in all mammalian tissues (Cotran et al., 1989). Hypoxia plays a role in the pathogenesis of stroke (for review see Hossmann, 1994), cardiac disease (Rumsey et al., 1994), cancer (for review see Adams, 1990), ocular disease (for example see Flammer, 1994), renal disease (Brezis and Rosen, 1995), rheumatoid arthritis (for review see Edmonds et al., 1995) and wound healing (for review see Hunt, 1988), among others. In many diseases, including cancer, the presence of hypoxia decreases the effectiveness of therapy and signals a poor prognosis (Brizel et al., 1996; Hockel et al.,

Oxygen Transport to Tissue XIX, edited by Harrison and Delpy
Plenum Press, New York, 1997

1993). In stroke and myocardial infarction, early determination of the extent and location of hypoxia could modify treatment and prognosis (for review see Nunn et al., 1995).

Potential markers that could be used to estimate tissue oxygen levels include blood vascular perfusion, acidosis (due to the production of lactate in ischemic tissue) or a direct marker of cellular pO_2. Positron emission tomographic (PET) and magnetic imaging spectroscopic (MRS) techniques have been developed to measure intracellular pH (Buxton et al., 1987; McCoy et al., 1995). Unfortunately, these techniques are complex and difficult to perform as routine procedures. In tumors, the highly variable blood flow results in a wide range of intracellular pH measurements. In any tissue, the measurement of pH is an indirect indicator of tissue hypoxia.

Blood vascular perfusion imaging depicts the delivery of plasma to the tissues; this measurement may not directly correspond to the delivery of oxygen to tissues. In tumor tissues, for example, elegant studies using a "window chamber model" have demonstrated the presence of plasma flow without red cells and flow in vessels which are devoid of oxygen (Secomb et al., 1993). Similarly, perfusion measurement do not address whether oxygen delivery is sufficient to meet the needs of the regional tissues, i.e. non-transmural cardiac infarction. None the less, many techniques have been developed to measure tissue perfusion, either directly or indirectly. Experimental techniques for measuring blood flow include direct measurements of blood flow using isolated organs or tumors (Evans, 1994) and *ex vivo* techniques (i.e. Hoechst dyes; see Koch et al, this conference). Clinically relevant techniques include Doppler blood flow, and washin or washout techniques. The regional distribution of radionuclide tracers (such as ^{99m}Tc-hexamethylpropylene amine oxime "HPOA" or thallium) have been used to study cerebral and cardiac ischemic disease. Although these methods can identify regions of low flow, they cannot differentiate between scar, functional parenchymal tissue, ischemia or necrosis.

An imaging agent or technique that specifically identifies hypoxic, viable (although perhaps dysfunctional) tissue is needed. Nitroimidazole agents are ideal for this function because they are preferentially bound in hypoxic viable tissues (first described by Varghese et al., 1976) but cannot be metabolized by regions of tissue necrosis. EF5 is a 2-nitroimidazole drug which binds to protein sulfhydryls in hypoxic cells (Lord et al., 1993; Koch et al., 1995). We have previously demonstrated that the adducts thus formed can be localized using specific monoclonal antibodies conjugated to the fluorescent dye, Cy3 (Lord et al., 1993). Using immunohistochemical imaging techniques, regions of hypoxia can be identified as areas of binding. Non-binding regions are either oxic tissues or necrotic tissues which cannot metabolize the EF5. Oxic viable tissues can be differentiated from necrotic tissues based upon re-staining of slides with hematoxalin and eosin or Hoechst 33342, which binds to intact DNA. One of the strengths of the EF5 binding technique is the large dynamic range of binding and the resultant ability to demonstrate a continuum of cellular pO_2 levels. This was first demonstrated using flow cytometric analyses (Koch, 1995) where the binding of 9L tumor cells incubated *in vitro* in 4% O_2 *vs.* N_2 varied by a factor of 50; moderate O_2 levels (1.2% O_2) show intermediate levels of EF5 binding. Using histogram analyses of immunohistochemical staining of tissue sections, this continuum of binding corresponding to cellular pO_2 can also be demonstrated (Laughlin, 1996).

The purpose of this paper is to demonstrate the wide range of applications of the EF5 technology in several disease processes: cancer, stroke and myocardial ischemia. Currently, the analysis of the presence and distribution of hypoxia by EF5 binding requires the acquisition of tissue for either immunohistochemistry or flow cytometric analyses. For analysis of hypoxia in neoplastic tissues, this is entirely appropriate because tumors are always biopsied or resected prior to definitive therapy. However, in other disease processes such as myocardial infarction and stroke, biopsy is neither feasible nor warranted. For these and other diseases, the ability to non-invasively image hypoxia is necessary. The design of EF5 with multiple fluorines on the side chain not only allowed Lord et al to raise specific monoclonal antibodies to this molecule, but allows the exchange of one or more of the fluorine molecules for ^{18}F and imaged by positron emission tomography (PET). Magnetic resonance imaging or spectroscopy of ^{19}F is also being developed.

MATERIALS AND METHODS

EF5 Immunohistochemistry

EF5, a pentafluorinated derivative of etanidazole [2-(2-nitro-1h-imidazol-1-yl)-n-(2,2,3,3,3-pentafluoro-propyl)acetamide] was synthesized by Dr. M. Tracy, SRI, Palo Alto, Ca. Monoclonal antibodies were made against radiochemically-produced adducts of EF5 and thiol-containing proteins as previously described (Lord, 1993). The antibodies used in the present study, ELK3–51 were conjugated with the fluorescent dye, Cy3 (Southwick et al., 1990). This dye is available in a form which reacts with secondary protein-amines (BDS, Pittsburgh, Pa.). The dye: protein ratio was about 4.

Animal and Tissue Studies

All animal use procedures were in accordance with the National Institutes of Health Guide for Care and Use of Laboratory Animals, and all protocols were approved by the Committees on Animal Research at the respective Universities. In all experimental protocols, EF5 was administered as an intravenous or intraperitoneal injection of 10mM EF5 prepared in 0.9% saline; the equivalent whole-body concentration was 100 μm in all studies except the human xenograft model wherein the final whole body concentration was 200 μM. One-10 hr following EF5 administration, the animal was anesthetized and tissue removed. The tissue was cooled, weighed and quick-frozen in isopentane at -50°, as previously described (Evans, 1995). Tissue sections were cut at 14 μm using an HM 505n cryostat and collected onto poly-l-lysine coated slides. Tissue sections were blocked, stained and rinsed (Koch et al, 1995). Residual unbound drug is removed during the fixation stage. Slides were photographed using a Nikon or Zeiss fluorescence microscope and the lamp was calibrated using standard dye solution (Laughlin, 1996).

Hypoxia-Ischemia in the Newborn Rat Brain

Sprague Dawley rat pups at postnatal day 7 (P7; Bantam Kingman, Fremont, Ca) were used as previously described (Rice et al., 1981; Ferriero et al., 1990) with some modifications. Thirty minutes after intraperitoneal injection of EF5 (0.1 mM/kg), rat pups were anesthetized with a gas mixture containing 1% halothane in 30%O_2 and 70% N_2O_2. They then underwent a right common carotid artery coagulation through a ventral midline neck incision. The wound was sutured and pups were returned to their dam for 90 min. of recovery. Pups were then placed in an 8% O_2/92% N_2 humidified atmosphere at a constant temperature of 37°C for 2.5 hours. Immediately after the hypoxia exposure, rats were decapitated and their brains quickly frozen at -70°C until further analysis. Controls included

EF5 injected rats receiving hypoxia alone with carotid occlusion; EF5 injected rats receiving neither hypoxia nor the carotid occlusion; saline injected rats receiving either hypoxia alone or hypoxia-ischemia.

Myocardial Infarction in Mice

Six week old C57Bl6 male mice (20–25 gm) were initially anesthetized with methoxyfluorane, maintained under general anesthesia with 0.5% isofluorane in 95% O_2 5% CO_2 and ventilated on a Harvard respirator at 115 breaths per minute; 0.5 ml tidal volume. 0.2 ml of 10mM EF5 was administered via a femoral vein catheter. Thirty minutes later, a median sternotomy was performed. A 9–0 silk suture was placed into the epicardial muscle layer between the pulmonary artery and the left atrium in order to ligate the left coronary artery. Control mice were sham operated wherein the suture was inserted, but not tied. One, 4 or 10 hours later, the heart was removed, retroperfused with cold phosphate buffered saline, frozen and prepared for EF5 immunohistochemistry.

Hypoxia in Human Colon Tumor Xenografts Grown in Nude Mice

Five million human colon carcinoma cells (HCT116) cells suspended in 0.1 ml Matrigel matrix (Collaborative Biomedical Products, Bedford, Mass.) were injected subcutaneously over the flank of nude mice. 10–12 days later 7–10 mm diameter tumors were palpable. EF5 was injected twice over 30 minutes; the first dose was given intravenously via the tail vein and the second dose was given intraperitoneally. Total whole body dose equivalent was 200 μm. Three hours following the second injection, the tumor was excised, frozen and prepared for EF5 immunohistochemistry.

RESULTS AND DISCUSSION

Cerebral Hypoxia/Ischemia

Perinatal asphyxia represents the most important cause of neurological morbidity in the full term, as well as the low birth weight, infant. It is often associated with severe neurological sequelae such as cerebral palsy, mental retardation and epileptic seizures (Shaywitz, 1987). Clues to potential neuroprotective strategies may depend on the development of reliable methods to detect early evidence of neonatal hypoxic-ischemic injury. This would also help predict the neurological outcome of the affected children. To study the effects of perinatal hypoxic-ischemic encephalopathy on the EF5 binding in the brain, P7 rat pups were subjected to right carotid permanent occlusion followed by exposure to a low oxygen environment (8%) for 2.5 hours. Hypoxia alone (no ligation) produced patchy areas of low intensity EF5 binding scattered throughout the brain tissue of the contralateral hemisphere (Fig. 1A, left hemisphere). In contrast, increased cellular EF5 binding was found throughout the ipsilateral (ligation and hypoxia) cortex, hippocampus, striatum, thalamus and corpus callosum (Fig. 1A), regions known to be most susceptible to neonatal hypoxic-ischemic delayed injury (Rice et al., 1981). Untreated controls and vehicle injected animals showed no EF5 binding activity (Fig. 1B).

In this animal model, whereas hypoxia exposure alone does not lead to any evidence of cellular damage, the combination of carotid occlusion (ischemia) and low oxygen exposure (hypoxia) usually results in delayed brain injury (Rice et al., 1981). However, evidence of

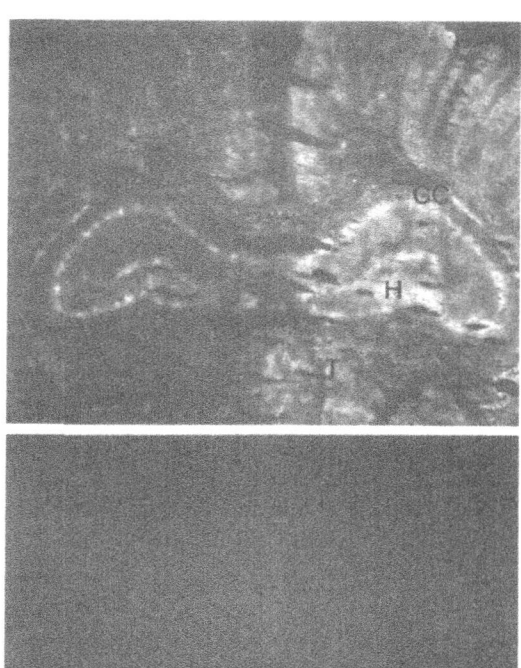

Figure 1. Photomicrographs of 14µm thick coronal sections showing the distribution profile of EF5 binding in newborn rat brain following perinatal hypoxia-ischemia. 1A: Upper panel: brain from an EF5 injected rat immediately after hypoxia ischemia. Whereas hypoxia alone (no ligation) led to a light and patchy EF5 binding pattern in the contralateral hemisphere (left side) combined hypoxia and ischemia produced a strong and diffuse EF5 binding throughout the hemisphere ipsilateral to the carotid occlusion (right side). 1B: Lower panel: brain from an EF5 injected control (normoxic) rat showing no constitutive EF5 binding activity. Abbreviations: C= Corpus collosum, H=hippocampus, Th= Thalamus.

brain damage and neuronal loss in the ipsilateral hemisphere is usually not seen for at least 12–24 hours after the end of the insult (Beilharz et al., 1995). In the present study, interhemispheric differences in the level of EF5 adducts were detectable immediately after the end of the insult, a time at which there is no reported cellular damage in ipsilateral hemispheric areas that have been shown to undergo subsequent delayed injury. Since the level of *in vivo* formation of macromolecular adducts of EF5 in neonatal rat brain depends on the degree of oxygen depletion in the tissue, our findings suggest that the level of intracellular EF5 binding may serve as a marker of cellular vulnerability to neonatal hypoxia-ischemic brain injury. This study describes a new sensitive method for monitoring the level and distribution of cellular brain hypoxia in a rat model of perinatal asphyxia which may provide the basis for the development of a non-invasive imaging method aimed at the early detection of brain regions of high vulnerability to ischemic injury in asphyxiated newborn infants. This would allow physicians to predict susceptibility early before actual cell death so that appropriate therapies can be instituted at the bedside, immediately after the insult.

Myocardial Infarction

Figure 2 demonstrates the EF5 binding in a sham treated (Fig. 2A) and infarcted (Fig. 2B) mouse heart. Homogeneous, minimal binding is seen in the sham treated heart. However, following infarction, 2/3 of the circumference of the left ventricle has regions of

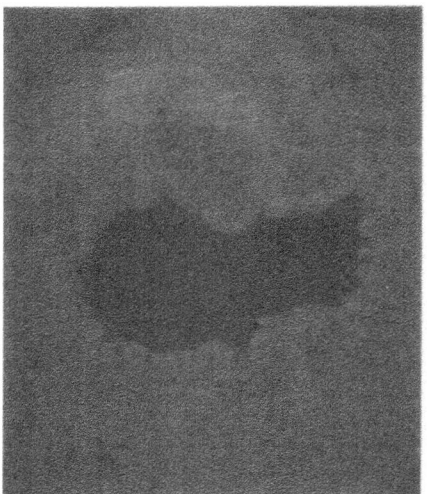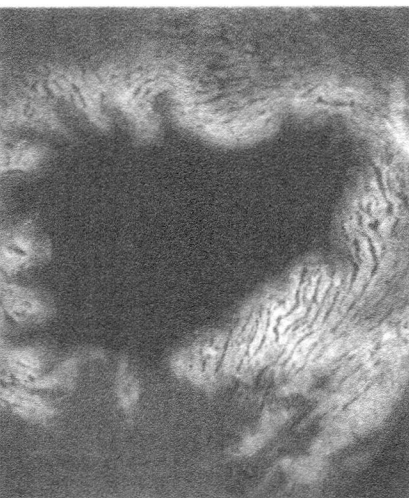

Figure 2. Photomicrographs of 14μm thick transverse sections showing the distribution profile of EF5 binding in mouse hearts following ligation of the left coronary artery and vein. Cross section through the mouse cardiac apex. 2A: EF5 control; sham infarction. Minimal and homogeneous binding is seen throughout the left ventricular wall. 2B: Left coronary artery infarction: EF5 binding is seen in the wall of the left ventricle. Minimal binding is seen for the first several cell layers of the endocardium. This is because these cells have access to the oxygenated blood in the ventricle. Heterogeneous EF5 binding is seen in the ventricular muscle wall. The relatively low binding in the mid-portion of the left ventricular wall may be due to intramural vascularization and/or drug (EF5) access limitations.

intense EF5 binding. For several cell layers adjacent to the ventricular lumen, minimal binding is seen, suggesting that oxygen can diffuse from the blood within the lumen into adjacent myocytes. Brighter binding is then seen, representing hypoxia in the myocytes at a distance from the lumen. The decreased binding in the middle portion of the ventricular wall may be the result of nutrient support from vessels within the muscular wall, or limitations in drug access. Similar considerations apply to the increased binding near the serosal margin of the left ventricular wall.

The ability to map regions of hypoxia is of use in delineating the role of this stimulus in myocyte dysfunction and death *in vivo*. As one example, we are using this approach to investigate the mechanism of myocyte apoptosis in a mouse model of myocardial infarction. Initial results demonstrate correlations between the regions of hypoxia and apoptosis and between the degrees of hypoxia and apoptosis at the time points studied (data not shown). These correlations suggest that hypoxia may play a mechanistic role in the pathogenesis of programmed cell death in this setting. It is likely that the ability to mark hypoxic myocardium could be exploited in the clinical setting. For example, such an approach may be useful in identifying whether ongoing hypoxia plays a role in residual areas of myocardial dysfunction in patients with ischemic heart disease, or following angioplasty of coronary bypass surgery.

Colon Carcinoma Xenografts

The importance of hypoxic cells in limiting the radiation response of animal tumors is very well recognized (for review see (Chapman et al., 1983; Moulder and Rockwell,

1984)). For many years, however, the presence and therefore the significance of hypoxia in the response of human tumors to radiation treatment was unclear. Urbach (Urbach, 1956) was the first to measure low oxygen tensions in malignant human skin tumors directly with pO2 polarography. Gatenby et al (Gatenby et al., 1988) demonstrated a significant relationship between low mean pO2 values in lymph node metastases of head and neck tumors and failure to respond to fractionated radiotherapy, similar to studies on cancer of the uterine cervix (Kolstad, 1968). Using the Eppendorf needle electrode in human tumors have raised the possibility that hypoxia may also have more far ranging implications for tumor behavior, e.g. prediction of poor surgical outcome and tumor metastasis (Hockel et al., 1993; Brizel et al., 1996). We have previously shown that EF5 binding can predict the radiation resistance in multicellular tumor spheroids (Woods et al., 1996) and individual rat (Laughlin et al., 1996) and mouse (Lee et al., 1996) tumors. The studies presented herein on the human colon xenografts represent the first *in vivo* use of EF5 in human tumors. Because of the short half life of EF5 in mice (58 minutes) and the lower absolute binding of EF5 in human vs. rodent cells (Koch, unpublished data), it is necessary to increase the whole body concentration of EF5 to 200 µM for xenograft studies. As expected from studies in rodent tumors, a tremendous heterogeneity of binding patterns is seen in the human colon carcinoma HCT 116 grown subcutaneously in nude mice (Fig. 3). Control studies demonstrated the absence of non-specific binding and autofluorescence in mice injected with saline alone (data not shown). Because of the reported relationship between hypoxia and gene induction (Pahl and Baeuerle, 1994; Mazure et al., 1996; Graeber et al., 1996) further studies to determine the presence and distribution of hypoxia, apoptosis and P53 are planned.

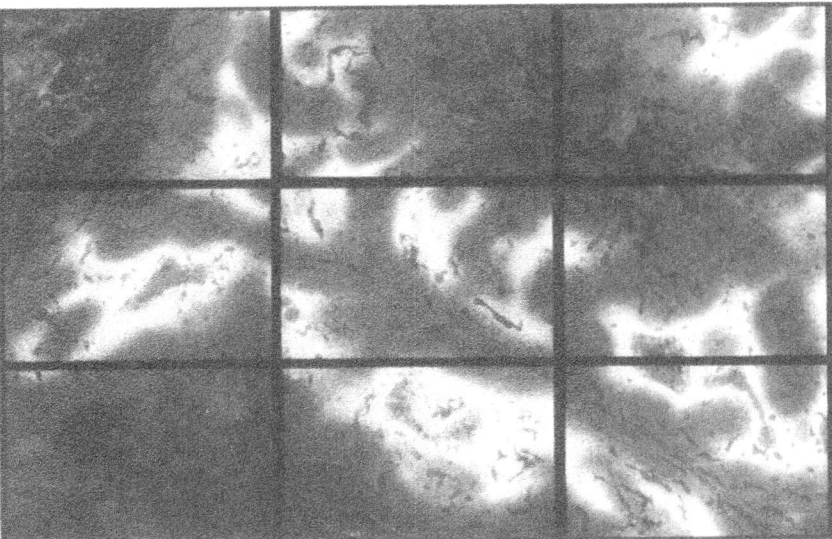

Figure 3. Photomicrographs of 14µm thick sections showing the distribution profile of EF5 binding in a human colon carcinoma grown in a nude mouse (3.4 x 2.6 mm). Section has been stained to demonstrate EF5 binding. Brightly binding regions represent hypoxic tissues. Areas of non-binding are either regions of tissue necrosis, blood vessels or devoid of tissue. Regions of low binding represent oxic tumor cells. Notice the heterogeneous patterns of hypoxia.

In conclusion, cardiovascular disease and cancer represent the leading causes of death in most developed countries. In these and many other disease processes, hypoxia plays a critical role in the pathogenesis, diagnosis and treatment. We have developed a biochemically activated probe for hypoxia, the 2-nitroimidazole, EF5 (2-(2-nitro-1H-imidazol-1-yl)-N-(2,2,3,3,3-pentafluoropropyl) acetamide). The nitroimidazole structure is preferentially reduced as the oxygen level decreases and, upon reduction, the drug binds to protein sulfhydryls in the activating cell. Drug adducts are detected by highly specific monoclonal antibody techniques. In this paper, we document the presence of high EF5 binding regions in several pathological conditions: including neonatal brain hypoxia/ischemia, myocardial infarction, irradiated skin and cancer. Moderate to high EF5 binding was found in ischemic brain and heart, and focal areas of very high binding were found in the keratinized layers of irradiated skin. Binding in tumors was highly heterogeneous, suggesting oxygen partial pressures ranging from near-normal to anoxia. Analysis of EF5 binding is useful as high resolution monitor of tissue hypoxia under pathological conditions characterized by poor oxygen delivery to tissue.

ACKNOWLEDGMENTS

The authors would like to express their appreciation to those who provided technical support for the studies shown herein: Tim Jenkins, Cecelia Chan, Kristine Laughlin, Reza Alavi, Albert Giandomenico, Isaac Sasso.

The research was supported by funds provided by the National Heart and Lung Institute (RNK: HL02699; DLG: HL44075), the National Cancer Institute (SME: R29 CA62331), the National Institute of Health (DMF: P20NS32553; FRS: RO1 NS28167) and the Medical Research Council of Canada (MB: Postdoctoral fellowship).

REFERENCES

1. Adams GE. The clinical relevance of tumor hypoxia. Eur. J. Cancer, *26*: 420–421, 1990.
2. Beilharz EJ, Williams CE, Dragunow M, et al. Mechanisms of delayed cell death following hypoxic-ischemic injury in the immature rat: evidence for apoptosis during selective neuronal loss. Molecular Brain Research, *29*: 1–14, 1995.
3. Brezis M and Rosen S. Hypoxia of the renal medulla - its implications for disease. N. Engl. J. Med., *332*: 647–655, 1995.
4. Brizel DM, Scully SP, Harrelson JM, et al. Tumor oxygenation predicts for the likelihood of distant metastases in human soft tissue sarcoma. Cancer Research, *56*:941–943, 1996.
5. Buxton RB, Alpert NM, Babikian V, et al. Evaluation of the 11CO2 positron emission tomographic method for measuring brain pH. I. pH changes measured in states of altered pCO2. J. Cereb Blood Flow Metab, *6*: 709–719, 1987.
6. Chapman JD, Franko AJ, Koch CJ. The fraction of hypoxic clonogenic cells in tumor populations. In: Biological Bases and Clinical Implications of Tumor Radioresistance, 61–73, 1983.
7. Cotran RS, Kumar V, Robbins SL. Cellular injury and adaptation. In: Robbins Pathologic Basis of Disease, 1–38, Eds: Cotran, Kumar, Robbins, 1989.
8. Edmonds SE, Ellis G, Gaffney K, et al. Hypoxia and the rheumatoid joint: immunological and therapeutic implications. Scand. J. of Rheum.-Supplement, *101*: 163–168, 1995.
9. Evans SM, and Koch CJ. Characterization of the 9L glioma as a tissue isolated epigastric implant. Radiat. Oncol. Invest., *2*: 134–143, 1994.
10. Evans SM, Koch CJ, Joiner B, et al: 2-Nitroimidazole Binding for Identification of Hypoxic Cell Fraction in Cells and Tissues of Epigastric 9L Tumors. Br. J. Cancer, *72*:875–882, 1995.
11. Ferriero DM, Soberano HQ, Simon RP, et al. Hypoxia-ischemia induces heat shock protein-like (HSP72) immunoreactivity in neonatal rat brain. Developmental Brain Research, *53*: 145–150, 1990.

12. Flammer J. The vascular concept of glaucoma. Survey of Ophthalmology, *38 Suppl*: S3–6, 1994.

13. Gatenby RA, Kessler HB, Rosenblum JS, et al. Oxygen distribution in squamous cell carcinoma metastases and its relationship to outcome of therapy. IJROBP, *14*: 831–838, 1988.

14. Graeber TG, Osmanian C, Jacks T, et al. Hypoxia mediated selection of cells with diminished apoptotic potential in solid tumors. Nature, *379*: 88–91, 1996.

15. Hockel M, Knoop C, Schlenger K, et al. Intratumor pO2 predicts survival in advanced cancer of the uterine cervix. Radiotherapy and Oncology, *26*: 45–50, 1993.

16. Hossmann KA. Viable thresholds and the penumbra of focal ischemia. Annals of Neurology, *36*: 557–565, 1994.

17. Hunt TH. Prospective: A retrospective perspective on the nature of wounds. Prog. Clin. Biol. Res., *266*: xiii-xx, 1988.

18. Koch CJ, Evans SM, Lord EM. Oxygen dependence of cellular uptake of EF5 [2-(2-nitro-1H-imidazol-1-yl)-N-(2,2,3,3,3-pentafluoropropyl)acetamide]: Analysis of drug adducts by fluorescent antibodies vs. bound radioactivity. Br. J. Cancer, *72*:869–874, 1995.

19. Kolstad P. Intercapillary distance, oxygen tension, and local recurrence in cervix cancer. Scand. J. Clin. Lab. Invest. Suppl., *106*: 145–147, 1968.

20. Laughlin KM, Evans SM, Lord EM et al. Biodistribution of EF5 [2-(2-nitro-iH-imidazole-1yl)-N-(2,2,3,3,3-pentafluoropropyl)acetamide] in BALB/c mice bearing EMT6 tumors; implications for oxygen measurements in normal and tumor tissues. J. Pharm. and Exp. Therapeutics, *277*:1049–1057, 1996.

21. Lee J, Siemann DW, Koch CJ et al. Direct Relationship between radiobiological hypoxia in tumors and monoclonal antibody detection of EF5 cellular adducts. Int. J. Cancer, *67*:372–378, 1996.

22. Lord EM, Harwell L, Koch CJ. Detection of hypoxic cells by monoclonal antibody recognizing 2-nitroimidazole adducts. Cancer Research, *53*: 5271–5276, 1993.

23. Mazure NM, Chen EY, Yeh P, et al Oncogenic transformation and hypoxia act to modulate vascular endothelial growth factor expression. Cancer Research, *56*: 3436–3440, 1996.

24. McCoy CL, Parkins CS, Chaplin DJ, et al. The effect of blood flow modification on intra- and extracellular pH measured by 31P MRS in murine tumors. Br. J. Cancer, *69*: 905–911, 1995.

25. Moulder JE, Rockwell SC. Hypoxic fractions of solid tumors: experimental techniques, methods of analysis and a survey of existing data. IJROBP, *10*: 695–712, 1984.

26. Nunn A, Linder K, Strauss HW. Nitroimidazoles and imaging hypoxia. Eur. J.Nuc. Med., *22*: 265–280, 1995.

27. Pahl HL, Baeuerle PA. Oxygen and the control of gene expression. BioEssays, *16*: 497–502, 1994.

28. Rice JE, Vannucci RC, Brierley JB. The influence of immaturity on hypoxic-ischemic brain damage in the rat. Annals of Neurology, *9*: 131–141, 1981.

29. Rumsey WL, Patel B, Kuczynski B, et al. Potential of nitroimidazoles as markers of hypoxia in heart. Adv. Exp. Med Biol., *315*: 263–270, 1994.

30. Secomb TW, Hsu R, Dewhirst MW, et al. Analysis of oxygen transport to tumor tissue by microvascular networks. IJROBP, *25*: 481–489, 1993.

31. Shaywitz BA. The sequelae of hypoxic ischemic encephalopathy. Sem. in Perinat., *11*: 180–190, 1987.

32. Southwick PL, Ernst LA, Tauriello EW, et al. Cyanine dye labeling reagents - carboxymethylindocyanine succinimidyl esters. Cytometry, *11*: 418–430, 1990.

33. Urbach F. Pathophysiology of malignancy: I. Tissue oxygenation of benign and malignant tumors of the skin. Proc. Soc. Exp. Biol. Med., *92*: 644–649, 1956.

34. Varghese AJ, Gulyas S, Mohindra JK. Hypoxia-dependent reduction of 1-(2-nitro-1-imidazolyl)-3-methoxy-2-propanol by Chinese hamster ovary cells and KHT tumor cells *in vitro* and *in vivo*. Cancer Research, *36*: 3761–3765, 1976.

35. Woods MR, Lord EM, Koch CJ. Prediction of hypoxic radioresistance by monoclonal antibody reactive with 2-nitroimidazole adducts. IJROBP., *34*: 93–101, 1996.

85

FLUORESCENCE LIFETIME IMAGING OF THE SKIN PO$_2$ SUPPLY

Instrumentation and Results

Paul Hartmann,[1] Werner Ziegler,[1] and Dietrich W. Lübbers[2]

[1]AVL List GmbH
Biomedical Research and Development
Kleiststraße 48, A-8020 Graz, Austria
[2]Max-Planck-Institut für molekulare Physiologie
Rheinlanddamm 201, D-44139 Dortmund, Germany

1. INTRODUCTION

The quality of the local blood microcirculation and the cutaneous "respiration" are important factors in the diagnosis of circulatory disturbances, which may arise as a consequence of a number of clinical diseases. Some of the more important are smokers leg, burns, ulcerations, or tumors of the skin.

Hitherto established methods of PO_2-measurements are subject to spatial limitations. The widely used electrodes in particular suffer from oxygen consumption, dependence on oxygen transport conditions, and interference of electromagnetic fields.

Luminescence spectroscopy has a great potential for clinical and analytical applications because of its low detection limits, and high temporal and spatial resolution. Lifetime sensors offer additional advantages over intensity measurements: they are independent of dye concentration and illumination conditions in a wide range, since the decay profile is an inherent molecular property. Recently developed 'fluorescence lifetime imaging' techniques (Morgan et al. 1990, Lakowicz et al. 1991) have enabled the measurement of luminescence lifetimes with high spatial resolution.

Time-resolved imaging of phosphorescence quenching by oxygen in tissues has been developed by Wilson and coworkers (Wilson et al. 1989, Wilson et al. 1992). They employ invasive techniques based on phosphorescent porphyrin-complexes incorporated into the blood vessels and take advantage of long-wavelength emission which is not absorbed by tissue. However, the available phosphorescence indicators are not applicable to humans.

The method presented in this work is completely non-invasive. It is based on lifetime imaging of luminescent polymer layer systems ('optodes') which cover selected areas of interest (from a few mm^2 to dozens of cm^2) on the human skin. The method

Oxygen Transport to Tissue XIX, edited by Harrison and Delpy
Plenum Press, New York, 1997

provides virtually calibration-free images of oxygen concentrations or fluxes (Lübbers et al. 1995, Holst et al. 1995). The results achieved with the described instrumentation suggest that quantitative fluorescence lifetime imaging of oxygen flux or $tcPO_2$ has the potential to open new possibilities for the assessment of the microcirculation and related problems in clinical practice.

2. EXPERIMENTAL

2.1. Oxygen Sensors

Tris(1,10'-phenanthroline) ruthenium(II) chloride (Aldrich, Steinheim, Germany; denoted RuPhen) was adsorbed at silica-gel (Whatman, Clifton, NJ) which was subsequently embedded in a filler-free polysiloxane (ABCR, Darmstadt, Germany) (Wolfbeis et al. 1986).

Based on this sensitive layer, the following two different sensor types were prepared.

2.1.1. tcPO_2-Sensors. Transcutaneous oxygen sensor membranes consist of a permeable optical isolation layer of black polysiloxane (Wacker, Burghausen, Germany or General Electric, Waterford, NY), the oxygen-sensitive layer (50µm wet layer thickness), and a transparent carrier foil (Mylar®, DuPont, Bad Homburg, Germany), which is practically impermeable to oxygen to terminate the oxygen diffusion through the sensor (see Fig. 1a).

2.1.2. Oxygen Flux Sensors. The flux optodes are designed with an oxygen permeable polymer (PFA 6510, Novofol, Siegsdorf, Germany) replacing the Mylar foil. A constant flux of oxygen through the sensor is maintained (see Fig. 1b).

2.2. Measurement System

An array of 8 blue LEDs (Nichia, Nürnberg, Germany) is periodically modulated by the output of a synthesized function generator (SRS DS345, Sunnyvale, CA) and a suitable driver circuit (Fig.2). The phase-shifted sensor luminescence is detected by an intensified relay optics (IRO, PCO, Kelheim, Germany). This device includes a photocathode to convert the detected photons into electrons, a micro channel plate (MCP) to amplify the photocurrent, and a phosphor-screen, where the electrons are reconverted into phosphorescence, which in turn is detected by a standard video-CCD camera with 756 x 581 pixels (VC 44, PCO). The IRO is modulated at the same frequency as the excitation source (homodyne detection). Thus, a time-invariant photocurrent is produced, generating a steady-

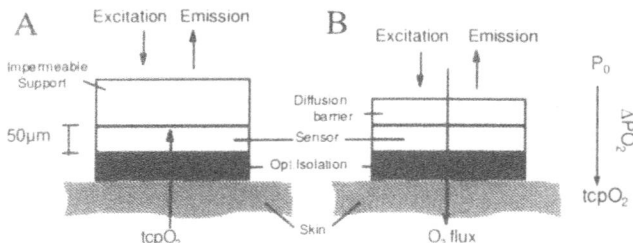

Figure 1. Cross sections of a transcutaneous sensor (A) and an oxygen flux sensor (B) applied to the skin.

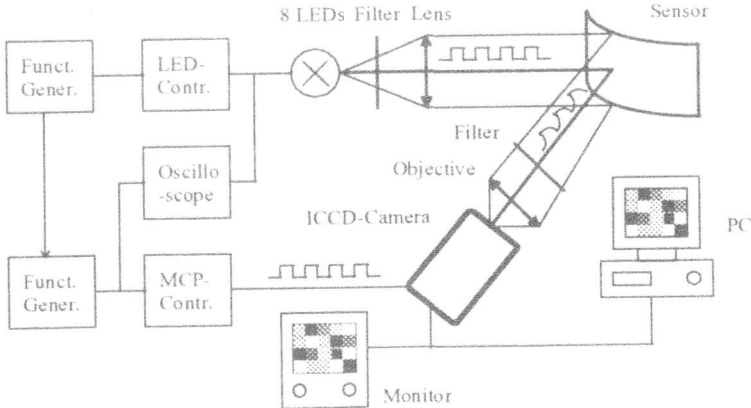

Figure 2. Schematic representation of the instrumental setup. LED...Light emitting diode, MCP...microchannel plate, ICCD...intensified charge-coupled device.

state image on the phosphor-screen. A conventional frame grabber board (Bitflow Raptor, Stemmer, Munich, Germany) and PC serve to control devices and signal processing.

2.3. Modulation Techniques

The LEDs were driven by a square wave output of the function generator.

The IRO was modulated by gating the photocathode (with a square wave from a second function generator, whose time base was coupled to the first device).

For square wave excitation the luminescent response of the sensor is a sum of sine functions according to the Fourier series of the actual excitation shape.

Synchronous gating of the photocathode (Wang et al. 1991) was achieved by applying a square wave to the pulse-input channel of the IRO.

We chose a measurement mode where the gated camera synchronously integrates a first image of the dye emission pixel-by-pixel for the period where the LEDs are on (rise period of the luminescence), and in a second picture for a period where the LEDs are off (decay period of the luminescence) (Khalil et al. 1991). Generally, 64 images were averaged to improve the signal-to-noise ratio (*SNR*) of the signal levels of the sensors.

After subtraction of the dark current of the CCD from each of the luminescence images, a ratio image was calculated, which is pixel-by-pixel related to the apparent lifetime of the luminescence of the sensing layer.

2.4. Determination of PO_2

The gray values of the ratio image can be transformed to PO_2-values by a calibration curve (Fig.3, for details see: Hartmann et al. 1996, 1997), which has to be determined once by in-factory calibration. This is done by placing a sensor membrane into a thermostated measurement chamber (T=38°C) supplied with gases of different oxygen content saturated with water vapor. The gray values of the image ratio was obtained by the measurement technique described above for a number of gas compositions, each one resulting in a point of the calibration curve. We took an averaged reading of all pixels within a sensor area of approximately 1 cm².

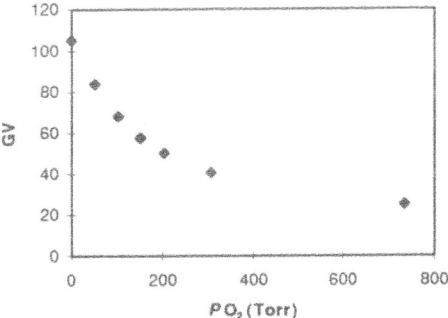

Figure 3. Calibration curve of a tc*PO*$_2$ sensor at T=38°C and 100% RH: 8 bit-gray values vs. *PO*$_2$ of the applied calibration gas.

3. RESULTS AND DISCUSSION

The sensors in principle respond to the oxygen concentration in the sensing layer. The Mylar-supported sensors measure the *PO*$_2$ of a surface under investigation (e.g., the tc*PO*$_2$).

The quantity measured by flux sensors is again the *PO*$_2$ corresponding to the oxygen concentration in the sensing layer, while the oxygen flux *JO*$_2$ between ambient air and the skin (Lübbers et al. 1995) is given by Fick's first law.

$$JO_2 = K(P_0 - PO_2)/d \tag{1}$$

$$K = \alpha D \tag{2}$$

α is the solubility constant, D the diffusion constant, and d the thickness of the diffusion barrier, respectively. P_0 is the oxygen partial pressure of the ambient air. It is assumed, that the permeability of the sensing layer is high compared to the one of the diffusion barrier. With the help of Fick's first law and the calibration curve of the sensor the flux can be related to the measurable decay parameters.

The suitability of the setup was demonstrated with the help of a flux sensor based on a PFA membrane, attached to a surface providing an artificial oxygen distribution. Fig.4 shows an image of this distribution in a sensor covering an area of 4x3 cm², which was obtained by the lifetime-selective method described above. Partial pressures up to one atmosphere can be resolved, but the measuring accuracy increases with decreasing *PO*$_2$ (see

Figure 4. Lifetime image of the luminescence of an oxygen sensor covering an area of 4 x 3 cm² of a surface with an artificial oxygen distribution, generated by leading gases of various oxygen content by tubes into a porous polymer dummy. Different 8 bit-gray values represent different oxygen concentrations in the sensing layer. Represented *PO*$_2$-range: 0–760 torr.

Table 1. Technical specifications for a
standard sensor area of 4x3 cm^2

Spatial resolution	500µm
Accuracy	
at PO_2=0 torr	0.4 torr
at PO_2=160 torr	5 torr
Frequency	150 kHz
Net acquisition time	2.5 sec.

Tab.1). In-factory calibration of the sensors will be sufficient, provided, that the calibration function remains stable for the time the sensor is in use. Clinical measurements are expected to be calibration-free.

Spatial resolution is limited by lateral diffusion of oxygen within the skin (which is higher than the respective quantity in the sensing membrane). Fig.5 shows a gradient of gray values of an lifetime image, appearing between sensor areas which stick tightly to the skin (and, thus, measure the actual tcPO_2) and sensor areas which were locally removed (lifted) from the skin and equilibrated with a gas stream of pure oxygen (PO_2=P$_0$ = 730 torr) which was applied from outside. From this graph it can be concluded, that a resolution of better than 500µm can easily be obtained for clinical measurements. This is also in agreement with model calculations based on the diffusion properties of the sensor layers and the skin. For many clinical applications this spatial resolution is sufficient (it is also comparable with established Laser Doppler Scanning methods).

After application of the sensor to the skin, equilibration of the luminescence signal took several minutes and varied from case to case, dependent on the diffusion properties of oxygen both in the membrane and the epidermis, as well as on temperature and humidity. However, this is not expected to be a major problem for clinical applications.

We compared the results of tcPO_2-measurements obtained by lifetime imaging of optical sensor membranes (optodes) with the readings of a conventional electrode (Hellige Oxykapnomonitor), applied to the forearms of various healthy volunteers. To induce significant circulation changes, two occlusions were set successively. Fig.6 shows a typical result of the optode (integrated over a sensing area of 1 cm^2) compared to the readings of the electrode. The absolute values do not coincide, probably due to the difference of the investigated skin areas.

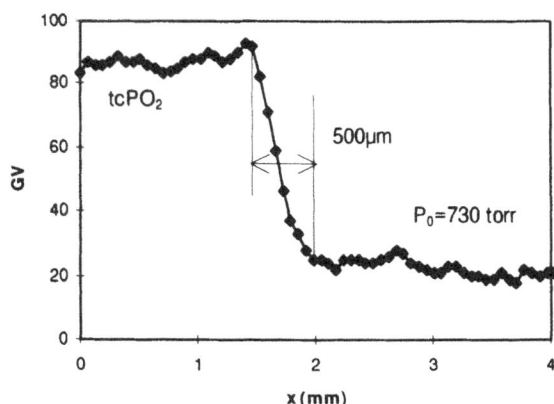

Figure 5. Gray values (GV) of luminescence in direction of a high PO_2-gradient in a tcPO_2 sensor applied to the skin of a volunteer. The low gray values (high PO_2) were realized by lifting the sensor locally and flooding the lifted parts with pure oxygen atmosphere. (T=38°C)

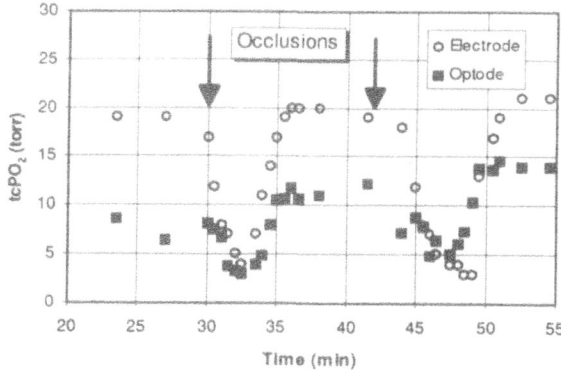

Figure 6. Comparison of lifetime-based PO_2-measurements with a tcPO_2 sensor applied to the forearm of a healthy volunteer and PO_2-readings of a standard electrode on two successive occlusions versus time after application of the sensor. All measurements at T=38°C.

The imaging capabilities of the new technique are shown in Fig.7. Before covering the skin with a tcPO_2-membrane, a strip of nicotinic acid was applied to a small skin area. The improved local circulation lead to higher tcPO_2-values, which became apparent by the higher gray values of the lifetime image (Fig.7).

The technical specifications for the breadboard status of the instrument are given in Table 1.

4. CONCLUSIONS

Lifetime imaging of luminescent oxygen sensors offers the possibility of non-invasive, quantitative and calibration-free imaging of distributions of tcPO_2 or the cutaneous "respiration" (oxygen flux). The method may play an important role in the diagnosis of, e.g., chronic venous insufficiency (estimated number of patients in Germany: >5 millions), ulcerations (1 million patients), wound healing, burns (vital and dead parts could be distinguished), amputations (locating the boundary to insufficiently supplied regions), or microcirculatory disturbances (e.g., smokers leg or gangrenes).

With this technique it should also be possible to evaluate quantitatively, e.g., the effects of drugs for improvement of circulation.

ACKNOWLEDGMENTS

We acknowledge gratefully the helpful assistance of R. Duschek, and stimulating discussions with Prof. E. Pilger and Dr. C. Schulze Bauer, Graz, and with Dr. G. Holst,

Strip of Nicotinic Acid

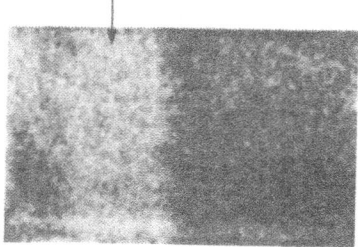

Figure 7. Effect of a strip of nicotinic acid (Thermo-Rheumon®, Tropon) on the microcirculation of the skin of a healthy volunteer, monitored by a lifetime image of an oxygen sensor foil of area 4x2.5cm². T=38°C. Higher gray values in the left part of the image indicate an increase of oxygen supply by microcirculation compared to the untreated skin.

Bremen. Financial support was received from Forschungsförderungsfonds der gewerblichen Wirtschaft.

REFERENCES

Hartmann P., and Ziegler W., (1996) Lifetime imaging of luminescent oxygen sensors based on all-solid-state technology, Anal. Chem., in press.

Hartmann P., Ziegler W., Holst G., and Lübbers D.W., (1997) Oxygen Flux Fluorescence Lifetime Imaging, Sensors and Actuators B, in press.

Holst G., Köster T., Voges E., and Lübbers D.W., (1995) FLOX- an oxygen-flux-measuring system using a phase-modulation method to evaluate the oxygen-dependent fluorescence lifetime, Sensors and Actuators, B 29, 231–239.

Khalil G., Gouterman M.P., and Green E., (1991) Methods and sensor for measuring oxygen concentration, US Patent, Nr. 5043286.

Lakowicz J.R., and Berndt K.W., (1991) Lifetime-selective fluorescence imaging using an rf phase-sensitive camera, Rev. Sci. Instrum., 62, 1727–1734.

Lübbers D.W., Köster T., and Holst G.A., (1995) Hybrid fiberoptical sensor for determining the oxygen partial pressure and the oxygen flux in biomedical applications, Proc. SPIE, 2388, 507–518.

Morgan C.G., Mitchell A.C., and Murray J.G., (1990) Fluorescence Lifetime imaging using an imaging photon detector with a radiofrequency photon correlation system, Proc. SPIE, 1204, 798–807.

Wang X.F., Uchida T., Coleman D.M., and Minami S., (1991) A two-dimensional fluorescence lifetime imaging system using a gated image intensifier, Appl. Spectrosc., 45, 360–366.

Wilson D.F., Rumsey W.L., and Vanderkooi, J.N., (1989) Oxygen distribution in isolated perfused liver observed by phosphorescence imaging, Adv. Exp. Med. Biol., 248, 109–115.

Wilson D.F., and Cerniglia G.J., (1992) Localization of tumors and evaluation of their state of oxygenation by phosphorescence imaging, Cancer Res., 52, 3988–3993.

Wolfbeis O.S., Leiner M.J.P., and Posch H.E., (1986) A new sensing material for optical oxygen measurement, with the indicator embedded in an aqueous phase, Mikrochim. Acta, 1986 III, 359–366.

86

EFFECT OF LIPOPOLYSACCHARIDE ON INTRA-PHAGOSOMAL OXYGEN CONCENTRATION AS MEASURED BY EPR OXIMETRY

Simon K. Jackson,[1] Philip E. James,[2] and Harold M. Swartz[2]

[1]Department of Medical Microbiology
University of Wales College of Medicine
Cardiff, Wales, United Kingdom
[2]EPR Center, Department of Radiology
Dartmouth Medical School
Hanover, New Hampshire

INTRODUCTION

Lipopolysaccharide (LPS) is the endotoxin from the outer membrane of Gram-negative bacteria that is associated with the high morbidity and mortality in patients with septic shock (Morrison, 1978). Our previous studies have shown that LPS can influence mitochondrial oxygen consumption in a variety of cell types (James and Jackson, 1995; Jackson, 1996) and alter the oxygen utilization of organs in experimental septic shock (James and Bacic, 1996). It is also suggested that LPS can augment the respiratory burst associated with phagocytosis in macrophages. We sought to investigate the effects of LPS on the respiratory burst and the distribution of oxygen within macrophages.

The intracellular concentration of oxygen $[O_2]$ reflects the balance between oxygen delivery and cellular consumption in viable cells and tissues. It usually is inferred rather than measured but it would be desirable to measure this parameter directly, preferably at specific intracellular loci. Although everyone agrees that there must be some gradient between the extracellular- and intracellular compartment, there is disagreement in the literature regarding the size of this gradient (Glockner, 1989; Tamura, 1978; Jones and Kennedy, 1982; Jones and Mason, 1978). But experiments to verify these findings have until now been hampered by experimental limitations (Ligeza, 1994) and the lack of suitable probes to measure selectively the oxygen concentration at particular locations.

A new EPR-based method was developed to obtain selective information on the $[O_2]$ within phagosomes. The method utilizes the selective incorporation of an oxygen-sensitive nitroxide into phagosomes of macrophages stimulated with zymosan. Addition of a second oxygen-sensitive probe whose EPR signal does not overlapp with that contained within the phagosomes enabled us to simultaneously monitor the $[O_2]$ in both the extracellular and phagosomal compartments (James and Grinberg, 1995).

We also used EPR spin-trapping (inclusion of a stable nitroxide compound to stabilize free radical species (Finkelstein, 1980)) to assess production of superoxide species from the NADPH-oxidase system within the phagosomes of these macrophages. The effect of LPS in sustaining the respiratory burst and the resulting effect on phagosomal $[O_2]$ was of particular interest, since it is important that the intracellular concentration of oxygen within macrophages be sufficient to supply and maintain mitochondrial oxygen consumption as well as production of reactive oxygen species (Root and Cohen, 1981). The latter forms the basis of their microbicidal and tumoricidal activity (Murray and Cohn, 1980), and release of these species is instrumental in bacterial destruction (Johnston, 1975). Membrane stimulation of human neutrophils results in activation of the membrane-bound NADPH-oxidase, which consumes a large quantity of oxygen and reduces it to various active oxygen species, including superoxide (O_2^-)(Pou, 1989), H_2O_2 (Root and Metcalff, 1977), and subsequently formed ·OH (Britigan,1986).

METHODS

Chemicals

Perdeuterated nitroxides 4-oxo-2,2,6,6-tetramethylpiperidine-d_{16}–1-oxyl (^{15}N PDT) and 2,2,6,6-tetramethyl,4-trimethylammonium iodide (d-Cat$_1$) were purchased from MSD Isotopes (St. Louis, MO). All other chemicals were obtained from Sigma Chemical Co.(St. Louis, MO) unless otherwise stated.

Cell Culture

The murine macrophage-like cell line RAW 264.7 was obtained from the American Type Culture Collection (Rockville, MD). Cells were grown in Dulbecco's modified Eagles medium (DMEM) supplemented with 10% heat-inactivated fetal bovine serum (Hyclone Laboratories, Logan, UT), penicillin (50 U/ml), and streptomycin (250 mg/ml). One day prior to the experiment, the cells were placed into 75 cm^3 tissue culture flasks (Corning, NY) at a density of 1–1.5x10^{5}cells/ml in 20 ml medium and allowed to adhere to the bottom of the flask by 24 hrs incubation at 37^{0}C and 95%O$_2$. Identical samples and samples for comparison were set up from the same solution, so that (after 24 hrs incubation) each flask contained 5x10^6 adherent macrophages.

EPR Measurements

The structures of the oxygen-sensitive probes used in this study are illustrated in Figure 1. The sensitivity of d-Cat$_1$ and ^{15}N PDT to oxygen has been discussed in previous communications (Glockner, 1989; Hu, 1992) and the technique of EPR oximetry has been used extensively to measure oxygen concentrations *in vitro* and *in vivo* (Swartz, 1992). The linewidth and shape obtained from EPR spectra of these two nitroxides varies with oxygen concentration, and can be calibrated accordingly. Oxygen concentration measurements were obtained from 100μl samples (5x10^{7}cells/ml) made up in serum free medium containing dextran (10% by weight) to avoid settling of the cells in the EPR cavity. Each

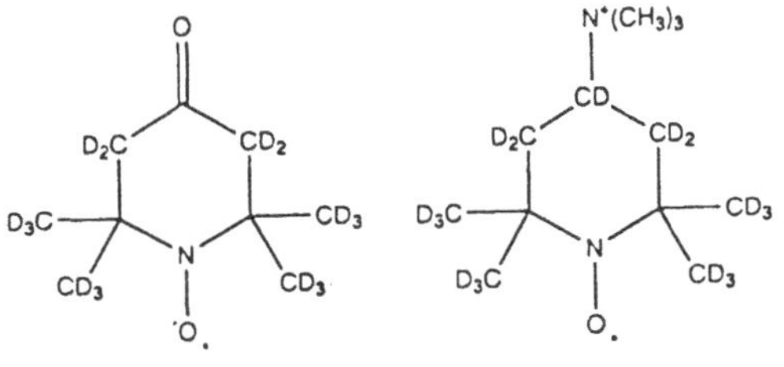

Figure 1. Chemical structures of the nitroxides used in these experiments. [15]N PDT is neutral and can equilibrate across the cell membrane. d-Cat[1] has a net positive charge and cannot traverse the bilayer.

PDT   d-CAT 1

aliquot was drawn into gas-permeable teflon tubing (Norell Inc., Mays Landing, NJ) which was folded in a 'W' conformation and put into a quartz EPR tube (3mm i.d.) which was open at both ends, allowing adequate aeration of the cell sample in the EPR cavity.

All spectra were recorded on a Bruker ER 220D-SRC ESR spectrometer fitted with a gas-flow variable temperature accessory. Samples were maintained at 37^0C and in a perfused gas flow rate of 80 L/hr. Spectrometer conditions varied according to the sample; incident microwave power was always kept at a level which avoided broadening due to power saturation effects (typically 0.5mW); amplitude of modulation was kept at less than one third of the linewidth (which varied according to the oxygen concentration) to avoid over-modulation of the EPR spectrum (typically 80mG for ^{15}N PDT and 200mG for both d-Cat[1] and AT); the center of the field was 3310 Gauss for d-Cat[1] and 3306 Gauss for ^{15}N PDT and was adjusted and maintained at this value by using the field frequency lock. Both ^{15}N PDT and d-Cat[1] could be studied in the same sample since their spectra do not overlap (Figure 2). The most intensive line was used; for ^{15}N PDT this was the low-field line and for d-Cat[1] the mid-field line.

Principle of Method to Obtain Selective Measurements of pO$_2$

We exploit a particularly useful characteristic of monocytes (macrophages), their ability to phagocytize or ingest foreign particles or bacteria; this capacity plays an important role in their tumoricidal and bactericidal actions (Murray, 1980). The method is based on the fact that incorporation of a charged nitroxide (d-Cat[1]) into the phagosome ensured that this probe remained in the intraphagosomal space and reported on the oxygen concentration therein, with little or no leakage into the cytoplasm or extracellular space during the time course of the experiment. d-Cat[1] (2.5mM) was incorporated into the phagosomes of activated macrophages by addition into the culture flask (20ml), and then Zymosan (0.25mg/ml) was added to induce phagocytosis and an oxidative burst in these macrophages. After a 25 min. incubation period at 37^0C, the cells were harvested by gentle scraping with a rubber policeman and washed four times with DMEM so as to remove any d-Cat[1] that was not incorporated into phagosomes. The cells were resuspended in medium containing 10% dextran for EPR measurement. In viable unstimulated cells the d-Cat[1] (positively charged) remained in the extracellular medium and could be removed from the cells by adequate washing.

To measure the extracellular oxygen concentration in the same sample (shown schematically in Figure 3) ^{15}N PDT was added to each 100μl aliquot (cell concentration 5×10^7/ml; final concentration 0.02mM): This gave the effective extracellular [O$_2$] because less than 5% of the total EPR signal arises from the intracellular compartment.

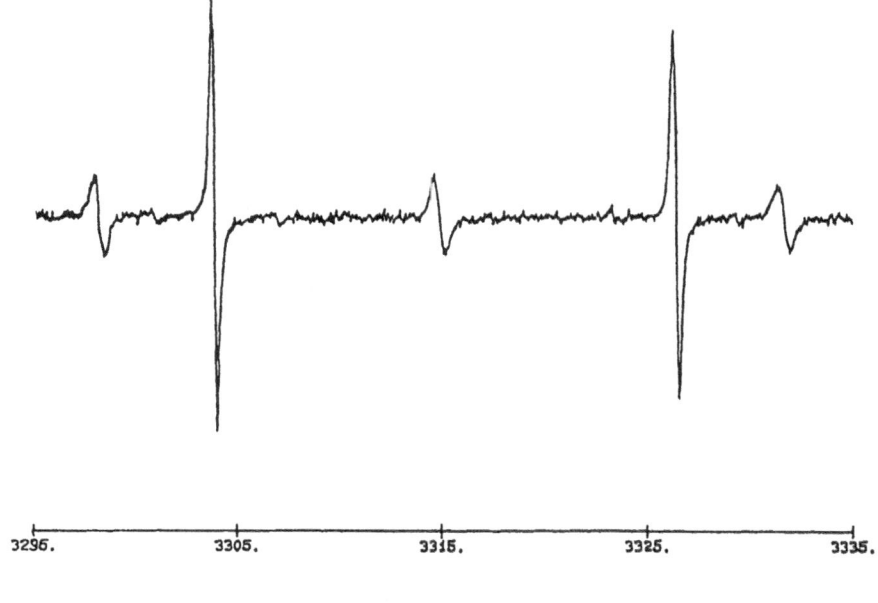

MAGNETIC FIELD (GAUSS)

Figure 2. A typical spectrum obtained from macrophages stimulated with zymosan in the presence of d-Cat$_1$, then washed and ^{15}N PDT added. The two more intensive lines arise from ^{15}N PDT (extracellular) and the three less intensive lines from d-Cat$_1$ (phagosomal). This allows simultaneous measurement of oxygen concentration within the same sample.

To calibrate the linewidths with [O$_2$] each of the above samples was duplicated and treated in exactly the same way except for the addition in the final solution calibration sample of sufficient NaCN (5mM) to inhibit cellular respiration. For each sample both ^{15}N PDT and d-Cat$_1$ were calibrated at various [O$_2$] being perfused through the cavity. Oxygen concentrations (µM) were calculated from the equation below (see also Glockner. 1989):

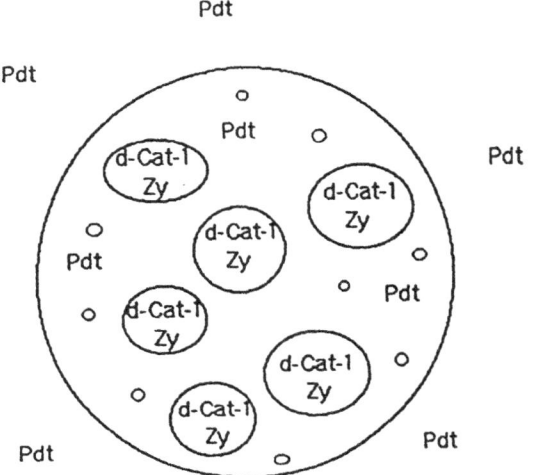

Figure 3. A schematic diagram of the macrophage stimulated with zymosan in the presence of d-Cat$_1$ (25 minutes incubation prior to washing), with subsequent addition of PDT. Large circles represent phagosomes.

$$[O_2] = 210 \times (LW_m - LW_n) / (LW_a - LW_n)$$

where 210µM is the oxygen concentration in media equilibrated in air at 37^0C; m represents the sample measurement; n or a are the calibration linewidths in 100% nitrogen or 100% air respectively.

For each sample, spectra were recorded three times and a mean value used in equation 1 (above) to calculate $[O_2]$ for this sample.

Spin Trapping of Superoxide

Reaction mixtures contained either 2.5 or 5×10^7 macrophages/ml, a spin trap (0.1M; 5,5-dimethyl-1-pyrroline-N-oxide; DMPO), zymosan (0.25mg/ml) to stimulate an oxidative burst, and DMEM (without added serum) which was supplemented with Diethylenetriaminepentacetic acid (DETAPAC; 0.1mM). Experiments were performed using different concentrations of oxygen in the gas flowing through the EPR cavity (varying between 60, 100 and 210µM). Immediately after addition of zymosan, 100µl aliquots were drawn into gas-permeable tubing and placed into the EPR cavity as described above. This ·procedure (with additional time for spectrometer tuning) took less than two minutes, after which time 20 second scans were repeated up to 80 mins. The amplitude of the low field peak of the EPR spectrum of the DMPO-OOH adduct (see Figure 4) was monitored and used as a measure of the concentration of radical present at a given time after stimulation. The use of such data for absolute quantitative comparisons is difficult because the observed intensity refelects a dynamic equilibrium between formation and destruction/conversion of the spin adduct and these rates may be affected by the condiyions in the sample. The decomposition of superoxide and hydroxide adducts of DMPO however, is independent of the concentration of oxygen (Samuni, 1989).

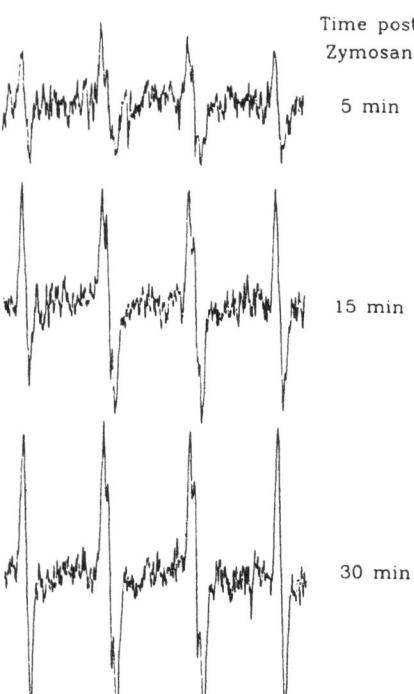

Time post
Zymosan

5 min

15 min

30 min

Figure 4. Progressive production of superoxide (–OOH) trapped by DMPO following zymosan stimulation of J774 macrophages. An arrow indicates the low field peak, the amplitude of which was usd as a measure of radical concentration at that time.

RESULTS

Linewidth measurements of d-Cat₁ inside phagosomes showed that under these experimental conditions an average value for the oxygen concentration within phagosomes was $163 \pm 13.6 \mu M$ and that in the extracellular compartment was $185\mu M$. For the above samples oxygen consumption (after zymosan stimulation) was calculated to be 1.9 ± 0.4 nmoles/min/10^6 cells. Measurements of intraphagosomal and extracellular $[O_2]$ in samples pre-treated for one hour with LPS showed the phagosomal $[O_2]$ was $90 +/- 21$ μM. When both d-Cat₁ and ^{15}N PDT were located in the extracellular compartment, the average oxygen concentration obtained from the linewidths of each respective probe compared closely and gave a similar value to that obtained from extracellular ^{15}N PDT when d-Cat₁ was inside the phagosomes.

Macrophages incubated in $210\mu M$ oxygen (air) showed an $[O_2]$ in the extracellular compartment of $185.5\mu M$. This decreased to $84.4\mu M$ at $100\mu M$ and 35.8 μM at $60\mu M$. The $[O_2]$ within phagosomes at these extracellular $[O_2]$ was 163.9 μM, 51.2 μM and 25.8 μM respectively.

Superoxide Production

Typical spectra of the -OOH radical adduct of DMPO were recorded following zymosan stimulation of macrophages (Figure 4; a 12 line spectrum, spectral hyperfine splittings were $a_N = 14.2$ G, $a_H = 11.3$ G, $a_{Hy} = 1.3$ G).

The amplitude of the low field peak of this spectrum was monitored with time and increased with sequential scans, indicating progressive free radical generation. The profile of superoxide production by macrophages incubated in $210\mu M$ oxygen is shown in Figure 5, and is compared with that of macrophages incubated at $100\mu M$ and $60\mu M$ extracellular oxygen. Superoxide production at the lower $[O_2]$ was 50% of that at the two higher $[O_2]$. We failed to observe trapped species in samples where the macrophages had previously been treated with an NADPH-oxidase blocker (2mM p-chloromercurobenzoate).

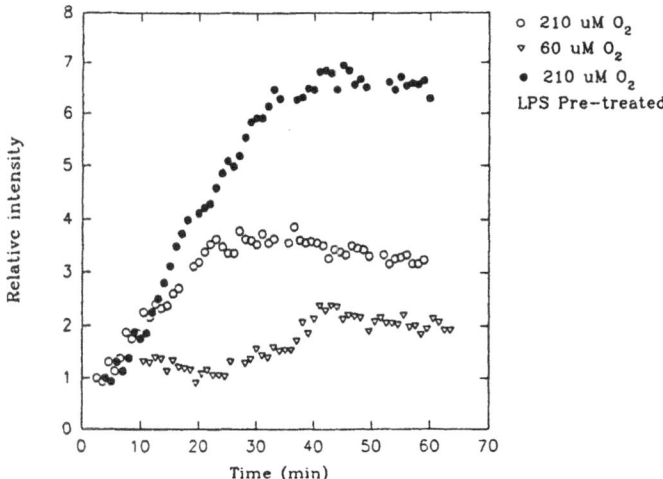

Figure 5. Progressive superoxide production at 210 and 60 μM extracellular $[O_2]$. Data following LPS pre-treatment are also shown for comparison.

Macrophages pre-treated for 1 hour with 10 μg/ml LPS showed an enhanced and sustained respiratory burst, above that induced by zymosan alone.

DISCUSSION

These results clearly demonstrate that production of reactive oxygen-centered radicals by the NADPH-oxidase system of macrophages can be affected by the availability of sufficient oxygen. LPS seems to have a dual effect on macrophage oxygen consumption; whilst inhibiting mitochondrial consumption, LPS also stimulates an enhanced and sustained production of reactive oxygen species (primarily superoxide).

Under our experimental conditions, with a constant supply of air (210 μM oxygen), zymosan-activated macrophages rapidly generated a DMPO-OOH spectrum. The $[O_2]$ in the extracellular compartment was 185 μM and that within phagosomes in these macrophages was 163 μM - sufficient to maintain maximal superoxide production. When the extracellular $[O_2]$ was reduced to 84 μM (by reducing the $[O_2]$ in the inflowing gas), the $[O_2]$ within phagosomes was 51μM, and at 35.8 μM extracellular $[O_2]$ was 25.8 μM within phagosomes. Consequently, production of superoxide species from within phagosomes was reduced at the lower level and was dependent on whether the $[O_2]$ within phagosomes was above the Km for oxygen of the NADPH-oxidase, which has been calculated at 50μM.

Macrophages pre-treated with LPS showed an enhanced and sustained production of superoxide. This was observed at all levels of extracellular $[O_2]$ studied. From these results, it appears that macrophages have some ability to control the consumption of oxygen between sites (mitochondria and/or phagosomal). We have previously shown that LPS inhibits mitochondrial oxygen consumption in macrophages (James, 1995), and together with the results presented here, conclude that this inhibition of mitochondrial consumption may be beneficial and essential to maintaining macrophage production of reactive oxygen species during phagocytosis, and has important implications for macrophage function in infections.

Our observation of a substantial difference between the $[O_2]$ in the extracellular and phagosomal compartments is consistent with previous findings from this laboratory (and others) showing significant differences between the average intra- and extracellular oxygen concentration in cells rapidly consuming oxygen that cannot be explained in terms of the free diffusion of oxygen alone (Glockner, 1989; Jones and Kennedy, 1982).

REFERENCES

Britigan, B.E., Rosen, G.M., Chai, Y., and Cohen, M.S. (1986) Do human neutrophils make hydroxyl radical? J.Biol.Chem., 261(10):4426–4431.

Finkelstein, E., Rosen, G.M., and Rauckman, E.J. (1980) Spin trapping of superoxide and hydroxyl radical: Practical aspects. Arch.Biochem.Biophys., 200(1):1–16.

Glockner, J.F., Swartz, H.M., and Pals, M.A. (1989) Oxygen gradients in CHO cells: Measurement and characterization by electron spin resonance. J.Cell.Physiol., 140:505–511.

Glover, R.E., Rowlands, C.C., Mile, B. and Jackson, S.K. (1996) Lipopolysaccharide decreases oxygen consumption in MonoMac 6 cells; an EPR oximetry study. Biochim. Biophys. acta, 1036: 5–9.

Hu, H., Sosnovsky, G., and Swartz, H.M. (1992) Simultaneous measurements of the intra- and extra-cellular oxygen concentration in viable cells. Biochim.Biophys.Acta, 1112:161–166.

James, P.E., Grinberg, O.Y., Michaels, G., and Swartz, H.M. (1995) Intraphagosomal oxygen in stimulated macrophages. J.Cell.Physiol., 163:241–247.

James, P.E., Jackson, S.K., Grinberg, O. and Swartz, H.M. (1995) The effects of endotoxin on oxygen consumption of various cell types in vitro: An EPR oximetry study. Free Rad.Biol.Med., 18:641–647.

James, P.E., Bacic, G., Grinberg, O.Y., Goda, F., Dunn, J.F., Jackson, S.K., and Swartz, H.M. (1996). Endotoxin-induced chnages in intrarenal pO_2 measured by in vivo electron paramagnetic resonance oximetry and magnetic resonance imaging. Free Rad. Biol.Med., 21:25–34.

Jones, D.P. and Kennedy, F.G. (1982) Intracellular oxygen supply during hypoxia Biochem. Biophys. Res. Commun. 105:419–424.

Jones, D.P. and Mason, H.S. (1978) Gradients of O_2 concentration in hepatocytes. J. Biol. Chem., 253:4874–4880.

Johnston, R.B.,Jr., Keele, B.B.,Jr., Misra, H.P., Lehmeyer, J.E., Webb, L.S., Baehner, R.L., and Rajagopalan, K.V. (1975) The role of superoxide anion generation in phagocytic bactericidal activity. studies with normal and chronic granulomatous disease leukocytes. J.Clin.Invest., 55:1357–1372.

Ligeza, A., Swartz, H.M. and Subczynski, W.K. (1994) Spin label oximetry in dense cell suspensions: Problems in closed and open chamber methods. Curr.Topics Biophys., 18(1):29–38.

Morrison, D.C., Ulevitch, R.J. (1978). The efects of bacterial endotoxin on host mediation systems. Am.J.Path., 93:527–536.

Murray, H.W., and Cohn, Z.A. (1980) Mononuclear phagocyte antimicrobial and antitumor activity: The role reactive oxygen intermediates. J.Invest.Dermatol., 74:285–288.

Pou, S., Cohen, M.S., Britigan, B.E., and Rosen, G.M. (1989) Spin-trapping and human neutrophils. J.Biol.Chem., 264(21):12299–12302.

Root, R.K., and Metcalff, J.A. (1977) H_2O_2 release from human granulocytes during phagocytosis. J.Clin.Invest., 60:1266–1279.

Root, R.K., and Cohen, M.S. (1981) Rev.Infect.Dis., 3:565–598.

Swartz, H.M., Boyer, S., Brown, D., Chang, K., Gast, P., Glockner, J.F., Hu, H., Lui, K.J., Moussavi, M., Nilges, M., Norby, S.W., Smirnov, A., Vahidi, N., Walczak, T., Wu, M., and Clarkson, R.B. (1992) The use of EPR for the measurement of the concentration of oxygen in vivo in tissues under physiologically pertinent conditions and concentration. In: Oxygen Transport to Tissue XIV. W.Erdmann and D.F.Bruley, eds. Plenum Press, New York, pp. 221–228.

Tamura, M., Oshino, N., Chance, B., and Silver, I.A. (1978) Optical measurements of intracellular oxygen concentration of rat heart in vitro. Arch.Biochem.Biophys., 191:8–22.

87

PROTEIN C DETECTION VIA FLUOROPHORE MEDIATED IMMUNO-OPTICAL BIOSENSOR

James O. Spiker,[1] Kyung A. Kang,[1] William Drohan,[2] and Duane F. Bruley[1]

[1]Department of Chemical and Biochemical Engineering
University of Maryland Baltimore County (UMBC)
Baltimore, Maryland 21228
[2]American Red Cross
15601 Crabbs Branch Way
Rockville, Maryland 20855

1. INTRODUCTION

Protein C

Hemostasis is a physiological state where the concentration of coagulants in a human body is well balanced with that of anticoagulants. For oxygen to be properly transported to the entire body, hemostasis must be maintained at all times. Protein C (PC) is one of only a few anticoagulating factors in the bloodstream. As a potent anticoagulant and antithrombotic, PC plays a key role in regulating hemostasis, ensuring the proper transport of oxygen throughout the body. When anticoagulants are not maintained at sufficient levels, the resulting inhibition of oxygen flow produces massive clotting problems. If PC deficiency is not diagnosed and properly treated immediately, an individual may experience debilitating trauma, thrombo-embolic insults in major organs, or death (Bruley and Drohan, 1990). This paper presents a continuation of the research initiated by Kang, et al., concerning the development of a PC biosensor. Background information regarding to the activation of PC, as well as the frequency of occurrence of PC deficiency, has been presented at the 23rd Annual ISOTT Conference (Kang, et al., in press).

Need for a Highly Sensitive, Specific, Portable, Real-Time PC Biosensor

The average concentration of PC in normal human plasma is approximately 4 μg/ml. Due to the low concentration, existing assays are unable to provide an efficient, accurate method of gauging PC levels in samples. Moreover, blood plasma contains an abundance of proteins, some of which are structurally homologous to PC molecules. This imposes the requirements of high specificity and sensitivity upon a proposed PC biosensor. Currently,

Oxygen Transport to Tissue XIX, edited by Harrison and Delpy
Plenum Press, New York, 1997

ELISA (enzyme-linked immunosorbent assay) is the assay of choice for acquiring PC concentration in plasma. However, besides being time consuming, ELISA is an expensive process that makes use of non-reusable test kits. Therefore, the development of a highly sensitive, specific, portable, and real-time biosensor is urgently needed for the diagnosis of PC deficiency as well as for other blood factor monitoring.

Proposed Immuno-Optical PC-Biosensor

The PC-biosensor under development involves immobilizing monoclonal antibody against PC (anti-PC) on the surface of an optical fiber. A PC sample conjugated with a fluorophore will adsorb to the surface of the fiber, being highly specific for PC, and allow fluorescent light to be generated. The emitted light intensity can be correlated with the PC concentration in the sample. Since monoclonal antibodies are covalently immobilized on the fiber, the biosensor can be reused after the adsorbed PC molecules are washed off, using a proper elution buffer (Kang, et al., 1992). In large part, selection of the elution buffer is determined by the binding characteristics of the monoclonal antibody in use. For example, our initial experiments involved a Ca^+ dependent monoclonal antibody that would release bound PC molecules in the presence of a buffer containing Ca^+. The current procedure for measuring PC concentration (for both calibration and actual sample measurements) using the proposed immuno-optical PC biosensor is as follows:

1. Equilibrate the anti-PC immobilized optical fiber with the equilibrium buffer.
2. Adsorb PC conjugated with fluorophore to the anti-PC on the optical fiber.
3. Wash the sensor with the washing buffer.
4. Apply the input light at the excitation wavelength and measure emitted intensity.
5. Elute the adsorbed PC with the elution buffer.
6. Regenerate the sensor using the regeneration buffer.
7. Re-equilibrate the fiber with the equilibrium buffer for the next measurement.

Aside from the aforementioned direct binding assay (i.e. PC is directly conjugated with the fluorophore), a method making use of a sandwich assay has been proposed. Here,

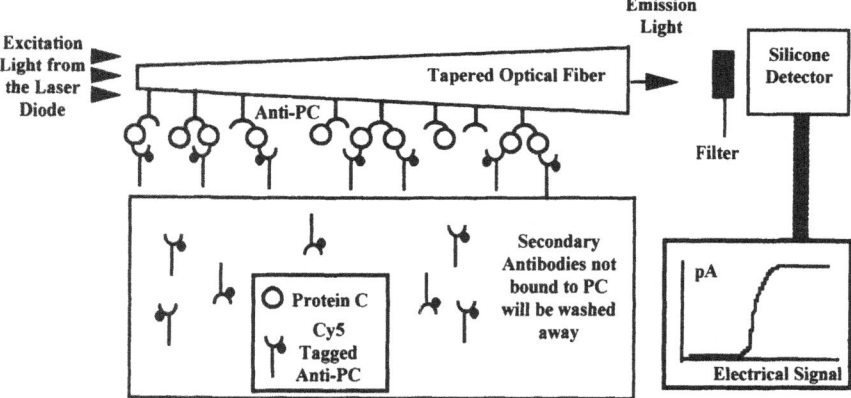

Figure 1. Schematic diagram of basic principles of an immuno-optical Protein C biosensor. Protein C (PC) in a sample adsorbs to monoclonal antibody against PC that has been adsorbed to the surface of a quartz fiber. The PC is then probed with a fluorophore tagged secondary antibody. Following a washing step, the amount of fluorescent light generated may be correlated to the concentration of PC in the sample.

native PC is adsorbed to the surface of the fiber and then probed with a fluorophore tagged secondary antibody against PC. A careful examination of each assay will reveal which method produces the most beneficial results. However, it is believed that the sandwich assay will more closely model the method in which such a biosensor is utilized in a clinical setting (Figure 1). The experimental procedure for PC concentration measurement using a sandwich assay is as follows:

1. Equilibrate the anti-PC immobilized optical fiber with the equilibrium buffer.
2. Adsorb PC in the sample to the anti-PC on the optical fiber.
3. Wash the sensor with washing buffer.
4. React the biosensor with fluorophore tagged secondary antibody against PC.
5. Wash the sensor with washing buffer.
6. Apply the input light at the excitation wavelength and measure intensity.
7. Elute the adsorbed PC with the elution buffer.
8. Regenerate the sensor using the regeneration buffer.
9. Re-equilibrate the fiber with the equilibrium buffer.

3. PRELIMINARY STUDY

3.1. Materials, Instruments, and Methods

Protein C, Monoclonal and Polyclonal Antibody Against Protein C

The University of Maryland Baltimore Country (UMBC) has an on-going research project (Protein C Project) with the collaboration of the American Red Cross (ARC), Rockville, Maryland. Both Protein C and monoclonal antibody are supplied by ARC as needed. Polyclonal antibody (goat fraction) against human PC is purchased from Sigma Chemical Company (Saint Louis, MO).

Immobilization of Anti-PC on a Quartz Fiber

Anti-PC is immobilized on the quartz optical fiber by the method developed by Bhatia, et al. (1989).

Conjugation of Fluorophore, Cy5, to PC and Secondary Antibody

The fluorophore, Cy5 (Biological Detection Systems, Inc., Pittsburgh, PA), is furnished in small vials capable of labeling 1 mg of protein. Up to 1 mg of PC or polyclonal antibody against PC (pAb-PC) are dissolved in 1 ml of 0.1 M sodium carbonate-sodium bicarbonate buffer (pH 9.3) and added to the dye vial. The reaction is incubated at room temperature for 30 minutes, with periodic mixing. Fluorophore tagged proteins are separated from unreacted Cy5 molecules via a 50 kd concentrator (Amicon Inc., Beverly, MA) that removes particles smaller than 50kd (Note: pAb-PC = 150 kd, PC = 62 kd, Cy5 = 1 kd).

Fluorometer

While this work represents the continuation of the PC biosensor project at UMBC, it also incorporates a number of changes in the experimental design as presented by Kang, et

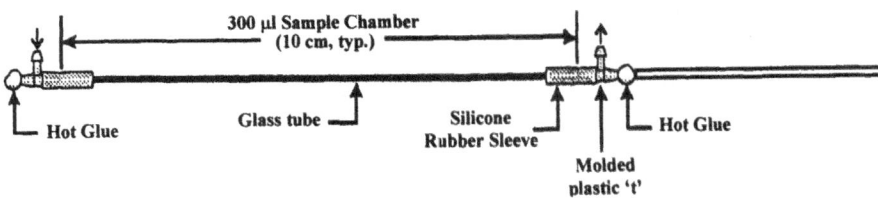

Figure 2. Schematic diagram of sample chamber assembly. A quartz fiber is inserted into a glass tube flanked by two plastic 'T' connectors. The connections are sealed with hot glue, creating a sample chamber size ~300 µl (diagram based on work done by Research International).

al., at the last ISOTT conference. The most significant change can be found in a departure from our previous fluorometer (Kang, et al., in press). Its sheer bulky size, cumbersome system of peristaltic pumps, and use of an outdated halogen lamp as a light source ultimately gave rise to the search for an alternative fluorometer. We have since adopted the Analyte 2000™, designed and built by Research International (Woodinville, Washington). In large part, the Analyte 2000™ has garnered attention due to its use in the detection of toxins, such as *Staphylococcal EnterotoxinB* (Ogert, et al., 1993; Templeman, et al., 1995). The Analyte 2000™ has been successfully used, for example, to detect 5 ng/ml concentrations of *Clostridium botulinum* toxins in under a minute (Beal, 1995). The PC biosensor project at UMBC marks one of the few times the Analyte 2000™ has been utilized for the detection of molecules as large as PC.

The physical size of the Analyte 2000™ is 20 cm X 8.8 cm x 11.2 cm, lending to great mobility. The fiber used for this experiment is a quartz fiber, 1 mm diameter with a tapered end (1 mm x 25 cm, Figure 2). Prior to antibody immobilization, the fiber is prepared with a chemical treatment. The fiber is then inserted into a glass tube flanked by two plastic 'T' connectors. The ends of the glass tube are sealed with hot glue, creating a ~300 µl sample chamber. Antibody immobilization may take place directly in the sample chamber. All sample additions occur via a syringe attached to the injection port. The small chamber size leads to an effective use of protein. This is of particular importance with this project due to the excessive cost of obtaining PC samples. Excitation light of 635 nm is produced via a 5 mW laser diode operated at a derated 1.5 mW level. By applying this excitaion light, Cy5 generates photons at a wavelength of 667 nm. A dichroic long-pass filter is placed before a silicone detector in order to distinguish emission light from excitation light. In the detector, photon energy is amplified and converted to electrical energy. The intensity of the arriving photons can be converted to a digital signal by a computer for display on a monitor.

3.2. Preliminary Results and Discussion

Direct Binding Assay

In all fiber preparations, antibody immobilization was performed with a solution of monoclonal antibody at a concentration of 50 µg/ml. After non-specific binding tests proved to be successful, attention was turned to direct binding assays in order to qualitatively prove the feasibility of this system. Here, PC directly conjugated with Cy5 was adsorbed to the surface of the optical fiber. Figure 3 shows the addition of 10, 5, 2.5, and 1.0 µg/ml of PC-Cy5 (each in .02M Sodium Citrate, 0.1% Tween 20, pH 7.2) to the sensor. The biosensor was able to detect down to 1.0 µg/ml PC-Cy5 at a satisfactory level, how-

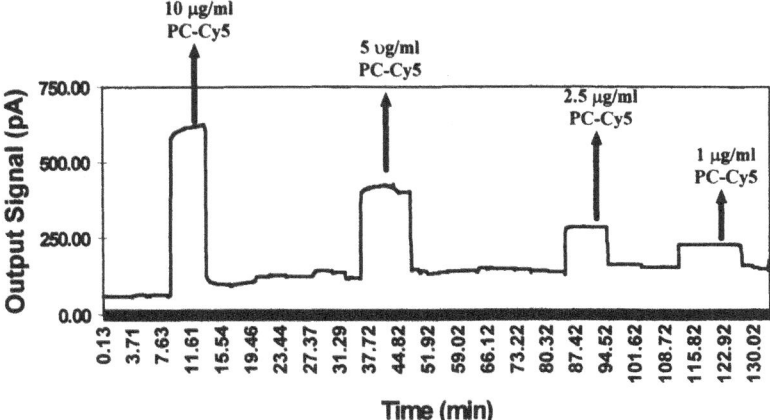

Figure 3. Direct Binding Assay. Fluorescent emissions representing the reaction between Cy5 tagged PC samples (10, 5, 2.5, 1.0 µg/ml samples in .02M Sodium Citrate, 0.1% Tween 20, pH 7.2) and monoclonal antibody on the fiber surface. The biosensor was able to detect 1.0 µg/ml PC-Cy5 concentrations at a satisfactory level.

ever, this degree of sensitivity will have to be greatly improved. Our goal is to reliably detect levels as low as 0.5 µg/ml.

Sandwich Assay

Cy5 tagged polyclonal antibody against PC (pAb-Cy5) was initially added to the sample chamber alone in order to discern if levels of non-specific binding would be detected. As Figure 4 shows, a modicum of non-specific binding was indeed detected after

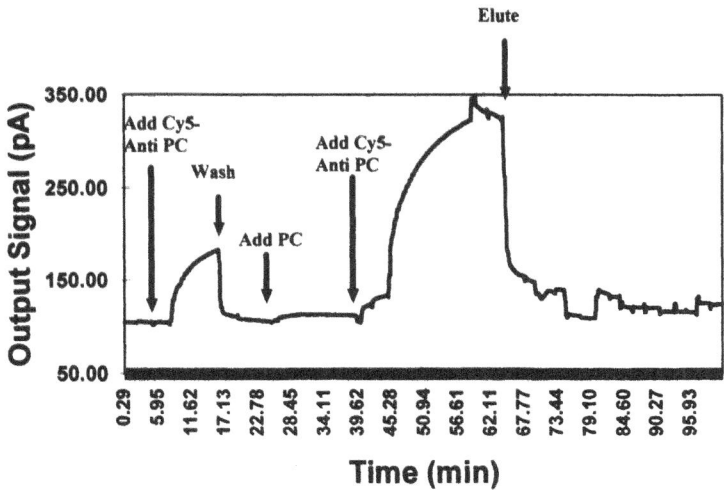

Figure 4. Sandwich Assay. Fluorescent emissions representing the reaction between PC (10 µg/ml in .02M Sodium Citrate, 0.1% Tween 20, pH 7.2) adsorbed to the surface of the optical fiber and Cy5 tagged polyclonal antibody against PC (100 µg/ml in .02M Sodium Citrate, 0.1% Tween 20, pH 7.2). pAb-Cy5 was initially added to the chamber alone. The output signal observed from this sample was eliminated after a washing step. The pAb-Cy5 was re-introduced to the PC sample, and a significantly higher output signal was achieved.

the addition of 100 µg/ml pAb-Cy5 (in .02M Sodium Citrate, 0.1% Tween 20, pH 7.2). However, the baseline was returned after a routine washing step. When the same pAb-Cy5 was reacted to an adsorbed sample of 10 µg/ml PC (also in .02M Sodium Citrate, 0.1% Tween 20, pH 7.2), a significantly higher output signal was observed. Although the probed PC sample was detected, the sensitivity of the reaction must nevertheless be increased. This will be achieved through a gradual refinement of the experimental protocol, including the discernment of optimal Cy5 to protein ratios and the optimal concentration of monoclonal antibody used to coat the optical fiber.

4. FUTURE WORK

Part of the goal in developing a PC biosensor is to provide a low-cost alternative to conventional measuring assays, such as ELISA. While the optical fibers may be reused several times, due to the fact that the monoclonal antibody is covalently bound, it is proposed to more thoroughly study the antibody activity after prolonged use. This will reveal an estimate for how long the fibers can be used before requiring a new coat of monoclonal antibodies. In the same vein, it is proposed to modify the monoclonal antibody immobilization procedure by incorporating an avidin-biotin system. It is believed that this method may alleviate the problem of lost antibody reactivity incurred during the immobilization step.

The experimental procedure should be optimized in order to make use of agents in their most optimal concentrations. This includes examining the best Cy5 to protein ratios, as well as learning the best concentration of monoclonal antibody used to coat the optical fiber.

Attention must be directed to characterizing the binding kinetics of the probe molecules. Research in this area may reveal a method to directly correlate an observed output signal with a specific PC concentration. In addition, knowledge of the binding kinetics may help to distinguish between specific and non-specific interactions.

In an esthetic note, the Analyte 2000™'s computer program may also be optimized to create a more user-friendly environment. It is the intent to make the process of gauging PC concentrations as efficient and effortless as possible.

5. CONCLUSIONS

The preliminary results hold promise for the development of a highly specific, sensitive, real-time, and portable Protein C biosensor. Besides being able to detect PC concentrations in a wide array of samples, the proposed biosensor may serve to expedite the detection of PC deficiency in a clinical setting. The same principles behind a PC biosensor may be applied to developing sensors for other blood factors, particularly since there is a lack of real-time biosensors for clotting and anti-clotting factors. Applications of highly specific and sensitive biosensors are only limited by the availability of probes for the plasma factor in question.

ACKNOWLEDGMENTS

This research is partly supported by The Whitaker Foundation and The American Red Cross. The authors would like to thank Research International for the use of their instrument and technical advice. The authors would also like to thank researchers at the Naval Research Laboratory of Washington D.C. for their technical support.

REFERENCES

Beal, Clifford, "An Invisible Enemy," *International Defense Review*, 3, 36–41, 1995.

Bhatia, S.K., Shriver-Laker, L.C., Prior, K.J., Georger, J.H., Calvert, J.M., Bredehorst, R., and Ligler F.S., "Use of thiol-terminal silanes and heterobifunctional crosslinkers for immobilization of antibodies on silica surfaces," *Analytical Biochemistry*, 178, 408–413, 1989.

Bruley, D.F. and Drohan, W.N., *Protein C and Related Anticoagulants - Advances in Applied Biotechnology Series*, 11, Gulf Publishing Company, Houston, 1990.

Kang, K.A., Ryu, D., Drohan, W.D., and Orthner, C.L., "Effect of Matrices on the Immunoaffinity Purification of Protein C," *Biotech. & Bioengr*, 39 (11), 1086–1098, 1992.

Kang, K.A., Anis, N.A., Eldefrawi, M.E., Drohan, W.D., and Bruley, D.F., "Reusable, Real-Time, Immuno-Optical Protein C Biosensor," ISOTT XVIII, Advances in Experimental Medicine and Biology, in press.

Ogert, R.A., Shriver, L.C., and Ligler, F.S., "Toxin Detection Using a Fiber Optic-based Biosensor," Proc. SPIE-Int. Soc. Opt. End. (1993), 1885 (Proceedings of Advances in Fluorescence Sensing Technology, 1993), 11–17.

Templeman, L.A., King, K.D., Anderson, G.P., and Ligler, F.S., "Quantitating *Staphylococcal Enterotoxin B* in Diverse Media using a Portable Fiber Optic Biosensor," *Analytical Biochemistry*, 233, 50–57, 1996.

88

PREDICTING OSCILLATION IN ARTERIAL SATURATION FROM CARDIORESPIRATORY VARIABLES

Implications for the Measurement of Cerebral Blood Flow with NIRS during Anaesthesia

A. T. Lovell,[1] H. Owen-Reece,[1] C. E. Elwell,[2] M. Smith,[1] and J. C. Goldstone[1]

[1]Division of Academic Anaesthesia
Department of Surgery
University College London Medical School
The Middlesex Hospital
Mortimer St., London, W1N 8AA, United Kingdom
[2]Department of Medical Physics and Bioengineering
University College London
Shropshire House, 11-20 Capper St.
London, WC1E 6JA, United Kingdom

1. INTRODUCTION

The potential of near infrared spectroscopy (NIRS) as a non invasive tissue oxygenation monitor was first outlined by Jöbsis (Jöbsis, 1977). Extension of the basic technique to measure tissue blood flow using a Fick technique was developed by Edwards and Reynolds (Edwards et al., 1988; Edwards et al., 1993). This uses a rapid change in arterial oxyhaemoglobin concentration to act as an intravascular tracer, avoiding the problem of recirculation of indicator.

$$Blood\ Flow\left(ml \cdot 100g^{-1} \cdot min^{-1}\right) = \frac{\Delta HbO_2 - \Delta Hb}{2 \cdot \left[tHb \cdot 10^{-2}\right] \cdot \int_0^t \Delta SpO_2\,dt} \cdot \frac{\left[MW_{Hb} \cdot 10^{-6}\right]}{\left[D_t \cdot 10\right]} \tag{1}$$

This technique has been successfully validated against ^{133}Xenon clearance for the measurement of cerebral blood flow (CBF) at flows upto 40 ml 100 g^{-1} min^{-1} in neonates

Oxygen Transport to Tissue XIX, edited by Harrison and Delpy
Plenum Press, New York, 1997

(Skov et al., 1991; Bucher et al., 1993), and against venous occlusion plethysmography in the adult forearm (Edwards et al., 1993).

In the last few years there has been increasing interest in the use of NIRS to monitor adult cerebral haemodynamics (Kirkpatrick et al., 1995; Mason et al., 1994; Villringer et al., 1993); Elwell et al., 1992; Owen-Reece et al., 1996). A major problem with the CBF technique is that with the induction of mild hypoxia, oscillation in arterial saturation and HbO_2 have been reported (Elwell et al., 1992). The finding that the dominant frequency of a fast fourier transform of both arterial saturation and HbO_2 were very strongly correlated with the breathing rate strongly suggested that these oscillations were of a respiratory nature (Elwell et al., 1994a).

The most likely explanation for the respiratory "artifact" was that the intrathoracic pressure changes during respiration were causing cyclical changes in cardiac output and that these were responsible for the oscillations in HbO_2. Elwell and colleagues have demonstrated that oscillations in HbO_2 occur whilst breathing room air, and that the magnitude of these oscillations increases with expiratory loading (Elwell et al., 1994b; Elwell et al., 1996). It was hypothesized that expiratory loading decreases the venous return and hence cardiac output during the respiratory cycle. However, the magnitude of the changes in cardiac output would be very unlikely to be sufficient to explain the changes in arterial saturation. Oscillations of the arterial partial pressure were first observed 50 years ago by Bjurstead. Consideration of the composition of gas within an alveolus would suggest that it should oscillate at the same frequency as respiration, due to the intermittent nature of gas flow, and this has been amply confirmed by studies of Yokota, Kreuzer and others during the 1970's (Yokota et al., 1973; (Yokota & Kreuzer, 1973; Folgering et al., 1978).

The measurement of CBF varies if the rise of HbO_2 and SpO_2 do not begin at the same time (Elwell et al., 1992). For example, a one second offset between HbO_2 and SpO_2 can change the measured CBF by 70%. Clearly if SpO_2 is oscillating determining the correct temporal relationship to HbO_2 can be difficult, and this may partially explain the previously high reported coefficient of variation of this technique.

Our model was developed to test the hypothesis that changes in ventilatory parameters might play a major role in the oscillations seen in arterial saturation.

2. THE MODEL

We have derived a two compartmental model. One compartment representing the dead space of the respiratory tract, in which gas mixing but no uptake of oxygen to the blood occurs, and the other representing the alveoli in which gas mixing and oxygen uptake occurs. The model assumes complete, and immediate, mixing of gas within each compartment.

The alveolar volume oscillates in a sinusoidal manner around the mean alveolar volume. During expiration, the volume of the alveolar compartment (V_A) is given by

$$V_A(t) = FRC + \frac{V_T}{2} + \cos\left(\frac{t}{k_1}\right) \cdot \frac{V_T}{2} = FRC + \frac{V_T}{2} \cdot \left\{1 + \cos\left(\frac{t}{k_1}\right)\right\} \tag{2}$$

where k_1 is a term used to set the inspiratory to expiratory ratio and respiratory rate, defined in Eq. 8 and 9 below. If a short time (δt) passes then the change in volume of the alveolar compartment (δV) is given by

$$\delta V = V_A(t) - V_A(t + \delta t) \tag{3}$$

substituting Eq. 2 in Eq. 3 yields

$$\delta V = \frac{V_T}{2} \cdot \left\{ \cos\left(\frac{t}{k_1}\right) - \cos\left(\frac{t + \delta t}{k_1}\right) \right\} \tag{4}$$

The fractional concentration of oxygen in the alveoli ($F_A O_2(t)$) will vary with time. Since no oxygen is added but oxygen consumption continues during expiration.

$$F_A O_2(t + \delta t) = \frac{F_A O_2(t) \cdot V_A(t) - \dot{v}o_2}{V_A(t)} \tag{5}$$

During expiration gas leaves the alveoli and enter the dead space, and an equal volume of gas is lost to the atmosphere from the dead space. If complete mixing of gas occurs then the fractional concentration of oxygen in the dead space ($F_D O_2(t)$) is given by

$$F_D O_2(t + \delta t) = \frac{F_D O_2(t) \cdot V_D - \{F_I O_2 - F_D O_2(t)\} \cdot \delta V}{V_D} \tag{6}$$

During inspiration the volume of the alveolar compartment (V_A) is given by

$$V_A(t) = FRC + \frac{V_T}{2} + \cos\left(\frac{\pi(k_2 - 1) - \pi(k_1 - 1) + t}{k_2}\right) \cdot \frac{V_T}{2}$$

$$= FRC + \frac{V_T}{2} \cdot \left\{ 1 + \cos\left(\frac{\pi(k_2 - k_1) + t}{k_2}\right) \right\} \tag{7}$$

where k_1 and k_2 are terms used to set the inspiratory to expiratory ratio and respiratory rate such that the duration of one breath is $2\pi(k_1 + k_2)$ seconds and the inspiratory to expiratory ratio is given by

$$\text{inspiratory to expiratory ratio} = \frac{k_2}{k_1} \tag{8}$$

and

$$k_1 + k_2 = \frac{30}{\pi(\text{breaths per minute})} \tag{9}$$

If a short time (δt) passes then during inspiration the change in volume of the alveolar compartment (δV) is given by substituting Eq. 7 in Eq. 3

$$\delta V = \frac{V_T}{2} \cdot \left\{ \cos\left(\frac{\pi(k_2 - k_1) + t}{k_2} \right) - \cos\left(\frac{\pi(k_2 - k_1) + t + \delta t}{k_2} \right) \right\}$$

(10)

During inspiration gas passes from the dead space into the alveoli and an equal volume of gas enter the dead space with a fractional concentration of oxygen of F_1O_2. If the gas entering the alveoli from the dead space and that entering the dead space is immediately mixed then

$$F_AO_2(t + \delta t) = \frac{F_AO_2(t) \cdot V_A(t) + F_DO_2(t) \cdot \delta V - \dot{v}o_2}{V_A(t)}$$

(11)

$$F_DO_2(t + \delta t) = \frac{F_DO_2(t) \cdot V_D + \{F_1O_2 - F_DO_2(t)\} \cdot \delta V}{V_D}$$

(12)

The model first solves Eq. 8 and 9 to yield k_1 and k_2. If the transition from inspiration to expiration is taken as t=0 then $F_AO_2(0)$ can be calculated using the alveolar gas equation and $F_DO_2(0)$ can be approximated by F_1O_2.

Because of the need to be able to produce an asymetrical "sine" wave the use of two separate waves was chosen. However, implicit with this technique is the fact that in order to synthesize a repeating waveform the value of t used in Eq. 2 and 7 must be set to zero on each transition from inspiration to expiration.

Conversion from alveolar oxygen concentration to arterial saturation is performed using the method of Farmery and Roe (Farmery & Roe, 1996) assuming complete equilibration of the mixed pulmonary venous blood with the alveoli and no diffusion barrier to oxygenation.

3. RESULTS

The results of a typical desaturation - resaturation run, for a 70 kg male, are shown in Figure 1. It is very clear that the magnitude of the oscillation increases as the arterial saturation falls, and that the magnitude of the oscillation is influenced by the tidal volume. The link between respiratory frequency and oscillatory frequency is clear. The shape of the saturation-time curve appears identical to that seen in clinical practice. Formal comparison with experimental measurements is in progress. The influence of the patient's oxygen consumption (VO_2) and alveolar volume (V_A) are shown in Figure 2. The importance of minimizing the reduction in FRC during anaesthesia is clearly shown. The effects of prolongation of the inspiratory to expiratory (I:E) time at differing values for VO_2 are shown in Table 1. The effects of even relatively small changes of the I:E ratio is quite marked at high VO_2. This effect is magnified by the use of larger tidal volumes. However, if small tidal volumes are used and the patient has a lower than normal VO_2, the effect of changes in I:E ratio can probably be ignored.

The effects of signal averaging upon the peak to peak magnitude of the oscillations in arterial saturation is shown in Table 2. It is clear that applying a moving average smoothing function can reduce the magnitude of these oscillations, but this also introduces a phase delay between the oscillation in the arterial signal and that actually measured.

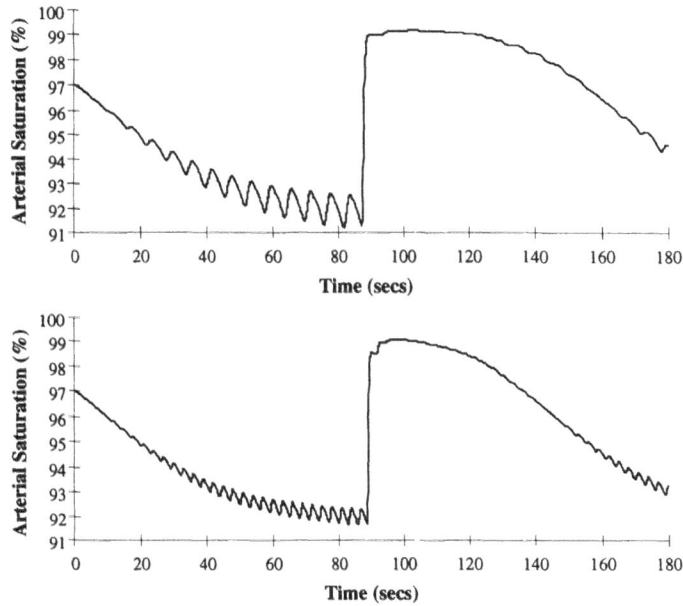

Figure 1. Computer predicted SpO$_2$ against time. V$_A$ 2500 ml. VO$_2$ 250 ml min^{-1}. F$_I$O$_2$ 0.12. At 90 seconds 3 breaths of 100% O$_2$ were delivered. Top panel V$_T$ 1000. 10 breaths min^{-1}. Bottom Panel V$_T$ 500. 20 breaths min^{-1}.

This time offset must be taken account of when performing measurments of CBF. In practice the effects of signal averaging will be to dramatically reduce the number of time points as well. This may further reduce the magnitude of the oscillations since the peaks and troughs would be unlikely to fall precisely at either end of one of the time periods. Given that the current generation of commercially available NIRS instruments have a maximum sampling rate of 2 Hz it is not surprising that these oscillations have not been consistently observed.

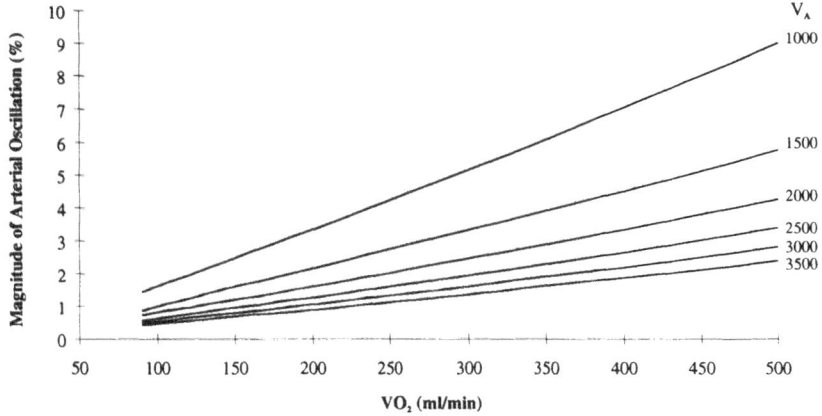

Figure 2. Effects of changes in alveolar volume (V$_A$) and minute oxygen consumption (VO$_2$) on the peak to peak magnitude of oscillation of the computer predicted arterial saturation when desaturated to a mean arterial saturation of 90%. V$_T$ 500 ml 14 breaths min^{-1} V$_D$/V$_T$ 0.3.

Table 1. Effects of changes in inspiratory:expiratory (I:E) ratio and minute oxygen consumption (VO_2) on the peak to peak magnitude of oscillation of the computer predicted arterial saturation. Mean predicted arterial saturation 90%. V_T 500 ml 14 breaths min^{-1} V_D/V_T 0.3

I : E Ratio	VO_2 (ml min^{-1})					
	100	150	200	250	300	400
4 : 1	.35	.53	.72	.91	1.10	1.50
3 : 1	.38	.57	.77	.99	1.18	1.61
2 : 1	.42	.64	.86	1.09	1.32	1.80
1 : 1	.52	.78	1.06	1.34	1.63	2.21
1 : 2	.63	.95	1.28	1.61	1.96	2.62
1 : 3	.69	1.04	1.39	1.76	2.14	2.90
1 : 4	.72	1.09	1.47	1.85	2.25	3.06

4. DISCUSSION

Oscillations in arterial saturation are likely to be a universal finding in adults subjected to a desaturation manoeuver. The magnitude of these oscillations increases dramatically as the saturation falls. At first sight this seems to be a contradiction with the work of Kreuzer and his colleagues (Folgering et al., 1978) who reported a much larger value for ΔPaO_2 at high values for PaO_2 than at low values. However, once it is realized that at the high values for PaO_2 used in their study the oxygen dissociation curve is flat, then it is clear that the saturation change would be very small. Previous studies (Nye, 1970) using a two compartment model under normoxic conditions have observed oscillations in alveolar PO_2 which changes with changes in the respiratory pattern. Because these studies were only conducted under normoxic conditions where the oxygen dissociation curve is essentially flat the equivalent change in arterial saturation was only 0.5% despite a 10 mmHg change in PO_2.

It is possible to dramatically reduce the magnitude of these oscillations by minimizing the tidal volume (V_T) used. However, reductions in the V_T limit the rate of rise of alveolar oxygen concentration and hence arterial saturation following exposure to 100%

Table 2. Effects of applying moving average smoothing to the peak to peak magnitude of the oscillations in the predicted arterial saturation. Mean predicted arterial saturation 90%. Simulations performed for a typical 70 kg person. Minute ventilation 7 l min^{-1}. V_A 2500 ml VO_2 250 ml min^{-1} V_D/V_T 0.3

Duration of moving average (sec)	Tidal volume (ml)		
	500	750	1000
Raw data	1.61	2.27	3.29
0.25	1.59	2.25	3.28
0.50	1.52	2.20	3.24
0.75	1.44	2.12	3.18
1.00	1.35	2.04	3.10
1.50	1.15	1.86	2.93
2.00	0.95	1.65	2.73

oxygen. It is important that this rate of rise is as fast as possible in order to maximize the input of tracer, oxyhaemoglobin, to the brain. An alternative strategy to minimize these oscillations is to maximize the alveolar volume (V_A), or minimize the VO_2. Under anaesthesia V_A generally falls by 20–30% from the awake value, although there are methods, such as the addition of PEEP, that can be applied to reduce this. The VO_2 generally falls under anaesthesia since active movement, the prime consumer of oxygen, generally ceases. It is possible to further reduce the VO_2 by the use of mechanical ventilation. The most likely explanation for the effects of changes in V_T and V_A is that the smaller the V_T:V_A ratio the smaller the proportion of alveolar gas that is exchanged each breath, and therefore the greater the ability for the alveolar gas to act as a buffer.

The other strategy for minimizing oscillation in arterial saturation is prolongation of the duration of active inspiration. With mechanical ventilation of patients this is fairly simple to achieve, but is all but impossible in conscious volunteers. The explanation behind the changes in oscillation with changes in the I:E ratio and VO_2 is that during expiration oxygen is being consumed from the alveolar gas, but is not being replaced. The cyclical nadir in P_AO_2 will be reached early in inspiration when gas that had resided in the dead space at the end of expiration reenters the alveoli. Thus the greater the proportion of the respiratory cycle spent in expiration, or the higher the VO_2, the greater the difference in oxygen concentration in the alveolar gas at the nadir compared to the end of inspiration. In this context, the addition of an inspiratory pause will lead to an increase in oscillation assuming complete and uniform equilibration of gas in the alveoli. This is because no further oxygen is delivered to the alveoli during this pause, but consumption is already proceeding.

The effects of the signal averaging built into many pulse oximeters is to markedly reduce the magnitude of these oscillations and to cause a considerable change in their phase angle compared to the true saturation of arterial blood. The averaging built in to the current generation of NIRS instruments when operated at their maximum capture rates causes only a very small reduction in magnitude of the oscillation.

Under anaesthesia it is often possible to select ventilatory parameters which minimize the introduction of respiratory "artifacts" on NIRS signals and thus facilitate the use of dynamic NIRS techniques. Furthermore this model suggests that there maybe categories of patients, such as those with a high VO_2 and low V_A where the oxygen CBF technique may be impractical due to respiratory artifacts.

It is possible to criticize the assumptions of our model on several grounds. The use of a two compartmental model is an oversimplification. Although there are millions of alveoli, in health they should all exchange oxygen in a relatively similar manner. It is possible to expand the dead space compartment into multiple smaller compartments, but if this approach is taken then very careful consideration has to be given to the partitioning of gas between these sub-compartments.

The assumption of continuous pulmonary blood flow was made to facilitate simplification of the model. In reality the pulmonary blood flow is pulsatile and will thus apply a convolution function to the arterial saturation signal. The currently commercially available NIRS instruments have a maximum sampling rate of 2 Hz. From Nyquist's theorem the maximum observed frequency of any convolution function is thus 1 Hz, which equates to a heart rate of 60 min^{-1}. Thus in clinical practice with the currently available devices the effect of the convolution will generally not be directly observable as an oscillating signal. Yokota and Kreuzer have shown that the effect of pulsatile blood flow is to cause a damping of the arterial oscillations (Yokota & Kreuzer, 1973). This is predominantly due to mixing within the left ventricle, with very little difference from aorta to the carotid (Folgering et al., 1978), or more distally in the arterial tree (Yokota & Kreuzer, 1973). Typically this damping leads to a reduction by approximately 50% of the oscillation seen in the left atrium, and predicted by our model. Furthermore, as the ratio of ventilation rate to heart rate decreases the attenuation of the oscillation decreases.

The idea that perfect equilibrium of end capillary blood with alveolar gas may seem at first sight extremely optimistic. However, in health blood traveling through the pulmonary capillaries is fully saturated by the time it has spent a third of its residence time. Thus there is a very considerable reserve before diffusion limitation for oxygenation will occur. In pathological states or circumstances with a very rapid pulmonary blood flow, such as exercise, this may not be true. The finding of a very strong correlation between fluctuations in end capillary and arterial saturations, although at reduced magnitude, supports this (Yokota & Kreuzer, 1973).

Gas flow, even in health, is usually turbulent which should produce near perfect mixing. However, if there are regions of the lungs with unequal time constants, such as in asthma, or very low gas flow rates are used, then this may not be true. Thus assuming complete and immediate mixing of gas in both compartments is probably reasonable. However, if the dead space compartment was further partitioned the exact location of the sub-compartments would have to determine whether this assumption could remain, or whether laminar flow considerations should be introduced.

We have assumed that the induction of hypoxia does not lead to an increase in total oxygen consumption (VO_2). If the VO_2 changes as hypoxia is induced then the rate that the alveolar oxygen reservoir is depleted during the expiratory and early inspiratory phase of each respiratory cycle will change. Over the range of saturations that the model has been designed to cope with this should be a reasonable approximation. The hypoxia chemoreceptors usually remain relatively silent until an arterial saturation well below 90%.

It is known that the dead space changes with lung volume. However, whether V_D/V_T remains constant throughout the respiratory cycle is less clear. In the extreme circumstance when the deflating lung approaches its residual volume (RV) during expiration this is clearly untrue. However, even allowing for the reductions in FRC that are known to occur during anaesthesia, alveolar volume is usually maintained well above RV.

5. CONCLUSIONS

In order to minimize the induction of respiratory "artifacts" when performing CBF NIRS studies using the oxygen method, care should be taken to use as high a baseline arterial saturation as possible. It is however inevitable that this limits the magnitude of the "tracer" input, and may lead to a rise in the signal to noise ratio. Attempts at minimizing the arterial oscillations by manipulation of either V_T or V_A involve an inevitable compromise if the oxygen CBF method is used. This is because of the fundamental requirement of the oxygen CBF technique for a rapid rise in arterial saturation which necessitates a high V_T:V_A ratio and directly conflicts with the requirements to minimize baseline arterial oscillations, namely a low V_T:V_A ratio. Prolongation of the duration of active inspiration, although effective is currently only possible with a small number of anaesthetic ventilators, and is impossible in the spontaneously breathing patient. Minimization of the patient's oxygen consumption, in order to minimize the oxygen consumption during the expiratory phase whilst making perfect logical sense, is only rarely amenable to manipulation.

Minimization of the induction of these respiratory "artifacts" in order to facilitate the measurement of CBF thus requires a carefully weighted compromise over the choice of respiratory parameters. Some clinical circumstances may preclude manipulation of some of the respiratory variables, and the resultant oscillation in the baseline arterial saturation may be such as to preclude the use of the oxygen CBF technique. For these patients other methods of measuring CBF will have to be applied.

ACKNOWLEDGMENT

This work has been supported by a grant from the Wellcome Trust.

REFERENCES

Bucher, HU; Edwards, AD; Lipp, AE; Duc, G (1993): Comparison between Near Infrared Spectroscopy and [133]Xenon Clearance for Estimation of Cerebral Blood Flow in Critically Ill Preterm Infants. *Pediatr Res* 33, 56–60.

Edwards, AD; Richardson, C; van der Zee, P; Elwell, C; Wyatt, JS; Cope, M; Delpy, DT; Reynolds, EOR (1993): Measurement of hemoglobin flow and blood flow by near-infrared spectroscopy. *J Appl Physiol* 75, 1884–1889.

Edwards, AD; Wyatt, JS; Richardson, C; Delpy, DT; Cope, M; Reynolds, EOR (1988): Cotside measurement of cerebral blood flow in ill newborn infants by near infrared spectroscopy. *Lancet* 2, 770–771.

Elwell, CE; Cope, M; Edwards, AD; Wyatt, JS; Reynolds, EOR; Delpy, DT (1992): Measurement of cerbral blood flow in adult humans using near infrared spectroscopy - methodology and possible errors. *Adv Exp Med Biol* 317, 235–245.

Elwell, CE; Cope, M; Edwards, AD; Wyatt, JS; Delpy, DT; Reynolds, EOR (1994a): Quantification of adult cerebral hemodynamics by near-infrared spectroscopy. *J Appl Physiol* 77, 2753–2760.

Elwell, CE; Owen-Reece, H; Cope, M; Edwards, AD; Wyatt, JS; Reynolds, EOR; Delpy, DT (1994b): Measurement of changes in cerebral haemodynamics during inspiration and expiration using near infrared spectroscopy. *Adv Exp Med Biol* 345, 619–626.

Elwell, CE; Owen-Reece, H; Wyatt, JS; Cope, M; Reynolds, EOR; Delpy, DT (1996): Influence of Respiration and Changes in Expiratory Pressure on Cerebral Haemoglobin Concentration Measured by Near Infrared Spectroscopy. *J Cereb Blood Flow Metab* 16, 353–357.

Farmery, AD; Roe, PG (1996): A model to describe the rate of oxyhaemoglobin desaturation during apnoea. *Br J Anaesth* 76, 284–291.

Folgering, H; Smolders, FDJ; Kreuzer, F (1978): Respiratory oscillations of the arterial PO_2 and their effects on the ventilatory controlling system of the cat. *Pflügers Arch* 375, 1–7.

Jöbsis, FF (1977): Noninvasive, infrared monitoring of cerebral and myocardial oxygen sufficiency and circulatory parameters. *Science* 198, 1264–1267.

Kirkpatrick, PJ; Smielewski, P; Whitfield, PC; Czosnyka, M; Menon, D; Pickard, JD (1995): An observational study of near-infrared spectroscopy during carotid endarterectomy. *J Neurosurg* 82, 756–763.

Mason, PF; Dyson, EH; Sellars, V; Beard, JD (1994): The assessment of cerebral oxygenation during carotid endarterectomy utilising near infrared spectroscopy. *Eur J Vasc Surg* 8, 590–594.

Nye Jr, RE (1970): Influence of the cyclical pattern of ventilatory flow on pulmonary gas exchange. *Respir Physiol* 10, 321–337.

Owen-Reece, H; Elwell, CE; Harkness, W; Goldstone, J; Delpy, DT; Wyatt, JS; Smith, M (1996): Use of near infrared spectroscopy to estimate cerebral blood flow in conscious and anaesthetized adult subjects. *Br J Anaesth* 76, 43–48.

Skov, L; Pryds, O; Greisen, G (1991): Estimating Cerebral Blood Flow in Newborn Infants: Comparison of Near Infrarred Spectroscopy and [133]Xe Clearance. *Pediatr Res* 30, 570–573.

Villringer, A; Planck, J; Hock, C; Schleinkofer, L; Dirnagl, U (1993): Near infrared spectroscopy (NIRS): a new tool to study hemodynamic changes during activation of brain function in human adults. *Neurosci Lett* 154, 101–104.

Yokota, H; Hoofd, LJC; Kreuzer, F (1973): Alveolar oxygen tension in anesthetized, artificially ventilated dogs. *Pflügers Arch* 340, 273–290.

Yokota, H; Kreuzer, F (1973): Alveolar to arterial transmission of oxygen fluctuations due to respiration in anesthetized dogs. *Pflügers Arch* 340, 291–306.

Appendix. Definition of variables and their units

Variable	Definition	Units
CBF	Cerebral blood flow	ml $100g^{-1}$ min^{-1}
D_t	Cerebral tissue density	g ml^{-1}
F_AO_2	Fractional alveolar oxygen concentration	—
F_DO_2	Fractional dead space oxygen concentration	—
F_IO_2	Fractional inspired oxygen concentration	—
FRC	Functional residual capacity of the lungs	litres
ΔHb	Change in cerebral deoxyhaemoglobin concentration	μmol $litre^{-1}$
ΔHbO_2	Change in cerebral oxyhaemoglobin concentration	μmol $litre^{-1}$
I:E	Inspiratory to expiratory ratio	—
k_1	Expiratory time factor	—
k_2	Inspiratory time factor	—
MW_{Hb}	Molecular weight of haemoglobin	g
SpO_2	Arterial saturation	%
t	Time	sec
tHb	Tissue haemoglobin concentration	g dl^{-1}
V_A	Alveolar volume	litres
V_D	Dead space volume	litres
V_T	Tidal volume	litres
$\dot{v}O_2$	Minute oxygen consumption	litres sec^{-1}

89

SEPARATION OF PROTEIN C FROM FRACTION IV OF THE COHN PROCESS USING IMMOBILIZED METAL AFFINITY CHROMATOGRAPHY

Sriram S. Tadepalli, Duane F. Bruley, Kyung A. Kang, and William Drohan

University of Maryland Baltimore County
The American Red Cross

1. INTRODUCTION

A book[1] which deals with most aspects of protein C has been of considerable help in this work. Protein C is azymogen precursor present in human blood.[2,14] Its absence could be harmful to the patient, and could ultimately lead to death. Hence, it has become important to provide protein C to all patients who need it at an affordable price. To achieve this, protein C must be produced in large quantities and at a low cost. Present methods to produce protein C are very expensive. Purification by immunoaffinity following the Cohn precipitation process produces protein C which can only be used as a model protein for recombinant protein C experiments'. It is too expensive to treat patients prophylactively. Hence, another approach might be to produce protein C using Immobilized Metal Affinity Chromatography (IMAC) since IMAC is inexpensive and its separation efficiency approaches that of immunoaffinity chromatography.[15]

1.1. Immunoaffinity as a Means of Separating Protein C From Fraction IV of the Cohn Paste

The first step is to obtain fraction IV of the Cohn paste. Protein C is then separated from the Cohn paste by the following steps. Barium citrate adsorption is used to separate the vitamin K dependent proteins. This step is repeated two to three times. This is followed by ammonium sulfate precipitation to achieve better separation of the proteins. The final step is the purification of protein C by immunoaffinity chromatography.[4] This is a very expensive step due to the cost of the monoclonal antibodies which are necessary to bind protein C with high specificity. Therefore, it is desirable to supplement this step of the purification process with IMAC.

Oxygen Transport to Tissue XIX, edited by Harrison and Delpy
Plenum Press, New York, 1997

1.2. Immobilized Metal Affinity Chromatography (IMAC)

Immobilized Metal Affinity Chromatography was discovered in 1975 in the pioneering work of Porath.[5] It was soon found that IMAC combined high specificity of binding with relatively low cost.[6] The principle behind IMAC lies in the attraction of proteins possessing histidine residues to the immobilized metal ions at neutral pH. Protein C, containing 17 histidine residues appears to be an ideal protein to be purified using IMAC.[2]

Much work has been done on IMAC since its discovery. The Temkin isotherm which is used in gas adsorption, is suitable for describing IMAC.[7,8] This isotherm fits heterogeneous adsorption which has been observed with proteins containing multiple histidine residues. The typical equilibration condition for IMAC is a neutral pH at which proteins containing histidine residues are easily attracted to the metal. Among the ligands used are Imino Di Acetic acid (IDA), Tetra Ethylene Di amine (TED)[9] and Nitrilo TriAcetic acid (NTA). For purposes of scale up, it is felt that IDA is the best ligand to be used since it is available commercially and on a large scale.[3]

1.2.1. Mechanism of IMAC[6]

(a) Adsorption. There are two steps here. First, the metal ion is immobilized by chelate formation. Second, the protein is adsorbed to the metal. The metal acts like a Lewis acid, accepting electrons. The two steps can be represented as:

$$Ch + M \rightarrow Ch\sim M \text{ (chelate formation)} \tag{1}$$

and

$$Ch\sim M + P \rightarrow Ch\sim M\cdots P \text{ (protein attachment)} \tag{2}$$

where Ch is the chelating ligand, M, the metal ion and P, the protein.

(b) Metal Ion transfer. Some proteins (metalloproteins) have a greater attraction for the metal ion than the chelate. This can be represented as:

$$Ch\sim M + P \rightarrow Ch + MP \tag{3}$$

This is called metal ion transfer.

(c) Desorption: This can occur by three mechanisms:

a.) When a competing ligate donates an electron (acting like a Lewis base).

$$Ch\sim M \cdots P + L \rightarrow ChML + P \tag{4}$$

$$Ch\sim M \cdots P + L \rightarrow Ch + LMP \tag{5}$$

b.) When a metal ion M* accepts an electron, thus acting as a Lewis acid. There are two ways by which this can occur:

$$Ch\sim M \cdots P + M^* \rightarrow Ch\sim M^* + MP \tag{6}$$

$$Ch\text{\textasciitilde}M \cdots P + M* \rightarrow Ch\text{\textasciitilde}M + M*P \qquad\qquad (7)$$

c.) By lowering the pH, positively charged hydrogen ions, called protons, can replace the immobilized metal ion.

$$Ch\text{\textasciitilde}M \cdots P + H^+ \rightarrow Ch\text{\textasciitilde} H + MP \qquad\qquad (8)$$

$$Ch\text{\textasciitilde}M \cdots P + H^+ \rightarrow Ch\text{\textasciitilde} M + H \qquad\qquad (9)$$

This type of desorption is called protonation, and was found to be the most effective method to recover protein C in the previous work.[3]

1.2.2. Advantages of IMAC. Probably the most important advantage of IMAC is its low cost as compared to immunoaffinity. It is only slightly less effective in terms of its affinity for proteins. Therefore, it approaches the efficiency of immunoaffinity chromatography. Also. a single packed column can be used repeatedly, approximately 25 times and great care need not be exercised to handle the columns.

1.2.3. Disadvantages of IMAC. An important disadvantage of IMAC is the **metal ion transfer**[10] which can occur when the immobilized metal ions have a greater affinity for the proteins than for the gel on which they are immobilized. This can result in the stripping of the metal by the metalloproteins. Among other disadvantages is the ligand leakage where the immobilized ligand leaks and metal catalyzed oxidation where the immobilized metal ion is reduced.[11]

2. PREVIOUS WORK DONE AND CHANGES INCORPORATED IN PRESENT WORK

Earlier work[3] to find the most optimal conditions for the separation of protein C from a mixture of proteins using IMAC has been presented. This work focused on determining the right system (ligand, metal ion etc.) with the appropriate buffers to give the most efficient process for separation. Copper was found to be the best metal, this the best buffer and protonation, the best means of elusion. However, this work suggested using the same buffer for both equilibration and elusion. Equilibration was done at pH 7.4 and elusion was done at pH 5.0. In the present work, two different buffers are being used for equilibration and elusion. This is because every buffer has a specific pH range to work in. The range is within pKa +1 pH units and pKa -1 pH units, where pKa is the thermodynamic dissociation constant of the weak acid of the buffer. At pKa+1 pH units and pKa-1 pH units, the buffering capacity falls to one third of its maximum capacity. Below pKa-1 pH units or above pKa+1 pH units, the buffering capacity is so small as to be of little value.[12] The pKa of tris buffer is 8.08. Therefore, the tris buffer can typically be used only between pH 7.1 and pH 9.1. If it is used below pH 7.1, it has little buffering capacity and most likely will not maintain the expected pH. Hence, it is recommended not to use the tris buffer to maintain a pH of 5.0. It is better to use the acetate buffer for a pH of 5.0. In the present study, phosphate buffer has been used at a pH of 7.06 and acetate buffer at a pH of 5.0.

3. WORK DONE

3.1. Materials and Methods

The goal of this work is to obtain the efficiency of separation of protein C from processed Cohn fraction IV of blood plasma using IMAC. The gel matrix chosen for IMAC is the Chelating Sepharose Fast Flow resin which is available from Pharmacia; very similar to the Hi Trap chelating gel which comes in prepacked columns. This gel matrix was chosen because it is available in bulk quantities which is important for scaleup. The Hi Trap Chelating column is useful for lab scale testing and its properties coincide with the Chelating Sepharose Fast Flow.[13] It was stated by the technical support of Pharmacia Company that both gels have very similar properties and that lab scale testing can be done with either gel with large scale testing being done with the Chelating Sepharose Fast Flow It was decided to use the Hi Trap gel in future experiments as it comes prepacked and in known volumes.

Copper was chosen as the metal ion to be immobilized for two reasons. First, copper is the best metal both in terms of its affinity for the gel and the protein. Second, previous work[3] showed copper to be much better than the other metals tested, viz., nickel and zinc. The IMAC set up consisted of a peristaltic pump (Pharmacia LKB pump P1) and a BioRad column of 0.4 cm ID and 10cm length. The flow rate was set constant at 30 ml/hr. The resulting chromatogram was recorded on a Servogor 120 chart recorder. An ELISA and an SDS PAGE were performed on the resulting fractions. A second run using changes incorporated in the previous work (see planned future work for more details) was performed using a Pharmacia FPLC system.

4. RESULTS AND DISCUSSIONS

Repeated ELISAs of the fractions obtained using the FPLC have shown that all protein C injected into the column was recovered in the first three minutes after injection. This led to the conclusion that protein C in the sample did not bind at all to the column. A stripping of the copper by the sample (Cohn paste) during the wash step following the sample injection into the column was also observed. The amount of copper stripped has not been quantified, but a sizable portion of the immobilized copper was visually observed to have been stripped. ELISAs of the wash and elusion fractions showed a much greater amount of protein C in the wash than in the elusion. In fact, practically all the protein C injected was recovered in the first fraction of the wash. Two explanations can be offered for this. One is that, the citrate buffer in which the sample is believed to be constituted is known to form a complex with the copper ions[13] which could result in the stripping of the immobilized copper ions. The second could be that metalloproteins are present in the sample which strip the copper. It is felt that protein C is not responsible for stripping of the metal (as was shown in the previous work[3]). An SDS PAGE was also performed for the samples obtained using the first set up. Bands were observed close to the 20kDa and the 42kDa standards in the SDS PAGE in both the wash and elusion fractions of the chromatographic run. These bands could have indicated protein C but this could not be ascertained as the Cohn paste was believed to contain vitamin K dependent proteins having molecular weights close to protein C, and present in far excess amount. Some of these proteins are believed to possess light and heavy chains and could split at almost the same molecular weight as protein C[16] in an SDS PAGE. Since the main focus in this work is to find the efficiency of separation of protein C from the Cohn paste using IMAC, it is thought that the SDS PAGE is not absolutely necessary to confirm the presence of protein C. An ELISA will definitely confirm the presence of protein C since it uses specific antibodies which bind only to protein C. An ELISA also gives a quantitative estimate of the amount of protein C present which is of more use in determining the efficiency of separation of protein C from the Cohn paste using IMAC. An Amidolytic activity assay was not performed to find the percentage activity of protein C. However, it is assumed that the protein C recovered is active since previous results obtained[3] showed protein C to be so. In the SDS PAGE results (Figure 1) lanes 5 and 6 represent the wash and elusion fractions analyzed, respectively. This clearly shows that the wash fraction contains a greater number of proteins than the elution This is a photograph of a silver stained gel.

5. PLANNED FUTURE WORK

Copper will be used, as it has been proven to be far superior to the other metals tested in the previous work[3] viz. nickel and zinc. Two loadings (viz., 100% metal loading and 66% metal loading which were used in the previous work[3]) will not be used. Instead only one metal loading percentage of 60% will be investigated, as suggested by Pharmacia.[13]

The prepacked Hi Trap Chelating column will be used instead of the Chelating Sepharose Fast flow because it is already packed and of known volume. It is planned to use a sample of known composition and if possible, to try and avoid citrate buffer. If, however, the metal continues to be stripped by the sample, then it is planned to strip and reload the metal after each run as some metal which is being stripped during the chromatographic process will not be available during the subsequent run. Two different buffers will be used, one for equilibration and one for elusion. A low salt concentration will be used as suggested in the previous work[3] because this will only introduce a small amount of foreign material into the system. A gradient elusion will be used instead of a step elusion as this method has been proven to be better than a step elusion in the previous work.[3] The

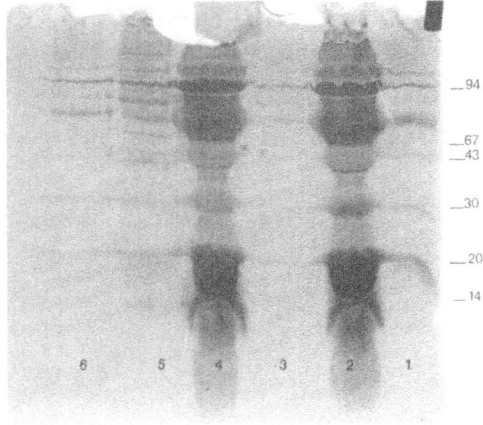

Figure 1. SDS PAGE results.

response as defined previously[3] will not be calculated and instead, the percentage efficiency of recovery in the wash and elusion stages will be calculated.

REFERENCES

1. D. F. Bruley and W. N. Drohan *Protein C and Related Anticoagulants,* Advances in Applied Biotechnology Series, Volume 11, 1990.
2. M. V. Chaubal, K. A. Kang, S. S. Tadepalli, W. N. Drohan and D. F. Bruley, *ISOTT(1995).*
3. J. C. Dalton, *M.S. Thesis,* University of Maryland Baltimore County (1996).
4. K. A. Kang, D. Ryu, W. N. Drohan, and C. L. Orthnew, *Biotechnology and Bioengineering, 39,* 1–11 (1992).
5. J. Porath, J. Carlsson, I. Olsson and G. Belfrage, *Nature,* 258, 598–599 (1975).
6. J. Porath, *Protein Expression and Purification,* 3, 263–281 (1992).
7. R. J. Todd, *Ph.D. Thesis,* California Institute of Technology (1993).
8. R. J. Todd, R. D. Johnson and F. H. Arnold, *Journal of Chromatography A* 662, 13–26 (1994).
9. T. Yip, T. W. Hutchens, *Methods in Molecular Biology,* Editors A. Kenney, S. Fowell, 2, 17–32 (1992).
10. E. Sulkowski, *Trends in Biotechnology,* 3, 1, 1–7 (1985).
11. R. Krishnamurthy, R. D. Madurawe, K. D. Bush and J. A. Lumpkin, *Biotechnology Progress, 11,* 643 650 (1995).
12. *Data for Biochemical Research,* 3rd edition, Oxford: Clarendon Press, p419, (1986).
13. Pharmacia Biotech, *Technical Support.*
14. W. H. Velander, T. Morcol, D. B. Clark, D. Gee, W. N. Drohan, *Protein C and Related Anticoagulants,* Eds. D. F. Bruley, W. N. Drohan, Gulf Publishing Company, 2, 11–28 (1990).
15. F. H. Arnold, *Biotechnology, 9,* 152–156 (1990).
16. Dr. Annemarie Ralston, Plasma Derivatives Laboratory, Holland Laboratories, American Red Cross, Rockville, Maryland.

90

BLOOD OXYGENATION

Heterogeneity of Hypoxic Tissues Monitored using Bold MR Imaging

Jeff F. Dunn and Harold M. Swartz

NMR and EPR Research Centers
Department of Radiology
Dartmouth-Hitchcock Medical Center
Hanover, New Hampshire 03755

1. INTRODUCTION

The capacity to monitor tissue oxygenation, both non-invasively and with high spatial and temporal resolution, would be of significant value in the study of tissue responses to hypoxia. While oxygen concentration at the mitochondrial level is desired for studies of metabolic regulation, very useful information often can be obtained by measuring oxygenation (pO_2 or content) in the vascular system.

Deoxygenated hemoglobin has a different susceptibility from oxygenated hemoglobin (Pauling et al., 1936), and so can be used as a magnetic resonance (MR) contrast agent (Belliveau et al., 1991; Ogawa et al., 1990; Ogawa et al., 1993; Thulborn et al., 1982; Turner et al., 1991). Appropriate use of MR sequences result in images that are sensitive to oxygenation, and have given rise to the technique termed blood oxygenation level dependent (BOLD) imaging. This has significant advantages for monitoring oxygenation in that it is sensitive over the same range of oxygen tensions as is the oxy-hemoglobin dissociation curve, and the contrast agent is a natural substance, making the technique non-invasive.

As a consequence of the differences in susceptibility, the relaxation properties of mobile protons are altered in proximity to deoxygenated hemoglobin (deoxyHb). The main influence is on the transverse (spin-spin) relaxation rate in the plane perpendicular to the main magnetic field (Bo). The "intrinsic" rate ($1/T_2$ or R_2) can be increased by factors such as inhomogeneities in the local magnetic field caused by variation in susceptibility. This faster rate is termed $1/T_2^*$ or R_2^*. The effect increases with increasing magnetic strength, making high field scanners more sensitive to the effect.

The field of BOLD imaging has rapidly expanded due to its use for determining areas of brain activation. In these studies, although arterial oxygen tension is constant, the

Oxygen Transport to Tissue XIX, edited by Harrison and Delpy
Plenum Press, New York, 1997

total content of deoxyHb changes with local cardiovascular responses to brain activation. The result is a decrease in R_2^* and R_2 at or near the area of activation. Although the causes are still being investigated, it is generally assumed that a decrease in deoxyHb is one of the main causes of this change in relaxation (perhaps by increasing flow resulting in an oversupply of oxygen, decreased oxygen extraction fraction, and an increase in venous saturation).

Conversely, in our experiments, we change the arterial oxygen tension and use BOLD contrast to study tissue responses to hypoxia. This paper shows the utility of such a technique for studying oxygenation in blood *in vitro*, and kidney, brain, and tumors *in vivo*.

2. METHODS

2.1. MR Imaging

BOLD imaging was performed at 7T using a SMIS spectrometer and a MAGNEX horizontal bore magnet. Standard gradient echo (GE) imaging was used as well as an echo planar imaging (EPI) sequence adapted to 7T, a multi-echo GE sequence designed to measure R_2^*, and a multi-echo sequence for measuring R_2.

For GE-EPI (Ding et al., 1996; Mansfield, 1977), the read gradients were ramped up within 150 μs with a 1/4 sine shape. Phase blips were applied during the ramp time of the read gradients with a half-sine lobe. Imaging time was 54 ms with 64 phase encoding steps and asymmetric k-space sampling. The effective echo time (TE) was 17 ms.

R_2 values were measured using a multi-echo sequence with a selective 90° pulse (90x), and a non-selective composite refocusing pulse (90x-180y-90x) (Levitt et al., 1981). Spoiler gradients along the slice direction before and after each refocusing pulse dephased signal from spins outside the slice. Stimulated echo formation was prevented by varying the spoiler gradient amplitudes (Poon et al., 1992). Acquisition parameters were: 50 KHz sweep width, 12 ms echo time, 8.8 ms echo spacing, 16 echoes, FOV = 30 mm, TR = 2 s, matrix size 128x128, slice thickness 1 mm.

R_2^* imaging was done with a standard FLASH sequence modified by oscillating the read gradient to generate a series of echoes. The TR/TE/echo spacing was 1000/12/4 ms. A reference was included in the FOV to monitor sequence stability. In most experiments, a scavenging mechanism was used to prevent oxygen from collecting in the magnet.

In each case, regions of interest (ROI's) were identified and mean pixel values at each time or echo were measured and used to calculate the relaxation rates by plotting ln(SI) vs. TE. The value of ΔR_2^*, when shown, was calculated from changes in SI compared to control using $\Delta R_2^* = -\ln(SIn/SIo)/TE$, where SIn = mean SI at each time point, SIo = mean SI of the control, and TE = echo time.

2.2. Biological Models

Packed human red cells (hct = 48–61%) were gassed with 5%CO_2, N_2 and O_2 to equilibrate with varying oxygen tensions. After equilibration, a sample was transferred to a quartz NMR tube and capped with oil to prevent contamination. The pO_2, pH, % saturation and deoxyhemoglobin content were measured in a second aliquot using a pO_2 meter and a co-oximeter (for measuring hemoglobin saturation and content). Samples were imaged with echo planar gradient echo (GE) or multi-echo SE sequences.

The data from mouse kidney were obtained by reprocessing of the data from a previously published study (James et al., 1996). GE imaging (TR/TE/flip angle = 200/10/45°) was undertaken in Balb/C mice after induction of endotoxemia by venous injection of LPS.

Imaging of the rat brain during hypoxia was done using EPI on rats positioned in a 6 cm birdcage coil. Animals were anesthetized with ketamine/xylaxine (80/11) and blood samples were drawn from a femoral artery cannula. A nose cone was used to deliver hydrated inspired gases at FiO_2s of approximately 0.30, 0.20, 0.15 and 0.10. Images were obtained over a time-course and the last 3 images at each FiO_2 were averaged to calculate R_2^*. These data have been modified from those published in abstract form (Ding et al., 1996).

Tumor imaging was done using multi-echo FLASH for quantification of ΔR_2^*. The tumor model was the CNS-1 rat astrocytoma (Kruse et al., 1994). Hyperoxygenation was induced with carbogen ($95\% O_2 / 5\% CO_2$) delivered via a nose cone with a scavenging device to remove the gases from the magnet.

Body temperatures of rats and mice were measured with a rectal probe and maintained with a pad containing circulating heated water.

3. RESULTS AND DISCUSSION

Intact human RBCs show a curvilinear relationship between R_2, and a linear relationship between ΔR_2^* and deoxyhemoglobin content (Fig. 1). The changes in R_2 are almost twice that previously reported for rat blood (Thulborn, et al., 1982), but are in the range of values reported by Meyer et al. (Meyer et al., 1995). The echo spacing influences this result, which is why we used a spacing that is achievable on imaging systems. Reduction in the spacing decreases the observed effect on R_2, and this probably explains some of the differences among reported studies.

The hemoglobin content itself can be a significant factor. It is important to note that the material influencing the susceptibility is deoxyhemoglobin, which is why we are presenting the data as deoxyHb content. Values reported as fractional saturation will change significantly with hematocrit.

The non-linearity of the R_2 vs. deoxyHb curve is in agreement with Thulborn et al. (Thulborn, et al., 1982), but differ from a more recent study indicating that R_2 varies line-

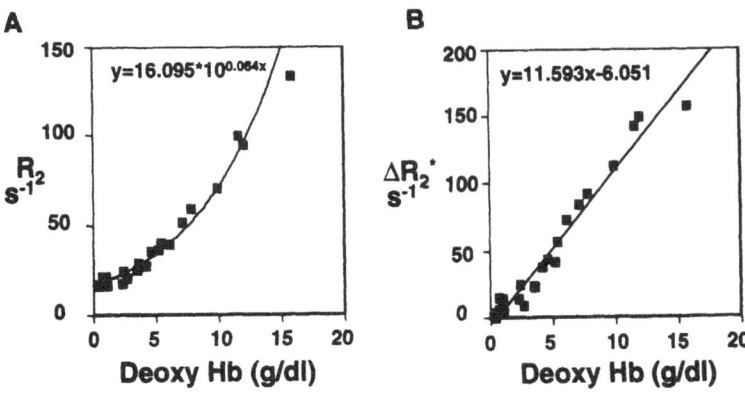

Figure 1. Proton relaxation rates in human red blood cell suspensions of varying deoxyhemoglobin content. A. R_2, measured using a multi-echo SE imaging sequence and, B. ΔR_2^*, calculated from signal intensity changes in GE-EPI images.

arly with fractional hemoglobin saturation (Meyer, et al., 1995). This discussion is not within the scope of this paper, but serves to indicate that the parameters used for the collection of the data can influence the results, and that the detailed aspects of the causes of these effects are still under active investigation.

The situation *in vivo* becomes even more complex. For a good discussion of the influence of local susceptibility variations on relaxation rates readers are referred to (Yablonskiy et al., 1994). One practical argument with respect to the interpretation and design of BOLD experiments relates to the use of GE or spin-echo (SE) sequences, and has been paraphrased from (Hoppel et al., 1993). In small vessels such as capillaries, water can diffuse across the blood cell, through the local inhomogeneity induced by deoxyHb and out into the surrounding tissue. This induces dephasing which will reduce signal in a SE image (sensitive to R_2). In larger vessels, due to surface/volume effects, there is relatively little extra-vessel water coming into proximity with the deoxyHb. Instead, there are static inhomogeneities with a different field in the vessel compared to outside the vessel. The result is that GE sequences sensitive to R_2* should be relatively more sensitive to deoxyHb changes in these larger vessels, while SE sequences will be more sensitive to smaller vessels.

Other factors which may influence the BOLD effect include: vessel orientation and image plane (Thulborn, et al., 1982), red cell shape and size (Gillis et al., 1995), and magnetic field (Thulborn, et al., 1982; Yablonskiy, et al., 1994).

Having identified some of these significant variables, we will show the potential of the technique for studying the heterogeneity of tissue responses to oxygenation.

The kidney has one of the largest blood contents in the body. With such a large hemoglobin content, there should be significant BOLD effects when oxygenation is varied. Acute renal failure is a common occurrence in endotoxemia, perhaps influenced by changes in renal perfusion and oxygenation. We measured cortex and medullary pO_2 using electron spin resonance, in parallel with MR studies of kidney response using the BOLD effect (James, et al., 1996). Figure 2 shows the data analyzed in a form that highlights changes in R_2* with time in the different regions. The magnitude of these changes is very large for a tissue, if one compares them to the changes observed in whole blood (Fig. 1), supporting the premise that GE imaging is very sensitive to changes in kidney oxygenation.

A key piece of evidence supporting the role of deoxyHb in this study were the parallel studies of pO_2. The cortex pO_2 declined and the medullary pO_2 increased within minutes of LPS infusion. This is matched by increases in the relaxation rate of the cortex (suggesting increased deoxyHb) and decreases in the medulla (Fig. 2). Although EPR measurements of pO_2 in the inner medulla were not undertaken, the decreases in R_2* indicate an increase in deoxyHb.

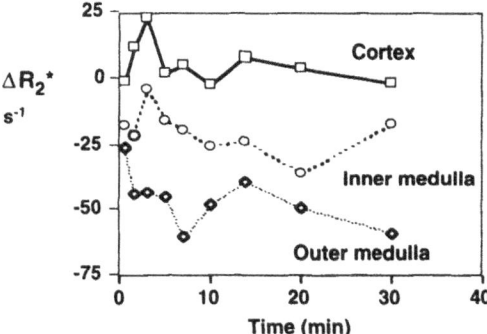

Figure 2. Regional changes in R_2* in mouse kidney medulla and cortex after induction of endotoxemia. Calculated from changes in gradient echo signal intensity. LPC was injected via tail vein at zero time. The different regions of interest were monitored using gradient echo MRI (TE = 10ms) at 7T. Parallel studies using EPR oximetry indicated that cortex pO_2 decreased transiently after LPS injection (induction of endotoxemia) while pO_2 in the outer medulla showed a transient increase (James, P.E. et al., 1996).

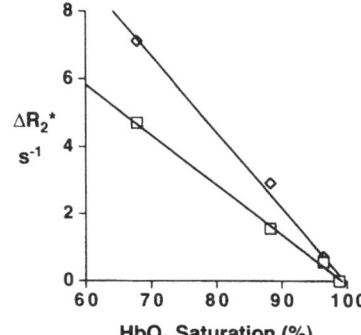

Figure 3. Regional changes in R_2* from rat brain vs. arterial pO_2. Calculated from changes in GE-EPI signal intensity. Squares = cortex, diamonds = caudate putamen.

GE imaging of brain during hypoxia was one of the first examples of the use of the BOLD effect *in vivo* (Ogawa, et al., 1990). We implemented a version of EPI for use at 7T, as this is a fast imaging technique and the high field will increase the sensitivity of the BOLD effect over studies done at whole body field strengths (up to 4T). We monitored changes in brain T_2* with declining arterial oxygen tensions (caused by reducing inspired oxygen tension). The changes in R_2* with arterial saturation were not equal between major brain structures, with the caudate putamen exhibiting a greater increase in relaxation rate than the cortex. As these changes are sensitive to deoxyHb, they represent a balance between blood volume and mean saturation. The blood volume of the caudate putamen is lower than the cortex before and during hypoxia (Bereczki et al., 1993). Therefore these observations indicate that the caudate putamen is able to extract more oxygen than the cortex during hypoxia, reducing the mean hemoglobin saturation.

The use of hyperoxygenation to oxygenate what might otherwise be hypoxic regions has been applied to tumor biology (Karczmar et al., 1994; Robinson et al., 1995). We have begun to quantify the changes in R_2* using a multi-echo FLASH sequence in an attempt to grade the tumors by their cardiovascular response to carbogen (95%O_2/5%CO_2). Tumor response to radiation therapy relates, in part, to the concentration of oxygen present. Carbogen has been used with varying success to increase radiation response (Falk et al., 1992). It may be possible to identify the causes of this variation by studying the effects of carbogen on the BOLD image contrast. Inferences gained about tumor oxygenation and local vascular control may, one day, also be useful for identifying the volume of oxygenated and growing tumor, from deoxygenated "dormant" or necrotic tumor.

Quantification of the BOLD effect in the CNS-1 rat brain tumor indicates that there are hypoxic regions which re-oxygenate upon administration of carbogen. Analysis of in-

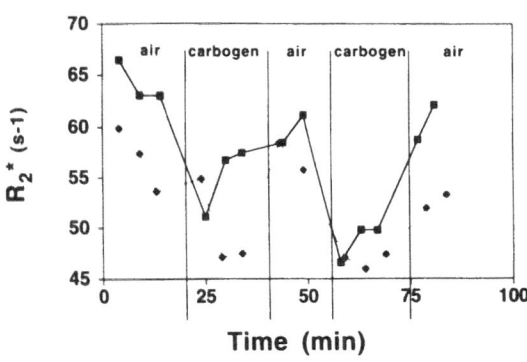

Figure 4. Using hyperoxygenated blood as a contrast agent in tumor imaging. Changes in R_2* from two rat CNS-1 astrocytomas implanted in basal ganglia measured using a multi-echo GE imaging sequence. Carbogen (95% O_2/5% CO_2) was alternated with air as the breathing gas where indicated.

dividual images indicates the presence of significant heterogeneity of response. Notice that the data is presented as absolute relaxation times as it was collected with a multi-echo Flash sequence. Such patterns of re-oxygenation may be useful for assessing vascularity or treatment response.

In summary, we have shown that studies of hypoxic tissues using the BOLD technique can provide significant information about oxygenation in various tissues and that these data can be used to make inferences as to oxygenation within the tissues themselves. Changing arterial blood saturation can provide a natural means of perturbing the system, allowing one to monitor the heterogeneity of response. The method can also be used to monitor tissue oxygenation during pathological perturbations such as endotoxemia in kidney. Because of the depth of penetration and the wide availability of NMR imaging systems, this approach may have widespread applicability. The value of this approach should increase further, as understanding of its physical and physiological bases improves, permitting more direct inferences to tissue oxygenation from the data on the oxygenation of the blood within the tissues.

ACKNOWLEDGMENTS

Thanks to Dr. Rolett for many helpful discussions, to Peter Carignan for helping with the blood data, Youssef Zaim Wadgheri for writing the multi-echo FLASH and collecting the tumor data, Shujun Ding for the EPI sequence and work on brain hypoxia. Thanks also to Drs. Hickey, Pikas and Hoopes for the CNS-1 tumors and to Phil James, Simon Jackson, Goran Bacic and Fuminori Goda for pioneering the kidney work.

REFERENCES

Belliveau, J.W., D.N. Kennedy, R.C. McKinstry, B.R. Buchbinder, R.M. Weisskoff, M.S. Cohen, J.M. Vevea, T.J. Brady and B.R. Rosen. *Science*, 1991; 254:716–717.

Bereczki, D., L. Wei, T. Otsuka, V. Acuff, K. Pettigrew, C. Patlak and J. Fenstermacher. *J Cereb Blood Flow Metab*, 1993; 13:475–86.

Ding, S., E.L. Rolett, J.B. Weaver and J.F. Dunn. Monitoring regional variation in cerebral microvascular response to hypoxia in rat using EPI at 7T, *Proc. Int. Soc. Magn. Reson. Med.* 1–524 (New York, 1996).

Falk, S., R. Ward and N. Bleehen. *Br J Cancer*, 1992; 66:919–924.

Gillis, P., S. Peto, F. Moiny, J. Mispelter and C.A. Cuenod. *Magn Reson Med*, 1995; 33:93–100.

Hoppel, B.E., R.M. Weisskoff, K.R. Thulborn, J.B. Moore, K.K. Kwong and B.R. Rosen. *Magn Reson Med*, 1993; 30:715–23.

James, P.E., G. Bacic, O.Y. Grinberg, F. Goda, J.F. Dunn, S.K. Jackson and H.M. Swartz. *Free Radical Biol Med*, 1996; 21:25–34.

Karczmar, G., J. River, J. Li, S. Vijayakumar, Z. Goldman and M. Lewis. *NMR in Biomedicine*, 1994; 7:3–11.

Kruse, C., M. Molleston, E. Parks, P. Schiltz, B. Kleinschmidt-DeMasters and W. Hickey. *J Neuro Oncol*, 1994; 22:191–200.

Levitt, M.H. and R. Freeman. *J Magn Reson*, 1981; 43:65–80.

Mansfield, P. *J Phys C*, 1977; 10:L55-L58.

Meyer, M.E., O. Yu, B. Eclancher, D. Grucker and J. Chambron. *Magn Reson Med*, 1995; 34:234–41.

Ogawa, S., T.-M. Lee, A. Nayak and P. Glynn. *Magn Reson Med*, 1990; 14:68–78.

Ogawa, S., T.M. Lee and B. Barrere. *Magn Reson Med*, 1993; 29:205–10.

Pauling, L. and C. Coryell. *Proc Natl Acad Sci*, 1936; 22:210.

Poon, C.S. and R.M. Henkelman. *JMRI*, 1992; 2:541–553.

Robinson, S., F. Howe and J. Griffiths. *Int J Radiation Oncology Biol Phys*, 1995; 33:855–859.

Thulborn, K., J. Waterton, P. Mathews and G. Radda. *Biochim Biophys Acta*, 1982; 714:265–270.

Turner, R., D. Le Bihan, C.T. Moonen, D. Despres and J. Frank. *Magn Reson Med*, 1991; 22:159–66.

Yablonskiy, D.A. and E.M. Haacke. *Magn Reson Med*, 1994; 32:749–63.

91

A NEW, WATER SOLUBLE, PHOSPHOR FOR OXYGEN MEASUREMENTS *IN VIVO*

Leu-Wei Lo, Sergei A. Vinogradov, Cameron J. Koch, and David F. Wilson

Departments of Biochemistry and Biophysics and of Radiation Oncology
Medical School
University of Pennsylvania
Philadelphia, Pennsylvania 19104

1. INTRODUCTION

Oxygen dependent quenching of phosphorescence is an optical method for non-invasive measurements of oxygen in the microvasculature of tissue (see [3–7,10,11]). It is unique among the oxygen measuring methods in that it is noninvasive and yet provides quantitative measurements with an excellent temporal resolution (milliseconds). Development of the technology has been rapid, and the phosphors available for oxygen measurements has been rapidly expanding. The phosphor which has been in greatest use is Pd-meso-tetra-(4-carboxyphenyl) porphine and this has been thoroughly calibrated[2]. The solubility of this phosphor in aqueous solutions at physiological pH is, however, low. In order to improve the aqueous solubility and to provide a phosphor environment which optimized for physiological measurements, the phosphor is dissolved in solutions containing high concentrations of serum albumin. Albumin binds the phosphor, providing both the necessary solubility in biological media and restricting the access of oxygen to the excited state phosphor. The latter is useful in that the restricted access to oxygen results in the the phosphorescence lifetime changing from near 650 μsec at zero oxygen to about 16 μsec at air saturation (38 °C). On the other hand, dissolving the phosphor in solutions containing high concentrations of albumin means that if these solutions are then injected into the blood of an animal, the animal is exposed to relatively large amounts of foreign protein as well as any endotoxin and other contaminants in the albumin. The necessity for injecting a foreign protein can be avoided by removing blood from the animal, preparing serum, dissolving the phosphor to the serum, and then reinjecting the serum with phosphor. Using a water soluble phosphor would allow this rather complicated procedure to be avoided, since the phosphor could be dissolved in physiological saline and the resulting solution injected.

Vinogradov and Wilson (these proceedings) have chemically altered the Pd-meso-tetra-(4-carboxyphenyl) porphyrin to make it fully water soluble at physiological pH values. The resulting phosphor has a molecular weight of 2,442 and a total of 16 carboxyl groups attached to the porphyrin. This new phosphor (designated Oxy-Phor R2), appears to be well suited for *in vivo* oxygen measurements, and may become the phosphor of choice when it is important to minimize possible side effects of phosphor injection.

2. MATERIALS AND METHODS

Oxy-Phor R2

The Oxy-Phor R2 was prepared as described in Vinogradov and Wilson (these proceedings) and is very soluble (to at least 10^{-2} M) at all pH values more alkaline than about 3.0. The spectra of the absorption and phosphorescence are the same as for the underivatized meso-tetra-(4-carboxyphenyl) porphine reported by Lo and coworkers[2].

Measurement of the Oxygen Dependence of the Phosphorescence Lifetime

The oxygen dependence of phosphorescence lifetime was determined using a thermostatted chamber (±0.1 C) constructed of glass and ceramic. The chamber (12 ml volume) was fitted with a high accuracy oxygen electrode and stirred with a magnetic stirring bar (see [2]). The oxygen in the solution in the chamber was slowly removed by including glucose (20 mM) in the solution and adding the enzyme glucose oxidase plus catalase. As oxygen was removed in the oxidation of glucose, continuous measurements were made of the oxygen electrode current and the phosphorescence lifetime. The data were then used to test the fit of the oxygen dependence of the phosphorescence lifetime to the Stern-Volmer relationship:

$$\frac{T^0}{T} = 1 + k_Q \cdot T^0 \cdot PO_2$$

Where T^0 and T are the phosphorescence lifetimes at zero oxygen pressure and at an oxygen pressure PO_2, respectively, and k_Q is the second order rate constant for quenching of phosphorescence.

The dependence of the lifetime at zero oxygen (T^0) and the quenching constant (k_Q) on serum albumin concentration, pH, temperature, and phosphor concentration were then determined.

Excretion of Oxy-Phor R2 in the Urine

Excretion of the phosphor by the kidney was measured following injection of approximately 3 x and 9 x the amount required for oxygen imaging measurements into mice. The urine was collected, placed in a volumetric flask, and diluted to the selected volume. The content of Oxy-Phor R2 in the collected urine was measured using a scanning spectrophotometer by the absorption maxima at 524 nm and 415 nm.

3. RESULTS

Dependence of the Phosphorescence Lifetime on Oxygen Pressure

The oxygen dependence of the phosphorescence of Oxy-Phor R2 was tested for fit to the Stern-Volmer relationship. For each of the different experimental conditions reported in this paper the fit to the linear form of the Stern-Volmer relationship had a correlation coefficient of greater than 0.998. As the concentration of bovine serum albumin was changed from zero to 4% by weight at pH 7.4 and 38° C, the values of the lifetime at zero oxygen pressure increased from 250 to 600 µsec. The albumin dependence is saturated at approximately 1% albumin and does not change significantly as the concentration is further increased. This means the oxygen quenching characteristics are essentially independent of albumin concentration throughout the range of from 3 to 5% found in blood serum, assuring saturation of phosphor after injection into the blood.

The Dependence of the Calibration Constants for Oxygen Dependent Quenching of Phosphorescence on pH and Temperature

The calibration constants have been measured at pH values from 6.4 to 7.8 and temperatures from 23° C to 38° C (Table 1) while the values at different albumin concentrations and ionic strength (NaCl concentrations) are given in Table 2. Each value is for an independent experiment and the values were obtained from fit of the complete oxygen dependence to the Stern-Volmer relationship. The quenching constant (k_Q) increased as the pH was made more alkaline, but this increase is only about 10% from pH 6.4 to 7.4. The value of the lifetime at zero oxygen is, however, unchanged over this pH range. There appears to be a slightly greater pH dependence in k_Q from 7.4 to 7.8, and use for oxygen measurements in this pH range would require the dependence be more completely measured.

Table 1. Quenching constants and lifetimes at zero oxygen pressure and different pH and temperature values

	Temp (°C)	K_q (torr^{-1} s^{-1})	τ^0 (µs)
pH 6.4	23	211	730
	28	241	688
	33	298	647
	38	360	607
pH 6.8	23	220	723
	28	251	689
	33	319	656
	38	408	605
pH 7.4	23	225	730
	28	274	686
	33	339	632
	38	409	601
pH 7.8	23	241	666
	28	308	629
	33	388	596
	38	455	580

The phosphorescence characteristics are dependent on the concentration of albumin in the medium, with significant decrease in the quenching constant occurring when albumin was added. The effect of albumin is largely saturated by 1% by weight and further increase has little effect. Similarly, the dependence on NaCl concentration is small even when the concentration is increased from 60 to 180 mM. The changes observed in physiology are at most a few mM, a range which would not significantly alter the value of k_Q. The value of the phosphorescence lifetime at zero oxygen pressure is not affected by changing the NaCl concentration from 60 to 180 mM.

Possible Toxicity and Excretion by the Kidneys

In preliminary experiments, the Oxy-Phor R2 was injected into 6 mice. In three mice, the amount injected was 3 times (77 mg/kg) that required for imaging phosphorescence and in the other three mice the amount injected was 9 times (231 mg/kg) the amount needed for imaging phosphorescence. At 1, 5, and 10 days after the injection the mice were euthanized and a histopathology examination performed by a veterinary pathologist. No evidence of tissue injury was found even at the highest concentration used. When rats or mice were injected with phosphor and kept in metabolic cages or the urine otherwise collected, measurements of the absorption spectra of the urine samples indicated recovery of from 40 to 60% of the amount injected within 12 hours. Most of this is excreted with a half time of 2–3 hours.

4. DISCUSSION

Oxy-Phor R2 is well suited to measurements of oxygen in vivo. The oxygen dependence has an excellent fit to the Stern-Volmer relationship throughout the temperature range tested (23 to 38 °C). The measured changes in lifetime are large, providing an excellent dynamic range for high accuracy oxygen measurements at all physiologically relevant values. The fact that Oxy-Phor R2 is very water soluble makes it easy to inject as a solution in physiological saline, minimizing the possibility of toxicity by eliminating the need for injecting with albumin. Excretion of much of the phosphor through the kidney further decreases possible toxicity, and also makes possible repetitive measurements in a single

Table 2. The calibration constants for Oxy-Phor R2 with different concentrations of albumin and NaCl in the solutions at pH 7.4 and 38°C

Albumin (% by weight)	K_q (torr^{-1} s^{-1})	τ^0 (μs)
0.0	2108	250
0.5	471	584
1.0	457	584
2.0	410	602
4.0	409	601
[NaCl] (mM)	K_q (torr^{-1} s^{-1})	τ^0 (μs)
0	452	548
60	427	605
120	409	601
180	385	615

animal. Thus, phosphor can be injected and the oxygen measurements made but by the next day the phosphor will have been eliminated in the urine and the procedure can be repeated. On the other hand, excretion means that if oxygen measurements are to be made over more than about 2 hours, it may be necessary to inject more phosphor. The changes in phosphor concentration in the blood will not affect the oxygen measurements, which are based on phosphorescence lifetime and independent of phosphor concentration. The phosphorescence signal strength will, however, decrease with decreasing phosphor concentration and additional phosphor may need to be injected to maintain adequate signal to noise in the measurements.

The principles used in construction of Oxy-Phor R2 can be also applied to the development of phosphores which absorb and phosphoresce in the near infra red region of the spectrum.[1,8,9] This will provide equally well behaved phosphors which can be used to measure oxygen through tissue thicknesses of up to several centimeters.

5. CONCLUSION

Oxy-Phor R2 appears to provide properties which can be of significant value in the measurements of oxygen in tissue. This is particularly true when it is important to minimize the amount of foreign material; such as albumin, which is injected.

ACKNOWLEDGMENT

This research was supported in part by grant CA-56679.

REFERENCES

1. Lo, L.-W., Jenkins, W.T., Vinogradov, S.A., Evans, S.M., and Wilson, D.F. (1996) Oxygen distribution in the vasculature of mouse tissue in vivo measured using a near infra red phosphor. *Adv. Exp. Med. Biol.* in press.
2. Lo, L.-W., Koch, C.J., and Wilson, D.F., (1996) Calibration of oxygen dependent quenching of the phosphorescence of Pd-meso-tetra (4-carboxyphenyl) porphine: a phosphor with general application for measuring oxygen concentration in biological systems. *Analy. Biochem.* 236, 153–160.
3. Pawlowski, M. and Wilson, D.F. (1992) Monitoring of the oxygen pressure in the blood of live animals using the oxygen dependent quenching of phosphorescence. *Adv. Exptl. Med. Biol.* 316: 179–185.
4. Rumsey, W.L., Vanderkooi, J.M., and Wilson, D.F. (1988) Imaging of phosphorescence: A novel method for measuring the distribution of oxygen in perfused tissue. *Science* 241, 1649–1651.
5. Rumsey, W.L., Pawlowski, M., Lejavardi, N., and Wilson, D.F. (1994) Oxygen pressure distribution in the heart in vivo and evaluation of the ischemic "border zone". *Am. J. Physiol.* 266: H1676–1680.
6. Vanderkooi, J.M., Maniara, G., Green, T.J., and Wilson, D.F. (1987) "An optical method for measurement of dioxygen concentration based on quenching of phosphorescence". *J. Biol. Chem.* 262, 5476–5482.
7. Vinogradov, S.A. and Wilson, D.F. (1994) Phosphorescence lifetime analysis with a quadratic programming algorithm for determining quencher distributions in heterogenous systems. *Biophys. J.* 67, 2048–2059.
8. Vinogradov, S.A. and Wilson, D.F. (1995) Metallotetrabenzoporphyrins. New phosphorescent probes for oxygen measurements. *J. Chem. Soc. Perkin Trans. II*, 2, 103–111.
9. Vinogradov, S.A., Lo, L.-W., Jenkins, W.T., Evans, S.M., Koch, C., and Wilson, D.F. (1996) Non invasive imaging of the distribution of oxygen in tissue in vivo using near infra-red phosphors. *Biophys. J.* 70, 1609–1617.

10. Wilson, D.F., Pastuszko, A., DiGiacomo, J.E., Pawlowski, M., Schneiderman, R. and Delivoria-Papadopoulos, M. (1991) Effect of hyperventilation on oxygenation of the brain cortex of newborn piglets. *J. Applied Physiol.* 70, 2691–2696.

11. Wilson, D.F., Rumsey, W.L., Green, T.J. and Vanderkooi, J.M. (1988) The oxygen dependence of mitochondrial oxidative phosphorylation measured by a new optical method for measuring oxygen. *J. Biol. Chem.* 263, 2712–2718.

92

"DENDRITIC" PORPHYRINS

New Protected Phosphors for Oxygen Measurements *in Vivo*

Sergei A. Vinogradov and David F. Wilson

Department of Biochemistry and Biophysics
School of Medicine
University of Pennsylvania
Philadelphia, Pennsylvania 19104

1. INTRODUCTION

It was defined previously that a phosphor molecule used for oxygen measurements *in vivo* can be described as a combination of two major components: *optical chromophor* (OC) and the *surrounding environment* (SE) (Vinogradov and Wilson, 1995). We have shown that extended porphyrins of Pd, Pt and Lu present an interesting class of *optical chromophors* with strong near IR absorption bands, high phosphorescence quantum yields, and lifetimes suitable for precise oxygen measurements (Vinogradov and Wilson, 1994a). The first primitive water soluble derivatives of such chromophors have been synthesized and tested *in vivo* for oxygen imaging through the thickness of tissue (Vinogradov et al., 1996). However, the primary limitation in application of the phosphorescence quenching technology remains the lack of optimal biologically compatible phosphors, namely phosphors with *surrounding environments* optimized for biological applications. An optimal *surrounding environment* is a ligand or a set of ligands linked to the central chromophor to provide high water solubility, no toxicity, good protection from chemically active components of the blood and tissue, and appropriate size for excretion through the kidney. Ideally, the surrounding environment forms an inert globular structure around the chromophor, through which only small molecules, such as oxygen, can penetrate (Figure 1).

In the present communication, we introduce a new approach to the design of surrounding environments for the phosphors, based on the use of *dendritic ligands. Dendritic polymer growth* is known as an effective way to build globular, three dimensional supramolecular structures around a functionalized center (Frechet, 1994). Dendritic polymers or *dendrimers* are combined of monomeric units which allow branching at each polymerization step. Two strategies are described for the construction of dendrimers: *convergent* and *divergent* growths. In the former polymer synthesis starts from the periphery and ends by linking the preformed

Oxygen Transport to Tissue XIX, edited by Harrison and Delpy
Plenum Press, New York, 1997

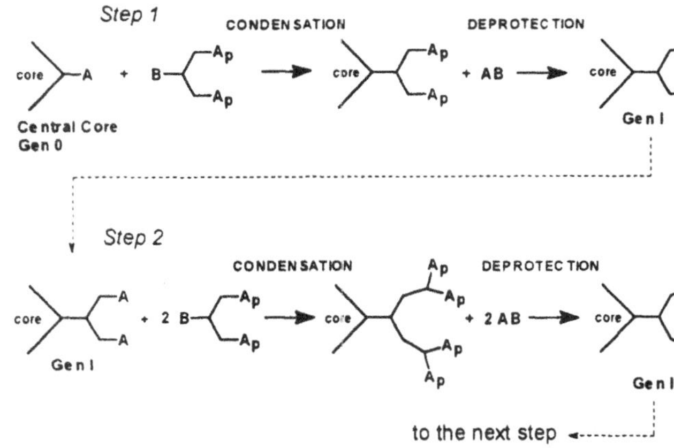

Scheme 1. A, B - reacting functionalities; A_p - protected group; Gen I, Gen II ... - dendrimers of the corresponding generations.

branched fragments to a central core (Hawker et al, 1992). The latter approach uses reverse order of synthesis, and so layers of monomeric units are added to a central core sequentially, until a desirable degree of branching is achieved (Tomalia et al, 1986).

It has been shown that dendrimers tend to form globular structures as the number of monomeric units increases and eventually cover the central core fragment (Bluemen and Schnorer, 1990; Tomalia et al, 1990). Recently dendrimers were used for modification of the porphyrin molecule (Jin et al, 1993). Measurements of the quenching of fluorescence of Zn porphyrin encapsulated into a large dendritic cage demonstrated that dendrimers can provide significant protection for the porphyrin core, serving as a barrier for the large molecules and a trap for the smaller species.

We have carried out computer simulations (Macromodel, Unix Version 3.5, MM2 force field) of porphyrins trapped inside dendritic cages. It appears that accessibility of the porphyrin core for small molecules, such as oxygen, decreases with increase of the size of dendrimer sphere. Furthermore, modification of the outer layer with functional groups of different hydrophobicity can substantially affect the 3D-structure of dendrimer in water solution. This can be directly quantitated by measuring the values of the phosphorescence quenching constants for oxygen (k_q) and the effects of the quenchers on the distributions of phosphorescence lifetimes (Vinogradov and Wilson, 1994b).

The present paper describes the synthesis of the first family of dendritic water soluble porphyrins and their phosphorescence quenching properties.

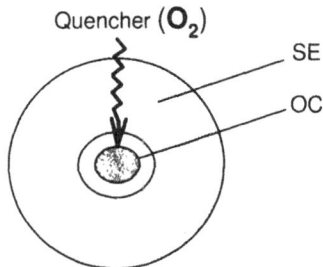

Figure 1. Schematic diagram of the phosphor for *in vivo* oxygen measurements.

2. MATERIALS AND METHODS

In this work Pd *meso*-tetra-(4-carboxyphenyl) porphyrin (PdTCPP) was used as a core fragment to initiate dendrimer growth. The "divergent" approach to the construction of dendritic ligands was used. This approach allowed more economical use of starting materials, particularly PdTCPP, and facilitated measurements of quenching properties of each new generation during the synthesis.

Based upon the natural occurrence of L-amino acids, we chose to use protected derivatives of L-glutamic acid (esters) as monomeric units for the dendrimer construction. The final structure, therefore can be described as *hyperbranched polyamide, composed of glutamates*. The well described condensation-deprotection strategy, widely used for polypeptide synthesis (Greenstein and Winitz, 1961), was applied. The modification of outer layers of dendritic porphyrins was carried out in the same fashion, using ester-protected L-leucine and 11-aminoundecanoic acid. All chemicals were purchased from the standard suppliers or synthesized as published previously. Newly synthesized compounds were characterized by standard chemical analysis, Size Exclusion Chromatography (SEC), ^{1}H NMR and UV-VIS spectroscopy. Earlier described phosphorimeters (Wilson and Vinogradov, 1992; Lo et al, 1996) and algorithms (Vinogradov and Wilson, 1994b) were used for precise characterization of the decay kinetics and determination of oxygen quenching constants.

3. RESULTS AND DISCUSSION

A series of hyperbranched dendritic porphyrins were synthesized using layer-by-layer "divergent growth" strategy. The general principals of the divergent growth are shown below (Scheme 1).

More specifically, the groups used in our particular synthesis were as follows: **Core = PdTCPP; A = COOH; B = NH$_2$; A$_p$ = COOAll or COOBz** (All - allyl, Bz - benzyl).

Two forms of each dendrimer were purified and characterized: the ester form, before the protecting groups were removed, and the free acid form, after ester hydrolysis. Some of the synthesized compounds are listed in Table 1. We suggest using the general formula **PdPGlu$_n$OR** for identification of dendritic polyglutamic porphyrins based on PdTCPP. In this formula **PdP** is PdTCPP; **Glu$_n$** shows number of glutamic layers attached to the initiating core, and thus **n** is a dendrimer generation number; **OR** is terminating group. For example, PdPGlu$_1$OAll is PdTCPP modified with one layer of glutamic acid diallyl esters (notice that the actual number of the glutamate molecules is 4, as there are 4 carboxyl groups on the initiating PdTCPP molecule). Accordingly, PdPGlu$_2$OH is PdTCPP sequentially modified with 2 layers of glutamates, which brings the total number of glutamate fragments to 12 (4 on the first layer and 8 on the second layer), and terminated by the 16 unprotected carboxyl groups.

First two generations of polyglutamic porphyrins, both esters and free acids, were structurally characterized by ^{1}H NMR, allowing detailed peak assignment. For higher dendrimer generations resonances which correspond to glutamic and porphyrinic parts of the molecule are relatively well resolved showing good quantitative agreement with general brutto formulas. However, due to the multiple conformations in solutions individual peaks were grouped into broad multiplets and detailed peak assignment was not possible.

In order to have control over the volume occupied by dendritic branches in solution, the outer layer of a dendrimer can be modified with groups of different hydrophobicity.

Table 1. Dendritic porphyrins based on PdTCPP

Unmodified dendrimers	Modified dendrimers
	Gen I
PdPGlu$_1$OAll	PdPGlu$_1$Leu$_1$OAll
PdPGlu$_1$OBz	PdPGlu$_1$Leu$_1$OH
PdPGlu$_1$OH	PdPGlu$_1$Leu$_1$OAll
	PdPGlu$_1$Leu$_1$OH
	PdPGlu$_1$UDA$_1$OAll
	PdPGlu$_1$UDA$_1$OH
	Gen II
PdPGlu$_2$OAll	PdPGlu$_2$Leu$_1$OAll
PdPGlu$_2$OBz	PdPGlu$_2$Leu$_1$OH
PdPGlu$_2$OH	PdPGlu$_2$UDA$_1$OAll
	PdPGlu$_2$UDA$_1$OH
	Gen III
PdPGlu$_3$OAll	
PdPGlu$_3$OH	
	Gen IV
PdPGlu$_4$OH	

PdTCPP - parent porphyrin

Indeed, in water media more hydrophobic groups would force dendritic branches to fold back closer to the parent hydrophobic core, making the whole structure more compact. The optimal choice of the outer layer groups thus would allow construction of well protected phosphors with relatively few layers of dendrimer growth. This is particularly important from the biological point of view when one of the necessary requirements for the phosphor is its relatively small size, allowing effective filtration through the kidney. We have used L-leucine (**Leu**) and 11-aminoundecanoic acid (**UDA**) to modify outer layers of dendritic porphyrins of early generations. The corresponding compounds are also listed in the table with names formed according to the rule described above.

Preliminary phosphorescence quenching measurements have been made for some of the dendritic porphyrins in DMF (dimethylformamide) and water solutions. The experimental setup for the measurements is described in the accompanying paper in this volume (*see* Lo et al, in this volume). Figure 2 demonstrates Stern-Volmer plots for the three chosen phosphors in DMF (A) and water (B) correspondingly.

As seen in the figure, oxygen quenching constants for all 3 porphyrins in DMF (A) solution are first, very high (about 9000 Torr^{-1}sec^{-1}) and, second, have very similar absolute values. This suggests that in organic media, particularly in dipolar aprotic solvents, dendrimer structures are extended and oxygen molecules are able to reach central porphyrin core with ease. Such result can be expected. Indeed, for the given generation numbers (II and III) dendritic branches are not yet large enough to form ball-like structures based on pure topological reasons (Blumen and Schnorer, 1990). On the contrary, negatively charged surface carboxyl groups tend to occupy larger space in order to lower electrostatic repulsion. Thus the whole molecule has a rather "spider-like" shape.

The situation changes dramatically when molecules are in an aqueous environment. The absolute values of the quenching constants decrease, and what is more important, the differences between quenching efficiencies for different size dendrimers become much more evident (Fig.2, B). The increase of the molecular size leads to more effective protec-

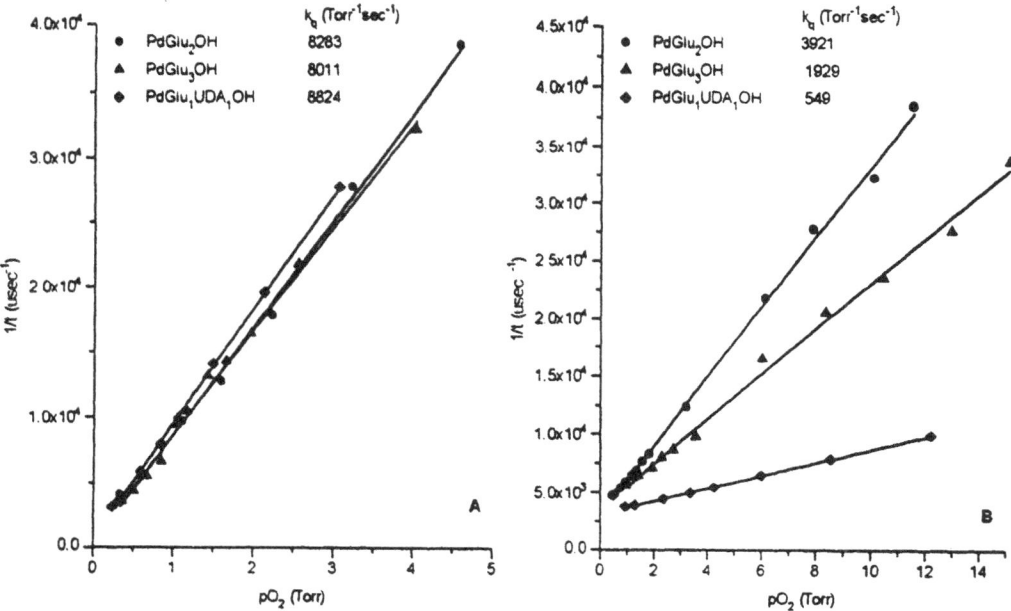

Figure 2. Stern-Volmer plots of the selected phosphors. A - in DMF solution; B - in water solution.

tion of the center and thus to a decrease of the quenching constants. The most distinct effect is seen in case of the dendrimer modified with long chain fatty acids. Hydrophobic interactions force the flexible hydrocarbon chains to fold back over the central core porphyrin, increasing its protection as seen in the decreased accessibility to oxygen.

4. CONCLUSION

The use of the dendrimers for the construction of the environment surrounding the phosphor allows control over the oxygen dependent quenching of phosphorescence. The dendritic porphyrins synthesized in this study form a group of phosphors which are promising agents for oxygen measurements in biological systems. (Supported by grants NS-31465 and CA-56679)

REFERENCES

Blumen, A. and H. Schnorer, 1990, Fractals and related hierarchical models in polymer science. *Angewandte Chemie, Int. Ed. Eng.* **29**: 113.

Frechet J. M. J., 1994, Functional polymers and dendrimers: Reactivity, molecular architecture, and interfacial energy, *Science*, **263**: 1710.

Greenstein J. and M. Winitz, 1961, *"Chemistry of Amino Acids"*, Vol. 1, John Wiley & Sons, New York.

Hawker, C.J. and J.M.J. Frechet. 1992, Unusual macromolecular architectures: The convergent growth approach to dendritic polyesters and novel block copolymers. *J. Am. Chem. Soc.* **114**: 8405.

Jin, R.H., T. Aida, and S. Inoue. 1993, 'Caged' porphyrin; the first dendritic molecule having a core photochemical functionality. *J. Chem. Soc. Chem. Commun.* 1260.

Lo, L.-W., Koch, C.J., and Wilson, D.F., 1996, Calibration of oxygen dependent quenching of the phosphorescence of Pd-meso-tetra (4-carboxyphenyl) porphine: a phosphor with general application for measuring oxygen concentration in biological systems, *Anal. Biochem.*, **236**: 153.

Tomalia, D., H. Baker, J. Dewald, M. Hall, G. Kallos, S. Martin, J. Roeck, J. Ryder, and P. Smith. 1986, Dendritic macromolecules: synthesis of starburst dendrimers. *Macromolecules.* **19**: 2466.

Tomalia, D.A., A.M. Naylor, and W.A. Goddard III. 1990, Starburst dendrimers: Molecular-level control of size, shape, surface chemistry, topology, and flexibility from atoms to macroscopic matter. *Angew. Chem., Int. Ed. Eng.* **29**: 138.

Vinogradov, S. A., and D. F. Wilson, 1994 (a), Metallotetrabenzoporphyrins. New phosphorescent probes for oxygen measurements. *J. Chem. Soc., Perkin Trans. II*, 103.

Vinogradov, S. A., and D. F. Wilson, 1994 (b), Phosphorescence lifetime analysis with quadratic programming algorithm for determining quencher distributions in heterogeneous systems. *Biophys. J.* **67**: 2048.

Vinogradov, S. A., and D. F. Wilson, 1995, Extended porphyrins: New IR phosphors for oxygen measurements, *Adv. Exptl. Med. Biol.*, in press.

Vinogradov, S.A., Lo, L.-W., Jenkins, W.T., Evans, S.M., Koch, C., and Wilson, D.F., 1996, Non invasive imaging of the distribution of oxygen in tissue in vivo using near infra-red phosphors. *Biophys. J.*, **70**: 1609.

Wilson, D.F. and S. A. Vinogradov, 1992, Recent advances in oxygen measurements using phosphorescent quenching, *Adv. Exptl. Med. Biol.*, **361**: 61.

93

WHAT DOES EPR OXIMETRY WITH SOLID PARTICLES MEASURE—AND HOW DOES THIS RELATE TO OTHER MEASURES OF PO$_2$?

Harold M. Swartz, Jeff Dunn, Oleg Grinberg, Julia O'Hara, and Ted Walczak

Department of Radiology
Dartmouth Medical School
Hanover, New Hampshire 03755

1. INTRODUCTION

1.1. Electron Paramagnetic Resonance (EPR or, Completely Equivalently, ESR) Oximetry

The technique of in vivo EPR oximetry has been more widely introduced only recently (Ferrari et al., 1994, Halpern et al., 1994, Swartz et al., 1994). Our laboratory has concentrated on spectroscopy at 1.2GHz, using particulate oxygen-sensitive paramagnetic materials placed at the site (or sites) of interest (Dunn et al., 1995, Gallez et al., 1996, Goda et al., 1996, Goda et al., 1995, Jiang et al., 1995, Liu et al., 1995, Liu et al., 1994a, Liu et al., 1994b, O'Hara et al., 1995a, O'Hara et al., 1995b, Swartz et al., 1992). This form of EPR oximetry has a number of features that seem potentially advantageous for making biologically useful measurements. These features include 1) the capability of making repeated non-invasive measurements from the same site without the need for anesthesia (after the placement of oxygen-sensitive paramagnetic materials at the site (or sites) of interest); 2) high accuracy and sensitivity, especially for lower levels of oxygen; 3) rapid response (seconds); 4) measurements which are not perturbed by factors such as pH, temperature, osmotic strength; 5) a high degree of stability and inertness in biological systems; 6) the availability of the oxygen-sensitive paramagnetic materials in a variety of forms ranging from a slurry of very small particles to a single macroscopic crystal; 7) the calibration is very stable; and 8) there is a very high degree of specificity of the measurements because there usually are no other EPR responsive materials present in sufficient concentrations to affect the measurements. EPR oximetry, of course, also has some potential limitations and uncertainties. These include 1) a certain level of invasiveness; for many uses the particulate oxygen-sensitive paramagnetic materials need to be physically placed at the site(s) of interest 2) availability; at the present time the instrumentation required for these measurements is not yet widely available 3) depth of penetration; the most sensitive approach, using an exciting frequency of 1 GHz, limits non-invasive measurements to a depth of about 10 mm; and 4) the nature of the parameter that is measured by the technique is not fully understood by many scientists. The purpose of this paper is to address the last question, doing so in a context that relates EPR oximetry to other methods.

1.2. The Need to Understand the Nature of Measurements of pO_2

While it is essential with all types of measurements of pO_2 that there is clear understanding of what is being measured so that the results can be interpreted accurately for the proposed application, this is especially important for a new method such as EPR. It is also important to understand what are likely to be the limitations and potential advantages of the methods so that the appropriate method for a given application can be chosen.

The seemingly simple question as to what does a measurement reflect in terms of the oxygenation of the biological system is, upon closer examination, quite complex and the answer depends very much on the physiological or pathophysiological question that is being posed (Kelleher et al., 1995). There are several types of information that are needed in order to address this question adequately:

1. The physical basis of the measurement
2. The spatial region that is being directly probed by the method;
3. The relationship of that spatial region to the volume and/or process that is being studied.

In this manuscript we examine each of these three areas, first from a general point of view and then with respect to EPR oximetry. In order to indicate the relationship of the EPR method to other methods we also briefly consider several other methods which have been suggested to provide a means to make quantitative measurements of oxygenation in tissues *in vivo* (see table 1): The methods for measuring oxygen have been recently reviewed (Chapman, 1991, Raleigh et al., 1996, Stone et al., 1993) and these reviews provide additional information on many of the methods discussed in this manuscript.

2. DISCUSSION

2.1. The Physical Basis of the Measurement of Oxygen

2.1.1 General Considerations. There are at least three conceptually different types of measurements that currently are in use with various methods:

2.1.1.1. A parameter that is directly proportional to the *partial pressure of oxygen* (pO_2) (e.g. most Clark electrodes). The pO_2 is independent of the medium in which the measurement is made and therefore from a physical-chemical point of view, it is a very fundamental parameter. If, however, the biological phenomenon that is being studied is proportional to the concentration of oxygen $[O_2]$, then it is necessary to know the solubility of oxygen in the compartment of interest in order to use the measurement of the pO_2 to obtain the needed information as to the $[O_2]$.

Table 1. Comparison of key properties of various quantitative methods to measure oxygen

METHOD	PARAMETER MEASURED	MINIMUM TIME REQUIRED FOR MEASUREMENT	REGION OF SAMPLES MEASURED DIRECTLY	APPROXIMATE RESOLUTION
CLARK ELECTRODE	pO_2 (current generated by the electrolytic decomposition of dioxygen)	Minutes[1]	Interstitial volume in contact with the tip of the electrode	2.5 torr
METHODS BASED ON HEME PROTEINS, ESPECIALLY HEMOGLOBIN	Physiological parameter (amount of dioxygen molecules bound to chromophoric ligands)	Instantaneous	Location of the proteins (in vascular system for Hb methods; intracellular for cytochromes)	2.5 μm
PHOSPHORESCENT AND FLUORESCENT METHODS BASED ON REDOX STATES OF INTERMEDIATES	Physiological parameter (ratio of reduced and oxidized states of redox couple)	Instantaneous	Sites of the redox intermediates (usually intracellular)	2.5 μm
PHOSPHORESCENT AND FLUORESCENT METHODS BASED ON QUENCHING BY OXYGEN	$[O_2]$	Instantaneous	Sites of the introduced probe molecules, usually intravascular	0.1 μm
NMR PERFLUOROCARBON RELAXATION	$[O_2]$ (effect on relaxation rates)	Minutes[2]	Sites of the emulsion	10 μm
SUBSTANCES WHICH LOCALIZE IN HYPOXIC AREAS (e.g., 2-nitroimidazoles)	Physiological parameter (combination of perfusion and $[O_2]$ at time of administration)	Minutes[3]	Tissues where substances localize	see footnote 4
EPR OXIMETRY BASED ON SOLID PARTICLES	pO_2 (effect on linewidth of EPR spectrum)	Seconds	Sites of the particles (usually interstitial; can be intracellular)	0.1 torr
EPR OXIMETRY BASED ON SOLUBLE MATERIALS	$[O_2]$ (effect on linewidth of EPR spectrum)	Instantaneous	Sites of the soluble molecules (usually throughout the tissues)	5 μm

[1] Time needed to obtain accurate measurement of the current
[2] Time required to obtain sufficient data to make relaxation calculations
[3] Currently requires biopsy; methods for attaching signaling molecules under development
[4] Comparative indication of amount of hypoxia in tissues

2.1.1.2. A parameter that is directly proportional to the concentration of oxygen ($[O_2]$) (e.g. quenching of phosphorescence of injected metalloporphyrins (Wilson and Cerniglia, 1992)). The $[O_2]$ often is the parameter for which the biological process of interest is proportional (e.g. rate of an enzymatic reaction). It should be noted, however, that most of this class of oxygen measuring techniques do not actually directly measure the $[O_2]$, but measure parameters which respond proportionately to the number of dioxygen molecules they encounter (the collision rate). The latter is a product of the diffusion rate and $[O_2]$ and therefore one needs to know *a priori* the diffusion rate of oxygen in the medium that is being measured (e.g. the intracellular compartment or the intravascular compartment), or make an independent measurement of that rate. Also, if this type of experimental measurement is not made in the medium of interest, then one needs to know the solubility of oxygen in both media (that in which the measurement is made and that of the phenomenon of interest) in order to calculate the $[O_2]$ at the site of interest.

2.1.1.3. A physiological parameter that has an expected relationship to either the pO_2 or the $[O_2]$ (e.g. methods based on the absorption of light by hemoglobin (Fenton and Gayeski, 1990) or the localization of drugs whose state is affected by the availability of oxygen (Urtasun et al., 1986)). Often such physiological parameters are affected by a variety of factors such as pH, 2–3-DPG, and pCO_2 for heme proteins and perfusion and reactivity for hypoxic localizing drugs. Therefore the relationship of these parameters to the $[O_2]$ or pO_2 is likely to rest on a number of assumptions whose validity may vary with the physiological state; when these assumptions are well supported, these methods can provide very useful data on oxygenation.

2.1.2. The Physical Basis of EPR Oximetry Based on Solid Particles. EPR oximetry which is based on solid particles such as India ink, lithium phthalocyanine, or carbon aggregates (coals or chars) directly reflects the pO_2. The effect of the pO_2 is to *broaden* the line width of the EPR signal of the particulate paramagnetic materials used with this approach and therefore this method is especially sensitive for *low* values of pO_2. The inherent time resolution for this technique depends on the rate of equilibration of the dissolved gases with the paramagnetic material and therefore will vary with the type of paramagnetic material that is used. Some materials, especially lithium phthalocyanine (LiPc), reach equilibration essentially instantly and therefore their inherent time resolution is very short. Due to their very extensive external and internal surface areas on which gases can absorb and/or the time required to displace gases which are already absorbed on them the carbon based materials such as coals and chars will have response times that vary with both the type of material and its physical configuration. The principal variable for the latter is the size of the particles of the material. The inherent time resolution with any specific particulate material can be determined fairly precisely in calibration experiments. The most commonly used materials such as fusinite and gloxy have effective time resolution in the order of seconds to minutes under most experimental conditions.

2.1.3. The Physical Basis of EPR Oximetry Which Is Based on Soluble Materials. The data obtained with EPR oximetry which is based on soluble materials such as nitroxides reflects the collision rate with molecular oxygen and hence respond to the product of the $[O_2]$ and the diffusion rate, similar to the situation for the quenching of phosphorescence or fluorescence. As is the case with EPR oximetry based on particulate materials, the effect of oxygen is to broaden the EPR spectra. The broadening effects from the encounters with oxygen differ, however, in that they are much smaller and have a similar change per unit of oxygen concentration throughout the whole range of oxygen concentrations. Therefore this technique has less sensitivity at lower concentrations of oxygen. The inherent time resolution is instantaneous.

2.2. The Spatial Region that is Being Directly Probed by the Method

2.2.1. General Considerations. This can vary considerably with the technique and therefore can have a very significant impact on the interpretation of the measurement for purposes of understanding the biological phenomena. There are a number of inter-related aspects that determine the spatial region that is being probed by the various methods including: the actual sites from which the information is being obtained (i.e. where does the oxygen interact with the measuring method, as described above and in the table), the spatial distribution of the oxygen-sensing probes, and the volume from which the information is interrogated.

The distribution of the oxygen sensing probes will vary, especially on whether naturally occurring or administered materials are used. The possibilities for administration include placement within the vascular compartment (e.g. the usual phosphorence methods), distributed into tissues via the vascular compartment (e.g. hypoxic localizing nitroimidazoles), or placed directly into the tissues (e.g. EPR oximetry with particulate materials).

The volume from which the information is obtained becomes more complex with those techniques that involve the recording of the emission or absorption of a radiation such as infrared light. In this type of technique-the volume that is subjected to the exciting radiation and the volume from which the responses are collected may have a major impact on the spatial region from which the data are actually obtained. The pertinent considerations include the distribution of the interrogating radiation as it passes through tissue and the distribution from which responses are received. The types of radiation that are used with these techniques can undergo considerable absorption and/or scattering, thereby affecting the volume from which the information is obtained and the efficiency with which information is obtained over the volume. There are a variety of techniques which can be employed to try to shape the volume from which the information is collected by controlling the volume which receives the exciting radiation or the volume from which the data are collected.

2.2.2. The Spatial Region That Is Being Directly Probed by EPR Oximetry Based on Solid Particles. **The basic information that is reflected in the EPR spectra for EPR oximetry based on solid particles is the pO2 that is on and in the paramagnetic materials.** Under most experimental conditions the particles are administered to a specific location and therefore the volume from which the data are obtained is defined by the administration of the material. If the paramagnetic materials are in a region with a homogeneous pO_2 then the interpretation is very straightforward, because the method does not consume or remove oxygen it must reflect the pO_2 in the immediately surrounding tissues. If the paramagnetic materials are in a region that is heterogeneous in regard to pO_2 the resulting EPR spectra will, in principle, proportionately reflect the full distribution of the various values of pO_2. While in the future it is quite possible that one will be able to extract out this full information from the spectra, at this time the data that are obtained usually consist of the average pO_2 over the volume of the paramagnetic material. *This is a direct reflection of the average pO2 of the biological materials (including intracellular and extracellular water) which are in direct contact with the paramagnetic material.*

EPR oximetry based on solid particles can provide information on a variety of different spatial volumes, depending on the technique that is used. If a single particle is used, it then reflects the pO_2 of the biological material that is in direct contact with it. The spatial resolution of that variant of the technique is determined by the amount of paramagnetic material that is required for these measurements. Lithium phthalocyanine currently is the most sensitive oxygen-sensitive paramagnetic particulate material and under typical conditions an adequate signal/noise uses a single particle of lithium phthalocyanine whose minimum volume is approximately 0.05 mm^3. It follows therefore that EPR oximetry cannot report on a smaller volume than .05 mm^3 and that it will provide an average pO_2 for that volume. If several macroscopic particles are used, then each particle will reflect the average pO_2 of the biological materials immediately around it. Under appropriate conditions, the pO_2 for each particle can be resolved separately (Smirnov et al (1993). If a slurry of particles is used, then the resulting spectra are a function of the pO_2 over the volume of biological material that is in contact with each of the particles. With current techniques, both an average pO_2 for the entire set of particles or the average pO_2 at two of the edges of the slurry can be measured.

2.2.3. The Spatial Region That Is Being Directly Probed by EPR Oximetry Based on Soluble Materials. EPR oximetry based on soluble materials such as nitroxides reflects the $[O_2]$ at each of the nitroxide molecules within the sensitive volume of the EPR spectrometer. The nitroxides usually are administered systemically and therefore the volume that is interrogated is defined by the characteristics of the detector of the EPR spectrometer: its sensitive volume and where it is placed. Considerable effort is in progress to carry out imaging with this type of EPR oximetry (Halpern et al (1994). The principal limitations to such imaging are the sensitivity of the nitroxides and of the spectrometer. If these limitations can be overcome, this technique could provide spatially resolved information on the $[O_2]$ in the form of 2 or 3 dimensional maps. Currently this approach is used principally to measure the average $[O_2]$ over relatively large volumes but the resolution seems to be improving.

2.3. The Relationship of the Spatial Region That Is Probed by a Method to the Biological Volume and/or Process That Is Being Studied

2.3.1 General Considerations. This relationship clearly can be quite complex and will, for all methods, crucially affect interpretation of the biological significance of the measurements of oxygen. The complexities arise principally because within functioning biological systems there almost always are quite large gradients in the pO_2 and $[O_2]$ due to the consumption of dioxygen by cellular processes and the way that oxygen is delivered by the vascular network. Virtually all of the methods provide measurements at particular sites within these heterogeneous systems and therefore some assumptions and/or extrapolations are required to infer to the pO_2 and/or $[O_2]$ at the sites of biological interest. The heterogeneity usually occurs on both macroscopic and microscopic scales and will differ in the vascular and extravascular compartments and probably also within the volume of individual cells. Depending on the information that is sought, some methods may offer particular advantages. In general more assumptions and/or extrapolations will be required with higher rates of consumption of oxygen and greater differences in the blood supply to the various regions of interest. As a general rule the biological usefulness of a measurement decreases in proportion to the distance from the oxygen-sensing molecules to the site of interest, and the amount of heterogeneity of oxygen that is present.

2.3.2. The Relationship of the Spatial Region That Is Probed to the Biological Volume and/or Process That Is Being Studied by EPR Oximetry Based on Solid Particles. The principal factors that may affect how these measurements relate to the processes of interest reside in the extent to which they sample the volume of interest, as discussed above in the section on general considerations. The use of slurries provides the widest sampling volume in proportion to the amount of paramagnetic material that is used.

Regardless of whether a solid particle or a slurry is used, however, with present methods of analysis this technique cannot measure microheterogeneity on a physical scale smaller than the size of the paramagnetic material required to provide an adequate signal to measure (currently 0.05 mm^3). It is likely that in the near future data analysis will progress such that it will be able to determine whether heterogeneity of pO_2 is present in the volume of biological material in contact with the paramagnetic material. Over a longer period of time, it may become possible to obtain a quantitative description of that heterogeneity, but not its precise spatial distribution.

2.3.3. The Spatial Region That Is Being Directly Probed by EPR Oximetry Based on Soluble Materials. Additional factors that may affect these measurements involve the occurrence of non-uniform distribution (due to the usual factors plus the possibility of metabolism to a non-paramagnetic state) which can affect the spectral parameters on which the oxygen measurements are based.

3. SUMMARY REGARDING MEASUREMENTS OF OXYGEN BY EPR OXIMETRY BASED ON PARTICULATE MATERIALS

The physical nature of the measurement is the average pO_2 (partial pressure of oxygen) in contact with the oxygen-sensitive paramagnetic material. The effects are based on the direct interaction of oxygen with the paramagnetic material and do not involve consumption or significant binding of the O_2. The effects on the EPR spectra are specific for diffusable paramagnetic molecules and inasmuch as there are no other paramagnetic molecules in tissues that occur at sufficient concentrations to affect the measurements, the method is specific for molecular oxygen.

The region of the sample that is measured directly is that immediately surrounding the paramagnetic material. If the paramagnetic material is a macroscopic particle, it reflects the pO_2 in the tissue in contact with the surface of the particle. If the paramagnetic material is a slurry of small particles, then it reports on the average pO_2 of the sum of the surfaces in contact with the individual components of the slurry.

The spatial resolution of the technique is limited by the amount of paramagnetic material that is required for these measurements (e.g. as a single particle of lithium phthalocyanine the minimum volume is approximately 0.05 mm^3). It follows therefore that EPR oximetry cannot report on a smaller volume than .05 mm^3 and that it will provide an average pO_2 for the volume.

For EPR oximetry as well as for all other techniques, the volume of the tissue (or other biological system) for which the measurements provide insights, depends greatly on the rates of consumption by the cells and the delivery of O_2. When the consumption of oxygen is high and/or the delivery of oxygen is limited, then there are likely to be significant gradients in the pO_2 in the tissue/sample. If there are only small gradients, then the measured pO_2 will reflect the average pO_2 for the entire sample. If there are large gradients of pO_2, the situation is complex and will depend in part on the volume sensed directly by the technique. If the gradients occur over a volume smaller than that directly measured by the technique and are similar throughout the sample (e.g. for EPR, if they arise from a Krough cylinder mechanism) then the measurement will be a reasonable approximation of the average pO_2 in the sample. If there are macroscopic gradients in the pO_2 (e.g. the presence of non-perfused areas) then the measurement will be valid only for the type of compartment in which the sensing element is located (e.g. necrotic area vs. viable cellular area).

ACKNOWLEDGMENT

This work was supported by NIH Grant GM 51630.

REFERENCES

Chapman, J. Measurement of tumor hypoxia by invasive and non-invasive procedures: a review of recent clinical studies. Radiother Oncol 1991; 20(Suppl 1):13–19.

Dunn, J.F., Ding, S., O'Hara, J.A., Liu, K.J., Rhodes, E., Weaver, J.B. and Swartz, H.M. The apparent diffusion constant measured by MRI correlates with pO_2 in a RIF-1 tumor. Magn. Reson. Med. 1995; 34:515–519.

Fenton, B.M. and Gayeski, T.E. Determination of microvascular oxyhemoglobin saturations using cryospectrophotometry. Am J Physiol 1990; 259(6 Pt 2):H1912–20.

Ferrari, M., Quaresima, V., Ursini, C.L., Alecci, M. and Sotgiu, A. In vivo electron paramagnetic resonance spectroscopy-imaging in experimental oncology: the hope and the reality. Int J Rad Oncol Biol Phys 1994; 29:421–425.

Gallez, B., Bacic, G., Goda, F., Jiang, J., O'Hara, J.A., Dunn, J.F. and Swartz, H.M. Use of nitroxides for assessing perfusion, oxygenation and viability of tissues: In vivo EPR and MRI studies. Magn Reson. Med. 1996; 35:97–106.

Goda, F., Bacic, G., O'Hara, J.A., Gallez, B., Swartz, H.M. and Dunn, J.F. The relationship between partial pressure of oxygen and perfusion in two murine tumors after x-ray irradiation: a combined gadopentetate dimeglumine dynamic magnetic resonance imaging and in vivo electron paramagnetic resonance study. Cancer Res 1996; 56:3344–3349.

Goda, F., O'Hara, J.A., Liu, K.J., Rhodes, E.S., Dunn, J.F. and Swartz, H.M. Comparisons of measurements of pO2 in tissue in vivo by EPR oximetry and micro-electrodes. Oxygen Transport to Tissue XVIII (In Press, 1995).

Halpern, H.J., Yu, C., Peric, M., Barth, E., D.J. and Teicher, B.A. Oximetry deep in tissues with low-frequency electron paramagnetic resonance. Proc Natl Acad Sci U S A 1994; 91:13047–51.

Jiang, J., Liu, K.J., Shi, X. and Swartz, H.M. Detection of short-lived free radicals by low-frequency electron paramagnetic resonance spin trapping in whole living animals. Arch Biochem Biophys 1995; 319:570–3.

Kelleher, D.K., Thews, O. and Vaupel, P.W. Oxygenation status of experimental rat tumors: Impact of duration of anesthesia and positioning of tumor. In Tumor Oxygenation (ed. P W Vaupel, D.K.K., M Gunderoth) 18–25 (Gustav Fischer Verlag, Stuttgart, 1995).

Liu, K.J., Bacic, G., Hoopes, P.J., Jiang, J., Du, H., Ou, L.C., Dunn, J.F. and Swartz, H.M. Assessment of cerebral pO2 by EPR oximetry in rodents: effects of anesthesia, ischemia, and breathing gas. Brain Res 1995; 685:91–8.

Liu, K.J., Grinstaff, M.W., Jiang, J., Suslick, K.S., Swartz, H.M. and Wang, W. In vivo measurement of oxygen concentration using sonochemically synthesized microspheres. Biophys J 1994a; 67:896–901.

Liu, K.J., Jiang, J., Swartz, H.M. and Shi, X. Low-frequency EPR detection of chromium(V) formation by chromium(VI) reduction in whole live mice. Arch Biochem Biophys 1994b; 313:248–52.

O'Hara, J.A., Goda, F., Dunn, J.F. and Swartz, H.M. The potential for EPR oximetry to guide treatment planning for tumors. Oxygen Transport to Tissue XVIII, In Press (1996).

O'Hara, J.A., Goda, F., Liu, K.J., Bacic, G.A., Hoopes, P.J. and Swartz, H.M. The pO$_2$ in a murine tumor after radiation: an in vivo electron paramagnetic resonance oximetry study. Radiat. Res. 1995b; 144:222–229.

Raleigh, J.A., Dewhirst, M.W. and Thrall, D.E. Measuring tumor hypoxia. in *Hypoxia and Its Clinical Significance* (ed. Raleigh, J.A.) 37–45 1996).

Smirnov, A.I., Norby, S.W., Clarkson, R.B., Walczak, T., and Swartz, H.M., "Simultaneous Multi Site EPR Spectroscopy *In Vivo*," Magn. Reson. Med., 30:213–220 (1993).

Stone, H.B., Brown, J.M., Phillips, T.L. and Sutherland, R.M. Oxygen in human tumors: correlations between methods of measurement and response to therapy. Summary of a workshop held November 19–20, 1992, at the National Cancer Institute, Bethesda, Maryland. Radiat. Res. 1993; 136:422–34.

Swartz, H.M., Bacic, G., Friedman, B., Goda, F., Grinberg, O., Hoopes, P.J., Jiang, J., Liu, K.J., Nakashima, T., O'Hara, J.A. and Walczak, T. Measurements of pO$_2$ in vivo, including human subjects, by electron paramagnetic resonance. in *Oxygen Transport to Tissue XVI* (ed. M.C. Hogan, et al) 119–128 (Plenum Press, New York, 1994).

Swartz, H.M., Boyer, S., Brown, D., Chang, K., Gast, P., Glockner, J.F., Hu, H., Liu, K.J., Moussavi, M., Nilges, M., Norby, S.W., Smirnov, A., Vahidi, N., Walczak, T., Wu, M. and Clarkson, R.B. The Use of EPR for the Measurement of the Concentration of Oxygen In Vivo In Tissues Under Physiologically Pertinent Conditions and Concentrations. in *Oxygen Transport to Tissue XIV* (eds. Erdmann, W. & Bruley, D.F.), pp 221–228. (Plenum Publishing Corp, New York, 1992).

Urtasun, R.C., Chapman, J.D., Raleigh, J.A. et al. Binding of ^{3}H-misonidazole to solid human tumors as a measure of tumor hypoxia. Int J Radiat Oncol Bio Phys 1986; 12:1263–1267.

Wilson, D.F. and Cerniglia, G.J. Localization of tumors and evaluation of their state of oxygenation by phosphorescence imaging. Cancer Res 1992; 52:3988–3993.

94

MATHEMATICAL MODELLING OF TUMOUR GROWTH AND ANGIOGENESIS

B. D. Sleeman

Department of Applied Mathematical Studies
University of Leeds
Leeds, LS2 9JT, United Kingdom

1. INTRODUCTION AND METHODS

Solid tumours are known to develop through two distinct phases of growth. The first phase is that of avascular growth and is well modelled by the growth of multicellur spheroids in vitro. By making use of the mathematical theory of non-linear elasticity we propose that the concept of a "strain-energy function" may be fundamentally important in attempting to understand the various processes involved in avascular growth.

The transition from the avascular phase to the vascular phase depends on the tumours ability to induce new blood vessels from the surrounding tissue to sprout towards and then gradually penetrate the tumour thus providing it with an adequate blood supply and microcirculation. A mathematical model based on reaction diffusion and chemotactic mechanisms is described. It provides an insight into the properties of angiogenesis which is well supported by experimental observations.

2. A MATHEMATICAL MODEL OF TUMOUR GROWTH BASED ON THE THEORY OF NONLINEAR ELASTICITY

Our consideration of mathematical models of tumour development concentrates on the early stages of growth and considers the tumour to be composed of a large central necrotic core surrounded by a layer of live proliferating cells on the tumour surface several cells thick. Initially the tumour is assumed to be spherical. All live tumour cells are assumed to be identical and each is considered to be an incompressible structure of fixed volume. The gross internal forces in the necrotic core are characterised by a pressure distribution P. Cell adhesion produces a surface tension force at the boundary. For a benign tumour there is a balance between nutrient supply and growth. Unstable development of the tumour arises when the internal pressure forces overcome the surface tension and inter-cellular adhesion. The instability is initially manifested as a pinching or a corrugation

Oxygen Transport to Tissue XIX, edited by Harrison and Delpy
Plenum Press, New York, 1997

of the boundary surface at the equator of the tumour and more elaborate instability configurations may lead to further subdivision or disintegration with subsequent invasion of the surrounding tissue.

The basic tool for our modelling is thin shell elasticity theory (see Landau & Lifschitz [1] and Fung [2]). We work in terms of averaged variables. That is we average over the reference thickness of the shell. Let R be the initial radius of the tumour and r its current radius and denote by H and h the corresponding thickness of the layer of live cells. The average deformation gradient, ie the measure of the deformation of the layer of live cells is expressed in terms of the averaged principal stretches $\bar{\lambda}_i$ ($i = 1,2,3$). These are measures of how much the layer of live cells stretches or compresses along each of the spherical polar directions corresponding to $i = 1,2$ and 3 respectively. Under the incompressibility assumption.

$$\bar{\lambda}_1 \bar{\lambda}_2 \bar{\lambda}_3 = 1$$

and in view of the symmetry of the deformation we can write

$$\bar{\lambda}_1 = \bar{\lambda}_2 = \lambda \text{ and } \bar{\lambda}_3 = \lambda^{-2}$$

As a consequence $r = \lambda R$, $h = \lambda^{-2}H$.

For a growing tumour λ 1 and so as the tumour increases in size the cell layer decreases in thickness. The next step in the modelling is to observe [3] that for an incompressible isotropic elastic material the principal average stresses $\bar{\sigma}_i$ are given by

$$\bar{\sigma}_i = \bar{\lambda}_i \frac{\partial \bar{W}}{\partial \bar{\lambda}_i}, i = 1,2,3$$

where the "strain energy" function $\bar{W} = \bar{W}(\bar{\lambda}_1, \bar{\lambda}_2, \bar{\lambda}_3)$ is symmetric in its arguments. $\bar{W}$ is a measure of the elastic potential energy stored in the layer of live proliferating cells at the tumour surface. Because of the assumed spherical symmetry we can rewrite $\bar{W}$ in the form $\bar{W} = \hat{W}(\lambda, \lambda, \lambda^{-2})$ From equilibrium considerations it can be shown [4] that the internal expansive pressure P is given by

$$P = \frac{2h}{r} \bar{\sigma}_{11} = \frac{H}{R\lambda^2} \frac{\partial \hat{W}}{\partial \lambda}.$$

The essential features of the initial phase of solid tumour growth are thus captured by the model. That is the gross internal forces are characterised by an internal expansive pressure distribution which is counter balanced by a surface tension at the outer boundary of the tumour. Of course the cause of growth is the uptake of nutrients by the cells, while cells in the interior die and necrosis takes place here due to the lack of vital nutrients. The role of surface tension and surface curvature is not fully understood. In [4] we have investigated the stability of the tumour under small deformations from the radially symmetric configuration. To this end the strain energy function is chosen to be of the form

$$\hat{W}(\lambda) = \mu_r \left(2\lambda^{\alpha_r} + \lambda^{-\alpha_r} - 3 \right) / \alpha_r$$

where μ_r, α_r are positive parameters and summation over a finite number of such terms is indicated by the repetition of the index r. The form of $\hat{W}$ was introduced by Haughton and Ogden [3] in their study of rubber like materials. The stability analysis suggests the following criteria for assessing the degree of malignancy of a tumour

benign tumour: $-\dfrac{3}{2} < \alpha_r \le -1$ or $2 \le \alpha_r < 3$;

malignant tumour: $-1 < \alpha_r < 1$.

For a full discussion of these ideas see Chaplain and Sleeman [4]. For the use of elasticity theory in modelling other biological tissues see Demirary [5], Gou [6] and Bogen [7].

3. TUMOUR ANGIOGENESIS

The transition from the dormant avascular state to the vascular state depends on the tumours ability to induce new blood vessels from the surrounding tissue to sprout towards and then to penetrate the tumour, thus providing it with an adequate blood supply and microcirculation. To accomplish neovascularisation tumours secrete diffusible chemical substances known collectively as tumour angiogenesis factor (TAF) into the surrounding tissue and extra cellular matrix (ECM).

Angiogenesis is a complex phenomena involving mechanical factors, energy imbalance due to hypoxia and inflammatory processes. Indeed the exact mechanisms for vacular growth are extremely complex and far from being understood. There are at least fifty factors which may be involved in angiogenesis, see Rakusan [8].

In order to develop at least a qualitative mathematical model of tumour angiogenesis we concentrate on the following essential events [9,10]

1. degradation of the basement membrane in neighbouring normal capillaries by enzymes secreted by the endothelial cells in response to TAF stimulus.
2. migration of endothelial cells.
3. proliferation of endothelial cells.

Let $n(x,t)$ denote the density of endothelial cells and $c(x,t)$ denote the concentration of TAF in the host tissue at the point x and time t. The model [11] then takes the form

Rate of increase in	=	diffusion of	−	loss due to	−	decay of,
$c(x,t)$		$c(x,t)$		$n(x,t)$		$c(x,t)$
Rate of increase in	=	migration of	+	mitotic generation of	−	cell loss.
$n(x,t)$		$n(x,t)$		$n(x,t)$		

We assume that the local uptake of TAF by endothelial cells is governed by Michael-Menten kinetics and that uptake rate depends linearly on the cell density and that the chemical decay is linear.

The resulting model for TAF concentration is

$$\frac{\partial c}{\partial t} = D_c \nabla^2 c - \frac{Qcn}{(Km+c)n_0} - dc,$$

where D_c, Q, Km, n_0 and d are parameters estimated from available data (see [12]).

To model the endothelial cell density we postulate a general conservation equation of the form

$$\frac{\partial n}{\partial t} + \nabla \cdot \underset{\sim}{J} = F(n)G(c) - H(n),$$

where $\underset{\sim}{J}$ is the cell flux assumed to consist of random motion together with chemotaxis. ie

$$\underset{\sim}{J} = -D_n \nabla n + n \, \chi_0 \nabla c,$$

where D_n is the diffusion coefficient and χ_0 is the chemotactic coefficient. $F(n)$ governs mitosis and is represented by a logistic law while $G(c)$ is a "switch" incorporating a threshold below which cell proliferation does not occur. Cell loss is assumed to be first order. The above equations together with appropriate initial and boundary conditions complete the formulation of the model (see [11,12] for details).

As observed in experiments performed on test animals [13,14,15] capillary networks grow in essentially a two-dimensional domain. To model this we consider an idealised but nevertheless realistic situation of a circular tumour situated centrally in a homogeneous ECM bounded by a circular limbus.

Groups of endothelial cells of fixed concentration able to respond to TAF are arranged around the edge of the ECM as shown in figure 1 in which the TAF concentration,

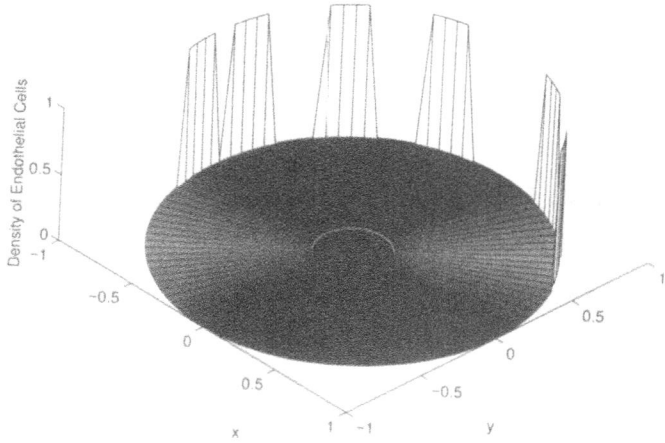

Figure 1. Initial endothelial cell distribution.

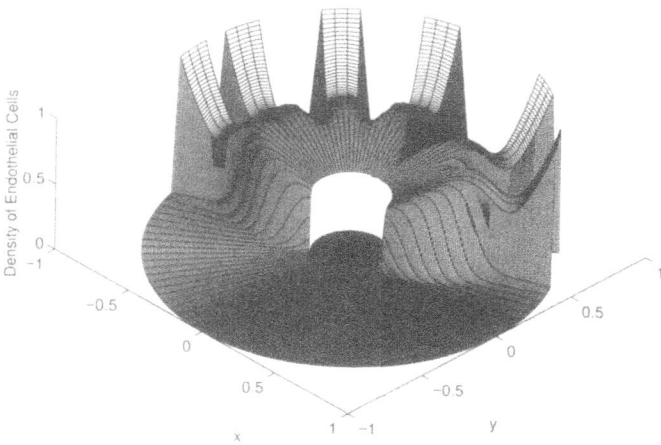

Figure 2. Endothelial cell distribution at about 8 days showing mitosis and complete anastomosis together with tumour vascularisation.

based on numerical evidence [16], is approximated by a steady state law. Figure 2 shows the evolution of endothelial cells after about 8 days. Here we clearly see an initial decrease in endothelial cell density followed by rapid mitosis and evidence of anastomosis within the ECM. The results shown in figure 2 and developed in more detail in [16] accord well with experimental evidence.

The ECM is a very heterogeneous medium consisting of interstitial tissue, collagen fibre, fibronectin fibrils and fribonectin as well as other components. Thus not only is the ECM an obstacle to EC growth but can also act as a tortuous medium guiding the growth of capillary sprouts. In order to take this into account we have taken the chemotactic constant χ_0 to be a random variable, an idea exploited earlier with Regalado et al [17] in the study of fungal mycelea growth and by Meinhardt [18] in modelling patterns on sea shells. Specifically a randomly generated multiplicative grid of chemotactic constants was imposed on the deterministic value. That is, within each grid element i of ECM a random value P_i drawn from a uniform distribution on [0, 1.125] was assigned. The chemotactic value within that particularly grid element was taken to be $\chi_i = P_i \chi_0$

Performing a numerical experiment under the same conditions as before resulted in This shows mitosis, anastomosis and tumour vascularisation. There is also evidence suggesting the development of capillary sprouts.

4. CONCLUSIONS

This paper describes how the essential characteristics of solid tumour growth may be modelled using ideas from nonlinear elasticity and how the resulting theory can be used to characterise the structure of solid tumours leading to mechanisms for distinguishing between benign and malignant tumours. The idea of a strain energy function plays a central role in the development of the theory. This is only the beginning of thinking about tumours in this way. Further development must be directed towards obtaining a greater insight in the modelling of tumour strain energy functions.

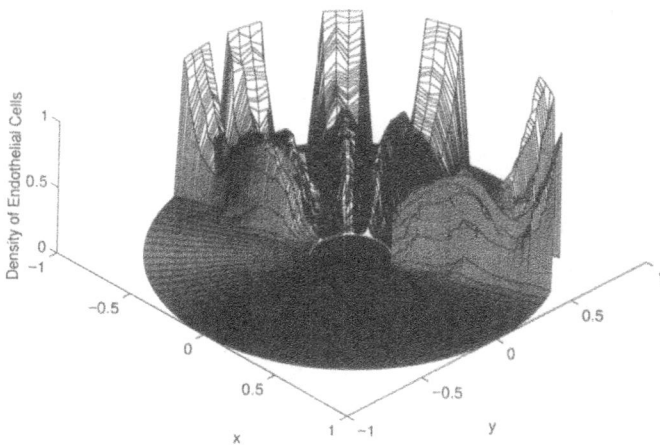

Figure 3. Endothelial cell distribution at about 8 days in the heterogeneous case showing mitosis, anastomosis and tumour vascularisation. There is also evidence of capillary sprouting.

The paper is also concerned with the angiogenic response of endothelial cells to the stimulus of TAF. By using reaction diffusion and chemotaxis, a model has been described which is able to demonstrate the important features of angiogenesis. By modelling the ECM as a heterogeneous random medium it is also possible to capture evidence of capillary branching and sprouting.

Chemotaxis is clearly an important factor influencing angiogenesis and anastomosis. However much work is needed to appropriately take account of the heterogeneity of the ECM and its role in enhancing or inhibiting capillary growth. Fibronectin is clearly important in enhancing EC adhesion to collagen [19]. Thus as well as responding to the chemotactic effect of TAF on EC, EC migration is also regulated by the haptotactic effect of fibronectin [20]. The idea of using a stochastic formulation is new and its biological significance for the study of angiogenesis is likely to have important implications.

REFERENCES

1. L.D. Landau & E.M. Lifschitz. Theory of Elasticity, Pergamon, London (1959).
2. Y.C. Fung. Biomechanics, Springer, New York (1981).
3. D.M. Haughton and R.W. Ogden. On the incremental equations in nonlinear elasticity. I Membrane theory. *J. Mech. Phys. Solids*, **26**, (1978), 93–110.
4. M.A.J. Chaplain & B.D. Sleeman. Modelling the growth of solid tumours and incorporating a method for their classification using nonlinear elasticity theory. *J. Math. Biol.*, **31**, (1993), 431–473.
5. H. Demirary. Large deformation analysis of some soft biological tissues. ASME. *J. Biomech. Eng.*, **103**, (1981), 73–78.
6. P.F. Gou. Strain energy functions for biological tissues. *J. Biomech.*, **3** (1970), 547–550.
7. D.K. Bogen. Strain energy description of biological swelling. I. Single fluid compartment models. ASME *J. Biomech. Eng.*, **109**, (1987), 252–256.
8. K. Rakusan. Coronary Angiogenesis: from morphometry to molecular biology and back. *Ann. New York Academy of Sciences*, **752**, (1995), 257–266.
9. J. Folkman. The vascularisation of tumours. *Sci. Am.*, **234**, (1976), 58–73.
10. N. Paweletz & M. Knierim. Tumour-related angiogenesis. *Crit. Rev. Oncol. Hematol.*, **9**, (1989), 197–242.
11. M.A.J. Chaplain & A.M. Stuart. A model mechanism for the chemotactic response of endothelial cells to tumour angiogenesis factor. *I.M.A. J. Math. Appl. Med. Biol.*, **10**, (1993), 149–168.

12. M.A.J. Chaplain, S.M. Giles, B.D. Sleeman & R.J. Jarvis. A mathematical analysis of a model for tumour angiogenesis. *J. Math. Biol.*, **33**, (1995), 744- 770.

13. M.A. Grinbrone, R.S. Cotran, S.B. Leapman & J. Folkman. Tumour growth and neovascularisation: an experimental model using the rabbit cornea. *J. Nat. Cancer Inst.,* **52**, (1974), 413–427.

14. D.F. Zawicki, R.J. Jain, G.W. Schmid-Schoenbeis & S. Chien. Dynamics of neovascularisation in normal tissue. *Microvas. Res.* (1981).

15. V.R. Mathukkaruppan, L. Kubai & R. Auerbach. Tumour induced neovascularisation in the mouse eye. *J. Nat. Cancer Inst.*, **69**, (1982), 699- 705.

16. B.D. Sleeman, D.S. Burn & R. Nimmo. Endothelial cell response to tumour angiogenesis. (1996) submitted.

17. C. Regalado, J.W. Crawford, K. Ritz & B.D. Sleeman. The origins of spatial heterogeneity in vegetative fungal mycella: a reaction diffusion model. *Mycological Research* (1996) to appear.

18. H. Meinhardt. The algorithmic beauty of sea shells. Springer-Verlag. Berlin, Heidelberg, (1995).

19. S.L. Schor, A.M. Schor & G.W. Brazill. The effects of fibronectin on the migration of human foreskin fibroblasts and syrian hamster melanoma cells into three-dimensional gels of lattice collagen fibres. *J. Cell Sci.*, **48**, (1981), 301–314.

20. M.M. Sholley, G.P. Ferguson, H.R. Seibel, J.L. Montour & J.D. Wilson. Mechanisms of neovascularisation. Vascular sprouting can occur without proliferation of endothelial cells. *Lab. Invest.*, **51**, (1984), 624–634.

95

EXTRACTION PRESSURES CALCULATED FOR RAT HEART AND DOG SKELETAL MUSCLE AND APPLICATION IN MODELS OF TISSUE OXYGENATION

L. Hoofd and C. Bos

Department of Physiology
University of Nijmegen
Geert Groote Plein Noord 21, 6525 EZ Nijmegen, The Netherlands

1. INTRODUCTION

The Extraction Pressure (EP) is a quantity that accounts for the loss of O_2 pressure in the immediate vicinity of a capillary. It can be thought of as a pressure drop due to the extraction of O_2 from the capillary before any further transport into the tissue occurs. EP is virtually absent for a capillary filled with homogeneous fluid but it will be manifest when individual erythrocytes as oxygen sources can be discerned. It originates from the locally steep O_2 gradients around these individual sources, in particular within the blood compartment itself. This means that the tissue experiences a pO_2 that is an amount EP lower than the actual capillary pO_2, or better, erythrocyte pO_2. For example, when blood pO_2 is 8 kPa and EP equals 1 kPa, the surrounding tissue experiences a capillary pO_2 of 7 kPa.

The phenomenon is a gradient phenomenon and thus depends on local O_2 delivery. In accordance, model calculations mostly show that EP is nicely proportional to the local O_2 supply area of a capillary. For example, an EP of 1 kPa for a supply area of 400 μm^2 would lead to a doubled EP of 2 kPa for a doubled area of 800 μm^2 (Bos et al., 1995). However, the EP also depends strongly on other capillary parameters, in particular blood flow and haematocrit.

The concept of an extraction pressure can be incorporated in tissue models encompassing many capillaries, by taking into account its value for each individual capillary. The global effect on tissue pO_2 can be estimated from a representative value of EP. For example, an EP value of 1 kPa for an 'average' capillary could be interpreted as globally lowering the tissue pO_2 by about 1 kPa. For local or specific effects, however, a complete tissue model has to be considered.

Data are presented here for EP as calculated by a numerical model for two types of muscle tissue, often modelled in the literature: rat heart and dog skeletal muscle. The resulting EP values were incorporated in a large-scale model to calculate the effects on tissue pO_2. The parameter range in the models covers muscle situations from rest up to heavy exercise.

2. MATERIALS AND METHODS

2.1. Calculation of the EPs

A model of a capillary part for one erythrocyte (RBC) and surrounded by a corresponding tissue portion was used to calculate EP for various input data. The model could be restricted to an elementary unit because it has been shown earlier that the resulting EP is essentially independent of the number of RBCs involved (Bos et al, 1995). Also, the elementary unit could be taken circularly symmetric because asymmetries in radial O_2 supply can be accounted for separately, in terms of a 'background' field. For these elementary units, EP was calculated numerically according to Bos et al. (1996), taking into account plasma motion in between the RBCs and non-equilibrium facilitated diffusion by hemoglobin (Hb) in the RBCs, for a variety of RBC velocities v_{RBC} and capillary haematocrits Hct, where Hct is the volume fraction of RBCs in the capillary.

2.2. Tissue Model

The resulting EPs were incorporated into a multi-capillary model as described earlier (Hoofd, 1995a). This models tissue oxygen pressure p in terms of capillary O_2 supply areas A_i of the ith capillary, where there are N capillaries. The A_i are calculated for any situation from the boundary conditions, total supply area A, of all the capillaries together, and capillary pressures p_{ck}, i.e., the apparent pO_2 at the capillary rim of each kth capillary. Incorporating EP, this capillary rim pressure is related to the pO_2 of the capillary blood, p_{bk}:

$$p_{ck} = p_{bk} - EP_k \quad : \quad EP_k = EP \frac{A_k}{\frac{1}{N} A} \tag{1}$$

where EP_k is the extraction pressure for the kth capillary taken to be proportional to the overall EP according to its supply area. This assumes the same conditions in each capillary (flow, haematocrit). Although it is perfectly possible to incorporate different situations for each capillary, the primary concern of this paper was investigation of the effects of EP and not of the capillary heterogeneity leading to heterogeneous EP distributions.

Subsequent layers form a block of tissue, where each p_{bk} of the next layer is calculated by subtracting the O_2 extracted in the layer (Hoofd, 1995b). For this, the Hill equation was used relating the O_2 pressure p and O_2 saturation of Hb sO_2Hb in the RBC:

$$sO_2Hb = \frac{p^n}{p^n + p_{50Hb}{}^n} \tag{2}$$

which yields an adequate description for the cases here ($sO_2Hb > 10\%$). Facilitation of O_2 transport by myoglobin (Mb) in the tissue is incorporated, through a facilitation pressure

p_F expressing the maximum contribution of myoglobin to (facilitated) O_2 diffusion (Hoofd, 1995a), and the O_2 saturation sO_2Mb of Mb, taken to be in equilibrium with tissue O_2 pressure p according to the hyperbolic description.

$$sO_2Mb = p/(p+p_{50Mb}).$$

2.3. Input Data

The basic input data were according to rat heart (Hoofd et al., 1990) and to dog gracilis muscle (Hoofd & Turek, 1994) and are listed in Table 1. Data include heterogeneity of capillary spacing, where capillary locations were taken from realistic tissue cross sections. For both tissues, capillaries were straight and parallel to one side of the tissue block. Capillaries were staggered in six groups; in three of the groups (**A**) all capillaries started with a blood pO_2 of p_a, in the other three groups (**V**) with a p_{bk} distribution that the **A** capillaries had halfway along their length. Consequently, the block length was half the capillary length.

The impact of the extraction pressure on tissue pO_2 is best shown when the O_2 delivery to the tissue is kept the same for each situation. This is done through a constant value of the product $v_{RBC} \times Hct$ as given in Table 1.

Table 1. Data used in the calculations. *Width* and *Height* are the dimensions of the cross sections of the tissue block with *Length* equal to half the capillary length, r_c is the capillary radius, V_{RBC} is the volume of one erythrocyte, $\dot{Q}$ is the tissue's O_2 consumption, D its oxygen diffusion coefficient and $\wp$ its oxygen permeability; for other symbols see text. Only for the tissue situations, the product of v_{RBC} and Hct is a constant

Dimensional data:	Width (μm)	Height (μm)	Length (μm)	N	r_c (μm)
Rat heart	177.5	140.4	250	71	2.4
Dog gracilis	250	145	456	35	2.3
Blood data:	p_a (kPa)	p_{50Hb} (kPa)	n	$v_{RBC} \times Hct$ (mm/sec)	V_{RBC} ((μm)^3)
Rat heart	13.3	4.93	2.7	0.4	61
Dog gracilis	8.1	3.52	2.65	0.9	66
Tissue data:	$\dot{Q}$ (mol·m^{-3}·s^{-1})	D (m^2·s^{-1})	$\wp$ (mol·m^{-1}·kPa^{-1}·s^{-1})	p_F (kPa)	p_{50Mb} (kPa)
Rat heart	0.3	$1.9\ 10^{-9}$	$1.18\ 10^{-11}$	1.87	0.71
Dog gracilis	0.0991	$1.16\ 10^{-9}$	$0.818\ 10^{-11}$	5.28	0.71

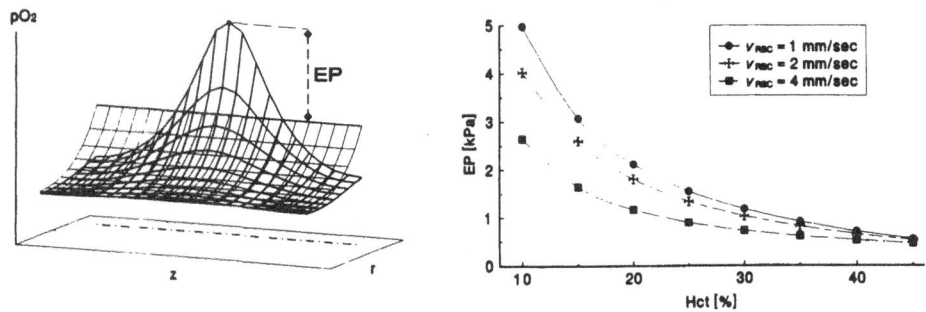

Figure 1. Principle of extraction pressure EP (left panel) and values calculated for rat heart (right panel) for a wide range of capillary haematocrit values Hct and RBC velocities as indicated.

3. RESULTS

First, EP values were calculated for average capillary supply area and for a variety of values of v_{RBC} and Hct. For rat heart, the results are shown in the right panel of figure 1. The left panel is a visualisation of EP; the tissue at some distance experiences a capillary pressure that is EP lower than the actual erythrocyte pressure. This facilitates the interpretation of the right panel. For dog gracilis muscle, EP values were only calculated as indicated below.

For each situation, pO_2 histograms were obtained by calculating pO_2 in the tissue block at a large number of locations (30000–130000), in a regular grid. In addition, characterising each histogram, mean $(\bar{p})$, standard deviation (σ_p) and skewness (τ_p) of the pO_2 distribution were calculated.

3.1. Rat Heart Tissue

The data for normal working rat heart were for a capillary blood velocity $v_{RBC}=1$ mm/sec and a capillary haematocrit Hct= 40%, yielding EP= 0.7 kPa for a capillary with an average supply area of 177.5x140.4/71= 351 μm^2. The resulting pO_2 histogram is shown in panel **b** of figure 2. Mean pO_2 was $\bar{p}$ = 3.79 kPa, standard deviation σ_p= 1.96 kPa and skewness τ_p= +1.78 kPa, which means that higher pO_2s are much more frequent than lower ones. Without EP (panel **a**), mean pO_2 would be 4.46 kPa with comparable values of σ_p= 1.99 kPa and τ_p= +1.81 kPa.

However, EP depends not only on O_2 delivery but on the actual combination of v_{RBC} and Hct. Doubling the first while halving the second leads to a much higher EP of 1.8 kPa and consequently to a different pO_2 histogram, in spite of unchanged O_2 delivery (panel **c**). A small amount (1.5%) of the tissue now has pO_2s calculated below zero which means that some of the tissue will be anoxic. Mean pO_2 has dropped to 2.84 kPa. Quadrupling v_{RBC} while lowering Hct to 10% leads to the histogram of panel **d**, EP= 2.6 kPa, $\bar{p}$ = 2.22 kPa and a further increase in anoxic tissue (5.4% calculated below zero). σ_p and τ_p remained similar for all cases.

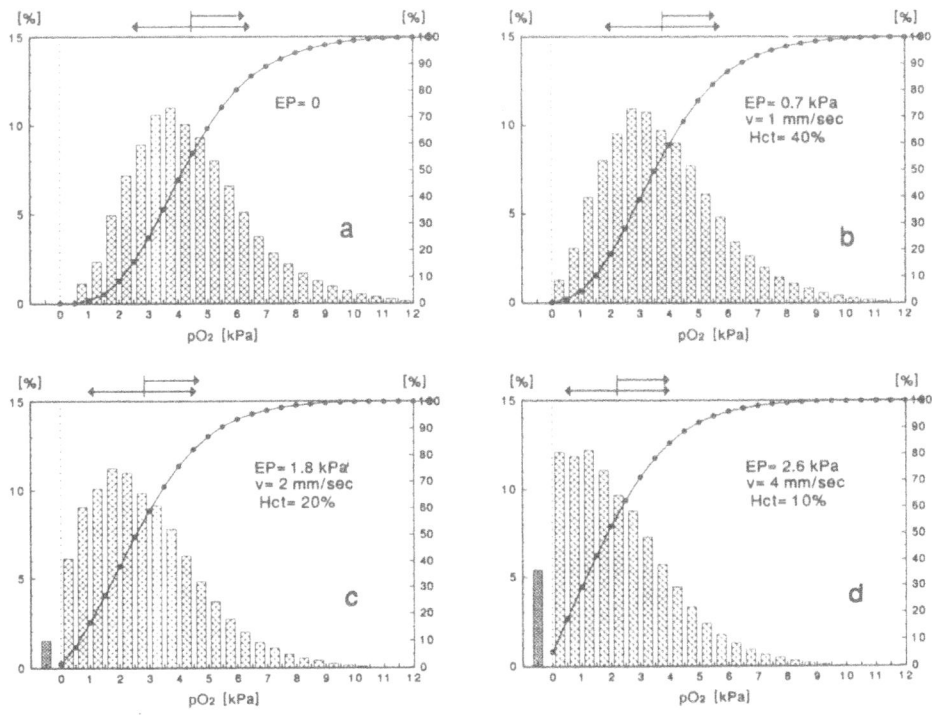

Figure 2. pO_2 histograms (bars; left axis) and cumulative histograms (dots connected by lines; right axis) calculated for rat heart tissue and for various situations as indicated in the panels. Arrows indicate mean ($\bar{p}$), standard deviation (σ_p) and skewness (τ_p) of the pO_2 distribution.

3.2. Dog Gracilis Tissue

These data are essentially for tetanic contraction (Groebe, 1990) and lead to EP= 0.73 kPa for a supply area of $250 \times 145/35 = 1036$ μm^2, for the combination $v_{RBC}= 2$ mm/sec and Hct= 45%. The resulting histograms for this EP and for EP= 0 are shown in figure 3. Without EP, only few anoxic areas arise (1.3% below zero), for a quite low mean tissue pO_2 of = 1.65 kPa, hardly larger than its standard deviation σ_p= 1.36 kPa so quite a skewed distribution τ_p= +1.35 kPa. Including EP results in an appreciable amount of anoxic regions (5.9% calculated below zero) and a mean tissue pO_2 of $\bar{p} = 1.25$ kPa with σ_p= 1.20 kPa and τ_p= +1.25 kPa. Doubling v_{RBC} while halving Hct resulted in EP= 1.02 kPa, somewhat increasing anoxic regions (8.9% calculated below zero) and lowering mean tissue pO_2 (p= 1.12 kPa), so the resulting pO_2 histogram is only slightly shifted to the left (σ_p= 1.13 kPa; τ_p= +1.21 kPa).

4. DISCUSSION AND CONCLUSIONS

Incorporation of locally steep gradients in O_2 transport in and directly around the capillaries can have an important effect on tissue pO_2. In both tissue models, EP can easily

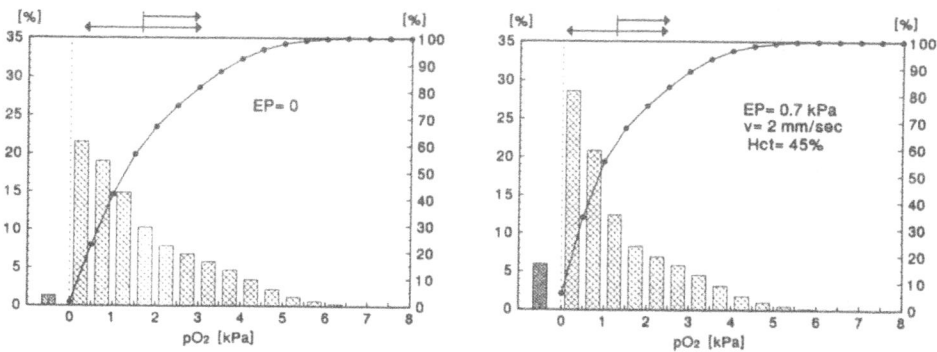

Figure 3. pO$_2$ histograms (bars; left axis) and cumulative histograms (dots connected by lines; right axis) calculated for dog gracilis situation without (left panel) and with (right panel) incorporating EP. Arrows indicate mean, standard deviation and skewness of the pO$_2$ distribution.

reach values comparable to the mean tissue pO$_2$. Since the EP depends on the same tissue parameters that govern the transport further into the tissue, this conclusion essentially does not depend on the choice of parameter values. The finding that pericapillary gradients can be important is not new and was sometimes referred to as the 'Capillary Barrier' (Honig et al., 1984; Turek et al., 1991). However, the quantification in terms of an extraction pressure is an easy way to understand its significance and provides a simple means of incorporation into large-scale tissue models.

The EP originates from steep pO$_2$ gradients around each RBC, so it will be minimal when RBCs are close together or when the blood is otherwise homogeneous. The latter will be the case in particular when the blood is replaced by or largely replaced by a homogeneous blood substitute; its homogenously carrying O$_2$ will be a property advantageous for oxygen supply to the tissue. However, such a situation will not come down to EP=0 as in the above examples. There still will be a gradient from the centre of the capillary to its periphery. Extending the Krogh model by a parabolic profile of p inside the capillary leads to:

$$p_{hk} = p_{ck} + \text{EP} \quad ; \quad \text{EP} = \frac{\dot{Q}}{8\wp}(R^2 - r_c^2) \tag{3}$$

where R is the Krogh cylinder radius ($\pi R^2 = A/N$). For the above situations, this yields extraction pressures of EP= 0.34 kPa for rat heart and EP= 0.49 kPa for the dog gracilis muscle, again for an 'average' capillary. However, also these EPs are expected to be decreased by the velocity profile inside the capillary, as based on general considerations of mass transport.

More importantly, current modelling predicts that O$_2$ delivery is no longer the sole determinant of tissue oxygenation. EP depends differently on blood flow and haematocrit, where the haematocrit effects are often found to dominate the blood flow effects. Mostly, an increase of the blood flow by twofold concomitantly halving the Hct results in unmistakably lower pO$_2$ values. This indicates that the particulate nature dictates that the blood flow should be increased more than twofold when the Hct is divided by two. In fact, EP represents a diffusional resistance from the erythrocytes to the tissue which is an increasing function of the erythrocyte spacing.

REFERENCES

Bos, C., Hoofd, L., Oostendorp, T. (1995). Mathematical model of erythrocytes as point-like sources, Math. Biosci. 125: 165–189.

Bos, C., Hoofd, L., Oostendorp, T. (1996). The effect of separate red blood cells on capillary tissue oxygenation calculated with a numerical model, IMA J. Math. Appl. Med. Biol., in press.

Groebe, K. (1990). A versatile model of steady state O_2 supply to tissue; application to skeletal muscle. Biophys. J. 57: 485–498.

Honig, C.R., Gayeski, T.E.J., Federspiel, W., Clark, A., Jr., Clark, P. (1984). Muscle O2 gradients from hemoglobin to cytochrome: new concepts, new complexities. In: Adv. Exper. Med. Biol. 169, Plenum Press, New York: 23–38.

Hoofd, L. (1995a). Calculation of oxygen pressures in tissue with anisotropic capillary orientation. I: Two-dimensional analytical solution for arbitrary capillary characteristics. Math. Biosci. 129: 1–23.

Hoofd, L. (1995b). Calculation of oxygen pressures in tissue with anisotropic capillary orientation. II: Coupling of two-dimensional planes. Math. Biosci. 129: 25–39.

Hoofd, L., Olders, J., Turek, Z. (1990). Oxygen pressures calculated in a tissue volume with parallel capillaries, In: Adv. Exper. Med. Biol. 277, Plenum Press, New York: 21–29.

Hoofd, L., Turek Z. (1994). Effect of realistic capillary spacing on pO_2 calculations in dog gracilis muscle, Med. Biol. Eng. Comput. 32: 436–439.

Turek, Z., Rakusan, K., Olders, J., Hoofd, L., Kreuzer, F. (1991). Computed myocardial Po2 histograms: effects of various geometrical and functional conditions. J. Appl. Physiol. 70: 1845–1853.

96

DIAMETER CONTROL IN THE ARTERIOLAR TREE BY CHANGES IN POST-CAPILLARY RESISTANCE

A Theoretical Study

Karlfried Groebe

Institut für Physiologie und Pathophysiologie
Johannes Gutenberg-Universität Mainz
Duesbergweg 6, D-55099 Mainz, Germany

1. INTRODUCTION

Presently, the mechanisms underlying local adjustment of organ perfusion to the metabolic needs of the tissue are not well understood. Even though a large number of vasoactive substances is known to be released upon variations in tissue metabolism,[24,28] the exact mechanisms of their action are unclear: Because diffusion is a slow and inefficient transport process for covering long distances, arterioles can only sense those changes in concentration and hence in release rate of these mediators with acceptable spatial and temporal resolution that originate from their immediate vicinity. On the other hand, perfusion needs to be regulated such that also the worst supplied tissue regions located remote from the arterioles ("lethal corners") receive sufficient supply of oxygen and nutrients. Thus, the question arises: how can the information about the metabolic status of these tissue regions be transmitted to the arteriolar tree in order to adjust blood flow to their needs? A number of suggestions has been advanced on how this gap in the signal chain might be bridged[9,13,26,29] all of which, however, fail to furnish satisfactory explanations: Tissue diffusion of vasoactive substances has been ruled out before. Upstream conduction of dilator signals along the capillary wall appears to be effective in controlling the feeding arteriole only if signals are generated within the proximal one third of the capillary[29] which excludes the "lethal corners". Diffusion of vasodilator substances from venules to their accompanying arterioles, despite being the best supported one of all presently proposed explanations,[9,13,26] is not adequate because venular-arteriolar distances are highly variable,[7] the smallest branches of the vascular tree are not arranged as such pairs at all, and in the larger vessel pairs, an arteriole often supplies a tissue microregion different from the one drained by its accompanying venule.[22]

Oxygen Transport to Tissue XIX, edited by Harrison and Delpy
Plenum Press, New York, 1997

In a recent publication on the mechanisms of perfusion control,[12] focus has been shifted to the venous system which has long been viewed as largely passive channels for returning the blood to the heart, and it has been proposed that perfusion rate is basically controlled by metabolite-induced changes in post-capillary resistance. There are indeed a number of reasons to believe in a more active role of the venular tree in regulating blood flow: *(i)* Venules are capable of reacting to changes in the concentrations of vasoactive substances with dilation/constriction=[2] *(ii)* Venules in skeletal muscle have been observed to actively dilate and (later) to constrict again following brief periods of electrical muscle stimulation.[21] *(iii)* During changes in perfusion pressure or perfusion rate, not only pre-capillary but also post-capillary resistance changes in a characteristic fashion.[24] *(iv)* Venules have been shown to exhibit active responses to transmural pressure and luminal flow.[4,18] *(v)* From their analysis of vascular resistance in arteriolar networks of cat sartorius muscle, Popel *et al.*[23] concluded "... that postarteriolar resistances may play a very important role in distribution of flow in the microvascular network". Moreover, also from a teleological point of view, the venous system would be an ideal site for perfusion regulation because the above mentioned problems of understanding signal transmission from the "lethal corners" to the flow controller would not apply. Nevertheless, in the adjustment of perfusion to the metabolic needs of the tissue, the largest conductance changes occur in the arterioles, and it is an essential requirement for each theory of perfusion control to explain how the latter may be brought about. Besides direct vasodilator action, there are two obvious ways of affecting arteriolar diameter and resistance which are myogenic response to changes in transmural pressure[15] as well as shear stress- and hence blood flow-induced vasodilation.[17] In[12] it has been proposed that any changes in pre-capillary conductance during metabolic perfusion regulation are mediated by arteriolar myogenic and flow-dependent diameter control and can be interpreted as responses of the arteriolar tree to changes in post-capillary flow conductance. In the present paper, the feasibility of this hypothesis is assessed by calculating the pressure courses expected to exist in arterioles in the presence of myogenic and shear stress-dependent mechanisms for ranges of arterial pressures and blood flow rates.

MODEL AND DATA

Calculations have been performed for the microvasculature in cat tenuissimus muscle representing one of the animal models for which the most complete morphometric and hemodynamic data sets are available. A symmetrically branching microvascular tree was considered which consists of four generations of arterial microvessels each exhibiting average dimensions and blood flow rates.

In the following, the structure of the model is briefly outlined; for a more detailed description see.[12] Pressure drop per unit length of vessel, dP/dl, at blood flow rate $\dot{Q}$ and vessel diameter d is governed by Hagen-Poiseuille's law

$$\frac{dP}{dl} = -\frac{128\dot{Q}\eta(d,v)}{\pi d^4} \tag{1}$$

in which apparent viscosity $\eta(d,v)$ may be expressed as a function of vessel diameter d and mean blood velocity v. Arterioles exhibit myogenic response to changes in wall tension, and wall shear stress-induced vasodilation, thus arteriolar diameter d_a may be represented as a function of transvascular pressure P and blood flow rate $\dot{Q}$: $d_a = d_a(d_{a0}, P, \dot{Q})$, where d_{a0} is the arteriolar reference diameter at zero pressure and flow for given levels of vascular tone and concentration of vasoactive substances. By substituting d and v in Eq. (1) by $d_a(d_{a0}, P, \dot{Q})$ and $4\dot{Q}/(\pi \cdot d_a(d_{a0}, P, \dot{Q})^2)$, respectively, and by rearranging one obtains

$$\frac{-\pi \, d_a(d_{a0}, P, \dot{Q})^4}{128\dot{Q}\eta\left(d_a(d_{a0}, P, \dot{Q}), \dfrac{4\dot{Q}}{\pi} d_a(d_{a0}, P, \dot{Q})^{-2}\right)} \, dP = dl$$

(2)

which is a first order ordinary differential equation with separated variables. Starting with a given pressure at the origin of the arteriolar tree, with given flow rates $\dot{Q}$, reference diameters d_{a0}, and functions $d_a(d_{a0}, P, \dot{Q})$ specific to the individual vascular generations, Eq. 2 may be integrated numerically to find the pressure course $P(l)$ along the vessels, from which the corresponding diameters $d(l) = d_a(d_{a0}, P(l), \dot{Q})$ may be computed.

Input data for four arteriolar generations have been selected from the literature to match the situation in cat tenuissimus muscle as closely as possible. Average vessel lengths have been estimated from.[7,8,22] Pressure in the organ artery and discharge hematocrit were $P_A = 105$ $mmHg$[11] and 0.4,[1] respectively. Blood viscosity in the arterioles as a function of vessel diameter, discharge hematocrit, and wall shear rate was taken from fits to experimental data in vitro[16] and in cat mesentery in vivo.[19] Resting flows in individual vessels were obtained from velocity-diameter data[7,11] according to,[20] and total number of vessels in each generation was derived from the parent/daughter flow ratios. Pressure-diameter relations for (active and passive) arterioles have been obtained by fitting suitable expression functions to experimental data by Davis.[3] As measurements had been made in vitro, data were converted to in vivo conditions by applying the ratios of in vivo to in vitro myogenic indices given in.[3] To account for flow-dependent diameter changes, arteriolar diameter at zero flow was multiplied by the non-dimensionalized diameter-flow relation given in[17] ($\ln(\dot{Q}) = 2.87 \cdot \ln(0.5d) - 5.47$). As in this formula d approaches 0 for $\dot{Q} \to 0$ — which is unrealistic — vessel diameter has been bounded from below by 90% the one at resting flow.[25] Reference arteriolar diameters d_{a0} at zero pressure and flow were chosen such that under resting conditions actual diameters, pressures, and velocities in arteriolar generations A1 to A4 approximated the ones given in.[30] In tenuissimus muscle at resting perfusion, $\dot{Q}_{rest}$, approximately one third of the capillaries are perfused.[10] At higher blood flows, perfused capillary fractions of 75% at $\dot{Q} = 2 \cdot \dot{Q}_{rest}$ or of 100% at $\dot{Q} \geq 4 \cdot \dot{Q}_{rest}$ were employed (cf.[14]).

RESULTS

Figs. 1 and 2 show calculated transmural pressures (top) and vessel diameters (bottom) as functions of position along the microvasculature for a range of arterial pressures at resting perfusion, $\dot{Q}_{rest}$, (Fig. 1); or for a range of blood flow rates (given as multiples of $\dot{Q}_{rest}$) at arterial pressure $P_A = 105$ $mmHg$ (Fig. 2). Fig. 1 illustrates that despite large variations in P_A, pressures at the capillary origin (upper panel, right edge of graph) are fairly constant, demonstrating the ability of the arteriolar system to autoregulate capillary pressure. E.g., for a 95 $mmHg$ increase in P_A (compared to normal $P_A = 105$ $mmHg$; dash-dot-dotted and solid traces), pressure at the capillary origin rises by no more than 6 $mmHg$. The largest change (-11 $mmHg$) occurs when P_A drops to 56 $mmHg$ (wide dots) indicating

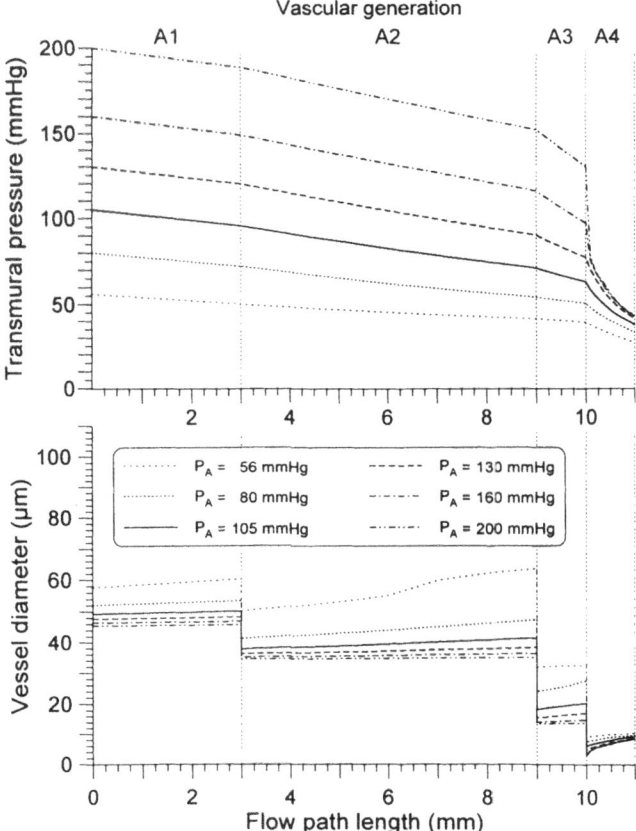

Figure 1. Calculated transmural pressures (top) and vessel diameters (bottom) as functions of position along the microvasculature for arterial pressures P_A ranging from 56 to 200 *mmHg* (see in-figure legends) at resting perfusion $\dot{Q}=\dot{Q}_{rest}$. Vertical dotted lines delineate vascular generations A1 to A4.

that the system approaches the limit of its autoregulatory range. Due to their dependence on wall tension, arteriolar diameters become smaller with increasing arterial pressure (bottom panel). This response is most pronounced in A2 and A3 vessels. Also, note the increases in diameters within branching orders at same P_A, corresponding to falling transmural pressures along the vessels.

As the bottom panel in Fig. 2 (variable $\dot{Q}$ at constant $P_A=105$ *mmHg*) shows, there is vasodilation with increasing $\dot{Q}$. For $\dot{Q}\geq\dot{Q}_{rest}$, this dilation is large enough to even *raise* arteriolar pressures despite larger flows (*cf.* Eq. 1). *E.g.*, pressure at the capillary origin rises from 38 *mmHg* at $\dot{Q}=\dot{Q}_{rest}$ to 54 *mmHg* at $\dot{Q}=8\cdot\dot{Q}_{rest}$. This effect is brought about by flow-induced vasodilation (which is partly counteracted by myogenic response). As can be seen from Eq. 1, reduction in pressure drops at augmented flows may occur if the increase in vessel diameter exceeds (increase in flow).[1/4] In working muscles, a rise in capillary perfusion pressure with blood flow rate is essential for achieving high performance levels at which capillary resistance may become limiting to perfusion: Owing to the small compliance of capillaries, their diameters and hence flow resistances hardly change, and recruitment of capillaries that are unperfused at rest could account for no more than a roughly three-fold increase in flow if pressure drop across the capillary bed were held constant. Hence, compared to resting conditions, a 2.5-fold increase in perfusion pressure is required to drive an eight times larger blood flow across the capillary bed. As post-capillary pressure drops are rather small to begin with, pre-capillary pressure drops must decrease to allow for the perfusion rates observed at maximum performance. Interestingly, vascular

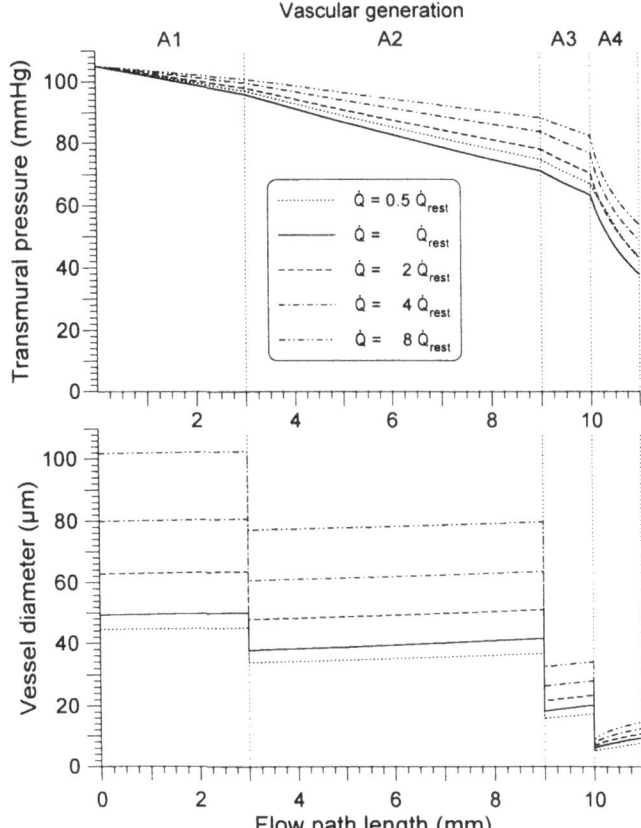

Figure 2. Calculated transmural pressures (top) and vessel diameters (bottom) as functions of position along the microvasculature for arterial pressure P_A=105 *mmHg* and blood flow rates $\dot{Q}$ ranging from $0.5 \cdot \dot{Q}_{rest}$ to $8 \cdot \dot{Q}_{rest}$ (which about equals blood flow capacity of cat tenuissimus muscle[6]). Vertical dotted lines delineate vascular generations A1 to A4.

pressures at $\dot{Q}=0.5\cdot\dot{Q}_{rest}$ (dotted) are also higher than at $\dot{Q}_{rest}$. This is because at resting flows the potential for shear-induced arteriolar *constriction* is very limited.[25]

With regard to the above hypothesis suggesting that changes in pre-capillary flow conductance during metabolic perfusion regulation can be understood as responses of the arteriolar tree to changes in post-capillary conductance, the results allow for the following conclusions: Flow-dependent diameter control in arterioles in combination with myogenic response brings about overproportionate decreases in arteriolar conductance as compared to the corresponding increases in blood flow. Consequently, pressure drops in arterioles are not merely kept constant at increasing perfusion rates, but rather arteriolar pressures become even higher — which is essential in muscles at highest performance for compensating for the enhanced pressure drops required to drive the blood stream across the capillary bed. These findings suggest that shear stress-dependent vasodilation in arterioles may be potent enough to account for the fall in pre-capillary resistance which is observed during metabolic perfusion regulation. As a consequence, perfusion rate may well be controlled by according changes in post-capillary resistance, and the variations in pre-capillary resistance may be caused by arteriolar responses to the corresponding hemodynamic changes in the venules.

DISCUSSION

Based on measured relations between microvascular diameters and luminal pressures and flows that are presently available, the above model calculations demonstrate that the changes in conductance of the arterial tree experimentally observed during muscular work at increasing performance, may be brought about by arteriolar responses to the corresponding (and much smaller) active reductions in post-capillary resistance. No direct vasodilator action on pre-capillary vessels is required; the entire vasculature can, in principle, be controlled by effects of vasoactive metabolites on venular diameters, which may resolve the problems of signal transmission between "lethal corners" and feeding arterioles.

The latter statement is not meant to imply that transmural pressure and luminal flow are the only factors affecting arteriolar diameters. Rather, the present hypothesis is intended to propose a plausible and realistic concept of a stable feedback loop for metabolic perfusion regulation, the individual components of which may, however, be modified in their actions by a variety of factors. In this way, the operation of the controller could be adapted to extraordinary conditions, its regulatory range could be shifted, *etc.*, while the feedback loop as such should remain intact and should only become modulated in its function. As an example of modulating factors, there is central control of the microvasculature *via* the autonomic nervous system and humoral factors as epinephrine, angiotensin, *etc.* that would be expected to change the "set points" of the local mechanisms (which may be represented by unstressed diameters at zero pressure and flow, d_{a0}, or by the functions describing P-d- or $\dot{Q}$-d-relations). Furthermore, there are conducted vasodilation in arterioles[27] (which has most likely contributed to the experimental $\dot{Q}$-d-relations[17] underlying the present simulations and therefore should, at least in part, be implicitly considered in the model), and even direct effects of vasoactive metabolites on arterioles may be present (which again may serve to establish new set points for the proposed mechanism to operate at greatly altered supply/demand ratios, and which should play a role at least under most critical supply conditions in which metabolite concentrations in tissue are expected to be high).

There is yet another problem to which the presently proposed conception of local perfusion regulation may offer a solution: Beyond adjusting blood flow to the metabolic needs of the tissue, the microvasculature has to fulfil a second task which is control of hydrostatic pressure in the capillary bed. Hydrostatic pressure is an important determinant of fluid filtration in exchange vessels and has to be kept low enough to avoid interstitial edema on the one hand, and high enough to allow for perfusion pressures that suffice to maintain the required blood flow on the other. Obviously, there is a potential conflict between pressure and perfusion control that necessitates coordination between the two, the physical basis and modalities of which are unknown. In the present conception, control of perfusion and hydrostatic pressure are largely uncoupled in such a way that perfusion is regulated by the venules, and the arteriolar tree may be viewed as a pressure controller that keeps capillary pressure (almost) constant at varying systemic pressures, and that adjusts it as needed for providing the blood flow preset by the venular flow regulator. The presence of two distinct mechanisms for perfusion and hydrostatic pressure control has also been concluded from experimental observations, *e.g.*, by Davis[5] who, furthermore, suggested that "... changes in R_v may be important".

In conclusion, the present paper shows that blood flow autoregulation by metabolite-mediated active changes in venular diameters in combination with control of capillary pressure by myogenic response and shear stress-induced dilation in arterioles may possess the potential to play a dominant role in the adjustment of perfusion to tissue metabolic needs. By this mechanism, a number of problems in current understanding of local microvascular control could be resolved: *(i)* The proposed conception is based on a well defined and unambiguous closed feedback loop for metabolic perfusion regulation which enables *(ii)* strictly local perfusion control and *(iii)* tight temporal coupling between demand and microvascular response. *(iv)* Pressure and perfusion controls are largely uncoupled, allowing for easy integration of their potentially conflicting requirements. *(v)* The flow of blood by itself is suggested to transport vasodilator information from downstream of the "lethal corners" to the arterioles which resolves the problem of long range signal transmission.

REFERENCES

1. P.L. Altman, D.S. Dittmer, "Biology Data Book", Federation of American Societies for Experimental Biology, Bethesda, 1972
2. D.N. Damon, B.R. Duling, Venular reactivity in the hamster cheek pouch and cremaster muscle, *Microvasc.Res.* 31:379–383 (1986)
3. M.J. Davis, Myogenic response gradient in an arteriolar network, *Am.J.Physiol.* 264:H2168–H2179 (1993)
4. M.J. Davis, Spontaneous contractions of isolated bat wing venules are inhibited by luminal flow, *Am.J.Physiol.* 264:H1174–H1186 (1993)
5. M.J. Davis, Control of bat wing capillary pressure and blood flow during reduced perfusion pressure, *Am.J.Physiol.* 255:H1114–H1129 (1988)
6. E. Eriksson, B. Lisander, Change in precapillary resistance in skeletal muscle vessels studied by intravital microscopy, *Acta Physiol.Scand.* 86:295–305 (1972)
7. E. Eriksson, M. Myrhage, Microvascular dimensions and blood flow in skeletal muscle, *Acta Physiol.Scand.* 86:211–222 (1972)
8. E. Eriksson, B. Lisander, Low flow states in the microvessels of skeletal muscle in cat, *Acta Physiol.Scand.* 86:202–210 (1972)
9. J.C. Falcone, H.G. Bohlen, EDRF from rat intestine and skeletal muscle venules causes dilation of arterioles, *Am.J.Physiol.* 258:H1515–H1523 (1990)
10. K. Fronek, B.W. Zweifach, Microvascular blood flow in cat tenuissimus muscle, *Microvasc.Res.* 14:181–189 (1977)
11. K. Fronek, B.W. Zweifach, Microvascular pressure distribution in skeletal muscle and the effect of vasodilation, *Am.J.Physiol.* 228:791–796 (1975)
12. K. Groebe, Pre-capillary servo control of blood pressure and post-capillary adjustment of flow to tissue metabolic status: A new paradigm for local perfusion regulation, *Circulation*, in press
13. R.L. Hester, Uptake of metabolites by postcapillary venules: Mechanism for the control of arteriolar diameter, *Microvasc.Res.* 46:254–261 (1993)
14. C.R. Honig, C.L. Odoroff, J.L. Frierson, Active and passive capillary control in red muscle at rest and in exercise, *Am.J.Physiol.* 243:H196–H206 (1982)
15. P.C. Johnson, The myogenic response, in: "Handbook of Physiology, Sect. 2: The Cardiovascular System, Vol. IV: Microcirculation", E.M. Renkin, C.C. Michel, eds., pp. 409–442, American Physiological Society, Bethesda, 1984
16. M.F. Kiani, A.G. Hudetz, A semi-empirical model of apparent blood viscosity as a function of vessel diameter and discharge hematocrit, *Biorheology* 28:65–73 (1991)
17. A. Koller, G. Kaley, Endothelial regulation of wall shear stress and blood flow in skeletal muscle microcirculation, *Am.J.Physiol.* 260:H862–H868 (1991)
18. L. Kuo, F. Arko, W.M. Chilian, M.J. Davis, Coronary venular responses to flow and pressure, *Circ.Res.* 72:607–615 (1993)
19. H.H. Lipowsky, Network hemodynamics and the shear rate dependency of blood viscosity, in: "Microvascular Networks: Experimental and Theoretical Studies", A.S. Popel, P.C. Johnson, eds., pp. 182–196, Karger, Basel, 1986
20. H.H. Lipowsky, B.J. Zweifach, Application of the "two-slit" photometric technique to the measurement of microvascular volumetric flow rates, *Microvasc.Res.* 15:93–101 (1978)
21. J.M. Marshall, H.C. Tandon, Direct observation of muscle arterioles and venules following contraction of skeletal muscle fibers in the rat, *J.Physiol.(London)* 350:447–459 (1984)
22. R. Myrhage, E. Eriksson, Vascular arrangements in hindlimb muscles of the cat, *J.Anat.* 131:1–17 (1980)
23. A.S. Popel, A. Liu, B. Dawant, A. Koller, P.C. Johnson, Distribution of vascular resistance in terminal arteriolar networks of cat sartorius muscle, *Am.J.Physiol.* 254:H1149–H1156 (1988)

24. E.M. Renkin, Control of microcirculation and blood-tissue exchange, **in:** "Handbook of Physiology, Sect. 2: The Cardiovascular System, Vol. IV: Microcirculation", E.M. Renkin, C.C. Michel, eds., pp. 627–687, American Physiological Society, Bethesda, 1984

25. Y. Saito, A. Eraslan, R.L. Hester, Role of EDRFs in the control of arteriolar diameter during increased metabolism of striated muscle, *Am.J.Physiol.* 267:H195–H200 (1994)

26. Y. Saito, A. Eraslan, R.L. Hester, Importance of venular flow in control of arteriolar diameter in hamster cremaster muscle, *Am.J.Physiol.* 265:H1294–H1300 (1993)

27. S.S. Segal, B.R. Duling, Flow control among microvessels coordinated by intercellular conduction, *Science* 234:868–870 (1986)

28. J.T. Shepherd, Circulation to skeletal muscle, **in:** "Handbook of Physiology, Sect. 2: The Cardiovascular System, Vol. III: Peripheral Circulation and Organ Blood Flow", J.T. Shepherd, F.M. Abboud, eds., pp. 319–370, American Physiological Society, Bethesda, 1983

29. H.S. Song, K. Tyml, Evidence for sensing and integration of biological signals by the capillary network, *Am.J.Physiol.* 265:H1235–H1242 (1993)

30. B.J. Zweifach, H.H. Lipowsky, Pressure-flow relations in blood and lymph microcirculation, **in:** "Handbook of Physiology, Sect. 2: The Cardiovascular System, Vol. IV: Microcirculation", E.M. Renkin, C.C. Michel, eds., pp. 251–307, American Physiological Society, Bethesda, 1984

97

ON THE CONTRIBUTION OF RESPIRATORY GASES TO CEREBRAL AUTOREGULATION

D. A. Wilson

Department of Anesthesiology
The Johns Hopkins University
Baltimore, Maryland 21287-4961

1. INTRODUCTION

Autoregulation has been defined as the occurrence of vasodilation when cerebral perfusion pressure (CPP) decreases and the occurrence of vasoconstriction when CPP increases.[2] Two main mechanisms, myogenic and metabolic, have been proposed to explain autoregulation. In the metabolic theory several metabolites have been proposed and each may contribute to the vascular readjustments, but no single metabolite appears to account for all of the changes that result from either a raising or a lowering of CPP. The principle event initiating a change in vasoactive metabolite level is in each case a disturbance in tissue oxygen (O_2) concentration.

The purpose of this work was to develop a general model of the metabolic theory and to utilize this model to investigate the roles of the respiratory gases, O_2 and carbon dioxide (CO_2), in cerebral autoregulation. The model evaluated here is principally based upon two hypotheses: that linear relationships exist between the cerebral venous blood gases and cerebrovascular resistance, and that homeostasis of tissue concentration of O_2 is the object of autoregulation.

2. METHODS

2.1. Model Derivation

The model is illustrated diagrammatically in Figure 1. All variables and dimensions are listed and defined in Table 1. For simplicity, variables regulated by outside systems are treated as independent variables (CPP, Pa_{O2}, Pa_{CO2}, CMR_{O2}, CMR_{CO2}, k_{CO2}, k_{O2}, and α_{CO2}). The model consists of two closed-loop feedback systems that function interdependently through the same controlling system. The principal control relationship is based upon a linear relationship between CVR and cerebral venous O_2 content. Hence, the governing equation may be defined as

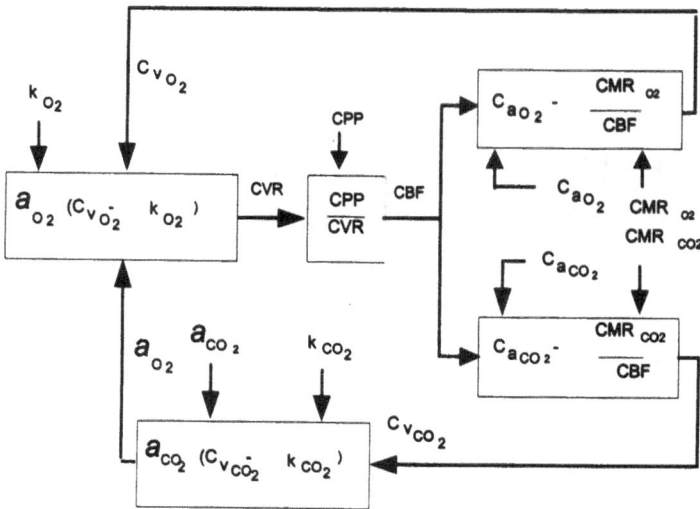

Figure 1. Cerebral chemostat dependency diagram.

$$CVR = \alpha_{O2}(Cv_{O2} - k_{O2})$$

$$(1)$$

The controlled systems are each mathematically defined according to the law of mass conservation,[3] where, for O_2,

$$Cv_{O2} = Ca_{O2} - \frac{CMR_{O2}}{CBF}$$

$$(2)$$

and for carbon dioxide,

Table 1. Definition of dependent and independent variables and their dimensions

Variable	Description	Dimensions
Independent		
Ca_{O2}	Arterial oxygen content	ml O_2/ml blood
Ca_{CO2}	Arterial carbon dioxide content	ml CO_2/ml blood
CPP	Cerebral perfusion pressure	mmHg
CMR_{O2}	Cerebral oxygen uptake	ml O_2/min/100g
CMR_{CO2}	Cerebral carbon dioxide production rate	ml CO_2/min/100g
α_{CO2}	Cerebrovascular reactivity to carbon dioxide	PRU/ml CO_2/ml blood
k_{O2}	O_2 setpoint	ml O_2/ml blood
k_{CO2}	CO_2 setpoint	ml CO_2/ml blood
Dependent		
CBF	Cerebral blood flow	ml/min/100g
CVR	Cerebrovascular resistance	PRU
α_{O2}	Cerebrovascular reactivity to oxygen	PRU/ml O_2/ml blood
Cv_{O2}	Cerebral venous oxygen content	ml O_2/ml blood
Cv_{CO2}	Cerebral venous carbon dioxide content	ml CO_2/ml blood

$$Cv_{CO2} = Ca_{CO2} - \frac{CMR_{CO2}}{CBF} \tag{3}$$

The model assumes that at steady state the tissue concentrations of both O_2 and CO_2 are in equilibrium with their respective cerebral venous blood concentrations. Thus, tissue concentration and cerebral venous concentration will be used interchangeably for modeling purposes. In equation 1, k_{O2} is a controller setpoint. α_{O2} expresses the sensitivity of the vessels to O_2. While equations 1 and 2 provide for feedback influence of O_2 via the incorporation of venous O_2 concentration in each equation, neither relationship incorporates a term to account for the known effects of CO_2 on cerebral vessels. [2,8] Theoretically, CO_2 could act on vascular resistance in a summated manner with O_2. Previous attempts to develop this type of model indicate such a model lacks responsiveness to individual blood gases due to counteracting influences of the two gases on vascular resistance. Massik et al.[4] found that hypercarbia causes the Ca_{O2}/CBF relationship to shift upward in a parallel manner. Such an effect is consistent with the hypothesis that increasing CO_2 concentration depresses apparent O_2 reactivity. Thus,

$$\alpha_{O2} = \alpha_{CO2}(Cv_{CO2} - k_{CO2}) \tag{4}$$

where, k_{CO2} is the setpoint of the CO_2 regulating mechanism and α_{CO2} describes vascular CO_2 reactivity. The regulator mechanism of the CO_2 controlling system operates, in effect, by determining the vascular response to O_2.

CBF is a convective process, dependent upon the product of CPP and vascular conductance (1/CVR). To relate equations 2, 3, and 1

$$CBF = \frac{CPP}{CVR} \tag{5}$$

Equations 1–5 can be solved graphically by rearranging CBF to terms of CVR and plotting the result as Figure 2 shows. Line A describes equation 2 and line B describes equation 1. The Cv_{O2} that satisfies both equations is the solution and lies at the point of intersection. Line B' illustrates the effect of CO_2 on apparent O_2 reactivity. A quantitative solution can be derived mathematically by substituting equation 1 into equation 5 which yields,

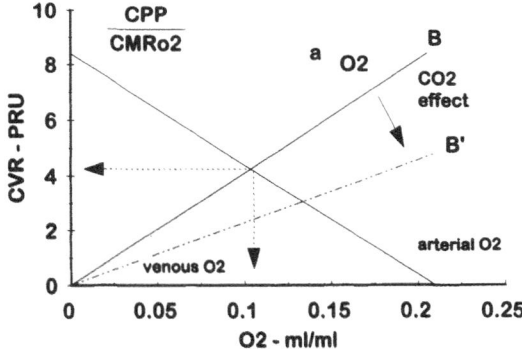

Figure 2. Graphical solution.

$$CBF = \frac{CPP}{\alpha_{O2}(Cv_{O2} - k_{O2})} \tag{6}$$

Substituting equations 2, 3 and 4 into 6, yields

$$CBF = \frac{CPP}{\alpha_{CO2}[(Ca_{CO2} - \frac{CMR_{CO2}}{CBF}) - k_{CO2}][(Ca_{O2} - \frac{CMR_{O2}}{CBF}) - k_{O2}]} \tag{7}$$

a unique equation defining CBF in terms of the independent variables CPP, Ca_{CO2}, Ca_{O2}, CMR_{CO2}, and CMR_{O2}. Equation 7 can be rearranged to

$$CBF^2[(Ca_{O2} - k_{O2})(Ca_{CO2} - k_{CO2})] - CBF[\frac{CPP}{\alpha_{CO2}} + CMR_{O2}(Ca_{CO2} - k_{CO2})$$

$$+ CMR_{CO2}(Ca_{O2} - k_{O2})] + CMR_{CO2}CMR_{O2} = 0 \tag{8}$$

an equation that follows the general form

$$aCBF^2 - bCBF + c = 0 \tag{9}$$

where

$$a = (Ca_{O2} - k_{O2})(Ca_{CO2} - k_{CO2}) \tag{9a}$$

$$b = \frac{CPP}{\alpha_{CO2}} + CMR_{O2}(Ca_{CO2} - k_{CO2}) + CMR_{CO2}(Ca_{O2} - k_{O2}) \tag{9b}$$

$$c = \frac{CMR_{CO2}}{CMR_{O2}} \tag{9c}$$

Thus, a numerical solution for CBF may be obtained using the quadratic form

$$CBF = -\frac{b}{2a} \pm \sqrt{\frac{b - 4ac}{2a}} \tag{10}$$

Since CPP, CMR_{O2}, CMR_{CO2}, Ca_{O2}, and Ca_{CO2} are readily measurable, the only remaining unknowns are α_{CO2}, k_{O2}, and k_{CO2}.

2.2. Experimental

To determine α_{CO_2}, k_{O_2}, and k_{CO_2}, eight dogs were anesthetized with sodium pentobarbital anesthesia and catheterized to measure CBF by the microsphere method. Catheters were placed in the femoral and omocervical arteries for blood pressure measurement and blood gas sample withdrawal. A catheter was placed in the superior sagittal sinus for sampling cerebral venous blood. After a suitable stabilization period, each animal was subjected to the following inspired gas mixtures for a 10 minute period: 10% CO_2, room air, 12% O_2, 10% O_2, 8% O_2, and 6% O_2. Cerebral blood flow and blood gas measurements were then made. At the end of the experiment the animals were killed by intravenous KCl injection, the brains autopsied, and blood flow computed. Cerebrovascular resistance was calculated as defined in equation 5. CO_2 content (ml/ml) was estimated as $P_{CO_2}/220$.

To determine the role of the respiratory gases in pressure/flow processes, data from thirteen dogs for which CBF and blood gas data were available, were fit retrospectively. In eight animals, arterial blood pressure was reduced from normotensive levels to approximately 80, 60, 40, and 20 mmHg during normoxic conditions. CBF and arterial and cerebral venous blood gases were measured at each pressure level. The remaining five dogs were ventilated on 12% inspired O_2 and a similar hemorrhage protocol repeated. Theoretical pressure/flow relationships were constructed using equation 10 and the values of α_{CO_2}, k_{CO_2}, and k_{O_2} derived using the CO_2 and O_2 protocol described previously.

3. RESULTS

Hemodynamic, blood gas, and metabolic measurements are shown in Table 2. From these values, α_{CO_2}, k_{CO_2}, and k_{O_2} were derived by determining the best least squares fit of CVR and the cerebral venous blood gases to the model: $CVR = \alpha_{CO_2}(Cv_{CO_2} - k_{CO_2})(Cv_{O_2} - k_{O_2})$. The average value of α_{CO_2} was -376 ± 45 PRU/ml/ml. k_{CO_2} averaged 0.429 ± 0.012 ml/ml and k_{O_2} averaged 0.0049 ± 0.0035 ml/ml. The overall test for goodness-of-fit is illustrated by plotting the predicted CBF for each measured condition against the true measured CBF (Figure 3). The correlation coefficient was 0.987. Generally, blood flow predictions during normocapnic/normoxic and hypercarbic conditions were more accurate (-2% and -1%, respectively) than were those made during hypoxic conditions (13% overestimate).

Table 2. Blood gas and hemodynamic changes induced by hypoxic and hypercapnic exposures

	Hypercarbia 10% CO_2	Room air	12% O2	Hypoxia 10% O_2	8% O_2	6% O_2
CPP (mmHg)	128 ±6	130 ±6	137 ±7	130 ±9	111 ±15	69 ±13
Pa_{CO_2} (mmHg)	85 ±4	32 ±1	31 ±1	32 ±1	33 ±2	30 ±1
Pcv_{CO_2} (mmHg)	90 ±3	50 ±1	45 ±1	41 ±1	42 ±2	38 ±1
Ca_{O_2} (vol%)	20.2 ±0.4	20.3 ±0.4	15.0 ± 1.2	10.3 ± 1.1	8.1 ±1.9	4.1 ±0.2
Cv_{O_2} (vol%)	17.4 ±0.7	8.3 ±0.6	5.9 ±0.6	3.5 ± 0.3	3.6 ±1.4	0.8 ±0.2
CBF (ml/min/100g)	160 ±12	23 ±2	34 ±5	58 ±13	66 ± 15	45 ±11
CMR_{O_2} (ml/min/100g)	4.31 ±0.61	2.70 ±0.26	2.77 ±0.21	3.06 ±0.26	2.32 ±0.12	1.40 ±0.16

Mean ± SE; n=8

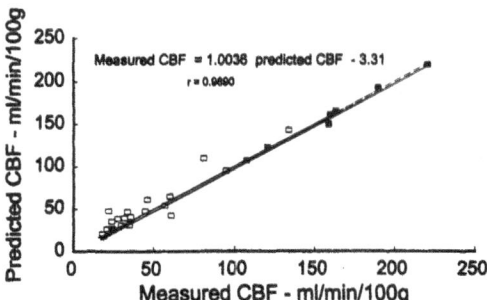

Figure 3. Test for goodness-of-fit of predicted CBF to measured CCBF during normoxic (+), hypoxic (□), and hypercapnic (■) conditions.

Figure 4 shows the results of predicting the autoregulation relationship using parameters derived by forcing arterial CO_2 and O_2. The two panels compare the predicted "autoregulatory" curves against the pressure/flow data from dogs breathing room air and 12% O_2, respectively. In both instances, when the predictions were based solely upon the level of metabolism present during normotension normoxic-normocapnic CBF was predicted accurately, but the overall fit was poor. However, when the level of metabolism at each pressure level was used to construct the curves, the predicted curves lie within the error of the measured curves and simulated the non-linear nature of the relationships.

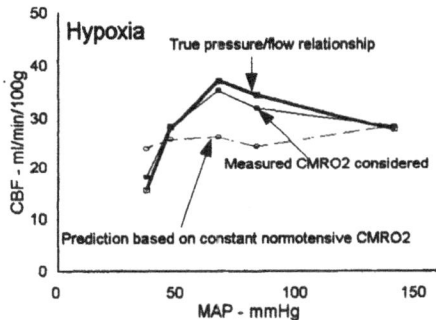

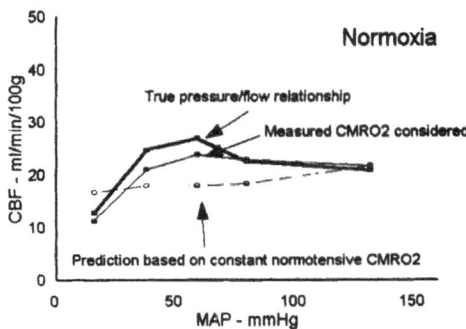

Figure 4. Comparison of measured and predicted autoregulation relationships in normoxic and hypoxic conditions.

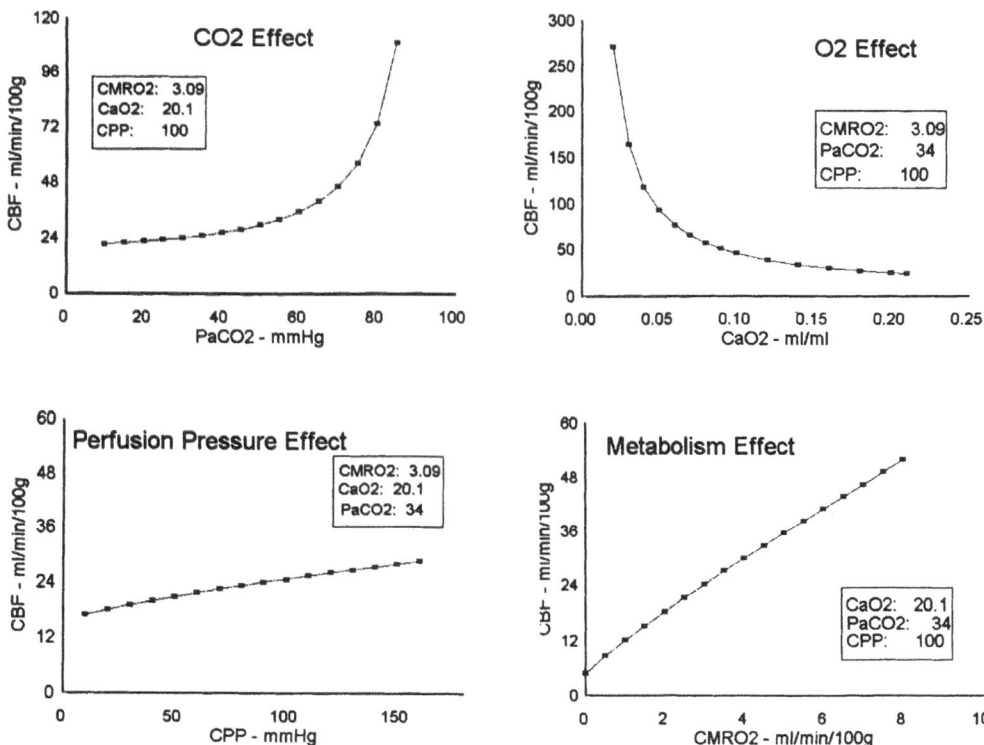

Figure 5. Predicted steady-state relationships. All predictions are based on values of α_{CO_2} = -376 PRU/ml/ml, k_{CO_2} = 0.429 ml/ml, and k_{O_2} = 0.0049 ml/ml.

4. DISCUSSION

The need for purposeful regulation of cerebral O_2 delivery is obvious. The cellular and molecular mechanisms responsible for this regulation have been the subjects of intense investigation. Less attention has focused upon the use of models to study the processes of feedback regulation and their role in the maintenance of tissue homeostasis.[6,7] The goal of this study was to investigate the role of feedback regulation of respiratory gases in cerebral pressure/flow autoregulation. To achieve this goal a mathematical model was developed and the parameters measured experimentally. The quantified parameters were then used to predict the manner in which CBF would be altered when perfusion pressure was reduced. The prediction was compared to experimentally determined pressure/flow relationships in two differing respiratory states. The findings strongly suggest that in the normal circulation, O_2 and CO_2 account for a substantial proportion of the observed changes in cerebrovascular resistance that maintain CBF relatively constant. More importantly, CPP/CBF autoregulation appears to be a special case of a more general autoregulator process that strives to acheive constancy in the tissue concentration of respiratory gases in a variety of abnormal respiratory and hemodynamic states.

Using parameters derived in response to changes in inspired O_2 and carbon dioxide, the model was able to predict the major characteristics of the cerebral pressure/flow relationship, the CO_2/CBF relationship, the O_2/CBF relationship, and the response to altered O_2 uptake (Figure 5). Although data are not shown for every combination, there exists families of relationships around each CO_2, O_2, $CMRO_2$, or CPP relationship that are determined by the status of the independent variables. For example, hypoxia was shown to shift the CPP/CBF relationship upward as has been experimentally demonstated previously.[5] Moreover, the model provides some insight into understanding why some animal models yield CPP/CBF relationships that are relatively without slope, while in other models the CPP/CBF relationships show a subtle correlation with CPP, or may, as in this study, be biphasic. Uncontrollable factors, such as hormonally induced changes in $CMRO_2$, or subtle changes in arterial blood gases, can significantly affect the characteristics of the autoregulation curve.[1] Additionally, any compensatory changes in the setpoint variables or in respiratory gas reactivity variables would also contribute unexplained variability to experimentally measured relationships. The model offers a means of quantifying this variance.

The model described was purposefully kept simple to study the most essential elements required to achieve respiratory gas control. One could, however, elaborate the feedback limbs to incorporate other variables. Such models would be useful in integrating multifactorial theories of regulation. The model also provides a means for quantifying system gain and for studying and interpreting potential mechanisms of compensation and adjustment.

The main characteristics of the normal canine cerebral circulation are qualitatively explained by a linear interaction between respiratory gases and cerebrovascular resistance, balanced against factors effecting gas delivery, clearance, and uptake/production. With respect to the cerebral pressure/flow relationship, these observations suggest that "autoregulation" is but a special case of a more general self-regulating process which attempts to maintain the concentrations of O_2 and carbon dioxide constant in a variety of conditions. As such, hypercapnic, hypoxic, and metabolism-induced changes in CBF can also be considered cerebral autoregulatory phenomena.

ACKNOWLEDGMENTS

This work was supported by National Institute of Neurological Disorders and Stroke Grant NS-20020.

REFERENCES

1. Chen RY, Fan FC, Schuessler GB, Simchon S, Kim S, Chien S. Regional cerebral blood flow and oxygen consumption of the canine brain during hemorrhagic hypotension. Stroke. 15:343–350, 1984.
2. Heistad DD, Kontos HA. Cerebral Circulation. In: Handbook of Physiology. Sect. 2, vol. 3, pt. 1, ch. 5. Eds, Shepherd JT and Abboud FM. American Physiological Society, Bethesda, pp 137–182, 1983.
3. Kety SS. Theory and applications of the exchange of inert gas at the lungs and tissues. Pharmacol. Rev. 3:1–41, 1951.
4. Massik J, Jones MD, Jr, Miyabe M, Tang YI, Hudak MI, Koehler RC, and Traystman RJ. Hypercapnia and response of cerebral blood flow to hypoxia in newborn lambs. J Appl. Physiol. 66:1065–1070, 1989.
5. Traystman RJ, Wilson DA Koehler RC, Borel C, McPherson RW, Jones MD Jr., Rogers MC. Effect of hypoxia on cerebral autoregulation. J Cerebral Blood Flow and Metabolism 5:S485-S486, 1985.
6. Ursino M, Giammarco PD, Belardinelli E. A mathematical model of cerebral blood flow chemical regulation - Part I: Diffusion processes. IEEE Tran Biomed Eng 36:183–191, 1989.
7. Ursino M, Giammarco PD, Belardinelli E. A mathematical model of cerebral blood flow chemical regulation - Part II: Reactivity of cerebral vascular bed. IEEE Tran Biomed Eng 36:192–202, 1989.
8. Wilson DA, Traystman RJ, and Rapela CE. Transient analysis of the canine cerebrovascular response to carbon dioxide. Circ. Res. 56:596–605, 1985.

98

TEMPORAL FLUCTUATIONS IN REGIONAL RED BLOOD CELL FLUX IN THE RAT BRAIN CORTEX IS A FRACTAL PROCESS

A. Eke,[1] P. Hermán,[1] J. B. Bassingthwaighte,[2] G. M. Raymond,[2] I. Balla,[1] and C. Ikrényi[1]

[1]Experimental Research Department II
Institute of Physiology
Semmelweis University of Medicine
P.O. Box 446, 1448 Budapest, Hungary
[2]National Simulation Resource
Center for Bioengineering
University of Washington
Seattle, Washington 98195

1. INTRODUCTION

Temporal fluctuations ever since it was observed and recognized as an elementary feature of blood flow in the microcirculation by Krogh in 1929 (Krogh, 1929) became the essence of concepts termed "vasomotion" and "flow motion". Such fluctuations in flow are common in various microcirculatory beds including that of the brain and can be continuously monitored by laser-Doppler flowmetry (Stern, 1975; Nilsson et al., 1980; Rosenblum et al., 1987; Fasano et al., 1988; Hudetz et al., 1992; Morita-Tsuzuki et al., 1992). When challenged, they often show slow (6–12 cycles/minute) oscillatory pattern, which allow for a relatively simple characterization if the higher frequencies are eliminated from the signal (Hudetz et al., 1992; Morita-Tsuzuki et al., 1992). Temporal variation in microcirculatory flow is however multifactorial and the interactions among these factors can manifest in a complex structuring of the time series recorded by high-resolution laser-Doppler flowmetry (LDF). With no apparent dominating frequency present (Fig. 3), they are much too complex to be analyzed in specific terms by conventional descriptive statistics, amplitude and frequency measures. Thus we have used fractal methods genuinely holistic in nature to gain statistical insight into random signals of this kind and to determine if under unchallenged control conditions they represent disorganized behavior or show long-range correlations (West and Goldberger, 1987; Bassingthwaighte, 1988; West and Shlesinger, 1990; Weibel, 1991; Bassingthwaighte et al., 1994).

Oxygen Transport to Tissue XIX, edited by Harrison and Delpy
Plenum Press, New York, 1997

2. METHODS

2.1. Fractals and Time Series

Fractals describe objects and events which are characterized by consistencies in the degree of correlation between neighboring structures or events (Bassingthwaighte, 1988; Bassingthwaighte et al., 1995). The correlation may be positive or negative and is given by

$$r_1 = 2^{2H-1} - 1 \tag{1}$$

where H is the Hurst coefficient (Hurst, 1951). H is a measure of "roughness" of the structure or the signal, H near 1.0 indicating a high degree of smoothness and strong positive correlation, H near 0 indicating a high degree of roughness and strong negative correlation. The Hurst coefficient is independent of the Euclidean dimension E, in which the object is embedded and relates to the fractal dimension, D as

$$D = E + H - 1 \tag{2}$$

Geometrical fractals are characterized by self-similarity of their structure viewed at varying scales of observation (Mandelbrot, 1983). Because the units in which a variable's change is measured and the units of time in which these changes are recorded are different, fractal time series are not self-similar but self-affine.

When a signal has self-affine structuring, then this is reflected in its Fourier power spectrum

$$|A|^2 \approx \frac{1}{f^{\beta}} \tag{3}$$

If the signal is composed of random independent events (random Gaussian noise, rGn), then $\beta = 0$, $\beta = 2H - 1$, $H = 0.5$ and $r_1 = 0$. When rGn is integrated to random Brownian motion (rBm) its $\beta = 2$, $\beta = 2H + 1$, $H = 0.5$ and $r_1 = 0$. Around the rGn and rBm, signals composed of correlated events are found, fractional Gaussian noises (fGn's) $(-1<\beta<1)$ and fractional Brownian motions (fBm's) $(1<\beta<3)$ (Mandelbrot and Ness, 1968).

2.2. Laser-Doppler Flowmetry in the Rat Brain Cortex

Urethane in a dose of 130 mg/100 g body weight was given i.p. for general anesthesia. The animals (n=8) were artificially ventilated by a gas mixture of 30% O_2 and 70% N_2. Ventilation was adjusted so that pO_2 in arterial blood samples be as close as possible to 100 mmHg. Core temperature was maintained by a Yellow Spring Body Temperature Controller switching an infra heating lamp at the threshold value of the rectal temperature of 37.5 °C. Mean arterial blood pressure (MABP) was measured by a Statham pressure transducer via a catheter introduced into the common carotid artery toward the heart. MABP and electrical activity of the brain cortex (ECoG) was recorded by a Grass polygraph. Red blood cell (RBC) flux was continuously measured with a time constant of 0.1 second under control conditions by a laser-Doppler Flowmeter (Moor Instruments,

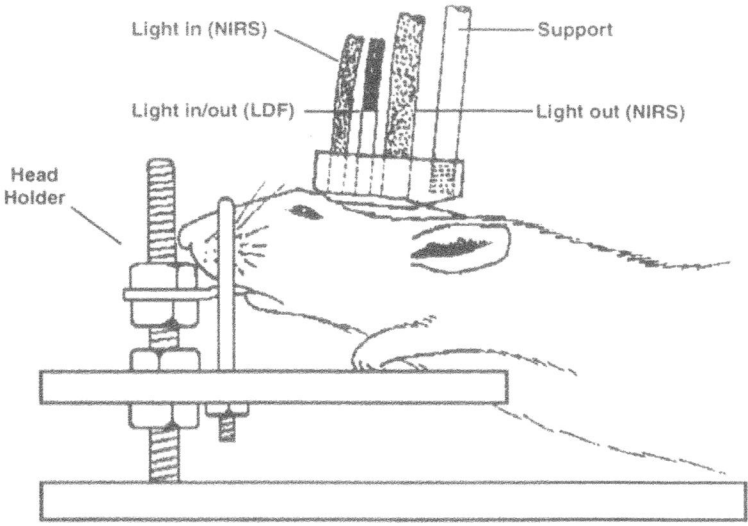

Figure 1. Experimental arrangement for laser-Doppler flometry (LDF) and Near Infrared Spectrophotometry (NIRS) in the brain cortex of the anesthetized rat.

MBF3D) in the brain cortex of anesthetized rats through the thinned (closed) calvarium (Figure 1).

In our polished skull preparation using an illumination fiber of 0.4 mm in diameter the laser light penetrates across the brain cortex and beyond into the subcortical white matter, and even beyond reaching as deep as the subcortical nuclei and the thalamus. Scattering of laser light is seen in all directions. Subcortical white matter due to its high scattering coefficient allows little intensity to penetrate thus appears as a darker band splitting the photon diffusion sphere. Photons would have equal difficulty in escaping

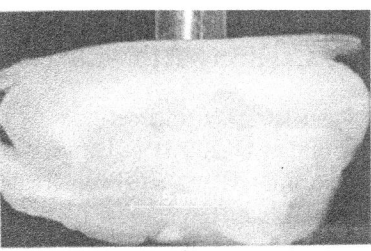

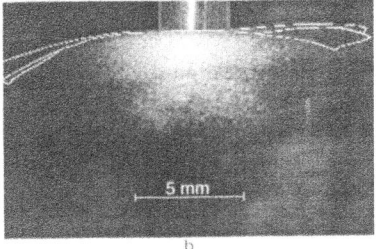

Figure 2. Assessment of the volume of measurement of laser-Doppler flometry in the rat brain cortex. Panel a: side view of a mid-line sagittal section of the rat brain in white light. Note the piece of the skull thinned for the RBC-flux measurement being interposed between the LDF-probe and the brain cortex. Panel b: Same view as in *panel a* with the Laser light being turned on and the white light turned off. The outline of the calvarium being trans-illuminated is shown.

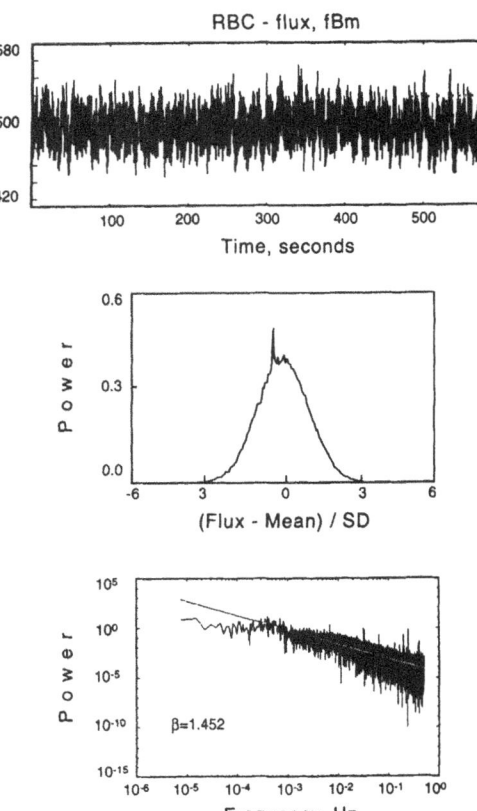

Figure 3. Schematics of the initial steps of data analysis. Upper panel: A record of RBC-flux in its original form. According to our signal classification, it is a fractional Brownian motion (fBm). Middle panel: Probability density funtion generated from the signal shown at the top. Lower panel: Fourier power-spectrum analysis. Power is plotted as a function of frequency on a log-log scale calculated from the time series shown at the top. The line of regression through the data is shown, which relates to the spectral index of the signal, ß as ß = -slope. Note, that the signal is fBm with ß>1.

and reentering the cortex across this boundary. Hence the volume actually sampled by LDF is sandwiched between the pial surface and the subcortical white matter. It has a disk shape with the thickness of the brain cortex (2.5 mm) and 5–6 mm in diameter. These dimensions correspond well with the theoretical model of Bonner and Nossal (Bonner and Nossal, 1981).

2.3. Creating RBC-flux Time Series

RBC-flux time series were created in a length of 2^{17} data points by sampling the signal output at 200 Hz@12 bit, representing 655 seconds each. All time series collected in the 8 animals were proven fractional Brownian motions (ß>1, Fig. 3 and Table I.) thus successive differences of the RBC-flux signal were taken to generate signals in fractional Gaussian noise format for analysis by the dispersional method for the Hurst coefficient H (Fig. 4.). The analyses were carried out on Sun workstations with a software developed by the group of Bassingthwaighte at the Center for Bioengineering, University of Washington (Bassingthwaighte and Raymond, 1995) operated via the Internet from Budapest. Additional analysis and data processing were carried out in Budapest on Macintosh Quadra 650 and 660AV computers by programs developed locally. Oxygen saturation and hemoglobin content of the volume of cerebro-cortical tissue under laser-Doppler measurement were also carried out by a Near Infrared Spectrophotometry (data are not shown and analyzed in this study). The latter did not show any interference with the laser-Doppler measurements

Table I. Characterization of RBC-flux time series with descriptive statistical and fractal tools

Time series	n	Mean	± SD	Skewness	Kurtosis
ts.1	131072	2488.843	32.638	-0.611	4.480
ts.2	131072	2461.350	15.306	0.187	3.003
ts.3	131072	2735.590	21.311	-0.174	3.649
ts.4	131072	2501.470	16.537	0.025	-2.941
ts.5	131072	2445.365	32.537	-0.247	2.710
ts.6	131072	2387.799	23.654	-0.011	2.900
ts.7	131072	2388.845	16.205	0.191	3.202
ts.8	131072	2456.693	22.038	0.675	4.113
Mean		2483.244	22.528	0.004	3.375
± SD		109.988	6.901	0.376	0.640

Time series	Fourier (done on fBm)			Dispersion (done on fGn)		
	β	H	r_l	r	H	r_l
ts.1	1.494	0.247	-0.296	-0.995	0.254	-0.289
ts.2	1.466	0.233	-0.309	-0.998	0.163	-0.373
ts.3	1.477	0.239	-0.304	-0.997	0.158	-0.378
ts.4	1.452	0.226	-0.316	-0.997	0.139	-0.394
ts.5	1.510	0.255	-0.288	-0.999	0.220	-0.322
ts.6	1.507	0.254	-0.289	-0.999	0.213	-0.328
ts.7	1.458	0.229	-0.313	-0.996	0.145	-0.389
ts.8	1.480	0.240	-0.303	-0.997	0.162	-0.374
Mean	1.481	0.240	-0.302	-0.997	0.182	-0.356
± SD	0.022	0.011	0.011	0.001	0.042	0.038

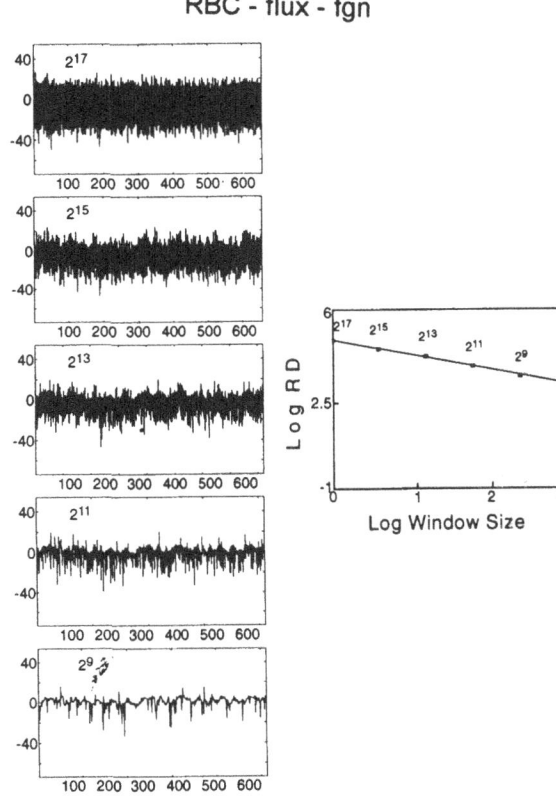

RBC - flux - fgn

Figure 4. Steps of dispersional analysis. First, successive differences of RBC-flux signal shown at the top of Fig. 3. is taken (upper left) because the dispersional analysis can only be applied to fractional Gaussian noises (fGn). Then, neighbouring elements are lumped in succession resulting in series of decreasing length (2^{17} to 2^{9} datapoints) and relative dispersion. The Hurst coefficient is found by a linear regression analysis on the log-log plot of relative dispersion and window size according to Eq. 4.

2.4. Fractal Analysis of RBC-flux Time Series

In the present study the dispersional analysis was chosen because of its good overall performance under a variety of signal quality (Bassingthwaighte and Raymond, 1995).This method is based on finding the variability in the average signal over windows m elements long. In an iterative process it finds the Hurst coefficient from the regression analysis of log RD(m) versus log m/m_o (Fig. 4.), where RD is relative dispersion (SD/mean) and m_o is a reference window size, thus

$$\log RD(m) = (H\text{-}1) \log(m/m_o) + \log RD(m_o) \tag{4}$$

Dispersional method requires signals be subjected to analysis in noise format. Thus the signals need first be tested by the Fourier spectral analysis on their ß before further processing and if they prove fBm, successive differences need to be taken to generate the fGn signal format (Fig. 3).

3. RESULTS

At a mean blood pressure of 101.25 ± 14.58 mm Hg the mean level of the LDF signal was 2483.244 ± 109.99 (mean ± SD in arbitrary units), the mean amplitude of LDF fluctuations within the time window of 655 seconds were 8.18 ± 2.55 % of the mean LDF signal. The mean skewness of the probability density functions (Fig. 2.) was 0.004 ± 0.376 indicating asymmetry of the distributions. Kurtosis>3 indicating high peakedness relative to Gaussian distribution were found in 5 of the 8 cases with one being marginal (3.375 ± 0.64). Thus probability density functions of the RBC-flux time series did not show a typical Gaussian distribution.

RBC-flux time series proved fractional Brownian motions (ß>1, Fig. 3 and Table I.). H of 0.182 ± 0.042 was found by dispersional analyis in the 8 rats studied. This value of H implies that the correlation r_1 between neighbouring elements in the series is negative, $r_1 = -0.356$ (Table I.). These results are also supported by H of 0.24 ± 0.011 given by the Fourier analyis. In none of the studied cases had the fractal model represented by Eq. 4 to be rejected because of a poor fit (see r in Table I).

4. DISCUSSION

These findings indicate that RBC-flux in the rat cerebro-cortical microvessels is a fractal anticorrelated process. The basis appears to be chaotic dynamical fluctuations in diameters of arteries in series along the flow path, with fractal interpeak intervals and amplitudes. The disparity in timings of resistance changes in nearby vessels can cause local "steal" whereby an increase in flow in one vessel occurs at the expense of a reduction in flow in a nearby vessel. Owing to the modular and hexagonal organization of the vasculature in the neocortex of the rat (Bär, 1980) the monitored hemispherical volume contains a few hundred perforating arterial units of different sizes. Each of these can exhibit vaso- and flowmotion of a given character. Their manifestation in the temporal structuring of the overall RBC flux time series, as our findings indicate, is not independent of each other. On a more local level, hanging up of red blood cells at vessel bifurcations can create a flip-flop effectively alternating flow in between competing pathways. Frequency components of our RBC-flux time series are higher than that of potential interfering sources (cardiac and pulmonary). The frequency components of our RBC-flux time series should reflect local rheology since frequencies of this magnitude would significantly attenuate over distances beyond the dimensions of the laser-Doppler measurement.

ACKNOWLEDGMENTS

Supported by OTKA Grants I/3 2040, E 012235, T 016953 and NIH Grant TW00442. The authors thank Dr. Péter Sándor for lending his LD Flowmeter.

REFERENCES

Bär, T. (1980). The Vascular System of the Cerebral Cortex. Adv. in Anat. Embriol. and Cell Biol. Berlin, Heidelberg, New York, Springer-Verlag.

Bassingthwaighte, J. B. (1988). "Physiological heterogeneity: Fractals link determinism and randomness in structures and function." News Physiol Sci 3: 5–10.

Bassingthwaighte, J. B., D. A. Beard, et al. (1995). Fractal Structures and Processes. Workshop on Chaos and Fractals. Mobile, Alabama, Published by the American Institute of Physics.

Bassingthwaighte, J. B., L. S. Liebovitch, et al. (1994). Fractal Physiology. New York, London, Oxford University Press.

Bassingthwaighte, J. B. and G. M. Raymond (1995). "Evaluation of the dispersional analysis method for fractal time series." Ann Biomed Eng 23: 491–505.

Bonner, R. and R. Nossal (1981) "Model for laser Doppler measurements of blood flow in tissue." Appl Opt 20(12): 2097–2107.

Fasano, V. A., R. Urciuoli, et al. (1988). "Intraoperative use of laser Doppler in the study of cerebral microvascular circulation." Acta Neurochir (Wien) 95: 40–48.

Hudetz, A. G., R. J. Roman, et al. (1992). "Spontaneous flow oscillations in the cerebral cortex during acute changes in mean arterial pressure." J Cereb Blood Flow and Metabol 12: 491–499.

Hurst, H. E. (1951). "Long-term storage capacity of reservoirs." Trans Amer Soc Civ Engrs 116: 770–808.

Krogh, A. (1929). The anatomy and physiology of capillaries. New Haven, CT, Yale University Press.

Mandelbrot, B. B. (1983). The Fractal Geometry of Nature. San Francisco, W.H. Freeman and Co.

Mandelbrot, B. B. and J. W. v. Ness (1968). "Fractional brownian motions, fractional noises and applications." SIAM Rev 10: 422–437.

Morita-Tsuzuki, Y., E. Bouskela, et al. (1992). "Vasomotion in the rat cerebral microcirculation recorded by laser-Doppler flometry." Acta Physiol Scand 146: 431–439.

Nilsson, G. E., T. Tenland, et al. (1980). "A new instrument for continuous measurement of tissue blood flow by light beating spectroscopy." IEEE Trans Biomed Eng :

Rosenblum, B. R., R. F. Bonner, et al. (1987). "Intraoperative measurement of cortical blood flow adjacent to cerebral AVM using laser Doppler velocimetry." J. Neurosurg. 66: 396–399.

Stern, M. D. (1975). "In vivo evaluation of microcirculation by coherent light scattering." Nature 254: 56–58.

Weibel, E. R. (1991). "Fractal geometry: a design principle for living organisms." Am. J. Physiol. 261: L361-L369.

West, B. J. and A. L. Goldberger (1987). "Physiology in fractal dimensions." Am. Sci. 75: 354–365.

West, B. J. and M. Shlesinger (1990). "The noise in natural phenomena." Am Sci 78: 40–45.

AUTHOR INDEX

SUBJECT INDEX

The manufacturer's authorised representative in the EU is Springer
Nature Customer Service Centre GmbH, Europaplatz 3, 69115 Heidelberg,
Germany. If you have any concerns regarding our products, please
contact ProductSafety@springernature.com

Printed and bound by CPI Group (UK) Ltd, Croydon, CR0 4YY

23/04/2026
02095623-0017